Basic College Mathematics

Seventh Edition

Margaret L. Lial
American River College

Stanley A. Salzman
American River College

Diana L. Hestwood
Minneapolis Community and
Technical College

PEARSON

Addison
Wesley

Boston • San Francisco • New York • London • Sydney
Tokyo • Singapore • Madrid • Mexico City • Paris
Cape Town • Hong Kong • Montreal

Publisher	Greg Tobin
Editor-in-Chief	Maureen O'Connor
Editorial Project Management	Lauren Morse
Editorial Assistant	Marcia Emerson
Managing Editor/Production Supervisor	Ron Hampton
Text and Cover Design	Dennis Schaefer
Supplements Production	Jason Miranda
Production Services	Elm Street Publishing Services, Inc.
Media Producers	Sharon Smith and Sara Anderson
Marketing Manager	Jay Jenkins
Marketing Coordinator	Tracy Rabinowitz
Prepress Services Buyer	Caroline Fell
Technical Art Supervisor	Joseph K. Vetere
First Print Buyer	Hugh Crawford
Composition Services	Pre-Press Company, Inc.
Cover Photo	© Daryl Benson/Masterfile
Cover Image	Walkway through field: Kouchibouguac National Park, New Brunswick, Canada

Photo Credits   All photos from PhotoDisc except the following: Stan Salzman, pp. 1, 61 left, 87; Michael S. Yamashita/Corbis, p. 2; Beth Anderson, pp. 8 top right, 8 bottom left, 61 right, 162 left, 192 right, 265 top, 265 bottom, 308, 378 right, 392 right, 401 bottom right, 402 right, 442 left, 490 right, 497 top, center, and bottom right; 498 right, 498 left, 499 top left, 500 bottom left, 520 bottom right, 520 top right, 527 bottom left, 574 bottom left, 704 right, 721 right, 732, 744 right, 757; Thinkstock (RF), pp. 8 bottom right, 258, 364 left, 371, 449 left, 462 right, 625, 746; Black and Decker, p. 17 left; Greg Fiume/NewSport/Corbis, p. 17 right; Kelly-Mooney Photography/Corbis, p. 12; Brand X Pictures (RF), p. 28 top left; Digital Vision, pp. 28 top right, 145 left, 145 right, 162 right, 209, 218 right, 237 right, 298 right, 299 bottom right, 313, 315, 379 left, 482 right, 718, 755; BSF/New Sport/Corbis (TL), p. 52; Dodge, p. 93; Courtesy of Jelly Belly, pp. 109 top and bottom; Reuters/Corbis, pp. 113, 289, 338, 345; Petmarket, p. 151 top; Bettmann/Corbis, p. 151 bottom; Corbis, pp. 154 left, 238 left, 266 top and bottom, 292 left and right, 299 left, 499 bottom left, 500 bottom right, 503 center and right, 528, 576, 623, 704 left, 711, 735, 760; Public domain, p. 164; photo and recipe courtesy of Jelly Belly Candy Co., p. 172; Cameron Campbell, p. 187; Bill Greenblatt/Liaison, p. 218 left; Crown Plumbing, p. 242 left; public domain and Comstock (RF), p. 290 left; NationBill/Corbis Sygma, p. 290 right; NASA, pp. 299 top right, 471, 574 bottom right; Bob Daemmrich/Image Works, p. 306 right; Dorling Kindersley, p. 314 center; Brand X Pictures (RF), p. 314 right; PictureQuest/Brand X Pictures, p. 322; PictureQuest/PhotoDisc, p. 324; Internet Product Shots, p. 331; Gary W. Carter/Corbis, p. 368; Newman's Own, 401 bottom left; Bruce Anderson Photo, p. 402 left; Robert Brenner/PhotoEdit, p. 413 left; Golden Horseshoe Assoc., p. 424 left; Comstock, p. 458 left; Sears, p. 462 left; Martin Harvey/Corbis, p. 479 left; Dwi Oblo/Stringer/Reuters/Corbis, p. 479 right; Brand X Pictures (RF), p. 480 left; Richard Carson/Reuters Corbis, p. 480 right; CMSP/PAL, p. 499 bottom right; Thierry Orban/Corbis Sygma, p. 509; Michael Newman/PhotoEdit, p. 520 top left; reprinted courtesy of Caterpillar, p. 522; PictureQuest/Thinkstock, p. 540; PictureQuest/Image Farm, p. 643; GE Healthcare, p. 632.

Library of Congress Cataloging-in-Publication Data

Lial, Margaret L.
    Basic college mathematics.—7th ed. / Margaret L. Lial, Stanley A. Salzman, Diana L. Hestwood.
        p.   cm.
    Includes index.
    ISBN 0-321-25780-4 (Student Edition)

    ISBN 0-321-29280-4 (Hardback Edition)
    1. Mathematics.   I. Salzman, Stanley A.   II. Hestwood, Diana L.   III. Title.
QA37.3.L48 2005
513'.1—dc22   2004052582

1 2 3 4 5 6 7 8 9 10   VHP   08 07 06 05 04

Contents

Index of Applications

Index of Focus on Real-Data Applications

Preface

The seventh edition of *Basic College Mathematics* continues our ongoing commitment to provide the best possible text and supplements package that will help instructors teach and students succeed. To that end, we have addressed the diverse needs of today's students through an attractive design, updated applications and graphs, helpful features, careful explanation of concepts, and an expanded package of supplements and study aids. We have also responded to the suggestions of users and reviewers and have added many new examples and exercises based on their feedback.

The text is designed to help students achieve success in a developmental mathematics program. It provides the necessary review and coverage of whole numbers, fractions, decimals, ratio and proportion, percent, and measurement, as well as an introduction to algebra and geometry, and a preview of statistics. This text is part of a series that also includes the following books:

- *Essential Mathematics,* Second Edition, by Lial and Salzman
- *Prealgebra,* Third Edition, by Lial and Hestwood
- *Introductory Algebra,* Eighth Edition, by Lial, Hornsby, and McGinnis
- *Intermediate Algebra,* Eighth Edition, by Lial, Hornsby, and McGinnis
- *Introductory and Intermediate Algebra,* Third Edition, by Lial, Hornsby, and McGinnis

HALLMARK FEATURES

We believe students and instructors will welcome the following helpful features.

▶ *Chapter Openers* New and updated chapter openers feature real-world applications of mathematics that are relevant to students and tied to specific material within the chapters. Examples of topics include finding the best buy on cell phone service, home improvements, recipes, weather, fishing, astronomy, and work/career applications. (See pp. 1, 109, and 315—Chapters 1, 2, and 5.)

▶ *Real-Life Applications* We are always on the lookout for interesting data to use in real-life applications. As a result, we have included many new or updated examples and exercises throughout the text that focus on real-life applications of mathematics. Students are often asked to find data in a table, chart, graph, or advertisement. (See pp. 279, 308, and 331.) These applied problems provide an up-to-date flavor that will appeal to and motivate students. A comprehensive Index of Applications appears at the beginning of the text.

▶ *Figures and Photos* Today's students are more visually oriented than ever. Thus, we have made a concerted effort to include mathematical figures, diagrams, tables, and graphs whenever possible. (See pp. 24, 279, and 551.) Many of the graphs use a style similar to that seen by students in today's print and electronic media. Photos have been incorporated to enhance applications in examples and exercises. (See pp. 266, 289, and 480.)

▶ *Emphasis on Problem Solving* Introduced at the end of Chapter 1, our six-step process for solving application problems is integrated throughout the text. The six steps, *Read, Plan,*

Estimate, Solve, State the Answer, and *Check,* are emphasized in boldface type and repeated in specific problem-solving examples in Chapters 1, 2, 3, 5, 6, 7, and 9. (See pp. 89, 149, and 347.)

▶ *Learning Objectives* Each section begins with clearly stated, numbered objectives, and the material within sections is keyed to these objectives so that students know exactly what concepts are covered. (See pp. 129, 281, and 425.)

▶ *Cautions and Notes* These color-coded and boxed comments, one of the most popular features of previous editions, warn students about common errors and emphasize important ideas throughout the exposition. (See pp. 157, 261, and 547.) There are more of these in the seventh edition than in the sixth, and the new text design makes them easier to spot; Cautions are highlighted in yellow and Notes are highlighted in purple.

▶ *Calculator Tips* These optional tips, marked with a calculator icon, offer basic information and instruction for students using calculators in the course. (See pp. 254, 276, and 569.) In addition, an Introduction to Calculators is included as an appendix.

▶ *Margin Problems* Margin problems, with answers immediately available on the bottom of the page, are found in every section of the text. (See pp. 81, 349, and 548.) This key feature allows students to immediately practice the material covered in the examples in preparation for the exercise sets. Based on reviewer feedback, we have added more margin exercises to the seventh edition.

▶ *Ample and Varied Exercise Sets* The text contains a wealth of exercises to provide students with opportunities to practice, apply, connect, and extend the skills they are learning. Numerous illustrations, tables, graphs, and photos have been added to the exercise sets to help students visualize the problems they are solving. Problem types include skill building, writing, estimation, and calculator exercises, as well as applications and correct-the-error problems. In the Annotated Instructor's Edition of the text, the writing and estimation exercises are marked with icons for writing ✎ and for estimation ≈ so that instructors may assign these problems at their discretion. Exercises suitable for calculator work are marked in both the student and instructor editions with a calculator icon ▦ . (See pp. 98–105, 271–274, and 287–290.)

▶ *Relating Concepts Exercises* These sets of exercises help students tie concepts together and develop higher level problem-solving skills as they compare and contrast ideas, identify and describe patterns, and extend concepts to new situations. (See pp. 74, 258, and 332.) These exercises make great collaborative activities for pairs or small groups of students.

▶ *Summary Exercises* There are now eight sets of in-chapter summary exercises: whole numbers, fraction basics, fraction computation, decimals, ratios/rates/proportions, percent, geometry concepts, and operations with signed numbers. These exercises provide students with the all-important *mixed* practice they need at these critical points in their skill development. (See pp. 135–136, 291–292, and 345–346.)

▶ *Study Skills Component* Poor study skills are a major reason why students do not succeed in math. A few generic tips sprinkled here and there are not enough to help students change their behavior. So, in this text, a desk-light icon at key points in the text directs students to one of 12 activities in a separate *Study Skills Workbook.* These carefully designed activities, correlated directly to the text material, cover note taking, homework, study cards, test preparation, test taking, preparing for final exams, and more. (See pp. 7, 95, and 243.) This unique workbook explains *how* the brain actually learns and remembers so students understand *why* the study skills activities will help them succeed in the course. Students are introduced to the workbook in a To the Student section at the beginning of the text.

▶ *Focus on Real-Data Applications* Each one-page activity presents a relevant and in-depth look at how mathematics is used in the real world. Designed to help instructors answer the often-asked question, "When will I ever use this stuff?," these activities ask students to read and interpret data from newspaper articles, the Internet, and other familiar, real-world sources. (See pp. 6, 112, and 360.) The activities are well-suited to collaborative work or they can be completed by individuals or used for open-ended class discussions. Instructor teaching notes and activity extensions are provided in the *Printed Test Bank and Instructor's Resource Guide.*

▶ *Test Your Word Power* This feature, incorporated into each chapter summary, helps students understand and master mathematical vocabulary. Key terms from the chapter are presented along with four possible definitions in a multiple-choice format. Answers and examples illustrating each term are provided. (See pp. 301, 355, and 605.)

▶ *Ample Opportunity for Review* Each chapter ends with a Chapter Summary featuring: Key Terms with definitions and helpful graphics, New Formulas, Test Your Word Power, and a Quick Review of each section's content with additional examples. Also included is a comprehensive set of Chapter Review Exercises keyed to individual sections, a set of Mixed Review Exercises, and a Chapter Test. Beginning with Chapter 2, each chapter concludes with a set of Cumulative Review Exercises. (See pp. 233–247, 301–314, and 355–370.)

▶ *Diagnostic Pretest* A diagnostic pretest is included on p. xxiii of the text and covers all the material in the book, much like a sample final exam. This pretest can be used to facilitate student placement in the correct chapter according to skill level. The pretest also exposes students to the scope of the course content.

WHAT'S NEW IN THIS EDITION

The scope and sequence of topics in *Basic College Mathematics* has stood the test of time and rates highly with our reviewers. Therefore, you will find the table of contents intact, making the transition to the new edition easier.

• Throughout the text, examples and exercises have been adjusted or replaced to reflect current data and practices. Applications have been updated and cover a wider variety of topics such as the fields of technology and health sciences. For example, Chapter 10, Statistics, includes current real-world graphs, tables, and charts to help show the relevance of the topic being discussed.

• There is an increased emphasis on readability. Many students who are in developmental-level mathematics courses also have difficulty with reading skills. Additionally, increasing numbers of students do not speak English as their first language. With these students in mind, we have increased our efforts to use explanations that are clear, complete, and written in an understandable style, while still maintaining mathematical accuracy. We have designed page layouts so that white space or divider lines separate concepts, key points are highlighted with color, and textual clues direct students to related material. When words at the end of a line are broken by a hyphen and continued on the next line, we've made sure that splitting the word does not create reading problems (for example, breaking the words "dig-its" or "an-swer" would be avoided).

• Four new sets of Summary Exercises have been added; eight chapters now include these helpful sets of review exercises. The Summary Exercises are positioned within the chapters to provide critically important mixed review so that students are better prepared to progress to new material.

• Chapter 8, Geometry, now provides opportunities for students to make sketches of geometric shapes as part of the problem-solving process (see pp. 549, 558, and 575). In addition, students are now asked to recognize and name the various shapes and solids within the mixed review exercises.

WHAT SUPPLEMENTS ARE AVAILABLE?

For a comprehensive list of the supplements and study aids that accompany *Basic College Mathematics,* Seventh Edition, see pages xiv and xv.

Student Supplements	Instructor Supplements

Student's Solutions Manual
- By Jeffrey A. Cole, *Anoka-Ramsey Community College*
- Provides detailed solutions to the odd-numbered section-level exercises and to all margin, Relating Concepts, Summary, Chapter Review, Chapter Test, and Cumulative Review Exercises
  ISBN: 0-321-27938-7

Study Skills Workbook
- By Diana Hestwood and Linda Russell
- The activities in the workbook teach students how to use the textbook effectively, plan their homework, take notes, make mind maps and study cards, manage study time, and prepare for and take tests
- A desk-light icon at key points in the text directs students to correlated activities
  ISBN: 0-321-27937-9

Videotape Series
- Features an engaging team of lecturers
- Provides comprehensive coverage of each section and topic in the text
  ISBN: 0-321-27941-7

Digital Video Tutor
- Complete set of digitized videos on CD-ROM for students to use at home or on campus
- Ideal for distance learning or supplemental instruction
  ISBN: 0-321-27949-2

New! Additional Skill and Drill Manual
- Provides additional practice and test preparation for students
  ISBN: 0-321-33168-0

Addison-Wesley Math Tutor Center
- Staffed by qualified mathematics instructors
- Provides tutoring on examples and odd-numbered exercises from the textbook through a registration number with a new textbook or purchased separately
- Accessible via toll-free telephone, toll-free fax, e-mail, or the Internet
  www.aw-bc/tutorcenter

Annotated Instructor's Edition
- Provides answers to all text exercises in color next to the corresponding problems
- Icons identify writing and calculator exercises
  ISBN: 0-321-26686-2

Instructor's Solutions Manual
- By Jeffrey A. Cole, *Anoka-Ramsey Community College*
- Provides complete solutions to all even-numbered section-level exercises
  ISBN: 0-321-27943-3

Answer Book
- By Jeffrey A. Cole, *Anoka-Ramsey Community College*
- Provides answers to all the exercises in the text
  ISBN: 0-321-27942-5

New! Adjunct Support Manual
- Includes resources designed to help both new and adjunct faculty with course preparation and classroom management
- Offers helpful teaching tips correlated to the sections of the text
  ISBN: 0-321-27950-6

Printed Test Bank and Instructor's Resource Guide
- By James J. Ball, *Indiana State University*
- The test bank contains two diagnostic pretests, six free-response and two multiple-choice test forms per chapter, and two final exams
- The resource guide contains teaching suggestions for each chapter, additional practice exercises for every objective of every section, a correlation guide from the sixth to the seventh edition, phonetic spellings for all key terms in the text, and teaching notes and extensions for the Focus on Real-Data Applications in the text
  ISBN: 0-321-27936-0

TestGen
- Enables instructors to build, edit, print, and administer tests
- Features a computerized bank of questions developed to cover all text objectives
- Available on a dual-platform Windows/Macintosh CD-ROM
  ISBN: 0-321-27940-9

MathXL® Tutorials on CD (ISBN: 0-321-27939-5) This interactive tutorial CD-ROM provides algorithmically generated practice exercises that are correlated at the objective level to the content of the text. Every exercise is accompanied by an example and a guided solution designed to involve students in the solution process. Selected exercises may also include a video clip to help students visualize concepts. The software recognizes common student errors and provides appropriate feedback; it also tracks student activity and scores and can generate printed summaries of students' progress.

MathXL®: www.mathxl.com MathXL is an online homework, tutorial, and assessment system that accompanies your Addison-Wesley textbook in mathematics or statistics. With MathXL, instructors can create, edit, and assign online homework and tests using algorithmically generated exercises correlated to your textbook. All student work is tracked in MathXL's online gradebook. Students can take chapter tests in MathXL and receive personalized study plans based on their test results. The study plan diagnoses weaknesses and links students directly to tutorial exercises for the objectives they need to study and retest. Students can also access supplemental video clips directly from selected exercises.

MyMathLab MyMathLab is a series of text-specific, easily customizable online courses for Addison-Wesley textbooks in mathematics and statistics. MyMathLab is powered by CourseCompass—Pearson Education's online teaching and learning environment—and by MathXL—our online homework, tutorial, and assessment system. MyMathLab gives instructors the tools they need to deliver all or a portion of their course online, whether students are in a lab setting or working from home. MyMathLab provides a rich and flexible set of course materials, featuring free-response exercises that are algorithmically generated for unlimited practice and mastery. Students can also use online tools, such as video lectures, animations, and a multimedia textbook, to independently improve their understanding and performance. Instructors can use MyMathLab's homework and test managers to select and assign online exercises correlated directly to the textbook, and they can import TestGen tests into MyMathLab for added flexibility. MyMathLab's online gradebook—designed specifically for mathematics and statistics—automatically tracks students' homework and test results and gives the instructor control over how to calculate final grades. Instructors can also add offline (paper-and-pencil) grades to the MathXL gradebook. MyMathLab is available to qualified adopters. For more information, visit our Web site at *www.mymathlab.com* or contact your Addison-Wesley sales representative.

ACKNOWLEDGMENTS

The comments, criticisms, and suggestions of users, nonusers, instructors, and students have positively shaped this textbook over the years, and we are most grateful for the many responses we have received. The feedback gathered for this revision of the text was particularly helpful, and we especially wish to thank the following individuals who provided invaluable suggestions for this and the previous edition:

George Alexander, *University of Wisconsin*
Sonya Armstrong, *West Virginia State College*
Vernon Bridges, *Durham Technical Community College*
Solveig R. Bender, *William Rainey Harper College*
Barbara Brown, *Anoka-Ramsey Community College*
Ernie Chavez, *Gateway Community College*
Terry Joe Collins, *Hinds Community College*
Martha Daniels, *Central Oregon Community College*
Matthew Flacche, *Camden Community College*
Donna Foster, *Piedmont Technical College*
Lourdes Gonzalez, *Miami-Dade Community College*
Joe Howe, *St. Charles County Community College*
Rose Kaniper, *Burlington County College*

Douglas Lewis, *Yakima Valley Community College*
Valerie Maley, *Cape Fear Community College*
Judy Mee, *Oklahoma City Community College*
Wayne Miller, *Lee College*
Kathy Peay, *Sampson Community College*
Thea Philliou, *College of Santa Fe*
Jane Roads, *Moberly Area Community College*
Richard D. Rupp, *Del Mar College*
Ellen Sawyer, *College of DuPage*
Lois Schuppig, *College of Mount St. Joseph*
Mary Lee Seitz, *Erie Community College—City Campus*
Kathryn Taylor, *Santa Ana College*
Sven Trenholm, *North Country Community College*
Bettie A. Truitt, *Black Hawk College*
Jackie Wing, *Angelina College*

Our sincere thanks go to these dedicated individuals at Addison-Wesley who worked long and hard to make this revision a success: Maureen O'Connor, Ron Hampton, Dennis Schaefer, Jay Jenkins, Beth Anderson, Sharon Smith, Sara Anderson, Tracy Rabinowitz, Lauren Morse, and Marcia Emerson. Gina Linko and Phyllis Crittenden of Elm Street Publishing Services provided their customary excellent production work. We are most grateful to Peg Crider for researching and writing several of the Focus on Real-Data Applications features; Bernice Eisen for her accurate and useful index; Becky Troutman for preparing the comprehensive Index of Applications; Abby Tanenbaum for writing the Diagnostic Pretest; and Janis Cimperman, Jon Becker, and Ellen Sawyer for accuracy checking the manuscript.

The ultimate measure of this textbook's success is whether it helps students master basic skills, develop problem-solving techniques, and increase their confidence in learning and using mathematics. In order for us, as authors, to know what to keep and what to improve for the next edition, we need to hear from you, the instructor, and you, the student. Please tell us what you like and where you need additional help by sending an e-mail to math@awl.com. We appreciate your feedback.

In appreciation of your lasting support and never-ending enthusiasm: family, colleagues, and more than a generation of motivated students.

Stan Salzman

This book is dedicated to my dad, who always told me when I was young that girls could learn math, and to my students at Minneapolis Community and Technical College, who keep me in touch with the real world.

Diana L. Hestwood

Feature Walk-Through

Chapter Openers New and updated chapter openers feature real-world applications of mathematics that are relevant to students and tied to specific material within the chapters.

Decimals

4

4.1 Reading and Writing Decimals
4.2 Rounding Decimals
4.3 Adding and Subtracting Decimals
4.4 Multiplying Decimals
4.5 Dividing Decimals
Summary Exercises on Decimals
4.6 Writing Fractions as Decimals

Nearly 49 million Americans go fishing at least once a year, making it America's fourth most popular recreational activity. (*Source:* National Sporting Goods Association.) In **Section 4.3**, Exercises 47–50, these friends will use decimal numbers when paying for new equipment. But will decimals help them catch their limit? (See **Section 4.1**, Exercises 59–62, and **Section 4.6**, Exercises 65–68.)

---

82 Chapter 1 Whole Numbers

2 Use the bar graph to find the approximate number of fans who picked each sport as their favorite.

(a) Pro football

(b) Pro baseball

(c) Pro basketball

(d) College basketball

(e) Golf

FAN APPEAL

Source: The Harris Poll.

EXAMPLE 2 Using a Bar Graph

Use the bar graph to find the number of fans who picked college football as their favorite sport.

Use a ruler or straightedge to line up the top of the bar labeled "College football," with the numbers on the left edge of the graph, labeled "Number of Adults Fans." We see that 10 out of 100 adult fans picked college football as their favorite sport.

◀◀◀ **Work Problem 2 at the Side.**

OBJECTIVE 3 Read and understand a line graph. A line graph is often used for showing a trend. The following line graph shows the U.S. Bureau of the Census predictions for U.S. population growth to the year 2100.

3 Use the line graph to find the predicted population of the United States for each year.

(a) 2050

(b) 2075

(c) 2100

ANOTHER CENTURY OF GROWTH

Source: U.S. Bureau of the Census.

EXAMPLE 3 Using a Line Graph

Figures and Photos Today's students are more visually oriented than ever. Thus, a concerted effort has been made to add mathematical figures, diagrams, tables, and graphs whenever possible. Many of the graphs use a style similar to that seen by students in today's print and electronic media. Photos have been incorporated to enhance applications in examples and exercises.

Relating Concepts These sets of exercises help students tie together topics and develop problem-solving skills as they compare and contrast ideas, identify and describe patterns, and extend concepts to new situations. These exercises make great collaborative activities for pairs or small groups of students.

---

74   Chapter 1 Whole Numbers

65. In Chicago, the Sears Tower tenants recycled 1,667,300 pounds of paper last year. Round this number to the nearest ten thousand, nearest hundred thousand, and nearest million. (*Source:* Trizec Properties.)

66. Round 621,999,652 to the nearest thousand, nearest ten thousand, and nearest hundred thousand.

67. American pharmaceutical companies spent $25,765,475,000 last year to develop new products. Round this amount to the nearest hundred thousand, nearest hundred million, and nearest billion. (*Source:* American Demographics.)

68. In one year the U.S. Federal Food Assistance Program paid out $18,915,762,568 in food stamps. Round this amount to the nearest hundred thousand, nearest hundred million, and nearest ten billion. (*Source:* U.S. Department of Agriculture.)

RELATING CONCEPTS (EXERCISES 69–75) For Individual or Group Work

To see how both rounding and front end rounding are used in solving problems, work Exercises 69–75 in order.

69. A number rounded to the nearest thousand is 72,000. What is the *smallest* whole number this could have been before rounding?

70. A number rounded to the nearest thousand is 72,000. What is the *largest* whole number this could have been before rounding?

71. When front end rounding is used, a whole number rounds to 8000. What is the *smallest* possible original number?

72. When front end rounding is used, a whole number rounds to 8000. What is the *largest* possible original number?

The graph below shows the greatest number of pitching appearances for active professional baseball pitchers.

ON THE MOUND
Active pitchers with the most appearances include the following:

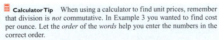

Jesse Orosco	John Franco	Dan Plesac	Mike Jackson	Mike Stanton
1155	998	979	937	794

Source: Major League Baseball.

73. Round the number of appearances given for each pitcher to the nearest ten.

74. Use front end rounding to round the number of appearances for each pitcher.

75. (a) What is one advantage of using front end rounding instead of rounding to the nearest ten?

(b) What is one disadvantage?

Focus on Real-Data Applications

'Til Debt Do You Part!

Ladies Home Journal recently released statistics showing the average wedding costs in the United States in 2004. In 2001, the cost of the average wedding was $20,357. The average number of wedding guests has grown to over 200.

Category	Average Cost in 2004
Misc. expenses (stationery, clergy, gifts, limousine)	$1583
Bouquets and other flowers	$ 950
Photography and videography	$1647
Music	$1042
Engagement and wedding rings (bride and groom)	$5099
Rehearsal dinner	$ 877
Bride's wedding dress and headpiece	$1242
Bridal attendants' dresses (average of 5 bridesmaids)	$ 992
Mother of the bride's apparel	$ 297
Men's formalwear (ushers, best man, groom)	$ 849
Wedding reception	$9590
Grand Total	

Source: *Ladies Home Journal;* Association of Bridal Consultants.

1. What is the grand total of expenses shown in the chart?
2. How much more expensive was a wedding in 2004 compared to 2001?
3. The groom pays for the bouquets and flowers, the rehearsal dinner, the bride's engagement and wedding rings ($4400), the clergy ($300), and the groom's formalwear ($125). What is the total amount spent by the groom?
4. If you budgeted $45 per person for the wedding reception and you invited 210 guests to a wedding in 2005, how much money would you have spent compared to the average 2004 wedding reception costs?
5. If you budget $7200 for the reception and the caterer charges $45 per person, how many guests can you invite? How much of your budget is left over?
6. If you budget $10,500 for the reception and the caterer charges $45 per person, how many guests can you invite? How much of your budget is left over?
7. What type of arithmetic problem did you work to get the answers to Problem 6? What is the mathematical term for the "left over" budget?

Focus on Real-Data Applications These one-page activities, found throughout the text, present even more relevant and in-depth looks at how mathematics is used in the real world. Designed to help instructors answer the often-asked question, "When will I ever use this stuff?," these activities ask students to read and interpret data from newspaper articles, the Internet, and other familiar, real sources. The activities are well-suited to collaborative work and can also be completed by individuals or used for open-ended class discussions.

Calculator Tip When using a calculator to find unit prices, remember that division is *not* commutative. In Example 3 you wanted to find cost per ounce. Let the *order* of the *words* help you enter the numbers in the correct order.

cost	**per**	**ounce**
Enter the cost.	*Per* means divide.	Enter number of ounces.
$2.73	÷	36	= 0.076 (rounded)

If you entered 36 ÷ 2.73 =, you'd get the number of *ounces* per *dollar*. How could you use that information to find the best buy? (*Answer:* The best buy would be to get the greatest number of ounces per dollar.)

Finding the best buy is sometimes a complicated process. Things that affect the cost per unit can include "cents off" coupons and differences in how much use you'll get out of each unit.

ANSWERS
3. (a) 4 quarts, at $1.62 per quart
   (b) 12 cans, at $0.291 per can (rounded)

Calculator Tips These optional tips, marked with calculator icons, offer basic information and instruction for students using calculators in the course.

End-of-Chapter Material One of the most admired features of the Lial textbooks is the extensive and well-thought-out end of chapter material. At the end of each chapter, students will find:

Key Terms are listed, defined, and referenced back to the appropriate section number.

Test Your Word Power To help students understand and master mathematical vocabulary, Test Your Word Power has been incorporated in each Chapter Summary. Students are quizzed on Key Terms from the chapter in a multiple-choice format. Answers and examples illustrating each term are provided.

Chapter 3
SUMMARY

KEY TERMS

3.1	like fractions	Fractions with the same denominator are called *like fractions*.
	unlike fractions	Fractions with different denominators are called *unlike fractions*.
3.2	least common multiple	Given two or more whole numbers, the least common multiple is the smallest whole number that is divisible by all the numbers.
	LCM	The abbreviation for *least common multiple* is LCM.
3.3	least common denominator	When unlike fractions are rewritten as like fractions having the least common multiple as the denominator, the new denominator is the least common denominator.
	LCD	The abbreviation for *least common denominator* is LCD.
3.4	carrying	Carrying is the method used in adding mixed numbers when the sum of the fractions is greater than 1. Carry from the fraction to the whole number.
	borrowing	Borrowing is the method used in subtracting mixed numbers when the fraction part of the minuend is less than the fraction part of the subtrahend.

TEST YOUR WORD POWER

See how well you have learned the vocabulary in this chapter. Answers follow the Quick Review.

1. **Like fractions** are
   A. fractions that are equivalent
   B. fractions that are not equivalent
   C. fractions that have the same numerator
   D. fractions that have the same denominator.

2. Two or more fractions are **unlike fractions** if
   A. they are not equivalent
   B. they have different numerators
   C. they have different denominators
   D. they are improper fractions.

3. The **least common multiple** is
   A. the smallest whole number that is divisible by each of two or more numbers
   B. the smallest numerator
   C. the smallest denominator
   D. the smallest whole number that is not divisible by a group of numbers.

4. The abbreviation **LCM** stands for
   A. the largest common multiple
   B. the longest common multiple
   C. the most likely common multiplier
   D. the least common multiple.

5. The **least common denominator** is
   A. needed when multiplying fractions
   B. needed when dividing fractions
   C. the least common multiple of the denominators in a fraction problem
   D. any denominator that is common to a group of fractions.

6. The abbreviation **LCD** stands for
   A. the largest common denominator
   B. the least common denominator
   C. the least common divisor
   D. the most likely common denominator.

233

A Chapter Test helps students practice for the real thing.

Study Skills Component A desk-light icon at key points in the text directs students to a separate *Study Skills Workbook* containing activities correlated directly to the text. This unique workbook explains how the brain actually learns, so students understand *why* the study tips presented will help them succeed in the course.

Test **243**

Chapter 3
TEST

Study Skills Workbook
Activity 10: Taking a Test

Add or subtract. Write answers in lowest terms.

1. $\frac{5}{8} + \frac{1}{8}$ 2. $\frac{1}{16} + \frac{7}{16}$

3. $\frac{7}{10} - \frac{3}{10}$ 4. $\frac{7}{12} - \frac{5}{12}$

Find the least common multiple of each set of numbers.

5. 2, 3, 4   6. 6, 3, 5, 15   7. 6, 9, 27, 36

Add or subtract. Write answers in lowest terms.

8. $\frac{3}{8} + \frac{1}{4}$ 9. $\frac{2}{9} + \frac{5}{12}$

1. _____
2. _____
3. _____
4. _____
5. _____
6. _____
7. _____
8. _____
9. _____
10.

---

234   Chapter 3 Adding and Subtracting Fractions

QUICK REVIEW

Concepts	Examples
3.1 Adding Like Fractions Add numerators and write in lowest terms.	$\frac{3}{4} + \frac{1}{4} + \frac{5}{4} = \frac{3 + 1 + 5}{4} = \frac{9}{4} = 2\frac{1}{4}$
3.1 Subtracting Like Fractions Subtract numerators and write in lowest terms.	$\frac{7}{8} - \frac{5}{8} = \frac{7 - 5}{8} = \frac{2 \div 2}{8 \div 2} = \frac{1}{4}$
3.2 Finding the Least Common Multiple (LCM) Method of using multiples of the larger number: List the first few multiples of the larger number. Check each one until you find the multiple that is divisible by the smaller number.	$\frac{1}{3} + \frac{1}{4}$ 4, 8, 12, 16, . . . ← Multiples of 4 First multiple divisible by 3 (12 ÷ 3 = 4) The least common multiple (LCM) of 3 and 4 is 12.
3.2 Finding the Least Common Multiple (LCM) Method of prime numbers: First find the prime factorization of each number. Then use the prime factors to build the least common multiple.	Factors of 9 $9 = 3 \cdot 3$ $15 = 3 \cdot 5$ $LCM = 3 \cdot 3 \cdot 5 = 45$ Factors of 15 The least common multiple (LCM) of 9 and 15 is 45.

Quick Review sections give students not only the main concepts from the chapter (referenced back to the appropriate section), but also an adjacent example of each concept.

Review Exercises are keyed to the appropriate sections so that students can refer to examples of that type of problem if they need help.

Chapter 3
REVIEW EXERCISES

[3.1] *Add or subtract. Write answers in lowest terms.*

1. $\frac{5}{7} + \frac{1}{7}$ 2. $\frac{4}{9} + \frac{3}{9}$ 3. $\frac{1}{8} + \frac{3}{8} + \frac{2}{8}$ 4. $\frac{5}{16} - \frac{3}{16}$

5. $\frac{5}{10} + \frac{3}{10}$ 6. $\frac{5}{12} - \frac{3}{12}$ 7. $\frac{36}{62} - \frac{10}{62}$ 8. $\frac{68}{75} - \frac{43}{75}$

Solve each application problem. Write answers in lowest terms.

9. Nurse Suzie Brasher screened $\frac{7}{16}$ of her patients in her first hour on duty and $\frac{5}{16}$ of her patients in the second hour. What fraction of her patients did she screen in the two hours?

10. The Koats for Kids committee members completed $\frac{5}{8}$ of their Web-page design in the morning and $\frac{3}{8}$ in the afternoon. How much less did they complete in the afternoon than in the morning?

Simplify by using the order of operations.

63. $8\left(\frac{1}{4}\right)^2$ 64. $12\left(\frac{3}{4}\right)^2$ 65. $\left(\frac{2}{3}\right)^2 \cdot \left(\frac{3}{8}\right)^2$

66. $\frac{7}{8} \div \left(\frac{1}{8} + \frac{3}{4}\right)$ 67. $\left(\frac{1}{2}\right)^2 \cdot \left(\frac{1}{4} + \frac{1}{2}\right)$ 68. $\left(\frac{1}{4}\right)^3 + \left(\frac{5}{8} + \frac{3}{4}\right)$

MIXED REVIEW EXERCISES

Simplify by using the order of operations as necessary. Write answers in lowest terms and as whole or as mixed numbers when possible.

69. $\frac{7}{8} - \frac{1}{8}$ 70. $\frac{7}{10} - \frac{3}{10}$ 71. $\frac{29}{32} - \frac{5}{16}$ 72. $\frac{1}{4} + \frac{1}{8} + \frac{5}{16}$

73. $\begin{array}{r} 6\frac{2}{3} \\ -4\frac{1}{2} \\ \hline \end{array}$ 74. $\begin{array}{r} 9\frac{1}{2} \\ +16\frac{3}{4} \\ \hline \end{array}$ 75. $\begin{array}{r} 7 \\ -1\frac{5}{8} \\ \hline \end{array}$ 76. $\begin{array}{r} 2\frac{3}{5} \\ 8\frac{5}{8} \\ +\frac{5}{16} \\ \hline \end{array}$

77. $\begin{array}{r} 32\frac{5}{12} \\ -17 \\ \hline \end{array}$ 78. $\frac{7}{22} + \frac{3}{22} + \frac{3}{11}$ 79. $\left(\frac{1}{4}\right)^2 \cdot \left(\frac{2}{5}\right)^3$

Mixed Review Exercises require students to solve problems without the help of section reference.

Cumulative Review Exercises
CHAPTERS 1-3

For each number, name the digit that has the given place value.

1. 871
   hundreds
   ones

2. 5,629,428
   millions
   thousands

Round each number to the nearest ten, nearest hundred, and nearest thousand.

	Ten	Hundred	Thousand
3. 1438	_____	_____	_____
4. 59,803			

≈ *Use front end rounding to estimate each answer. Then add, subtract, multiply, or divide to find the exact answer.*

5. *Estimate:*   *Exact:*
   Rounds to   2361
   Rounds to   386
   Rounds to   47,304
   Rounds to   + 29,728
   +

6. *Estimate:*   *Exact:*
       24,276
   −   − 9 887

7. *Estimate:*   *Exact:*
       4468
   ×   × 280

8. *Estimate:*   *Exact:*
   $\overline{)}$ $35\overline{)112,385}$

Add, subtract, multiply, or divide as indicated.

9. $\begin{array}{r} 4 \\ 8 \\ 5 \\ +9 \\ \hline \end{array}$ 10. $\begin{array}{r} 375,899 \\ 521,742 \\ +357,968 \\ \hline \end{array}$ 11. $\begin{array}{r} 1687 \\ -1096 \\ \hline \end{array}$ 12. $\begin{array}{r} 3,896,502 \\ -1,094,807 \\ \hline \end{array}$

13. $3 \times 8 \times 5$ 14. $6 \cdot 3 \cdot 7$ 15. $5(8)(4)$

Cumulative Review Exercises gather various types of exercises from preceding chapters to help students remember and retain what they are learning throughout the course.

To the Student: Success in Mathematics

There are two main reasons students have difficulty with mathematics:

- Students start in a course for which they do not have the necessary background knowledge.

- Students don't know how to study mathematics effectively.

Your instructor can help you decide whether this is the right course for you. We can give you some study tips listed below.

Studying mathematics *is* different from studying subjects like English and history. The key to success is regular practice. This should not be surprising. After all, can you learn to play the piano or ski well without a lot of regular practice? The same is true for learning mathematics. Working problems nearly every day is the key to becoming successful. Here are some ideas to help you succeed in studying mathematics.

1. *Attend class regularly.* Pay attention to what your instructor says and does in class, and take careful notes. In particular, note the problems the instructor works on the board and copy the complete solutions. Keep these notes separate from your homework to avoid confusion when you review them later.

2. Don't hesitate to *ask questions in class.* It is not a sign of weakness but of strength. There are always other students with the same question who are too shy to ask.

3. *Read your text carefully.* Many students read only enough to get by, usually only the examples. Reading the complete section will help you solve the homework problems. Most exercises are keyed to specific examples or objectives that will explain the procedures for working them.

4. Before you start on your homework assignment, *rework the problems the teacher worked in class.* This will reinforce what you have learned. Many students say, "I understand it perfectly when you do it in class, but I get stuck when I try to do the homework problems by myself."

5. Do your homework assignment only *after reading the text* and reviewing your notes from class. Check your homework against the answers in the back of the book. If you get a problem wrong and are unable to understand why, mark that problem and ask your instructor about it. Then practice working additional problems of the same type to reinforce what you have learned.

6. *Work as neatly as you can.* Write your symbols clearly, and make sure the problems are clearly separated from each other. Working neatly will help you to think clearly and also make it easier to review the homework before a test.

7. After you complete a homework assignment, *look over the text again.* Try to identify the main ideas that are in the lesson. Often they are clearly highlighted or boxed in the text.

8. *Use the chapter test at the end of each chapter as a practice test.* Work through the problems under test conditions, without referring to the text or the answers until you are finished. You may want to time yourself to see how long it takes you. When you finish, check your answers against those in the back of the book, and rework the problems you missed.

9. *Keep all quizzes and tests that are returned to you,* and use them when you study for future tests and the final exam. These quizzes and tests indicate what concepts your instructor considers to be most important. Be sure to correct any problems that you missed on these tests, so you will have the corrected work to study.

10. *Don't worry if you do not understand a new topic right away.* As you read more about it and work through the problems, you will gain understanding. Each time you review a topic you will understand it a little better. Few people understand each topic completely right from the start.

NOTE

Reading a list of study tips is a good start, but you may need some help actually *applying* the tips to your work in this math course.

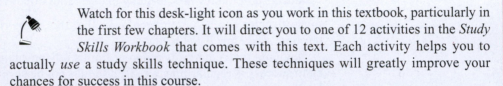 Watch for this desk-light icon as you work in this textbook, particularly in the first few chapters. It will direct you to one of 12 activities in the *Study Skills Workbook* that comes with this text. Each activity helps you to actually *use* a study skills technique. These techniques will greatly improve your chances for success in this course.

- Find out *how your brain learns new material.* Then use that information to set up effective ways to learn math.

- Find out *why short-term memory is so short* and what you can do to help your brain remember new material weeks and months later.

- Find out *what happens when you "blank out" on a test* and simple ways to prevent it from happening.

All the activities in the *Study Skills Workbook* are brain-friendly ways to enjoy and succeed at math. Whether you need help with note taking, managing homework, taking tests, or preparing for a final exam, you'll find specific, clearly explained ideas that really work because they're based on research about how the brain learns and remembers.

1392

Diagnostic Pretest

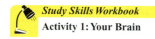

[Chapter 1]

1. Use digits to write eighty-nine million, twenty-three thousand, five hundred seven.

1. $89,023,507$

2. Subtract.
$$\begin{array}{r} 7009 \\ -\ 2678 \\ \hline 4331 \end{array}$$

2. 4331

3. Divide. $20,213 \div 29$

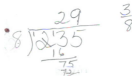

3. 697

4. Round 89,658 to the nearest thousand.

$90,000$

4. $90,000$

[Chapter 2]

5. Write $\dfrac{235}{8}$ as a mixed number.

$$8\overline{)235}\quad \frac{3}{8}$$
$$\begin{array}{r}29\\ \underline{16}\\ 75\\ \underline{72}\end{array}$$

5. $29\frac{3}{8}$

6. Write the prime factorization of 392 using exponents.

6. _____

7. A cake recipe calls for $2\frac{1}{4}$ cups of flour. How much flour is needed to make 5 cakes?

$2\frac{1}{4} \cdot 5 \qquad \frac{9}{4} \times \frac{5}{1} = \frac{45}{4}\quad 11\frac{1}{4}$

7. $11\frac{1}{4}$

≈ 8. First estimate the answer. Then find the exact answer. Write the exact answer as a mixed number.

$$5\frac{3}{4} \cdot 2\frac{1}{8}$$

8. *Estimate:* _____

Exact: _____

[Chapter 3]

9. Find the least common multiple of 5, 8, 12, and 30.

$2\times5\times2\times3\times2$
$10\times2\times3\times2$
$20\times3\times2$
60×2
120

9. 120

10. Solve. Write your answer in lowest terms.

$$\frac{9}{10} - \frac{1}{15}$$

10. $\dfrac{5}{6}$

≈ 11. First estimate the answer. Then find the exact answer. Write the exact answer as a mixed number.

$$8\frac{2}{9} + 12\frac{5}{6}$$

11. *Estimate:* _____

Exact: _____

12. Use the order of operations to simplify.

$$\frac{9}{16} + \frac{1}{6}\left(\frac{3}{4}\right)^2 + \left(\frac{5}{6} - \frac{2}{3}\right)$$

12. $\dfrac{35}{48}$

13. _____1.4_____

14. _____

15. _____

16. _____

17. _____

18. _____

19. _____

20. _____

21 _____

22. _____

23. _____

24. _____

[Chapter 4]

13. Round $1.3852 to the nearest cent.

14. Write $6\frac{5}{9}$ as a decimal. Round to the nearest thousandth if necessary.

15. Use the order of operations to simplify.
$$4.5^2 - 3.2 + 0.6(12)$$
PEMDAS

16. Find the cost (to the nearest cent) of 5.3 pounds of chicken at $1.59 per pound.

[Chapter 5]

17. Write the ratio "50 minutes to 4 hours" as a fraction in lowest terms. Change to the same units if necessary.

18. Determine whether the proportion is true or false.
$$\frac{16.2}{23.6} = \frac{5.4}{8}$$

19. Find the unknown number in the proportion. Write your answer as a mixed number.
$$\frac{2\frac{1}{2}}{x} = \frac{\frac{3}{4}}{8}$$

20. Martha earns $59.50 in 7 hours. How much will she earn in 40 hours?

[Chapter 6]

Write each number as a percent.

21. 0.582

22. $8\frac{3}{4}$

23. The price of a cell phone is $128 plus 6% sales tax. Find the total cost of the phone including sales tax.

24. Roberto's annual salary was $24,500. After his first year on the job, he received a raise to $26,215. Find the percent of increase.

[Chapter 7]
Convert each measurement.

25. (a) 36 quarts to gallons

    (b) $4\frac{3}{4}$ pounds to ounces

26. (a) 0.675 meter to centimeters

    (b) 4528 grams to kilograms

27. Jonathan weighed 3 kg 170 g at birth. When he was 5 months old, he weighed 6 kg 90 g. How much weight had he gained, in kilograms?

28. Write the most reasonable metric unit in each blank. Choose from km, m, cm, mm, L, mL, kg, g, and mg.

    (a) Leanne took a 325 _____ pill for her headache.

    (b) The cover of this textbook is about 21 _____ wide.

25. (a) _____

    (b) _____

26. (a) _____

    (b) _____

27. _____

28. (a) _____

    (b) _____

[Chapter 8]

29. Find the perimeter.

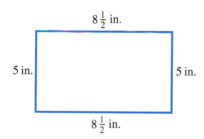

30. Find the area.

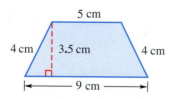

31. Find the volume. Use 3.14 as the approximate value for π.

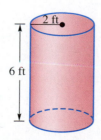

29. _____

30. _____

31. _____

32. Find the unknown length.

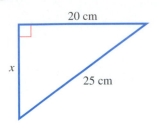

20 cm

x

25 cm

32. _____

[Chapter 9]

33. _____

33. Simplify. $12 - (-13)$

34. _____

34. Simplify. $(-0.4)(11.5)$

35. _____

35. Find the value of $7p - 8q$ if $p = -3$ and $q = -4$.

36. _____

36. Solve. $8r + 9 = 2r - 3$

[Chapter 10]

37. _____

37. The circle graph to the right shows what parents say stresses them out the most about the beginning of a new school year. If the total number of parents in the survey was 6800, how may said "peer influence on clothing choices" was most stressful?

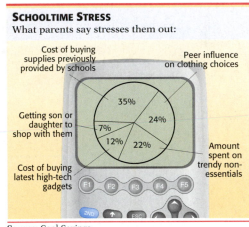

SCHOOLTIME STRESS
What parents say stresses them out:

Cost of buying supplies previously provided by schools

Peer influence on clothing choices

Getting son or daughter to shop with them

Cost of buying latest high-tech gadgets

Amount spent on trendy non-essentials

35%   24%   7%   12%   22%

Source: Cool Savings.

38. _____

38. Find the mean age of these members of an extended family. Round your answer to the nearest tenth, if necessary.

15, 52, 16, 51, 29, 55, 26, 85, 29, 55

39. _____

39. Find the weighted mean.

Round your answer to the nearest tenth, if necessary.

Quiz Score	Frequency
10	3
9	8
8	10
7	9
6	4
5	1

40. _____

40. Find the median for these hourly wages:

$7.50, $6.25, $9.80, $8.10, $13.75, $11.00

Whole Numbers

1

The little red wagon that many of us remember from childhood was probably a Radio Flyer. The company, known today as Radio Flyer, Inc., first began making wagons in 1917. Children are still enjoying the 50 different products manufactured by the company today. The products range from a 40-inch wood wagon with air tires that sells for $150 to a 25-cent key chain. Each day, one hundred employees at the Chicago plant are capable of making 8000 wagons. In this chapter, we discuss whole numbers, which are used daily in all our lives. (See Exercise 39 in **Section 1.6** and Example 2 and Exercises 1 and 2 in **Section 1.10**.)

1.1 Reading and Writing Whole Numbers

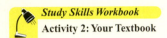

Study Skills Workbook
Activity 2: Your Textbook

1 Identify the place value of the 4 in each whole number.

(a) 341

(b) 714

(c) 479

Knowing how to read and write numbers is an important step in learning mathematics.

OBJECTIVE 1 **Identify whole numbers.** The **decimal system** of writing numbers uses the ten digits

$$0, 1, 2, 3, 4, 5, 6, 7, 8, 9$$

to write any number. For example, these digits can be used to write **the whole numbers:**

$$0, 1, 2, 3, 4, 5, 6, 7, 8, 9, 10, 11, 12, 13 \ldots$$

The three dots indicate that the list goes on forever.

OBJECTIVE 2 **Give the place value of a digit.** Each digit in a whole number has a **place value,** depending on its position in the whole number. The following place value chart shows the names of the different places used most often and has the whole number 554,170 entered.

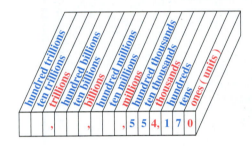

China has 554,170 fast-food restaurants, more than any other country, and three times as many as the United States. (*Source:* Euromonitor.)

China has 554,170 fast-food restaurants, the greatest number of any country in the world. Each of the 5s in 554,170 represents a different amount because of its position, or **place value** within the number. The **place value** of the 5 on the left is 5 hundred thousands (500,000). The **place value** of the 5 on the right is 5 ten thousands (50,000).

EXAMPLE 1 **Identifying Place Values**

Identify the place value of 8 in each whole number.

(a) 28   **(b)** 85   **(c)** 869

    8 ones   8 tens   8 hundreds

Notice that the value of 8 in each number is different, depending on its location (place) in the number.

Work Problem 1 at the Side.

EXAMPLE 2 **Identifying Place Values**

Identify the place value of each digit in the number 725,283.

7 2 5 , 2 8 3
                    3 ones
                    8 tens
                    2 hundreds
                    5 thousands
                    2 ten thousands
                    7 hundred thousands

ANSWERS

1. **(a)** tens   **(b)** ones   **(c)** hundreds

Notice the comma between the hundreds and thousands position in the number 725,283 in Example 2.

Work Problem 2 at the Side. ▶▶▶

Using Commas

Commas are used to separate each group of three digits, starting from the right. This makes numbers easier to read. (An exception: Commas are frequently omitted in four-digit numbers such as 9748 or 1329.) Each three-digit group is called a **period.** Some instructors prefer to just call them **groups.**

EXAMPLE 3 Knowing the Period or Group Names

Write the digits in each period of 8,321,456,795.

8,321,456,795

8 billions ←
321 millions ←
456 thousands ←
795 ones ←

Work Problem 3 at the Side. ▶▶▶

Use the following rule to read a number with more than three digits.

Writing Numbers in Words

Start at the left when writing a number in words or saying it aloud. Write or say the digit names in each period (group), followed by the name of the period, except for the period name "ones," which is *not* used.

OBJECTIVE 3 **Write a number in words or digits.** The following examples show how to write names for whole numbers.

EXAMPLE 4 Writing Numbers in Words

Write each number in words.

(a) 57

This number means 5 tens and 7 ones, or 50 ones and 7 ones. Write the number as

fifty-seven.

Continued on Next Page

---

2 Identify the place value of each digit.

(a) 14,218

(b) 460,329

3 In the number 3,251,609,328 identify the digits in each period (group).

(a) billions period

(b) millions period

(c) thousands period

(d) ones period

ANSWERS

2. **(a)** 1 : ten thousands
   4 : thousands
   2 : hundreds
   1 : tens
   8 : ones
   (b) 4 : hundred thousands
   6 : ten thousands
   0 : thousands
   3 : hundreds
   2 : tens
   9 : ones
3. **(a)** 3   **(b)** 251   **(c)** 609   **(d)** 328

4 Write each number in words.

(a) 18

(b) 36

(c) 418

(d) 902

5 Write each number in words.

(a) 3104

(b) 95,372

(c) 100,075,002

(d) 11,022,040,000

6 Rewrite each number using digits.

(a) one thousand, four hundred thirty-seven

(b) nine hundred seventy-one thousand, six

(c) eighty-two million, three hundred twenty-five

(b) 94

> ninety-four

(c) 874

> eight hundred seventy-four

(d) 601

> six hundred one

‹‹‹ Work Problem 4 at the Side.

CAUTION
The word *and* should never be used when writing whole numbers. You will often hear someone say "five hundred *and* twenty-two," but the use of "and" is not correct since "522" is a whole number. When you work with decimal numbers, the word *and* is used to show the position of the decimal point. For example, 98.6 is read as "ninety-eight *and* six tenths." Practice with decimal numbers is the topic of **Section 4.1.**

EXAMPLE 5 **Writing Numbers in Words by Using Period Names**

Write each number in words.

(a) 725,283

seven hundred twenty-five **thousand,** two hundred eighty-three

Number in period | Name of period | Number in period (not necessary to write "ones")

(b) 7252

seven **thousand,** two hundred fifty-two

Name of period | No period name needed

(c) 111,356,075

one hundred eleven **million,** three hundred fifty-six **thousand,** seventy-five

(d) 17,000,017,000

seventeen **billion,** seventeen **thousand**

‹‹‹ Work Problem 5 at the Side.

EXAMPLE 6 **Writing Numbers in Digits**

Rewrite each number using digits.

(a) six **thousand,** twenty-two

6022

(b) two hundred fifty-six **thousand,** six hundred twelve

256,612

(c) nine **million,** five hundred fifty-nine

9,000,559

Zeros indicate there are no thousands.

‹‹‹ Work Problem 6 at the Side.

▦ **Calculator Tip**   Does your calculator show a comma between each group of three digits? Probably not, but try entering a long number such as 34,629,075. Notice that there is no key with a comma on it, so you do not enter commas. A few calculators may show the position of the commas *above* the digits, like this ⬇ ⬇

> **34'629'075**

Most of the time you will have to write in the commas where needed.

OBJECTIVE **4** **Read a table.**   A common way of showing number values is by using a **table.** Tables organize and display facts so that they are more easily understood. The following table shows some past facts and future predictions for the United States.

NUMBERS FOR THE 21ST CENTURY
These estimated numbers give us a glimpse of what we can expect in the 21st century.

Year	1990s	2005	2020
U.S. population	261 million	296 million	338 million
Births	16 million	14 million	13 million
Household income	$42,936	$47,808	$53,375
Average salary	$21,129	$24,861	$28,050

Source: Family Circle magazine; World Almanac.

If you read from left to right along the row labeled "U.S. population," you find that the population in 1990 was 261 million, then the population in 2005 was 296 million, and the estimated population for 2020 is 338 million.

EXAMPLE 7 **Reading a Table**

Use the table to find each number, and write the number in words.

(a) The predicted household income in the year 2020
    Read from left to right along the row labeled "Household income" until you reach the 2020 column and find $53,375.

> Fifty-three thousand, three hundred seventy-five dollars

(b) The average salary in 1990.
    Read from left to right along the row labeled "Average salary." In the 1990 column you find $21,129.

> Twenty-one thousand, one hundred twenty-nine dollars.

Work Problem 7 at the Side. ▶▶▶

NOTE
Notice in Example 7 that hyphens are used when writing numbers in words. A hyphen is used when writing the numbers 21 through 99 (twenty-one through ninety-nine), except for numbers ending in zero (20, 30, 40, ... 90).

7 Use the table to find each number, and write the number in digits, or write the number in words.

(a) The number of births in 2005

(b) The predicted number of births in 2020

(c) Household income in 1990

(d) The predicted average salary in 2020

Real-Data Applications

'Til Debt Do You Part!

Ladies Home Journal recently released statistics showing the average wedding costs in the United States in 2004. In 2001, the cost of the average wedding was $20,357. The average number of wedding guests has grown to over 200.

Category	Average Cost in 2004
Misc. expenses (stationery, clergy, gifts, limousine)	$1583
Bouquets and other flowers	$ 950
Photography and videography	$1647
Music	$1042
Engagement and wedding rings (bride and groom)	$5099
Rehearsal dinner	$ 877
Bride's wedding dress and headpiece	$1242
Bridal attendants' dresses (average of 5 bridesmaids)	$ 992
Mother of the bride's apparel	$ 297
Men's formalwear (ushers, best man, groom)	$ 849
Wedding reception	$9590
Grand Total	

Source: Ladies Home Journal; Association of Bridal Consultants.

1. What is the grand total of expenses shown in the chart?

2. How much more expensive was a wedding in 2004 compared to 2001?

3. The groom pays for the bouquets and flowers, the rehearsal dinner, the bride's engagement and wedding rings ($4400), the clergy ($300), and the groom's formalwear ($125). What is the total amount spent by the groom?

4. If you budgeted $45 per person for the wedding reception and you invited 210 guests to a wedding in 2005, how much money would you have spent compared to the average 2004 wedding reception costs?

5. If you budget $7200 for the reception and the caterer charges $45 per person, how many guests can you invite? How much of your budget is left over?

6. If you budget $10,500 for the reception and the caterer charges $45 per person, how many guests can you invite? How much of your budget is left over?

7. What type of arithmetic problem did you work to get the answers to Problem 6? What is the mathematical term for the "left over" budget?

1.1 Exercises

FOR EXTRA HELP

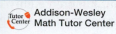 Addison-Wesley Math Tutor Center

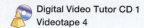 MathXL

 Digital Video Tutor CD 1 Videotape 4

Student's Solutions Manual

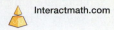 MyMathLab

Interactmath.com

Study Skills Workbook
Activity 3: Homework

*Write the digit for the given **place value** in each whole number. See Examples 1 and 2.*

1. 3065
   thousands *3*
   tens *6*

2. 4681
   thousands *4*
   ones *1*

3. 18,015
   ten thousands *1*
   hundreds *0*

4. 86,332
   ten thousands
   ones

5. 7,628,592,183
   millions *8 2 8*
   thousands *2*

6. 1,700,225,016
   billions *700*
   millions *225*

*Write the digits for the given **period** (group) in each whole number. See Example 3.*

7. 3,561,435
   millions *3*
   thousands *561*
   ones *435*

8. 28,785,203
   millions
   thousands
   ones

9. 60,000,502,109
   billions *60*
   millions *0*
   thousands *502*
   ones *109*

10. 100,258,100,006
   billions
   millions
   thousands
   ones

11. Do you think the fact that humans have four fingers and a thumb on each hand explains why we use a number system based on ten digits? Explain.

Yes—there is evidence that this is true. It is common to use fingers when counting.

12. The decimal system uses ten digits. Fingers and toes are often referred to as digits. In your opinion, is there a relationship here? Explain.

Write each number in words. See Examples 4 and 5.

13. 23,115

14. 37,886

15. 346,009

16. 218,033

17. 25,756,665

18. 999,993,000

Write each number using digits. See Example 6.

19. sixty-three thousand, one hundred sixty-three

20. ninety-five thousand, one hundred eleven

21. ten million, two hundred twenty-three

22. one hundred million, two hundred

Write the numbers from each sentence using digits. See Example 6.

23. There are three million, two hundred thousand parachute jumps in the United States each year. (*Source:* History Channel.)

24. The United States Postal Service set a record of two hundred eighty million, four hundred eighty-nine thousand postmarked pieces of mail on a single day. (*Source:* U.S. Postal Service.)

25. The Binney & Smith Company in Pennsylvania makes about two billion Crayola crayons each year. (*Source:* Binney & Smith Company.)

26. Walt Disney Company spent one billion, seven hundred fifty-seven million, five hundred thousand dollars on advertising in one year. (*Source:* Crain Communications Inc.; *Advertising Age.*)

27. There are eight hundred fifty-four thousand, seven hundred ninety-five boxes of Jell-O gelatin sold each day. (*Source: USA Today.*)

28. Americans consume three hundred fifty-three million cups of coffee each day. (*Source: Parade* magazine.)

29. Rewrite eight hundred trillion, six hundred twenty-one million, twenty thousand, two hundred fifteen by using digits.

800,000,621,020,215
T   B   M   Th Tens

30. Rewrite 70,306,735,002,102 in words

The table at the right shows various ways people get to work. Use the table to answer Exercises 31–34. See Example 7.

31. Which method of transportation is least used? Write the number in words.

32. Which method of transportation is most used? Write the number in words.

33. Find the number of people who walk to work or work at home, and write it in words.

34. Find the number of people who carpool, and write in words.

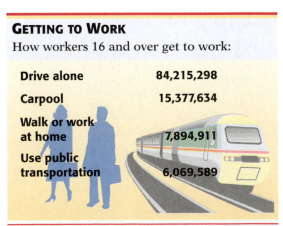

GETTING TO WORK
How workers 16 and over get to work:

Drive alone	84,215,298
Carpool	15,377,634
Walk or work at home	7,894,911
Use public transportation	6,069,589

Source: U.S. Bureau of the Census.

Add + Add = Sum

1.2 Adding Whole Numbers

There are four dollar bills at the left and two at the right. In all, there are six dollar bills.

The process of finding the total is called **addition.** Here 4 and 2 were added to get 6. Addition is written with a + sign, so that

$$4 + 2 = 6.$$

OBJECTIVE 1 Add two single-digit numbers.   In addition, the numbers being added are called **addends,** and the resulting answer is called the **sum** or **total.**

$$
\begin{array}{r}
4 \quad \leftarrow \text{Addend} \\
+\ 2 \quad \leftarrow \text{Addend} \\
\hline
6 \quad \leftarrow \text{Sum (total)}
\end{array}
$$

Addition problems can also be written horizontally, as follows.

$$
\begin{array}{ccccc}
4 & + & 2 & = & 6 \\
\uparrow & & \uparrow & & \uparrow \\
\text{Addend} & \text{Addend} & & \text{Sum}
\end{array}
$$

Commutative Property of Addition

By the **commutative property of addition,** changing the order of the addends in an addition problem does not change the sum.

For example, the sum of $4 + 2$ is the same as the sum of $2 + 4$. This allows the addition of the same numbers in a different order.

EXAMPLE 1 Adding Two Single-Digit Numbers

Add, and then change the order of numbers to write another addition problem.

(a) $5 + 3 = 8$ and $3 + 5 = 8$

(b) $7 + 8 = 15$ and $8 + 7 = 15$

(c) $8 + 3 = 11$ and $3 + 8 = 11$

(d) $8 + 8 = 16$

Work Problem 1 at the Side. ▶▶▶

Associative Property of Addition

By the **associative property of addition,** changing the grouping of the addends in an addition problem does not change the sum.

For example, the sum of $3 + 5 + 6$ may be found as follows.

$$(3 + 5) + 6 = 8 + 6 = 14 \qquad \text{Parentheses tell what to do first.}$$

Another way to add the same numbers is

$$3 + (5 + 6) = 3 + 11 = 14.$$

Either grouping gives the answer 14.

OBJECTIVES

1 Add two single-digit numbers.

2 Add more than two numbers.

3 Add when carrying is not required.

4 Add with carrying.

5 Solve application problems with carrying.

6 Check the answer in addition.

❶ Add, and then change the order of numbers to write another addition problem.

(a) $2 + 6$

(b) $9 + 5$

(c) $7 + 8$

(d) $6 + 9$

ANSWERS
1. **(a)** $8; 6 + 2 = 8$ **(b)** $14; 5 + 9 = 14$
    (c) $15; 8 + 7 = 15$ **(d)** $15; 9 + 6 = 15$

2 Add each column of numbers.

(a)
```
   3
   8
   5
   4
 + 6
```

(b)
```
   5
   6
   3
   2
 + 4
```

(c)
```
   9
   6
   8
   7
 + 3
```

(d)
```
   3
   8
   6
   4
 + 8
```

OBJECTIVE 2 **Add more than two numbers.**   To add several numbers, first write them in a column. Add the first number to the second. Add this sum to the third digit; continue until all the digits are used.

EXAMPLE 2 **Adding More Than Two Numbers**

Add 2, 5, 6, 1, and 4.

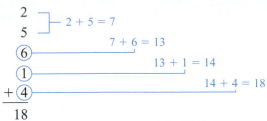

Work Problem 2 at the Side.

NOTE
By the commutative and associative properties of addition, numbers may also be added starting at the bottom of a column. Adding from the top or adding from the bottom will give the same answer.

OBJECTIVE 3 **Add when carrying is not required.**   If numbers have two or more digits, you must arrange the numbers in columns so that the ones digits are in the same column, tens are in the same column, hundreds are in the same column, and so on. Next, you add column by column starting at the right.

EXAMPLE 3 **Adding without Carrying**

Add $511 + 23 + 154 + 10$.

First line up the numbers in columns, with the ones column at the right.

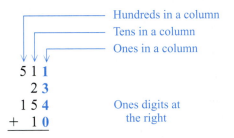

Now start at the right and add the ones digits. Add the tens digits next, and finally, the hundreds digits.

```
   5 1 1
     2 3
   1 5 4
 +   1 0
   6 9 8
```

Sum of ones
Sum of tens
Sum of hundreds

The sum of the four numbers is 698.

Work Problem 3 at the Side. ▶▶▶

OBJECTIVE **4** **Add with carrying.** If the sum of the digits in any column is more than 9, use **carrying.**

EXAMPLE 4 **Adding with Carrying**

Add 47 and 29.

Add ones.

$$\begin{array}{r} 47 \\ +\ 29 \\ \end{array}$$

Sum of ones is 16.

Because 16 is 1 ten plus 6 ones, write 6 in the ones column and carry 1 to the tens column.

$$\begin{array}{r} 1 \\ 47 \\ +\ 29 \\ \hline 6 \\ \end{array} \qquad 7 + 9 = 16$$

Add the tens column, including the carried 1.

$$\begin{array}{r} 1 \\ 47 \\ +\ 29 \\ \hline 76 \\ \end{array}$$

Sum of digits in tens column

Work Problem 4 at the Side. ▶▶▶

EXAMPLE 5 **Adding with Carrying**

Add 324 + 7855 + 23 + 7 + 86.

Add the digits in the ones column.

$$\begin{array}{r} 2 \\ 324 \\ 7855 \\ 23 \\ 7 \\ +\ 86 \\ \hline 5 \\ \end{array}$$

Carry 2 to the tens column.

Sum of the ones column is 25.

Write 5 in the ones column.

In 25, the 5 represents 5 ones and is written in the ones column, while 2 represents 2 tens and is carried to the tens column.

Now add the digits in the tens column, including the carried 2.

$$\begin{array}{r} 1\ 2 \\ 324 \\ 7855 \\ 23 \\ 7 \\ +\ 86 \\ \hline 95 \\ \end{array}$$

Carry 1 to the hundreds column.

Sum of the tens column is 19. Write 9 in the tens column.

Continued on Next Page

---

3 Add.

(a) $\begin{array}{r} 26 \\ +\ 73 \end{array}$

(b) $\begin{array}{r} 534 \\ +\ 265 \end{array}$

(c) $\begin{array}{r} 42{,}305 \\ +\ 11{,}563 \end{array}$

4 Add by carrying.

(a) $\begin{array}{r} 66 \\ +\ 27 \end{array}$

(b) $\begin{array}{r} 58 \\ +\ 33 \end{array}$

(c) $\begin{array}{r} 56 \\ +\ 37 \end{array}$

(d) $\begin{array}{r} 34 \\ +\ 49 \end{array}$

ANSWERS
3. (a) 99 (b) 799 (c) 53,868
4. (a) 93 (b) 91 (c) 93 (d) 83

5 Add by carrying as necessary.

(a)
```
    42
   651
   396
 +  87
_____
```

(b)
```
   162
  4271
   372
 + 8976
_____
```

(c)
```
    57
     4
   392
   804
    51
 +  27
_____
```

(d)
```
  7821
   435
    72
   305
 + 1693
_____
```

6 Add by carrying mentally.

(a)
```
   816
   363
    17
     2
     5
 + 7654
_____
```

(b)
```
  3305
   650
   708
    29
    40
     6
 +   3
_____
```

(c)
```
 15,829
    765
     78
     15
      9
      7
 + 13,179
_____
```

Add the hundreds column, including the carried 1.

```
  1 12
   324
  7855
    23
     7
 +  86
_____
   295
```

Carry 1 to the thousands column.

Sum of the hundreds column is 12.

Write 2 in the hundreds column.

Add the thousands column, including the carried 1.

```
  112
   324
  7855
    23
     7
 +  86
_____
  8295
```

Sum of the thousands column is 8.

Finally, 324 + 7855 + 23 + 7 + 86 = 8295.

Work Problem 5 at the Side.

NOTE

For additional speed, try to carry mentally. Do not write the carried number, but just remember it as you move to the top of the next column. Try this method. If it works for you, use it.

Work Problem 6 at the Side.

OBJECTIVE 5 Solve application problems with carrying. In **Section 1.10** we will describe how to solve application problems in more detail. The next two examples are application problems that require adding.

EXAMPLE 6 Applying Addition Skills

On this map of the Walt Disney World area in Florida, the distance in miles from one location to another is written alongside of the road. Find the shortest route from Altamonte Springs to Clear Lake.

Continued on Next Page

Approach   Add the mileage along various routes to determine the distances from Altamonte Springs to Clear Lake. Then select the shortest route.

Solution   One way from Altamonte Springs to Clear Lake is through Orlando. Add the mileage numbers along this route.

8	Altamonte Springs to Pine Hills
5	Pine Hills to Orlando
+ 8	Orlando to Clear Lake
21 →	miles from Altamonte Springs to Clear Lake, going through Orlando

Another way is through Castleberry, Bertha, and Winter Park. Add the mileage numbers along this route.

5	Altamonte Springs to Castleberry
6	Castleberry to Bertha
7	Bertha to Winter Park
+ 7	Winter Park to Clear Lake
25 →	miles from Altamonte Springs to Clear Lake through Bertha and Winter Park

The shortest route from Altamonte Springs to Clear Lake is through Orlando.

Work Problem 7 at the Side. ▶▶▶

EXAMPLE 7 **Finding a Total Distance**

Use the map in Example 6 to find the total distance from Shadow Hills to Castleberry to Orlando and back to Shadow Hills.

Approach   Add the mileage from Shadow Hills to Castleberry to Orlando and back to Shadow Hills to find the total distance.

Solution   Use the numbers from the map.

9	Shadow Hills to Bertha
6	Bertha to Castleberry
5	Castleberry to Altamonte Springs
8	Altamonte Springs to Pine Hills
5	Pine Hills to Orlando
8	Orlando to Clear Lake
+ 11	Clear Lake to Shadow Hills
52 →	miles from Shadow Hills to Castleberry to Orlando and back to Shadow Hills

Work Problem 8 at the Side. ▶▶▶

EXAMPLE 8 **Finding a Perimeter**

Find the number of feet of fencing needed to enclose the college parking lot shown below.

The short way to write *feet* is *ft*.

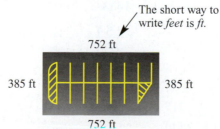

752 ft

385 ft   385 ft

752 ft

Approach   Find the **perimeter,** or total distance around the lot, by adding the lengths of all the sides.

Continued on Next Page

---

7 Use the map in Example 6 to find the shortest route from Lake Buena Vista to Conway.

8 The road is closed between Orlando and Clear Lake, so this route cannot be used. Use the map in Example 6 to find the next shortest route from Orlando to Clear Lake.

9 Solve the problem. Find the number of feet of fencing needed to enclose the park shown.

818 ft

348 ft   348 ft

818 ft

Solution   Use the lengths shown.

$$752$$
$$385$$
$$752$$
$$+\ 385$$
$$\overline{2274 \text{ ft}}$$

The amount of fencing needed is 2274 ft, which is the perimeter of (distance around) the lot.

◄◄◄ Work Problem 9 at the Side.

OBJECTIVE 6   Check the answer in addition.   Checking the answer is an important part of problem solving. A common method for checking addition is to re-add from bottom to top. This is an application of the commutative and associative properties of addition.

10 Check the following additions. If an answer is incorrect, find the correct answer.

(a)   63
        4
        9
    + 28
    ──────
      104

(b)   927
        395
        64
    + 251
    ──────
       1637

(c)   79
        218
        7
    + 639
    ──────
       953

(d)   21,892
        11,746
    + 43,925
    ──────────
       79,563

EXAMPLE 9   Checking Addition

Check the following addition.

Add down.

$$\mathbf{1428} \nwarrow$$
$$738$$
$$63$$
$$125$$
$$17$$
$$+\ 485$$
$$\overline{\mathbf{1428}}$$

Adding down and adding up should give the same answer.

Add up.
Check.

Here the answers agree, so the sum is probably correct.

EXAMPLE 10   Checking Addition

Check the following additions. Are they correct?

(a)

$$785$$ $$\mathbf{1033} \nwarrow$$
$$63$$ $$785$$
$$+\ 185$$ $$63$$
$$\overline{1033}$$ $$+\ 185$$
              $$\overline{\mathbf{1033}}$$

Correct, because both answers are the same.

Add up.
Check.

(b)

$$635$$ $$\mathbf{2454} \nwarrow$$
$$73$$ $$635$$
$$831$$ $$73$$
$$+\ 915$$ $$831$$
$$\overline{2444}$$ $$+\ 915$$
              $$\overline{\mathbf{2444}}$$

Error, because the answers are different.

Add up.
Check.

Re-add to find that the correct sum is 2454.

◄◄◄ Work Problem 10 at the Side.

ANSWERS

9. 2332 ft
10. **(a)** correct   **(b)** correct
    (c) incorrect; should be 943
    (d) incorrect; should be 77,563

1.2 Exercises

FOR EXTRA HELP

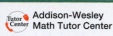 Addison-Wesley Math Tutor Center

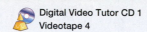 MathXL

 Digital Video Tutor CD 1 Videotape 4

Student's Solutions Manual

MyMathLab MyMathLab

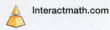 Interactmath.com

Add. See Examples 1–3.

1. 43 + 54	**2.** 18 + 11	**3.** 56 + 33	**4.** 83 + 15	**5.** 317 + 572

6. 574 + 325	**7.** 318 151 + 420	**8.** 135 253 + 410	**9.** 6310 252 + 1223	**10.** 121 5705 + 3163

11. 932 + 44 + 613

12. 517 + 131 + 250

13. 1251 + 4311 + 2114

14. 3241 + 1513 + 2014

15. 12,142 + 43,201 + 23,103

16. 41,124 + 12,302 + 23,500

17. 3213 + 5715

18. 6344 + 1655

19. 38,204 + 21,020

20. 63,251 + 36,305

Add, carrying as necessary. See Examples 4 and 5.

21. 87 + 63	**22.** 19 + 92	**23.** 86 + 69	**24.** 37 + 85	**25.** 47 + 74

26. 97 + 79	**27.** 67 + 78	**28.** 96 + 47	**29.** 73 + 29	**30.** 68 + 37

31. 746 + 905	**32.** 621 + 359	**33.** 306 + 848	**34.** 798 + 206	**35.** 278 + 135

36. 172 + 156	**37.** 928 + 843	**38.** 686 + 726	**39.** 526 + 884	**40.** 116 + 897

41.	3574 + 2817	42.	6871 + 7528	43.	7896 + 3728	44.	9382 + 7586	45.	9625 + 7986

46.	5718 5623 + 7436	47.	9056 78 6089 + 731	48.	4022 709 8621 + 37	49.	18 708 9286 + 636	50.	1708 321 61 + 8926

51.	422 6074 435 + 8663	52.	6505 173 7044 + 168	53.	321 9603 8 21 + 1604	54.	7631 5983 7 36 + 505	55.	2109 63 16 3 + 9887

56.	322 6508 93 745 18 + 2005	57.	553 97 2772 437 63 + 328	58.	3187 810 527 76 2665 + 317	59.	413 85 9919 602 31 + 1218	60.	576 7934 60 781 5968 + 371

Check each addition. If an answer is incorrect, find the correct answer. See Examples 9 and 10.

61. ____	62. ____	63. ____	64. ____	65. ____
832 468 + 791 2091	326 852 + 679 1857	179 214 + 376 759	17 296 713 + 94 1220	4713 28 615 + 64 5420

66. _____	67. _____	68. _____	69. _____	70. _____
3 628 72 564 + 7 319 11,583	678 7 952 56 718 + 2 173 11,377	516 8 760 24 189 + 1 723 11,212	4 714 27 77 8 878 + 636 14,332	6 715 283 9 617 13 + 81 16,719

71. Explain the commutative property of addition in your own words. How is this used when checking an addition problem?

72. Explain the associative property of addition. How can this be used when adding columns of numbers?

For Exercises 73–76, use the map to find the shortest route between each pair of cities.
See Examples 6 and 7.

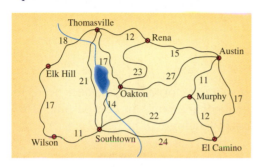

73. Southtown and Rena

74. Elk Hill and Oakton

75. Thomasville and Murphy

76. Murphy and Thomasville

Solve each application problem.

77. The Twin Lakes Food Bank raised $3482 at a flea market and $12,860 at their annual auction. Find the total amount raised at these two events.

78. A clothing store ordered 75 tops and 52 pairs of shorts. How many items were ordered?

79. There are 413 women and 286 men on the sales staff. How many people are on the sales staff?

80. One department in an office building has 283 employees while another department has 218 employees. How many employees are in the two departments?

81. A leaf blower costs $114 and a weed trimmer costs $98. Find the total cost for both tools.

82. The attendance at one NBA game is 23,821 fans. The next game has an attendance of 17,974 fans. Find the total attendance for the two games.

Solve each problem involving perimeter. See Example 8.

83. A concrete curb is to be built around a parking lot. How many feet of curbing will be needed? (Disregard the driveways.)

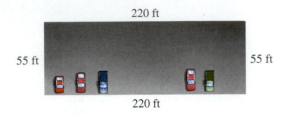

220 ft

55 ft   55 ft

220 ft

84. Because of heavy snowfall this winter, Maria needs to put new rain gutters around her entire roof. How many feet of gutters will she need?

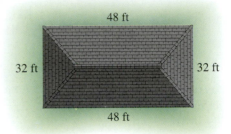

48 ft

32 ft   32 ft

48 ft

85. Martin plans to frame his back patio with redwood lumber. How many feet of lumber will he need?

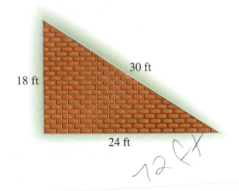

30 ft

18 ft

24 ft

86. Due to a recent tornado, Carl Jones needs to replace all the fencing around his cow pasture. How many meters of fencing will he need?

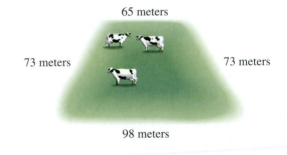

65 meters

73 meters   73 meters

98 meters

RELATING CONCEPTS (EXERCISES 87–94) For Individual or Group Work

Recall the place values of digits discussed in Section 1.1 and **work Exercises 87–94 in order.**

87. Write the largest four-digit number possible using the digits 4, 1, 9, and 2. Use each digit once.

88. Using the digits 4, 1, 9, and 2, write the smallest four-digit number possible. Use each digit once.

89. Write the largest five-digit number possible using the digits 6, 2, and 7. Use each digit at least once.

90. Using the digits 6, 2, and 7, write the smallest five-digit number possible. Use each digit at least once.

91. Write the largest seven-digit number possible using the digits 4, 3, and 9. Use each digit at least twice.

92. Using the digits 4, 3, and 9, write the smallest seven-digit number possible. Use each digit at least twice.

93. Explain your rule or procedure for writing the largest number in Exercise 91.

94. Explain your rule or procedure for writing the smallest number in Exercise 92.

1.3 Subtracting Whole Numbers

Suppose you have $9, and you spend $2 for parking. You then have $7 left. There are two different ways of looking at these numbers.

As an addition problem:

$$\$2 \quad + \quad \$7 \quad = \quad \$9$$

Amount spent   Amount left   Original amount

As a subtraction problem:

$$\$9 \quad - \quad \$2 \quad = \quad \$7$$

Original amount   Subtraction symbol   Amount spent   Amount left

OBJECTIVES

1. Change addition problems to subtraction and subtraction problems to addition.
2. Identify the minuend, subtrahend, and difference.
3. Subtract when no borrowing is needed.
4. Check subtraction answers by adding.
5. Subtract by borrowing.
6. Solve application problems with subtraction.

OBJECTIVE 1 Change addition problems to subtraction and subtraction problems to addition. As shown in the preceding box, an addition problem can be changed to a subtraction problem and a subtraction problem can be changed to an addition problem.

EXAMPLE 1 Changing Addition Problems to Subtraction

Change each addition problem to a subtraction problem.

(a) $4 + 1 = 5$

Two subtraction problems are possible:

$$5 - 1 = 4 \quad \text{or} \quad 5 - 4 = 1$$

These figures show each subtraction problem.

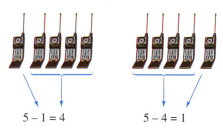

$$5 - 1 = 4 \qquad\qquad 5 - 4 = 1$$

(b) $8 + 7 = 15$

$$15 - 7 = 8 \quad \text{or} \quad 15 - 8 = 7$$

Work Problem 1 at the Side. ▶▶▶

EXAMPLE 2 Changing Subtraction Problems to Addition

Change each subtraction problem to an addition problem.

(a) $8 - 3 = 5$

$$8 = 3 + 5$$

It is also correct to write $8 = 5 + 3$.

— **Continued on Next Page**

1 Write two subtraction problems for each addition problem.

(a) $5 + 3 = 8$

$$8 - 3 = 5$$
$$8 - 5 = 3$$

(b) $7 + 4 = 11$

(c) $15 + 22 = 37$

(d) $23 + 55 = 78$

2 Write an addition problem for each subtraction problem.

(a) $7 - 5 = 2$

(b) $9 - 4 = 5$

(c) $21 - 15 = 6$

(d) $58 - 42 = 16$

3 Subtract.

(a)
$$\begin{array}{r} 74 \\ -\ 43 \\ \hline \end{array}$$

(b)
$$\begin{array}{r} 68 \\ -\ 24 \\ \hline \end{array}$$

(c)
$$\begin{array}{r} 429 \\ -\ 318 \\ \hline \end{array}$$

(d)
$$\begin{array}{r} 3927 \\ -\ 2614 \\ \hline \end{array}$$

(e)
$$\begin{array}{r} 5464 \\ -\ 324 \\ \hline \end{array}$$

(b) $18 - 13 = 5$

$18 = 13 + 5$ or $18 = 5 + 13$

(c) $29 - 13 = 16$

$29 = 13 + 16$ or $29 = 16 + 13$

▶▶◀ Work Problem 2 at the Side.

OBJECTIVE 2 Identify the minuend, subtrahend, and difference. In subtraction, as in addition, the numbers in a problem have names. For example, in the problem $8 - 5 = 3$, the number 8 is the **minuend,** 5 is the **subtrahend,** and 3 is the **difference** or answer.

$$8 \quad - \quad 5 \quad = \quad 3 \leftarrow \text{Difference (answer)}$$

$\uparrow$ Minuend $\uparrow$ Subtrahend

$$\begin{array}{r} 8 \leftarrow \text{Minuend} \\ -\ 5 \leftarrow \text{Subtrahend} \\ \hline 3 \leftarrow \text{Difference} \end{array}$$

OBJECTIVE 3 Subtract when no borrowing is needed. Subtract two numbers by lining up the numbers in columns so the digits in the ones place are in the same column. Next, subtract by columns, starting at the right with the ones column.

EXAMPLE 3 Subtracting Two Numbers

Subtract:

Ones digits are lined up in the same column.

(a)
$$\begin{array}{r} 53 \\ -\ 21 \\ \hline 32 \end{array}$$

$3 - 1 = 2$
$5 - 2 = 3$

Ones digits are lined up.

(b)
$$\begin{array}{r} 385 \\ -\ 161 \\ \hline 224 \end{array}$$

$\leftarrow 5 - 1 = 4$
$8 - 6 = 2$
$3 - 1 = 2$

(c)
$$\begin{array}{r} 9431 \\ -\ 210 \\ \hline 9221 \end{array}$$

$\leftarrow 1 - 0 = 1$
$3 - 1 = 2$
$4 - 2 = 2$
$9 - 0 = 9$

▶▶◀ Work Problem 3 at the Side.

OBJECTIVE 4 Check subtraction answers by adding. Use addition to check your answer to a subtraction problem. For example, check $8 - 3 = 5$ by *adding* 3 and 5.

$$3 + 5 = 8, \quad \text{so} \quad 8 - 3 = 5 \quad \text{is correct.}$$

EXAMPLE 4 Checking Subtraction by Using Addition

Check each answer.

(a)

$$\begin{array}{r} 89 \\ - 47 \\ \hline 42 \end{array}$$

Rewrite as an addition problem, as shown in Example 2.

Subtraction problem
$$\left.\begin{array}{r} 89 \\ - 47 \\ \hline 42 \\ \hline \mathbf{89} \end{array}\right\}$$
Addition problem
$$\begin{array}{r} \mathbf{47} \\ + \mathbf{42} \\ \hline \mathbf{89} \end{array}$$

Because $47 + 42 = 89$, the subtraction was done correctly.

(b) $72 - 41 = 21$

Rewrite as an addition problem.

$$72 = 41 + 21$$

But, $41 + 21 = 62$, not 72, so the subtraction was done incorrectly. Rework the original subtraction to get the correct answer, 31.

(c)

$$\begin{array}{r} 374 \\ - 141 \\ \hline 233 \end{array} \longleftarrow \text{Match}$$

$$141 + 233 = 374$$

The answer checks.

Work Problem 4 at the Side. ▷▷▷

OBJECTIVE 5 Subtract by borrowing. When a digit in the minuend is less than the one directly below it, we subtract by using **borrowing.**

EXAMPLE 5 Subtracting with Borrowing

Subtract 19 from 57.

Write the problem.

$$\begin{array}{r} 57 \\ - 19 \end{array}$$

In the ones column, 7 is **less** than 9, so to subtract, we must **borrow a 10** from the 5 (which represents 5 tens, or 50).

$$50 - 10 = 40 \longrightarrow \begin{array}{cc} 4 & 17 \\ \cancel{5} & \cancel{7} \\ - 1 & 9 \end{array} \longleftarrow 10 + 7 = 17$$

Now we can subtract $17 - 9$ in the ones column and then $4 - 1$ in the tens column.

$$\begin{array}{cc} 4 & 17 \\ \cancel{5} & \cancel{7} \\ - 1 & 9 \\ \hline 3 & 8 \end{array} \quad \text{Difference}$$

Finally, $57 - 19 = 38$. Check by adding 19 and 38; you should get 57.

Work Problem 5 at the Side. ▷▷▷

---

4 Use addition to determine whether each answer is correct. If incorrect, what should it be?

(a)
$$\begin{array}{r} 76 \\ 45 \\ \hline 31 \end{array}$$

(b)
$$\begin{array}{r} 53 \\ - 22 \\ \hline 21 \end{array}$$

(c)
$$\begin{array}{r} 374 \\ - 251 \\ \hline 113 \end{array}$$

(d)
$$\begin{array}{r} 7531 \\ - 4301 \\ \hline 3230 \end{array}$$

5 Subtract.

(a)
$$\begin{array}{r} 58 \\ - 19 \end{array}$$

(b)
$$\begin{array}{r} 86 \\ - 38 \end{array}$$

(c)
$$\begin{array}{r} 41 \\ - 27 \end{array}$$

(d)
$$\begin{array}{r} 863 \\ - 47 \end{array}$$

(e)
$$\begin{array}{r} 762 \\ - 157 \end{array}$$

ANSWERS
4. (a) correct (b) incorrect; should be 31
    (c) incorrect; should be 123 (d) correct
5. (a) 39 (b) 48 (c) 14 (d) 816
    (e) 605

6 Subtract.

(a)   927
      − 43

(b)   675
      − 86

(c)   477
      − 389

(d)   1437
      − 988

(e)   8739
      − 3892

EXAMPLE 6   Subtracting with Borrowing

Subtract by borrowing as necessary.

(a)   7856
      − 137

There is no need to borrow, as
4 is greater than 3.

$$10 + 6 = 16$$
                4 16
      7 8 $\not{5}$ $\not{6}$
      − 1 3 7
      7 7 1 9   Difference

(b)   635
      − 546

$600 - 100 = 500$ $100 + 20 = 120$ (12 tens = 120)
                                        $10 + 5 = 15$
                5 12 15
          $\not{6}$ $\not{3}$ $\not{5}$
      −   5 4 6
              8 9   Difference

(c)   3648
      − 1769

      2 15 13 18
      $\not{3}$ $\not{6}$ $\not{4}$ $\not{8}$
    − 1 7 6 9
      1 8 7 9

Work Problem 6 at the Side.

Sometimes a minuend has 0s (zeros) in some of the positions. In such cases, borrowing may be a little more complicated than what we have shown so far.

EXAMPLE 7   Borrowing with Zeros

Subtract.

       4607   *Hundreds position*
    − 3168

It is not possible to borrow from the tens position. Instead we must first borrow from the hundreds position.

$600 - 100 = 500$ $100 + 0 = 100$
                5 10
      4 $\not{6}$ $\not{0}$ 7
    − 3 1 6 8

Now we may borrow from the tens position.

                    9          $100 - 10 = 90$
              5 $\not{10}$ 17 $10 + 7 = 17$
      4 $\not{6}$ $\not{0}$ $\not{7}$
    − 3 1 6 8
                    9

Continued on Next Page

Complete the problem.

$$
\begin{array}{r}
\overset{9}{}\\
5\ \overset{\cancel{10}}{\cancel{6}}\ 17\\
4\ \cancel{6}\ \cancel{0}\ \cancel{7}\\
-\ 3\ 1\ 6\ 8\\
\hline
1\ 4\ 3\ 9\quad\text{Difference}
\end{array}
$$

Check by adding 1439 and 3168; you should get 4607.

Work Problem 7 at the Side. ▶▶▶

EXAMPLE 8 Borrowing with Zeros

Subract.

(a) 708
 − 149

$100 + 0 = 100$ ⎯ $100 − 10 = 90$ (9 tens = 90)
$700 − 100 = 600$ $10 + 8 = 18$

$$
\begin{array}{r}
\overset{9}{}\\
6\ \overset{\cancel{10}}{\cancel{}}\ 18\\
\cancel{7}\ \cancel{0}\ \cancel{8}\\
-\ 1\ 4\ 9\\
\hline
5\ 5\ 9
\end{array}
$$

(b) 380
 − 276

$80 − 10 = 70$ (7 tens = 70) $10 + 0 = 10$

$$
\begin{array}{r}
7\ 10\\
3\ \cancel{8}\ \cancel{0}\\
-\ 2\ 7\ 6\\
\hline
1\ 0\ 4
\end{array}
$$

(c) 9000
 − 6999

$$
\begin{array}{r}
\overset{9}{}\ \overset{9}{}\\
8\ 10\ 10\ 10\\
\cancel{9}\ \cancel{0}\ \cancel{0}\ \cancel{0}\\
-\ 6\ 9\ 9\ 9\\
\hline
2\ 0\ 0\ 1
\end{array}
$$

Work Problem 8 at the Side. ▶▶▶

As we have seen, an answer to a subtraction problem can be checked by adding.

EXAMPLE 9 Checking Subtraction by Using Addition

Use addition to check each answer.

Check

(a) **613**   275
 − 275   *Match* + 338
 338   **613**   Correct

Continued on Next Page

---

7 Subtract.

(a)   206
 − 177

(b)   703
 − 415

(c)   7024
 − 2632

8 Subtract.

(a)   308
 − 159

(b)   570
 − 368

(c)   1570
 − 983

(d)   7001
 − 5193

(e)   4000
 − 1782

9 Use addition to check each answer. If the answer is incorrect, find the correct answer.

(a)
$$357$$
$$-\ 168$$
$$\overline{189}$$

(b)
$$570$$
$$-\ 328$$
$$\overline{252}$$

(c)
$$14,726$$
$$-\ 8\ 839$$
$$\overline{5\ 887}$$

10 Use the table from Example 10 to find, on average,

(a) how much more a person with a Bachelor's degree earns each year than a person with an Associate of Arts degree.

$$62,458$$
$$39,902$$

(b) how much more a person with an Associate of Arts degree earns each year than a person who is not a high school graduate.

Check

(b)
$$1915$$ $$1635$$
$$-\ 1635$$ $$Match$$ $$+\ 280$$
$$\overline{280}$$ $$\overline{1915}$$ Correct

Check

(c)
$$15,803$$ $$7\ 325$$
$$-\ 7\ 325$$ $$No\ Match$$ $$+\ 8\ 578$$
$$\overline{8\ 578}$$ $$\overline{15,903}$$ Error

Rework the original problem to get the correct answer, 8478.

◀◀◀ Work Problem 9 at the Side.

OBJECTIVE 6 Solve application problems with subtraction. As shown in the next example, subtraction can be used to solve an application problem.

EXAMPLE 10 Applying Subtraction Skills

Use the table to find how much more, on average, a person with an Associate of Arts degree earns each year than a high school graduate.

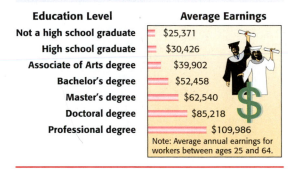

EDUCATION PAYS

The more education adults get, the higher their annual earnings.

Education Level	Average Earnings
Not a high school graduate	$25,371
High school graduate	$30,426
Associate of Arts degree	$39,902
Bachelor's degree	$52,458
Master's degree	$62,540
Doctoral degree	$85,218
Professional degree	$109,986

Note: Average annual earnings for workers between ages 25 and 64.

Source: U.S. Bureau of the Census.

Approach   The average earnings for a person with an Associate of Arts degree is $39,902 each year and the average for a high school graduate is $30,426. Find how much more a college graduate earns by subtracting $30,426 from $39,902.

Solution
$$\$39,902 \leftarrow \text{Associate of Arts degree}$$
$$-\ 30,426 \leftarrow \text{High school graduate}$$
$$\overline{\$9\ 476} \leftarrow \text{More earnings}$$

On average, a person with an Associate of Arts degree earns $9476 more each year than a high school graduate.

◀◀◀ Work Problem 10 at the Side.

1.3 Exercises

FOR EXTRA HELP | Addison-Wesley Math Tutor Center | 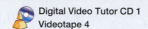 MathXL | Digital Video Tutor CD 1 Videotape 4 | Student's Solutions Manual | MyMathLab | 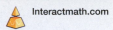 Interactmath.com

Work each subtraction problem. Use addition to check each answer. See Examples 3 and 4.

1.   48
   − 32

2.   17
   − 13

3.   86
   − 53

4.   78
   − 35

5.   77
   − 60

6.   87
   − 63

7.   335
   − 122

8.   602
   − 301

9.   552
   − 451

10.   888
   − 215

11.   7352
   − 241

12.   4420
   − 310

13.   5546
   − 2134

14.   1875
   − 1362

15.   6259
   − 4148

16.   9654
   − 4323

17.   24,392
   − 11,232

18.   57,921
   − 34,801

19.   46,253
   − 5 143

20.   75,904
   − 3 702

Use addition to check each subtraction problem. If an answer is not correct, find the correct answer. See Example 4.

21.   54
   − 42
   ‾‾‾‾
   12

22.   87
   − 43
   ‾‾‾‾
   44

23.   89
   − 27
   ‾‾‾‾
   63

24.   47
   − 35
   ‾‾‾‾
   13

25.   382
   − 261
   ‾‾‾‾
   131

26.   754
   − 342
   ‾‾‾‾
   412

27.   4683
   − 3542
   ‾‾‾‾
   1141

28.   5217
   − 4105
   ‾‾‾‾
   1132

29.   8643
   − 1421
   ‾‾‾‾
   7212

30.   9428
   − 3124
   ‾‾‾‾
   6324

Subtract by borrowing as necessary. See Examples 5–8.

31.   75
   − 37

32.   86
   − 28

33.   94
   − 49

34.   68
   − 39

35.   57
   − 38

36.   47
   − 29

37.   828
   − 547

38.   916
   − 618

39.   771
   − 252

40.   973
   − 788

41.   7538
       $-$ 479

42.   5863
       $-$ 1295

43.   9988
       $-$ 2399

44.   3576
       $-$ 1658

45.   38,335
       $-$ 29,476

46.   82,731
       $-$ 14,826

47.   40
       $-$ 37

48.   80
       $-$ 73

49.   60
       $-$ 37

50.   70
       $-$ 27

51.   308
       $-$ 289

52.   600
       $-$ 599

53.   4041
       $-$ 1208

54.   4602
       $-$ 2063

55.   9305
       $-$ 1530

56.   7120
       $-$ 6033

57.   1580
       $-$ 1077

58.   3068
       $-$ 2105

59.   2006
       $-$ 1850

60.   8203
       $-$ 5365

61.   8240
       $-$ 6056

62.   7050
       $-$ 6045

63.   8503
       $-$ 2816

64.   16,004
       $-$ 5 087

65.   80,705
       $-$ 61,667

66.   81,000
       $-$ 55,456

67.   66,000
       $-$ 34,444

68.   77,000
       $-$ 65,308

69.   20,080
       $-$ 13,496

70.   80,056
       $-$ 23,869

Use addition to check each subtraction problem. If an answer is incorrect, find the correct answer. See Example 9.

71.   9428
       $-$ 4509

       4919

72.   1671
       $-$ 1325

       1346

73.   2548
       $-$ 2278

       270

74.   5274
       $-$ 1130

       4144

75.   93,758
       $-$ 52,869

       40,889

76.   82,357
       $-$ 14,396

       68,961

77.   36,778
       $-$ 17,405

       19,373

78.   34,821
       $-$ 17,735

       17,735

79. An addition problem can be changed to a subtraction problem and a subtraction problem can be changed to an addition problem. Give two examples of each to demonstrate this.

80. Can you use the commutative and the associative properties in subtraction? Explain.

Solve each application problem. See Example 10.

81. A man burns 187 calories during 60 minutes of sitting at a computer while a woman burns 140 calories at the same activity. How many fewer calories does a woman burn than a man? (*Source:* www .cookinglight.com)

187
140
47 calories

82. A woman burns 302 calories during 60 minutes of walking, while a man burns 403 calories doing the same activity. How many more calories does a man burn than a woman? (*Source:* www.cookinglight .com)

83. Toronto's skyline is dominated by the CN Tower, which rises 1821 ft. The Sears Tower in Chicago is 1454 ft high. Find the difference in height between the two structures. (*Source:* Trizec Properties; *World Almanac.*)

1821 ft
d
1454 ft

CN Tower   Sears Tower

84. The fastest animal in the world, the peregrine falcon, dives at 217 miles per hour (mph). A Boeing 747 cruises at 580 mph. How much faster is the plane?

Diving peregrine
217 mph

Boeing 747
580 mph

85. A cruise ship has 1815 passengers. When in port at Grand Cayman, 1348 passengers go ashore for the day while the others remain on the ship. How many passengers remain on the ship?

1348
-1815
467 passengers

86. In a recent three-month period there were 81,465 Ford Explorers and 70,449 Jeep Grand Cherokees sold. Which vehicle had greater sales? By how much? (*Source:* J. D. Power and Associates.)

87. On Tuesday, 5822 people went to a soccer game, and on Friday, 7994 people went. Which day had more people at the game? How many more?

88. In 1964, its first year on the market, the Ford Mustang sold for $2500. In 2006, the Ford Mustang sold for $29,398. Find the increase in price. (*Source: Consumer Reports.*)

89. Patriot Flag Company manufactured 14,608 U.S. flags and sold 5069. How many flags remain unsold?

90. Eye exams have been given to 14,679 children in the school district. If there are 23,156 students in the school district, how many have not received eye exams?

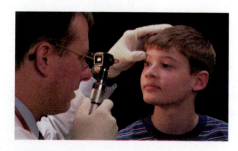

91. The Jordanos now pay rent of $650 per month. If they buy a house, their housing expense will be $913 per month. How much more will they pay per month if they buy a house?

92. A retired couple who used to receive a Social Security payment of $1479 per month now receives $1568 per month. Find the amount of the monthly increase.

93. On Monday, 11,594 people visited Arcade Amusement Park, and 12,352 people visited the park on Tuesday. Which day had more people visit the park? How many more?

94. The highest annual food cost per household was $7442 in San Francisco-Oakland-San Jose, California. The lowest annual food cost was $4589 in Tampa-St. Petersburg-Clearwater, Florida. How much more was the annual household food cost in the highest metropolitan area than the lowest? (*Source:* U.S. Bureau of the Census)

Solve each application problem. Add or subtract as necessary.

95. A survey of large hotels found that the average salary for a general manager of a deluxe spa and tennis resort is one hundred one thousand, five hundred dollars per year, while spa and tennis directors earn $44,000. How much more does a general manager earn than a spa and tennis director?

96. There are 24 million business enterprises in the United States. If only 7000 of these business are large businesses having 500 or more employees, how many are small and midsize businesses?

The table shows the deliveries made by Diana Lopez, a United Parcel Service driver. Use the table to answer Exercises 97–100.

PACKAGE DELIVERY (LOPEZ)

Day	Number of Deliveries
Monday	137
Tuesday	126
Wednesday	119
Thursday	89
Friday	147

97. How many more deliveries did Diana make on Monday than on Thursday?

98. How many more deliveries did Diana make on Friday than on Tuesday?

99. Find the total deliveries made on the two busiest days.

100. Find the total deliveries made on the two slowest days.

1.4 Multiplying Whole Numbers

Suppose we want to know the total number of exercise bicycles available at the gym. The bicycles are arranged in four rows with three stations in each row. Adding the number 3 a total of 4 times gives 12.

$$3 + 3 + 3 + 3 = 12$$

This result can also be shown with a figure.

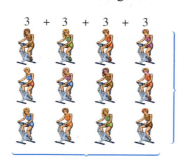

3 + 3 + 3 + 3

3 bicycles in each row

4 rows

OBJECTIVE 1 Identify the parts of a multiplication problem. Multiplication is a shortcut for repeated addition. In the exercise bicycle example, instead of *adding* $3 + 3 + 3 + 3$ to get 12, we can *multiply* 3 by 4 to get 12. The numbers being multiplied are called **factors**. The answer is called the **product**. For example, the product of 3 and 4 can be written with the symbol $\times$, a raised dot, or parentheses, as follows.

$$3 \leftarrow \text{Factor (also called } multiplicand)$$
$$\times\ 4 \leftarrow \text{Factor (also called } multiplier)$$
$$\overline{12} \leftarrow \text{Product (answer)}$$

$$3 \times 4 = 12 \quad or \quad 3 \cdot 4 = 12 \quad or \quad (3)(4) = 12 \quad or \quad 3(4) = 12$$

Work Problem 1 at the Side.

Commutative Property of Multiplication

By the **commutative property of multiplication,** the product (answer) remains the same when the order of the factors is changed. For example,

$$3 \times 5 = 15 \quad and \quad 5 \times 3 = 15.$$

CAUTION
Recall that addition also has a commutative property. For example, $4 + 2$ gives the same sum as $2 + 4$. Subtraction, however, is ***not*** commutative.

EXAMPLE 1 Multiplying Two Numbers

Multiply. (Remember that a raised dot or parentheses means to multiply.)

(a) $3 \times 4 = 12$

(b) $6 \cdot 0 = 0$ (The product of any number and 0 is 0; if you give no money to each of 6 relatives, you give no money.)

(c) $4(8) = 32$

Continued on Next Page

---

1 Identify the factors and the product in each multiplication problem.

(a) $8 \times 5 = 40$

(b) $6(4) = 24$

(c) $7 \cdot 6 = 42$

(d) $(3)(9) = 27$

2 Multiply.

(a) 7×4

(b) 0×9

(c) $8(5)$

(d) $6 \cdot 5$

(e) $(3)(8)$

3 Multiply.

(a) $3 \times 2 \times 5$

(b) $4 \cdot 7 \cdot 1$

(c) $(8)(3)(0)$

You may want to review this multiplication table.

Multiplication Table

×	1	2	3	4	5	6	7	8	9
1	1	2	3	4	5	6	7	8	9
2	2	4	6	8	10	12	14	16	18
3	3	6	9	12	15	18	21	24	27
4	4	8	12	16	20	24	28	32	36
5	5	10	15	20	25	30	35	40	45
6	6	12	18	24	30	36	42	48	54
7	7	14	21	28	35	42	49	56	63
8	8	16	24	32	40	48	56	64	72
9	9	18	27	36	45	54	63	72	81

◀◀◀ Work Problem 2 at the Side.

OBJECTIVE 2 Do chain multiplications. Some multiplications contain more than two factors.

Associative Property of Multiplication

By the **associative property of multiplication,** grouping the factors differently does not change the product.

EXAMPLE 2 Multiplying Three Numbers

Multiply $2 \times 3 \times 5$.

$$(2 \times 3) \times 5 \qquad \text{Parentheses tell what to do first.}$$
$$6 \quad \times 5 = 30$$

Also,

$$2 \times (3 \times 5)$$
$$2 \times \quad 15 = 30$$

Either grouping results in the same product.

◀◀◀ Work Problem 3 at the Side.

Calculator Tip The calculator approach to Example 2 uses chain calculations.

$$2 \;\times\; 3 \;\times\; 5 \;=\; 30$$

A problem with more than two factors, such as the one in Example 2, is called a **chain multiplication.**

OBJECTIVE **3** **Multiply by single-digit numbers.** Carrying may be needed in multiplication problems with larger factors.

EXAMPLE 3 **Multiplying with Carrying**

Multiply.

(a) $\begin{array}{r} 53 \\ \times\ 4 \end{array}$

Start by multiplying in the ones column.

$\begin{array}{r} \mathbf{1} \\ 53 \\ \times\ 4 \\ \hline 2 \end{array}$ $4 \times 3 = \mathbf{12}$ Carry the 1 to the tens column.
Write 2 in the ones column.

Next, multiply 4 ones and 5 tens.

$\begin{array}{r} \mathbf{1} \\ 53 \\ \times\ 4 \\ \hline 2 \end{array}$ $4 \times 5 = \mathbf{20}$ tens

Add the 1 that was carried to the tens column.

$\begin{array}{r} \mathbf{1} \\ 53 \\ \times\ 4 \\ \hline 212 \end{array}$ $20 + 1 = \mathbf{21}$ tens

(b) $\begin{array}{r} 724 \\ \times\ 5 \end{array}$

Work as shown.

$\begin{array}{r} 12 \\ 724 \\ \times\ 5 \\ \hline 3620 \end{array}$ ← $5 \times 4 = \mathbf{20}$ ones; write 0 ones and carry 2 tens.

$5 \times 2 = \mathbf{10}$ tens; add the 2 tens to get 12 tens;
write 2 tens and carry 1 hundred.

$5 \times 7 = \mathbf{35}$ hundreds; add the
1 hundred to get 36 hundreds.

Work Problem 4 at the Side. ❱❱❱

OBJECTIVE **4** **Use multiplication shortcuts for numbers ending in zeros.** The product of two whole number factors is also called a **multiple** of either factor. For example, since $4 \cdot 2 = 8$, the whole number 8 is a multiple of both 4 and 2. *Multiples of 10* are very useful when multiplying. A **multiple of 10** is a whole number that ends in 0, such as 10, 20, or 30; 100, 200, or 300; 1000, 2000, or 3000. There is a short way to multiply by these multiples of 10. Look at the following examples.

$$26 \times 1 = 26$$
$$26 \times 1\mathbf{0} = 26\mathbf{0}$$
$$26 \times 1\mathbf{00} = 26\mathbf{00}$$
$$26 \times 1\mathbf{000} = 26,\mathbf{000}$$

Do you see a pattern? These examples suggest the rule on the next page.

4 Multiply.

(a) $\begin{array}{r} 53 \\ \times\ 5 \end{array}$

(b) $\begin{array}{r} 79 \\ \times\ 0 \end{array}$

(c) $\begin{array}{r} 758 \\ \times\ 8 \end{array}$

(d) $\begin{array}{r} 2831 \\ \times\ 7 \end{array}$

(e) $\begin{array}{r} 4714 \\ \times\ 8 \end{array}$

ANSWERS
4. **(a)** 265 **(b)** 0 **(c)** 6064
   (d) 19,817 **(e)** 37,712

5 Multiply.

(a) 63×10

(b) 305×100

(c) 714×1000

Multiplying by Multiples of 10

To multiply a whole number by 10, 100, or 1000, attach one, two, or three zeros to the right of the whole number.

EXAMPLE 4 Using Multiples of 10 to Multiply

Multiply.

(a) $59 \times 10 = 590$ — Attach 0.

(b) $74 \times 100 = 7400$ — Attach 00.

(c) $803 \times 1000 = 803,000$ ← Attach 000.

Work Problem 5 at the Side.

You can also find the product of other multiples of 10 by attaching zeros.

EXAMPLE 5 Using Multiples of 10 to Multiply

Multiply.

(a) 75×3000
Multiply 75 by 3, and then attach three zeros.

$$75 \times 3000 = 225,000$$

$$\begin{array}{r} 75 \\ \times\ 3 \\ \hline 225 \end{array}$$ Attach 000.

6 Multiply.

(a) 16×50

(b) 73×400

(b) 150×70
Multiply 15 by 7, and then attach two zeros.

$$150 \times 70 = 10,500 \longleftarrow \text{Attach 00.}$$

$$\begin{array}{r} 15 \\ \times\ 7 \\ \hline 105 \end{array}$$

(c) $\begin{array}{r} 180 \\ \times\ 30 \end{array}$

Work Problem 6 at the Side.

OBJECTIVE 5 Multiply by numbers having more than one digit. The next example shows multiplication when both factors have more than one digit.

(d) $\begin{array}{r} 4200 \\ \times\ 80 \end{array}$

(e) $\begin{array}{r} 800 \\ \times\ 600 \end{array}$

EXAMPLE 6 Multiplying with More Than One Digit

Multiply 46 and 23.

First multiply 46 by 3.

Carrying is needed here.

$$\begin{array}{r} \overset{1}{4}6 \\ \times\ 3 \\ \hline 138 \end{array} \longleftarrow 46 \times 3 = 138$$

Continued on Next Page

Now multiply 46 by 20.

$$
\begin{array}{r}
1 \\
46 \\
\times\ 20 \\
\hline
920
\end{array}
$$
$\leftarrow$ 46 × 20 = 920

Add the results.

$$
\begin{array}{r}
46 \\
\times\ 23 \\
\hline
138 \\
+\ 920 \\
\hline
1058
\end{array}
$$
138 $\leftarrow$ 46 × 3
920 $\leftarrow$ 46 × 20

Add.

Both 138 and 920 are called **partial products.** To save time, the 0 in 920 is usually not written.

$$
\begin{array}{r}
46 \\
\times\ 23 \\
\hline
138 \\
92 \\
\hline
1058
\end{array}
$$
$\leftarrow$ { 0 not written. Be very careful to place the 2 in the tens column.

Work Problem 7 at the Side.))))

EXAMPLE 7 Using Partial Products

Multiply.

(a)
$$
\begin{array}{r}
2\,3\,3 \\
\times\ 1\,3\,2 \\
\hline
4\,6\,6 \\
6\,9\,9 \\
2\,3\,3 \\
\hline
3\,0,7\,5\,6
\end{array}
$$
(Tens lined up)
(Hundreds lined up)
$\leftarrow$ Product

(b)
$$
\begin{array}{r}
538 \\
\times\ 46
\end{array}
$$

First multiply by 6.

$$
\begin{array}{r}
2\,4 \\
538 \\
\times\ \ \ 46 \\
\hline
3228
\end{array}
$$
$\leftarrow$ Carrying is needed here.

Now multiply by 4, being careful to line up the tens.

$$
\begin{array}{r}
1\ 3 \\
2\ 4 \\
5\ 3\ 8 \\
\times\ \ \ \ 4\ 6 \\
\hline
3\ 2\ 2\ 8 \\
2\ 1\ 5\ 2 \\
\hline
2\ 4,7\ 4\ 8
\end{array}
$$
Finally, add the partial products.

Work Problem 8 at the Side.))))

7 Complete each multiplication.

(a)
$$
\begin{array}{r}
35 \\
\times\ 54 \\
\hline
140 \\
175
\end{array}
$$

(b)
$$
\begin{array}{r}
76 \\
\times\ 49 \\
\hline
684 \\
304
\end{array}
$$

8 Multiply.

(a)
$$
\begin{array}{r}
52 \\
\times\ 16
\end{array}
$$

(b)
$$
\begin{array}{r}
81 \\
\times\ 49
\end{array}
$$

(c)
$$
\begin{array}{r}
75 \\
\times\ 63
\end{array}
$$

(d)
$$
\begin{array}{r}
234 \\
\times\ \ 73
\end{array}
$$

(e)
$$
\begin{array}{r}
835 \\
\times\ 189
\end{array}
$$

ANSWERS

7. (a) 1890 **(b)** 3724
8. (a) 832 **(b)** 3969 **(c)** 4725
    (d) 17,082 **(e)** 157,815

9 Multiply.

(a)   28
    × 60

(b)   728
    × 50

(c)   562
    × 109

(d)   3526
   × 6002

10 Find the total cost of the following items.

(a) 289 redwood planters at $12 per planter

(b) 205 DVD/CD players at $89 each

(c) 12 delivery vans at $24,300 per van

24,300
× 12
291,600

When 0 appears in the multiplier, be sure to move the partial products to the left to account for the position held by the 0.

EXAMPLE 8 Multiplying with Zeros

Multiply.

(a)   1 3 7
      × 3 0 6
        8 2 2
        0 0 0   (Tens lined up)
      4 1 1   (Hundreds lined up)
      4 1,9 2 2

(b)   1 4 0 6
      × 2 0 0 1
         1 4 0 6
       0 0 0 0   ← (Zeros to line up tens)
      0 0 0 0   ← (Zeros to line up hundreds)
      2 8 1 2
      2,8 1 3,4 0 6

       1 4 0 6
    × 2 0 0 1
       1 4 0 6
    2 8 1 2 0 0   ← Zeros are written so this partial product starts in the thousands column.
    2,8 1 3,4 0 6

NOTE
In Example 8(b) in the alternative method on the right, zeros were inserted so that thousands were placed in the thousands column. This is a commonly used shortcut.

◀◀◀ **Work Problem 9 at the Side.**

OBJECTIVE 6 Solve application problems with multiplication. The next example shows how multiplication can be used to solve an application problem.

EXAMPLE 9 Applying Multiplication Skills

Find the total cost of 36 cellular phones priced at $89 each.

Approach To find the cost of all the cellular phones, multiply the number of phones (36) by the cost of one phone ($89).

Solution Multiply 36 by 89.

           36
        × 89
          324
          288
         3204

The total cost of the cellular phones is $3204.

Calculator Tip If you are using a calculator for Example 9, you will do this calculation.

36 ⊗ 89 ⊜ 3204

◀◀◀ **Work Problem 10 at the Side.**

1.4 Exercises

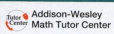

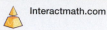

Work each chain multiplication. See Example 2.

1. $2 \times 6 \times 2$

2. $3 \times 5 \times 3$

3. $8 \times 6 \times 1$

4. $2 \times 4 \times 5$

5. $7 \cdot 8 \cdot 0$

6. $9 \cdot 0 \cdot 5$

7. $4 \cdot 1 \cdot 6$

8. $1 \cdot 5 \cdot 7$

9. $(4)(5)(2)$

10. $(4)(1)(9)$

11. $(3)(0)(7)$

12. $(0)(9)(4)$

13. Explain in your own words the commutative property of multiplication. How do the commutative properties of addition and multiplication compare to each other? *They are the same. You can add & multiply factors in any order - Answer will be the same*

14. Explain in your own words the associative property of multiplication. How do the associative properties of addition and multiplication compare to each other?

Multiply. See Example 3.

15. $\begin{array}{r} 35 \\ \times\ 6 \\ \hline \end{array}$

16. $\begin{array}{r} 53 \\ \times\ 7 \\ \hline \end{array}$

17. $\begin{array}{r} 34 \\ \times\ 7 \\ \hline \end{array}$

18. $\begin{array}{r} 76 \\ \times\ 5 \\ \hline \end{array}$

19. $\begin{array}{r} 642 \\ \times\ \ 5 \\ \hline \end{array}$

20. $\begin{array}{r} 472 \\ \times\ \ 4 \\ \hline \end{array}$

21. $\begin{array}{r} 624 \\ \times\ \ 3 \\ \hline \end{array}$

22. $\begin{array}{r} 852 \\ \times\ \ 7 \\ \hline \end{array}$

23. $\begin{array}{r} 2153 \\ \times\ \ \ 4 \\ \hline \end{array}$

24. $\begin{array}{r} 1137 \\ \times\ \ \ 3 \\ \hline \end{array}$

25. $\begin{array}{r} 2521 \\ \times\ \ \ 4 \\ \hline \end{array}$

26. $\begin{array}{r} 2544 \\ \times\ \ \ 3 \\ \hline \end{array}$

27. $\begin{array}{r} 2561 \\ \times\ \ \ 8 \\ \hline \end{array}$

28. $\begin{array}{r} 7326 \\ \times\ \ \ 5 \\ \hline \end{array}$

29. $\begin{array}{r} 36,921 \\ \times\ \ \ \ 7 \\ \hline \end{array}$

30. $\begin{array}{r} 28,116 \\ \times\ \ \ \ 4 \\ \hline \end{array}$

Multiply. See Examples 4 and 5.

31. $\begin{array}{r} 40 \\ \times\ 7 \\ \hline \end{array}$

32. $\begin{array}{r} 20 \\ \times\ 7 \\ \hline \end{array}$

33. $\begin{array}{r} 80 \\ \times\ 6 \\ \hline \end{array}$

34. $\begin{array}{r} 70 \\ \times\ 5 \\ \hline \end{array}$

35. $\begin{array}{r} 740 \\ \times\ \ 3 \\ \hline \end{array}$

36. $\begin{array}{r} 400 \\ \times\ \ 8 \\ \hline \end{array}$

37. $\begin{array}{r} 600 \\ \times\ \ 6 \\ \hline \end{array}$

38. $\begin{array}{r} 860 \\ \times\ \ 7 \\ \hline \end{array}$

39. $\begin{array}{r} 125 \\ \times\ 30 \\ \hline \end{array}$

40. $\begin{array}{r} 246 \\ \times\ 50 \\ \hline \end{array}$

41. 1635
× 40

42. 7311
× 50

43. 900
× 300

44. 400
× 700

45. 43,000
× 2 000

46. 11,000
× 9 000

47. 970 · 50

48. 730 · 40

49. 800 · 900

50. 850 · 700

51. 9700 · 200

52. 10,050 · 300

Multiply. See Examples 6–8.

53. 28
× 17

54. 16
× 34

55. 75
× 32

56. 82
× 32

57. 83
× 45

58. (75)(21)

59. (58)(41)

60. (82)(67)

61. (67)(92)

62. (26)(33)

63. (28)(564)

64. (58)(312)

65. (619)(35)

66. (681)(47)

67. (55)(286)

68. 286
× 574

69. 735
× 112

70. 621
× 415

71. 538
× 342

72. 3228
× 751

73. 9352
× 264

74. 528
× 106

75. 215
× 307

76. 218
× 106

77. 428
× 201

78. 3706
× 208

79. 6310
× 3078

80. 3533
× 5001

81. 2195
× 1038

82. 1502
× 2009

83. A classmate of yours is not clear on how to use a shortcut to multiply a whole number by 10, by 100, or by 1000. Write a short note explaining how this can be done.

84. Show two ways to multiply when a 0 is in the factor that is multiplying. Use the problem 291 × 307 to show this.

Solve each application problem. See Example 9.

85. Carepanian Company, a health care supplier, purchased 300 cartons of Thera Bond Gym Balls. If there are 10 balls in each carton, find the total number of balls purchased.

86. A medical supply house has 30 bottles of vitamin C tablets, with each bottle containing 500 tablets. Find the total number of vitamin C tablets in the supply house.

87. There are 12 tomato plants to a flat. If a gardener buys 18 flats, find the total number of tomato plants he bought. *12 × 18 = 216 plants*

88. A hummingbird's wings beat about 65 times per second. How many times do the hummingbird's wings beat in 30 seconds?

89. A new Toyota Highlander gets 22 miles per gallon on the highway. How many miles can it go on 16 gallons of gas? (*Source:* Toyota Motor Company.)

90. Squid are being hauled out of the Santa Barbara Channel by the ton. They are then processed, renamed calamari, and exported. Last night 27 fishing boats each hauled out 40 tons of squid. What was the total catch for the night? (*Source: Santa Barbara News Press.*) *27 × 40 = 1080*

Find the total cost of the following items. See Examples 7–9.

91. 75 first-aid kits at $8 per kit

92. 38 employees at $64 per day

93. 65 rebuilt alternators at $24 per alternator

94. 62 wheelchair cushions at $44 per cushion

95. 206 desktop computers at $548 per computer

96. 520 printers at $219 per printer

 Multiply.

97. 21 · 43 · 56

98. (600)(8)(75)(40)

Use addition, subtraction, or multiplication to solve each application problem.

99. In a forest-planting project, trees are planted 450 trees to an acre. Find the number of trees needed to plant 85 acres.

100. The largest living land mammal is the African elephant, and the largest mammal of all time is the blue whale. An African elephant weighs 15,225 pounds and a blue whale weighs 28 times that amount. Find the weight of the blue whale.

101. The average number of miles Americans drove annually in 1980 was 8813 miles. In 2000 the average was 11,988 miles. How many more miles were driven by the average American in 2000 than in 1980? (*Source: Energy Information Administration Monthly Review.*)

102. As part of a Low Income Home Energy Assistance Program, the state of Florida will receive $1,100,000, while Vermont will receive $505,551. How much more will Florida receive than Vermont? (*Source:* U.S. Department of Energy.)

103. A medical center purchased six laptop computers at $880 each, six printers at $235 each, and five fax machines at $140 each. Find the total cost of this equipment.

104. A motorcycle club traveled 640 miles on Saturday, 438 miles on Sunday, and 535 miles on Monday. Find the total number of miles traveled.

RELATING CONCEPTS (EXERCISES 105–114) For Individual or Group Work

Work Exercises 104–113 in order.

105. Add.

(a) $189 + 263$

(b) $263 + 189$

106. Your answers to Exercise 105(a) and (b) should be the same. This shows that the order of numbers in an addition problem does not change the sum. This is known as the _____ property of addition.

107. Add. Recall that parentheses tell you what to do first.

(a) $(65 + 81) + 135$

(b) $65 + (81 + 135)$

108. Since the answers to Exercise 107(a) and (b) are the same, we see that grouping the addition of numbers in any order does not change the sum. This is known as the _____ property of addition.

109. Multiply.

(a) 220×72

(b) 72×220

110. Since the answers to Exercise 109(a) and (b) are the same, we see that the product remains the same when the order of the factors is changed. This is known as the _____ property of multiplication.

111. Multiply. Recall that parentheses tell you what to do first.

(a) $(26 \times 18) \times 14$

(b) $26(18 \times 14)$

112. Since the answers to Exercise 111(a) and (b) are the same, we see that the grouping of numbers in any order when multiplying gives the same product. This is known as the _____ property of multiplication.

113. Do the commutative and associative properties apply to subtraction? Explain your answer using several examples.

114. Do you think that the commutative and associative properties will apply to division? Explain your answer using several examples.

1.5 Dividing Whole Numbers

Suppose the cost of a fast-food lunch is $12 and is to be divided equally by three friends. Each person would pay $4, as shown here.

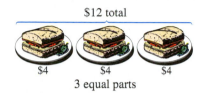

$12 total

$4 $4 $4

3 equal parts

OBJECTIVE 1 Write division problems in three ways. Just as $3 \cdot 4$, 3×4, and $(3)(4)$ are different ways of indicating the multiplication of 3 and 4, there are several ways to write 12 divided by 3.

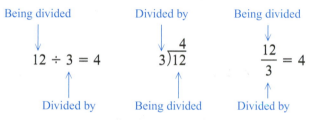

Being divided   Divided by   Being divided

$$12 \div 3 = 4 \qquad 3\overline{)12}^{\,4} \qquad \frac{12}{3} = 4$$

Divided by   Being divided   Divided by

We will use all three division symbols, $\div$, $\overline{)}\,$, and —. In courses such as algebra, a slash symbol, /, or a fraction bar, —, is most often used.

EXAMPLE 1 Using Division Symbols

Write each division problem using two other symbols.

(a) $12 \div 4 = 3$

This division can also be written as shown below.

$$4\overline{)12}^{\,3} \quad \text{or} \quad \frac{12}{4} = 3$$

(b) $\dfrac{15}{5} = 3$

$$15 \div 5 = 3 \quad \text{or} \quad 5\overline{)15}^{\,3}$$

(c) $5\overline{)20}^{\,4}$

$$20 \div 5 = 4 \quad \text{or} \quad \frac{20}{5} = 4$$

Work Problem 1 at the Side. ▶▶▶

OBJECTIVE 2 Identify the parts of a division problem. In division, the number being divided is the **dividend,** the number divided by is the **divisor,** and the answer is the **quotient.**

$$\textcolor{green}{\textbf{dividend}} \div \textcolor{blue}{\textbf{divisor}} = \textcolor{red}{\textbf{quotient}}$$

$$\textcolor{blue}{\textbf{divisor}}\overline{)\textcolor{green}{\textbf{dividend}}}^{\,\textcolor{red}{\textbf{quotient}}} \qquad \frac{\textcolor{green}{\textbf{dividend}}}{\textcolor{blue}{\textbf{divisor}}} = \textcolor{red}{\textbf{quotient}}$$

1 Write each division problem using two other symbols.

(a) $24 \div 6 = 4$

(b) $9\overline{)36}^{\,4}$

(c) $48 \div 6 = 8$

(d) $\dfrac{42}{6} = 7$

ANSWERS

1. (a) $6\overline{)24}^{\,4}$ and $\dfrac{24}{6} = 4$

(b) $36 \div 9 = 4$ and $\dfrac{36}{9} = 4$

(c) $6\overline{)48}^{\,8}$ and $\dfrac{48}{6} = 8$

(d) $6\overline{)42}^{\,7}$ and $42 \div 6 = 7$

2 Identify the dividend, divisor, and quotient.

(a) $15 \div 3 = 5$

(b) $18 \div 6 = 3$

(c) $\dfrac{28}{7} = 4$

(d) $9\overline{)27}$ with quotient 3

3 Divide.

(a) $0 \div 5$

(b) $\dfrac{0}{9}$

(c) $\dfrac{0}{24}$

(d) $37\overline{)0}$

EXAMPLE 2 Identifying the Parts of a Division Problem

Identify the dividend, divisor, and quotient.

(a) $35 \div 7 = 5$

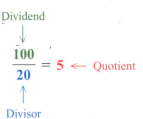

$35 \div 7 = 5 \leftarrow$ Quotient

Dividend   Divisor

(b) $\dfrac{100}{20} = 5$

Dividend

$\dfrac{\mathbf{100}}{\mathbf{20}} = \mathbf{5} \leftarrow$ Quotient

Divisor

(c) $8\overline{)72}$ with quotient 9

$9 \leftarrow$ Quotient
$8\overline{)72} \leftarrow$ Dividend

Divisor

Work Problem 2 at the Side.

OBJECTIVE 3 Divide 0 by a number. If no money, or \$0, is divided equally among five people, each person gets \$0. The general rule for dividing 0 follows.

Dividing 0 by a Number

The number **0** divided by any nonzero number is **0**.

EXAMPLE 3 Dividing 0 by a Number

Divide.

(a) $0 \div 12 = 0$

(b) $0 \div 1728 = 0$

(c) $\dfrac{0}{375} = 0$

(d) $129\overline{)0}$ with quotient 0

Work Problem 3 at the Side.

Just as a subtraction such as $8 - 3 = 5$ can be written as the addition $8 = 3 + 5$, any division can be written as a multiplication. For example, $12 \div 3 = 4$ can be written as

$$3 \times 4 = 12 \quad \text{or} \quad 4 \times 3 = 12.$$

EXAMPLE 4 **Changing Division Problems to Multiplication**

Change each division problem to a multiplication problem.

(a) $\dfrac{20}{4} = 5$ becomes $4 \cdot 5 = 20$ or $5 \cdot 4 = 20$

(b) $8\overline{)48}^{\,6}$ becomes $8 \cdot 6 = 48$ or $6 \cdot 8 = 48$

(c) $72 \div 9 = 8$ becomes $9 \cdot 8 = 72$ or $8 \cdot 9 = 72$

Work Problem 4 at the Side. ▶▶▶

OBJECTIVE **4** **Recognize that a number cannot be divided by 0.**
Division of any number by 0 cannot be done. To see why, try to find

$$9 \div 0 = ?$$

As we have just seen, any division problem can be converted to a multiplication problem so that

divisor · quotient = dividend.

If you convert the preceding problem to its multiplication counterpart, it reads as follows.

$$0 \cdot ? = 9$$

You already know that 0 times any number must always be 0. Try any number you like to replace the "**?**" and you'll aways get 0 instead of 9. Therefore, the division probem $9 \div 0$ cannot be done. Mathematicians say it is *undefined* and have agreed never to divide by 0. However, $0 \div 9$ *can* bc done. Check by rewriting it as a multiplication problem.

$$0 \div 9 = 0 \quad \text{because} \quad 0 \cdot 9 = 0 \text{ is true.}$$

Dividing a Number by 0
Since dividing any number by 0 cannot be done, we say that division by **0** is *undefined*. It is impossible to compute an answer.

EXAMPLE 5 **Dividing Numbers by 0**

All the following are undefined.

(a) $\dfrac{6}{0}$ is undefined.

(b) $0\overline{)8}$ is undefined.

(c) $18 \div 0$ is undefined.

(d) $\dfrac{0}{0}$ is undefined.

4 Write each division problem as a multiplication problem.

(a) $5\overline{)15}^{\,3}$

(b) $\dfrac{32}{4} = 8$

(c) $48 \div 8 = 6$

5 Divide. If the division is not possible, write "undefined."

(a) $\dfrac{4}{0}$ $0\overline{)4}$ ∪

(b) $\dfrac{0}{4}$ ∅

(c) $0\overline{)36}$

36 ÷ 0 = undefined

(d) $36\overline{)0}$

0 ÷ 36 = 0

(e) $100 \div 0$

∪

(f) $0 \div 100$

0

6 Divide.

(a) $8 \div 8$

(b) $15\overline{)15}$

(c) $\dfrac{37}{37}$

Division Involving 0

$$0 \div \text{nonzero number} = 0 \quad \text{and} \quad \frac{0}{\text{nonzero number}} = 0$$

but

$$\text{nonzero number} \div 0 \quad \text{and} \quad \frac{\text{nonzero number}}{0} \text{ are undefined.}$$

CAUTION
When 0 is the divisor in a problem, you write "undefined" as the answer. Never divide by 0.

◀◀◀ **Work Problem 5 at the Side.**

Calculator Tip Try these two problems on your calculator. Jot down your answers.

9 ÷ 0 = _____   0 ÷ 9 = _____

When you try to divide by 0, the calculator cannot do it, so it shows the word "Error" or the letter "E" (for error) in the display. But, when you divide 0 by 9 the calculator displays 0, which is the correct answer.

OBJECTIVE 5 **Divide a number by itself.** What happens when a number is divided by itself? For example, what is $4 \div 4$ or $97 \div 97$?

Dividing a Number by Itself
Any *nonzero* number divided by itself is **1**.

EXAMPLE 6 **Dividing a Nonzero Number by Itself**

Divide.

(a) $16 \div 16 = 1$

(b) $32\overline{)32}$ (with 1 above)
$$32\overset{1}{\overline{)32}}$$

(c) $\dfrac{57}{57} = 1$

◀◀◀ **Work Problem 6 at the Side.**

OBJECTIVE 6 **Divide a number by 1.** What happens when a number is divided by 1? For example, what is $5 \div 1$ or $86 \div 1$?

Dividing a Number by 1
Any number divided by 1 is itself.

EXAMPLE 7 Dividing Numbers by 1

Divide.

(a) $5 \div 1 = 5$

(b) $1{\overline{\smash{)}26}}$ with quotient 26

(c) $\dfrac{41}{1} = 41$

Work Problem 7 at the Side. ▶▶▶

7 Divide.

(a) $9 \div 1$

(b) $1{\overline{\smash{)}18}}$

(c) $\dfrac{43}{1}$

OBJECTIVE 7 Use short division. **Short division** is a method of dividing a number by a one-digit divisor.

EXAMPLE 8 Using Short Division

Divide: $3{\overline{\smash{)}96}}$.

First, divide 9 by 3.

$$3{\overline{\smash{)}96}} \quad \text{quotient } 3 \quad \leftarrow \dfrac{9}{3} = 3$$

Next, divide 6 by 3.

$$3{\overline{\smash{)}96}} \quad \text{quotient } 32 \quad \leftarrow \dfrac{6}{3} = 2$$

Work Problem 8 at the Side. ▶▶▶

8 Divide.

(a) $2{\overline{\smash{)}24}}$

(b) $3{\overline{\smash{)}93}}$

(c) $4{\overline{\smash{)}88}}$

(d) $2{\overline{\smash{)}624}}$

When two numbers do not divide exactly, the leftover portion is called the **remainder.**

EXAMPLE 9 Using Short Division with a Remainder

Divide 147 by 4.

Write the problem.

$$4{\overline{\smash{)}147}}$$

Because 1 cannot be divided by 4, divide 14 by 4. Notice that the 3 is placed over the 4 in 14.

$$4{\overline{\smash{)}14^{2}7}} \quad \text{quotient } 3 \qquad \dfrac{14}{4} = 3 \text{ with 2 left over}$$

Next, divide 27 by 4. The final number left over is the remainder. Use **R** to indicate the remainder, and write the remainder to the side.

$$4{\overline{\smash{)}14^{2}7}} \quad \text{quotient } 3\,6\ \mathbf{R3} \qquad \dfrac{27}{4} = 6 \text{ with 3 left over}$$

Work Problem 9 at the Side. ▶▶▶

9 Divide.

(a) $2{\overline{\smash{)}125}}$

(b) $3{\overline{\smash{)}215}}$

(c) $4{\overline{\smash{)}538}}$

(d) $\dfrac{819}{5}$

⑩ Divide.

(a) $4\overline{)523}$

EXAMPLE 10 Dividing with a Remainder

Divide 1809 by 7.

Divide 7 into 18.

$$7\overline{)18^{4}09} \qquad \frac{18}{7} = 2 \text{ with 4 left over}$$

Divide 7 into 40.

$$7\overline{)18^{4}0^{5}9} \qquad \frac{40}{7} = 5 \text{ with 5 left over}$$

Divide 7 into 59.

$$7\overline{)18^{4}0^{5}9} \text{ R3} \qquad \frac{59}{7} = 8 \text{ with 3 left over}$$

◄◄◄ Work Problem 10 at the Side.

(b) $\dfrac{515}{7}$

NOTE
Short division takes practice but is useful in many situations.

(c) $3\overline{)1885}$

OBJECTIVE **8** **Use multiplication to check the answer to a division problem.**   **Check** the answer to a division problem as follows.

Checking Division

$$(\text{divisor} \times \text{quotient}) + \text{remainder} = \text{dividend}$$

Parentheses tell you what to do first: Multiply the divisor by the quotient, then add the remainder.

(d) $6\overline{)1415}$

EXAMPLE 11 Checking Division by Using Multiplication

Check each answer.

(a) $5\overline{)458}$ quotient 91 R3

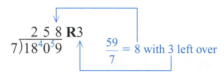

(divisor × quotient) + remainder = dividend

$$(5 \times 91) + 3$$
$$455 + 3 = 458$$

Matches original dividend, so the division was done correctly.

Continued on Next Page

$$\begin{array}{r} 239\ \mathbf{R}4 \\ \textbf{(b)}\ 6\overline{)1437} \end{array}$$

$$(\text{divisor} \times \text{quotient}) + \text{remainder} = \text{dividend}$$

$$(6 \ \times \ 239) \ + \ 4$$

$$1434 \quad + \quad 4 \quad = 1438$$

Does not match original dividend.

The answer does not check. Rework the original problem to get the correct answer, 239 **R**3.

CAUTION

A common error when checking division is to forget to add the remainder. Be sure to add any remainder when checking a division problem.

Work Problem 11 at the Side. ▷▷▷

O B J E C T I V E 9 Use tests for divisibility.   It is often important to know whether a number is *divisible* by another number. You will find this useful in Chapter 2 when writing fractions in lowest terms.

Divisibility

One whole number is **divisible** by another if the remainder is 0.

Use the following tests to decide whether one number is divisible by another number.

Tests for Divisibility

A number is divisible by

2	if it ends in 0, 2, 4, 6, or 8. These are the even numbers.
3	if the sum of its digits is divisible by 3.
4	if the last two digits make a number that is divisible by 4.
5	if it ends in 0 or 5.
6	if it is divisible by both 2 and 3.
7	has no simple test.
8	if the last three digits make a number that is divisible by 8.
9	if the sum of its digits is divisible by 9.
10	if it ends in 0.

The most commonly used tests are those for 2, 3, 5, and 10.

---

11 Use multiplication to check each division. If an answer is incorrect, give the correct answer.

$$\textbf{(a)}\ 2\overline{)65} \quad \begin{array}{c}32\ \mathbf{R}1\end{array}$$

$$\textbf{(b)}\ 7\overline{)586} \quad \begin{array}{c}83\ \mathbf{R}4\end{array}$$

$$\textbf{(c)}\ 3\overline{)1223} \quad \begin{array}{c}407\ \mathbf{R}2\end{array}$$

$$\textbf{(d)}\ 5\overline{)2383} \quad \begin{array}{c}476\ \mathbf{R}3\end{array}$$

12 Which numbers are divisible by 2?

(a) 258

(b) 307

(c) 4216

(d) 73,000

Divisibility by 2

A number is divisible by **2** if the number ends in 0, 2, 4, 6, or 8. All even numbers are divisible by 2.

EXAMPLE 12   **Testing for Divisibility by 2**

Are the following numbers divisible by 2?

(a) 986   Ends in 6

Because the number ends in 6, which is an even number, the number 986 is divisible by 2.

(b) 3255 is not divisible by 2.   Ends in 5, and not in 0, 2, 4, 6, or 8

▶◀ **Work Problem 12 at the Side.**

Divisibility by 3

A number is divisible by **3** if the sum of its digits is divisible by **3**.

EXAMPLE 13   **Testing for Divisibility by 3**

Are the following numbers divisible by 3?

(a) 4251

    Add the digits.

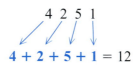

$$4 + 2 + 5 + 1 = 12$$

Because 12 is divisible by 3, the number 4251 is divisible by 3.

(b) 29,806

    Add the digits.

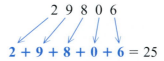

$$2 + 9 + 8 + 0 + 6 = 25$$

Because 25 is not divisible by 3, the number 29,806 is not divisible by 3.

13 Which numbers are divisible by 3?

(a) 743

(b) 5325

(c) 374,214

(d) 205,633

CAUTION

Be careful when testing for divisibility by adding the digits. This method works only for the numbers 3 and 9.

▶◀ **Work Problem 13 at the Side.**

ANSWERS

12. all but b
13. b and c

Divisibility by 5 and by 10

A number is divisible by **5** if it ends in 0 or 5.
A number is divisible by **10** if it ends in 0.

EXAMPLE 14 Testing for Divisibility by 5

Are the following numbers divisible by 5?

(a) 12,900 ends in 0 and is divisible by 5.

(b) 4325 ends in 5 and is divisible by 5.

(c) 392 ends in 2 and is not divisible by 5.

Work Problem 14 at the Side. ▶▶▶

EXAMPLE 15 Testing for Divisibility by 10

Are the following numbers divisible by 10?

(a) 700 and 9140 both end in 0 and are divisible by 10.

(b) 355 and 18,743 do not end in 0 and are not divisible by 10.

Work Problem 15 at the Side. ▶▶▶

14 Which numbers are divisible by 5?

(a) 180

    36

(b) 635

    127

(c) 8364

    1672.8

(d) 206,105

    41221

15 Which numbers are divisible by 10?

(a) 270

    27

(b) 495

    49:5

(c) 5030

    503

(d) 14,380

    1438

ANSWERS
14. all but c
15. all but b

Real-Data Applications

Study Time

As a college student, you need to plan a schedule to accommodate many responsibilities: attending class, preparing for class, studying for exams, traveling to and from college, part-time work, family responsibilities, and personal time. First, you must calculate the amount of time dedicated to each of your obligations.

As an example, suppose that you are a full-time student enrolled in 12 credit hours of class: 4 credits of biology, 4 credits of computer science, 3 credits of mathematics, and 1 credit of physical education. Biology and computer science each have 3 hours of lecture and a 3-hour lab each week. Your biology and mathematics instructors recommend that you spend an additional 2 hours per week of study time for each hour of lecture time. Your physical education class is only 1 credit, but you are in class 3 hours each week.

Activity	Hours per Week
Class time	12
Lab time	12
Study time for mathematics and biology	15
Travel time to and from college	5
Part-time work (including travel time)	25
Sleep (8 hours per day)	56
Meals (3 hours per day)	21
Hygiene (baths, dressing, etc.)	7
Other (housecleaning, laundry, etc.)	14

1. How many total hours are in one week? 168

2. Fill in the table entries for the number of hours in a week spent on class time, lab time, study time for biology and mathematics, sleep, and meals.

3. How many hours per week are spent on college-related activities?

4. How much more time is required for personal time (sleeping, eating, hygiene) than for college-related activities?

5. Based on the table data, how many hours per week are spent on the activities listed?

6. How many hours per week are available for other activities, such as dating, shopping, family responsibilities, and so on?

7. List two additional activities that are not listed in the table. Estimate the amount of time per week required for each activity.

8. Based on your analysis of the time constraints of this schedule, what changes would you make to your schedule and other obligations?

Handwritten margin notes:
Biology — 3 hours of lecture time (total=6)
3 hour-lab
4 bio
4 cs
3 math
1 PE = 12 credit hours
3+3+3 math

1.5 Exercises

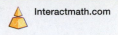
Write each division problem using two other symbols. See Example 1.

1. $24 \div 4 = 6$

2. $36 \div 3 = 12$

3. $\dfrac{45}{9} = 5$

4. $\dfrac{56}{8} = 7$

5. $2\overline{)16}$ with quotient 8

6. $8\overline{)48}$ with quotient 6

Dividing ÷ 0 = U

Divide. If the division is not possible, write "undefined." See Examples 3, 5, 6, and 7.

7. $9 \div 9$

8. $36 \div 9$

9. $\dfrac{14}{2}$

10. $\dfrac{10}{0}$ $0\overline{)10}$ = U

$10 \div U = U$

11. $22 \div 0$

12. $6 \div 6$

13. $\dfrac{24}{1}$

14. $\dfrac{12}{1}$

15. $15\overline{)0}$

16. $\dfrac{0}{12}$

17. $0\overline{)43}$

18. $\dfrac{8}{0}$

19. $\dfrac{15}{1}$

20. $\dfrac{0}{0}$

21. $\dfrac{8}{1}$

22. $\dfrac{0}{5}$

Divide by using short division. Use multiplication to check each answer. See Examples 8, 9, and 10.

23. $3\overline{)75}$

24. $5\overline{)85}$

25. $7\overline{)126}$

26. $6\overline{)168}$

27. $4\overline{)1216}$

28. $5\overline{)2305}$

29. $4\overline{)2509}$

30. $8\overline{)1335}$

31. $6\overline{)9137}$

32. $9\overline{)8371}$

33. $6\overline{)1854}$

34. $8\overline{)856}$

35. 12,020 ÷ 4   **36.** 8012 ÷ 4   **37.** 30,036 ÷ 6   **38.** 32,008 ÷ 8

39. 2434 ÷ 3   **40.** 5993 ÷ 7   **41.** 12,947 ÷ 5   **42.** 33,285 ÷ 9

43. 29,298 ÷ 4   **44.** 17,937 ÷ 6   **45.** 12,630 ÷ 4   **46.** 46,560 ÷ 7

47. 21,040 ÷ 8   **48.** $\dfrac{8199}{9}$ **49.** $\dfrac{74,751}{6}$ **50.** $\dfrac{72,543}{5}$

51. $\dfrac{71,776}{7}$ **52.** $\dfrac{77,621}{3}$ **53.** $\dfrac{128,645}{7}$ **54.** $\dfrac{172,255}{4}$

Use multiplication to check each answer. If an answer is incorrect, find the correct answer. See Example 11.

55. $5\overline{)1877}$ 375 **R2**   **56.** $3\overline{)1282}$ 427 **R1**   **57.** $3\overline{)5725}$ 1908 **R2**   **58.** $5\overline{)2158}$ 432 **R3**

1908 × 3 + 2 =
5725

59. $7\overline{)4692}$ 650 **R2**   **60.** $9\overline{)5974}$ 663 **R5**   **61.** $6\overline{)21,409}$ 3 568 **R2**   **62.** $6\overline{)3192}$ 532

63. $8\overline{)16,019}$ 2 002 **R3**   **64.** $8\overline{)33,664}$ 4 208   **65.** $6\overline{)69,140}$ 11,523 **R2**   **66.** $3\overline{)82,598}$ 27,532 **R1**

67. $9\overline{)86,655}$ 9 628 **R7**   **68.** $7\overline{)50,809}$ 7 258 **R4**   **69.** $8\overline{)222,576}$ 27,822   **70.** $4\overline{)311,216}$ 77,804

71. Explain in your own words how to check a division problem using multiplication. Be sure to include what must be done if the quotient includes a remainder.

72. Describe the three divisibility rules that you feel might be most useful to you and tell why.

Solve each application problem.

73. The Carnival Cruise Line has 2624 linen napkins. If it takes eight napkins to set each table, find the number of tables that can be set. (*Source: USA Today.*)

2624 napkins
8
328 tables

74. A school district will distribute 1620 new science books equally among 12 schools. How many books will each school receive?

75. Last year in Seattle, 56,000 people went to the circus over a 5-day period. If the same number of people attended each day, what was the daily attendance? (*Source:* Feld Entertainment, Vienna, Virginia.)

76. One gallon of orange juice will serve nine people. How many gallons are needed for 3483 people?

77. Lottery winnings of $436,500 are divided equally among nine Starbucks employees. Find the amount received by each employee.

78. How many 5-pound bags of organic whole wheat flour can be filled from a 17,175-pound bin of flour?

79. If 8 gallons of fertilizer are needed for each acre of land, find the number of acres that can be fertilized with 1080 gallons of fertilizer.

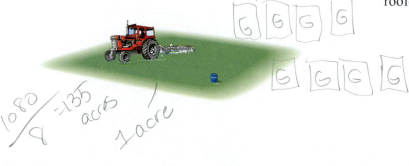

Fertilizer

1080/8 = 135 acres
1 acre

you have 1080 gallons of F

80. A roofing contractor has purchased 1134 squares of roofing material. If each cabin needs 9 squares of material, find the number of cabins that can be roofed. (1 square measures 10 ft by 10 ft.)

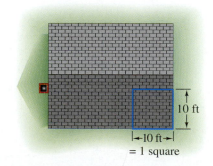

10 ft
|←10 ft→|
= 1 square

81. A class-action lawsuit settlement of $6,825,000 will be divided evenly among six injured people. Find the amount received by each person.

82. A 12,000-square foot condominium at the edge of Central Park in Manhattan sold for a record $45,000,000. If the buyer pays for the condominium in eight equal payments, find the amount of each payment. (*Source: USA Today.*)

83. Kaci Salmon, a supervisor at Albany Electric, earns $36,540 per year. Find the amount of her earnings in a 3-month period.

84. When Alex Rodriguez played shortstop for the Texas Rangers, he earned $25,000,000 in one year. If he was paid in four equal checks, find the amount of each check.

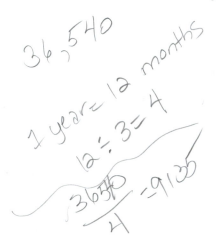

Put a ✓ mark in the blank if the number at the left is divisible by the number at the top. Put an X in the blank if the number is not divisible by the number at the top.
See Examples 12–15.

	2	3	5	10			2	3	5	10
85. 60	✓	✓	✓	✓	**86.** 35	___	___	___	___
87. 92	___	___	___	___	**88.** 96	___	___	___	___
89. 445	___	___	___	___	**90.** 897	___	___	___	___
91. 903	___	___	___	___	**92.** 500	___	___	___	___
93. 5166	___	___	___	___	**94.** 8302	___	___	___	___
95. 21,763	___	___	___	___	**96.** 32,472	___	___	___	___

1.6 Long Division

If the total cost of 42 external cable modems is $3066, we can find the cost of each modem using **long division**. Long division is used to divide by a number with more than one digit.

OBJECTIVE 1 Do long division. In long division, estimate the various numbers by using a **trial divisor**, which is used to get a **trial quotient**.

OBJECTIVES

1 Do long division.

2 Divide numbers ending in 0 by numbers ending in 0.

3 Use multiplication to check division answers.

EXAMPLE 1 Using a Trial Divisor and a Trial Quotient

Divide: $42\overline{)3066}$.

Because 42 is closer to 40 than to 50, use the first digit of the divisor as a trial divisor.

$$42$$
↑
└────── Trial divisor

Try to divide the first digit of the dividend by 4. Since 3 cannot be divided by 4, use the first *two* digits, 30.

$$\frac{30}{4} = 7 \text{ with remainder 2}$$

$$7 \leftarrow \text{Trial quotient}$$
$$42\overline{)3066}$$
↑
└────── 7 goes over the 6, because

$$\frac{306}{42} \text{ is about 7.}$$

Multiply 7 and 42 to get 294; next, subtract 294 from 306.

$$
\begin{array}{r}
7 \\
42\overline{)3066} \\
294 \quad \leftarrow 7 \times 42 \\
\overline{12} \quad \leftarrow 306 - 294
\end{array}
$$

Bring down the 6 at the right.

$$
\begin{array}{r}
7 \\
42\overline{)3066} \\
294\downarrow \\
\overline{126} \quad \leftarrow 6 \text{ brought down}
\end{array}
$$

Use the trial divisor, 4.

First two digits of 126 → $\frac{12}{4} = 3$ The cost of each modem is $73.

$$
\begin{array}{r}
73 \\
42\overline{)3066} \\
294 \\
\overline{126} \\
126 \quad \leftarrow 3 \times 42 = 126 \\
\overline{0}
\end{array}
$$

Check the answer by multiplying 42 and 73. The product should be 3066.

1 Divide.

(a) $28\overline{)2296}$

(b) $16\overline{)1024}$

(c) $61\overline{)8784}$

(d) $\dfrac{2697}{93}$

2 Divide.

(a) $24\overline{)1344}$

(b) $72\overline{)4472}$

(c) $65\overline{)5416}$

(d) $89\overline{)6649}$

> **CAUTION**
> The *first digit* of the quotient in long division must be placed in the proper position over the dividend.

◀◀◀ Work Problem 1 at the Side.

EXAMPLE 2 **Dividing to Find a Trial Quotient**

Divide: $58\overline{)2730}$.

Use 6 as a trial divisor, since 58 is closer to 60 than to 50.

First two digits of dividend ⟶ $\dfrac{27}{6} = 4$ with 3 left over

4 ← Trial quotient

$58\overline{)2730}$
$\underline{232}$ ← $4 \times 58 = 232$
41 ← $273 - 232 = 41$ (smaller than 58, the divisor)

Bring down the 0.

4
$58\overline{)2730}$
$\underline{232}{\downarrow}$
410 ← 0 brought down

First two digits of 410 ⟶ $\dfrac{41}{6} = 6$ with 5 left over

46 ← Trial quotient

$58\overline{)2730}$
$\underline{232}$
410
$\underline{348}$ ← $6 \times 58 = 348$
62 ← Greater than 58

The remainder, 62, is greater than the divisor, 58, so 7 should be used instead of 6.

47 **R4** ←
$58\overline{)2730}$
$\underline{232}$
410
$\underline{406}$ ← $7 \times 58 = 406$
4 ← $410 - 406$

◀◀◀ Work Problem 2 at the Side.

Sometimes it is necessary to insert a 0 in the quotient.

EXAMPLE 3 **Inserting Zeros in the Quotient**

Divide: $42\overline{)8734}$.

Start as above.

$$
\begin{array}{r}
2 \\
42\overline{)8734} \\
84 \quad \leftarrow 2 \times 42 = 84 \\
\hline
3 \quad \leftarrow 87 - 84 = 3
\end{array}
$$

Bring down the 3.

$$
\begin{array}{r}
2 \\
42\overline{)8734} \\
84\downarrow \\
\hline
33 \quad \leftarrow 3 \text{ brought down}
\end{array}
$$

Since 33 cannot be divided by 42, place a 0 in the quotient as a placeholder.

$$
\begin{array}{r}
20 \quad \leftarrow 0 \text{ in quotient} \\
42\overline{)8734} \\
84 \\
\hline
33
\end{array}
$$

Bring down the final digit, the 4.

$$
\begin{array}{r}
20 \\
42\overline{)8734} \\
84\downarrow \\
\hline
334 \quad \leftarrow 4 \text{ brought down}
\end{array}
$$

Complete the problem.

$$
\begin{array}{r}
207\ \mathbf{R}40 \\
42\overline{)8734} \\
84 \\
\hline
334 \\
294 \\
\hline
40
\end{array}
$$

The answer is 207 **R**40.

CAUTION
There **must be a digit** in the quotient (answer) above every digit in the dividend once the answer has begun. Notice in Example 3 that a **0** was used to assure an answer digit above every digit in the dividend.

Work Problem 3 at the Side.

OBJECTIVE 2 **Divide numbers ending in 0 by numbers ending in 0.**
When the divisor and dividend both contain zeros at the far right, recall that these numbers are multiples of 10. As with multiplication, there is a short way to divide these multiples of 10. Look at the following examples.

$$26{,}000 \div 1 = 26{,}000$$
$$26{,}000 \div 10 = 2600$$
$$26{,}000 \div 100 = 260$$
$$26{,}000 \div 1000 = 26$$

Do you see a pattern? These examples suggest the following rule.

3 Divide.

(a) $17\overline{)1823}$

(b) $23\overline{)4791}$

(c) $39\overline{)15{,}933}$

(d) $78\overline{)23{,}462}$

4 Divide.

(a) $70 \div 10$

(b) $2600 \div 100$

(c) $505,000 \div 1000$

Dividing a Whole Number by 10, 100, or 1000
Divide a whole number by 10, 100, or 1000 by dropping the appropriate number of zeros from the whole number.

EXAMPLE 4 Dividing by Multiples of 10

Divide.

One 0 in divisor.

(a) $60 \div 10 = 6$

0 dropped

Two zeros in divisor

(b) $3500 \div 100 = 35$

00 dropped

Three zeros in divisor

(c) $915,000 \div 1000 = 915$

000 dropped

◄◄◄ Work Problem 4 at the Side.

In the next example, we find the quotient for other multiples of 10 by dropping zeros.

5 Divide.

(a) $50\overline{)6250}$

(b) $130\overline{)131,040}$

(c) $3400\overline{)190,400}$

EXAMPLE 5 Dividing by Multiples of 10

Divide:

(a) $40\overline{)11,000}$ Drop one zero from the divisor and the dividend.

$$
\begin{array}{r}
275 \\
4\overline{)1100} \\
8 \\
\hline
30 \\
28 \\
\hline
20 \\
20 \\
\hline
0
\end{array}
$$

Since $1100 \div 4$ is 275, then $11,000 \div 40$ is also 275.

(b) $3500\overline{)31,500}$ Drop two zeros from the divisor and the dividend.

$$
\begin{array}{r}
9 \\
35\overline{)315} \\
315 \\
\hline
0
\end{array}
$$

Since $315 \div 35$ is 9, then $31,500 \div 3500$ is also 9.

NOTE
Dropping zeros when dividing by multiplies of 10 **does not** change the quotient (answer).

◄◄◄ Work Problem 5 at the Side.

Answers
4. (a) 7 (b) 26 (c) 505
5. (a) 125 (b) 1008 (c) 56

OBJECTIVE **3** **Use multiplication to check division answers.**
Answers in long division can be checked just as answers in short division were checked.

EXAMPLE 6 **Checking Division by Using Multiplication**

Check each answer.

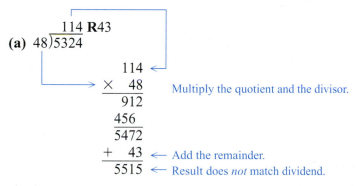

(a) 48)5324 114 **R**43

$$\begin{array}{r} 114 \\ \times\ \ 48 \\ \hline 912 \\ 456\ \ \\ \hline 5472 \\ +\ \ \ 43 \\ \hline 5515 \end{array}$$

Multiply the quotient and the divisor.

← Add the remainder.
← Result does *not* match dividend.

The answer does ***not*** check. Rework the original problem to get 110 **R**44.

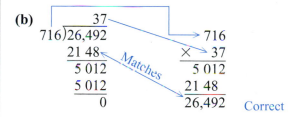

(b)
$$\begin{array}{r} 37 \\ 716)\overline{26,492} \\ 21\ 48\ \ \\ \hline 5\ 012 \\ 5\ 012 \\ \hline 0 \end{array}$$
Matches

$$\begin{array}{r} 716 \\ \times\ \ \ 37 \\ \hline 5\ 012 \\ 21\ 48\ \ \\ \hline 26,492 \end{array}$$
Correct

Calculator Tip To check the answer to Example 6(a), don't forget to add the remainder.

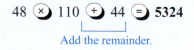

48 ⊗ 110 ⊕ 44 ⊜ **5324**

Add the remainder.

CAUTION
When checking a division problem, first multiply the quotient and the divisor. Then be sure to ***add any remainder*** before checking it against the original dividend.

Work Problem 6 at the Side. ▶▶▶

6 Decide whether each answer is correct. If the answer is incorrect, find the correct answer.

(a) 16)608
$$\begin{array}{r} 38 \\ 16)\overline{608} \\ 48\ \ \\ \hline 128 \\ 128 \\ \hline 0 \end{array}$$

(b) 426)19,170
$$\begin{array}{r} 42\ R178 \\ 426)\overline{19,170} \\ 17\ 04\ \ \\ \hline 1\ 130 \\ 952 \\ \hline 178 \end{array}$$

(c) 514)29,316
$$\begin{array}{r} 57\ R18 \\ 514)\overline{29,316} \\ 25\ 70\ \ \\ \hline 3\ 616 \\ 3\ 598 \\ \hline 18 \end{array}$$

Real-Data Applications

Sharing Travel Expenses

Sharing costs for meals, entertainment, and travel can be a challenging mathematical problem, especially if the costs are incurred over several days and paid for by different people.

As an example, suppose that Donna and Jerry (couple), Peg and Richard (couple), and Jo (single) decide to spend a week touring Wales and Ireland. Donna has booked the accommodations, and each person will be responsible for their own meals and hotel expenses. However, to save on transportation costs, Richard (who lives in England) will use his personal car, and the group will share the costs for insurance, gasoline, parking, and the ferry. At the end of the trip, the costs were those listed in the table below. All the costs are rounded to the nearest U.S. dollar, although the travelers actually paid the amounts in British pounds and Irish punts.

Expense	Paid By	Cost (in U.S. dollars)
Supplemental car insurance	Richard	21
Ferry (prebooked)	Donna	220
Ferry (upgrade to fast boat)	Richard	40
Parking (Dublin, Ireland)	Richard	21
Gasoline (Wales)	Richard	46
Gasoline (Ireland)	Richard	13
Gasoline (Ireland)	Richard	22
Gasoline (Ireland)	Richard	24
Gasoline (Wales)	Richard	46
Gasoline (Wales)	Richard	21

1. What was the total amount spent on transportation?

2. To compute the shared costs, divide the total transportation costs by the number of travelers.

   (a) How many whole dollars does each traveler owe for transportation costs? (The division does not come out evenly; there is a remainder.)

   (b) How many dollars are left over (the remainder)? What should be done with the remainder costs to ensure a fair division of the costs?

   (c) If the total transportation costs are rounded up to the nearest multiple of 5, then how much would each traveler owe?

3. Using your answer from Problem 2(c), how much is Donna and Jerry's share of the costs as a couple? Do they owe money or are they owed money? If they owe money, how much and to whom should it be paid? If money is owed to them, who owes the money and how much is owed?

4. Using your answer from Problem 2(c), how much is Richard and Peg's share of the costs as a couple? Do they owe money or are they owed money? If they owe money, how much and to whom should it be paid? If money is owed to them, who owes the money and how much is owed?

5. Did each traveler pay the same amount toward transportation costs?

1.6 Exercises

 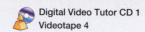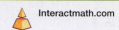
Decide where the first digit in the quotient would be located. Then without finishing the division, you can tell which of the three choices is the correct answer. Circle your choice. See Examples 1 and 2.

1. 50)2650

    5   53   530

2. 14)476

    3   34   304

3. 18)4500

    2   25   250

4. 35)5600

    16   160   1600

5. 86)10,327

    12   120 **R7**   1200

6. 46)24,026

    5   52   522 **R14**

7. 26)28,735

    11   110   1105 **R5**

8. 12)116,953

    974 **R2**   9746 **R1**   97,460

9. 21)149,826

    71   713   7134 **R12**

10. 64)208,138

    325 **R2**   3252 **R10**   32,521

11. 523)470,800

    9 **R100**   90 **R100**   900 **R100**

12. 230)253,230

    11   110   1101

Divide by using long division. Use multiplication to check each answer. See Examples 1–3.

13. 18)1319

14. 58)3654

15. 23)10,963

16. 83)39,692

17. 26)62,583

18. 28)84,249

19. 74)84,819

20. 238)186,948

21. 153)509,725

22. 308)26,796

23. 420)357,000

24. 900)153,000

Use multiplication to check each answer. If an answer in incorrect, find the correct answer.
See Example 6.

25.
$$\begin{array}{r} 101 \text{ R4} \\ 35\overline{)3549} \end{array}$$

26.
$$\begin{array}{r} 42 \text{ R26} \\ 64\overline{)2712} \end{array}$$

27.
$$\begin{array}{r} 658 \text{ R9} \\ 28\overline{)18,424} \end{array}$$

28.
$$\begin{array}{r} 239 \text{ R121} \\ 145\overline{)34,776} \end{array}$$

29.
$$\begin{array}{r} 62 \text{ R3} \\ 614\overline{)38,068} \end{array}$$

30.
$$\begin{array}{r} 174 \text{ R368} \\ 557\overline{)97,286} \end{array}$$

31. Describe in your own words a shortcut you can use to divide multiples of 10 by 10, by 100, or by 1000. Write an example problem and solve it.

32. Suppose you have a division problem with a remainder in the answer. Explain how to check your answer by writing an example problem that has a remainder.

Solve each application problem by using addition, subtraction, multiplication, or division as needed. See Examples 3–5.

33. A cross-country cyclist will ride 90 miles each day. If the coast-to-coast distance that the rider travels is 3150 miles, how many days will it take to cross the country?

34. A new bridge from Owensboro, Kentucky, to Rockport, Indiana, is 2200 feet long and cost $55,998,800. Find the construction cost per foot. (*Source:* Federal Highway Administration.)

35. Don Gracey, the Mountain Timesmith, has serviced and repaired 636 clocks this year. He has worked on 272 wall clocks and 308 table clocks. The remainder were standing floor clocks. Find the number of floor clocks he worked on this year.

36. Two separated parents each share some of the $3718 education costs of their child. If one parent paid $1880, how much did the other pay?

37. To complete her college education, Judy Martinez received education loans of $34,080 including interest. Find her monthly payment if the loan is to be paid off in 96 months (8 years).

38. A consultant charged $13,050 for evaluating a school's compliance with the Americans with Disabilities Act. If the consultant worked 225 hours, find the rate charged per hour.

39. Radio Flyer, Inc. manufactures 1000 wagons each hour. How many wagons can be manufactured in one month of 22 workdays of 8 hours per day? (*Source:* Costco Connection.)

1000 hr
X
176

176,000

40. There are two conveyor lines at Marton Salt Company, each of which packages 240 sacks of salt per hour. If the lines operate for 8 hours, find the total number of sacks of salt packaged by the two lines.

41. The average U.S. household of 2.5 people spent $2028 eating away from home last year. Find the average weekly household cost of eating away from home. (*Hint:* 1 year equals 52 weeks.) (*Source:* U.S. Bureau of Labor Statistics consumer expenditure surveys.)

$39 a week

$\frac{2028}{52}$

42. Former professional basketball player Junior Bridgeman now owns 120 Wendy's restaurants with 4080 employees. Find the average number of employees at each restaurant. (*Source:* National Basketball Retired Players Association.)

RELATING CONCEPTS (EXERCISES 43–50) For Individual or Group Work

Knowing and using the rules of divisibility is necessary in problem solving.
Work Exercises 43–50 in order.

43. If you have $0 and you divide this amount among three people, how much will each receive?

44. When 0 is divided by any nonzero number, the result is _____.

45. Divide.

8 ÷ 0

46. We say that division by 0 is *undefined* because it is

_____ to compute the answer. Give an
(possible/impossible)
example involving cookies that will support your answer.

47. Divide.

(a) 14 ÷ 1

(b) $1\overline{)17}$

(c) $\dfrac{38}{1}$

48. Any number divided by 1 is the number itself. Is this also true when multiplying by 1? Give three examples that support your answer.

49. Divide.

(a) 32,000 ÷ 10

(b) 32,000 ÷ 100

(c) 32,000 ÷ 1,000

50. Write a rule that explains the shortcut for doing divisions like the ones in Exercise 49.

Real-Data Applications

Post Office Facts

The United States Postal Service posts a Web page on the Internet that gives a list of facts about their service. Some of those facts are given in the table below.

Resource/Service	Number
1. Mail collection boxes	326,000
2. Post offices	38,019
3. Delivery points	138 million
4. Pieces of First Class mail delivered each year	200 billion
5. Processing plants sorting and shipping the mail	331
6. Pounds of mail carried on commercial airline flights annually	2.7 billion
7. Commercial airline flights per day	15,000
8. Miles driven to move the mail annually	1.15 billion
9. Vehicles to pick up, transport, and deliver the mail	215,530
10. Customers a day who transact business at the post offices	7 million

Source: United States Postal Service, www.usps.com

1. Write the number of delivery points entirely in digits. (Do not use the word "million" in your answer.)

2. Write the number of miles driven annually to move the mail entirely in digits.

3. Estimate the number of post offices, rounded to the nearest thousand.

4. Use your estimate from Problem 3 to compute the average number of customers who transact business each day per post office. Round the answer to the nearest person. (*Hint:* To find the average number of customers per day who transact business at each post office, divide the total number of customers per day by the number of post offices. Recall that you can use a shortcut to simplify a division problem by crossing out the same number of zeros in the divisor and the dividend.)

5. Find the total number of commercial airline flights per year that carry postal freight. Write the result entirely in words. (*Hint:* Assume that planes fly 365 days per year.)

6. Find the average number of pounds of mail per commercial airline flight, rounded to the nearest 10 pounds. (*Hint:* Divide the total number of pounds of mail carried annually by the number of flights per year, calculated in Problem 5. Use the division shortcut described in Problem 4.)

7. Mail is sorted and shipped from processing plants to the post offices. Which of the following is a rough estimate of the number of post offices served by each processing plant: 10, 100, or 1000? Explain your choice.

Summary Exercises on Whole Numbers

Write the digit for the given ***place value*** *in each whole number.*

1. 631,548

| ten thousands | millions | hundred millions |

| tens | hundred thousands | thousands |

2. 76,047,309   **3.** 9,181,576,423   *Write each*

Write each number in words.

4. 86,002   **5.** 425,208,733

Add or subtract as indicated.

6. 46   **7.** 166 + 739   **8.** 82   **9.** 798
    + 51   − 61   − 389

10. 6382 + 4062 + 7129   **11.** 75 + 81,579 + 506 + 4

12. 1704   **13.** 55,000   **14.** 70,552
     − 1027   − 17,326   − 34,663

Multiply using the shortcut for multiples of 10.

15. 56 × 10   **16.** 140 × 40   **17.** 500   **18.** 3600
                                                       × 700   ×   70

Write the numbers in each sentence using digits.

19. Last year General Motors sold eight million, six hundred forty-two thousand, one hundred eighty vehicles worldwide. (*Source:* Associated Press.)

20. Worldwide sales for Toyota last year were six million, seven hundred eighty-three thousand, four hundred thirty vehicles. (*Source:* Associated Press.)

Multiply or divide as indicated. If the division is not possible, write "undefined."

21. $8 \div 8$

22. $0 \div 9$

23. $\dfrac{12}{0}$

24. $\dfrac{15}{1}$

25. 7×8

26. $(6)(0)(5)$

27. $8 \cdot 4 \cdot 3$

28. $\dfrac{608}{2}$

29. $3\overline{)8252}$

30. $4569 \div 6$

31. $\begin{array}{r} 65 \\ \times\ 52 \\ \hline \end{array}$

32. $\begin{array}{r} 507 \\ \times\ 435 \\ \hline \end{array}$

33. $(28)(72)$

34. $(41)(36)$

35. $25\overline{)1950}$

36. $18\overline{)3780}$

37. $\begin{array}{r} 3602 \\ \times\ 5008 \\ \hline \end{array}$

38. $62\overline{)31{,}400}$

39. $630\overline{)32{,}760}$

40. $351\overline{)424{,}011}$

1.7 Rounding Whole Numbers

One way to get a rough check on an answer is to *round* the numbers in the problem. **Rounding** a number means finding a number that is close to the original number, but easier to work with.

For example, the county planning commissioner might be discussing the need for more affordable housing. To demonstrate this, she probably would not need to say that the county is in need of 8235 more affordable housing units— she probably could say that the county needs 8200 or even 8000 housing units.

OBJECTIVE **1** **Locate the place to which a number is to be rounded.** The first step in rounding a number is to locate the *place to which the number is to be rounded.*

EXAMPLE 1 **Finding the Place to Which a Number Is to Be Rounded**

Locate and draw a line under the place to which each number is to be rounded.

(a) Round 83 to the nearest ten. Is 83 closer to 8<u>0</u> or to 9<u>0</u>?

83 is closer to 80.

Tens place

(b) Round 54,702 to the nearest thousand. Is it closer to 5<u>4</u>,000 or to 5<u>5</u>,000?

54,702 is closer to 55,000.

Thousands place

(c) Round 2,806,124 to the nearest hundred thousand. Is it closer to 2,<u>8</u>00,000 or to 2,<u>9</u>00,000?

2,806,124 is closer to 2,800,000.

Hundred thousands place

Work Problem 1 at the Side.

OBJECTIVE **2** **Round numbers.**   Use the following rules for rounding whole numbers.

Rounding Whole Numbers

Step 1   Locate the **place** to which the number is to be rounded. Draw a line under that place.

Step 2(a)   Look only at the next digit to the right of the one you underlined. If it is **5 or more, increase** the underlined digit by 1.

Step 2(b)   If the next digit to the right is **4 or less, do not change** the digit in the underlined place.

Step 3   **Change** all digits to the right of the underlined place to zeros.

EXAMPLE 2 **Using Rounding Rules for 4 or Less**

Round 349 to the nearest hundred.

Step 1   Locate the place to which the number is being rounded. Draw a line under that place.

3<u>4</u>9

Hundreds place

Continued on Next Page

OBJECTIVES

1 Locate the place to which a number is to be rounded.

2 Round numbers.

3 Round numbers to estimate an answer.

4 Use front end rounding to estimate an answer.

1 Locate and draw a line under the place to which each number is to be rounded. Then answer the question.

(a) 373 (nearest ten)

Is it closer to 370 or to 380?

_____

(b) 1482 (nearest thousand)

Is it closer to 1000 or to 2000?

_____

(c) 89,512 (nearest hundred)

Is it closer to 89,500 or to 89,600?

_____

(d) 546,325 (nearest ten thousand)

Is it closer to 540,000 or to 550,000?

_____

2 Round to the nearest ten.

(a) <u>6</u>2   *10*

60

(b) 94

90

(c) 134

130

(d) 7543

7540

3 Round to the nearest thousand.

(a) <u>3</u>683   *1000*

4000

(b) <u>6</u>502

7000

(c) 84,<u>6</u>21

85,000

(d) <u>5</u>5,960

56,000

Step 2   Because the next digit to the right of the underlined place is 4, which is 4 or less, do *not* change the digit in the underlined place.

Next digit is 4 or less.

34<u>9</u>

3 remains 3.

Step 3   Change all digits to the right of the underlined place to zeros.

3<u>4</u>9 rounded to the nearest hundred is 300.

In other words, 349 is closer to 300 than to 400.

◀◀◀ **Work Problem 2 at the Side.**

EXAMPLE 3   **Using Rounding Rules for 5 or More**

Round 36,833 to the nearest thousand.

Step 1   Find the place to which the number is to be rounded and draw a line under that place.

3<u>6</u>,833

Thousands

Step 2   Because the next digit to the right of the underlined place is 8, which is 5 or more, add 1 to the underlined place.

Next digit is 5 or more.

3<u>6</u>,833

Change 6 to 7.

Step 3   Change all digits to the right of the underlined place to zeros.

Change to 0.

36,833 rounded to the nearest thousand is 37,000.

Change 6 to 7.

In other words, 36,833 is closer to 37,000 than to 36,000.

◀◀◀ **Work Problem 3 at the Side.**

EXAMPLE 4   **Using Rounding Rules**

(a) Round 2382 to the nearest ten.

Step 1   2382

Tens place

Step 2   The next digit to the right is 2, which is 4 or less.

Next digit is 4 or less.

2382

Leave 8 as 8.

Step 3   23**82**   Change to 0.

2382 rounded to the nearest ten is 2380.
In other words, 2382 is closer to 2380 than to 2390.

Continued on Next Page

(b) Round 13,961 to the nearest hundred.

Step 1   13,961
↑ Hundreds place

Step 2   The next digit to the right is 6.

Next digit is 5 or more.
13,961
Change 9 to 10; write 0 and carry 1 into thousands place.
3 + carried 1 = 4
Change to 0

Step 3   14,061

13,961 rounded to the nearest hundred is 14,000.
In other words, 13,961 is closer to 14,000 than to 13,900.

> **NOTE**
> In Step 2 of Example 4(b), notice that the first three digits increased from 139 to 140 when we added 1 to the hundreds place.
>
> (13,9)61 rounded to (14,0)00

Work Problem 4 at the Side. ▶▶▶

EXAMPLE 5 **Rounding Large Numbers**

(a) Round 37,892 to the nearest ten thousand.

Step 1   37,892
↑ Ten thousands place.

Step 2   The next digit to the right is 7.

Next digit is 5 or more.
37892
↑ Change 3 to 4.
Change 0.

Step 3   47,892

37,892 rounded to the nearest ten thousand is 40,000.

(b) Round 528,498,675 to the nearest million.

Step 1   528,498,675
↑ Millions place
Next digit is 4 or less.

Step 2   528,498,675
↑ Leave 8 as 8
Change to 0.

Step 3   528,498,675

528,498,675 rounded to the nearest million is 528,000,000.

Work Problem 5 at the Side. ▶▶▶

4 Round each number as indicated.

(a) 3458 to the nearest ten

(b) 6448 to the nearest hundred

(c) 73,077 to the nearest hundred

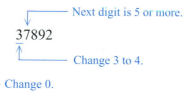
73,100

(d) 85,972 to the nearest hundred

5 Round each number as indicated.

(a) 14,598 to the nearest ten thousand

10,000

(b) 724,518,715 to the nearest million

725,000,000

6 Round each number to the nearest ten and to the nearest hundred.

(a) 458

(b) 549

(c) 9308

Sometimes a number must be rounded to different places.

EXAMPLE 6 Rounding to Different Places

Round 648 **(a)** to the nearest ten and **(b)** to the nearest hundred.

(a) to the nearest ten

┌─ Next digit is 5 or more.
648
└─ Tens place (4 + 1 = 5)

648 rounded to the nearest ten is 650.

(b) to the nearest hundred

┌─ Next digit is 4 or less.
648
└─ Hundreds place stays the same.

648 rounded to the nearest hundred is 600.

Notice that if 648 is rounded to the nearest ten (650), and then 650 is rounded to the nearest hundred, the result is 700. If, however, 648 is rounded directly to the nearest hundred, the result is 600 (not 700).

◀◀ **Work Problem 6 at the Side.**

CAUTION
Before rounding to a different place, always go back to the *original, unrounded* number.

EXAMPLE 7 Applying Rounding Rules

Round each number to the nearest ten, nearest hundred, and nearest thousand.

(a) 4358
First round 4358 to the nearest ten.

┌─ Next digit is 5 or more.
4358
└─ Tens place (5 + 1 = 6)

4358 rounded to the nearest ten in 4360.

Now go back to 4358, the *original* number, before rounding to the nearest hundred.

┌─ Next digit is 5 or more.
4358
└─ Hundreds place (3 + 1 = 4)

4358 rounded to the nearest hundred is 4400.

Again, go back to the *original* number before rounding to the nearest thousand.

┌─ Next digit is 4 or less.
4358
└─ Thousands place stays the same.

4358 rounded to the nearest thousand is 4000.

─── **Continued on Next Page**

(b) 680,914

First, round to the nearest ten.

Next digit is 4 or less.

680,91**4**

Tens place stays the same.

680,914 rounded to the nearest ten is 680,910

   Go back to 680,914, the *original* number, to round to the nearest hundred.

Next digit is 4 or less.

680,**9**14

Hundreds place stays the same.

680,914 rounded to the nearest hundred is 680,900.

   Go back to the *original* number to round to the nearest thousand.

Next digit is 5 or more.

680,**9**14

Thousands place (0 + 1 = 1)

680,914 rounded to the nearest thousand is 681,000.

Work Problem 7 at the Side. ▶▶▶

OBJECTIVE 3 Round numbers to estimate an answer. Numbers may be rounded to **estimate** an answer. An estimated answer is one that is close to the exact answer and may be used as a check when the exact answer is found. The "≈" sign is often used to show that an answer has been rounded or estimated and is almost equal to the exact answer, ≈ means "approximately equal to."

EXAMPLE 8 **Using Rounding to Estimate an Answer**

Estimate each answer by rounding to the nearest ten.

(a)

$$
\begin{array}{r}
76 \longrightarrow 80 \\
53 \longrightarrow 50 \\
38 \longrightarrow 40 \\
+\,91 \longrightarrow +\,90 \\
\hline
260
\end{array}
$$
Rounded to the nearest ten

260   Estimated answer

(b)

$$
\begin{array}{r}
27 \qquad 30 \\
-\,14 \qquad -\,10 \\
\hline
20
\end{array}
$$
Rounded to the nearest ten

20   Estimated answer

(c)

$$
\begin{array}{r}
16 \qquad 20 \\
\times\,21 \qquad \times\,20 \\
\hline
400
\end{array}
$$
Rounded to the nearest ten

400   Estimated answer

Work Problem 8 at the Side. ▶▶▶

7 Round each number to the nearest ten, nearest hundred, and nearest thousand.

(a) 4078

(b) 46,364

(c) 268,328

8 Estimate each answer by ≈ rounding to the nearest ten.

(a)
$$
\begin{array}{r}
16 \\
74 \\
58 \\
+\,31 \\
\hline
\end{array}
$$

(b)
$$
\begin{array}{r}
53 \\
-\,19 \\
\hline
\end{array}
$$

(c)
$$
\begin{array}{r}
46 \\
\times\,74 \\
\hline
\end{array}
$$

ANSWERS

7. **(a)** 4080; 4100; 4000
   (b) 46,360; 46,400; 46,000
   (c) 268,330; 268,300; 268,000
8. **(a)** 20 + 70 + 60 + 30 = 180
   (b) 50 − 20 = 30
   (c) 70 × 50 = 3500

9 Estimate each answer by
≈ rounding to the nearest
hundred.

(a)   358
         743
         822
       + 978

(b)   842
       − 475

(c)   723
       × 478

10 Use front end rounding to
≈ estimate each answer.

(a)   36
        3852
         749
       + 5474

(b)   2583
       − 765

(c)   648
       × 67

EXAMPLE 9 Using Rounding to Estimate an Answer

Estimate each answer by rounding to the nearest hundred.

(a)   152 ⟶ 200
       749 ⟶ 700   } Rounded to the nearest hundred
       576 ⟶ 600
     + 819 ⟶ + 800
              2300   Estimated answer

(b)   780   800
     − 536   − 500   } Rounded to the nearest hundred
                  300   Estimated answer

(c)   664   700
     × 834   × 800   } Rounded to the nearest hundred
              560,000   Estimated answer

◀◀◀ **Work Problem 9 at the Side.**

OBJECTIVE 4 Use front end rounding to estimate an answer. A convenient way to estimate an answer is to use *front end rounding*. With **front end rounding,** we round to the highest possible place so that all the digits become 0 except the first one. For example, suppose you want to buy a big-screen television for $949, a home entertainment center for $759, and a big, comfortable chair for $525. Using front end rounding, you can estimate the total cost of these purchases.

Television $949 ⟶ $900
Entertainment center $759 ⟶   800
Big chair $525 ⟶ + 500
                                  $2200 ⟵ Estimated total cost

EXAMPLE 10 Using Front End Rounding to Estimate an Answer

Estimate each answer using front end rounding.

(a)   3825   4000
        72   70   } All digits changed to
       565   600   } 0 except first digit,
     + 2389   + 2000 } which is rounded
                 6670   Estimated answer

(b)   6712   7000 } First digit rounded and
     − 825   − 800 } all others changed to 0.
                 6200   Estimated answer

(c)   725   700
     × 86   ×   90
              63,000   Estimated answer

NOTE
When using front end rounding, all the digits become 0 except the highest-place digit (the first digit).

◀◀◀ **Work Problem 10 at the Side.**

1.7 Exercises

FOR EXTRA HELP

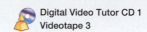 Addison-Wesley Math Tutor Center

 Math XP MathXL

Digital Video Tutor CD 1 Videotape 3

Student's Solutions Manual

MyMathLab MyMathLab

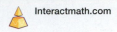 Interactmath.com

Round each number as indicated. See Examples 1–5.

1. 624 to the nearest ten 620

2. 509 to the nearest ten

3. 855 to the nearest ten 860

4. 946 to the nearest ten 950

5. 6771 to the nearest hundred 6800

6. 5847 to the nearest hundred 5800

7. 86,813 to the nearest hundred 86,800

8. 17,211 to the nearest hundred 17,200

9. 28,472 to the nearest hundred 28,500

10. 18,273 to the nearest hundred 18,300

11. 5996 to the nearest hundred

12. 4452 to the nearest hundred 4500

13. 15,758 to the nearest thousand

14. 28,465 to the nearest thousand 28,000

15. 78,499 to the nearest thousand

16. 14,314 to the nearest thousand 14,000

17. 7760 to the nearest thousand 8000

18. 49,706 to the nearest thousand 50,000

19. 12,987 to the nearest ten thousand

20. 6599 to the nearest ten thousand

21. 595,008 to the nearest ten thousand 600,000

22. 725,182 to the nearest ten thousand 730,000

23. 4,860,220 to the nearest million 5,000,000

24. 13,713,409 to the nearest million 14,000,000

Round each number to the nearest ten, nearest hundred, and nearest thousand. See Examples 6 and 7.

	Ten	Hundred	Thousand			Ten	Hundred	Thousand
25. 4476	_____	_____	_____		**26.** 6483	_____	_____	_____
27. 3374	_____	_____	_____		**28.** 7632	_____	_____	_____
29. 6048	_____	_____	_____		**30.** 7065	_____	_____	_____

	Ten	Hundred	Thousand			Ten	Hundred	Thousand
31. 5343	_____	_____	_____		**32.** 7456	_____	_____	_____
33. 19,539	_____	_____	_____		**34.** 59,806	_____	_____	_____
35. 26,292	_____	_____	_____		**36.** 78,519	_____	_____	_____
37. 93,706	_____	_____	_____		**38.** 84,639	_____	_____	_____

39. Write in your own words the three steps that you would use to round a number when the digit to the right of the place to which you are rounding is 5 or more.

40. Write in your own words the three steps that you would use to round a number when the digit to the right of the place to which you are rounding is 4 or less.

≈ *Estimate each answer by rounding to the nearest ten. Then find the exact answer. See Example 8.*

41. *Estimate:*   *Exact:*

Rounds to
$$
\begin{array}{r}
25 \\
63 \\
47 \\
+\ 84 \\
\hline
\end{array}
$$
+ _____

42. *Estimate:*   *Exact:*

$$
\begin{array}{r}
56 \\
24 \\
85 \\
+\ 71 \\
\hline
\end{array}
$$
+ _____

43. *Estimate:*   *Exact:*

$$
\begin{array}{r}
78 \\
-\ 43 \\
\hline
\end{array}
$$
− _____

44. *Estimate:*   *Exact:*

$$
\begin{array}{r}
57 \\
-\ 24 \\
\hline
\end{array}
$$
− _____

45. *Estimate:*   *Exact:*

$$
\begin{array}{r}
67 \\
\times\ 34 \\
\hline
\end{array}
$$
× _____

46. *Estimate:*   *Exact:*

$$
\begin{array}{r}
53 \\
\times\ 75 \\
\hline
\end{array}
$$
× _____

≈ *Estimate each answer by rounding to the nearest hundred. Then find the exact answer. See Example 9.*

47. *Estimate:*   *Exact:*

Rounds to
$$
\begin{array}{r}
863 \\
735 \\
438 \\
+\ 792 \\
\hline
\end{array}
$$
+ _____

48. *Estimate:*   *Exact:*

$$
\begin{array}{r}
623 \\
362 \\
189 \\
+\ 736 \\
\hline
\end{array}
$$
+ _____

49. *Estimate:*   *Exact:*

$$\begin{array}{r} 883 \\ -\ 448 \\ \hline \end{array}$$

 − _____

50. *Estimate:*   *Exact:*

$$\begin{array}{r} 614 \\ -\ 276 \\ \hline \end{array}$$

 − _____

51. *Estimate:*   *Exact:*

$$\begin{array}{r} 752 \\ \times\ 375 \\ \hline \end{array}$$

 × _____

52. *Estimate:*   *Exact:*

$$\begin{array}{r} 845 \\ \times\ 396 \\ \hline \end{array}$$

 × _____

≈ *Estimate each answer using front end rounding. Then find the exact answer. See Example 10.*

53. *Estimate:*   *Exact:*

Rounds to

$$\begin{array}{r} 8215 \\ 56 \\ 729 \\ +\ 3605 \\ \hline \end{array}$$

 + _____

54. *Estimate:*   *Exact:*

$$\begin{array}{r} 2685 \\ 73 \\ 592 \\ +\ 7183 \\ \hline \end{array}$$

 + _____

55. *Estimate:*   *Exact:*

$$\begin{array}{r} 687 \\ -\ 529 \\ \hline \end{array}$$

 − _____

56. *Estimate:*   *Exact:*

$$\begin{array}{r} 543 \\ -\ 174 \\ \hline \end{array}$$

 − _____

57. *Estimate:*   *Exact:*

$$\begin{array}{r} 939 \\ \times\ 29 \\ \hline \end{array}$$

 × _____

58. *Estimate:*   *Exact:*

$$\begin{array}{r} 864 \\ \times\ 74 \\ \hline \end{array}$$

 × _____

59. The number 3492 rounded to the nearest hundred is 3500, and 3500 rounded to the nearest thousand is 4000. But when 3492 is rounded directly to the nearest thousand it becomes 3000. Why is this true? Explain.

60. The use of rounding is helpful when estimating the answer to a problem. Why is this true? Give an example using either addition, subtraction, multiplication, or division to show how this works.

61. In 1900, the population of the United States was 76 million. Today it's 296 million. Round each of these numbers to the nearest ten million. (*Source:* Reiman Publications.)

62. In 1900, the average workweek in the United States was 59 hours. Today it's 38 hours. Round each of these numbers to the nearest ten. (*Source:* Reiman Publications.)

63. In 1900, the life expectancy in the United States was 47 years. Today it's 76 years. Round these numbers to the nearest ten. (*Source:* Reiman Publications.)

64. In 1900, the average age of the population in the United States was 23. Today it's 35. Round these numbers to the nearest ten. (*Source:* Reiman Publications.)

65. In Chicago, the Sears Tower tenants recycled 1,667,300 pounds of paper last year. Round this number to the nearest ten thousand, nearest hundred thousand, and nearest million. (*Source:* Trizec Properties.)

1,670,000
1,700,000
2,000,000

66. Round 621,999,652 to the nearest thousand, nearest ten thousand, and nearest hundred thousand.

67. American pharmaceutical companies spent $25,765,475,000 last year to develop new products. Round this amount to the nearest hundred thousand, nearest hundred million, and nearest billion. (*Source:* American Demographics.)

68. In one year the U.S. Federal Food Assistance Program paid out $18,915,762,568 in food stamps. Round this amount to the nearest hundred thousand, nearest hundred million, and nearest ten billion. (*Source:* U.S. Department of Agriculture.)

RELATING CONCEPTS (EXERCISES 69–75)   For Individual or Group Work

To see how both rounding and front end rounding are used in solving problems, work Exercises 69–75 in order.

69. A number rounded to the nearest thousand is 72,000. What is the *smallest* whole number this could have been before rounding?

71,500

70. A number rounded to the nearest thousand is 72,000. What is the *largest* whole number this could have been before rounding?

71. When front end rounding is used, a whole number rounds to 8000. What is the *smallest* possible original number?

7500

72. When front end rounding is used, a whole number rounds to 8000. What is the *largest* possible original number?

The graph below shows the greatest number of pitching appearances for active professional baseball pitchers.

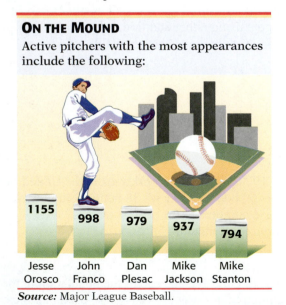

ON THE MOUND
Active pitchers with the most appearances include the following:

Jesse Orosco	John Franco	Dan Plesac	Mike Jackson	Mike Stanton
1155	998	979	937	794

Source: Major League Baseball.

73. Round the number of appearances given for each pitcher to the nearest ten.

1160

74. Use front end rounding to round the number of appearances for each pitcher.

75. (a) What is one advantage of using front end rounding instead of rounding to the nearest ten?

(b) What is one disadvantage?

1.8 Exponents, Roots, and Order of Operations

OBJECTIVE 1 Identify an exponent and a base. The product $3 \cdot 3$ can be written as 3^2 (read as "3 squared"). The small raised number 2, called an **exponent**, says to use 2 factors of 3. The number 3 is called the **base**. Writing 3^2 as 9 is called *simplifying the expression*.

EXAMPLE 1 Simplifying Expressions

Identify the exponent and the base, and then simplify each expression.

(a) 4^3

$$\text{Base} \rightarrow 4^3 \leftarrow \text{Exponent} \qquad 4^3 = 4 \times 4 \times 4 = 64$$

(b) $2^5 = 2 \times 2 \times 2 \times 2 \times 2 = 32$

The base is 2 and the exponent is 5.

Work Problem 1 at the Side.

OBJECTIVE 2 Find the square root of a number. Because $3^2 = 9$, the number 3 is called the **square root** of 9. The square root of a number is one of two identical factors of that number. Square roots of numbers are written with the symbol $\sqrt{}$.

$$\sqrt{9} = 3$$

Square Root

$$\sqrt{\text{number} \cdot \text{number}} = \sqrt{\text{number}^2} = \text{square root}$$

For example: $\sqrt{36} = \sqrt{6 \cdot 6} = \sqrt{6^2} = 6$ is the square root of 36.

To find the square root of 64 ask, "What number can be multiplied by itself (that is, *squared*) to give 64?" The answer is 8, so

$$\sqrt{64} = \sqrt{8 \cdot 8} = \sqrt{8^2} = 8.$$

A **perfect square** is a number that is the square of a *whole number*. The first few perfect squares are listed here.

Perfect Squares Table

$0 = 0^2$	$16 = 4^2$	$64 = 8^2$	$144 = 12^2$
$1 = 1^2$	$25 = 5^2$	$81 = 9^2$	$169 = 13^2$
$4 = 2^2$	$36 = 6^2$	$100 = 10^2$	$196 = 14^2$
$9 = 3^2$	$49 = 7^2$	$121 = 11^2$	$225 = 15^2$

EXAMPLE 2 Using Perfect Squares

Find each square root.

(a) $\sqrt{16}$ Because $4^2 = 16$, $\sqrt{16} = 4$. **(b)** $\sqrt{49} = 7$

(c) $\sqrt{0} = 0$ **(d)** $\sqrt{169} = 13$

Work Problem 2 at the Side.

OBJECTIVE 3 Use the order of operations. Frequently problems may have parentheses, exponents, and square roots, and may involve more than one operation. Work these problems by following the **order of operations**.

OBJECTIVES

1 Identify an exponent and a base.

2 Find the square root of a number.

3 Use the order of operations.

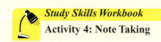
Study Skills Workbook
Activity 4: Note Taking

1 Identify the exponent and the base, and then simplify each expression.

(a) 4^2

(b) 5^3

(c) 3^4

(d) 2^6

2 Find each square root.

(a) $\sqrt{4}$ 2

(b) $\sqrt{25}$ 5

(c) $\sqrt{36}$ 6

(d) $\sqrt{225}$ 15

(e) $\sqrt{1}$ 1

ANSWERS
1. (a) 2; 4; 16 (b) 3; 5; 125
   (c) 4; 3; 81 (d) 6; 2; 64
2. (a) 2 (b) 5 (c) 6 (d) 15 (e) 1

(handwritten at top) $\sqrt{}$ $\times$ $\div$ $+$ $-$
PEMDAS

3 Simplify each expression.

(a) $4 + 5 + 2^2$

(handwritten) $9 + \;\; 4 = 13$

(b) $3^2 + 2^3$ *(handwritten)* $2 \cdot 2 \cdot 2$

(handwritten) $9 + 8 = 17$

(c) $4 \cdot 6 \div 12 - 2$

(handwritten) $4 \cdot \; .5 - 2$
$2 - 2 = 0$

(d) $60 \div 6 \div 2$

(handwritten) 5

(e) $8 + 6(14 \div 2)$

(handwritten) $8 + 6(7)$
$8 + 42 = 50$

4 Simplify each expression.

(a) $12 - 6 + 4^2$

(handwritten) $12 - 6 + 16$
$12 - 22 = -10$

(handwritten) $2 \cdot 2 \cdot 2$ (b) $2^3 + 3^2 - (5 \cdot 3)$

(handwritten) $8 + 9 - 15$
$17 - 15 = 2$

(c) $2 \cdot \sqrt{64} - 5 \cdot 3$

(handwritten) $2 \cdot 8 - 15$
$16 - 15 = 1$

(d) $20 \div 2 + (7 - 5)$

(handwritten) $10 + 2 = 12$

(e) $15 \cdot \sqrt{9} - 8 \cdot \sqrt{4}$

(handwritten) $15 \cdot 3 - 8 \cdot 2$
$45 - 16 = 29$

Order of Operations

1. Do all operations inside **parentheses** or **other grouping symbols.**
2. Simplify any expressions with **exponents** and find any **square roots.**
3. **Multiply** or **divide,** proceeding from left to right.
4. **Add** or **subtract,** proceeding from left to right.

EXAMPLE 3 **Understanding the Order of Operations**

Use the order of operations to simplify each expression.

(a) $8^2 + 5 + 2$

$$8^2 \quad + 5 + 2$$
$$8 \cdot 8 + 5 + 2 \qquad \text{Evaluate exponent first; } 8^2 \text{ is } 8 \cdot 8.$$
$$64 \quad + 5 + 2 \qquad \text{Add from left to right.}$$
$$69 \quad + 2 = 71$$

(b) $35 \div 5 \cdot 6$ — Divide first (start at left).
$\quad\; 7 \quad \cdot 6 = 42$ — Multiply.

(c) $9 + (20 - 4) \cdot 3$ — Work inside parentheses first.
$\quad 9 + \quad 16 \cdot 3$ — Multiply.
$\quad 9 + \quad\;\; 48 \quad = 57$ — Add last.

(d) $12 \cdot \sqrt{16} - 8(4)$ — Find the square root first.
$\quad 12 \cdot \quad 4 \quad - 8(4)$ — Multiply from left to right.
$\quad\;\; 48 \quad\; - \quad 32 \quad = 16$ — Subtract last.

◀◀◀ Work Problem 3 at the Side.

EXAMPLE 4 **Using the Order of Operations**

Use the order of operations to simplify each expression.

(a) $15 - 4 + 2$ — Subtract first (start at left).
$\quad\;\; 11 \quad + 2 = 13$ — Add.

(b) $8 + (7 - 3) \div 2$ — Work inside parentheses first.
$\quad 8 + \quad 4 \quad \div 2$ — Divide.
$\quad 8 + \quad\quad 2 \quad = 10$ — Add last.

(c) $4^2 \cdot 2^2 + (7 + 3) \cdot 2$ — Parentheses first.
$\quad 4^2 \cdot 2^2 + \quad 10 \quad \cdot 2$ — Evaluate exponents.
$\quad 16 \cdot 4 + \quad 10 \quad \cdot 2$ — Multiply from left to right.
$\quad\;\; 64 \quad + \quad\quad 20 \quad = 84$ — Add last.

(d) $4 \cdot \sqrt{25} - 7 \cdot 2 + \dfrac{0}{5}$ — Find the square root first.

$\quad 4 \cdot \quad 5 \quad - 7 \cdot 2 + \dfrac{0}{5}$ — Multiply or divide from left to right.

$\quad\quad 20 \quad\;\; - \quad 14 \quad + 0 = 6$ — Add or subtract last.

NOTE
Getting a correct answer always depends on following the order of operations.

◀◀◀ Work Problem 4 at the Side.

1.8 **Exercises**

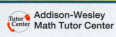
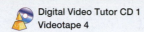

Identify the exponent and the base, and then simplify each expression. See Example 1.

1. 3^2

2. 2^3

3. 5^2

4. 4^2

5. 8^2

6. 10^3

7. 15^2

8. 11^3

Use the Perfect Squares Table on page 75 to find each square root. See Example 2.

9. $\sqrt{16}$
  4

10. $\sqrt{25}$
  5

11. $\sqrt{64}$

12. $\sqrt{36}$

13. $\sqrt{100}$

14. $\sqrt{49}$

15. $\sqrt{144}$
  12

16. $\sqrt{225}$

Fill in each blank. See Example 2.

17. $6^2 = \underline{6 \times 6}$ so $\sqrt{36} = 6$
  36

18. $9^2 = \underline{\hspace{2cm}}$ so $\sqrt{\hphantom{0}} = 9$

19. $20^2 = \underline{\hspace{2cm}}$ so $\sqrt{\hphantom{0}} = 20$

20. $30^2 = \underline{\hspace{2cm}}$ so $\sqrt{\hphantom{0}} = 30$

21. $35^2 = \underline{\hspace{2cm}}$ so $\sqrt{\hphantom{0}} = 35$

22. $38^2 = \underline{\hspace{2cm}}$ so $\sqrt{\hphantom{0}} = 38$

23. $25^2 =$ _____ so $\sqrt{} = 25$

24. $50^2 =$ _____ so $\sqrt{} = 50$

25. $100^2 =$ _____ so $\sqrt{} = 100$

26. $60^2 =$ _____ so $\sqrt{} = 60$

27. Describe in your own words a perfect square. Of the two numbers 25 and 50, identify which is a perfect square and explain why.

28. Use the following list of words and phrases to write the four steps in the order of operations.

add	square root
exponents	subtract
multiply	divide

parentheses or other grouping symbols

Simplify each expression by using the order of operations. See Examples 3 and 4.

29. $3^2 + 8 - 5$

3×3

$9 + 8 - 5$ $17 - 5 = \boxed{12}$

30. $5^2 + 5 - 6$

$25 + 5 - 6$

$30 - 6 = 24$

31. $3 \cdot 7 - 6$

32. $5 \cdot 7 - 7$

$35 - 7 = 28$

33. $8 \cdot 5 \div 10$

34. $6 \cdot 8 \div 8$

35. $25 \div 5(8 - 4)$

$25 \div 5(4)$

$25 \div 20 = 1.25$

36. $36 \div 18(7 - 3)$

$18(4)$

$36 \div 72 = .5$

37. $5 \cdot 3^2 + \dfrac{0}{8}$

$5 \cdot 9 + 0$

45

38. $8 \cdot 3^2 - \dfrac{10}{2}$

$8 \cdot 9 - \frac{10}{2}$

$72 - 5 = 14.4$

39. $4 \cdot 1 + 8(9 - 2) + 3$

$4 + 8(7) + 3 = 63$

56

40. $3 \cdot 2 + 7(3 + 1) + 5$

41. $2^2 \cdot 3^3 + (20 - 15) \cdot 2$

$4 \cdot 27 * 5 \cdot 2$

$108 + 10 = 118$

42. $4^2 \cdot 5^2 + (20 - 9) \cdot 3$

43. $5\sqrt{36} - 2(4)$

$5\cdot6 - 8 = 22$

44. $2 \cdot \sqrt{100} - 3(4)$

45. $8(2) + 3 \cdot 7 - 7 =$

$16 + 21 - 7$

46. $10(3) + 6 \cdot 5 - 20$

47. $2^3 \cdot 3^2 + 3(14 - 4)$

$8 \quad 9 \quad 3\cdot10$

48. $3^2 \cdot 4^2 + 2(15 - 6)$

49. $7 + 8 \div 4 + \dfrac{0}{7}$

0

$\dfrac{8 \div 4}{7 + 2} \qquad 0 \div 7$

50. $6 + 8 \div 2 + \dfrac{0}{8}$

51. $3^2 + 6^2 + (30 - 21) \cdot 2$

52. $4^2 + 5^2 + (25 - 9) \cdot 3$

53. $7 \cdot \sqrt{81} - 5 \cdot 6$

54. $6 \cdot \sqrt{64} - 6 \cdot 5$

55. $8 \cdot 2 + 5(3 \cdot 4) - 6$

56. $5 \cdot 2 + 3(5 + 3) - 6$

57. $4 \cdot \sqrt{49} - 7(5 - 2)$

58. $3 \cdot \sqrt{25} - 6(3 - 1)$

59. $7(4 - 2) + \sqrt{9}$

60. $5(4 - 3) + \sqrt{9}$

61. $7^2 + 3^2 - 8 + 5$

62. $3^2 - 2^2 + 3 - 2$

63. $5^2 \cdot 2^2 + (8 - 4) \cdot 2$

$25 \cdot 4 + \quad 4 \cdot 2$

$100 + \qquad 8$

64. $5^2 \cdot 3^2 + (30 - 20) \cdot 2$

65. $5 + 9 \div 3 + 6 \cdot 3$

66. $8 + 3 \div 3 + 6 \cdot 3$

67. $8 \cdot \sqrt{49} - 6(9 - 4)$

$8 \cdot 7 - 6(5)$

$56 - 30 = 26$

68. $8 \cdot \sqrt{49} - 6(5 + 3)$

69. $5^2 - 4^2 + 3 \cdot 6$

70. $3^2 + 6^2 - 5 \cdot 8$

71. $8 + 8 \div 8 + 6 + \dfrac{5}{5}$

72. $3 + 14 \div 2 + 7 + \dfrac{8}{8}$

73. $6 \cdot \sqrt{25} - 7(2)$

74. $8 \cdot \sqrt{36} - 4(6)$

75. $9 \cdot \sqrt{16} - 3\sqrt{25}$

76. $6 \cdot \sqrt{81} - 3 \cdot \sqrt{49}$

77. $7 \div 1 \cdot 8 \cdot 2 \div (21 - 5)$

78. $12 \div 4 \cdot 5 \cdot 4 \div (15 - 13)$

79. $15 \div 3 \cdot 2 \cdot 6 \div (14 - 11)$

80. $9 \div 1 \cdot 4 \cdot 2 \div (11 - 5)$

81. $6 \cdot \sqrt{25} - 4 \cdot \sqrt{16}$

82. $10 \cdot \sqrt{49} - 4 \cdot \sqrt{64}$

83. $5 \div 1 \cdot 10 \cdot 4 \div (17 - 9)$

84. $15 \div 3 \cdot 8 \cdot 9 \div (12 - 8)$

85. $8 \cdot 9 \div \sqrt{36} - 4 \div 2 + (14 - 8)$

86. $3 - 2 + 5 \cdot 4 \cdot \sqrt{144} \div \sqrt{36}$

87. $2 + 1 - 2 \cdot \sqrt{1} + 4 \cdot \sqrt{81} - 7 \cdot 2$

88. $6 - 4 + 2 \cdot 9 - 3 \cdot \sqrt{225} \div \sqrt{25}$

89. $5 \cdot \sqrt{36} \cdot \sqrt{100} \div 4 \cdot \sqrt{9} + 8$

90. $9 \cdot \sqrt{36} \cdot \sqrt{81} \div 2 + 6 - 3 - 5$

1.9 Reading Pictographs, Bar Graphs, and Line Graphs

We have all heard the saying "A picture is worth a 1000 words," and there may be some truth in this. Today, so much information and data are being presented in the form of pictographs, circle graphs, bar graphs, and line graphs that it is important to be able to read and understand these tools.

OBJECTIVES

1 Read and understand a pictograph.

2 Read and understand a bar graph.

3 Read and understand a line graph.

OBJECTIVE 1 Read and understand a pictograph. A **pictograph** is a graph that uses pictures or symbols. It displays information that can be compared easily. However, since a symbol is used to represent a certain quantity, it can be difficult to determine the amount represented by a fraction of a symbol.

The pictograph below compares the number of U.S. students studying in colleges and universities in foreign countries. In this pictograph, it is difficult to determine what fractional amount is represented by the partial pencils.

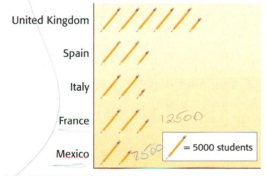

TRAVELING FOR EDUCATION
Top-five host countries for U.S. students studying abroad.

United Kingdom

Spain

Italy

France   12500

Mexico   7500   / = 5000 students

Source: Institute of International Education.

5000 ÷ 2 = 2500

1 Use the pictograph to answer each question.

(a) Which country has the greatest number of U.S. students attending its colleges and universities?

UK

(b) Approximately how many more U.S. students are studying in France than in Mexico?

5000

EXAMPLE 1 Using a Pictograph

Use the pictograph to answer each question.

(a) Which country has the lowest number of U.S. students attending its colleges and universities?

The row representing Mexico has the fewest symbols. This means that the lowest number of U.S. students is in Mexico.

(b) Approximately how many more U.S. students are attending colleges and universities in the United Kingdom than in Spain?

The row representing the United Kingdom has three more symbols than that for Spain. This means that 3 • 5000 or 15,000 more U.S. students are attending colleges and universities in the United Kingdom than in Spain.

Work Problem 1 at the Side. ▶▶▶

OBJECTIVE 2 Read and understand a bar graph. **Bar graphs** are useful for showing comparisons. For example, the following bar graph shows how many people out of every 100 fans surveyed chose each sport as their favorite.

ANSWERS
1. **(a)** United Kingdom
   (b) about 5000 more students

2 Use the bar graph to find the approximate number of fans who picked each sport as their favorite.

(a) Pro football

(b) Pro baseball

(c) Pro basketball

(d) College basketball

(e) Golf

3 Use the line graph to find the predicted population of the United States for each year.

(a) 2050

(b) 2075

(c) 2100

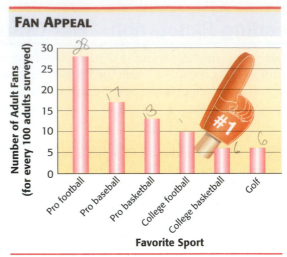

FAN APPEAL

Source: The Harris Poll.

EXAMPLE 2 Using a Bar Graph

Use the bar graph to find the number of fans who picked college football as their favorite sport.

Use a ruler or straightedge to line up the top of the bar labeled "College football," with the numbers on the left edge of the graph, labeled "Number of Adults Fans." We see that 10 out of 100 adult fans picked college football as their favorite sport.

◀◀◀ **Work Problem 2 at the Side.**

OBJECTIVE 3 Read and understand a line graph. A **line graph** is often used for showing a trend. The following line graph shows the U.S. Bureau of the Census predictions for U.S. population growth to the year 2100.

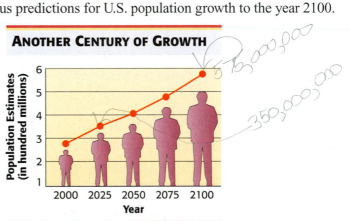

ANOTHER CENTURY OF GROWTH

Source: U.S. Bureau of the Census.

EXAMPLE 3 Using a Line Graph

Use the line graph to answer each question.

(a) What trend or pattern is shown in the graph?
The population will continue to increase.

(b) What is the estimated population for 2025?
Use a ruler or straightedge to line up the dot above the year labeled 2025 on the horizontal line with the numbers along the left edge of the graph. Notice that the label on the left side says "in hundred millions." Since the 2025 dot is halfway between 3 and 4, the population in 2025 is halfway between 3 • 100,000,000 and 4 • 100,000,000 or 300,000,000 and 400,000,000. That means that the predicted population in 2025 is about 350,000,000 people.

◀◀◀ **Work Problem 3 at the Side.**

ANSWERS

2. (a) 28 out of 100   (b) 17 out of 100
   (c) 13 out of 100   (d) 6 out of 100
   (e) 6 out of 100
3. (a) 400,000,000   (b) 475,000,000
   (c) 575,000,000

1.9 Exercises

FOR EXTRA HELP

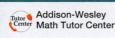 Addison-Wesley Math Tutor Center

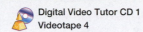 MathXL

 Digital Video Tutor CD 1 Videotape 4

Student's Solutions Manual

MyMathLab MyMathLab

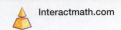 Interactmath.com

The following pictograph shows the number of retail stores for the seven companies with the greatest number of outlets. Use the pictograph to answer Exercises 1–6. See Example 1.

SOMETHING IN STORE
While Wal-Mart has the greatest amount of sales, it trails other chains in number of stores.

Dollar General
7-Eleven
Family Dollar
CVS
Walgreens
Rite-Aid
Wal-Mart

= 500 stores

Source: T. D. Linx.

1. Find the number of Family Dollar retail stores.

$$5500$$
$$+\ 250$$
$$\overline{5750}$$

 2. Approximately how many retail stores does 7-Eleven have?

3. Which company has the greatest number of retail stores?

 Dollar General

4. Which companies have the least number of retail stores?

5. Dollar General has 5750 retail stores. How many more is this than Family Dollar?

6. How many more retail stores does Walgreens have than Wal-Mart?

The following bar graph shows the results of a survey that was taken of 100 working adults to determine how they chose their careers. Use the bar graph to answer Exercises 7–12. See Example 2.

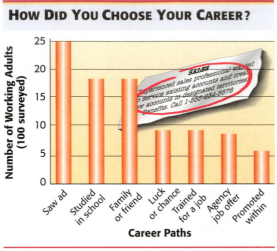

How Did You Choose Your Career?

Number of Working Adults (100 surveyed)

Career Paths: Saw ad, Studied in school, Family or friend, Luck or chance, Trained for a job, Agency job offer, Promoted within

Source: Market Facts/TeleNation for Career
Education Corporation.

7. How many people found their careers as a result of training for a job?

8. How many people found their careers because they studied for the career in school?

9. (a) Which career path was taken by the greatest number of people?

(b) How many people used this path?

10. (a) Which career path was taken by the least number of people?

(b) How many people used this path?

11. How many more people found their careers as a result of "Studied in school" than "Luck or chance"?

12. Find the total number of people who found their careers as a result of either "Studied in school" or "Trained for a job."

A local real estate association collected home sales data for their area and prepared the following line graph. Remembering that the sales data is shown in thousands, use the line graph to answer Exercises 13–16. See Example 3.

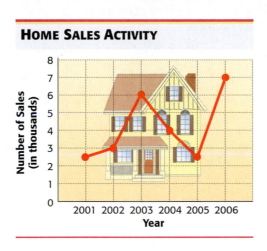

HOME SALES ACTIVITY

13. Which year had the greatest number of home sales? How many sales were there?

14. Which two years had the least number of home sales? How many sales were there in each of those years?

15. Find the increase in the number of home sales from 2005 to 2006.

16. Find the decrease in the number of home sales from 2004 to 2005.

17. Give three possible explanations for the decrease in home sales in 2004.

18. Give three possible explanations for the increase in home sales in 2006.

Getting a correct answer in mathematics always depends on following the order of operations. Insert grouping symbols (parentheses) so that each given expression evaluates to the given number. **Work Exercises 19–23 in order.**

19. $7 - 2 \cdot 3 - 6$; evaluate to 9

20. $4 + 2 \cdot 5 + 1$; evaluate to 36

21. $36 \div 3 \cdot 3 \cdot 4$; evaluate to 16

22. $56 \div 2 \cdot 2 \cdot 2 + \dfrac{0}{6}$; evaluate to 7

23. A fencing contractor is building 12 corrals like the one shown below for a rancher.

  (a) Use the order of operations to write an expression for the amount of fencing needed for one corral.

  (b) How many feet of fencing are needed for 12 corrals?

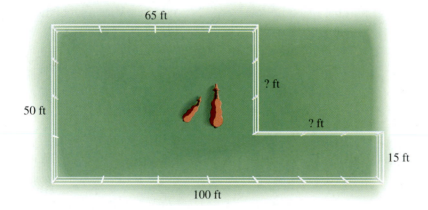

1.10 Solving Application Problems

Most problems involving applications of mathematics are written out in sentence form. You need to read the problem carefully to decide how to solve it.

OBJECTIVE 1 Find indicator words in application problems. As you read an application problem, look for **indicator words** that help you determine the necessary operations—either addition, subtraction, multiplication, or division. Some of these indicator words are shown here.

Addition	Subtraction	Multiplication	Division	Equals
plus	less	product	divided by	is
more	subtract	double	divided into	the same as
more than	subtracted from	triple	quotient	equals
added to	difference	times	goes into	equal to
increased by	less than	of	divide	yields
sum	fewer	twice	divided equally	results in
total	decreased by	twice as much	per	are
sum of	loss of			
increase of	minus			
gain of	take away			

CAUTION

The word *and* does not always indicate addition, so it does not appear as an indicator word in the preceding table. Notice how the "and" shows the location of several different operation signs below.

The sum of 6 *and* 2 is 6 + 2.
The difference of 6 *and* 2 is 6 − 2.
The product of 6 *and* 2 is 6 · 2.
The quotient of 6 *and* 2 is 6 ÷ 2.

OBJECTIVE 2 Solve application problems. Solve application problems by using the following six steps.

Solving an Application Problem

Step 1 **Read** the problem carefully and be certain you *understand* what the problem is asking. It may be necessary to read the problem several times.

Step 2 Before doing any calculations, **work out a plan** and try to visualize the problem. Draw a sketch if possible. Know which facts are given and which must be found. Use *indicator words* to help decide on the *plan* (whether you will need to add, subtract, multiply, or divide).

Step 3 **Estimate** a *reasonable answer* by using rounding.

Step 4 **Solve** the problem by using the facts given and your plan.

Step 5 **State the answer.**

Step 6 **Check** your work. If the answer does not seem reasonable, begin again by reading the problem.

1 Pick the most reasonable ≈ answer for each problem.

(a) A grocery clerk's hourly wage: $1.40; $14; $140

(b) The total length of five sport-utility vehicles: 8 ft; 18 ft; 80 ft; 800 ft

(c) The cost of heart bypass surgery: $1000; $100,000; $10,000,000

CAUTION

Be careful not to begin solving a problem before you understand what the problem is asking. Be certain that you know what the problem is asking before you try to solve it.

OBJECTIVE 3 Estimate an answer.   The six problem-solving steps give a systematic approach for solving word problems. Each of the steps is important, but special emphasis should be placed on Step 3, estimating a *reasonable answer.* Many times an "answer" just does not fit the problem.

What is a reasonable answer? Read the problem and try to determine the approximate size of the answer. Should the answer be part of a dollar, a few dollars, hundreds, thousands, or even millions of dollars? For example, if a problem asks for the cost of a man's shirt, would an answer of $20 be reasonable? $2000? $2? $200?

CAUTION

Always estimate the answer, then look at your final result to be sure it fits your estimate and is reasonable. This step will give greater success in problem solving.

Work Problem 1 at the Side.

2 Solve each problem.

(a) On a recent geology field trip, 84 fossils were collected. If the fossils are divided equally among John, Sean, Jenn, and Kara, how many fossils will each receive?

(b) This week there are 408 children attending a winter sports camp. If 12 children are assigned to each camp counselor, how many counselors are needed?

EXAMPLE 1 Applying Division

At a recent garage sale, the total sales were $584. If the money was divided equally among Tom, Rosetta, Maryann, and José, how much did each person get?

Step 1   **Read.** A reading of the problem shows that the four members in the group divided $584 equally.

Step 2   **Work out a plan.** The indicator words, ***divided equally,*** show that the amount each received can be found by dividing $584 by 4.

Step 3   **Estimate.** Round $584 to $600. Then $600 ÷ 4 = $150, so a reasonable answer would be a little less than $150 each.

Step 4   **Solve.** Find the actual answer by dividing $584 by 4.

$$\begin{array}{r} 146 \\ 4\overline{)584} \end{array}$$

Step 5   **State the answer.** Each person got $146.

Step 6   **Check.** The answer $146 is reasonable, as $146 is close to the estimated answer of $150. Is the answer $146 correct? Check by multiplying.

$146 ← Amount received by each person

× 4 ← Number of people

$584 ← Total sales; matches number given in problem

Work Problem 2 at the Side.

EXAMPLE 2 Applying Addition

One week, Maureen O'Connor, operations manager, decided to total the wagon production at Radio Flyer, Inc. The daily production figures were 7642 wagons on Monday, 8150 wagons on Tuesday, 7916 wagons on Wednesday, 8419 wagons on Thursday, and 7704 wagons on Friday. Find the total production for the week.

Step 1 **Read.** In this problem, the production for each day is given and the total production for the week must be found.

Step 2 **Work out a plan.** Add the daily production figures to arrive at the weekly total.

Step 3 **Estimate.** Because the production was about 8000 wagons per day for a week of 5 days, a reasonable estimate would be $5 \cdot 8000 = 40,000$ wagons.

Step 4 **Solve.** Find the actual answer by adding the production for the 5 days.

$$
\begin{array}{r}
39,831 \quad \text{Check by adding up.} \\
7642 \\
8150 \\
7916 \\
8419 \\
+\ 7704 \\
\hline
39,831 \leftarrow \text{Number of wagons for the week}
\end{array}
$$

Step 5 **State the answer.** O'Connor's total production figure for the week was 39,831 wagons.

Step 6 **Check.** This answer of 39,831 wagons is close to the estimate of 40,000 wagons, so it is reasonable. Add up the columns to check the exact answer.

Calculator Tip The calculator solution to Example 2 uses chain calculations.

$$7642 \ \oplus \ 8150 \ \oplus \ 7916 \ \oplus \ 8419 \ \oplus \ 7704 \ \ominus \ \mathbf{39{,}831}$$

> **Work Problem 3 at the Side.** ▶▶▶

EXAMPLE 3 Determining whether Subtraction Is Necessary

The number of students enrolled in Chabot College this year is 4084 fewer than the number enrolled last year. Enrollment last year was 21,382. Find the enrollment this year.

Step 1 **Read.** In this problem, the enrollment has decreased from last year to this year. The enrollment last year and the decrease in enrollment are given. This year's enrollment must be found.

Step 2 **Work out a plan.** The indicator word, *fewer,* shows that subtraction must be used to find the number of students enrolled this year.

Step 3 **Estimate.** Because the enrollment was about 21,000 students, and the decrease in enrollment is about 4000 students, a reasonable estimate would be $21,000 - 4000 = 17,000$ students.

Continued on Next Page

---

3 Solve each problem.

(a) During the semester, Cindy received the following points on examinations and quizzes: 92, 81, 83, 98, 15, 14, 15, and 12. Find her total points for the semester.

(b) Stephanie Dixon works at the telephone order desk of a catalog sales company. One week she had the following number of customer contacts: Monday, 78; Tuesday, 64; Wednesday, 118; Thursday, 102; and Friday, 196. How many customer contacts did she have that week?

ANSWERS
3. (a) 410 points (b) 558 customer contacts

4 Solve each problem.

(a) A home has a living area of 1450 square feet, while an apartment has 980 square feet. Find the difference in the number of square feet of the two living areas.

470

(b) The Antique Military Vehicle Collectors (AMVC) had $14,863 in their club treasury bank account. After writing a check for $1180 to rent a display hall, find the amount remaining in the club account.

13,683

5 Solve each problem.

(a) Brenda is paid $685 for each kitchen remodeling job that she sells. If she sold 6 remodeling jobs and had $320 in sales expense deducted, how much did she make?

(b) An Internet book company had sales of 12,628 books with a profit of $6 for each book sold. If 863 books are returned, how much profit remains?

Step 4   **Solve.** Find the exact answer by subtracting 4084 from 21,382.

$$
\begin{array}{r}
21{,}382 \\
-\ 4\ 084 \\
\hline
17{,}298
\end{array}
$$

Step 5   **State the answer.** The enrollment this year is 17,298 students.

Step 6   **Check.** The answer 17,298 is reasonable, as it is close to the estimate of 17,000. Check by adding.

$$
\begin{array}{r}
17{,}298 \quad \leftarrow \text{Enrollment this year} \\
+\ 4\ 084 \quad \leftarrow \text{Decrease in enrollment} \\
\hline
21{,}382 \quad \leftarrow \text{Enrollment last year; matches number given in problem}
\end{array}
$$

⟨⟨⟨ Work Problem 4 at the Side.

EXAMPLE 4   Solving a Two-Step Problem

In May, a landlord received $720 from each of eight tenants. After paying $2180 in expenses, how much rent money did the landlord have left?

Step 1   **Read.** The problem asks for the amount of rent remaining after expenses have been paid.

Step 2   **Work out a plan.** The wording *from each of eight tenants* indicates that the eight rents must be totaled. Since the rents are all the same, first, use multiplication to find the total rent received. Finally, subtract expenses.

Step 3   **Estimate.** The amount of rent is about $700, making the total rent received about $700 • 8 = $5600. The expenses are about $2000. A reasonable estimate of the amount remaining is $5600 − $2000 = $3600.

Step 4   **Solve.** Find the exact amount by first multiplying $720 by 8 (the number of tenants).

$$
\begin{array}{r}
\$\ 720 \\
\times \quad\ 8 \\
\hline
\$5760
\end{array}
$$

Finally, subtract the $2180 in expenses from $5760.

$$
\begin{array}{r}
\$5760 \\
-\ \$2180 \\
\hline
\$3580
\end{array}
$$

Step 5   **State the answer.** The amount remaining is $3580.

Step 6   **Check.** The answer of $3580 is reasonable, since it is close to the estimated answer of $3600. Check by adding the expenses to the amount remaining and then dividing by 8.

$$
\$3580 + \$2180 = \$5760
$$

$$
\overset{\$720}{8)\overline{5760}} \quad \leftarrow \begin{array}{l}\text{Matches the rent amount} \\ \text{given in the problem}\end{array}
$$

Work Problem 5 at the Side. ⟩⟩⟩

1.10 Exercises

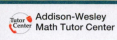

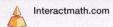

≈ *Solve each application problem. First use front end rounding to estimate the answer.*
Then find the exact answer. See Examples 1–4.

1. The billing department at Radio Flyer, Inc. sent out 80 invoices on Monday, 75 invoices on Tuesday, 135 invoices on Wednesday, 40 invoices on Thursday, and 52 invoices on Friday. How many invoices were sent out in this 5-day period?

Estimate:

Exact:

2. During a recent week, Radio Flyer, Inc. manufactured 32,815 Model #18 wagons, 4875 steel mini-wagons, 1975 wood 40-inch wagons, 15,308 scooters, and 9815 new-design plastic wagons. Find the total number of units manufactured.

Estimate:

Exact:

The graph shows the average credit card debt for college students by grade level.

3. On average how much more does a junior owe on credit card debt than a sophomore?

Estimate:

Exact:

4. How much more does a senior owe on credit card debt than a freshman on average?

Estimate:

Exact:

STUDENTS IN CHARGE

The percentage of college students using credit cards increased from 67% in 1998 to 83% in 2001. Average credit card debt, by grade level:

Source: Nellie Mae.

5. A packing machine can package 236 first-aid kits each hour. At this rate, find the number of first-aid kits packaged in 24 hours.

Estimate:

Exact:

6. If 450 admission tickets to a classic car show are sold each day, how many tickets are sold in a 12-day period?

Estimate:

Exact:

7. Ted Slauson, coordinator of Toys for Tots, has collected 2628 toys. If his group can give the same number of toys to each of 657 children, how many toys will each child receive?

Estimate:

Exact:

8. If profits of $680,000 are divided evenly among a firm's 1000 employees, how much money will each employee receive?

Estimate:

Exact:

9. The number of boaters and campers at the lake was 8392 on Friday. If this was 4218 more than the number of people at the lake on Wednesday, how many were there on Wednesday?

Estimate:

Exact:

10. The community has raised $52,882 for the homeless shelter. If the total amount needed for the shelter is $75,650, find the additional amount needed.

Estimate:

Exact:

11. Turn down the thermostat in the winter and you can save money and energy. In the upper Midwest, setting back the thermostat from 68° to 55° at night can save $34 per month on fuel. Find the amount of money saved in five months.

Estimate:

Exact:

12. The cost of tuition and fees at a community college is $785 per quarter. If Gale Klein has five quarters remaining, find the total amount that she will need for tuition and fees.

Estimate:

Exact:

≈ *The table shows the average starting salaries for selected occupations. Refer to the table to answer Exercises 13–16. First use front end rounding to estimate the answer. Then find the exact answer.*

13. How much more does an executive chef earn than a jet mechanic?

Estimate:

Exact:

14. How much more does a physical therapist earn than a social worker?

Estimate:

Exact:

FIRST PAYCHECKS

Typical annual starting salaries for different occupations.

Jet pilot	$82,400
Executive chef	$54,300
Physical therapist	$49,900
Jet mechanic	$45,000
Paralegal	$38,600
Librarian	$35,200
Social worker	$33,000
Accountant	$32,800
Realtor	$29,000
Firefighter	$25,500

Source: U.S. Department of Labor, Bureau of Labor Statistics.

15. Mr. Garrett is an accountant and Mrs. Garrett is a librarian. Mr. Harcos is a social worker and Mrs. Harcos is a physical therapist.

(a) Which couple has higher earnings?

Estimate:

Exact:

(b) Find the difference in the earnings.

Estimate:

Exact:

16. Mr. Gonzalez is a paralegal and Mrs. Gonzalez is a realtor. Mr. Horton is a firefighter and Mrs. Horton is a jet pilot.

(a) Which couple has higher earnings?

Estimate:

Exact:

(b) Find the difference in the earnings.

Estimate:

Exact:

17. Carolyn Phelps decides to establish a budget. She will spend $695 for rent, $340 for food, $435 for child care, $240 for transportation, $180 for other expenses, and she will put the remainder in savings. If her monthly take-home pay is $2240, find her monthly savings.

Estimate:

Exact:

18. Jared Ueda had $2874 in his checking account. He wrote checks for $308 for auto repairs, $580 for child support, and $778 for an insurance payment. Find the amount remaining in his account.

Estimate:

Exact:

19. There are 43,560 square feet in one acre. How many square feet are there in 138 acres?

Estimate:

Exact:

43,560 ×138

20. The number of gallons of water polluted each day in an industrial area is 209,670. How many gallons of water are polluted each year? (Use a 365-day year.)

Estimate:

Exact:

≈ *The Internet was used to find the following minivan optional features and the price of each feature. Use this information to answer Exercises 21–24.*

Safety and Security Options		Convenience and Comfort Options	
Option	Cost	Option	Cost
Bench seat with two child seats	$225	Power sliding door	$400
Supplemental side air bags	$390	8-way power seat	$370
Antilock brakes	$565	Roof rack	$250
Rear-window defroster	$195	Power door locks	$315
Power adjustable pedal	$185	Keyless entry	$150
Full-size spare tire	$160	AM/FM cassette and CD	$225

Source: www.edmunds.com

21. Find the total cost of all Safety and Security Options listed.

Estimate:

Exact:

22. Find the total cost of all Convenience and Comfort Options listed.

Estimate:

Exact:

23. A new-car dealer offers an option value package that includes a bench seat with two child seats, antilock brakes, keyless entry, and a power sliding door at a cost of $1220. If Jill buys the value package instead of paying for each option separately, how much will she save?

Estimate:

Exact:

24. A new-car dealer offers an option package that includes a bench seat with two child seats, antilock brakes, full-size spare tire, power door locks, 8-way power seat, and a roof rack for a total of $1750. How much will Samuel save if he buys the option package instead of paying for each option separately?

Estimate:

Exact:

25. The Enabling Supply House purchased 6 wheelchairs at $1256 each and 15 speech compression recorder-players at $895 each. Find the total cost.

Estimate:

Exact:

26. A college bookstore buys 17 desktop computers at $506 each and 13 printers at $482 each. Find the total cost.

Estimate:

Exact:

27. Being able to identify indicator words is helpful in determining how to solve an application problem. Write three indicator words for each of these operations: add, subtract, multiply, and divide. Write two indicator words that mean equals.

28. Identify and explain the six steps used to solve an application problem. You may refer to the text if you need help, but use your own words.

29. Write in your own words why it is important to estimate a reasonable answer. Give three examples of what might be a reasonable answer to a math problem from your daily activities.

30. First estimate by rounding to thousands, then find the exact answer to the following problem.

$$7438 + 6493 + 2380$$

Do the two answers vary by more than 1000? Why? Will estimated answers always vary from exact answers?

Estimate:

Exact:

Solve each application problem. See Examples 1–4.

31. A package of 3 undershirts costs $12, and a package of 6 pairs of socks costs $15. Find the total cost of 30 undershirts and 18 pairs of socks.

$10 \times 12 = 120$
$+45$
$3 \times 15 = 45$

32. Mark earned $11 per hour for 36 hours of work. Maria earned $13 per hour for 38 hours of work. Find their total combined income.

33. A car weighs 2425 pounds. If its 582-pound engine is removed and replaced with a 634-pound engine, what will the car weigh?

34. Barbara has $2324 in her preschool operating account. She spends $734 from this account, and then the class parents raise $568 in a rummage sale. Find the balance in the account after she deposits the money from the rummage sale.

35. In a recent survey of Reno/Lake Tahoe hotels, the cost per night at Harrah's in Reno was $59, while the cost at Harrah's in Lake Tahoe was $159 per night. Find the amount saved on a 5-night stay at Harrah's in Reno instead of staying at Harrah's in Lake Tahoe. (*Source:* Harrah's Casinos and Hotels, weekday rates.)

36. The most expensive hotel room in a recent study was the Ritz-Carlton at $495 per night, while the least expensive was Motel 6 at $66 per night. Find the amount saved in a 4-night stay at Motel 6 instead of staying at the Ritz-Carlton. (*Source:* Ritz-Carlton/Motel 6.)

37. A youth soccer association raised $7588 through fund-raising projects. After expenses of $838 were paid, the balance of the money was divided evenly among the 18 teams. How much did each team receive?

$7588 - 838 =$
$\div 18 =$

38. Feather Farms Egg Ranch collected 3545 eggs in the morning and 2575 eggs in the afternoon. If the eggs are packed in flats containing 30 eggs each, find the number of flats needed for packing.

39. A theater owner wants to provide enough setting for 1250 people. The main floor has 30 rows of 25 seats in each row. If the balcony has 25 rows, how many seats must be in each balcony row to satisfy the owner's seating requirements?

40. Jennie makes 24 grapevine wreaths per week to sell to gift shops. She works 40 weeks a year and packages six wreaths per box. If she ships equal quantities to each of five shops, find the number of boxes each store will receive.

Chapter 1

SUMMARY

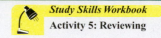

Study Skills Workbook
Activity 5: Reviewing

KEY TERMS

1.1	whole numbers	The whole numbers are 0, 1, 2, 3, 4, 5, 6, 7, 8, and so on.
	place value	The place value of each digit in a whole number is determined by its position in the whole number.
	table	A table is a display of facts in rows and columns.
1.2	addition	The process of finding the total is addition.
	addends	The numbers being added in an addition problem are addends.
	sum (total)	The answer in an addition problem is called the sum.
	commutative property of addition	The commutative property of addition states that the order of numbers in an addition problem can be changed without changing the sum.
	associative property of addition	The associative property of addition states that grouping the addition of numbers differently does not change the sum.
	carrying	The process of carrying is used in an addition problem when the sum of the digits in an column is greater than 9.
	perimeter	The perimeter is the distance around the outside edges of a figure.
1.3	minuend	The number from which another number (the subtrahend) is being subtracted is the minuend.
	subtrahend	The subtrahend is the number being subtracted in a subtraction problem.
	difference	The answer in a subtraction problem is called the difference.
	borrowing	The method of borrowing is used in subtraction if a digit is less than the one directly below it.
1.4	factors	The numbers being multiplied are called factors. For example, in $3 \times 4 = 12$, both 3 and 4 are factors.
	product	The answer in a multiplication problem is called the product.
	commutative property of multiplication	The commutative property of multiplication states that the product in a multiplication problem remains the same when the order of the factors is changed.
	associative property of multiplication	The associative property of multiplication states that grouping the numbers differently does not change the product.
	multiple	The product of two whole number factors is a multiple of those numbers.
1.5	dividend	The number being divided by another number in a division problem is the dividend.
	divisor	The divisor is the number doing the dividing in a division problem.
	quotient	The answer in a division problem is called the quotient.
	short division	A method of dividing a number by a one-digit divisor is short division.
	remainder	The remainder is the number left over when two numbers do not divide exactly.
1.6	long division	The process of long division is used to divide by a number with more than one digit.
1.7	rounding	Rounding is used to find a number that is close to the original number, but easier to work with. Use the $\approx$ sign, which means "approximately equal to."
	estimate	An estimated answer is one that is close to the exact answer.
	front end rounding	Rounding to the highest possible place so that all the digits become zeros except the first one is front end rounding.
1.8	square root	The square root of a whole number is the number that can be multiplied by itself to produce the given number.
	perfect square	A number that is the square of a whole number is a perfect square.
	order of operations	For problems or expressions with more than one operation, the order of operations tells what to do first, second, and so on to get the correct answer.
1.9	pictograph	A graph that uses pictures or symbols to show data is a pictograph.
	bar graph	A graph that uses bars of various heights to show quantity is a bar graph.
	line graph	A graph that uses dots connected by lines to show trends is a line graph.
1.10	indicator words	Words in a problem that indicate the necessary operations—addition, subtraction, multiplication, or division—are indicator words.

TEST YOUR WORD POWER

See how well you have learned the vocabulary in this chapter. Answers follow the Quick Review.

1. When using **addends** you are performing
   - A. division
   - B. subtraction
   - C. addition
   - D. multiplication.

2. The **subtrahend** is the
   - A. number being multiplied
   - B. number being subtracted
   - C. number being added
   - D. answer in division.

3. A **factor** is
   - A. the answer in an addition problem
   - B. one of two or more numbers being added
   - C. one of two or more numbers being multiplied
   - D. one of two or more numbers being divided.

4. The **divisor** is
   - A. the number being rounded
   - B. the number being multiplied
   - C. always the largest number
   - D. the number doing the dividing.

5. We use **rounding** to
   - A. avoid solving a problem
   - B. purely guess at the answer
   - C. help estimate a reasonable answer
   - D. find the remainder.

6. A **perfect square** is
   - A. the square of a whole number
   - B. the same as square root
   - C. similar to a perfect triangle
   - D. used when rounding.

QUICK REVIEW

Concepts

Examples

1.1 Reading and Writing Whole Numbers

Do not use the word *and* when writing a whole number. Commas help divide the periods or groups for ones, thousands, millions, and billions. A comma is not needed when a number has four digits or fewer.

795 is written *seven hundred ninety-five.*
9,768,002 is written *nine million, seven hundred sixty-eight thousand, two*

1.2 Adding Whole Numbers

Add from top to bottom, starting with the ones column and working left. To check, add from bottom to top.

(Add up to check.)

$$
\begin{array}{r}
1\ 1\ 4\ 0 \\
6\ 8\ 7 \\
2\ 6 \\
9 \\
+\ 4\ 1\ 8 \\
\hline
1\ 1\ 4\ 0
\end{array}
$$

Addends

Sum

1.2 Commutative Property of Addition

Changing the order of the addends in an addition problem does not change the sum.

$2 + 4 = 6$

$4 + 2 = 6$

By the commutative property, the sum is the same.

1.2 Associative Property of Addition

Grouping the addends differently when adding does not change the sum.

$(2 + 3) + 4 = 9$

$2 + (3 + 4) = 9$

By the associative property, the sum is the same.

1.3 Subtracting Whole Numbers

Subtract the subtrahend from the minuend to get the difference by borrowing when necessary. To check, add the difference to the subtrahend to get the minuend.

Problem

$$
\begin{array}{r}
6\ 12\ 18 \\
4\ 7\ 3\ 8 \\
-\ \ \ 6\ 4\ 9 \\
\hline
4\ 0\ 8\ 9
\end{array}
$$

Minuend
Subtrahend
Difference

Check

$$
\begin{array}{r}
4\ 0\ 8\ 9 \\
+\ \ \ 6\ 4\ 9 \\
\hline
4\ 7\ 3\ 8
\end{array}
$$

Concepts	Examples

1.4 Multiplying Whole Numbers

Use $\times$, $\cdot$ (a raised dot), or parentheses to indicate multiplication.

The numbers being multiplied are called *factors*. The multiplicand is being multiplied by the multiplier, giving the product. When the multiplier has more than one digit, partial products must be used and added to find the product.

3×4 or $3 \cdot 4$ or $(3)(4)$ or $3(4)$

$$
\begin{array}{r}
78 \\
\times \quad 24 \\
\hline
312 \\
156 \\
\hline
1872
\end{array}
$$

78 — Multiplicand ⎫ Factors
24 — Multiplier ⎭
312 — Partial product
156 — Partial product (move one position left)
1872 — Product

1.4 Commutative Property of Multiplication

The product in a multiplication problem remains the same when the order of the factors is changed.

$$3 \times 4 = 12$$
$$4 \times 3 = 12$$

By the commutative property, the product is the same.

1.4 Associative Property of Multiplication

The grouping of factors differently when multiplying does not change the product.

$$(2 \times 3) \times 4 = 24$$
$$2 \times (3 \times 4) = 24$$

By the associative property, the product is the same.

1.5 Dividing Whole Numbers

$\div$ and $\overline{)}$ mean divide.
Also a —, as in $\frac{25}{5}$, means to divide the top number (dividend) by the bottom number (divisor).

Divisor → $4\overline{)88}$ 22 ← Quotient, 88 ← Dividend

$$
\begin{array}{r}
22 \\
4\overline{)88} \\
88 \\
\hline
0
\end{array}
$$

$88 \div 4 = 22$

Dividend | Quotient
Divisor

$$\frac{88}{4} = 22 \leftarrow \text{Quotient}$$

1.7 Rounding Whole Numbers

Rules for Rounding:
Step 1 Locate the place to be rounded, and draw a line under it.
Step 2 If the next digit to the right is 5 or more, increase the underlined digit by 1. If the next digit is 4 or less, do not change the underlined digit.
Step 3 Change all digits to the right of the underlined place to zeros.

Round 726 to the nearest ten.

Next digit is 5 or more.

7<u>2</u>6

Tens place increases by 1 (2 + 1 = 3).

7<u>2</u>6 rounds to 7<u>3</u>0.

Round 1,498,586 to the nearest million.

Next digit is 4 or less.

<u>1</u>,498,586

Millions place does not change.

1,498,586 rounds to 1,000,000.

1.7 Front End Rounding

Front end rounding is rounding to the highest possible place so that all the digits become 0 except the first digit.

Round each number using front end rounding.

76 rounds to 80.

348 rounds to 300.

6512 rounds to 7000.

23,751 rounds to 20,000.

652,179 rounds to 700,000.

Concepts	Examples

1.8 Order of Operations

Problems may have several operations. Work these problems using the order of operations.

1. Do all operations inside parentheses or other grouping symbols.
2. Simplify any expressions with exponents and find any square roots $\left(\sqrt{} \right)$.
3. Multiply or divide proceeding from left to right.
4. Add or subtract proceeding from left to right.

Simplify, using the order of operations.

$7 \cdot \sqrt{9} - 4 \cdot 5$ Find the square root.

$7 \cdot 3 - 4 \cdot 5$ Multiply from left to right.

$\underbrace{21} - \underbrace{20} = 1$ Subtract.

1.9 Reading Pictographs, Bar Graphs, and Line Graphs

A *pictograph* uses pictures or symbols to show data.

A *bar graph* uses bars of various heights to show quantity.

A *line graph* uses dots connected by lines to show trends.

When reading a pictograph, be certain that you determine the quantity represented by each picture or symbol.

When reading a bar graph, use a straightedge to line up the top of the bar with the numbers along the left edge of the graph.

When reading a line graph, use a straightedge to line up the dot with the numbers along the left edge of the graph.

1.10 Application Problems

Steps for Solving an Application Problem

Step 1 **Read** the problem carefully, perhaps several times.

Step 2 **Work out a plan** before starting. Draw a sketch if possible.

Step 3 **Estimate** a reasonable answer.

Step 4 **Solve** the problem.

Step 5 **State the answer.**

Step 6 **Check** your work. If the answer is not reasonable, start over.

Manuel earns $118 on Sunday, $87 on Monday, and $63 on Tuesday. Find his total earnings for the 3 days.

Step 1 The earnings for each day are given, and the total for the 3 days must be found.

Step 2 Add the daily earnings to find the total.

Step 3 Since the earnings were about $100 + $90 + $60 = $250, a reasonable estimate would be approximately $250.

Step 4   **$268**   Check by adding up
      $118
        87
      + 63
      ———
      $268   Total earnings

Step 5 Manuel's total earnings are $268.

Step 6 The answer is reasonable, because it is close to the estimate of $250.

ANSWERS TO TEST YOUR WORD POWER

1. C. *Example:* In $2 + 3 = 5$, the 2 and the 3 are addends.
2. B. *Example:* In $5 - 4 = 1$, the 4 is the subtrahend.
3. C. *Example:* In $3 \times 5 = 15$, the numbers 3 and 5 are factors.
4. D. *Example:* In $8 \div 4 = 2$, $\frac{8}{4} = 2$, and $4\overline{)8}$, the 4 is the divisor.
5. C. *Example:* We can use rounding to estimate our answer and then determine whether the exact answer is reasonable.
6. A. *Example:* 25 is a perfect square because $5^2 = 25$ and 5 is a whole number.

Chapter **1**

R E V I E W   E X E R C I S E S

If you need help with any of these Review Exercises, look in the section indicated in brackets.

[1.1] *Write the digits for the given period or group in each number.*

1. 6573

thousands

ones

2. 36,215

thousands

ones

3. 105,724

thousands

ones

4. 1,768,710,618

billions

millions

thousands

ones

Rewrite each number in words.

5. 728

6. 15,310

7. 319,215

8. 62,500,005

Rewrite each number in digits.

9. ten thousand, eight

10. two hundred million, four hundred fifty-five

[1.2] *Add.*

11.   72
      + 38

12.   54
      + 67

13.   807
       4606
      +   51

14.   8215
          9
      + 7433

15.   2130
           453
          8107
      +   296

16.   5684
           218
          2960
      +   983

17.   5 732
         11,069
            37
          1 595
      + 22,169

18.   3 451
         12,286
            43
          1 291
      + 32,784

[1.3] *Subtract.*

19. $\begin{array}{r} 64 \\ -28 \end{array}$	**20.** $\begin{array}{r} 46 \\ -19 \end{array}$	**21.** $\begin{array}{r} 375 \\ -186 \end{array}$	**22.** $\begin{array}{r} 573 \\ -389 \end{array}$

23. $\begin{array}{r} 7416 \\ -567 \end{array}$	**24.** $\begin{array}{r} 5210 \\ -883 \end{array}$	**25.** $\begin{array}{r} 2210 \\ -1986 \end{array}$	**26.** $\begin{array}{r} 99{,}704 \\ -73{,}838 \end{array}$

[1.4] *Multiply.*

27. $\begin{array}{r} 7 \\ \times 7 \end{array}$	**28.** $\begin{array}{r} 8 \\ \times 0 \end{array}$	**29.** $8(4)$	**30.** $8(8)$

31. $(5)(9)$	**32.** $(6)(7)$	**33.** $7 \cdot 8$	**34.** $9 \cdot 9$

Work each chain multiplication.

35. $5 \times 4 \times 2$	**36.** $9 \times 1 \times 5$	**37.** $4 \times 4 \times 3$	**38.** $2 \times 2 \times 2$

39. $(6)(0)(8)$	**40.** $(7)(1)(6)$	**41.** $6 \cdot 1 \cdot 8$	**42.** $7 \cdot 7 \cdot 0$

Multiply.

43. $\begin{array}{r} 28 \\ \times 3 \end{array}$	**44.** $\begin{array}{r} 46 \\ \times 8 \end{array}$	**45.** $\begin{array}{r} 58 \\ \times 9 \end{array}$	**46.** $\begin{array}{r} 98 \\ \times 1 \end{array}$

47. $\begin{array}{r} 625 \\ \times 8 \end{array}$	**48.** $\begin{array}{r} 374 \\ \times 8 \end{array}$	**49.** $\begin{array}{r} 1349 \\ \times 4 \end{array}$	**50.** $\begin{array}{r} 9163 \\ \times 5 \end{array}$

51. $\begin{array}{r} 7456 \\ \times 2 \end{array}$	**52.** $\begin{array}{r} 2880 \\ \times 7 \end{array}$	**53.** $\begin{array}{r} 93{,}105 \\ \times 5 \end{array}$	**54.** $\begin{array}{r} 21{,}873 \\ \times 8 \end{array}$

55.   35
   × 25

56.   74
   × 32

57.   98
   × 12

58.   68
   × 75

59.   472
   × 33

60.   392
   × 77

61.   4051
   × 219

62.   1527
   × 328

Find each total cost.

63. 30 scientific calculators at \$12 per calculator

64. 76 subscribers at \$14 per subscription

65. 318 drill bit sets at \$64 per set

66. 114 earplugs at \$6 per earplug

Multiply by using the shortcut for multiples of 10.

67.   280
   × 50

68.   340
   × 70

69.   517
   × 400

70.   637
   × 500

71.   16,000
   × 8 000

72.   43,000
   × 2 100

[1.5] *Divide. If the division is not possible, write "undefined."*

73. $20 \div 4$

74. $35 \div 5$

75. $42 \div 7$

76. $18 \div 9$

77. $\dfrac{54}{9}$

78. $\dfrac{36}{9}$

79. $\dfrac{49}{7}$

80. $\dfrac{0}{6}$

81. $\dfrac{148}{0}$

82. $\dfrac{0}{23}$

83. $\dfrac{64}{8}$

84. $\dfrac{81}{9}$

[1.5–1.6] *Divide.*

85. $4\overline{)328}$

86. $3\overline{)294}$

87. $6\overline{)26,532}$

88. $76\overline{)26,752}$

89. $2704 \div 18$

90. $15,525 \div 125$

[1.7] *Round as indicated.*

91. 817 to the nearest ten

92. 15,208 to the nearest hundred

93. 20,643 to the nearest thousand

94. 67,485 to the nearest ten thousand

Round each number to the nearest ten, nearest hundred, and nearest thousand. Remember to round from the original number.

	Ten	Hundred	Thousand
95. 3487	_____	_____	_____
96. 20,065	_____	_____	_____
97. 98,201	_____	_____	_____
98. 352,118	_____	_____	_____

[1.8] *Find each square root by using the Perfect Squares Table on page 75.*

99. $\sqrt{16}$ **100.** $\sqrt{49}$ **101.** $\sqrt{144}$ **102.** $\sqrt{196}$

Identify the exponent and the base, and then simplify each expression.

103. 7^3 **104.** 3^6 **105.** 5^3 **106.** 4^5

Simplify each expression by using the order of operations.

107. $7^2 - 15$ **108.** $6^2 - 10$ **109.** $2 \cdot 3^2 \div 2$

PEMDAS

110. $9 \div 1 \cdot 2 \cdot 2 \div (11 - 2)$ **111.** $\sqrt{9} + 2(3)$ **112.** $6 \cdot \sqrt{16} - 6 \cdot \sqrt{9}$

$9 \div 1 \cdot 2 \cdot 2 \div 9$ $3 + 6 = 9$ 6 $L \cdot A - b \div 3$
$9 \quad 2 \cdot 2 \qquad \cdot 2$ $24 - 18 = 6$

[1.9] *The bar graph shows the number of parents out of 100 surveyed who nag their children about performing certain household chores.*

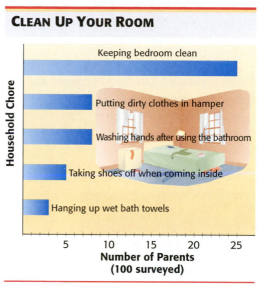

CLEAN UP YOUR ROOM

Household Chore

- Keeping bedroom clean
- Putting dirty clothes in hamper
- Washing hands after using the bathroom
- Taking shoes off when coming inside
- Hanging up wet bath towels

Number of Parents
(100 surveyed)

5   10   15   20   25

Source: Opinion Research Corporation for the Soap and Detergent Association.

113. How many parents nagged their children about washing hands after using the bathroom?

114. Find the number of parents who nagged their children about taking shoes off when coming inside.

115. Which household chore was nagged about by the greatest number of parents?

116. Which household chore was nagged about by the least number of parents?

≈ **[1.10]** *Solve each application problem. First use front end rounding to estimate the answer. Then find the exact answer.*

117. Find the cost of 65 chefs hats at $18 per hat.

Estimate:

Exact:

118. A pulley on an evaporative cooler turns 1400 revolutions per minute. How many revolutions will the pulley turn in 60 minutes?

Estimate:

Exact:

119. There are 144 plastic forks in a box. Find the number of plastic forks in 15 boxes.

Estimate:

Exact:

120. A drum contains 6000 brackets. How many brackets are in 30 drums?

Estimate:

Exact:

121. It takes 2000 hours of work to build one home. How many hours of work are needed to build 12 homes?

Estimate:

Exact:

122. A Japanese bullet train travels 80 miles in 1 hour. Find the number of miles traveled in 5 hours.

Estimate:

Exact:

123. The Kennedy Space Center charges $33 for each adult admission and $23 for each child. Find the total cost to admit a group of 22 adults and 25 children. (*Source:* Kennedy Space Center.)

Estimate:

Exact:

124. A newspaper carrier has 62 customers who take the paper daily and 21 customers who take the paper on weekends only. A daily customer pays $16 per month and a weekend-only customer pays $7 per month. Find the total monthly collections.

Estimate:

Exact:

125. Home Depot displays and sells stainless steel kitchen sinks that range in price from $39 to $388. Find the difference in price between the most and the least expensive sinks. (*Source:* Home Depot.)

Estimate:

Exact:

127. A food canner uses 1 pound of pork for every 175 cans of pork and beans. How many pounds of pork are needed for 8750 cans?

Estimate:

Exact:

129. Nitrogen sulfate is used in farming to enrich nitrogen-poor soil. If 625 pounds of nitrogen sulfate are spread per acre, how many acres can be spread with 32,500 pounds of nitrogen sulfate?

Estimate:

Exact:

126. A stamping machine produces 986 license plates each hour. How long will it take to produce 32,538 license plates?

Estimate:

Exact:

128. Rachel Leach writes a $520 check for rent and a $385 check for her car payment. If she started with $1924 in her checking account, how much remains in her account.

Estimate:

Exact:

130. Each home in a subdivision requires 180 feet of fencing. Find the number of homes that can be fenced with 5760 feet of fencing material.

Estimate:

Exact:

MIXED REVIEW EXERCISES*

Perform the indicated operations.

131. 4(83)

132. 7(64)

133.
$$309 - 56$$

134.
$$835 - 247$$

135.
$$662 + 379$$

136.
$$789 + 872$$

137.
$$38,140 - 6\,078$$

138.
$$29,156 - 4\,209$$

139. $21 \div 7$

140. $\dfrac{42}{6}$

141.
$$\begin{array}{r} 7\,218 \\ 3 \\ 18 \\ 1\,791 \\ 82,623 \\ + 1\,982 \end{array}$$

142.
$$\begin{array}{r} 3\,812 \\ 5 \\ 22 \\ 1\,836 \\ 75,134 \\ + 2\,369 \end{array}$$

143. $\dfrac{9}{0}$

144. $\dfrac{7}{1}$

145. $27,600 \div 4$

---

* The order of exercises in this final group does not correspond to the order in which topics occur in the chapter. This random ordering should help you prepare for the chapter test in yet another way.

146. 18,480 ÷ 8

147.
$$\begin{array}{r} 8430 \\ \times \quad 128 \\ \hline \end{array}$$

148.
$$\begin{array}{r} 21,702 \\ \times \quad 6 \\ \hline \end{array}$$

149. $34\overline{)3672}$

150. $68\overline{)14,076}$

151. Rewrite 376,853 in words.

152. Rewrite 408,610 in words.

153. Round 8749 to the nearest hundred.

154. Round 400,503 to the nearest thousand.

Find each square root.

155. $\sqrt{64}$

156. $\sqrt{81}$

Find each total cost.

157. 308 pairs of knee guards at $18 per pair

158. 84 dishwashers at $370 per dishwasher

159. 208 baseball hats at $11 per hat

160. 607 boxes of avocados at $26 per box

Solve each application problem.

161. There are 52 playing cards in a deck. How many cards are there in nine decks?

162. Your college bookstore receives textbooks packed 20 books per carton. How many textbooks are received in a delivery of 180 cartons?

163. Push-type gasoline-powered lawn mowers cost $100 less than self-propelled mowers that you walk behind. If a self-propelled mower costs $380, find the cost of a push-type mower.

164. The Country Day School wants to raise $218,450 to construct and equip a computer lab. If $103,815 has already been raised, how much more is needed?

American River Raft Rentals lists the following daily raft rental fees. Notice that there is an additional $2 launch fee payable to the park system for each raft rented. Use this information to solve Exercises 165 and 166.

AMERICAN RIVER RAFT RENTALS

Size	Rental Fee	Launch Fee
4-person	$28	$2
6-person	$38	$2
10-person	$70	$2
12-person	$75	$2
16-person	$85	$2

Source: American River Raft Rentals.

165. On a recent Tuesday the following rafts were rented: 6 4-person; 15 6-person; 10 10-person; 3 12-person; and 2 16-person. Find the total receipts, including the $2 per-raft launch fee.

166. On the 4th of July the following rafts were rented: 38 4-person; 73 6-person; 58 10-person; 34 12-person; and 18 16-person. Find the total receipts, including the $2 per-raft launch fee.

New York plans to build the world's tallest structure. Broadcasters are proposing to build a $200 million television tower to replace the antenna lost when the World Trade Center was destroyed by terrorists. The pictogram below shows the height of buildings around the world. Use this information to solve Exercises 167–170.

167. How much taller are the Petronas Towers in Malaysia than the Eiffel Tower in Paris?

168. How much taller is the Empire State Building in New York than the Canadian National Tower in Toronto?

169. (a) What is the combined height of the four existing buildings? Do not include the World Trade Center tower or the proposed antenna.

(b) Is the combined height of these four buildings greater or less than a mile? How much greater or less than a mile? (Hint: 1 mile = 5280 ft)

170. The estimated cost of building the proposed antenna is $200 million. What is the cost per foot?

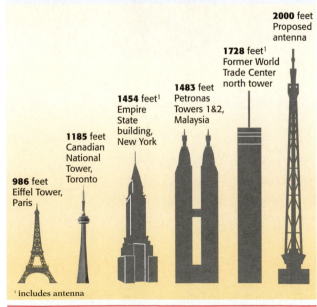

THE WORLD'S TALLEST STRUCTURE

Broadcasters are proposing to build a $200 million TV tower to replace the antenna lost when the World Trade Center towers collapsed.

2000 feet Proposed antenna

1728 feet[1] Former World Trade Center north tower

1483 feet Petronas Towers 1&2, Malaysia

1454 feet[1] Empire State building, New York

1185 feet Canadian National Tower, Toronto

986 feet Eiffel Tower, Paris

[1] includes antenna

Source: USA Today research; rendition by Peter Coe.

Chapter 1

TEST

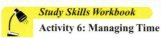

Study Skills Workbook
Activity 6: Managing Time

Write each number in words.

1. 9205   **2.** 25,065

3. Use digits to write four hundred twenty-six thousand, five.

Add.

4.
```
    853
     66
   4022
 + 3589
```

5.
```
  17,063
       7
      12
   1 505
  93,710
 +   333
```

Subtract.

6.
```
   9009
 − 7964
```

7.
```
   9075
 − 2869
```

Multiply

8. $7 \times 6 \times 4$ **9.** $57 \cdot 3000$

10. 85(19)   **11.**
```
   7381
 × 603
```

Divide. If the division is not possible, write "undefined."

12. $16\overline{)112{,}752}$ **13.** $\dfrac{835}{0}$

14. $19{,}241 \div 42$ **15.** $280\overline{)44{,}800}$

Round as indicated.

16. 6347 to the nearest ten

17. 76,502 to the nearest thousand

1. _____

2. _____

3. _____

4. _____

5. _____

6. _____

7. _____

8. _____

9. _____

10. _____

11. _____

12. _____

13. _____

14. _____

15. _____

16. _____

17. _____

18. _____

19. _____

20. *Estimate:* _____

    Exact: _____

21. *Estimate:* _____

    Exact: _____

22. *Estimate:* _____

    Exact: _____

23. *Estimate:* _____

    Exact: _____

24. _____

25. _____

Simplify each expression.

18. $5^2 + 8(2)$

19. $7 \cdot \sqrt{64} - 14 \cdot 2$

≈ *Solve each application problem. First use front end rounding to estimate the answer. Then find the exact answer.*

20. Amy collects the following monthly rents from the tenants in her fourplex: $485, $500, $515, and $425. After she pays expenses of $785, how much does she have left?

21. Hewlett-Packard assembles 542 computer printers each day. Find the number of workdays it would take to assemble 67,750 computer printers.

22. Barb Weaks paid $528 for tires, $195 for brakes, and $235 for a timing belt. If this money was withdrawn from her checking account, which had a balance of $1906, find her new balance.

23. An electronics manufacturer assembles 208 digital cameras each hour for 4 hours and 238 camcorders each hour for the next 4 hours. Find the number of both types of camera assembled in the 8-hour period.

24. Explain in your own words the rules for rounding numbers. Give an example of rounding a number to the nearest ten thousand.

25. List the six steps for solving application problems.

Multiplying and Dividing Fractions

2

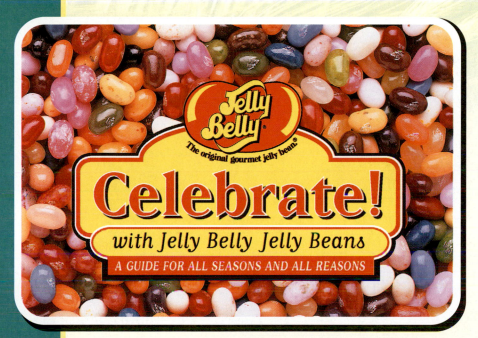

NEST COOKIES

$1\frac{1}{2}$ cups all purpose
   flour
1 tsp. baking powder
$\frac{1}{2}$ tsp. salt
$\frac{1}{2}$ cup shortening
$\frac{1}{2}$ cup sugar
1 egg white

1 tsp. vanilla
2 Tbs. milk
2 cups shredded
   coconut
5 oz. Jelly Belly
   jelly beans
   assorted flavors

Heat oven to 375° F. Sift together flour, baking powder, and salt and set aside. In large bowl beat shortening, sugar, egg white, and vanilla until well blended. Add flour mixture and milk until blended. Stir in coconut.

Roll dough into a ball, divide in half. Roll 15 one-inch balls from each half and place on ungreased baking sheet. Make thumb print depression in center of each ball to form nest. Bake 6 minutes. Remove from oven and place 4 Jelly Belly beans in center of each cookie. Return to oven and bake 5 more minutes. Transfer cookies to wire rack to cool. Makes 30 cookies.

Most recipes include ingredients that use common fractions and mixed numbers in their measurements. The recipe shown here uses Jelly Belly jelly beans and will make $2\frac{1}{2}$ dozen (30) cookies. But suppose you wanted to make 3 dozen, 10 dozen, or even $3\frac{1}{2}$ dozen cookies? You would have to multiply or divide each of the ingredients to arrive at the proper amounts needed. In this chapter we discuss multiplication and division of fractions, which you need to know when cooking or baking. (See Exercise 39 in **Section 2.7** and Exercises 25–28 in **Section 2.8**.)

2.1   Basics of Fractions

In Chapter 1 we discussed whole numbers. Many times, however, we find that parts of whole numbers are considered. One way to write parts of a whole is with **fractions.** Another way is with decimals, which is discussed in Chapter 4.

OBJECTIVE 1   Use a fraction to show which part of a whole is shaded. The number $\frac{1}{8}$ is a fraction that represents 1 of 8 equal parts. Read $\frac{1}{8}$ as "one eighth."

EXAMPLE 1   Identifying Fractions

Use fractions to represent the shaded portions and the unshaded portions of each figure.

(a)   The figure on the left has 6 equal parts. The 1 shaded part is represented by the fraction $\frac{1}{6}$. The *un*shaded part is $\frac{5}{6}$.

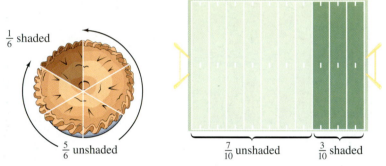

(b)   The 3 shaded parts of the 10-part figure on the right are represented by the fraction $\frac{3}{10}$. The *un*shaded part is $\frac{7}{10}$.

> **Work Problem 1 at the Side.**

Fractions can be used to show more than one whole object.

EXAMPLE 2   Representing Fractions Greater Than 1

Use a fraction to represent the shaded part of each figure.

(a)
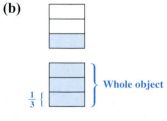

An area equal to 5 of the $\frac{1}{4}$ parts is shaded. Write this as $\frac{5}{4}$.

(b)

An area equal to 4 of the $\frac{1}{3}$ parts is shaded, so $\frac{4}{3}$ is shaded.

> **Work Problem 2 at the Side.**

OBJECTIVE 2   Identify the numerator and denominator.   In the fraction $\frac{2}{3}$, the number 2 is the **numerator** and 3 is the **denominator**. The bar between the numerator and the denominator is the *fraction bar.*

Fraction bar → $\dfrac{2}{3}$ ← Numerator
            ← Denominator

---

① Write fractions for the shaded portions and the unshaded portions of each figure.

(a)
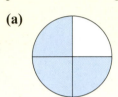

(b)

(c)
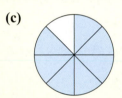

② Write fractions for the shaded portions of each figure.

(a)

(b)
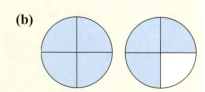

Numerators and Denominators

The **denominator** of a fraction shows the number of equivalent parts in the whole, and the **numerator** shows how many parts are being considered.

> **NOTE**
> Remember that a bar, —, is one of the division symbols, and that division by 0 is undefined. A fraction with a denominator of 0 is also undefined.

EXAMPLE 3 Identifying Numerators and Denominators

Identify the numerator and denominator in each fraction.

(a) $\dfrac{3}{4}$ (b) $\dfrac{8}{5}$

$\dfrac{3}{4}$ ← Numerator
   ← Denominator

$\dfrac{8}{5}$ ← Numerator
   ← Denominator

Work Problem 3 at the Side.

OBJECTIVE 3 Identify proper and improper fractions.   Fractions are sometimes called *proper* or *improper* fractions.

Proper and Improper Fractions

If the numerator of a fraction is *smaller* than the denominator, the fraction is a **proper fraction.** A proper fraction is less than 1 whole.

If the numerator is *greater than or equal to* the denominator, the fraction is an **improper fraction.** An improper fraction is greater than or equal to 1 whole.

Proper Fractions	Improper Fractions
$\dfrac{5}{8}$ $\dfrac{3}{5}$ $\dfrac{23}{24}$	$\dfrac{6}{5}$ $\dfrac{10}{10}$ $\dfrac{115}{112}$

EXAMPLE 4 Classifying Types of Fractions

(a) Identify all proper fractions in this list.

$$\frac{3}{4} \quad \frac{5}{9} \quad \frac{17}{5} \quad \frac{9}{7} \quad \frac{3}{3} \quad \frac{12}{25} \quad \frac{1}{9} \quad \frac{5}{3}$$

Proper fractions have a numerator that is smaller than the denominator. The proper fractions are shown below.

$$\frac{3}{4} \quad \text{← 3 is smaller than 4.} \qquad \frac{5}{9} \quad \frac{12}{25} \quad \frac{1}{9}$$

(b) Identify all improper fractions in the list in part (a).
   Improper fractions have a numerator that is equal to or greater than the denominator. The improper fractions are shown below.

$$\frac{17}{5} \quad \text{← 17 is greater than 5.} \qquad \frac{9}{7} \quad \frac{3}{3} \quad \frac{5}{3}$$

Work Problem 4 at the Side.

---

3 Identify the numerator and the denominator. Draw a picture with shaded parts to show each fraction. Your drawings may vary, but they should have the correct number of shaded parts.

(a) $\dfrac{2}{3}$

(b) $\dfrac{1}{4}$

(c) $\dfrac{8}{5}$

(d) $\dfrac{5}{2}$

4 From the following group of fractions:

$$\frac{2}{3} \quad \frac{4}{3} \quad \frac{3}{4} \quad \frac{8}{8} \quad \frac{3}{1} \quad \frac{1}{3}$$

(a) list all proper fractions;

(b) list all improper fractions.

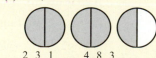

Real-Data Applications

Quilt Patterns

People who make quilts often base their designs on a block that is cut into a grid of 4, 9, 16, or 25 squares. The quilter chooses various colors for the pieces. Each quilt design shown was selected from an Archive of American Quilt Designs.

1. Identify the makeup of the block as 4, 9, 16, etc. Each color is what fractional part of the block?

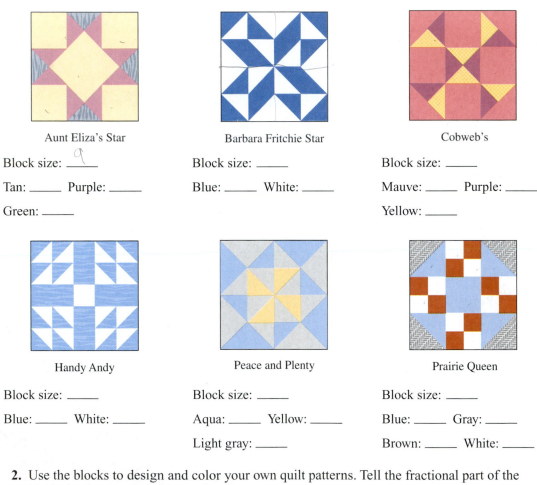

Aunt Eliza's Star

Block size: ___9___

Tan: _____ Purple: _____

Green: _____

Barbara Fritchie Star

Block size: _____

Blue: _____ White: _____

Cobweb's

Block size: _____

Mauve: _____ Purple: _____

Yellow: _____

Handy Andy

Block size: _____

Blue: _____ White: _____

Peace and Plenty

Block size: _____

Aqua: _____ Yellow: _____

Light gray: _____

Prairie Queen

Block size: _____

Blue: _____ Gray: _____

Brown: _____ White: _____

2. Use the blocks to design and color your own quilt patterns. Tell the fractional part of the block that is represented by each color.

3. Find the next two numbers in this pattern: 4, 9, 16, 25, _____, _____.

4. Explain how the pattern works:

2.1   **Exercises**

Write fractions to represent the shaded and unshaded portions of each figure. See Examples 1 and 2.

1.  $\dfrac{3}{4}$

2. 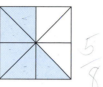 $\dfrac{5}{8}$

3. 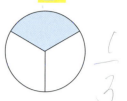 $\dfrac{1}{3}$

4. 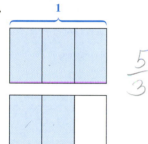 $\dfrac{5}{3}$

5. $\dfrac{7}{5}$

6. $\dfrac{11}{6}$

7. What fraction of these 6 bills has a life span of 2 years or greater? $\dfrac{5}{6}$

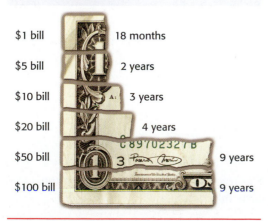

A BILL'S LIFE

A $1 bill lasts about 18 months as compared with the average lifespan of other denominations:

$1 bill	18 months
$5 bill	2 years
$10 bill	3 years
$20 bill	4 years
$50 bill	9 years
$100 bill	9 years

Source: Federal Reserve System; Bureau of Engraving and Printing.

8. What fraction of these 6 recording artists are women?

9. In an American Sign Language (ASL) class of 25 students, 8 are hearing impaired. What fraction of the students are hearing impaired? $\frac{8}{25}$

10. Of 98 bicycles in a bike rack, 67 are mountain bikes. What fraction of the bicycles are not mountain bikes?

11. Of the 218 trees in the park, 63 are oak trees. What fraction of the trees are not oak trees?

12. A women's community college basketball team has 18 members. If 7 of the players are sophomores and the rest are freshmen, find the fraction of the players who are freshmen. $\frac{11}{8}$

Identify the numerator and denominator. See Example 3.

	Numerator	**Denominator**		**Numerator**	**Denominator**
13. $\frac{4}{5}$	_____	_____	14. $\frac{5}{6}$	_____	_____
15. $\frac{9}{8}$	_____	_____	16. $\frac{7}{5}$	_____	_____

List the proper and improper fractions in each group. See Example 4.

		Proper	**Improper**
17.	$\frac{8}{5}$ $\frac{1}{3}$ $\frac{5}{8}$ $\frac{6}{6}$ $\frac{12}{2}$ $\frac{7}{16}$	_____	_____
18.	$\frac{1}{3}$ $\frac{3}{8}$ $\frac{16}{12}$ $\frac{10}{8}$ $\frac{6}{6}$ $\frac{3}{4}$	_____	_____
19.	$\frac{3}{4}$ $\frac{3}{2}$ $\frac{5}{5}$ $\frac{9}{11}$ $\frac{7}{15}$ $\frac{19}{18}$	_____	_____
20.	$\frac{12}{12}$ $\frac{15}{11}$ $\frac{13}{12}$ $\frac{11}{8}$ $\frac{17}{17}$ $\frac{19}{12}$	_____	_____

21. Write a fraction of your own choice. Label the parts of the fraction and write a sentence describing what each part represents. Draw a picture with shaded parts showing your fraction.

22. Give one example of a proper fraction and one example of an improper fraction. What determines whether a fraction is proper or improper? Draw pictures with shaded parts showing these fractions.

Fill in the blanks to complete each sentence.

23. The fraction $\frac{3}{8}$ represents ___3___ of the ___8___ equal parts into which a whole is divided.

24. The fraction $\frac{7}{16}$ represents _____ of the _____ equal parts into which a whole is divided.

25. The fraction $\frac{5}{24}$ represents _____ of the _____ equal parts into which a whole is divided.

26. The fraction $\frac{24}{32}$ represents _____ of the _____ equal parts into which a whole is divided.

2.2 Mixed Numbers

Suppose you had three whole cases of soft drink cans and half of another case. You would state this as a whole number and a fraction.

OBJECTIVE 1 Identify mixed numbers. When a whole number and a fraction are written together, the result is a **mixed number.** For example, the mixed number

$$3\frac{1}{2} \quad \text{represents} \quad 3 + \frac{1}{2},$$

or 3 wholes and $\frac{1}{2}$ of a whole. Read $3\frac{1}{2}$ as "three and one-half." As this figure shows, the mixed number $3\frac{1}{2}$ is equal to the improper fraction $\frac{7}{2}$.

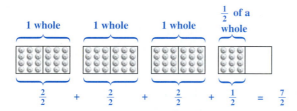

| 1 whole | 1 whole | 1 whole | $\frac{1}{2}$ of a whole |

$$\frac{2}{2} + \frac{2}{2} + \frac{2}{2} + \frac{1}{2} = \frac{7}{2}$$

Work Problem 1 at the Side. ▶▶▶

OBJECTIVE 2 Write mixed numbers as improper fractions. Use the following steps to write $3\frac{1}{2}$ as an improper fraction without drawing a figure.

Step 1   Multiply 3 and 2.

$$3\frac{1}{2} \qquad 3 \cdot 2 = 6$$

Step 2   Add 1 to the product.

$$3\frac{1}{2} \qquad 6 + 1 = 7$$

Step 3   Use 7, from Step 2, as the numerator and 2 as the denominator.

$$3\frac{1}{2} = \frac{7}{2}$$

— Same denominator

In summary, use the following steps to *write a mixed number as an improper fraction.*

Writing a Mixed Number as an Improper Fraction $3\frac{1}{2} = \frac{7}{2}$

Step 1   *Multiply* the denominator of the fraction and the whole number.

Step 2   *Add* to this product the numerator of the fraction.

Step 3   Write the result of Step 2 as the *numerator* and the original denominator as the *denominator.*

OBJECTIVES

1. Identify mixed numbers.
2. Write mixed numbers as improper fractions.
3. Write improper fractions as mixed numbers.

1 **(a)** Use these diagrams to write $1\frac{2}{3}$ as an improper fraction.

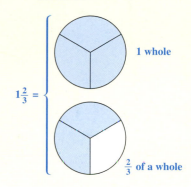

$1\frac{2}{3} =$
1 whole
$\frac{2}{3}$ of a whole

(b) Use these diagrams to write $2\frac{1}{4}$ as an improper fraction.

$2\frac{1}{4} =$
2 wholes
$\frac{1}{4}$ of a whole

ANSWERS

1. (a) $\frac{5}{3}$ (b) $\frac{9}{4}$

2 Write as improper fractions.

(a) $6\frac{1}{2}$ 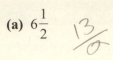 $\frac{13}{0}$

(b) $7\frac{3}{4}$ $\frac{31}{4}$

(c) $4\frac{7}{8}$

(d) $8\frac{5}{6}$

EXAMPLE 1   Writing a Mixed Number as an Improper Fraction

Write $7\frac{2}{3}$ as an improper fraction (numerator greater than denominator).

Step 1 $7\frac{2}{3}$ $7 \cdot 3 = 21$ Multiply 7 and 3.

Step 2 $7\frac{2}{3}$ $21 + 2 = 23$ Add 2. The numerator is 23.

Step 3 $7\frac{2}{3} = \frac{23}{3}$ Use the same denominator.

◀◀◀ **Work Problem 2 at the Side.**

OBJECTIVE 3   Write improper fractions as mixed numbers.   Write an improper fraction as a mixed number as follows.

> **Writing an Improper Fraction as a Mixed Number**
>
> Write an **improper fraction** as a mixed number by dividing the numerator by the denominator. The quotient is the whole number (of the mixed number), the remainder is the numerator of the fraction part, and the denominator remains unchanged.

EXAMPLE 2   Writing Improper Fractions as Mixed Numbers

Write each improper fraction as a mixed number.

(a) $\frac{17}{5}$

Divide 17 by 5.

$$\begin{array}{r} 3 \leftarrow \textbf{Whole number part} \\ 5\overline{)17} \\ 15 \\ \hline 2 \leftarrow \textbf{Remainder} \end{array}$$

The quotient **3** is the whole number part of the mixed number. The remainder **2** is the numerator of the fraction, and the denominator remains as **5**.

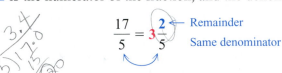 $\frac{17}{5} = 3\frac{2}{5}$ ← Remainder / Same denominator

We can verify this by using a diagram in which $\frac{17}{5}$ is shaded.

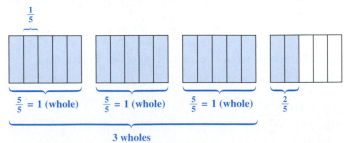

$\frac{1}{5}$

$\frac{5}{5} = 1$ (whole) $\frac{5}{5} = 1$ (whole) $\frac{5}{5} = 1$ (whole) $\frac{2}{5}$

3 wholes

Continued on Next Page

(b) $\dfrac{24}{4}$

Divide 24 by 4.

$$\begin{array}{r} 6 \\ 4\overline{)24} \\ \underline{24} \\ 0 \end{array} \quad \text{so} \quad \dfrac{24}{4} = 6$$

$\leftarrow$ No remainder

Verify this using a diagram in which $\dfrac{24}{4}$ is shaded.

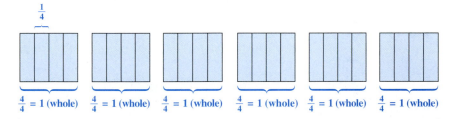

$\frac{1}{4}$

$\frac{4}{4} = 1$ (whole) $\frac{4}{4} = 1$ (whole) $\frac{4}{4} = 1$ (whole) $\frac{4}{4} = 1$ (whole) $\frac{4}{4} = 1$ (whole) $\frac{4}{4} = 1$ (whole)

6 wholes

NOTE

Recall that a proper fraction has a value that is less than 1. An improper fraction has a value that is greater than or equal to 1.

Work Problem 3 at the Side. ▶▶▶

3 Write as whole or mixed numbers.

(a) $\dfrac{6}{5}$ $1\frac{1}{5}$

(b) $\dfrac{9}{4}$ $2\frac{1}{4}$

(c) $\dfrac{35}{5}$

(d) $\dfrac{78}{7}$

2.2 **Exercises**

FOR
EXTRA
HELP

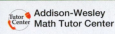 Addison-Wesley
Math Tutor Center

 MathXL

 Digital Video Tutor CD 1
Videotape 4

Student's
Solutions
Manual

MyMathLab
MyMathLab

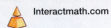 Interactmath.com

Write each mixed number as an improper fraction. See Example 1.

1. $1\frac{1}{4}$ $\frac{5}{4}$

2. $2\frac{1}{2}$

3. $4\frac{3}{5}$

4. $5\frac{2}{3}$

5. $6\frac{1}{2}$

6. $5\frac{3}{5}$

7. $8\frac{1}{4}$

8. $8\frac{1}{2}$

9. $1\frac{7}{11}$

10. $4\frac{3}{7}$

11. $6\frac{1}{3}$

12. $8\frac{2}{3}$

13. $10\frac{1}{8}$

14. $12\frac{2}{3}$

15. $10\frac{3}{4}$

16. $5\frac{4}{5}$

17. $3\frac{3}{8}$

18. $2\frac{8}{9}$

19. $8\frac{3}{5}$

20. $3\frac{4}{7}$

21. $4\frac{10}{11}$

22. $11\frac{5}{8}$

23. $32\frac{3}{4}$

24. $15\frac{3}{10}$

25. $18\frac{5}{12}$

26. $19\frac{8}{11}$

27. $17\frac{14}{15}$

28. $9\frac{5}{16}$

29. $7\frac{19}{24}$

30. $9\frac{7}{12}$

Write each improper fraction as a whole or mixed number. See Example 2.

31. $\frac{4}{3}$

32. $\frac{11}{9}$

33. $\frac{9}{4}$

34. $\frac{7}{2}$

35. $\frac{48}{6}$

36. $\frac{64}{8}$

37. $\frac{38}{5}$

38. $\frac{33}{7}$

39. $\frac{39}{8}$

40. $\frac{40}{9}$

41. $\frac{27}{3}$

42. $\frac{78}{6}$

43. $\frac{63}{4}$

44. $\frac{19}{5}$

45. $\frac{47}{9}$

46. $\frac{65}{9}$

47. $\frac{65}{8}$

48. $\frac{37}{6}$

49. $\frac{84}{5}$

50. $\frac{92}{3}$

51. $\frac{112}{4}$

52. $\frac{117}{9}$

53. $\frac{183}{7}$

54. $\frac{212}{11}$

55. Your classmate asks you how to change a mixed number to an improper fraction. Write a couple of sentences and give an example to show her how this is done.

$$6\tfrac{1}{2} = \tfrac{13}{2}$$

56. Explain in a sentence or two how to change an improper fraction to a mixed number. Give an example to show how this is done.

Write each mixed number as an improper fraction.

57. $250\tfrac{1}{2}$

58. $185\tfrac{3}{4}$

59. $333\tfrac{1}{3}$

60. $138\tfrac{4}{5}$

61. $522\tfrac{3}{8}$

62. $622\tfrac{1}{4}$

Write each improper fraction as a whole or mixed number.

63. $\dfrac{617}{4}$

64. $\dfrac{760}{8}$

65. $\dfrac{2565}{15}$

66. $\dfrac{2915}{16}$

67. $\dfrac{3917}{32}$

68. $\dfrac{5632}{64}$

Knowing the basics of fractions is necessary in problem solving. **Work Exercises 69–74 in order.**

69. Which of these fractions are proper fractions?

$$\frac{2}{3} \quad \frac{4}{5} \quad \frac{8}{5} \quad \frac{3}{4} \quad \frac{6}{6} \quad \frac{7}{10}$$

70. (a) The proper fractions in Exercise 69 are the ones where the ___hum.___ is smaller than the ___den___.

(b) Draw a picture with shaded parts to show each proper fraction in Exercise 69.

(c) The proper fractions in Exercise 69 are all _____ than 1.
(less/greater)

71. Which of these fractions are improper fractions?

$$\frac{5}{5} \quad \frac{3}{4} \quad \frac{10}{3} \quad \frac{2}{3} \quad \frac{5}{6} \quad \frac{6}{5}$$

72. (a) The improper fractions in Exercise 71 are the ones where the _____ is greater than or equal to the _____.

(b) Draw a picture with shaded parts to show each mixed number in Exercise 71.

(c) The improper fractions in Exercise 71 are all equal to or _____ than 1.
(less/greater)

73. Identify which of these fractions can be written as whole or mixed numbers and then write them as whole or mixed numbers.

$$\frac{5}{3} \quad \frac{7}{8} \quad \frac{7}{7} \quad \frac{11}{6} \quad \frac{4}{5} \quad \frac{15}{16}$$

74. (a) The fractions that can be written as whole or mixed numbers in Exercise 73 are _____ fractions, and their value is
(proper/improper)
always _____ 1.
(less than/greater than or equal to)

(b) Draw a picture with shaded parts to show each whole or mixed number in Exercise 73.

(c) Explain how to write an improper fraction as a whole or mixed number.

2.3 Factors

OBJECTIVE **1** **Find factors of a number.** You will recall that numbers multiplied to give a product are called **factors.** Because $2 \cdot 5 = 10$, both 2 and 5 are factors of 10. The numbers 1 and 10 are also factors of 10, because

$$1 \cdot 10 = 10$$

The various tests for divisibility show that 1, 2, 5, and 10 are the only whole number factors of 10. The products $2 \cdot 5$ and $1 \cdot 10$ are called **factorizations** of 10.

> **NOTE**
> The tests to decide whether one number is divisible by another number were shown in **Section 1.5.** You might want to review these. The ones that you will want to remember are those for 2, 3, 5, and 10.

EXAMPLE 1 **Using Factors**

Find all possible two-number factorizations of each number.

(a) 12

$$1 \cdot 12 = 12 \qquad 2 \cdot 6 = 12 \qquad 3 \cdot 4 = 12$$

The factors of 12 are 1, 2, 3, 4, 6, and 12.

(b) 60

$$1 \cdot 60 = 60 \qquad 2 \cdot 30 = 60$$
$$3 \cdot 20 = 60 \qquad 4 \cdot 15 = 60$$
$$5 \cdot 12 = 60 \qquad 6 \cdot 10 = 60$$

The factors of 60 are 1, 2, 3, 4, 5, 6, 10, 12, 15, 20, 30, and 60.

> **Work Problem 1 at the Side.** ▶▶▶

Composite Numbers
A number with a factor other than itself or 1 is called a **composite number.**

EXAMPLE 2 **Identifying Composite Numbers**

Which of the following numbers are composite?

(a) 6

Because 6 has factors of **2** and **3**, numbers other than 6 or 1, the number 6 is composite.

(b) 11

The number 11 has only two factors, 11 and 1. It is not composite.

(c) 25

A factor of 25 is **5**, so 25 is composite.

> **Work Problem 2 at the Side.** ▶▶▶

OBJECTIVE **2** **Identify prime numbers.** Whole numbers that are not composite are called **prime numbers,** except 0 and 1, which are neither prime nor composite.

Prime Numbers
A **prime number** is a whole number that has exactly *two different* factors, *itself* and *1.*

OBJECTIVES

1 Find factors of a number.
2 Identify prime numbers.
3 Find prime factorizations.

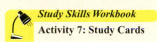
Study Skills Workbook
Activity 7: Study Cards

1 Find all the whole number factors of each number.

(a) 12 1, 2, 3, 4, 6, 12

(b) 16 1, 2, 4, 8, 16

(c) 36

(d) 80

2 Which of these numbers are composite?

2, 4, 5, 6, 8, 10, 11, 13, 19, 21, 27, 28, 33, 36, 42

ANSWERS
1. **(a)** 1, 2, 3, 4, 6, 12   **(b)** 1, 2, 4, 8, 16
   (c) 1, 2, 3, 4, 6, 9, 12, 18, 36
   (d) 1, 2, 4, 5, 8, 10, 16, 20, 40, 80
2. 4, 6, 8, 10, 21, 27, 28, 33, 36, 42

3 Which of the following are prime?

4, 7, 9, 13, 17, 19, 29, 33

The number 3 is a prime number, since it can be divided evenly only by itself and 1. The number 8 is not a prime number (it is composite), since 8 can be divided evenly by 2 and 4, as well as by itself and 1.

> **CAUTION**
> A prime number has *only two* different factors, itself and 1. The number 1 is not a prime number because it does not have *two different* factors; the only factor of 1 is 1.

EXAMPLE 3 Finding Prime Numbers

Which of the following numbers are prime?

$$2 \quad 5 \quad 11 \quad 15 \quad 27$$

The number 15 can be divided by 3 and 5, so it is not prime. Also, because 27 can be divided by 3 and 9, 27 is not prime. The other numbers in the list, 2, 5, and 11, are divisible by only themselves and 1, so they are prime.

▶ Work Problem 3 at the Side.

OBJECTIVE 3 Find prime factorizations. For reference, here are the prime numbers smaller than 50.

2	3	5	7	11
13	17	19	23	29
31	37	41	43	47

4 Find the prime factorization of each number.

(a) 8

> **CAUTION**
> All prime numbers are odd numbers except the number 2. Be careful, though, because *all odd numbers are not prime numbers*. For example, 9, 15, and 21 are odd numbers, but are *not* prime numbers.

(b) 28

The **prime factorization** of a number can be especially useful when we are adding or subtracting fractions and need to find a common denominator or write a fraction in lowest terms.

> **Prime Factorization**
> A **prime factorization** of a number is a factorization in which every factor is a *prime number*.

(c) 18

EXAMPLE 4 Determining the Prime Factorization

Find the prime factorization of 12.
  Try to divide 12 by the first prime, 2.

$$12 \div 2 = 6,$$

First prime

so

(d) 40

$$12 = 2 \cdot 6$$

Try to divide 6 by the prime, 2.

$$6 \div 2 = 3,$$

so

$$12 = 2 \cdot 2 \cdot 3$$

Factorization of 6

Because all factors are prime, the prime factorization of 12 is

$$2 \cdot 2 \cdot 3$$

▶ Work Problem 4 at the Side.

ANSWERS
3. 7, 13, 17, 19, 29
4. **(a)** $2 \cdot 2 \cdot 2$ **(b)** $2 \cdot 2 \cdot 7$
   (c) $2 \cdot 3 \cdot 3$ **(d)** $2 \cdot 2 \cdot 2 \cdot 5$

EXAMPLE 5 Factoring by Using the Division Method

Find the prime factorization of 48.

$$2\overline{)48}$$ Divide 48 by 2 (first prime).

$$2\overline{)24}$$ Divide 24 by 2.

All prime factors $2\overline{)12}$ Divide 12 by 2.

$$2\overline{)6}$$ Divide 6 by 2.

$$3\overline{)3}$$ Divide 3 by 3.

1   Continue to divide until the quotient is 1.

Because all factors (divisors) are prime, the prime factorization of 48 is

2 • 2 • 2 • 2 • 3

In Chapter 1, we wrote $2 \cdot 2 \cdot 2 \cdot 2$ as 2^4, so the prime factorization of 48 can be written, using exponents, as

$$2 \cdot 2 \cdot 2 \cdot 2 \cdot 3 = 2^4 \cdot 3$$

> **Work Problem 5 at the Side.** ▶▶▶

NOTE
When using the division method of factoring, the last quotient found is 1. The "1" is never used as a prime factor because 1 is neither prime nor composite. Besides, 1 times any number is the number itself.

EXAMPLE 6 Using Exponents with Prime Factorization

Find the prime factorization of 225.

$$3\overline{)225}$$ 225 is not divisible by 2; use 3.

$$3\overline{)75}$$ Divide 75 by 3.

All prime factors $5\overline{)25}$ 25 is not divisible by 3; use 5.

$$5\overline{)5}$$ Divide by 5.

1   Continue to divide until the quotient is 1.

Write the prime factorization.

$$3 \cdot 3 \cdot 5 \cdot 5,$$

with exponents as

$$3^2 \cdot 5^2$$

> **Work Problem 6 at the Side.** ▶▶▶

---

5 Find the prime factorization of each number. Write the factorization with exponents.

(a) 36

(b) 54

(c) 60

(d) 81

---

6 Write the prime factorization of each number using exponents.

(a) 48

(b) 44

(c) 90

(d) 120

(e) 180

ANSWERS
5. (a) $2^2 \cdot 3^2$ (b) $2 \cdot 3^3$ (c) $2^2 \cdot 3 \cdot 5$
   (d) 3^4
6. (a) $2^4 \cdot 3$ (b) $2^2 \cdot 11$ (c) $2 \cdot 3^2 \cdot 5$
   (d) $2^3 \cdot 3 \cdot 5$ (e) $2^2 \cdot 3^2 \cdot 5$

7 Complete each factor tree and give the prime factorization.

(a) 28

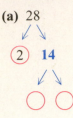

(b) 35

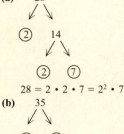

(c) 78

Another method of factoring is a *factor tree*.

EXAMPLE 7 Factoring by Using a Factor Tree

Find the prime factorization of each number.

(a) 30

Try to divide by the first prime, 2. Write the factors under the 30. Circle the 2, since it is a prime.

Since 15 cannot be divided evenly by 2, try the next prime, 3.

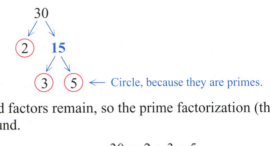

No uncircled factors remain, so the prime factorization (the circled factors) has been found.

$$30 = 2 \cdot 3 \cdot 5$$

(b) 24

Divide by 2.

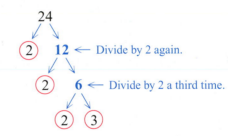

$24 = 2 \cdot 2 \cdot 2 \cdot 3$ or, using exponents, $24 = 2^3 \cdot 3$

(c) 45

Because 45 cannot be divided by 2, try 3.

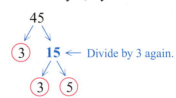

$45 = 3 \cdot 3 \cdot 5$ or, using exponents, $45 = 3^2 \cdot 5$

NOTE
The diagrams used in Example 7 look like tree branches, and that is why this method is referred to as using a *factor tree*.

Work Problem 7 at the Side.

ANSWERS
7. **(a)** $28 = 2 \cdot 2 \cdot 7 = 2^2 \cdot 7$
(b) $35 = 5 \cdot 7$
(c) $78 = 2 \cdot 3 \cdot 13$

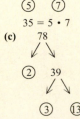

2.3 **Exercises**

FOR EXTRA HELP   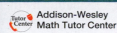 Addison-Wesley Math Tutor Center   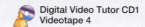 MathXl   Digital Video Tutor CD1 Videotape 4   📖 Student's Solutions Manual   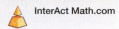 MyMathLab   ⚠ InterAct Math.com

Find all the factors of each number. See Example 1.

1. 8   2,4,8

2. 12

3. 15

4. 28

5. 48

6. 30

7. 36

8. 20

9. 40

10. 60

11. 64

12. 84

Decide whether each number is prime *or* composite. *See Examples 2 and 3.*

13. 6

14. 9

15. 5

16. 7

17. 16

18. 10

19. 13

20. 65

21. 19

22. 17

23. 25

24. 26

25. 48

26. 47

27. 45

28. 53

Find the prime factorization of each number. Write answers with exponents when repeated factors appear. See Examples 4–7.

29. 8   2·4
2 · 2·2

30. 6

31. 20

32. 40

33. 36   2·18 $2^2 \cdot 3^2$
9·2
2·3·3·2

34. 18

35. 25

36. 56

37. 68

38. 70

39. 72

40. 64

41. 44

42. 104

43. 100

44. 112

45. 125

46. 135

47. 180

48. 300

49. 320

50. 480

51. 360

52. 400   200   100·2 $2^4 \cdot 5^2$
2   50·2
2·25
5·5

53. Give a definition in your own words of both a composite number and a prime number. Give three examples of each. Which whole numbers are neither prime nor composite?

Composite   not composite

6   *3*

2·3

3·1

54. With the exception of the number 2, all prime numbers are odd numbers. Nevertheless, all odd numbers are not prime numbers. Explain why these statements are true.

55. Explain the difference between finding all possible factors of 24 and finding the prime factorization of 24.

56. Use the division method to find the prime factorization of 36. Can you divide by 3s before you divide by 2s? Does the order of division change the answers?

Find the prime factorization of each number. Write answers using exponents.

57. 350

58. 640

59. 960

60. 1000

61. 1560

62. 2000

63. 1260

64. 2200

RELATING CONCEPTS (EXERCISES 65–70) For Individual or Group Work

An understanding of factors and factorization will be needed to solve fraction problems.
Work Exercises 65–70 in order.

65. A prime number is a whole number that has exactly two different factors, itself and 1. List all prime numbers smaller than 50.

66. Explain what it is about the numbers in Exercise 65 that makes them prime.

67. With the exception of the number 2, all prime numbers are odd numbers. Why is this true?

68. Can a multiple of a prime number be prime (for example, 6, 9, 12, and 15 are multiples of 3)? Explain.

69. Find the prime factorization of 2100. Do not use exponents in your answer.

70. Write the answer to Exercise 69 using exponents for repeated factors.

2.4 Writing a Fraction in Lowest Terms

When working problems involving fractions, we must often compare two fractions to determine whether they represent the same portion of a whole. Look at the figures below.

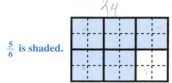

$\frac{5}{6}$ is shaded. $\frac{20}{24}$ is shaded.

The figures show areas that are $\frac{5}{6}$ shaded and $\frac{20}{24}$ shaded. Because the shaded areas are equivalent (the same portion), the fractions $\frac{5}{6}$ and $\frac{20}{24}$ are **equivalent fractions.** They represent the same portions of the whole.

$$\frac{5}{6} = \frac{20}{24}$$ 4 (common factor)

Because the numbers 20 and 24 both have 4 as a factor, 4 is called a **common factor** of the numbers. Other common factors of 20 and 24 are 1 and 2.

> **Work Problem 1 at the Side.** ▶▶▶

OBJECTIVE 1 Tell whether a fraction is written in lowest terms.

The fraction $\frac{5}{6}$ is written in *lowest terms* because the numerator and denominator have no common factor other than 1. However, the fraction $\frac{20}{24}$ is *not* in lowest terms because its numerator and denominator have common factors of 4, 2, and 1.

Writing a Fraction in Lowest Terms

A fraction is written in **lowest terms** when the numerator and denominator have no common factor other than 1.

EXAMPLE 1 Understanding Lowest Terms

Are the following fractions in lowest terms?

(a) $\frac{3}{8}$

The numerator and denominator have no common factor other than 1, so the fraction is in lowest terms.

(b) $\frac{21}{36}$

The numerator and denominator have a common factor of 3, so the fraction is not in lowest terms.

> **Work Problem 2 at the Side.** ▶▶▶

OBJECTIVE 2 Write a fraction in lowest terms using common factors. There are two common methods for writing a fraction in lowest terms. These methods are shown in the next examples. The first method works best when the numerator and denominator are small numbers.

1 Decide whether the given factor is a common factor of both numbers.

(a) 6, 12; 2

(b) 32, 64; 8

(c) 32, 56; 16

(d) 75, 81; 1

2 Are the following fractions in lowest terms?

(a) $\frac{4}{5}$

(b) $\frac{6}{18}$

(c) $\frac{9}{15}$

(d) $\frac{17}{46}$

❸ Write in lowest terms.

(a) $\dfrac{8}{16}$

(b) $\dfrac{9}{12}$

(c) $\dfrac{28}{42}$

(d) $\dfrac{30}{80}$

(e) $\dfrac{16}{40}$

EXAMPLE 2 **Writing Fractions in Lowest Terms**

Write each fraction in lowest terms.

(a) $\dfrac{20}{24}$

The largest common factor of 20 and 24 is 4. Divide both numerator and denominator by **4**.

$$\frac{20}{24} = \frac{20 \div 4}{24 \div 4} = \frac{5}{6}$$

(b) $\dfrac{30}{50} = \dfrac{30 \div 10}{50 \div 10} = \dfrac{3}{5}$ Divide both numerator and denominator by 10.

(c) $\dfrac{24}{42} = \dfrac{24 \div 6}{42 \div 6} = \dfrac{4}{7}$ Divide both numerator and denominator by 6.

(d) $\dfrac{60}{72}$

Suppose we made an error and thought 4 was the largest common factor of 60 and 72. Dividing by 4 would give

$$\frac{60}{72} = \frac{60 \div 4}{72 \div 4} = \frac{15}{18}$$

But $\frac{15}{18}$ is not in lowest terms, because 15 and 18 have a common factor of 3. So we divide by 3.

$$\frac{15}{18} = \frac{15 \div 3}{18 \div 3} = \frac{5}{6}$$

The fraction $\frac{60}{72}$ could have been written in lowest terms in one step by dividing by 12, the largest common factor of 60 and 72.

$$\frac{60}{72} = \frac{60 \div 12}{72 \div 12} = \frac{5}{6}$$

> **NOTE**
> Dividing the numerator and denominator by the same number results in an equivalent fraction.

In Example 2, we wrote fractions in lowest terms by dividing by a common factor. This method is summarized in the following steps.

> **The Method of Dividing by a Common Factor**
>
> *Step 1*   Find the largest number that will divide evenly into both the numerator and denominator. This number is a *common factor*.
>
> *Step 2*   *Divide* both numerator and denominator by the common factor.
>
> *Step 3*   *Check* to see whether the new fraction has any common factors (besides 1). If it does, repeat Steps 2 and 3. If the only common factor is 1, the fraction is in lowest terms.

ANSWERS

1. (a) $\dfrac{1}{2}$ (b) $\dfrac{3}{4}$ (c) $\dfrac{2}{3}$ (d) $\dfrac{3}{8}$ (e) $\dfrac{2}{5}$

◀◀◀ **Work Problem 3 at the Side.**

OBJECTIVE 3 Write a fraction in lowest terms using prime factors. The method of writing a fraction in lowest terms by division works well for fractions with small numerators and denominators. For larger numbers, when common factors are not obvious, use the method of *prime factors,* which is shown in the next example.

EXAMPLE 3 Using Prime Factors

Write each fraction in lowest terms.

(a) $\dfrac{24}{42}$

Write the prime factorization of both numerator and denominator. See **Section 2.3** for help.

$$\frac{24}{42} = \frac{2 \cdot 2 \cdot 2 \cdot 3}{2 \cdot 3 \cdot 7}$$

Just as with the other method, divide both numerator and denominator by any common factors. Write a **1** by each factor that has been divided.

$$\frac{24}{42} = \frac{\overset{1}{\cancel{2}} \cdot 2 \cdot 2 \cdot \overset{1}{\cancel{3}}}{\underset{1}{\cancel{2}} \cdot \underset{1}{\cancel{3}} \cdot 7}$$

Multiply the remaining factors in both numerator and denominator.

$$\frac{24}{42} = \frac{1 \cdot 2 \cdot 2 \cdot 1}{1 \cdot 1 \cdot 7} = \frac{4}{7}$$

Finally, $\frac{24}{42}$ written in lowest terms is $\frac{4}{7}$.

(b) $\dfrac{162}{54}$

Write the prime factorization of both numerator and denominator.

$$\frac{162}{54} = \frac{2 \cdot 3 \cdot 3 \cdot 3 \cdot 3}{2 \cdot 3 \cdot 3 \cdot 3}$$

Now divide by the common factors. ***Do not forget to write the 1s.***

$$\frac{162}{54} = \frac{\overset{1}{\cancel{2}} \cdot \overset{1}{\cancel{3}} \cdot \overset{1}{\cancel{3}} \cdot \overset{1}{\cancel{3}} \cdot 3}{\underset{1}{\cancel{2}} \cdot \underset{1}{\cancel{3}} \cdot \underset{1}{\cancel{3}} \cdot \underset{1}{\cancel{3}}}$$

$$= \frac{1 \cdot 1 \cdot 1 \cdot 1 \cdot 3}{1 \cdot 1 \cdot 1 \cdot 1} = \frac{3}{1} = 3$$

(c) $\dfrac{18}{90}$

$$\frac{18}{90} = \frac{\overset{1}{\cancel{2}} \cdot \overset{1}{\cancel{3}} \cdot \overset{1}{\cancel{3}}}{\underset{1}{\cancel{2}} \cdot \underset{1}{\cancel{3}} \cdot \underset{1}{\cancel{3}} \cdot 5} = \frac{1 \cdot 1 \cdot 1}{1 \cdot 1 \cdot 1 \cdot 5} = \frac{1}{5}$$

CAUTION

In Example 3(c) above, all factors of the numerator were divided. But $1 \cdot 1 \cdot 1$ is still 1, so the final answer is $\frac{1}{5}$ (***not*** 5).

4 Use the method of prime factors to write each fraction in lowest terms.

(a) $\dfrac{12}{36}$

(b) $\dfrac{32}{56}$

(c) $\dfrac{74}{111}$

(d) $\dfrac{124}{340}$

5 Is each pair of fractions equivalent?

(a) $\dfrac{24}{48}$ and $\dfrac{36}{72}$

(b) $\dfrac{45}{60}$ and $\dfrac{50}{75}$

(c) $\dfrac{20}{4}$ and $\dfrac{110}{22}$

(d) $\dfrac{120}{220}$ and $\dfrac{180}{320}$

In Example 3, we wrote fractions in lowest terms using prime factors. This method is summarized as follows.

The Method of Prime Factors

Step 1   Write the *prime factorization* of both numerator and denominator.

Step 2   *Divide* both numerator and denominator by any common factors.

Step 3   *Multiply* the remaining factors in the numerator and denominator.

◀◀◀ Work Problem 4 at the Side.

O B J E C T I V E 4 Determine whether two fractions are equivalent. The next example shows how to decide whether two fractions are equivalent.

EXAMPLE 4 Determining whether Two Fractions Are Equivalent

Determine whether each pair of fractions is equivalent. In other words, do both fractions represent the same part of a whole?

(a) $\dfrac{16}{48}$ and $\dfrac{24}{72}$

Use the method of prime factors to write each fraction in lowest terms.

$$\frac{16}{48} = \frac{\overset{1}{\cancel{2}} \cdot \overset{1}{\cancel{2}} \cdot \overset{1}{\cancel{2}} \cdot \overset{1}{\cancel{2}}}{\underset{1}{\cancel{2}} \cdot \underset{1}{\cancel{2}} \cdot \underset{1}{\cancel{2}} \cdot \underset{1}{\cancel{2}} \cdot 3} = \frac{1 \cdot 1 \cdot 1 \cdot 1}{1 \cdot 1 \cdot 1 \cdot 1 \cdot 3} = \frac{1}{3}$$

Equivalent $\left(\frac{1}{3} = \frac{1}{3}\right)$

$$\frac{24}{72} = \frac{\overset{1}{\cancel{2}} \cdot \overset{1}{\cancel{2}} \cdot \overset{1}{\cancel{2}} \cdot \overset{1}{\cancel{3}}}{\underset{1}{\cancel{2}} \cdot \underset{1}{\cancel{2}} \cdot \underset{1}{\cancel{2}} \cdot \underset{1}{\cancel{3}} \cdot 3} = \frac{1 \cdot 1 \cdot 1 \cdot 1}{1 \cdot 1 \cdot 1 \cdot 1 \cdot 3} = \frac{1}{3}$$

(b) $\dfrac{32}{52}$ and $\dfrac{64}{112}$

$$\frac{32}{52} = \frac{\overset{1}{\cancel{2}} \cdot \overset{1}{\cancel{2}} \cdot 2 \cdot 2 \cdot 2}{\underset{1}{\cancel{2}} \cdot \underset{1}{\cancel{2}} \cdot 13} = \frac{2 \cdot 2 \cdot 2}{1 \cdot 1 \cdot 13} = \frac{8}{13}$$

Not equivalent $\left(\frac{8}{13} \neq \frac{4}{7}\right)$

$$\frac{64}{112} = \frac{\overset{1}{\cancel{2}} \cdot \overset{1}{\cancel{2}} \cdot \overset{1}{\cancel{2}} \cdot \overset{1}{\cancel{2}} \cdot 2 \cdot 2}{\underset{1}{\cancel{2}} \cdot \underset{1}{\cancel{2}} \cdot \underset{1}{\cancel{2}} \cdot \underset{1}{\cancel{2}} \cdot 7} = \frac{1 \cdot 1 \cdot 1 \cdot 1 \cdot 2 \cdot 2}{1 \cdot 1 \cdot 1 \cdot 1 \cdot 7} = \frac{4}{7}$$

(c) $\dfrac{75}{15}$ and $\dfrac{60}{12}$

$$\frac{75}{15} = \frac{\overset{1}{\cancel{3}} \cdot \overset{1}{\cancel{5}} \cdot 5}{\underset{1}{\cancel{3}} \cdot \underset{1}{\cancel{5}}} = \frac{1 \cdot 1 \cdot 5}{1 \cdot 1} = 5$$

Equivalent $(5 = 5)$

$$\frac{60}{12} = \frac{\overset{1}{\cancel{2}} \cdot \overset{1}{\cancel{2}} \cdot \overset{1}{\cancel{3}} \cdot 5}{\underset{1}{\cancel{2}} \cdot \underset{1}{\cancel{2}} \cdot \underset{1}{\cancel{3}}} = \frac{1 \cdot 1 \cdot 1 \cdot 5}{1 \cdot 1 \cdot 1} = 5$$

◀◀◀ Work Problem 5 at the Side.

ANSWERS

1. (a) $\dfrac{1}{3}$ (b) $\dfrac{4}{7}$ (c) $\dfrac{2}{3}$ (d) $\dfrac{31}{85}$

5. (a) equivalent   (b) not equivalent
   (c) equivalent   (d) not equivalent

2.4 Exercises

 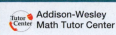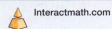

Put a ✓ mark in the blank if the number at the left is divisible by the number at the top. Put an X in the blank if the number is not divisible by the number at the top. (For help, see Section 1.5.)

	2	3	5	10			2	3	5	10
1. 60	___	___	___	___		**2.** 90	___	___	___	___
3. 48	___	___	___	___		**4.** 36	___	___	___	___
5. 160	___	___	___	___		**6.** 175	___	___	___	___
7. 138	___	___	___	___		**8.** 150	___	___	___	___

Write each fraction in lowest terms. See Example 2.

9. $\dfrac{6}{8}$ **10.** $\dfrac{6}{12}$ **11.** $\dfrac{3}{12}$ **12.** $\dfrac{4}{12}$

13. $\dfrac{15}{25}$ **14.** $\dfrac{32}{48}$ **15.** $\dfrac{36}{42}$ **16.** $\dfrac{22}{33}$

17. $\dfrac{56}{64}$ **18.** $\dfrac{21}{35}$ **19.** $\dfrac{180}{210}$ **20.** $\dfrac{72}{80}$

21. $\dfrac{72}{126}$ **22.** $\dfrac{73}{146}$ **23.** $\dfrac{12}{600}$ **24.** $\dfrac{8}{400}$

25. $\dfrac{96}{132}$ **26.** $\dfrac{165}{180}$ **27.** $\dfrac{60}{108}$ **28.** $\dfrac{112}{128}$

Write the numerator and denominator of each fraction as a product of prime factors and divide by the common factors. Then write the fraction in lowest terms. See Example 3.

29. $\dfrac{18}{24}$ $\quad \dfrac{2 \cdot \cancel{3} \cdot 3}{\cancel{2} \cdot 2 \cdot \cancel{3} \cdot 2 \cdot 4} \quad \dfrac{3}{4}$

30. $\dfrac{16}{64}$

31. $\dfrac{35}{40}$

32. $\dfrac{20}{32}$

33. $\dfrac{90}{180}$

34. $\dfrac{36}{48}$

35. $\dfrac{36}{12}$

36. $\dfrac{192}{48}$

37. $\dfrac{72}{225}$

38. $\dfrac{65}{234}$

Determine whether each pair of fractions is equivalent or not equivalent. See Example 4.

39. $\dfrac{3}{6}$ and $\dfrac{18}{36}$ $\dfrac{2 \cdot 3 \cdot 3}{2 \cdot 3 \cdot 2 \cdot 3}$ $\dfrac{1}{2}$
   equiv.

40. $\dfrac{3}{8}$ and $\dfrac{27}{72}$

41. $\dfrac{10}{24}$ and $\dfrac{12}{30}$

42. $\dfrac{15}{35}$ and $\dfrac{18}{40}$

43. $\dfrac{15}{24}$ and $\dfrac{35}{52}$

44. $\dfrac{21}{33}$ and $\dfrac{9}{12}$

45. $\dfrac{14}{16}$ and $\dfrac{35}{40}$

46. $\dfrac{27}{90}$ and $\dfrac{24}{80}$

47. $\dfrac{48}{6}$ and $\dfrac{72}{8}$

48. $\dfrac{33}{11}$ and $\dfrac{72}{24}$

49. $\dfrac{25}{30}$ and $\dfrac{65}{78}$

50. $\dfrac{24}{72}$ and $\dfrac{30}{90}$

51. What does it mean when a fraction is expressed in lowest terms? Give three examples.

52. Explain what equivalent fractions are and give an example of a pair of equivalent fractions. Show that they are equivalent (represent the same portion of a whole).

Write each fraction in lowest terms.

53. $\dfrac{160}{256}$

54. $\dfrac{363}{528}$

55. $\dfrac{238}{119}$

56. $\dfrac{570}{95}$

Summary Exercises on Fraction Basics

Write fractions to represent the shaded and unshaded portions of each figure.

1.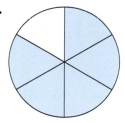

2.

3.

Identify the numerator and denominator.

	Numerator	Denominator		Numerator	Denominator
4. $\frac{3}{4}$	_____	_____	**5.** $\frac{8}{5}$	_____	_____

List the proper and improper fractions.

	Proper	Improper
6. $\frac{8}{2}$ $\frac{3}{5}$ $\frac{16}{7}$ $\frac{8}{8}$ $\frac{4}{25}$ $\frac{1}{32}$	_____	_____

Use the table to answer Exercises 7 and 8.

TYPES OF STORES IN THE SUNRISE MALL	
Clothing	8
Cards/gifts/accessories	6
Shoes	4
Movie theater	1
Food/restaurant	5
Other	7
Total	31

7. What fraction of the stores are cards, gifts, accessories stores? $\frac{6}{31}$

8. What fraction of the stores are not clothing or shoe stores?

Write each improper fraction as a mixed number or whole number.

9. $\frac{5}{2}$

10. $\frac{11}{8}$

11. $\frac{9}{7}$

12. $\frac{8}{3}$

13. $\frac{64}{8}$

14. $\frac{45}{9}$

15. $\frac{36}{5}$

16. $\frac{47}{10}$

Write each mixed number as an improper fraction.

17. $3\dfrac{1}{3}$

18. $5\dfrac{3}{8}$

19. $6\dfrac{4}{5}$

20. $10\dfrac{3}{5}$

21. $12\dfrac{3}{4}$

22. $4\dfrac{10}{13}$

23. $11\dfrac{5}{6}$

24. $23\dfrac{5}{8}$

Write the prime factorization of each number using exponents.

25. 10

26. 55

27. 36

28. 81

29. 280

30. 360

Write each fraction in lowest terms.

31. $\dfrac{4}{12}$

32. $\dfrac{7}{14}$

33. $\dfrac{4}{16}$

34. $\dfrac{15}{25}$

35. $\dfrac{18}{24}$

36. $\dfrac{30}{36}$

37. $\dfrac{56}{64}$

38. $\dfrac{6}{300}$

39. $\dfrac{25}{200}$

40. $\dfrac{125}{225}$

41. $\dfrac{88}{154}$

42. $\dfrac{70}{126}$

Write the numerator and denominator of each fraction as a product of prime factors and divide by the common factors. Then write the fraction in lowest terms.

43. $\dfrac{24}{36}$

44. $\dfrac{80}{160}$

45. $\dfrac{126}{42}$

46. $\dfrac{96}{112}$

2.5 Multiplying Fractions

OBJECTIVE 1 Multiply fractions.   Suppose that you give $\frac{1}{2}$ of your Energy Bar to your kickboxing partner Jennifer. Then Jennifer gives $\frac{1}{2}$ of her share to Tony. How much of the Energy Bar does Tony get to eat?

Start with a sketch showing the Energy Bar cut in half (2 equal pieces).

OBJECTIVES

1 Multiply fractions.

2 Use a multiplication shortcut.

3 Multiply a fraction and a whole number.

4 Find the area of a rectangle.

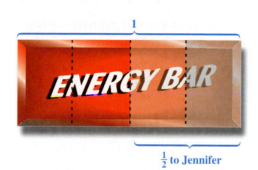

1

$\frac{1}{2}$ to Jennifer

$\frac{1}{2} \times \frac{1}{2} = \frac{1}{4}$

Next, take $\frac{1}{2}$ of the shaded area. (Here we are dividing $\frac{1}{2}$ into 2 equal parts and shading one darker than the other.)

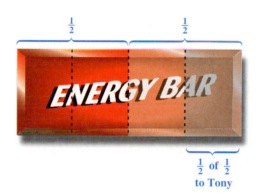

$\frac{1}{2}$ $\frac{1}{2}$

$\frac{1}{2}$ of $\frac{1}{2}$
to Tony

The sketch shows that Tony gets $\frac{1}{4}$ of the Energy Bar.

1 Use these figures to find $\frac{1}{4}$ of $\frac{1}{2}$.

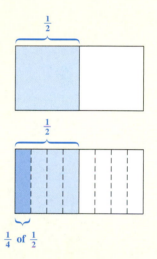

$\frac{1}{2}$

$\frac{1}{2}$

$\frac{1}{4}$ of $\frac{1}{2}$

Tony gets $\frac{1}{2}$ of $\frac{1}{2}$ of the Energy Bar. When used between two fractions, the word **of** tells us to multiply.

$$\frac{1}{2} \text{ of } \frac{1}{2} \quad \text{means} \quad \frac{1}{2} \cdot \frac{1}{2}$$

Tony's share of the Energy Bar is

$$\frac{1}{2} \cdot \frac{1}{2} = \frac{1}{4}$$

Work Problem 1 at the Side. ▶▶▶

of = ✗

The rule for multiplying fractions follows.

> ### Multiplying Fractions
> Multiply two fractions by multiplying the numerators and multiplying the denominators.

ANSWER

1. $\frac{1}{8}$

2 Multiply. Write answers in lowest terms.

(a) $\dfrac{1}{2} \cdot \dfrac{3}{4}$

(b) $\dfrac{3}{5} \cdot \dfrac{1}{3}$

(c) $\dfrac{5}{6} \cdot \dfrac{1}{2} \cdot \dfrac{1}{8}$

(d) $\dfrac{1}{2} \cdot \dfrac{3}{4} \cdot \dfrac{3}{8}$

Use this rule to find the product of $\frac{2}{3}$ and $\frac{1}{3}$ (multiply $\frac{2}{3}$ by $\frac{1}{3}$).

$$\frac{2}{3} \cdot \frac{1}{3} = \frac{2 \cdot 1}{3 \cdot 3} \quad \leftarrow \text{Multiply numerators.}$$
$$\leftarrow \text{Multiply denominators.}$$

Finish multiplying.

$$\frac{2}{3} \cdot \frac{1}{3} = \frac{2 \cdot 1}{3 \cdot 3} = \frac{2}{9} \quad \begin{array}{l} \leftarrow 2 \cdot 1 = 2 \\ \leftarrow 3 \cdot 3 = 9 \end{array}$$

Check that the final result is in lowest terms. $\frac{2}{9}$ is in lowest terms because 2 and 9 have no common factor other than 1.

EXAMPLE 1 Multiplying Fractions

Multiply. Write answers in lowest terms.

(a) $\dfrac{5}{8} \cdot \dfrac{3}{4}$

Multiply the numerators and multiply the denominators.

$$\frac{5}{8} \cdot \frac{3}{4} = \frac{5 \cdot 3}{8 \cdot 4} = \frac{15}{32}$$

Notice that 15 and 32 have no common factors other than 1, so the answer is in lowest terms.

(b) $\dfrac{4}{7} \cdot \dfrac{2}{5}$

$$\frac{4}{7} \cdot \frac{2}{5} = \frac{4 \cdot 2}{7 \cdot 5} = \frac{8}{35}$$

(c) $\dfrac{5}{8} \cdot \dfrac{3}{4} \cdot \dfrac{1}{2}$

$$\frac{5}{8} \cdot \frac{3}{4} \cdot \frac{1}{2} = \frac{5 \cdot 3 \cdot 1}{8 \cdot 4 \cdot 2} = \frac{15}{64}$$

Work Problem 2 at the Side.

OBJECTIVE 2 Use a multiplication shortcut. A multiplication shortcut that can be used with fractions is shown in Example 2.

EXAMPLE 2 Using the Multiplication Shortcut

Multiply $\frac{5}{6}$ and $\frac{9}{10}$. Write the answer in lowest terms.

$$\frac{5}{6} \cdot \frac{9}{10} = \frac{5 \cdot 9}{6 \cdot 10} = \frac{45}{60} \qquad \text{Not in lowest terms}$$

The numerator and denominator have a common factor other than 1, so write the prime factorization of each number.

$$\frac{5}{6} \cdot \frac{9}{10} = \frac{5 \cdot 9}{6 \cdot 10} = \frac{5 \cdot 3 \cdot 3}{2 \cdot 3 \cdot 2 \cdot 5}$$

Continued on Next Page

Next, divide by the common factors of 5 and 3.

$$\frac{5}{6} \cdot \frac{9}{10} = \frac{5 \cdot 9}{6 \cdot 10} = \frac{\overset{1}{\cancel{5}} \cdot \overset{1}{\cancel{3}} \cdot 3}{2 \cdot \underset{1}{\cancel{3}} \cdot 2 \cdot \underset{1}{\cancel{5}}}$$

Finally, multiply the remaining factors in the numerator and in the denominator.

$$\frac{5}{6} \cdot \frac{9}{10} = \frac{\mathbf{1 \cdot 1 \cdot 3}}{2 \cdot \mathbf{1} \cdot 2 \cdot \mathbf{1}} = \frac{3}{4} \qquad \text{Lowest terms}$$

As a shortcut, instead of writing the prime factorization of each number, find the product of $\frac{5}{6}$ and $\frac{9}{10}$ as follows.

First, divide by 5, a common factor of both 5 and 10.
$$\frac{\overset{1}{\cancel{5}}}{6} \cdot \frac{9}{\underset{2}{\cancel{10}}}$$

Next, divide by 3, a common factor of both 6 and 9.
$$\frac{\overset{1}{\cancel{5}}}{\underset{2}{\cancel{6}}} \cdot \frac{\overset{3}{\cancel{9}}}{\underset{2}{\cancel{10}}}$$

Finally, multiply.
$$\frac{1 \cdot 3}{2 \cdot 2} = \frac{3}{4}$$

CAUTION

When using the multiplication shortcut, you are dividing a numerator and a denominator by a common factor. Be certain that you divide a numerator and a denominator **by the same number.** If you do all possible divisions, your answer will be in lowest terms.

EXAMPLE 3 Using the Multiplication Shortcut

Use the multiplication shortcut to find each product. Write the answers in lowest terms and as mixed numbers where possible.

(a) $\dfrac{6}{11} \cdot \dfrac{7}{8}$

Divide both 6 and 8 by their common factor of 2. Notice that 7 and 11 have no common factor. Finally, multiply.

$$\frac{\overset{3}{\cancel{6}}}{11} \cdot \frac{7}{\underset{4}{\cancel{8}}} = \frac{3 \cdot 7}{11 \cdot 4} = \frac{21}{44} \qquad \text{Lowest terms}$$

(b) $\dfrac{7}{10} \cdot \dfrac{20}{21}$

Divide 7 and 21 by 7, and divide 10 and 20 by 10.

$$\frac{\overset{1}{\cancel{7}}}{\underset{1}{\cancel{10}}} \cdot \frac{\overset{2}{\cancel{20}}}{\underset{3}{\cancel{21}}} = \frac{1 \cdot 2}{1 \cdot 3} = \frac{2}{3} \qquad \text{Lowest terms}$$

Continued on Next Page

3 Use the multiplication shortcut to find each product.

(a) $\dfrac{3}{4} \cdot \dfrac{2}{3}$

(b) $\dfrac{6}{11} \cdot \dfrac{33}{21}$

(c) $\dfrac{20}{4} \cdot \dfrac{3}{40} \cdot \dfrac{1}{3}$

(d) $\dfrac{18}{17} \cdot \dfrac{1}{36} \cdot \dfrac{2}{3}$

(c) $\dfrac{35}{12} \cdot \dfrac{32}{25}$

$$\dfrac{\overset{7}{\cancel{35}}}{\underset{3}{\cancel{12}}} \cdot \dfrac{\overset{8}{\cancel{32}}}{\underset{5}{\cancel{25}}} = \dfrac{7 \cdot 8}{3 \cdot 5} = \dfrac{56}{15} \quad \text{or} \quad 3\dfrac{11}{15} \qquad \text{Mixed number}$$

(d) $\dfrac{2}{3} \cdot \dfrac{8}{15} \cdot \dfrac{3}{4}$

$$\dfrac{\overset{1}{\cancel{2}}}{\underset{1}{\cancel{3}}} \cdot \dfrac{\overset{4}{\cancel{8}}}{15} \cdot \dfrac{\overset{1}{\cancel{3}}}{\underset{1}{\underset{2}{\cancel{4}}}} = \dfrac{1 \cdot 4 \cdot 1}{1 \cdot 15 \cdot 1} = \dfrac{4}{15} \qquad \text{Lowest terms}$$

This shortcut is especially helpful when the fractions involve large numbers.

> **NOTE**
> There is no specific order that must be used when dividing numerators and denominators as long as both numerator and denominator are divided by the same number.

◀◀◀ **Work Problem 3 at the Side.**

OBJECTIVE 3 Multiply a fraction and a whole number. The rule for multiplying a fraction and a whole number follows.

> **Multiplying a Whole Number and a Fraction**
> Multiply a whole number and a fraction by writing the whole number as a fraction with a denominator of 1.

For example, write the whole numbers 8, 10, and 25 as follows.

$$8 = \dfrac{8}{1} \qquad 10 = \dfrac{10}{1} \qquad 25 = \dfrac{25}{1}$$

EXAMPLE 4 Multiplying by Whole Numbers

Multiply. Write answers in lowest terms and as whole numbers where possible.

(a) $8 \cdot \dfrac{3}{4}$

Write 8 as $\frac{8}{1}$ and multiply.

$$8 \cdot \dfrac{3}{4} = \dfrac{\overset{2}{\cancel{8}}}{1} \cdot \dfrac{3}{\underset{1}{\cancel{4}}} = \dfrac{2 \cdot 3}{1 \cdot 1} = \dfrac{6}{1} = 6$$

— **Continued on Next Page**

ANSWERS

3. (a) $\dfrac{\overset{1}{\cancel{3}}}{\underset{2}{\cancel{4}}} \cdot \dfrac{\overset{1}{\cancel{2}}}{\underset{1}{\cancel{3}}} = \dfrac{1}{2}$ (b) $\dfrac{\overset{2}{\cancel{6}}}{\underset{1}{\cancel{11}}} \cdot \dfrac{\overset{3}{\cancel{33}}}{\underset{7}{\cancel{21}}} = \dfrac{6}{7}$

(c) $\dfrac{\overset{1}{\cancel{20}}}{4} \cdot \dfrac{\overset{1}{\cancel{3}}}{\underset{2}{\cancel{40}}} \cdot \dfrac{1}{\cancel{3}} = \dfrac{1}{8}$

(d) $\dfrac{\overset{1}{\cancel{18}}}{17} \cdot \dfrac{1}{\underset{2}{\cancel{36}}} \cdot \dfrac{\overset{1}{\cancel{2}}}{\underset{1}{\cancel{3}}} = \dfrac{1}{51}$

(b) $12 \cdot \dfrac{5}{6}$

$$12 \cdot \frac{5}{6} = \frac{\overset{2}{\cancel{12}}}{1} \cdot \frac{5}{\underset{1}{\cancel{6}}} = \frac{2 \cdot 5}{1 \cdot 1} = \frac{10}{1} = 10$$

Work Problem 4 at the Side. ▶▶▶

OBJECTIVE **4** **Find the area of a rectangle.** To find the area of a rectangle (the amount of surface inside the rectangle), use the following formula.

> **Area of a Rectangle**
> The area of a rectangle is equal to the length multiplied by the width.
> $$\textbf{Area} = \textbf{length} \cdot \textbf{width}$$

For example, the rectangle shown here has an area of 12 square feet (ft^2).

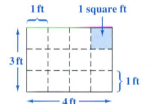

Area = length • width
Area = 4 ft • 3 ft
Area = 12 ft^2

Other units for measuring area are square inches (in.^2), square yards (yd^2), and square miles (mi^2). (See **Section 8.3** for more information on area.)

EXAMPLE 5 **Applying Fraction Skills**

Find the area of each rectangle.

(a) Find the area of each floor tile.

$\boxed{W}$ $\frac{3}{4}$ ft |← $\frac{11}{12}$ ft →| L LONG

Area = length • width

$$\text{Area} = \frac{11}{12} \cdot \frac{3}{4}$$

$$= \frac{11}{\underset{4}{\cancel{12}}} \cdot \frac{\overset{1}{\cancel{3}}}{4} \quad \text{Divide numerator and denominator by 3.}$$

$$= \frac{11}{16} \text{ ft}^2$$

Continued on Next Page

---

4 Multiply. Write answers in lowest terms.

(a) $8 \cdot \dfrac{1}{8}$

(b) $\dfrac{3}{4} \cdot 5 \cdot \dfrac{5}{3}$

(c) $\dfrac{3}{5} \cdot 40$

(d) $\dfrac{3}{25} \cdot \dfrac{5}{11} \cdot 99$

ANSWERS

4. **(a)** 1   **(b)** $6\frac{1}{4}$ **(c)** 24   **(d)** $\frac{27}{5}$ or $5\frac{2}{5}$

5 Find the area of each rectangle.

(a)
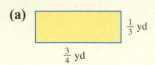
$\frac{1}{3}$ yd

$\frac{3}{4}$ yd

(b) a community college campus that is $\frac{3}{8}$ mile by $\frac{1}{3}$ mile

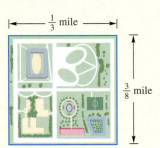

$\frac{1}{3}$ mile

$\frac{3}{8}$ mile

(c) a nature area that is $\frac{7}{5}$ mile by $\frac{5}{8}$ mile

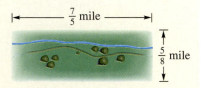

$\frac{7}{5}$ mile

$\frac{5}{8}$ mile

(b) Find the area of this wallpaper sample.

$\frac{7}{9}$ yd

$\frac{3}{14}$ yd

Multiply the length and width.

$$\text{Area} = \frac{7}{9} \cdot \frac{3}{14}$$

$$= \frac{\overset{1}{\cancel{7}}}{\underset{3}{\cancel{9}}} \cdot \frac{\overset{1}{\cancel{3}}}{\underset{2}{\cancel{14}}} \qquad \text{Divide numerator and denominator.}$$

$$= \frac{1}{6} \text{ square yard (yd}^2)$$

Work Problem 5 at the Side.

2.5 Exercises

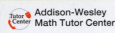

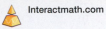
Multiply. Write answers in lowest terms. See Examples 1–3.

1. $\dfrac{1}{3} \cdot \dfrac{3}{4}$

2. $\dfrac{2}{5} \cdot \dfrac{3}{4}$

3. $\dfrac{2}{7} \cdot \dfrac{1}{5}$

4. $\dfrac{2}{3} \cdot \dfrac{1}{2}$

5. $\dfrac{8}{5} \cdot \dfrac{15}{32}$

6. $\dfrac{5}{9} \cdot \dfrac{4}{3}$

7. $\dfrac{2}{3} \cdot \dfrac{7}{12} \cdot \dfrac{9}{14}$

8. $\dfrac{7}{8} \cdot \dfrac{16}{21} \cdot \dfrac{1}{2}$

9. $\dfrac{3}{4} \cdot \dfrac{5}{6} \cdot \dfrac{2}{3}$

10. $\dfrac{2}{5} \cdot \dfrac{3}{8} \cdot \dfrac{2}{3}$

11. $\dfrac{9}{22} \cdot \dfrac{11}{16}$

12. $\dfrac{5}{12} \cdot \dfrac{7}{10}$

13. $\dfrac{5}{8} \cdot \dfrac{16}{25}$

14. $\dfrac{6}{11} \cdot \dfrac{22}{15}$

15. $\dfrac{14}{25} \cdot \dfrac{65}{48} \cdot \dfrac{15}{28}$

16. $\dfrac{35}{64} \cdot \dfrac{32}{15} \cdot \dfrac{27}{72}$

17. $\dfrac{16}{25} \cdot \dfrac{35}{32} \cdot \dfrac{15}{64}$

18. $\dfrac{39}{42} \cdot \dfrac{7}{13} \cdot \dfrac{7}{24}$

Multiply. Write answers in lowest terms and as whole or mixed numbers where possible. See Example 4.

19. $5 \cdot \dfrac{4}{5}$

20. $20 \cdot \dfrac{3}{4}$

21. $\dfrac{5}{8} \cdot 64$

22. $\dfrac{5}{6} \cdot 24$

23. $36 \cdot \dfrac{2}{3}$

24. $30 \cdot \dfrac{3}{10}$

25. $36 \cdot \dfrac{5}{8} \cdot \dfrac{9}{15}$

26. $35 \cdot \dfrac{3}{5} \cdot \dfrac{1}{2}$

27. $100 \cdot \dfrac{21}{50} \cdot \dfrac{3}{4}$

28. $400 \cdot \dfrac{7}{8}$

29. $\dfrac{2}{5} \cdot 200$

30. $\dfrac{6}{7} \cdot 245$

31. $142 \cdot \dfrac{2}{3}$

32. $\dfrac{12}{25} \cdot 430$

33. $\dfrac{28}{21} \cdot 640 \cdot \dfrac{15}{32}$

34. $\dfrac{21}{13} \cdot 520 \cdot \dfrac{7}{20}$

35. $\dfrac{54}{38} \cdot 684 \cdot \dfrac{5}{6}$

36. $\dfrac{76}{43} \cdot 473 \cdot \dfrac{5}{19}$

Find the area of each rectangle. See Example 5.

37.
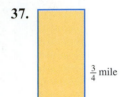
$\frac{3}{4}$ mile

$\frac{1}{3}$ mile

38.

$\frac{1}{4}$ ft

$\frac{7}{8}$ ft

39.
$\frac{3}{4}$ meter

12 meters

40.

$\frac{3}{8}$ in.

8 in.

$8 \cdot \frac{3}{8} = 3$

41.
$\frac{3}{14}$ in.

$\frac{7}{5}$ in.

42.

$\frac{9}{16}$ yd

$\frac{14}{15}$ yd

43. Write in your own words the rule for multiplying fractions. Make up an example problem to show how this works.

44. A useful shortcut when multiplying fractions is to divide a numerator and a denominator by the same number. Describe how this works and give an example.

Find the area of each rectangle in these application problems. Write answers in lowest terms and as whole or mixed numbers where possible. See Example 5.

45. Find the area of a heating-duct grill having a length of 2 yd and a width of $\frac{3}{4}$ yd.

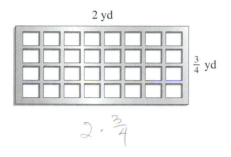

2 yd

$\frac{3}{4}$ yd

$2 \cdot \frac{3}{4}$

46. Find the area of the top of an office desk having a length of 2 yd and a width of $\frac{7}{8}$ yd.

2 yd

$\frac{7}{8}$ yd

47. A speedway property measures $\frac{3}{4}$ mile by 2 miles. Find the total area of the parcel.

48. A motorcycle racecourse is $\frac{2}{3}$ mile wide by 4 miles long. Find the area of the racecourse.

49. The Bel Air Grocery parking lot is $\frac{1}{4}$ mile long and $\frac{3}{16}$ mile wide, while the Raley's Market parking lot is $\frac{3}{8}$ mile long and $\frac{1}{8}$ mile wide. Which parking lot has the larger area?

50. The Rocking Horse Ranch is $\frac{3}{4}$ mile long and $\frac{2}{3}$ mile wide. The Silver Spur Ranch is $\frac{5}{8}$ mile long and $\frac{4}{5}$ mile wide. Which ranch has the larger area?

RELATING CONCEPTS (EXERCISES 51–56) For Individual or Group Work

Front end rounding can be used to estimate an answer when multiplying fractions.
Work Exercises 51–56 in order.

The table shows the five largest cable television systems in the U.S.A.

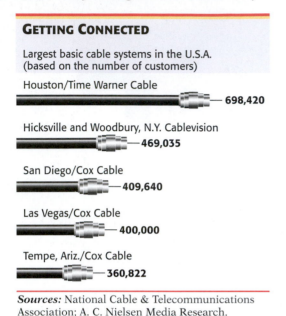

GETTING CONNECTED

Largest basic cable systems in the U.S.A.
(based on the number of customers)

Houston/Time Warner Cable
━━━━━━ 698,420

Hicksville and Woodbury, N.Y. Cablevision
━━━━ 469,035

San Diego/Cox Cable
━━━ 409,640

Las Vegas/Cox Cable
━━━ 400,000

Tempe, Ariz./Cox Cable
━━━ 360,822

Sources: National Cable & Telecommunications
Association; A. C. Nielsen Media Research.

51. Use front end rounding to estimate the total customers for the five largest systems.

52. Find the exact total number of customers of the five largest systems.

53. If $\frac{3}{5}$ of the Hicksville and Woodbury, New York, Cablevision customers subscribe to the premium channels, use front end rounding to estimate, and then find the exact number of customers who receive the premium channels.

54. If $\frac{3}{7}$ of the Tempe, Arizona/Cox Cable customers subscribe to the movie channels, use front end rounding to estimate, and then find the exact number of customers who receive the movie channels. Round to the nearest whole number.

Compare the estimated answers and the exact answers in Exercises 53 and 54. How can you get an estimated answer that is closer to the exact answer? Try rounding the number of customers to some multiple of the denominator in the fraction.

55. Refer to Exercise 53. Round the number of Hicksville and Woodbury, New York, Cablevision customers to some multiple of the denominator in $\frac{3}{5}$. Now estimate the answer, showing your work.

56. Refer to Exercise 54. Round the number of Tempe, Arizona/Cox Cable customers to some multiple of the denominator in $\frac{3}{7}$. Now estimate the answer, showing your work.

2.6 Applications of Multiplication

OBJECTIVE 1 Solve fraction application problems using multiplication. Many application problems are solved by multiplying fractions. Use the following indicator words for multiplication.

> product
> double
> triple
> times
> of
> twice
> twice as much

Look for these indicator words in the following examples.

OBJECTIVE

1 Solve fraction application problems using multiplication.

❶ Solve each problem.

(a) Shafali Patel pays $\frac{1}{4}$ of her monthly salary as a house payment. If her monthly salary is $4320, find her monthly house payment.

$$\frac{1}{4} \times \frac{4320}{1}$$

$$1080$$

EXAMPLE 1 Applying Indicator Words

Lois Stevens gives $\frac{1}{10}$ of her income to her church. One month she earned $2980. How much did she give to the church that month?

Step 1 **Read** the problem. The problem asks us to find the amount of money given to the church.

Step 2 **Work out a plan.** The indicator word is *of*: Stevens gave $\frac{1}{10}$ *of* her income. The word *of* indicates multiplication, so find the amount given to the church by multiplying $\frac{1}{10}$ and $2980.

Step 3 **Estimate** a reasonable answer. Round the income of $2980 to $3000. Then divide $3000 by 10 to find $\frac{1}{10}$ of the income (one of 10 equal parts). Our estimate is $3000 \div 10 = 300$. (Recall the shortcut for dividing by 10; drop one 0 from the dividend.)

$$\frac{1}{10} \times \frac{\overset{298}{\cancel{2980}}}{1}$$

Step 4 **Solve** the problem.

$$\text{amount} = \frac{1}{\underset{1}{\cancel{10}}} \cdot \frac{\overset{298}{\cancel{2980}}}{1} = \frac{298}{1} = 298$$

Step 5 **State the answer.** Stevens gave $298 to her church that month.

Step 6 **Check.** The exact answer, $298, is close to our estimate of $300.

(((Work Problem 1 at the Side.

(b) A retiring firefighter will receive $\frac{5}{8}$ of her highest annual salary as retirement income. If her highest annual salary is $62,504, how much will she receive as retirement income?

$$\frac{5}{8} \times \frac{\overset{7813}{\cancel{62504}}}{1} =$$

$$39,065$$

EXAMPLE 2 Solving a Fraction Application Problem

Of the 39 students in Carol Dixon's high school biology class, $\frac{2}{3}$ plan to go to college. How many plan to go to college?

Step 1 **Read** the problem. The problem asks us to find the number of students who plan to go to college.

Step 2 **Work out a plan.** Reword the problem to read

$$\frac{2}{3} \text{ of the students plan to go to college.}$$

Indicator word for multiplication

Continued on Next Page

2 At Village Drug, $\frac{5}{16}$ of the prescriptions are paid by a third party (insurance company). If 3696 prescriptions are filled, find the number paid by a third party.

Use the six problem-solving steps.

Step 3 **Estimate** a reasonable answer. Round the number of students in the class from 39 to 40. Then, $\frac{1}{2}$ of 40 is 20. Since $\frac{2}{3}$ is more than $\frac{1}{2}$, our estimate is that "more than 20 students" plan to go to college.

Step 4 **Solve** the problem. Find the number who plan to go to college by multiplying $\frac{2}{3}$ and 39.

$$\text{number who plan to go} = \frac{2}{3} \cdot 39$$

$$= \frac{2}{\overset{}{\underset{1}{3}}} \cdot \frac{\overset{13}{\cancel{39}}}{1} = \frac{26}{1} = 26$$

Step 5 **State the answer.** 26 students plan to go to college.

Step 6 **Check.** The exact answer, 26, fits our estimate of "more than 20."

◀◀◀ **Work Problem 2 at the Side.**

EXAMPLE 3 **Finding a Fraction of a Fraction**

In her will, a woman divides her estate into 6 equal parts. Five of the 6 parts are given to relatives. Of the sixth part, $\frac{1}{3}$ goes to the Salvation Army. What fraction of her total estate goes to the Salvation Army?

Step 1 **Read** the problem. The problem asks for the fraction of an estate that goes to the Salvation Army.

Step 2 **Work out a plan.** Reword the problem to read

the Salvation Army gets $\frac{1}{3}$ **of** $\frac{1}{6}$.

↑
Indicator word for multiplication

3 In a certain community, $\frac{1}{3}$ of the residents speak a foreign language. Of those speaking a foreign language, $\frac{3}{4}$ speak Spanish. What fraction of the residents speak Spanish?

Use the six problem-solving steps.

Step 3 **Estimate** a reasonable answer. If the estate is divided into 6 equal parts and each of these parts was divided into 3 equal parts, we would have 6 · 3 = 18 equal parts. Our estimate is $\frac{1}{18}$.

Step 4 **Solve** the problem. The Salvation Army gets $\frac{1}{3}$ **of** $\frac{1}{6}$.

↑
Indicator word

To find the fraction that the Salvation Army is to receive, multiply $\frac{1}{3}$ and $\frac{1}{6}$.

$$\text{fraction to Salvation Army} = \frac{1}{3} \cdot \frac{1}{6}$$

$$= \frac{1}{18}$$

Step 5 **State the answer.** The Salvation Army gets $\frac{1}{18}$ of the total estate.

Step 6 **Check.** The exact answer, $\frac{1}{18}$, matches our estimate.

◀◀◀ **Work Problem 3 at the Side.**

ANSWERS

2. 1155 prescriptions

3. $\frac{1}{4}$ speak Spanish

EXAMPLE 4 **Using Fractions with a Circle Graph**

The circle graph, or pie chart, shows where children 8 to 17 years of age make food purchases when away from home. If 2500 children were in the survey, find the number of children who buy food in the school cafeteria.

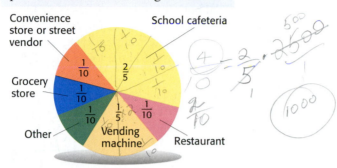

FINDING FOOD

Children ages 8 to 17 made food purchases at the following locations:

Source: Pursuant Inc. for American Dietetic Association Foundation.

Step 1 **Read** the problem. The problem asks for the number of children who buy food in the school cafeteria.

Step 2 **Work out a plan.** Reword the problem to read

$\frac{2}{5}$ **of** 2500 children buy food in the school cafeteria.

↑

Indicator word for multiplication

Step 3 **Estimate** a reasonable answer. $\frac{1}{2}$ of 2500 people is 1250 people. $\frac{2}{5}$ is less than $\frac{1}{2}$, so our estimate is "less than 1250 people."

Step 4 **Solve** the problem. Find the number who buy food in the school cafeteria by multiplying $\frac{2}{5}$ and 2500.

$$\text{number in school cafeteria} = \frac{2}{5} \cdot 2500$$

$$= \frac{2}{5} \cdot \frac{2500}{1}$$

$$= \frac{2}{\cancel{5}_{1}} \cdot \frac{\cancel{2500}^{500}}{1} \qquad \text{Divide both numerator and denominator by 5.}$$

$$= \frac{1000}{1} = 1000$$

Step 5 **State the answer.** 1000 children buy food in the school cafeteria.

Step 6 **Check.** The exact answer, 1000 children, fits our estimate of "less than 1250 children."

Work Problem 4 at the Side. ▶▶▶

④ Solve each problem using the six problem-solving steps. Use the circle graph in Example 4.

(a) What fraction of the children buy food from vending machines?

(b) What number of children buy food from vending machines?

(c) What fraction of the children buy food from a convenience store or street vendor?

(d) What number of children buy food from a convenience store or street vendor?

Real-Data Applications

Heart-Rate Training Zone

Performing aerobic exercise is beneficial both for improving aerobic fitness and for burning fat. For best results, you should keep your heart rate within the training zone for a minimum of 12 minutes. If you train at the higher end of the training zone, you will burn glycogen and improve aerobic fitness. Training for longer periods at the lower end of the training zone results in your body using fat reserves for energy.

Example: The training zone (TZ) is based on your heart rate (HR) for one minute. To see if you are in the training zone, measure your heart rate for 15 seconds. Compare it to the 15-second training zone. Fine the exact answer, and then round to the nearest whole number.

Instruction	Calculation	Example (age 22)
Calculate maximum heart rate (MHR)	$220 -$ your age	$220 - 22 = 198$
Calculate lower limit of training zone (TZ)	$\frac{3}{5} \times$ (MHR)	$\frac{3}{5} \times (198) = \frac{594}{5} = 118\frac{4}{5}$
Calculate upper limit of training zone (TZ)	$\frac{4}{5} \times$ (MHR)	$\frac{4}{5} \times (198) = \frac{792}{5} = 158\frac{2}{5}$
Calculate the exact 15-second training zone. Round the results to the nearest whole number.	$\left(\frac{1}{4} \times \text{lower TZ}, \frac{1}{4} \times \text{Upper TZ} \right)$	$\frac{1}{4} \times \frac{594}{5} = 29\frac{7}{10} ; \frac{1}{4} \times \frac{792}{5} = 39\frac{3}{5}$ HR between $29\frac{7}{10}$ and $39\frac{3}{5}$ HR between 30 and 40

Age	MHR	Lower Limit of TZ	Upper Limit of TZ	15-Second TZ (exact)	15-Second TZ (rounded)
18					
25					
30					
40					
50					
60					

1. Suppose you work in a physical fitness center and decide to design a poster to remind the clients of the training zone for their age. Compute the exact 15-second training zone for people of each of the following ages. Write fractions in lowest terms. Then round the answers to the nearest whole number.

2. Explain why the lower and upper training zones (TZ) are multiplied by $\frac{1}{4}$.

3. Explain why the 15-second training zone is lower for a person aged 50 in comparison to a person aged 20.

4. Based on the chart, what would you tell a 45-year-old person about their 15-second training zone?

2.6 Exercises

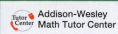
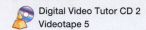

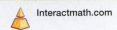

Solve each application problem. Look for indicator words. See Examples 1–4.

1. A file cabinet top is $\frac{3}{4}$ yd by $\frac{2}{3}$ yd. Find its area.

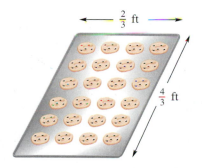

2. A kitty bed for Snuggles is $\frac{3}{8}$ yd by $\frac{7}{9}$ yd. Find its area.

3. A cookie sheet is $\frac{4}{3}$ ft by $\frac{2}{3}$ ft. Find its area.

4. One-third of the players elected to the Baseball Hall of Fame were pitchers. If 183 players are in the Hall of Fame, how many were pitchers? (*Source: National Baseball Hall of Fame.*)

$$\frac{1}{3} \times 183$$

5. Pete is helping Colin make a mahogany lamp table for Carolyn's birthday. Find the area of the top of the table if it is $\frac{4}{5}$ yd long by $\frac{3}{8}$ yd wide.

6. A convenience store sells 1650 items, of which $\frac{12}{25}$ are classified as junk food. How many of the items are junk food?

$$\frac{12}{25} \times \frac{\overset{66}{\cancel{1650}}}{1} = 792$$

7. Dan Crump had expenses of $6848 during one semester of college. His part-time job provided $\frac{3}{8}$ of the amount he needed. How much did he earn on his job?

8. Erin Hernandez produces $5680 in profits for her employer. If her personal earnings are $\frac{2}{5}$ of these profits, find the amount of her earnings.

9. A school gave scholarships to $\frac{5}{24}$ of its 1800 freshmen. How many freshman students received scholarships?

10. Erich Means estimates that it will cost him $12,200, including living expenses, to attend college full time for one year. If he must earn $\frac{3}{5}$ of the cost and borrow the balance, find the amount that he must earn.

11. At the Garlic Festival Fun Run, $\frac{7}{12}$ of the runners are women. If there are 1560 runners, how many are women?

12. A hotel has 408 rooms. Of these rooms, $\frac{9}{17}$ are for nonsmokers. How many rooms are for nonsmokers?

$$\frac{9}{17} \times \frac{408}{1} \quad 24 = 216$$

In a recent year, Americans purchased one billion books. The circle graph shows the purchase of these books by age group over a one-year period. Use this information to work Exercises 13–18.

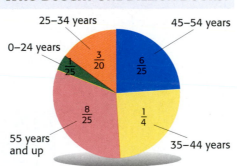

WHO BOUGHT ONE BILLION BOOKS?

Source: Book Industry Study Group.

13. Which age group purchased the least number of books? How many did this group purchase?

14. Which age group purchased the greatest number of books? How many did this group purchase?

15. Find the number of books purchased by those in the 35–44 year age group.

16. Find the number of books purchased by those in the 45–54 age group.

17. Without actually adding the fractions given for all the age groups, explain why their sum has to be 1.

18. Refer to Exercise 17. Suppose you added all the fractions for the age groups and did not get 1 as an answer. List some possible explanations.

The table shows the earnings for the Gomes family last year and the circle graph shows how they spent their earnings. Use this information to answer Exercises 19–24.

Month	Earnings	Month	Earnings
January	$4575	July	$5540
February	$4312	August	$3732
March	$4988	September	$4170
April	$4530	October	$5512
May	$4320	November	$4965
June	$4898	December	$6458

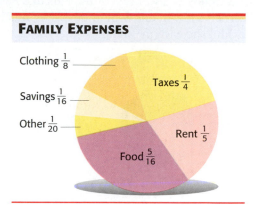

FAMILY EXPENSES

Clothing $\frac{1}{8}$

Savings $\frac{1}{16}$

Other $\frac{1}{20}$

Taxes $\frac{1}{4}$

Rent $\frac{1}{5}$

Food $\frac{5}{16}$

19. Find the Gomes family's total income for the year.

20. How much of their annual earnings went to taxes?

21. Find the amount of their rent for the year.

22. How much did they spend for food during the year?

23. Find their annual savings.

24. How much of their annual income was spent on clothing?

25. Here is how one student solved a multiplication problem. Find the error and solve the problem correctly.

$$\frac{9}{10} \times \frac{20}{21} = \frac{\overset{3}{\cancel{9}}}{\underset{1}{\cancel{10}}} \times \frac{\overset{2}{\cancel{20}}}{\underset{3}{\cancel{21}}} = \frac{6}{3} = 2$$

26. When two whole numbers are multiplied, the product is always larger than the numbers being multiplied. When two proper fractions are multiplied, the product is always smaller than the numbers being multiplied. Are these statements true? Why or why not?

Solve each application problem.

27. Pamela Denny is a waitress at the Pancake Parade and earned $128 in one 8-hour day. How much money did she earn in 3 hours?

16 per hour

48

28. Tori Buckley jogged 40 miles in 8 hours. How far did she jog in 5 hours?

25

29. LaDonna Washington is running for city council. She needs to get $\frac{2}{3}$ of her votes from senior citizens and 27,000 votes in all to win. How many votes does she need from voters other than the senior citizens?

30. The start-up cost of a Subs and Sandwich Shop is $32,000. If the bank will loan you $\frac{9}{16}$ of the start up and you must pay the balance, how much more will you need to open a shop?

31. A will states that $\frac{7}{8}$ of an estate is to be divided among relatives. Of the remaining estate, $\frac{1}{4}$ goes to the American Cancer Society. What fraction of the estate goes to the American Cancer Society?

32. A couple has $\frac{2}{5}$ of their total investments in real estate. Of the remaining investment, $\frac{1}{3}$ is invested in bonds. What fraction of the total investments is in bonds?

2.7 Dividing Fractions

OBJECTIVE 1 Find the reciprocal of a fraction. To divide fractions, we need to know how to find the **reciprocal** of a fraction.

OBJECTIVES

1. Find the reciprocal of a fraction.
2. Divide fractions.
3. Solve application problems in which fractions are divided.

Reciprocal of a Fraction

Two numbers are reciprocals of each other if their product is 1. To find the reciprocal of a fraction, interchange the numerator and denominator.

For example, the reciprocal of $\frac{3}{4}$ is $\frac{4}{3}$.

Fraction $\frac{3}{4}$ ✕ $\frac{4}{3}$ Reciprocal

> **NOTE**
> Notice that you invert, or "flip" a fraction to find its reciprocal.

EXAMPLE 1 Finding Reciprocals

Find the reciprocal of each fraction.

(a) The reciprocal of $\frac{1}{4}$ is $\frac{4}{1}$ because $\frac{1}{4} \cdot \frac{4}{1} = \frac{4}{4} = 1$

(b) The reciprocal of $\frac{2}{3}$ is $\frac{3}{2}$ because $\frac{2}{3} \cdot \frac{3}{2} = \frac{6}{6} = 1$

(c) The reciprocal of $\frac{3}{5}$ is $\frac{5}{3}$ because $\frac{3}{5} \cdot \frac{5}{3} = \frac{15}{15} = 1$

(d) The reciprocal of 8 is $\frac{1}{8}$ because $\frac{8}{1} \cdot \frac{1}{8} = \frac{8}{8} = 1$ Think of 8 as $\frac{8}{1}$.

> **Work Problem 1 at the Side.** ▶▶▶

> **NOTE**
> Every number has a reciprocal except 0. The number 0 has no reciprocal because there is no number that can be multiplied by 0 to get 1.
>
> $$0 \cdot (\text{reciprocal}) = 1$$
> ↑
> There is no number to use here that will give an answer of 1. When you multiply by 0, you always get 0.

1 Find the reciprocal of each fraction.

(a) $\frac{3}{4}$

(b) $\frac{3}{8}$

(c) $\frac{9}{4}$

(d) 16

In **Chapter 1,** we saw that the division problem $12 \div 3$ asks how many 3s are in 12. In the same way, the division problem $\frac{2}{3} \div \frac{1}{6}$ asks how many $\frac{1}{6}$s are in $\frac{2}{3}$. The figure illustrates $\frac{2}{3} \div \frac{1}{6}$.

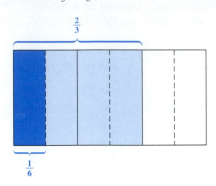

The figure shows that there are 4 of the $\frac{1}{6}$ pieces in $\frac{2}{3}$, or

$$\frac{2}{3} \div \frac{1}{6} = 4$$

OBJECTIVE **2** **Divide fractions.**   We will use reciprocals to divide fractions.

> **Dividing Fractions**
>
> To divide two fractions, multiply the first fraction by the reciprocal of the second fraction.

EXAMPLE 2   **Dividing One Fraction by Another**

Divide. Write answers in lowest terms and as mixed numbers where possible.

(a) $\dfrac{7}{8} \div \dfrac{15}{16}$

The reciprocal of $\dfrac{15}{16}$ is $\dfrac{16}{15}$

Reciprocals

$$\frac{7}{8} \div \frac{15}{16} = \frac{7}{8} \cdot \frac{16}{15}$$

Change division to multiplication.

$$= \frac{7}{\overset{}{\underset{1}{8}}} \cdot \frac{\overset{2}{16}}{15} \qquad \text{Divide the numerator and denominator by 8.}$$

$$= \frac{7 \cdot 2}{1 \cdot 15} \qquad \text{Multiply.}$$

$$= \frac{14}{15} \qquad \text{Lowest terms}$$

Continued on Next Page

(b) $\dfrac{\frac{4}{5}}{\frac{3}{10}}$

$$\dfrac{\frac{4}{5}}{\frac{3}{10}} = \dfrac{4}{5} \div \dfrac{3}{10}$$ Rewrite by using the $\div$ symbol for division.

$$= \dfrac{4}{\underset{1}{\cancel{5}}} \cdot \dfrac{\overset{2}{\cancel{10}}}{3}$$ The reciprocal of $\frac{3}{10}$ is $\frac{10}{3}$. Change "$\div$" to "$\cdot$". Divide the numerator and denominator by 5.

$$= \dfrac{4 \cdot 2}{1 \cdot 3}$$ Multiply.

$$= \dfrac{8}{3} = 2\dfrac{2}{3}$$ Rewrite the answer as a mixed number.

CAUTION
Be certain that the divisor fraction is changed to its reciprocal *before* you divide numerators and denominators by common factors.

Work Problem 2 at the Side. ▶▶▶

EXAMPLE 3 Dividing with a Whole Number

Divide. Write all answers in lowest terms and as whole or mixed numbers where possible.

(a) $5 \div \dfrac{1}{4}$

Write 5 as $\frac{5}{1}$. Next, use the reciprocal of $\frac{1}{4}$, which is $\frac{4}{1}$.

$$5 \div \dfrac{1}{4} = \dfrac{5}{1} \cdot \dfrac{4}{1}$$ Reciprocal of $\frac{1}{4}$ is $\frac{4}{1}$.

Reciprocals

$$= \dfrac{5 \cdot 4}{1 \cdot 1}$$ Multiply.

$$= \dfrac{20}{1} = 20$$ Rewrite the answer as a whole number.

Continued on Next Page

---

2 Divide. Write answers in lowest terms and as mixed numbers where possible.

(a) $\dfrac{1}{4} \div \dfrac{2}{3}$

(b) $\dfrac{3}{8} \div \dfrac{5}{8}$

(c) $\dfrac{\frac{2}{3}}{\frac{4}{5}}$

(d) $\dfrac{\frac{5}{6}}{\frac{7}{12}}$ $\dfrac{5}{\cancel{6}} \times \dfrac{\overset{2}{\cancel{12}}}{7} = \dfrac{10}{7}$

$1\dfrac{3}{7}$

3 Divide. Write answers in lowest terms and as whole or mixed numbers where possible.

(a) $10 \div \dfrac{1}{2}$

(b) $6 \div \dfrac{6}{7}$

(c) $\dfrac{4}{5} \div 6$

(d) $\dfrac{3}{8} \div 4$

4 Solve each problem using the six problem-solving steps.

(a) How many $\frac{3}{4}$-quart leafblower fuel tanks can be filled from 15 quarts of fuel?

(b) How many $\frac{2}{3}$-gallon garden sprayers can be filled from 36 gallons of insect spray?

(b) $\dfrac{2}{3} \div 6$

Write 6 as $\frac{6}{1}$. The reciprocal of $\frac{6}{1}$ is $\frac{1}{6}$.

$$\dfrac{2}{3} \div \dfrac{\mathbf{6}}{\mathbf{1}} = \dfrac{2}{3} \cdot \dfrac{\mathbf{1}}{\mathbf{6}}$$

$$\underbrace{}_{\text{Reciprocals}}$$

$$= \dfrac{\overset{1}{\cancel{2}}}{3} \cdot \dfrac{1}{\underset{3}{\cancel{6}}} \qquad \text{Divide the numerator and denominator by 2, then multiply.}$$

$$= \dfrac{1 \cdot 1}{3 \cdot 3} = \dfrac{1}{9} \qquad \text{Lowest terms}$$

◀◀◀ Work Problem 3 at the Side.

OBJECTIVE 3 Solve application problems in which fractions are divided. Many application problems require division of fractions. Recall that typical indicator words for division are *goes into, per, divide, divided by, divided equally,* and *divided into.*

EXAMPLE 4 Applying Fraction Skills

Goldie, the manager of the Burnside Deli, must fill a 10-gallon dill pickle crock with salt brine. She has only a $\frac{2}{3}$-gallon container to use. How many times must she fill the $\frac{2}{3}$-gallon container and empty it into the 10-gallon crock?

Step 1 **Read** the problem. We need to find the number of times Goldie needs to use a $\frac{2}{3}$-gallon container in order to fill a 10-gallon crock.

Step 2 **Work out a plan.** We can solve the problem by finding the number of times 10 can be divided by $\frac{2}{3}$.

Step 3 **Estimate** a reasonable answer. Round $\frac{2}{3}$ gallon to 1 gallon. In order to fill the 10-gallon container, she would have to use the 1-gallon container 10 times, so our estimate is 10.

Step 4 **Solve** the problem.

$$\overset{\text{Reciprocals}}{10 \div \dfrac{2}{3} = \dfrac{10}{1} \cdot \dfrac{3}{2}} \qquad \text{The reciprocal of } \tfrac{2}{3} \text{ is } \tfrac{3}{2}. \\ \text{Change "} \div \text{" to "} \cdot \text{."}$$

$$= \dfrac{\overset{5}{\cancel{10}}}{1} \cdot \dfrac{3}{\underset{1}{\cancel{2}}} \qquad \text{Divide the numerator and denominator by 2 and then multiply.}$$

$$= \dfrac{15}{1} = 15$$

Step 5 **State the answer.** Goldie must fill the container 15 times.

Step 6 **Check.** The exact answer, 15 times, is close to our estimate of 10 times.

◀◀◀ Work Problem 4 at the Side.

ANSWERS

3. **(a)** 20 **(b)** 7 **(c)** $\dfrac{2}{15}$ **(d)** $\dfrac{3}{32}$

4. **(a)** 20 fuel tanks **(b)** 54 sprayers

EXAMPLE 5 **Applying Fraction Skills**

At the Happi-Time Day Care Center, $\frac{6}{7}$ of the total budget goes to classroom operation. If there are 18 classrooms, what fraction of the classroom operating amount does each classroom receive?

Step 1 **Read** the problem. Since $\frac{6}{7}$ of the total budget must be split into 18 parts, we must find the fraction of the classroom operating amount received by each classroom.

Step 2 **Work out a plan.** We must divide the fraction of the total budget going to classroom operation ($\frac{6}{7}$) by the number of classrooms (18).

Step 3 **Estimate** a reasonable answer. Round $\frac{6}{7}$ to 1. If all of the operating expenses (1 whole) were divided between 18 classrooms, each classroom would receive $\frac{1}{18}$ of the operating expenses, our estimate.

Step 4 **Solve** the problem. We solve by dividing $\frac{6}{7}$ by 18.

$$\frac{6}{7} \div 18 = \frac{6}{7} \div \frac{18}{1}$$

$$= \frac{\overset{1}{\cancel{6}}}{7} \cdot \frac{1}{\underset{3}{\cancel{18}}} \qquad \text{The reciprocal of } \frac{18}{1} \text{ is } \frac{1}{18}. \text{ Change "} \div \text{" to "} \cdot \text{."}$$
$$\text{Divide the numerator and denominator by 6.}$$

$$= \frac{1}{21} \qquad \text{Multiply.}$$

Step 5 **State the answer.** Each classroom receives $\frac{1}{21}$ of the total budget.

Step 6 **Check.** The exact answer, $\frac{1}{21}$, is close to our estimate of $\frac{1}{18}$.

Work Problem 5 at the Side. ▶▶▶

5 Solve each problem using the six problem-solving steps.

(a) The top 12 employees at Mayfield Manufacturing will divide $\frac{3}{4}$ of the annual bonus money. What fraction of the bonus money will each employee receive?

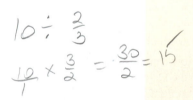

(b) A winning lottery ticket was purchased by 8 Starbucks Coffee employees. They will divide $\frac{4}{5}$ of the total lottery prize money. What fraction of the prize money will each receive?

Real-Data Applications

Hotel Expenses

Mathematics teachers attending conferences in New Orleans, Louisiana, and San Jose, California, found the following information about hotel rates on the organizations' Internet Web sites.

Hotel	Single	Double	Triple	Quad	Suites
New Orleans (January 2005)					
Marriott	$116	$121	$121	$121	$603
Sheraton (Club)	$157	$169	$194	$219	$598
San Jose (February 2005)					
Crowne Plaza	$157	$157	$167	$177	—
Hilton Towers	$167	$187	$207	$227	—
Hyatt Sainte Claire	$156	$176	$196	$216	—

1. The double rate is for two people sharing a room. What fractional part does each person pay?

2. (a) Multiply the double rate at the Hilton Towers in San Jose by $\frac{1}{2}$. What is the result?

 (b) Divide the double rate at the Hilton Towers in San Jose by 2. What is the result?

 (c) Explain what happened. How much money would one person owe if he or she shared a double room at the Hilton Towers in San Jose?

3. The triple rate is for three people sharing a room. What fractional part does each person pay?

4. How much money would one person owe if he or she shared a triple room at the Hyatt Sainte Claire in San Jose? How much money would each person save if they could book a triple room at the Crowne Plaza instead of the Hyatt Sainte Claire? Find your answer using two different methods, based on your observations in Problem 2.

5. The quad rate is for four people sharing a room. What fractional part does each person pay?

6. How much money would one person owe if he or she shared a quad room at the Sheraton in New Orleans? Find your answer using two different methods, based on your observations in Problem 2.

7. How many people would have to share a suite at the Sheraton in New Orleans for the cost per person to be less than sharing a quad room at the same hotel? Round the answer to the nearest whole number. (*Hint:* Estimate the cost per person for a quad room first. Then estimate the number of people needed to share the cost of the suite. Check your work using actual values.)

8. Suppose you have a travel allotment of $500 that can be spent on transportation, hotel, food, and registration fees. You and a colleague decide to attend the 3-day New Orleans conference and plan to share a room for three nights at the Marriott. Registration costs $150; the flight costs $129 round-trip; the taxi ride to the hotel costs $20 per person each way; and you budget $35 per day for meals. How much out-of-pocket expense will you have to pay? How much would you save if you could recruit a third person to share the room?

2.7 **Exercises**

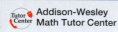

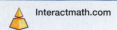
Find the reciprocal of each number. See Example 1.

1. $\dfrac{3}{8}$

2. $\dfrac{2}{5}$

3. $\dfrac{5}{6}$

4. $\dfrac{12}{7}$

5. $\dfrac{8}{5}$

6. $\dfrac{13}{20}$

7. 4

8. 10

Divide. Write answers in lowest terms and as whole or mixed numbers where possible. See Examples 2 and 3.

9. $\dfrac{1}{2} \div \dfrac{3}{4}$

10. $\dfrac{5}{8} \div \dfrac{7}{8}$

11. $\dfrac{7}{8} \div \dfrac{1}{3}$

12. $\dfrac{7}{8} \div \dfrac{3}{4}$

13. $\dfrac{3}{4} \div \dfrac{5}{3}$

14. $\dfrac{4}{5} \div \dfrac{9}{4}$

15. $\dfrac{7}{9} \div \dfrac{7}{36}$

16. $\dfrac{5}{8} \div \dfrac{5}{16}$

17. $\dfrac{15}{32} \div \dfrac{5}{64}$

18. $\dfrac{7}{12} \div \dfrac{14}{15}$

19. $\dfrac{\frac{13}{20}}{\frac{4}{5}}$

20. $\dfrac{\frac{9}{10}}{\frac{3}{5}}$

21. $\dfrac{\frac{5}{6}}{\frac{25}{24}}$

22. $\dfrac{\frac{28}{15}}{\frac{21}{5}}$

23. $12 \div \dfrac{2}{3}$

24. $7 \div \dfrac{1}{4}$

25. $\dfrac{\frac{18}{3}}{4}$

26. $\dfrac{\frac{12}{3}}{4}$

27. $\dfrac{\frac{4}{7}}{8}$

28. $\dfrac{\frac{7}{10}}{3}$

Solve each application problem by using division. See Examples 4 and 5.

29. Veterinarian Jasmine Cato has $\frac{8}{9}$ liter of medication. She wishes to prescribe this medication for 4 cats in her pet hospital. If she divides the medication evenly, how much will each cat receive?

$$\frac{8}{9} \div 4$$

$$\overset{2}{\frac{8}{9}} \times \frac{1}{\underset{1}{4}} = \frac{2}{9}$$

30. Harold Pishke, barber, has 15 quarts of conditioning shampoo. If he wants to put this shampoo into $\frac{3}{8}$ quart containers, how many containers can be filled?

31. Some college roommates want to make pancakes for their neighbors. They need 5 cups of flour, but have only a $\frac{1}{3}$-cup measuring cup. How many times will they need to fill their measuring cup?

$$5 \div \frac{1}{3}$$

$$\frac{5}{1} \times \frac{3}{1} = 15$$

32. The Sweepstakes Lottery pays out $\frac{7}{8}$ of the total revenue to 14 top winners. What fraction of the total revenue does each winner receive?

33. How many $\frac{1}{8}$-ounce eye drop dispensers can be filled with 11 ounces of eye drops?

$$11 \div \frac{1}{8}$$

$$\frac{11}{1} \times \frac{8}{1} = 88$$

34. It is estimated that each guest at a party will eat $\frac{5}{16}$ pound of peanuts. How many guests may be served with 10 pounds of peanuts?

35. Pam Trizlia had a small pickup truck that could carry $\frac{2}{3}$-cord of firewood. Find the number of trips needed to deliver 40 cords of wood.

36. Manuel Servin has a 200-yard roll of weather stripping material. Find the number of pieces of weather stripping $\frac{5}{8}$ yard in length that may be cut from the roll.

37. Your classmate is confused on how to divide by a fraction. Write a short note telling him how this should be done.

38. If you multiply positive proper fractions, the product is smaller than the fractions multiplied. When you divide by a proper fraction, is the quotient smaller than the numbers in the problem? Prove your answer with examples.

Solve each application problem using multiplication or division.

39. The recipe for the Jelly Belly Express loafcake calls for $\frac{3}{4}$ pound of Jelly Belly jelly beans in assorted colors. If you want to make 16 cakes, how many pounds of Jelly Belly jelly beans will you need?

40. An upholsterer wants to reupholster 24 barstools. If each barstool needs $\frac{7}{8}$ pound of brass tacks, how many pounds of brass tacks are needed?

41. Broadly Plumbing finds that $\frac{3}{4}$ can of pipe joint compound is needed to plumb each new home. How many homes can be plumbed with 156 cans of compound?

42. A mechanic uses on average $\frac{2}{3}$ gallon of gear lube to service each tractor differential. Find the number of tractors that can be serviced with 28 gallons of gear lube.

43. Mike and Charlie have completed $\frac{4}{5}$ of their river rafting excursion down the Colorado River. If the trip is a total of 380 miles, how many miles have they gone?

44. Laura has been working on a job that will require 81 hours to complete. If she has completed $\frac{7}{9}$ of the job, how many hours has she worked?

45. A dish towel manufacturer requires $\frac{3}{8}$ yard of cotton fabric for each towel. Find the number of dish towels that can be made from 912 yards of fabric.

46. Each patient will receive $\frac{7}{10}$ vial of medication. How many patients can be treated with 3150 vials of medication?

Many application problems are solved using multiplication and division of fractions.
Work Exercises 47–52 in order.

47. Perhaps the most common indicator word for multiplication is the word *of.* Circle the words in the list below that are also indicator words for multiplication.

more than	per
double	twice
times	product
less than	difference
equals	twice as much

48. Circle the words in the list below that are indicator words for division.

fewer	sum of
goes into	divide
per	quotient
equals	double
loss of	divided by

49. To divide two fractions, multiply the first fraction by the _____ of the second fraction.

50. Find the reciprocals for each number.

$$\frac{3}{4} \qquad \frac{7}{8} \qquad 5 \qquad \frac{12}{19}$$

The size of a square postage stamp is shown here. Use this to answer Exercises 51 and 52.

$\longmapsto \frac{15}{16}$ in. $\longmapsto$

$\frac{15}{16}$ in.

51. Find the perimeter (distance around the outside edges) of the postage stamp using multiplication. Explain how to find the perimeter of any regular 3-, 4-, 5-, or 6-sided figure using multiplication.

52. Find the area of the postage stamp. Explain how to find the area of any square.

2.8 Multiplying and Dividing Mixed Numbers

In Section 2.2 we worked with mixed numbers—a whole number and a fraction written together. Many of the fraction problems you encounter in everyday life involve mixed numbers.

OBJECTIVE 1 Estimate the answer and multiply mixed numbers.
When multiplying mixed numbers, it is a good idea to estimate the answer first. Then multiply the mixed numbers by using the following steps.

> ### Multiplying Mixed Numbers
>
> *Step 1*   *Change* each mixed number to an improper fraction.
>
> *Step 2*   *Multiply* as fractions.
>
> *Step 3*   Simplify the answer, which means to write it in *lowest terms,* and change it to a mixed number or whole number where possible.

To estimate the answer, round each mixed number to the nearest whole number. If the numerator is *half* of the denominator or *more,* round up the whole number part. If the numerator is *less* than half the denominator, leave the whole number as it is.

$$1\frac{5}{8} \quad \begin{matrix} \leftarrow \text{5 is more than 4.} \\ \leftarrow \text{Half of 8 is 4.} \end{matrix} \Bigg\} \quad 1\frac{5}{8} \text{ rounds up to 2}$$

$$3\frac{2}{5} \quad \begin{matrix} \leftarrow \text{2 is less than } 2\frac{1}{2}. \\ \leftarrow \text{Half of 5 is } 2\frac{1}{2} \end{matrix} \Bigg\} \quad 3\frac{2}{5} \text{ rounds to 3}$$

> **Work Problem 1 at the Side.** ▶▶▶

EXAMPLE 1 Multiplying Mixed Numbers

First estimate the answer. Then multiply to get an exact answer. Simplify your answers.

(a) $2\frac{1}{2} \cdot 3\frac{1}{5}$

Estimate the answer by rounding the mixed numbers.

$$2\frac{1}{2} \text{ rounds to 3} \quad \text{and} \quad 3\frac{1}{5} \text{ rounds to 3}$$

$$3 \cdot 3 = 9 \qquad \text{Estimated answer}$$

To find the exact answer, change each mixed number to an improper fraction.

$$\textit{Step 1} \qquad 2\frac{1}{2} = \frac{5}{2} \quad \text{and} \quad 3\frac{1}{5} = \frac{16}{5}$$

── **Continued on Next Page**

1 Round each mixed number to the nearest whole number.

(a) $4\frac{2}{3}$

4.6

(b) $3\frac{2}{5}$ 3.4

③

(c) $5\frac{3}{4}$  ⑥

(d) $4\frac{7}{12}$ 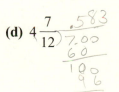 4.583

⑤

(e) $1\frac{1}{2}$

(f) $8\frac{4}{9}$

2 First estimate the answer. ≈ Then multiply to find the exact answer. Simplify your answers.

(a) $3\frac{1}{2} \quad \cdot \quad 6\frac{1}{3}$

$\downarrow \qquad \downarrow$

$\underline{\hspace{1.5cm}} \cdot \underline{\hspace{1.5cm}}$

$= \underline{\hspace{1.5cm}}$ *estimate*

(b) $4\frac{2}{3} \quad \cdot \quad 2\frac{3}{4}$

$\downarrow \qquad \downarrow$

$\underline{\hspace{1.5cm}} \cdot \underline{\hspace{1.5cm}}$

$= \underline{\hspace{1.5cm}}$ *estimate*

(c) $3\frac{3}{5} \quad \cdot \quad 4\frac{4}{9}$

$\downarrow \qquad \downarrow$

$\underline{\hspace{1.5cm}} \cdot \underline{\hspace{1.5cm}}$

$= \underline{\hspace{1.5cm}}$ *estimate*

(d) $5\frac{1}{4} \quad \cdot \quad 3\frac{2}{5}$

$\downarrow \qquad \downarrow$

$\underline{\hspace{1.5cm}} \cdot \underline{\hspace{1.5cm}}$

$= \underline{\hspace{1.5cm}}$ *estimate*

ANSWERS

1. (a) *Estimate:* $4 \cdot 6 = 24$; *Exact:* $22\frac{1}{6}$

(b) *Estimate:* $5 \cdot 3 = 15$; *Exact:* $12\frac{5}{6}$

(c) *Estimate:* $4 \cdot 4 = 16$; *Exact:* 16

(d) *Estimate:* $5 \cdot 3 = 15$; *Exact:* $17\frac{17}{20}$

Next, multiply.

Step 1   *Step 2*   *Step 3*

$$2\frac{1}{2} \cdot 3\frac{1}{5} = \frac{5}{2} \cdot \frac{16}{5} = \frac{\overset{1}{\cancel{5}}}{\underset{1}{\cancel{2}}} \cdot \frac{\overset{8}{\cancel{16}}}{\underset{1}{\cancel{5}}} = \frac{1 \cdot 8}{1 \cdot 1} = \frac{8}{1} = 8$$

The estimated answer is 9 and the exact answer is 8. The exact answer is reasonable.

(b) $3\frac{5}{8} \cdot 4\frac{4}{5}$

$3\frac{5}{8}$ rounds to 4   and $4\frac{4}{5}$ rounds to 5

$4 \cdot 5 = 20$ Estimated answer

Now find the exact answer.

Step 1   *Step 2*

$$3\frac{5}{8} \cdot 4\frac{4}{5} = \frac{29}{8} \cdot \frac{24}{5} = \frac{29}{\underset{1}{\cancel{8}}} \cdot \frac{\overset{3}{\cancel{24}}}{5} = \frac{29 \cdot 3}{1 \cdot 5} = \frac{87}{5}$$

Simplify the answer.

Step 3

$$\frac{87}{5} = 17\frac{2}{5} \qquad \text{Simplified answer}$$

The estimate was 20, so the exact answer of $17\frac{2}{5}$ is reasonable.

(c) $1\frac{3}{5} \cdot 3\frac{1}{3}$

$1\frac{3}{5}$ rounds to 2   and $3\frac{1}{3}$ rounds to 3

$2 \cdot 3 = 6$ Estimated answer

The exact answer is shown below.

$$1\frac{3}{5} \cdot 3\frac{1}{3} = \frac{8}{\underset{1}{\cancel{5}}} \cdot \frac{\overset{2}{\cancel{10}}}{3} = \frac{8 \cdot 2}{1 \cdot 3} = \frac{16}{3} = 5\frac{1}{3}$$

The estimate was 6, so the exact answer of $5\frac{1}{3}$ is reasonable.

◀◀◀ Work Problem 2 at the Side.

OBJECTIVE **2** **Estimate the answer and divide mixed numbers.** Just as you did when multiplying mixed numbers, it is also a good idea to estimate the answer when dividing mixed numbers. To divide mixed numbers, use the following steps.

Dividing Mixed Numbers

Step 1 *Change* each mixed number to an improper fraction.

Step 2 Use the *reciprocal* of the second fraction (divisor).

Step 3 *Multiply.*

Step 4 Simplify the answer, which means to write it in *lowest terms,* and change it to a mixed number or whole number where possible.

NOTE
Recall that the reciprocal of a fraction is found by interchanging the numerator and the denominator.

EXAMPLE 2 **Dividing Mixed Numbers**

First estimate the answer. Then divide to find the exact answer. Simplify your answers.

(a) $2\frac{2}{5} \div 1\frac{1}{2}$

First estimate the answer by rounding each mixed number to the nearest whole number.

$$2\frac{2}{5} \div \quad 1\frac{1}{2}$$

Rounded

$$2 \quad \div \quad 2 = 1 \quad \text{Estimated answer}$$

To find the exact answer, first change each mixed number to an improper fraction.

Step 1

$$2\frac{2}{5} \div 1\frac{1}{2} = \frac{12}{5} \div \frac{3}{2}$$

Next, use the reciprocal of the second fraction and multiply.

Step 2 *Step 3* *Step 4*

$$\frac{12}{5} \div \frac{3}{2} = \frac{\overset{4}{\cancel{12}}}{5} \cdot \frac{2}{\underset{1}{\cancel{3}}} = \frac{4 \cdot 2}{5 \cdot 1} = \frac{8}{5} = 1\frac{3}{5} \quad \text{Simplified answer}$$

Reciprocals

The estimate was 1, so the exact answer of $1\frac{3}{5}$ is reasonable.

Continued on Next Page

3 First estimate the answer.
≈ Then divide to find the exact answer. Simplify all answers.

(a) $3\frac{1}{8}$ · $6\frac{1}{4}$
            ↓   ↓

_____ · _____

= _____ *estimate*

(b) $5\frac{1}{3}$ · $1\frac{1}{4}$
            ↓   ↓

_____ · _____

= _____ *estimate*

(c) 8 ÷ $5\frac{1}{3}$
            ↓   ↓

_____ · _____

= _____ *estimate*

(d) $13\frac{1}{2}$ ÷ 18
            ↓   ↓

_____ · _____

= _____ *estimate*

3. (a) *Estimate:* $3 \div 6 = \frac{1}{2}$; *Exact:* $\frac{1}{2}$
   (b) *Estimate:* $5 \div 1 = 5$; *Exact:* $4\frac{4}{15}$
   (c) *Estimate:* $8 \div 5 = 1\frac{3}{5}$; *Exact:* $1\frac{1}{2}$
   (d) *Estimate:* $14 \div 18 = \frac{7}{9}$; *Exact:* $\frac{3}{4}$

(b) $8 \div 3\frac{3}{5}$

$$
\begin{array}{ccc}
8 & \div & 3\frac{3}{5}\\
\downarrow & \text{Rounded} & \downarrow\\
8 & \div & 4 = 2 \qquad \text{Estimate}
\end{array}
$$

Now find the exact answer.

Reciprocals

$$8 \div 3\frac{3}{5} = \frac{8}{1} \div \frac{18}{5} = \frac{8}{1} \cdot \frac{5}{\underset{9}{\cancel{18}}} = \frac{20}{9} = 2\frac{2}{9}$$

Write 8 as $\frac{8}{1}$.

The estimate was 2, so the exact answer of $2\frac{2}{9}$ is reasonable.

(c) $4\frac{3}{8} \div 5$

$$
\begin{array}{ccc}
4\frac{3}{8} & \div & 5
\end{array}
$$

Rounded

Reciprocals

$$4 \div 5 = \frac{4}{1} \div \frac{5}{1} = \frac{4}{1} \cdot \frac{1}{5} = \frac{4}{5} \qquad \text{Estimate}$$

The exact answer is shown below.

Reciprocals

$$4\frac{3}{8} \div 5 = \frac{35}{8} \div \frac{5}{1} = \frac{\overset{7}{\cancel{35}}}{8} \cdot \frac{1}{\cancel{5}} = \frac{7}{8}$$

Write 5 as $\frac{5}{1}$.

The estimate was $\frac{4}{5}$, so the exact answer of $\frac{7}{8}$ is reasonable.

Work Problem 3 at the Side.

OBJECTIVE 3 Solve application problems with mixed numbers. The next two examples show how to solve application problems involving mixed numbers.

EXAMPLE 3 **Applying Multiplication Skills**

The local Habitat for Humanity chapter is looking for 11 contractors who will each donate $3\frac{1}{4}$ days of labor to a community building project. How many days of labor will be donated in all?

Step 1 **Read** the problem. The problem asks for the total days of labor donated by the 11 contractors.

Step 2 **Work out a plan.** Multiply the number of contractors (11) and the amount of labor that each donates ($3\frac{1}{4}$ days).

Continued on Next Page

Step 3   **Estimate** a reasonable answer. Round $3\frac{1}{4}$ days to 3 days. Multiply 3 days by 11 contractors (3 • 11) to get an estimate of 33 days.

Step 4   **Solve** the problem. Find the exact answer.

$$11 \cdot 3\frac{1}{4} = 11 \cdot \frac{13}{4}$$

$$= \frac{11}{1} \cdot \frac{13}{4} = \frac{143}{4} = 35\frac{3}{4}$$

Step 5   **State the answer.** The community building project will receive $35\frac{3}{4}$ days of donated labor.

Step 6   **Check.** The answer, $35\frac{3}{4}$ days, is close to our estimate of 33 days.

> **Work Problem 4 at the Side.** ▶▶▶

EXAMPLE 4   **Applying Division Skills**

A dome tent for backpacking requires $7\frac{1}{4}$ yards of nylon material. How many tents can be made from $65\frac{1}{4}$ yards of material?

Step 1   **Read** the problem. The problem asks how many tents can be made from $65\frac{1}{4}$ yards of material.

Step 2   **Work out a plan.** Divide the number of yards of cloth ($65\frac{1}{4}$ yd) by the number of yards needed for one tent ($7\frac{1}{4}$ yd).

Step 3   **Estimate** a reasonable answer.

$$65\frac{1}{4} \div 7\frac{1}{4}$$

↓   Rounded   ↓

$$65 \div 7 \approx 9 \text{ tents} \qquad \text{Estimate}$$

Step 4   **Solve** the problem.

$$65\frac{1}{4} \div 7\frac{1}{4} = \frac{261}{4} \div \frac{29}{4}$$

$$= \frac{\overset{9}{\cancel{261}}}{\underset{1}{\cancel{4}}} \cdot \frac{\overset{1}{\cancel{4}}}{\underset{1}{\cancel{29}}} = \frac{9}{1} = 9 \qquad \text{Matches estimate}$$

Step 5   **State the answer.** 9 tents can be made from $65\frac{1}{4}$ yards of cloth.

Step 6   **Check.** The exact answer, 9, matches our estimate.

> **Work Problem 5 at the Side.** ▶▶▶

NOTE
When rounding mixed numbers to estimate the answer to a problem, the estimated answer usually varies somewhat from the exact answer. However, the importance of the estimated answer is that it will show you whether your exact answer is reasonable or not.

---

4 ≈ Use the six problem-solving steps. Simplify all answers.

(a) If one automobile requires $2\frac{5}{8}$ quarts of paint, find the number of quarts needed to paint 15 cars.

(b) Clare earns $\$9\frac{1}{4}$ per hour. How much would she earn in $6\frac{1}{2}$ hours? Write the answer as a mixed number.

5 ≈ Use the six problem-solving steps. Simplify all answers.

(a) The manufacture of one outboard engine propeller requires $4\frac{3}{4}$ pounds of brass. How many propellers can be manufactured from 57 pounds of brass?

(b) Campus student help is paid $\$8\frac{3}{4}$ per hour. Find the number of hours of student help that can be paid for with $315.

ANSWERS

4. (a) *Estimate:* 3 • 15 = 45; *Exact:* $39\frac{3}{8}$ quarts

   (b) *Estimate:* 9 • 7 = 63; *Exact:* $\$60\frac{1}{8}$

5. (a) *Estimate:* 57 ÷ 5 ≈ 11;
      Exact: 12 propellers
   (b) *Estimate:* 300 ÷ 9 ≈ 33; *Exact:* 36 hours

Real-Data Applications

Recipes

The side of the corn starch box shown has useful recipes for Fun-Time Dough and for Great Gravy. Suppose you have made the recipe before and know from experience that 1 pound makes enough Fun-Time Dough for three children.

ARGO

CORN STARCH

FAVORITE RECIPES

Fun-Time Dough

$1\frac{1}{2}$ cups Argo Corn Starch
$\frac{1}{2}$ cup flour
2 cups water
2 tsp cream of tartar
1 cup salt
1 T. vegetable oil

Mix all ingredients together in saucepan. Cook over medium heat, stirring constantly, until mixture gathers on the stirring spoon and forms dough. This will take about 6 minutes. Dump onto waxed paper until cool enough to handle and knead to form a pliable mass. Store in covered container or plastic bag. Food coloring may be added to make different colors.
Makes about 2 lbs. of Fun-Time Dough.

Great Gravy

3 T. bacon fat or meat drippings
2 T. Argo Corn Starch
$1\frac{1}{2}$ cups water
$\frac{3}{4}$ tsp. salt
$\frac{1}{8}$ tsp. pepper

Blend fat and Argo Corn Starch over low heat until it is a rich brown color, stirring constantly. Gradually add water, salt, and pepper. Heat to boiling over direct heat and then boil gently 2 minutes, stirring constantly.
Makes $1\frac{1}{2}$ cups.

— **Satisfaction Guaranteed** —

1. If you make the recipe as written, you will have enough Fun-Time Dough for how many children?

2. Suppose you work in a day care center. How many pounds of Fun-Time Dough will you need for nine children?

3. If you double the recipe, you would have 4 pounds of dough. By what fraction should you multiply each ingredient amount to make 3 pounds of dough?

4. Fill in the blanks with the ingredient amounts needed to make 3 pounds of Fun-Time Dough. Show your work.

   Corn starch: _____   Flour: _____

   Water: _____   Cream of tartar: _____

   Salt: _____   Vegetable oil: _____

You decide to make Great Gravy for Thanksgiving dinner.

5. How much gravy does the recipe make?

6. Suppose you decide to double the recipe. By what factor will you change the ingredient amounts?

7. You did not make enough! Suppose you decide to halve the recipe to make more gravy. Now, by what factor will you change the ingredient amounts?

8. If you need 4 cups of gravy, by what factor must you multiply each ingredient amount? Explain how you determined your answer.

2.8 Exercises

 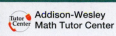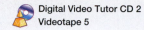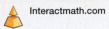
≈ *First estimate the answer. Then multiply to find the exact answer. Simplify all answers.*
See Example 1.

1. *Exact:*

$4\dfrac{1}{2} \cdot 1\dfrac{3}{4}$

Estimate:

____ • ____ = ____

2. *Exact:*

$2\dfrac{1}{2} \cdot 2\dfrac{1}{4}$

Estimate:

____ • ____ = ____

3. *Exact:*

$1\dfrac{2}{3} \cdot 2\dfrac{7}{10}$

Estimate:

____ • ____ = ____

4. *Exact:*

$4\dfrac{1}{2} \cdot 2\dfrac{1}{4}$

Estimate:

____ • ____ = ____

5. *Exact:*

$3\dfrac{1}{9} \cdot 1\dfrac{2}{7}$

Estimate:

____ • ____ = ____

6. *Exact:*

$6\dfrac{1}{4} \cdot 3\dfrac{1}{5}$

Estimate:

____ • ____ = ____

7. *Exact:*

$8 \cdot 6\dfrac{1}{4}$

Estimate:

____ • ____ = ____

8. *Exact:*

$6 \cdot 2\dfrac{1}{3}$

Estimate:

____ • ____ = ____

9. *Exact:*

$4\dfrac{1}{2} \cdot 2\dfrac{1}{5} \cdot 5$

Estimate:

____ • ____ • ____ = ____

10. *Exact:*

$5\dfrac{1}{2} \cdot 1\dfrac{1}{3} \cdot 2\dfrac{1}{4}$

Estimate:

____ • ____ • ____ = ____

11. *Exact:*

$3 \cdot 1\dfrac{1}{2} \cdot 2\dfrac{2}{3}$

Estimate:

____ • ____ • ____ = ____

12. *Exact:*

$\dfrac{2}{3} \cdot 3\dfrac{2}{3} \cdot \dfrac{6}{11}$

Estimate:

____ • ____ • ____ = ____

≈ *First estimate the answer. Then divide to find the exact answer. Simplify all answers.*
See Example 2.

13. *Exact:*

$1\dfrac{1}{4} \div 3\dfrac{3}{4}$

Estimate:

____ ÷ ____ = ____

14. *Exact:*

$1\dfrac{1}{8} \div 2\dfrac{1}{4}$

Estimate:

____ ÷ ____ = ____

15. *Exact:*

$2\dfrac{1}{2} \div 3$

Estimate:

____ ÷ ____ = ____

16. *Exact:*

$2\dfrac{3}{4} \div 2$

Estimate:

____ ÷ ____ = ____

17. *Exact:*

$9 \div 2\dfrac{1}{2}$

Estimate:

____ ÷ ____ = ____

18. *Exact:*

$5 \div 1\dfrac{7}{8}$

Estimate:

____ ÷ ____ = ____

19. *Exact:*

$$\frac{5}{8} \div 1\frac{1}{2}$$

Estimate:

____ ÷ ____ = ____

20. *Exact:*

$$\frac{3}{4} \div 2\frac{1}{2}$$

Estimate:

____ ÷ ____ = ____

21. *Exact:*

$$1\frac{7}{8} \div 6\frac{1}{4}$$

Estimate:

____ ÷ ____ = ____

22. *Exact:*

$$8\frac{2}{5} \div 3\frac{1}{2}$$

Estimate:

____ ÷ ____ = ____

23. *Exact:*

$$5\frac{2}{3} \div 6$$

Estimate:

____ ÷ ____ = ____

24. *Exact:*

$$5\frac{3}{4} \div 2$$

Estimate:

____ ÷ ____ = ____

≈ *For Exercises 25–42, first estimate the answer. Then solve each application problem by using the six problem-solving steps. Simplify all answers. See Examples 3 and 4. Use the recipe for Carrot Cake Cupcakes to work Exercises 25–28.*

CARROT CAKE CUPCAKES

12 paper bake cups
1¾ cups flour
1 cup packed brown sugar
1 tsp. baking powder
1 tsp. baking soda
1 tsp. ground cinnamon
½ tsp. salt
1 cup shredded carrots
¾ cup applesauce

⅓ cup vegetable oil
1 large egg
½ tsp. vanilla extract
1 container (16 oz.) ready-to-
   spread cream cheese frosting
Shredded coconut,
   tinted green
3 oz. Jelly Belly jelly beans,
   Orange Sherbet flavor

Preheat oven to 350° F. Place 12 bake cups in a muffin pan; set aside. In a large bowl, using a wire whisk, stir together flour, brown sugar, baking powder, baking soda, cinnamon, and salt. In a medium bowl, combine carrots, applesauce, oil, egg, and vanilla until blended. Add carrot mixture to flour mixture, stir well. Spoon batter into bake cups, filling ⅔ full. Bake until toothpick inserted in center comes out clean,

20–25 minutes. Cool cupcakes in pan 10 minutes; remove to wire rack and cool completely. Frost with cream cheese frosting. To make carrot design on cupcakes, place gourmet Jelly Belly jelly beans in a carrot shape on each cupcake, top with green coconut for carrot top. Makes 12 cupcakes.

25. If 30 cupcakes are baked (2½ times the recipe), find the amount of each ingredient.

   (a) Applesauce

     Estimate:

     Exact:

   (b) Salt

     Estimate:

     Exact:

   (c) Flour

     Estimate:

     Exact:

26. If 18 cupcakes are baked (1½ times the recipe), find the amount of each ingredient.

   (a) Flour

     Estimate:

     Exact:

   (b) Applesauce

     Estimate:

     Exact:

   (c) Vegetable oil

     Estimate:

     Exact:

27. How much of each ingredient is needed if you bake one-half of the recipe?

  (a) Vanilla extract

  Estimate:

  Exact:

  (b) Applesauce

  Estimate:

  Exact:

  (c) Flour

  Estimate:

  Exact:

28. How much of each ingredient is needed if you bake one-third of the recipe?

  (a) Flour

  Estimate:

  Exact:

  (b) Salt

  Estimate:

  Exact:

  (c) Applesauce

  Estimate:

  Exact:

29. Each home in a housing development needs $109\frac{1}{2}$ yards of prefinished baseboard. How many homes can be fitted if there are 1314 yards of baseboard available?

  Estimate:

  Exact:

30. How many $2\frac{1}{4}$-ounce vials of nose drops can be filled from a 198-ounce container?

  Estimate:

  Exact:

31. A manufacturer of floor jacks is ordering steel tubing to make the handles for the jack shown below. How much steel tubing is needed to make 135 of these jacks? The symbol for inch is "**″**" (*Source:* Harbor Freight Tools.)

  Estimate:

  Exact:

32. A wheelbarrow manufacturer uses handles made of hardwood. Find the amount of wood that is necessary to make 182 handles. The longest dimension shown in the advertisement below is the handle length. (*Source:* Harbor Freight Tools.)

  Estimate:

  Exact:

2-TON COMPACT FLOOR JACK
LOT NO. 36119
4000 LB. CAPACITY

- $19^1/_2$" handle
- Lifts 5" to $15^1/_4$"
- 21" x $9^1/_2$" x 6"
- Compact size & lightweight for portability—perfect for the trunk

6.0 CUBIC FT. WHEEL BARROW
LOT NO. 46852

- Steel construction with hardwood handles
- 14" tubeless pneumatic tire
- Fully rolled edge for added tray strength
- Overall dimensions: $61^1/_2$" L x 27" W x 24.9" H

33. Write the three steps for multiplying mixed numbers. Use your own words.

34. Refer to Exercise 33. In your own words, write the additional step that must be added to the rule for multiplying mixed numbers to make it the rule for dividing mixed numbers.

35. A tire manufacturer uses $20\frac{3}{4}$ pounds of rubber to make a tire. Find the number of tires that can be manufactured with 51,460 pounds of rubber.

Estimate:

Exact:

36. To remodel the apartments in her complex, Joleen needs $62\frac{1}{2}$ square yards of carpet for each apartment unit. How many apartment units can she remodel with 6750 square yards of carpet?

Estimate:

Exact:

37. A manufacturer of bird feeders cuts spacers from a tube that is $9\frac{3}{4}$ inches long. How many spacers can be cut from the tube if each spacer must be $\frac{3}{4}$-inch thick?

Estimate:

Exact:

38. A building contractor must move 12 tons of sand. If his truck can carry $\frac{3}{4}$ ton of sand, how many trips must he make to move the sand?

Estimate:

Exact:

39. A grape grower uses $6\frac{3}{4}$ gallons of a chemical for each acre of grapes. If she has $25\frac{1}{2}$ acres of grapes, how many gallons of the chemical are needed?

Estimate:

Exact:

40. A flooring contractor needs $24\frac{2}{7}$ boxes of tile to cover a kitchen floor. If there are 24 homes in a subdivision, how many boxes of tile are needed to cover all of the floors?

Estimate:

Exact:

Chapter 2
SUMMARY

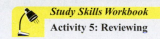

KEY TERMS

2.1	**numerator**	The number above the fraction bar in a fraction is called the numerator. It shows how many of the equivalent parts are being considered.
	denominator	The number below the fraction bar in a fraction is called the denominator. It shows the number of equal parts in a whole.
	proper fraction	In a proper fraction, the numerator is smaller than the denominator. The fraction is less than 1.
	improper fraction	In an improper fraction, the numerator is greater than or equal to the denominator. The fraction is equal to or greater than 1.
2.2	**mixed number**	A mixed number includes a fraction and a whole number written together.
2.3	**factors**	Numbers that are multiplied to give a product are factors.
	composite number	A composite number has at least one factor other than itself and 1.
	prime number	A prime number is a whole number other than 0 and 1 that has exactly two factors, itself and 1.
	factorizations	The numbers that can be multiplied to give a specific number (product) are factorizations of that number.
	prime factorization	In a prime factorization, every factor is a prime number.
2.4	**equivalent fractions**	Two fractions are equivalent when they represent the same portion of a whole.
	common factor	A common factor is a number that can be divided into two or more whole numbers.
	lowest terms	A fraction is written in lowest terms when its numerator and denominator have no common factor other than 1.
2.5	**multiplication shortcut**	When multiplying or dividing fractions, the process of dividing a numerator and denominator by a common factor can be used as a shortcut.
2.7	**reciprocal**	Two numbers are reciprocals of each other if their product is 1. To find the reciprocal of a fraction, interchange the numerator and the denominator.

NEW FORMULA

Area of a rectangle: Area = length • width

See how well you have learned the vocabulary in this chapter. Answers follow the Quick Review.

1. A **numerator** is
   A. a number greater than 5
   B. the number above the fraction bar in a fraction
   C. any number
   D. the number below the fraction bar in a fraction.

2. A **proper fraction**
   A. has a value less than 1
   B. has a whole number and a fraction
   C. has a value greater than 1
   D. is equal to 1.

3. A **mixed number** is
   A. equal to 1
   B. less than 1
   C. a whole number and a fraction written together
   D. a number multiplied by another number.

4. A **factor** is
   A. one of two or more numbers that are added to get another number
   B. the answer in division
   C. one of two or more numbers that are multiplied to get another number
   D. the answer in multiplication.

5. A whole number greater than 1 is **prime** if
   A. it cannot be factored
   B. it has just one factor
   C. it has only itself and 1 as factors
   D. it has at least two different factors.

6. A **common factor** can
   A. only be divided by itself and 1
   B. be divided into two or more whole numbers
   C. never be divided by 2
   D. only be divided by the numbers 5 and 10.

7. A fraction is in **lowest terms** when
   A. it cannot be divided
   B. it is a common fraction
   C. its numerator and denominator have no common factor other than 1
   D. it has a value less than 1.

8. To find the **reciprocal** of a fraction.
   A. multiply it by itself
   B. interchange the numerator and the denominator
   C. change it to an improper fraction
   D. change it to lowest terms.

QUICK REVIEW

Concepts	*Examples*
2.1 *Types of Fractions*	$\dfrac{2}{3}\quad \dfrac{3}{4}\quad \dfrac{15}{16}\quad \dfrac{1}{8}$ Proper fractions
Proper Numerator smaller than denominator; a value less than 1	
Improper Numerator equal to or greater than denominator; a value equal to or greater than 1	$\dfrac{17}{8}\quad \dfrac{19}{12}\quad \dfrac{11}{2}\quad \dfrac{5}{3}\quad \dfrac{7}{7}$ Improper fractions
2.2 *Converting Fractions*	$7\dfrac{2}{3} = \dfrac{23}{3} \longleftarrow 3 \times 7 + 2$
Mixed to Improper Multiply denominator by whole number, add numerator, and place over denominator.	Same denominator
Improper to Mixed Divide numerator by denominator and place remainder over denominator.	$\dfrac{17}{5} = 3\dfrac{2}{5}$ $5\overline{)17}$ $\leftarrow$ Divide numerator by denominator. $\dfrac{15}{2}$ Same denominator

Concepts	**Examples**
2.3 Prime Numbers Determine whether a whole number is evenly divisible only by itself and 1. (By definition, 0 and 1 are not prime.)	The prime numbers less than 100 are 2, 3, 5, 7, 11, 13, 17, 19, 23, 29, 31, 37, 41, 43, 47, 53, 59, 61, 67, 71, 73, 79, 83, 89, and 97.

2.3 Finding the Prime Factorization of a Number
Divide each factor by a prime number using a diagram that forms the shape of tree branches.

Find the prime factorization of 30. Use a factor tree.

Prime factors are circled.
$30 = 2 \cdot 3 \cdot 5$

2.4 Writing Fractions in Lowest Terms
Divide the numerator and denominator by the greatest common factor.

Write $\frac{30}{42}$ in lowest terms.

$$\frac{30}{42} = \frac{30 \div 6}{42 \div 6} = \frac{5}{7}$$

2.5 Multiplying Fractions
1. Multiply numerators by numerators and denominators by denominators.
2. Write answers in lowest terms if the multiplication shortcut was not used.

Multiply.

$$\frac{6}{11} \cdot \frac{7}{8} = \frac{\overset{3}{\cancel{6}}}{11} \cdot \frac{7}{\underset{4}{\cancel{8}}} = \frac{3 \cdot 7}{11 \cdot 4} = \frac{21}{44}$$

2.7 Finding the Reciprocal
To find the reciprocal of a fraction, interchange the numerator and denominator.

Find the reciprocal of each fraction.

$\frac{3}{4}$ The reciprocal of $\frac{3}{4}$ is $\frac{4}{3}$.

$\frac{8}{5}$ The reciprocal of $\frac{8}{5}$ is $\frac{5}{8}$.

9 The reciprocal of 9 is $\frac{1}{9}$.

2.7 Dividing Fractions
Use the reciprocal of the second fraction (divisor) and multiply as fractions.

Divide.

$$\frac{25}{36} \div \frac{\mathbf{15}}{\mathbf{18}} = \frac{\overset{5}{\cancel{25}}}{\underset{2}{\cancel{36}}} \cdot \frac{\overset{1}{\cancel{18}}}{\underset{3}{\cancel{15}}} = \frac{5 \cdot 1}{2 \cdot 3} = \frac{5}{6}$$

Reciprocals

Concepts	Examples

2.8 Multiplying Mixed Numbers

First estimate the answer. Then follow these steps.

Step 1 *Change* each mixed number to an improper fraction.

Step 2 *Multiply.*

Step 3 Simplify the answer, which means to write it in *lowest terms,* and change it to a mixed number or whole number where possible.

First estimate the answer. Then multiply to get the exact answer.

Estimate: *Exact:*

$$1\frac{3}{5} \quad \bullet \quad 3\frac{1}{3}$$

↓ Rounded ↓

$$2 \quad \bullet \quad 3 = 6$$

$$1\frac{3}{5} \cdot 3\frac{1}{3} = \frac{8}{\cancel{5}_{1}} \cdot \frac{\cancel{10}^{2}}{3}$$

$$= \frac{8 \cdot 2}{1 \cdot 3}$$

$$= \frac{16}{3} = 5\frac{1}{3}$$

Close to estimate

2.8 Dividing Mixed Numbers

First estimate the answer. Then follow these steps.

Step 1 *Change* each mixed number to an improper fraction.

Step 2 Use the *reciprocal* of the second fraction (divisor).

Step 3 *Multiply.*

Step 4 Simplify the answer, which means to write it in *lowest terms,* and change it to a mixed number or whole number where possible.

First estimate the answer. Then divide to get the exact answer.

Estimate: *Exact:*

$$3\frac{5}{9} \quad \div \quad 2\frac{2}{5}$$

↓ Rounded ↓

$$4 \quad \div \quad 2 = 2$$

$$3\frac{5}{9} \div 2\frac{2}{5} = \frac{32}{9} \div \frac{12}{5}$$

$$= \frac{\cancel{32}^{8}}{9} \cdot \frac{5}{\cancel{12}_{3}} = \frac{40}{27}$$

$$= 1\frac{13}{27}$$

Close to estimate

ANSWERS TO TEST YOUR WORD POWER

1. B. *Example:* In $\frac{3}{8}$, the numerator is 3.

2. A. *Example:* $\frac{1}{2}, \frac{3}{4}$, and $\frac{7}{8}$ are all proper fractions with a value less than 1.

3. C. *Example:* $2\frac{3}{8}$ and $5\frac{3}{4}$ are mixed numbers.

4. C. *Example:* Since $3 \cdot 5 = 15$, the numbers 3 and 5 are factors of 15.

5. C. *Example:* 3, 5, and 11 are prime numbers; 4, 8, and 12 are composite numbers.

6. B. *Example:* 3 is a common factor of both 6 and 9 because it can be evenly divided into each of them.

7. C. *Example:* $\frac{3}{8}, \frac{4}{5}$, and $\frac{5}{6}$ are in lowest terms but $\frac{6}{8}, \frac{3}{6}$, and $\frac{2}{4}$ are not.

8. B. *Example:* The reciprocal of $\frac{3}{8}$ is $\frac{8}{3}$, the reciprocal of $\frac{25}{4}$ is $\frac{4}{25}$, and the reciprocal of 6 or $\frac{6}{1}$ is $\frac{1}{6}$.

Chapter 2
REVIEW EXERCISES

[2.1] *Write the fraction that represents each shaded portion.*

1.

2.

3.

List the proper and improper fractions in each group.

	Proper	**Improper**
4. $\dfrac{1}{8}$ $\dfrac{4}{3}$ $\dfrac{5}{5}$ $\dfrac{3}{4}$ $\dfrac{2}{3}$	_____	_____
5. $\dfrac{6}{5}$ $\dfrac{15}{16}$ $\dfrac{16}{13}$ $\dfrac{1}{8}$ $\dfrac{5}{3}$	_____	_____

[2.2] *Write each mixed number as an improper fraction. Write each improper fraction as a mixed number.*

6. $4\dfrac{3}{4}$ **7.** $9\dfrac{5}{6}$ **8.** $\dfrac{27}{8}$ **9.** $\dfrac{63}{5}$

[2.3] *Find all factors of each number.*

10. 6 **11.** 24 **12.** 55 **13.** 90

Write the prime factorization of each number by using exponents.

14. 27 **15.** 150 **16.** 420

Simplify each expression.

17. 5^2 **18.** $6^2 \cdot 2^3$ **19.** $8^2 \cdot 3^3$ **20.** $4^3 \cdot 2^5$

[2.4] *Write each fraction in lowest terms.*

21. $\dfrac{16}{24}$ **22.** $\dfrac{36}{45}$ **23.** $\dfrac{75}{80}$

Write the numerator and denominator of each fraction as a product of prime factors. Then, write the fraction in lowest terms.

24. $\dfrac{25}{60}$ **25.** $\dfrac{384}{96}$

Decide whether each pair of fractions is equivalent or not equivalent, using the method of prime factors.

26. $\dfrac{3}{4}$ and $\dfrac{48}{64}$

27. $\dfrac{3}{4}$ and $\dfrac{42}{58}$

[2.5–2.8] *Multiply. Write answers in lowest terms, and as mixed numbers or whole numbers where possible.*

28. $\dfrac{4}{5} \cdot \dfrac{3}{4}$

29. $\dfrac{3}{10} \cdot \dfrac{5}{8}$

30. $\dfrac{70}{175} \cdot \dfrac{5}{14}$

31. $\dfrac{44}{63} \cdot \dfrac{3}{11}$

32. $\dfrac{5}{16} \cdot 48$

33. $\dfrac{5}{8} \cdot 1000$

Divide. Write answers in lowest terms, and as mixed numbers or whole numbers where possible.

34. $\dfrac{2}{3} \div \dfrac{1}{2}$

35. $\dfrac{5}{6} \div \dfrac{1}{2}$

36. $\dfrac{\dfrac{15}{18}}{\dfrac{10}{30}}$

37. $\dfrac{\dfrac{3}{4}}{\dfrac{3}{8}}$

38. $7 \div \dfrac{7}{8}$

39. $18 \div \dfrac{3}{4}$

40. $\dfrac{5}{8} \div 3$

41. $\dfrac{2}{3} \div 5$

42. $\dfrac{\dfrac{12}{13}}{3}$

Find the area of each rectangle.

43.

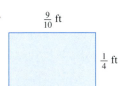

$\frac{9}{10}$ ft

$\frac{1}{4}$ ft

44.

$\frac{2}{3}$ in.

$\frac{7}{8}$ in.

45. Find the area of a rectangular piano top having a length of 5 ft and a width of $\frac{5}{6}$ ft.

46. Find the area of a rectangle having a length of 48 yd and a width of $5\frac{3}{4}$ yd.

First estimate the answer. Then multiply or divide to find the exact answer. Simplify all answers.

47. *Exact:*

$5\dfrac{1}{2} \cdot 1\dfrac{1}{4}$

Estimate:

_____ • _____ = _____

48. *Exact:*

$2\dfrac{1}{4} \cdot 7\dfrac{1}{8} \cdot 1\dfrac{1}{3}$

Estimate:

_____ • _____ • _____ = _____

49. *Exact:*

$$15\frac{1}{2} \div 3$$

Estimate:

_____ ÷ _____ = _____

50. *Exact:*

$$4\frac{3}{4} \div 6\frac{1}{3}$$

Estimate:

_____ ÷ _____ = _____

Solve each application problem by using the six problem-solving steps.

51. Blue Diamond Almonds has 320 tons of almonds. How many $\frac{5}{8}$ ton bins will be needed to store the almonds?

52. An estate is divided so that each of 5 children receives equal shares of $\frac{2}{3}$ of the estate. What fraction of the total estate will each receive?

≈ 53. How many window-blind pull cords can be made from $157\frac{1}{2}$ yards of cord if $4\frac{3}{8}$ yards of cord are needed for each blind? First estimate, and then find the exact answer.

Estimate:

Exact:

≈ 54. Working as a bookkeeper, Neta Fitzgerald is paid $8\frac{1}{2}$ per hour. Find her earnings for a week in which she works 38 hours. First estimate, and then find the exact answer.

Estimate:

Exact:

55. Ebony Wilson purchased 100 pounds of rice at the food co-op. After selling $\frac{1}{4}$ of this to her neighbor, she gives $\frac{2}{3}$ of the remaining rice to her parents. How many pounds of rice does she have left?

56. Tiffany Crowder received her Social Security check for $1275. After paying $\frac{1}{3}$ of this amount for rent, she paid $\frac{2}{5}$ of the remaining amount for food, utilities, and transportation. How much money does she have left?

57. The Citrus Heights Park District will divide $\frac{3}{4}$ of its budget among 8 local parks. What fraction of the total budget will each park receive?

58. In a morning of deep-sea fishing, 5 fishermen catch $\frac{4}{5}$ ton of salmon. If they divide the fish evenly, how much will each receive?

MIXED REVIEW EXERCISES

Multiply or divide as indicated. Simplify all answers.

59. $\dfrac{1}{2} \cdot \dfrac{3}{4}$

60. $\dfrac{2}{3} \cdot \dfrac{3}{5}$

61. $12\dfrac{1}{2} \cdot 2\dfrac{1}{2}$

62. $12\dfrac{1}{2} \cdot 2\dfrac{1}{4}$

63. $\dfrac{\frac{4}{5}}{8}$

64. $\dfrac{\frac{5}{8}}{4}$

65. $\dfrac{15}{31} \cdot 62$

66. $3\dfrac{1}{4} \div 1\dfrac{1}{4}$

Write each mixed number as an improper fraction. Write each improper fraction as a mixed number.

67. $\dfrac{8}{5}$

68. $\dfrac{153}{4}$

69. $5\dfrac{2}{3}$

70. $38\dfrac{3}{8}$

Write the numerator and denominator of each fraction as a product of prime factors; then write the fraction in lowest terms.

71. $\dfrac{8}{12}$

72. $\dfrac{108}{210}$

Write each fraction in lowest terms.

73. $\dfrac{75}{90}$

74. $\dfrac{48}{72}$

75. $\dfrac{44}{110}$

76. $\dfrac{87}{261}$

Solve each application problem.

77. The directions on a can of fabric glue say to apply ≈ $3\dfrac{1}{2}$ ounces of glue to each square yard of fabric. How many ounces are needed for $43\dfrac{5}{9}$ square yards? First estimate, and then find the exact answer.

Estimate:

Exact:

≈ **78.** Jack Anderson Trucking purchased some diesel fuel additive. The instructions say to use $7\dfrac{1}{4}$ quarts of additive for each tank of fuel. How many quarts are needed for $25\dfrac{1}{2}$ tanks? First estimate, and then find the exact answer.

Estimate:

Exact:

79. A postage stamp is $1\dfrac{3}{4}$ in. by $\dfrac{7}{8}$ in. Find its area.

80. A section of marble countertop is $\dfrac{1}{2}$ yard by $\dfrac{5}{8}$ yard. What is its area?

Chapter **2**

T E S T

Study Skills Workbook
Activity 8: Test Preparation

Write a fraction to represent each shaded portion.

1.

2.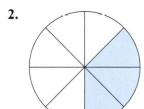

1. _____

2. _____

3. Identify all the proper fractions in this list:

$$\frac{2}{3} \quad \frac{4}{4} \quad \frac{6}{7} \quad \frac{5}{2} \quad \frac{1}{4} \quad \frac{5}{8} \quad \frac{30}{18}$$

3. _____

4. Write $3\frac{3}{8}$ as an improper fraction.

4. _____

5. Write $\frac{123}{4}$ as a mixed number.

5. _____

6. Find all factors of 18.

6. _____

Find the prime factorization of each number. Write the answers using exponents.

7. 45

7. _____

8. 144

8. _____

9. 500

9. _____

Write each fraction in lowest terms.

10. $\frac{36}{48}$ **11.** $\frac{60}{72}$

10. _____

12. The method of prime factors is used to write a fraction in lowest terms. Briefly explain how this is done. Use the fraction $\frac{56}{84}$ to show how this works.

11. _____

12. _____

13. _____

13. Explain how to multiply fractions. What additional step must be taken when dividing fractions?

14. _____

Multiply or divide. Write answers in lowest terms, and as mixed numbers or whole numbers where possible.

14. $\dfrac{3}{4} \cdot \dfrac{4}{9}$

15. _____

15. $54 \cdot \dfrac{2}{3}$

16. _____

16. Find the area of a rectangular x-ray machine measuring $\dfrac{7}{8}$ yard by $\dfrac{3}{4}$ yard.

17. _____

17. There are 3820 real estate agents in the county. If $\frac{3}{5}$ of these agents are women and the rest are men, find the number of agents who are men.

18. _____

18. $\dfrac{3}{4} \div \dfrac{5}{6}$

19. $\dfrac{\dfrac{7}{4}}{9}$

19. _____

20. _____

20. The Lincoln Fun Run committee has acquired 120 quarts of Sports Drink to sell in $\dfrac{3}{5}$-quart sports bottles at the race. Find the number of sports bottles that can be filled.

21. *Estimate:* _____

   Exact: _____

≈ *First estimate the answer. Then either multiply or divide to find the exact answer. Simplify all answers.*

22. *Estimate:* _____

21. $4\dfrac{1}{8} \cdot 3\dfrac{1}{2}$

22. $1\dfrac{5}{6} \cdot 4\dfrac{1}{3}$

   Exact: _____

23. *Estimate:* _____

   Exact: _____

23. $9\dfrac{3}{5} \div 2\dfrac{1}{4}$

24. $\dfrac{8\dfrac{1}{2}}{1\dfrac{2}{3}}$

24. *Estimate:* _____

   Exact: _____

25. *Estimate:* _____

25. A new vaccine is synthesized at the rate of $2\frac{1}{2}$ grams per day. How many grams can be synthesized in $12\frac{1}{4}$ days?

   Exact: _____

Cumulative Review Exercises

CHAPTERS 1–2

Name the digit that has the given place value in each number.

1. 783
   hundreds
   tens

2. 8,621,785
   millions
   ten thousands

Add, subtract, multiply, or divide as indicated.

3.
$$\begin{array}{r} 71 \\ 23 \\ 47 \\ +\ 36 \end{array}$$

4.
$$\begin{array}{r} 82{,}121 \\ 5\ 468 \\ 316 \\ +\ 61{,}294 \end{array}$$

5.
$$\begin{array}{r} 6537 \\ -\ 2085 \end{array}$$

6.
$$\begin{array}{r} 4{,}819{,}604 \\ -\ 1{,}597{,}783 \end{array}$$

7.
$$\begin{array}{r} 83 \\ \times\ 9 \end{array}$$

8. $9 \cdot 4 \cdot 2$

9.
$$\begin{array}{r} 3784 \\ \times\ \ \ 573 \end{array}$$

10.
$$\begin{array}{r} 563 \\ \times\ \ \ 800 \end{array}$$

11. $\dfrac{63}{7}$

12. $18\overline{)136{,}458}$

13. $33{,}886 \div 4$

14. $492\overline{)10{,}850}$

Round each number to the nearest ten, nearest hundred, and nearest thousand.

	Ten	**Hundred**	**Thousand**
15. 6583	_____	_____	_____
16. 76,271	_____	_____	_____

Simplify each expression by using the order of operations.

17. $2^5 - 6(4)$

18. $\sqrt{36} - 2 \cdot 3 + 5$

Solve each application problem using the six problem-solving steps.

19. The manager of Starbucks purchased 8 cases of small-size coffee cups for $46 per case and 12 cases of large-size coffee cups for $52 per case. Find the total cost of the 20 cases of coffee cups.

20. Sounds louder than 80 decibels may harm a person's hearing. A loud rock concert produces 150 decibels, while a lawn mower, shop tool, truck traffic, or subway produces 90 decibels. How many more decibels does a rock concert produce than a lawn mower? (*Source:* Ear Foundation and Plantronics.)

21. A typical adult loses 100 hairs a day out of approximately 120,000 hairs. If the lost hairs were not replaced, find the number of hairs remaining after two years. (1 year = 365 days.)

22. A group of 22 health care workers will divide 4136 work hours evenly this month. Find the number of hours each will work.

23. The glass face for a picture frame measures $\frac{11}{12}$ ft by $\frac{3}{4}$ ft. Find its area.

24. The Municipal Utility District says that the cost of using a gas oven is $\frac{1}{6}$ ¢ per minute. Find the cost of roasting a turkey for 6 hours.

Write proper *or* improper *for each fraction.*

25. $\frac{2}{3}$

26. $\frac{6}{6}$

27. $\frac{9}{18}$

Write each mixed number as an improper fraction. Write each improper fraction as a whole or mixed number.

28. $3\frac{3}{8}$

29. $6\frac{2}{5}$

30. $\frac{14}{7}$

31. $\frac{103}{8}$

Find the prime factorization of each number. Write answers using exponents.

32. 72

33. 126

34. 350

Simplify each expression.

35. $4^2 \cdot 2^2$

36. $2^3 \cdot 6^2$

37. $2^3 \cdot 4^2 \cdot 5$

Write each fraction in lowest terms.

38. $\frac{42}{48}$

39. $\frac{24}{36}$

40. $\frac{30}{54}$

Multiply or divide as indicated. Simplify all answers.

41. $\frac{1}{2} \cdot \frac{3}{4}$

42. $30 \cdot \frac{2}{3} \cdot \frac{3}{5}$

43. $7\frac{1}{2} \cdot 3\frac{1}{3}$

44. $\frac{3}{5} \div \frac{5}{8}$

45. $\frac{7}{8} \div 1\frac{1}{2}$

46. $3 \div 1\frac{1}{4}$

Adding and Subtracting Fractions

3

When Bryan Berg was a small boy, his grandfather taught him how to build a simple house of cards. Berg has set seven Guinness world records for the tallest freestanding card structures—no glue, no tape, no hidden supports, just the force of gravity and a very steady hand. While a high school student, he set his first record of 14 feet, 6 inches tall. That house of cards required 4 days, 208 decks of cards, and a scaffolds to build. Today, Berg is an architect and a student at Harvard University and holds the current record of 25 feet, $3\frac{1}{2}$ inches. As an architect and when *Stacking the Deck* (the title of his book), Berg will be adding and subtracting fractions. (See Exercises 49 and 64 in **Section 3.4.**) (*Source: Reader's Digest.*)

3.1 Adding and Subtracting Like Fractions

OBJECTIVES

1. Define like and unlike fractions.
2. Add like fractions.
3. Subtract like fractions.

In **Chapter 2** we looked at the basics of fractions and then practiced with multiplication and division of fractions and mixed numbers. In this chapter we will work with addition and subtraction of fractions and mixed numbers.

OBJECTIVE 1 Define like and unlike fractions. Fractions with the same denominators are **like fractions**. Fractions with different denominators are **unlike fractions**.

1 Next to each pair of fractions write *like* or *unlike*.

(a) $\dfrac{2}{5}$ $\dfrac{3}{5}$ _____

EXAMPLE 1 Identifying Like and Unlike Fractions

(a) $\dfrac{3}{4}, \dfrac{1}{4}, \dfrac{5}{4}, \dfrac{6}{4},$ and $\dfrac{4}{4}$ are **like** fractions.

All denominators are the same.

(b) $\dfrac{7}{12}$ and $\dfrac{12}{7}$ are **unlike** fractions.

Denominators are different.

(b) $\dfrac{2}{3}$ $\dfrac{3}{4}$ _____

> **NOTE**
> Like fractions have the *same* denominator.

◀◀◀ **Work Problem 1 at the Side.**

(c) $\dfrac{7}{12}$ $\dfrac{11}{12}$ _____

OBJECTIVE 2 Add like fractions. The figures below show you how to add the fractions $\frac{2}{7}$ and $\frac{4}{7}$.

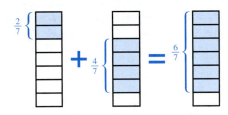

As the figures show,

(d) $\dfrac{3}{8}$ $\dfrac{3}{16}$ _____

$$\frac{2}{7} + \frac{4}{7} = \frac{6}{7}$$

Add like fractions as follows.

> **Adding Like Fractions**
>
> *Step 1* Add the numerators to find the numerator of the sum.
>
> *Step 2* Write the denominator of the like fractions as the denominator of the sum.
>
> *Step 3* Write the answer in lowest terms.

ANSWERS
1. **(a)** like **(b)** unlike **(c)** like **(d)** unlike

EXAMPLE 2 Adding Like Fractions

Add and write the answer in lowest terms.

(a) $\dfrac{1}{5} + \dfrac{2}{5}$

$$\dfrac{1}{5} + \dfrac{2}{5} = \overbrace{\dfrac{1+2}{5}}^{\text{Add numerators.}} = \dfrac{3}{5} \leftarrow \text{Same denominator}$$

(b) $\dfrac{1}{12} + \dfrac{7}{12} + \dfrac{1}{12}$

Step 1 $\overbrace{\dfrac{1+7+1}{12}}^{\text{Add numerators.}}$

Step 2 $= \dfrac{9}{12} \begin{array}{l} \leftarrow \text{Sum} \\ \leftarrow \text{Same denominator} \end{array}$

Step 3 $= \dfrac{9 \div \mathbf{3}}{12 \div \mathbf{3}} = \dfrac{3}{4}$ In lowest terms

CAUTION
Fractions may be added **only** if they have like denominators.

Work Problem 2 at the Side. ▶▶▶

OBJECTIVE 3 **Subtract like fractions.** The figures below show $\frac{7}{8}$ broken into $\frac{4}{8}$ and $\frac{3}{8}$.

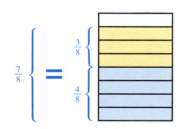

Subtracting $\frac{3}{8}$ from $\frac{7}{8}$ gives the answer $\frac{4}{8}$, or

$$\dfrac{7}{8} - \dfrac{3}{8} = \dfrac{4}{8}$$

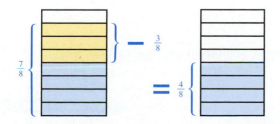

2 Add and write the answers in lowest terms.

(a) $\dfrac{3}{8} + \dfrac{1}{8}$

(b) $\begin{array}{r} \dfrac{2}{9} \\ + \dfrac{5}{9} \\ \hline \end{array}$

(c) $\dfrac{3}{16} + \dfrac{1}{16}$

(d) $\dfrac{3}{10} + \dfrac{1}{10} + \dfrac{4}{10}$

3 Subtract and simplify.

(a) $\dfrac{7}{8} - \dfrac{3}{8}$

Write $\frac{4}{8}$ in lowest terms.

$$\frac{7}{8} - \frac{3}{8} = \frac{4 \div 4}{8 \div 4} = \frac{1}{2}$$

The steps for subtracting like fractions are very similar to those for adding like fractions.

Subtracting Like Fractions

Step 1   Subtract the numerators to find the numerator of the difference.

Step 2   Write the denominator of the like fractions as the denominator of the difference.

Step 3   Write the answer in lowest terms.

(b)
$$\begin{array}{r} \dfrac{15}{16} \\[2mm] -\dfrac{3}{16} \\ \hline \end{array}$$

EXAMPLE 3   **Subtracting Like Fractions**

Subtract and simplify the answer.

(a) $\dfrac{15}{16} - \dfrac{3}{16}$ Subtract numerators.

Step 1 $\dfrac{15}{16} - \dfrac{3}{16} = \dfrac{\overbrace{15 - 3}}{16}$

Step 2 $= \dfrac{12}{16}$ ← Difference
   ← Same denominator

(c) $\dfrac{15}{3} - \dfrac{5}{3}$

Step 3 $= \dfrac{12 \div 4}{16 \div 4} = \dfrac{3}{4}$ In lowest terms

(b) $\dfrac{13}{4} - \dfrac{6}{4}$

   Subtract numerators.

$$\dfrac{13}{4} - \dfrac{6}{4} = \dfrac{\overbrace{13 - 6}}{4}$$ ← Same denominator

$$= \dfrac{7}{4}$$

(d)
$$\begin{array}{r} \dfrac{25}{32} \\[2mm] -\dfrac{6}{32} \\ \hline \end{array}$$

To simplify the answer, write $\frac{7}{4}$ as a mixed number.

$$\frac{7}{4} = 1\frac{3}{4}$$

CAUTION
Fractions may be subtracted *only* if they have like denominators.

◀◀◀ Work Problem 3 at the Side.

3.1 Exercises

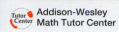

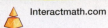

Add and simplify the answer. See Example 2.

1. $\dfrac{3}{8} + \dfrac{2}{8}$

2. $\dfrac{1}{5} + \dfrac{3}{5}$

3. $\dfrac{2}{6} + \dfrac{3}{6}$

4. $\dfrac{9}{11} + \dfrac{1}{11}$

5. $\dfrac{1}{4} + \dfrac{1}{4}$

6. $\dfrac{1}{14} + \dfrac{1}{14}$

7. $\begin{array}{r} \dfrac{9}{10} \\ + \dfrac{3}{10} \\ \hline \end{array}$

8. $\begin{array}{r} \dfrac{13}{12} \\ + \dfrac{5}{12} \\ \hline \end{array}$

9. $\begin{array}{r} \dfrac{2}{9} \\ + \dfrac{1}{9} \\ \hline \end{array}$

10. $\dfrac{7}{12} + \dfrac{3}{12}$

11. $\dfrac{6}{20} + \dfrac{4}{20} + \dfrac{3}{20}$

12. $\dfrac{1}{7} + \dfrac{2}{7} + \dfrac{3}{7}$

13. $\dfrac{4}{15} + \dfrac{2}{15} + \dfrac{5}{15}$

14. $\dfrac{5}{11} + \dfrac{1}{11} + \dfrac{4}{11}$

15. $\dfrac{3}{8} + \dfrac{7}{8} + \dfrac{2}{8}$

16. $\dfrac{4}{9} + \dfrac{1}{9} + \dfrac{7}{9}$

17. $\dfrac{2}{54} + \dfrac{8}{54} + \dfrac{12}{54}$

18. $\dfrac{7}{64} + \dfrac{15}{64} + \dfrac{20}{64}$

Subtract and simplify the answer. See Example 3.

19. $\dfrac{7}{8} - \dfrac{4}{8}$

20. $\dfrac{2}{3} - \dfrac{1}{3}$

21. $\dfrac{10}{11} - \dfrac{4}{11}$

22. $\dfrac{4}{5} - \dfrac{3}{5}$

23. $\dfrac{9}{10} - \dfrac{3}{10}$

24. $\dfrac{7}{14} - \dfrac{3}{14}$

25. $\begin{array}{r} \dfrac{31}{21} \\ - \dfrac{7}{21} \\ \hline \end{array}$

26. $\begin{array}{r} \dfrac{43}{24} \\ - \dfrac{13}{24} \\ \hline \end{array}$

27. $\begin{array}{r} \dfrac{27}{40} \\ - \dfrac{19}{40} \\ \hline \end{array}$

28. $\begin{array}{r} \dfrac{38}{55} \\ - \dfrac{16}{55} \\ \hline \end{array}$

29. $\dfrac{47}{36} - \dfrac{5}{36}$

30. $\dfrac{76}{45} - \dfrac{21}{45}$

31. $\dfrac{73}{60} - \dfrac{7}{60}$

32. $\dfrac{181}{100} - \dfrac{31}{100}$

33. In your own words, write an explanation of how to add like fractions. Consider using three steps in your explanation.

34. Describe in your own words the difference between *like* fractions and *unlike* fractions. Give three examples of each type.

Solve each application problem. Write answers in lowest terms.

35. In June, Jose and Juanita Romero had saved $\frac{2}{5}$ of the amount needed for a down payment on their first home. By November, they had saved another $\frac{2}{5}$ of the amount needed. What fraction of the amount needed have they saved?

36. After an initial payment to a lotto winner, the state lottery commission still owed the lotto winner $\frac{7}{10}$ of his total winnings. If the state pays the lotto winner another $\frac{3}{10}$ of the winnings, what fraction is still owed?

37. The Gerards owe $\frac{5}{9}$ of a loan for last year's vacation. If they pay $\frac{2}{9}$ of it this month, what fraction of the loan will they still owe?

38. Phil Fravesi, general contractor, completes $\frac{3}{12}$ of a hobby and toy train addition to his home in April. In May, he completes another $\frac{5}{12}$ of the project. What portion of the project has he completed?

39. An organic farmer purchased $\frac{9}{10}$ acre of land one year and $\frac{3}{10}$ acre the next year. She then planted carrots on $\frac{7}{10}$ acre of the land and squash on the remainder. How much land is planted in squash?

40. A forester planted $\frac{5}{12}$ acre in seedlings in the morning and $\frac{11}{12}$ acre in the afternoon. If $\frac{7}{12}$ acre of seedlings were destroyed by frost, how many acres remained?

3.2 Least Common Multiples

Only *like* fractions can be added or subtracted. Because of this, we must rewrite *unlike* fractions as *like* fractions before we can add or subtract them.

OBJECTIVE 1 Find the least common multiple.   We can rewrite unlike fractions as like fractions by finding the *least common multiple* of the denominators.

> **Least Common Multiple (LCM)**
>
> The **least common multiple (LCM)** of two whole numbers is the smallest whole number divisible by both those numbers.

EXAMPLE 1 Finding the Least Common Multiple

Find the least common multiple of 6 and 9.
     This list shows the multiples of 6.

$$6 \cdot 1 \quad 6 \cdot 2 \quad 6 \cdot 3 \quad 6 \cdot 4 \quad 6 \cdot 5 \quad 6 \cdot 6 \quad 6 \cdot 7 \quad 6 \cdot 8$$
$$6, \quad 12, \quad 18, \quad 24, \quad 30, \quad 36, \quad 42, \quad 48, \ldots$$

(The three dots at the end of the list show that the list continues in the same pattern without stopping.) The next list shows multiples of 9.

$$9 \cdot 1 \quad 9 \cdot 2 \quad 9 \cdot 3 \quad 9 \cdot 4 \quad 9 \cdot 5 \quad 9 \cdot 6 \quad 9 \cdot 7 \quad 9 \cdot 8$$
$$9, \quad 18, \quad 27, \quad 36, \quad 45, \quad 54, \quad 63, \quad 72, \ldots$$

The smallest number found in *both* lists is 18, so 18 is the **least common multiple** of 6 and 9; the number 18 is the smallest whole number divisible by both 6 and 9.

Multiples of 6:   6, 12, **18**, 24, 30, 36, 42, 48, . . .

Multiples of 9:   9, **18**, 27, 36, 45, 54, 63, 72, . . .

18 is the smallest number found in both lists. **18** is the least common multiple (LCM) of 6 and 9.

Work Problem 1 at the Side.)))

OBJECTIVE 2 Find the least common multiple using multiples of the largest number.   There are several ways to find the least common multiple. If the numbers are small, the least common multiple can often be found by inspection. Can you think of a number that can be divided evenly by both 3 and 4? What about 6 or 8, or perhaps 10 or 12? The number 12 will work; it is the least common multiple of the numbers 3 and 4. A method that works well to find the least common multiple is to write multiples of the larger number.
     In this case, 4 is larger than 3, so write the multiples of 4.

$$4, 8, 12, 16, 20, \ldots$$

Now, check each multiple of 4 to see if it is divisible by 3.

4 is *not* divisible by 3

8 is *not* divisible by 3

12 *is* divisible by 3

The first multiple of 4 that is divisible by 3 is 12, so 12 is the least common multiple of 3 and 4.

OBJECTIVES

1 Find the least common multiple.

2 Find the least common multiple using multiples of the largest number.

3 Find the least common multiple using prime factorization.

4 Find the least common multiple using an alternative method.

5 Write a fraction with an indicated denominator.

1 (a) List the multiples of 5.

5, _____, _____, _____,

_____, _____, _____,

_____, . . .

(b) List the multiples of 8.

8, _____, _____, _____,

_____, _____, _____, . . .

(c) Find the least common multiple of 5 and 8.

ANSWERS
1. (a) 10, 15, 20, 25, 30, 35, 40, . . .
   (b) 16, 24, 32, 40, 48, 56, . . .
   (c) 40

2 Use multiples of the larger number to find the least common multiple in each set of numbers.

(a) 2 and 5

(b) 3 and 9

(c) 6 and 8

(d) 4 and 7

3 Use prime factorization to find the LCM for each pair of numbers.

(a) 15 and 18

(b) 12 and 20

2. (a) 10 **(b)** 9 **(c)** 24 **(d)** 28

3. (a) 15 = 3 • 5   LCM = 2 • 3 • 3 • 5 = 90
18 = 2 • 3 • 3

15
∧
18

(b) 12 = 2 • 2 • 3 LCM = 2 • 2 • 3 • 5 = 60
20 = 2 • 2 • 5

12
∧
20

EXAMPLE 2 Finding the Least Common Multiple

Use multiples of the larger number to find the least common multiple of 6 and 9.

We start by writing the first few multiples of 9.

Multiples of 9

9, 18, 27, 36, 45, 54, . . .

Now, we check each multiple of 9 to see if it is divisible by 6. The first multiple of 9 that is divisible by 6 is 18.

9, **18**, 27, 36, 45, 54, . . .

First multiple divisible by 6 because 18 ÷ 6 = 3

The least common multiple of the numbers 6 and 9 is 18.

Work Problem 2 at the Side.

OBJECTIVE 3 Find the least common multiple using prime factorization. Example 2 shows how to find the least common multiple of two numbers by making a list of the multiples of the *larger* number. Although this method works well if both numbers are fairly small, it is usually easier to find the least common multiple for larger numbers by using *prime factorization*, as shown in the next example.

EXAMPLE 3 Applying Prime Factorization Knowledge

Use prime factorization to find the least common multiple of 9 and 12.
We start by finding the prime factorization of each number.

Factors of 9

9 = 3 • 3
12 = 2 • 2 • 3   LCM = 3 • 3 • 2 • 2 = 36

Factors of 12

Check to see that 36 is divisible by 9 (yes) and by 12 (yes). The smallest whole number divisible by both 9 and 12 is 36.

CAUTION
Notice that we did **not** have to repeat the factors that 9 and 12 have in common. In this case, the **3** in 2 • 2 • **3** = 12 was **not** used because 3 is already included in 3 • 3 = 9.

Work Problem 3 at the Side.

EXAMPLE 4 Using Prime Factorization

Find the least common multiple of 12, 18, and 20.

Continued on Next Page

Find the prime factorization of each number. Then use the prime factors to build the LCM.

$$12 = 2 \cdot 2 \cdot 3$$
$$18 = 2 \cdot 3 \cdot 3 \qquad LCM = 2 \cdot 2 \cdot 3 \cdot 3 \cdot 5 = 180$$
$$20 = 2 \cdot 2 \cdot 5$$

Check to see that 180 is divisible by 12 (yes) and by 18 (yes) and by 20 (yes). This smallest whole number divisible by 12, 18, and 20 is 180.

Note: We did *not* repeat the factors that 12, 18, and 20 have in common.

Work Problem 4 at the Side.

EXAMPLE 5 **Finding the Least Common Multiple**

Find the least common multiple for each set of numbers.

(a) 5, 6, 35
Find the prime factorization for each number.

$$5 = 5$$
$$6 = 2 \cdot 3 \qquad LCM = 2 \cdot 3 \cdot 5 \cdot 7 = 210$$
$$35 = 5 \cdot 7$$

The least common multiple of 5, 6, and 35 is 210.

(b) 10, 20, 24
Find the prime factorization for each number.

$$10 = 2 \cdot 5$$
$$20 = 2 \cdot 2 \cdot 5 \qquad LCM = 2 \cdot 2 \cdot 2 \cdot 3 \cdot 5 = 210$$
$$24 = 2 \cdot 2 \cdot 2 \cdot 3$$

The least common multiple of 10, 20, and 24 is 120.

Work Problem 5 at the Side.

OBJECTIVE 4 **Find the least common multiple using an alternative method.** Some people like the following *alternative method* for finding the least common multiple for larger numbers. Try both methods, and *use the one you prefer.* As a review, a list of the first few prime numbers follows.

$$2, 3, 5, 7, 11, 13, 17$$

---

4 Find the least common multiple of the denominators in each set of fractions.

(a) $\dfrac{3}{8}$ and $\dfrac{6}{5}$

8, 16, 24, 32, 40

5, 10, 15, 20, 25, 30, 35

40

(b) $\dfrac{5}{6}$ and $\dfrac{1}{14}$

6, 12, 18, 24, 30, 36, 42

14, 28, 42, 54

42

(c) $\dfrac{4}{9}, \dfrac{5}{18},$ and $\dfrac{7}{24}$

5 Find the least common multiple for each set of numbers.

(a) 6, 8

(b) 3, 6, 8

(c) 9, 36, 48

(d) 15, 20, 30, 40

ANSWERS

4. (a) $8 = 2 \cdot 2 \cdot 2$ $LCM = 2 \cdot 2 \cdot 2 \cdot 5 = 40$
$5 = 1 \cdot 5$

(b) $6 = 2 \cdot 3$ $LCM = 2 \cdot 3 \cdot 7 = 42$
$14 = 2 \cdot 7$

(c) $9 = 3 \cdot 3$ $LCM = 2 \cdot 2 \cdot 2 \cdot 3 \cdot 3 = 72$
$18 = 2 \cdot 3 \cdot 3$
$24 = 2 \cdot 2 \cdot 2 \cdot 3$

5. (a) 24   (b) 24   (c) 144   (d) 120

6 In the following problems, the divisions have already been worked out. Multiply the prime numbers on the left to find the least common multiple.

(a) 2 |6   1̸5̸
   3 |3   15
   5 |1̸   5
      1   1

(b) 2 |20   36
   2 |10   18
   3 |5̸   9
   3 |5̸   3
   5 |5   1̸
      1   1

EXAMPLE 6 Alternative Method for Finding the Least Common Multiple

Find the least common multiple for each set of numbers.

(a) 14 and 21

Start by trying to divide 14 and 21 by the first prime number in the list of prime numbers: 2, 3, 5, 7, 11, 13, and 17. Use the following shortcut. Divide by 2, the first prime.

$$2 \,\lfloor 14 \quad 2\!\!\!/1$$
$$\quad\quad 7 \quad 21$$

Because 21 cannot be divided evenly by 2, cross out 21 and bring it down. Divide by 3, the second prime.

$$2 \,\lfloor 14 \quad 2\!\!\!/1$$
$$3 \,\lfloor \;7 \quad 21$$
$$\quad\quad 7 \quad 7$$

Since 7 cannot be divided evenly by the third prime, 5, skip 5 and divide by the next prime, 7. Divide by 7, the fourth prime.

$$2 \,\lfloor 14 \quad 2\!\!\!/1$$
$$3 \,\lfloor \;7 \quad 21$$
$$7 \,\lfloor \;7 \quad 7$$
$$\quad\quad 1 \quad 1 \quad \text{All quotients are 1.}$$

When all quotients are 1, multiply the prime numbers on the left side.

$$\text{least common multiple} = \mathbf{2 \cdot 3 \cdot 7 = 42}$$

The least common multiple of 14 and 21 is 42.

(b) 6, 15, 18

Divide by 2.

$$2 \,\lfloor 6 \quad 1\!\!\!/5 \quad 18$$
$$\quad\quad 3 \quad 15 \quad 9 \quad \text{Cross out 15 and bring it down.}$$

Divide by 3.

$$2 \,\lfloor 6 \quad 1\!\!\!/5 \quad 18$$
$$3 \,\lfloor 3 \quad 15 \quad 9$$
$$\quad\quad 1 \quad 5 \quad 3$$

Divide by 3 again, since the remaining 3 can be divided.

$$2 \,\lfloor 6 \quad 1\!\!\!/5 \quad 18$$
$$3 \,\lfloor 3 \quad 15 \quad 9$$
$$3 \,\lfloor 1\!\!\!/ \quad 5\!\!\!/ \quad 3$$
$$\quad\quad 1 \quad 5 \quad 1$$

Finally, divide by 5.

$$2 \,\lfloor 6 \quad 1\!\!\!/5 \quad 18$$
$$3 \,\lfloor 3 \quad 15 \quad 9$$
$$3 \,\lfloor 1\!\!\!/ \quad 5\!\!\!/ \quad 3$$
$$5 \,\lfloor 1\!\!\!/ \quad 5 \quad 1\!\!\!/$$
$$\quad\quad 1 \quad 1 \quad 1 \quad \text{All quotients are 1.}$$

Multiply the prime numbers on the left side.

$$\mathbf{2 \cdot 3 \cdot 3 \cdot 5 = 90} \leftarrow \text{Least common multiple}$$

◀◀◀ Work Problem 6 at the Side.

ANSWERS
6. (a) 30 (b) 180

Work Problem 7 at the Side. 〉〉〉

OBJECTIVE 5 Write a fraction with an indicated denominator. Before we can add or subtract unlike fractions, we must find the least common multiple, which is then used as the denominator of the fractions.

EXAMPLE 7 **Writing a Fraction with an Indicated Denominator**

Write the fraction $\frac{2}{3}$ with a denominator of 15
   Find a numerator, so that these fractions are equivalent.

$$\frac{2}{3} = \frac{?}{15}$$

To find the new numerator, first divide **15** by **3**.

$$\frac{2}{3} = \frac{?}{15} \qquad 15 \div 3 = 5$$

Multiply both numerator and denominator of the fraction $\frac{2}{3}$ by 5

$$\frac{2}{3} = \frac{2 \cdot 5}{3 \cdot 5} = \frac{10}{15}$$

   This process is just the opposite of writing a fraction in lowest terms. Check the answer by writing $\frac{10}{15}$ in lowest terms; you should get $\frac{2}{3}$ again.

EXAMPLE 8 **Writing Fractions with a New Denominator**

Rewrite each fraction with the indicated denominator.

(a) $\frac{3}{8} = \frac{?}{48}$

   Divide 48 by 8, getting 6. Now multiply both numerator and denominator of $\frac{3}{8}$ by 6

$$\frac{3}{8} = \frac{3 \cdot 6}{8 \cdot 6} = \frac{18}{48} \qquad \text{Multiply numerator and denominator by 6.}$$

That is, $\frac{3}{8} = \frac{18}{48}$. As a check, write $\frac{18}{48}$ in lowest terms; you should get $\frac{3}{8}$ again.

(b) $\frac{5}{6} = \frac{?}{42}$

   Divide 42 by 6, getting 7. Next, multiply both numerator and denominator of $\frac{5}{6}$ by 7.

$$\frac{5}{6} = \frac{5 \cdot 7}{6 \cdot 7} = \frac{35}{42} \qquad \text{Multiply numerator and denominator by 7.}$$

This shows that $\frac{5}{6} = \frac{35}{42}$. As a check, write $\frac{35}{42}$ in lowest terms. Did you get $\frac{5}{6}$ again?

—— **Continued on Next Page**

7 Find the least common multiple of each set of numbers. Use whichever method you prefer.

(a) 3, 6, 10

(b) 15 and 40

(c) 9, 24

(d) 8, 21, 24

ANSWERS
7. (a) 30 **(b)** 120 **(c)** 72 **(d)** 168

8 Rewrite each fraction with the indicated denominator.

(a) $\dfrac{1}{4} = \dfrac{?}{16}$ 4

NOTE
In Example 7, on the previous page, the fraction $\frac{2}{3}$ was multiplied by $\frac{5}{5}$. In Example 8, the fraction $\frac{3}{8}$ was multiplied by $\frac{6}{6}$ and the fraction $\frac{5}{6}$ was multiplied by $\frac{7}{7}$. The fractions, $\frac{5}{5}$, $\frac{6}{6}$, and $\frac{7}{7}$ are all equal to 1.

$$\frac{5}{5} = 1 \qquad \frac{6}{6} = 1 \qquad \frac{7}{7} = 1$$

Recall that any number multiplied by 1 is the number itself.

◀◀◀ Work Problem 8 at the Side.

(b) $\dfrac{2}{3} = \dfrac{?}{15}$ 10

(c) $\dfrac{7}{16} = \dfrac{?}{32}$ 14

(d) $\dfrac{6}{11} = \dfrac{?}{33}$ 18

ANSWERS

8. **(a)** $\dfrac{4}{16}$ **(b)** $\dfrac{10}{15}$ **(c)** $\dfrac{14}{32}$ **(d)** $\dfrac{18}{33}$

3.2 **Exercises**

FOR EXTRA HELP   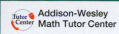 Addison-Wesley Math Tutor Center    MathXL   Digital Video Tutor CD 2 Videotape 6   Student's Solutions Manual   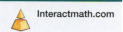 MyMathLab   Interactmath.com

Use multiples of the larger number to find the least common multiple in each set of numbers. See Examples 1 and 2.

1. 3 and 6

2. 2 and 4

3. 3 and 5

4. 3 and 7

5. 4 and 9

6. 4 and 10

7. 2 and 7

8. 6 and 8

9. 6 and 10

10. 12 and 16

11. 20 and 50

12. 25 and 75

Find the least common multiple of each set of numbers. Use any method. See Examples 2–6.

13. 4, 10

14. 6, 10

15. 12, 20

16. 9 and 15

17. 6, 9, 12

18. 20, 24, 30

19. 4, 6, 8, 10

20. 8, 9, 12, 18

21. 12, 15, 18, 20

22. 6, 9, 27, 36

23. 8, 12, 16, 36

24. 5, 6, 25, 30

Rewrite each fraction with a denominator of 24. See Examples 7 and 8.

25. $\dfrac{2}{3} =$

26. $\dfrac{3}{8} =$

27. $\dfrac{3}{4} =$

28. $\dfrac{5}{12} =$

29. $\dfrac{5}{6} =$

30. $\dfrac{7}{8} =$

Rewrite each fraction with the indicated denominator.

31. $\dfrac{1}{2} = \dfrac{}{6}$

32. $\dfrac{2}{3} = \dfrac{}{9}$

33. $\dfrac{3}{4} = \dfrac{}{16}$

34. $\dfrac{7}{10} = \dfrac{}{30}$

35. $\dfrac{7}{8} = \dfrac{}{32}$

36. $\dfrac{5}{12} = \dfrac{}{48}$

37. $\dfrac{3}{16} = \dfrac{}{64}$

38. $\dfrac{7}{8} = \dfrac{}{96}$

39. $\dfrac{8}{5} = \dfrac{}{20}$

40. $\dfrac{5}{8} = \dfrac{}{40}$

41. $\dfrac{9}{7} = \dfrac{}{56}$

42. $\dfrac{3}{2} = \dfrac{}{64}$

43. $\dfrac{7}{4} = \dfrac{}{48}$

44. $\dfrac{7}{6} = \dfrac{}{120}$

45. $\dfrac{8}{11} = \dfrac{}{132}$

46. $\dfrac{4}{15} = \dfrac{}{165}$

47. $\dfrac{3}{16} = \dfrac{}{144}$

48. $\dfrac{7}{16} = \dfrac{}{112}$

49. There are several methods for finding the least common multiple (LCM). Do you prefer the method using multiples of the largest number or the method using prime factorizations? Why? Would you ever use the other method?

50. Explain in your own words how to write a fraction with an indicated denominator. As part of your explanation, show how to change $\frac{3}{4}$ to a fraction having 12 as a denominator.

▦ *Find the least common multiple of the denominators of each pair of fractions.*

51. $\dfrac{25}{400}, \dfrac{38}{1800}$

52. $\dfrac{53}{600}, \dfrac{115}{4000}$

53. $\dfrac{109}{1512}, \dfrac{23}{392}$

54. $\dfrac{61}{810}, \dfrac{37}{1170}$

RELATING CONCEPTS (EXERCISES 55–62) For Individual or Group Work

Most people think that addition and subtraction of fractions is more difficult than multipli-cation and division of fractions. This is probably because a common denominator must be used. **Work Exercises 55–62 in order.**

55. Fractions with the same denominators are _____ fractions and fractions with different denominators are _____ fractions.

56. To subtract like fractions, first find the numerator of the difference by subtracting the _____. Write the denominator of the like fractions as the _____ of the difference. The answer is then written in _____ terms.

57. The _____ common multiple (LCM) of two numbers is the _____ whole number
(smallest/largest)
divisible by both those numbers.

58. The following shows the common multiples for both 8 and 10. What is the least common multiple for these two numbers?

Multiples of 8:   8, 16, 24, 32, 40, 48, 56, 64, 72, 80, 88, . . .

Multiples of 10:   10, 20, 30, 40, 50, 60, 70, 80, 90, . . .

Find the least common multiple for each set of numbers.

59. 5, 7, 14, 10

60. 25, 18, 30, 5

61. Explain why the least common multiple for 8, 3, 5, 4, and 10 is not 240. Find the least common multiple.

62. Demonstrate that the least common multiple of 55 and 1760 is 1760.

3.3 Adding and Subtracting Unlike Fractions

OBJECTIVE 1 Add unlike fractions. In this section, we add and subtract unlike fractions. To add unlike fractions, we must first change them to like fractions (fractions with the same denominator). For example, the figures below show $\frac{3}{8}$ and $\frac{1}{4}$.

OBJECTIVES

1 Add unlike fractions.
2 Add fractions vertically.
3 Subtract unlike fractions.

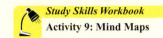

Study Skills Workbook
Activity 9: Mind Maps

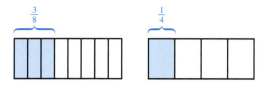

These fractions can be added by changing them to like fractions. Make like fractions by changing $\frac{1}{4}$ to the equivalent fraction $\frac{2}{8}$.

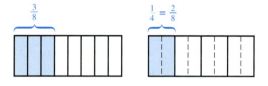

Next, add.

Becomes

$$\frac{3}{8} + \frac{1}{4} = \frac{3}{8} + \frac{2}{8} = \frac{5}{8}$$

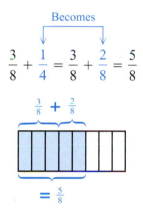

Use the following steps to add or subtract unlike fractions.

Adding or Subtracting Unlike Fractions

Step 1 Rewrite the *unlike fractions* as *like fractions* with the least common multiple as their new denominator. This new denominator is called the **least common denominator (LCD).**

Step 2 Add or subtract as with like fractions.

Step 3 Simplify the answer by writing it in lowest terms and as a whole or mixed number where possible.

EXAMPLE 1 Adding Unlike Fractions

Add $\frac{2}{3}$ and $\frac{1}{9}$.

The least common multiple of 3 and 9 is 9, so first write the fractions as like fractions with a denominator of 9. This is the *least common denominator* of 3 and 9.

Continued on Next Page

1 Add.

(a) $\dfrac{1}{2} + \dfrac{3}{8}$

(b) $\dfrac{3}{4} + \dfrac{1}{8}$

(c) $\dfrac{3}{5} + \dfrac{3}{10}$

(c) $\dfrac{1}{12} + \dfrac{5}{6}$

2 Add. Simplify all answers.

(a) $\dfrac{3}{10} + \dfrac{1}{5}$

(b) $\dfrac{5}{8} + \dfrac{1}{3}$

(c) $\dfrac{1}{10} + \dfrac{1}{3} + \dfrac{1}{6}$

ANSWERS

1. (a) $\dfrac{7}{8}$ (b) $\dfrac{7}{8}$ (c) $\dfrac{9}{10}$ (d) $\dfrac{11}{12}$

2. (a) $\dfrac{1}{2}$ (b) $\dfrac{23}{24}$ (c) $\dfrac{3}{5}$

Step 1
$$\dfrac{2}{3} = \dfrac{?}{9}$$

Divide 9 by 3, getting 3. Next, multiply numerator and denominator by 3.

$$\dfrac{2}{3} = \dfrac{2 \cdot 3}{3 \cdot 3} = \dfrac{6}{9}$$

Now, add the like fractions $\dfrac{6}{9}$ and $\dfrac{1}{9}$.

Becomes

Step 2
$$\dfrac{2}{3} + \dfrac{1}{9} = \dfrac{6}{9} + \dfrac{1}{9} = \dfrac{6 + 1}{9} = \dfrac{7}{9}$$

Step 3   Step 3 is not needed because $\dfrac{7}{9}$ is already in lowest terms.

▶◀ Work Problem 1 at the Side.

EXAMPLE 2 Adding Fractions

Add each pair of fractions using the three steps. Simplify all answers.

(a) $\dfrac{1}{3} + \dfrac{1}{6}$

The least common multiple of 3 and 6 is 6. Rewrite both fractions as fractions with a least common denominator of 6.

Rewritten as like fractions

Step 1
$$\dfrac{1}{3} + \dfrac{1}{6} = \dfrac{2}{6} + \dfrac{1}{6}$$

Add numerators.

Step 2
$$\dfrac{2}{6} + \dfrac{1}{6} = \dfrac{2 + 1}{6} = \dfrac{3}{6}$$

Step 3
$$\dfrac{3}{6} = \dfrac{1}{2} \quad \longleftarrow \text{In lowest terms}$$

(b) $\dfrac{6}{15} + \dfrac{3}{10}$

The least common multiple of 15 and 10 is 30, so rewrite both fractions with a least common denominator of 30.

Rewritten as like fractions

Step 1
$$\dfrac{6}{15} + \dfrac{3}{10} = \dfrac{12}{30} + \dfrac{9}{30}$$

Add numerators.

Step 2
$$\dfrac{12}{30} + \dfrac{9}{30} = \dfrac{12 + 9}{30} = \dfrac{21}{30}$$

Step 3
$$\dfrac{21}{30} = \dfrac{7}{10} \quad \longleftarrow \text{In lowest terms}$$

▶◀ Work Problem 2 at the Side.

OBJECTIVE 2 **Add fractions vertically.**   Fractions can also be added vertically (up and down).

EXAMPLE 3 **Vertical Addition of Fractions**

Add the following fractions vertically.

(a)

$$\frac{3}{8} = \frac{3 \cdot 3}{8 \cdot 3} = \frac{9}{24}$$

$$+\frac{7}{12} = \frac{7 \cdot 2}{12 \cdot 2} = \frac{14}{24}$$

Rewritten as like fractions

$$\frac{23}{24} \leftarrow \text{Add the numerators.}$$

(b)

$$\frac{2}{9} = \frac{2 \cdot 4}{9 \cdot 4} = \frac{8}{36}$$

$$+\frac{1}{4} = \frac{1 \cdot 9}{4 \cdot 9} = \frac{9}{36}$$

Rewritten as like fractions

$$\frac{17}{36} \leftarrow \text{Add the numerators.}$$

> **Work Problem 3 at the Side.**))))

OBJECTIVE 3 **Subtract unlike fractions.**   The next example shows subtraction of unlike fractions.

EXAMPLE 4 **Subtracting Unlike Fractions**

Subtract. Simplify all answers.

As with addition, rewrite unlike fractions with a least common denominator.

(a) $\dfrac{3}{4} - \dfrac{3}{8}$

Rewritten as like fractions

Step 1 $\dfrac{3}{4} - \dfrac{3}{8} = \dfrac{6}{8} - \dfrac{3}{8}$

Subtract numerators.

Step 2 $\dfrac{6}{8} - \dfrac{3}{8} = \dfrac{6-3}{8} - \dfrac{3}{8}$

Step 3   Not needed because $\frac{3}{8}$ is in lowest terms.

Continued on Next Page

---

3 Add the following fractions vertically.

(a) $\begin{array}{r} \frac{5}{8} \quad \frac{30}{48} \\ +\frac{1}{12} \quad \frac{4}{48} \\ \hline \frac{34}{48} = \frac{17}{24} \end{array}$

(b) $\begin{array}{r} \frac{7}{16} \\ +\frac{1}{4} \\ \hline \end{array}$

④ Subtract. Simplify all answers.

(a) $\dfrac{5}{8} - \dfrac{1}{4}$

(b) $\dfrac{4}{5} - \dfrac{3}{4}$

(c) $\quad \dfrac{7}{8}$

$\quad -\dfrac{2}{3}$

(d) $\quad \dfrac{5}{6}$

$\quad -\dfrac{1}{12}$

(b) $\dfrac{3}{4} - \dfrac{7}{12}$

Rewritten as like fractions

Step 1 $\qquad \dfrac{3}{4} - \dfrac{7}{12} = \dfrac{9}{12} - \dfrac{7}{12}$

Subtract numerators.

Step 2 $\qquad \dfrac{9}{12} - \dfrac{7}{12} = \dfrac{9-7}{12} - \dfrac{2}{12}$

Step 3 $\qquad \dfrac{2}{12} = \dfrac{1}{6}$ ← Lowest terms

(c) $\dfrac{3}{7} - \dfrac{5}{12}$

Rewritten as like fractions

Step 1 $\qquad \dfrac{3}{7} - \dfrac{5}{12} = \dfrac{36}{84} - \dfrac{35}{84}$

Subtract numerators.

Step 2 $\qquad \dfrac{36}{84} - \dfrac{35}{84} = \dfrac{36-35}{84} = \dfrac{1}{84}$

Step 3 $\qquad$ Not needed because $\frac{1}{84}$ is in lowest terms.

◀◀◀ Work Problem 4 at the Side.

3.3 Exercises

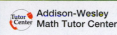

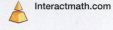
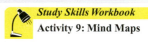

Add the following fractions. Simplify all answers. See Examples 1–3.

1. $\frac{3}{4} + \frac{1}{8}$

2. $\frac{1}{6} + \frac{2}{3}$

3. $\frac{2}{3} + \frac{2}{9}$

4. $\frac{3}{7} + \frac{1}{14}$

5. $\frac{9}{20} + \frac{3}{10}$

6. $\frac{5}{8} + \frac{1}{4}$

7. $\frac{3}{5} + \frac{3}{8}$

8. $\frac{5}{7} + \frac{3}{14}$

9. $\frac{2}{9} + \frac{5}{12}$

10. $\frac{5}{8} + \frac{1}{12}$

11. $\frac{1}{3} + \frac{3}{5}$

12. $\frac{2}{5} + \frac{3}{7}$

13. $\frac{1}{4} + \frac{2}{9} + \frac{1}{3}$

14. $\frac{3}{7} + \frac{2}{5} + \frac{1}{10}$

15. $\frac{3}{10} + \frac{2}{5} + \frac{3}{20}$

16. $\frac{1}{3} + \frac{3}{8} + \frac{1}{4}$

17. $\frac{4}{15} + \frac{1}{6} + \frac{1}{3}$

18. $\frac{5}{12} + \frac{2}{9} + \frac{1}{6}$

19. $\frac{1}{4} + \frac{1}{8}$

20. $\frac{7}{12} + \frac{1}{8}$

21. $\frac{5}{12} + \frac{1}{16}$

22. $\frac{3}{7} + \frac{1}{3}$

Subtract the following fractions. Simplify all answers. See Example 4.

23. $\dfrac{5}{6} - \dfrac{1}{3}$

24. $\dfrac{3}{4} - \dfrac{5}{8}$

25. $\dfrac{2}{3} - \dfrac{1}{6}$

26. $\dfrac{3}{4} - \dfrac{5}{8}$

27. $\dfrac{2}{3} - \dfrac{1}{5}$

28. $\dfrac{5}{6} - \dfrac{7}{9}$

29. $\dfrac{5}{12} - \dfrac{1}{4}$

30. $\dfrac{5}{7} - \dfrac{1}{3}$

31. $\dfrac{8}{9} - \dfrac{7}{15}$

32. $\begin{array}{r} \dfrac{4}{5} \\ -\dfrac{1}{3} \\ \hline \end{array}$

33. $\begin{array}{r} \dfrac{7}{8} \\ -\dfrac{4}{5} \\ \hline \end{array}$

34. $\begin{array}{r} \dfrac{5}{8} \\ -\dfrac{1}{3} \\ \hline \end{array}$

35. $\begin{array}{r} \dfrac{5}{12} \\ -\dfrac{1}{16} \\ \hline \end{array}$

36. $\begin{array}{r} \dfrac{7}{12} \\ -\dfrac{1}{3} \\ \hline \end{array}$

Solve each application problem.

Use the newspaper advertisement for this 4-piece chisel set to answer Exercises 37–38. (*Source:* Harbor Freight Tools.)

37. Find the difference in the cutting-edge width of the two chisels with the widest blades. The symbol " is for inches.

$1 - \dfrac{3}{4}$

38. Find the difference in the cutting-edge width of the two chisels with the narrowest blades. The symbol " is for inches.

$\dfrac{1}{4} \quad \dfrac{1}{2}$

39. The Weiner Works has $\frac{3}{4}$ acre of land. If $\frac{1}{6}$ acre must remain as a green belt and the remainder is buildable, find the amount of land that is buildable.

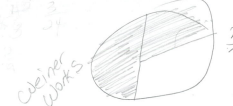

40. Della Daniel wants to open a day care center and has saved $\frac{2}{5}$ of the amount needed for start-up costs. If she saves another $\frac{1}{8}$ of the amount needed and then $\frac{1}{6}$ more, find the total portion of the start-up costs she has saved.

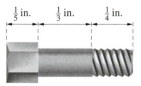

41. When installing cabinets, Cecil Feathers must be certain that the proper type and size of mounting screw is used. Find the total length of the screw shown.

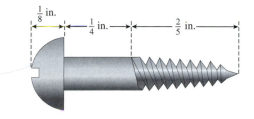

42. When installing a computer chassis, Carolyn Phelps must be certain that the proper type and size of bolt is used. Find the total length of the bolt shown.

$\frac{1}{5}$ in. $\frac{1}{3}$ in. $\frac{1}{4}$ in.

43. The hydraulic system on a forklift contains $\frac{7}{8}$ gallon of hydraulic fluid. A cracked seal resulted in a loss of $\frac{1}{6}$ gallon of fluid in the morning and another $\frac{1}{3}$ gallon in the afternoon. Find the amount of fluid remaining.

44. Adrian Ortega drives a tanker for the British Petroleum Company. He leaves the refinery with his tanker filled to $\frac{7}{8}$ of capacity. If he delivers $\frac{1}{4}$ of the tanker's capacity at the first stop and $\frac{1}{3}$ of the tanker's capacity at the second stop, find the fraction of the tanker's capacity remaining.

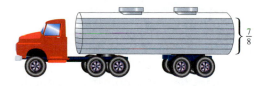

45. Step 1 in adding or subtracting unlike fractions is to rewrite the fractions so they have the least common multiple as a denominator. Explain in your own words why this is necessary.

46. Briefly list the three steps used for addition and subtraction of unlike fractions.

Refer to the circle graph to answer Exercises 47–50.

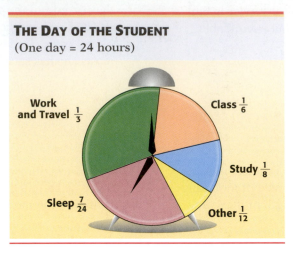

THE DAY OF THE STUDENT
(One day = 24 hours)

Work and Travel $\frac{1}{3}$

Class $\frac{1}{6}$

Study $\frac{1}{8}$

Sleep $\frac{7}{24}$

Other $\frac{1}{12}$

47. What fraction of the day was spent in class and study?

48. What fraction of the day was spent in work and travel and other?

49. In which activity was the greatest amount of time spent? How many hours did this activity take?

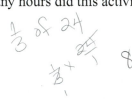

50. In which activity was the least amount of time spent? How many hours did this activity take?

51. Find the diameter of the hole in the mounting bracket shown. (The diameter is the distance across the center of the hole.)

Diameter

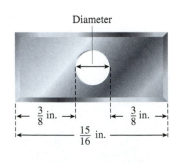

$\frac{3}{8}$ in.　$\frac{3}{8}$ in.

$\frac{15}{16}$ in.

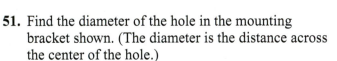

52. Chakotay is fitting a turquoise stone into a bear claw pendant. Find the diameter of the hole in the pendant. (The diameter is the distance across the center of the hole.)

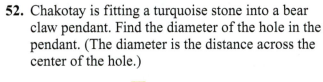

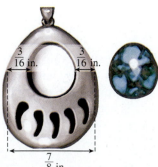

$\frac{3}{16}$ in.　$\frac{3}{16}$ in.

$\frac{7}{8}$ in.

3.4 Adding and Subtracting Mixed Numbers

Recall that a mixed number is the sum of a whole number and a fraction. For example,

$$3\frac{2}{5} \quad \text{means} \quad 3 + \frac{2}{5}$$

OBJECTIVE 1 Estimate an answer, then add or subtract mixed numbers. Add or subtract mixed numbers by adding or subtracting the fraction parts and then the whole number parts. It is a good idea to estimate the answer first, as we did when multiplying and dividing mixed numbers in **Section 2.8.**

Work Problem 1 at the Side. ▶▶▶

OBJECTIVES

1 Estimate an answer, then add or subtract mixed numbers.

2 Estimate an answer, then subtract mixed numbers by borrowing.

3 Add or subtract mixed numbers using an alternate method.

EXAMPLE 1 Adding and Subtracting Mixed Numbers

First estimate the answer. Then add or subtract to find the exact answer.

(a) $16\frac{1}{8} + 5\frac{5}{8} \approx$

Estimate: *Exact:*

$$16 \xleftarrow{\text{Rounds to}} \left\{ 16\frac{1}{8} \right.$$

$$+ \; 6 \xleftarrow{\text{Rounds to}} \left\{ + \; 5\frac{5}{8} \right.$$

$$22 \qquad\qquad 21\frac{6}{8} = 21\frac{3}{4} \leftarrow \text{Lowest terms}$$

Sum of whole numbers ⟵⎯⎯⎯⎯⎯⎯⎯ ⟶ Sum of fractions

In lowest terms $\frac{6}{8}$ *is* $\frac{3}{4}$, so the exact answer is $21\frac{3}{4}$ in lowest terms. This is *reasonable* because it is close to the estimate of 22.

(b) $8\frac{5}{8} - 3\frac{1}{12}$

Estimate: *Exact:*

$$9 \xleftarrow{\text{Rounds to}} \left\{ 8\frac{5}{8} = 8\frac{15}{24} \right.$$

$$- \; 3 \xleftarrow{\text{Rounds to}} \left\{ - 3\frac{1}{12} = -3\frac{2}{24} \right. \quad \text{Least common denominator}$$

$$6 \qquad\qquad 5\frac{13}{24}$$

Subtract whole numbers. ⟵⎯⎯⎯⎯⎯ ⟶ Subtract fractions.

The exact answer of $5\frac{13}{24}$ is *reasonable* because it is close to the estimated answer of 6. Just as before, check by adding $5\frac{13}{24}$ and $3\frac{1}{12}$; the sum should be $8\frac{5}{8}$.

Continued on Next Page

1 As a review of mixed numbers, write each mixed number as an improper fraction and each improper fraction as a mixed number.

(a) $\frac{9}{2}$

(b) $\frac{8}{3}$

(c) $4\frac{3}{4}$

(d) $3\frac{7}{8}$

② First estimate, and then add
≈ or subtract to find the exact answer.

(a) *Estimate:*   *Exact:*

$$7 \xleftarrow{\text{Rounds to}} \left\{ 6\frac{7}{8} = 6\frac{7}{8} \right.$$

$$+ 2 \xleftarrow{\text{Rounds to}} \left\{ + 2\frac{1}{4} = 2\frac{2}{8} \right.$$

(b) $25\frac{3}{5} + 12\frac{3}{10}$

Estimate:

_____ + _____ = _____

Exact:

_____ + _____ = _____

(c) *Estimate:*   *Exact:*

$$\xleftarrow{\hspace{1cm}} \left\{ 4\frac{7}{9} \right.$$

$$- \xleftarrow{\hspace{1cm}} \left\{ - 2\frac{2}{3} \right.$$

③ First estimate, and then add
≈ to find the exact answer.

(a) *Estimate:*   *Exact:*

$$\xleftarrow{\text{Rounds to}} \left\{ 9\frac{3}{4} \right.$$

$$+ \xleftarrow{\text{Rounds to}} \left\{ + 7\frac{1}{2} \right.$$

(b) *Estimate:*   *Exact:*

$$\xleftarrow{\text{Rounds to}} \left\{ 15\frac{4}{5} \right.$$

$$+ \xleftarrow{\text{Rounds to}} \left\{ + 12\frac{2}{3} \right.$$

ANSWERS

2. (a) $7 + 2 = 9; 9\frac{1}{8}$

(b) $26 + 12 = 38; 37\frac{9}{10}$

(c) $5 - 3 = 2; 2\frac{1}{9}$

3. (a) $10 + 8 = 18; 17\frac{1}{4}$

(b) $16 + 13 = 29; 28\frac{7}{15}$

> **NOTE**
> When estimating, if the numerator is *half* of the denominator or *more*, round up the whole number part. If the numerator is *less* than *half* the denominator, leave the whole number part as it is.

⫷⫷⫷ Work Problem 2 at the Side.

When you add the fraction parts of mixed numbers, the sum may be greater than 1. If this happens, **carrying** from the fraction column to the whole number column is the best procedure. (You cannot leave a whole number along with an improper fraction as the answer.)

EXAMPLE 2 Carrying When Adding Mixed Numbers

≈
First estimate, and then add $9\frac{5}{8} + 13\frac{7}{8}$.

Estimate: *Exact:*

$$10 \xleftarrow{\text{Rounds to}} \left\{ 9\frac{5}{8} \right.$$

$$+ 14 \xleftarrow{\text{Rounds to}} \left\{ + 13\frac{7}{8} \right.$$

$$24 \qquad\qquad 22\frac{12}{8}$$

Sum of whole numbers ⎯⎯⎯ Sum of fractions

The improper fraction $\frac{12}{8}$ can be written in lowest terms as $\frac{3}{2}$. Because $\frac{3}{2} = 1\frac{1}{2}$, the simplified sum is

Becomes

$$22\frac{12}{8} = 22 + \frac{12}{8} = 22 + 1\frac{1}{2} = 23\frac{1}{2}.$$

The estimate was 24, so the exact answer of $23\frac{1}{2}$ is reasonable.

> **NOTE**
> When adding mixed numbers, first add the fraction parts, then add the whole number parts. Then combine the two answers, simplifying the fraction part when necessary.

⫷⫷⫷ Work Problem 3 at the Side.

OBJECTIVE ② **Estimate an answer, then subtract mixed numbers by borrowing.** When subtracting mixed numbers, **borrowing** is necessary when the fraction part of the first number is less than the fraction part of the second number.

EXAMPLE 3 Borrowing When Subtracting Mixed Numbers

First estimate, and then subtract.
≈

(a) $7 - 2\frac{5}{6}$

Continued on Next Page

Estimate: *Exact:*

$$7 \xleftarrow{\text{Rounds to}} \begin{cases} 7 \end{cases} \xleftarrow{\substack{\text{There is no fraction here} \\ \text{from which to subtract } \frac{5}{6}}}$$

$$-3 \xleftarrow{\text{Rounds to}} \begin{cases} -2\dfrac{5}{6} \end{cases}$$

$$\overline{4}$$

It is not possible to subtract $\frac{5}{6}$ without borrowing from the whole number **7** first.

$$\overset{\text{Borrow 1.}}{\downarrow}$$

$$7 = 6 + \mathbf{1}$$

$$\overset{1 = \frac{6}{6}}{\downarrow}$$

$$= 6 + \frac{\mathbf{6}}{\mathbf{6}}$$

$$= 6\frac{6}{6}$$

Now, subtract.

$$7 = 6\frac{6}{6}$$

$$-2\frac{5}{6} = -2\frac{5}{6}$$

$$\overline{4\frac{1}{6}}$$

The estimate was 4, so the exact answer of $4\frac{1}{6}$ is reasonable.

(b) $8\dfrac{1}{3} - 4\dfrac{3}{5}$

Estimate: *Exact:*

$$8 \xleftarrow{\text{Rounds to}} \begin{cases} 8\dfrac{1}{3} = 8\dfrac{5}{15} \end{cases} \xleftarrow{}$$

$$-5 \xleftarrow{\text{Rounds to}} \begin{cases} -4\dfrac{3}{5} = -4\dfrac{9}{15} \end{cases} \xleftarrow{\text{Least common denominator}}$$

$$\overline{3}$$

It is not possible to subtract $\frac{9}{15}$ from $\frac{5}{15}$, so borrow from the whole number **8**.

$$\overset{\text{Borrow 1.}}{\downarrow}$$

$$8\frac{5}{15} = 8 + \frac{5}{15} = 7 + \mathbf{1} + \frac{5}{15}$$

$$\overset{1 = \frac{15}{15}}{\downarrow}$$

$$= 7 + \frac{\mathbf{15}}{\mathbf{15}} + \frac{5}{15}$$

$$= 7 + \frac{\mathbf{20}}{\mathbf{15}} \xleftarrow{\frac{15}{15} + \frac{5}{15}}$$

$$= 7\frac{20}{15}$$

Continued on Next Page

④ First estimate and then
subtract to find the exact
answer.

(a) *Estimate:*   *Exact:*

$$\underleftarrow{\text{Rounds to}}\ \left\{7\frac{1}{3}\right.$$

$$-\ \ \underleftarrow{\text{Rounds to}}\ \left\{-4\frac{5}{6}\right.$$

(b) *Estimate:*   *Exact:*

$$\underleftarrow{\qquad}\ \left\{4\frac{5}{8}\right.$$

$$-\ \ \underleftarrow{\qquad}\ \left\{-2\frac{15}{16}\right.$$

(c) *Estimate:*   *Exact:*

$$\underleftarrow{\qquad}\ \{15$$

$$-\ \ \underleftarrow{\qquad}\ \left\{-6\frac{4}{9}\right.$$

⑤ Add or subtract by changing
mixed numbers to improper
fractions. Write answers as
mixed numbers in lowest
terms.

(a) $3\frac{3}{8}$

$+\ 2\frac{1}{2}$

(b) $6\frac{3}{4}$

$-\ 4\frac{2}{3}$

Now, subtract.

$$8\frac{1}{3} = 8\frac{5}{15} = 7\frac{20}{15}$$

$$-4\frac{3}{5} = 4\frac{9}{15} = 4\frac{9}{15}$$

$$\rule{3cm}{0.4pt}$$

$$3\frac{11}{15}$$

The exact answer is $3\frac{11}{15}$ (lowest terms), which is reasonable because it is close
to the estimate of 3.

◀◀◀ Work Problem 4 at the Side.

**OBJECTIVE 3 Add or subtract mixed numbers using an alternate
method.** An alternate method for adding or subtracting mixed numbers is
to first change the mixed numbers to improper fractions. Then rewrite the unlike fractions as like fractions. Finally, add or subtract the numerators and
write the answer in lowest terms.

EXAMPLE 4 Adding or Subtracting Mixed Numbers

Add or subtract.

(a) $2\frac{3}{8} = \frac{19}{8} = \frac{19}{8}$

$+\ 3\frac{3}{4} = \frac{15}{4} = \frac{30}{8}$ ———— Least common denominator

$\dfrac{49}{8} = 6\frac{1}{8}$ Answer as mixed number

Change to
improper fractions.

(b) $4\frac{2}{3} = \frac{14}{3} = \frac{70}{15}$

$-\ 2\frac{1}{5} = \frac{11}{5} = \frac{33}{15}$ ———— Least common denominator

$\dfrac{37}{15} = 2\frac{7}{15}$ Answer as mixed number

Improper
fractions

◀◀◀ Work Problem 5 at the Side.

NOTE
The advantage of this alternate method of adding or subtracting mixed
numbers is that it eliminates the need to carry when adding or to borrow
when subtracting. It is also the most useful method for working with algebraic fractions. (See **Section 9.7.**) However, if the mixed numbers are
large, then the numerators of the improper fractions may become so
large that they are difficult to work with. In such cases, you may want to
keep the numbers as mixed numbers.

3.4 Exercises

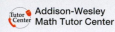

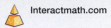
≈ *First estimate the answer. Then add to find the exact answer. Write answers as mixed numbers. See Examples 1 and 2.*

1. *Estimate:*   *Exact:*

Rounds to ⟵ $\left\{ 5\dfrac{1}{2} \right.$

$+$ Rounds to ⟵ $\left\{ +\,3\dfrac{1}{3} \right.$

___   _____

2. *Estimate:*   *Exact:*

$6\dfrac{3}{5}$

$+$ $+\,7\dfrac{1}{10}$

___   _____

3. *Estimate:*   *Exact:*

$7\dfrac{1}{3}$

$+$ $+\,4\dfrac{1}{6}$

___   _____

4. *Estimate:*   *Exact:*

$10\dfrac{1}{4}$

$+$ $+\,5\dfrac{5}{8}$

___   _____

5. *Estimate:*   *Exact:*

$\dfrac{5}{8}$

$+$ $+\,3\dfrac{7}{12}$

___   _____

6. *Estimate:*   *Exact:*

$12\dfrac{4}{5}$

$+$ $+\,\dfrac{7}{10}$

___   _____

7. *Estimate:*   *Exact:*

$24\dfrac{5}{6}$

$+$ $+\,18\dfrac{5}{6}$

___   _____

8. *Estimate:*   *Exact:*

15

$14\dfrac{6}{7}$

$+\,15.5 = 16$ $+\,15\dfrac{1}{2}$

___   _____

9. *Estimate:*   *Exact:*

$33\dfrac{3}{5}$

$+$ $+\,18\dfrac{1}{2}$

___   _____

10. *Estimate:*   *Exact:*

$18\dfrac{5}{8}$

$+$ $+\,6\dfrac{2}{3}$

___   _____

11. *Estimate:*   *Exact:*

$22\dfrac{3}{4}$

$+$ $+\,15\dfrac{3}{7}$

___   _____

12. *Estimate:*   *Exact:*

$7\dfrac{1}{4}$

$+$ $+\,25\dfrac{7}{8}$

___   _____

13. *Estimate:*   *Exact:*

$$12\frac{8}{15}$$

$$18\frac{3}{5}$$

$+$ $+\ 14\frac{7}{10}$

---

14. *Estimate:*   *Exact:*

$$14\frac{9}{10}$$

$$8\frac{1}{4}$$

$+$ $+\ 13\frac{3}{5}$

≈ *First estimate the answer. Then subtract to find the exact answer. Simplify all answers.*
See Examples 1 and 3.

15. *Estimate:*   *Exact:*

$$14\frac{7}{8}$$

$-$ $-\ 12\frac{1}{4}$

16. *Estimate:*   *Exact:*

$$14\frac{3}{4}$$

$-$ $-\ 11\frac{3}{8}$

17. *Estimate:*   *Exact:*

$$12\frac{2}{3}$$

$-$ $-\ 1\frac{1}{5}$

---

18. *Estimate:*   *Exact:*

$$11\frac{9}{20}$$

$-$ $-\ 4\frac{3}{5}$

19. *Estimate:*   *Exact:*

$$28\frac{3}{10}$$

$-$ $-\ 6\frac{1}{15}$

20. *Estimate:*   *Exact:*

$$15\frac{7}{20}$$

$-$ $-\ 6\frac{1}{8}$

---

21. *Estimate:*   *Exact:*

$$17$$

$-$ $-\ 6\frac{5}{8}$

22. *Estimate:*   *Exact:*

$$22$$

$-$ $-\ 4\frac{5}{8}$

23. *Estimate:*   *Exact:*

$$18\frac{3}{4}$$

$-$ $-\ 5\frac{4}{5}$

24. *Estimate:*   *Exact:*   **25.** *Estimate:*   *Exact:*   **26.** *Estimate:*   *Exact:*

$$14\frac{5}{8}$$ $$19\frac{2}{3}$$ $$20\frac{3}{5}$$

$-$ $$-3\frac{2}{3}$$ $-$ $$-11\frac{3}{4}$$ $-$ $$-12\frac{7}{15}$$

_____   _____   _____   _____   _____   _____

Add or subtract by changing mixed numbers to improper fractions. Write answers as mixed numbers when possible. See Example 4.

27. $\quad 7\frac{5}{8}$ **28.** $\quad 8\frac{3}{4}$ **29.** $\quad 4\frac{2}{3}$ **30.** $\quad 7\frac{5}{12}$

$\quad +1\frac{1}{2}$ $\quad +1\frac{5}{8}$ $\quad +6\frac{5}{6}$ $\quad +6\frac{2}{3}$

_____   _____   _____   _____

31. $\quad 2\frac{2}{3}$ **32.** $\quad 4\frac{1}{2}$ **33.** $\quad 3\frac{1}{4}$ **34.** $\quad 2\frac{4}{5}$

$\quad +1\frac{1}{6}$ $\quad +2\frac{3}{4}$ $\quad +3\frac{2}{3}$ $\quad +5\frac{1}{3}$

_____   _____   _____   _____

35. $\quad 1\frac{3}{8}$ **36.** $\quad 1\frac{5}{12}$ **37.** $\quad 3\frac{1}{2}$ **38.** $\quad 4\frac{1}{4}$

$\quad +6\frac{3}{4}$ $\quad +1\frac{7}{8}$ $\quad -2\frac{2}{3}$ $\quad -3\frac{7}{12}$

_____   _____   _____   _____

39. $8\frac{3}{4}$
$-\ 5\frac{7}{8}$

40. $12\frac{2}{3}$
$-\ 7\frac{11}{12}$

41. $7\frac{1}{4}$
$-\ 4\frac{2}{3}$

42. $4\frac{1}{10}$
$-\ 3\frac{7}{8}$

43. $9\frac{1}{5}$
$-\ 3\frac{3}{4}$

44. $10\frac{2}{7}$
$-\ 5\frac{5}{14}$

45. $\overset{5}{\cancel{6}}\frac{3}{7}\ \frac{9}{21}\ \overset{+21}{}=\frac{30}{21}$
$-\ 2\frac{2}{3}\ \frac{14}{21}\quad -\frac{14}{21}$
$\overline{}$
$\qquad 3\ \frac{16}{21}$

46. $\overset{7}{\cancel{8}}\frac{2}{15}\ \frac{4}{30}\ \overset{+30}{}\ \overset{34}{}$
$-\ 6\frac{1}{2}\ \frac{16}{30}$

47. In your own words, explain the steps you would take to add two large mixed numbers.

48. When subtracting mixed numbers, explain when you need to borrow. Explain how to borrow using your own example.

≈ *First estimate the answer. Then solve each application problem.*

49. At the beginning of this chapter you read about Bryan Berg, who builds houses of cards. While in high school, Bryan built a house of cards $14\frac{1}{2}$ ft tall, setting his first world record. Today his current world record is $25\frac{1}{4}$ ft tall. How much taller is his current world record than his first world record? (*Source: Guinness Book of World Records.*)

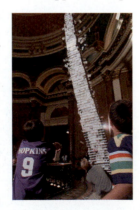
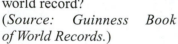

Estimate:

Exact:

50. The heaviest marine mammal in the world is the blue whale at $143\frac{3}{10}$ tons. The sixth heaviest is the humpback whale at $29\frac{1}{5}$ tons. Find the difference in their weights. (*Source: Top 10 of Everything.*)

Estimate:

Exact:

Use the newspaper advertisement for this 6-piece jumbo wrench set to answer Exercises 51–54. The symbol " is for inches.
(*Source:* Harbor Freight Tools.)

6-Piece Jumbo Wrench Set

• Sizes:
$1\frac{3}{8}$" to 2"

• Length:
$17\frac{3}{4}$" to $22\frac{3}{4}$"

$22\frac{3}{4}$"
$22\frac{1}{4}$"
21"
$20\frac{3}{4}$"
$18\frac{5}{8}$"
$17\frac{3}{4}$"

SALE!
$27⁹⁹

REGULAR PRICE $39.99

51. How much longer is the longest wrench than the second-to-shortest wrench?

Estimate:

Exact:

$$22\frac{3}{4} - 18\frac{5}{8}$$

$$22\frac{3}{4}\frac{6}{8}$$
$$-18\frac{5}{8}\frac{5}{8}$$
$$4\quad\frac{1}{8}$$

52. Find the difference in length between the shortest wrench and the second-to-longest wrench.

Estimate:

Exact:

53. What is the total length of the three longest wrenches?

Estimate:

Exact:

54. What is the total length of the three shortest wrenches?

Estimate:

Exact:

55. Andrea Abriani, a college student, works part-time at the Cyber Coffeehouse. She worked $3\frac{3}{8}$ hours on Monday, $5\frac{1}{2}$ hours on Tuesday, $4\frac{3}{4}$ hours on Wednesday, $3\frac{1}{4}$ hours on Thursday, and 6 hours on Friday. How many hours did she work altogether?

Estimate:

Exact:

56. On a recent vacation to Canada, Erin Gavin drove for $7\frac{3}{4}$ hours on the first day, $5\frac{1}{4}$ hours on the second day, $6\frac{1}{2}$ hours on the third day, and 9 hours on the fourth day. How many hours did she drive altogether?

Estimate:

Exact:

57. A craftsperson must attach a lead strip around all four sides of a stained glass window before it is installed. Find the length of lead stripping needed.

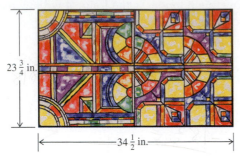

$23\frac{3}{4}$ in.

$34\frac{1}{2}$ in.

Estimate:

Exact:

58. To complete a custom order, Zak Morten of Home Depot must find the number of inches of brass trim needed to go around the four sides of the lamp base plate shown. Find the length of brass trim needed.

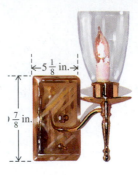

$5\frac{1}{8}$ in.

$\frac{7}{8}$ in.

Estimate:

Exact:

59. A cement-truck driver has $8\frac{7}{8}$ cubic yards of concrete in a truck. If he unloads $2\frac{1}{2}$ cubic yards at the first stop, 3 cubic yards at the second stop, and $1\frac{3}{4}$ cubic yards at the third stop, how much concrete remains in the truck?

Estimate:

Exact:

60. Marv Levenson bought 15 yards of Italian silk fabric. He made two shirts with $3\frac{3}{4}$ yards of the material, a suit for his wife with $4\frac{1}{8}$ yards, and a jacket with $3\frac{7}{8}$ yards. Find the number of yards of material remaining.

Estimate:

Exact:

61. The exercise yard at the correction center has four sides and is surrounded by $527\frac{1}{24}$ ft of security fencing. If three sides of the yard measure $107\frac{2}{3}$ ft, $150\frac{3}{4}$ ft, and $138\frac{5}{8}$ ft, find the length of the fourth side.

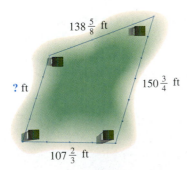

$138\frac{5}{8}$ ft

$150\frac{3}{4}$ ft

? ft

$107\frac{2}{3}$ ft

Estimate:

Exact:

62. Three sides of a parking lot are $108\frac{1}{4}$ ft, $162\frac{3}{8}$ ft, and $143\frac{1}{2}$ ft. If the distance around the lot is $518\frac{3}{4}$ ft, find the length of the fourth side.

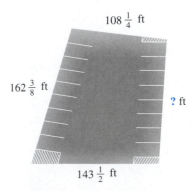

$108\frac{1}{4}$ ft

$162\frac{3}{8}$ ft

? ft

$143\frac{1}{2}$ ft

Estimate:

Exact:

63. A truck trailer is to be loaded with personal computers weighing $2\frac{5}{8}$ tons, computer monitors weighing $6\frac{1}{2}$ tons, computer printers weighing $1\frac{5}{6}$ tons, and assorted accessories weighing $3\frac{1}{4}$ tons. If the truck trailer weighs $7\frac{3}{8}$ tons empty, find the total weight after it has been loaded.

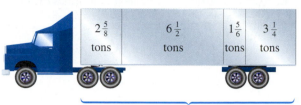

$2\frac{5}{8}$ tons $6\frac{1}{2}$ tons $1\frac{5}{6}$ tons $3\frac{1}{4}$ tons

Weight of empty trailer = $7\frac{3}{8}$ tons

Estimate:

Exact:

64. Bryan Berg, from the opening page of this chapter, built houses of cards reaching heights of $14\frac{1}{2}$ ft, $19\frac{3}{8}$ ft, $23\frac{5}{12}$ ft, and $25\frac{1}{4}$ ft (his current world record). Find the total height of these four houses of cards. (*Source: Reader's Digest.*)

Estimate:

Exact:

Find the unknown length, labeled with a question mark, in each figure.

65.

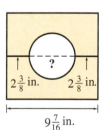

?

$2\frac{3}{8}$ in. $2\frac{3}{8}$ in.

$9\frac{7}{16}$ in.

66.

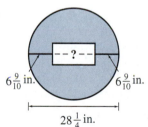

- - ? - -

$6\frac{9}{10}$ in. $6\frac{9}{10}$ in.

$28\frac{1}{4}$ in.

67.

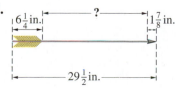

$6\frac{1}{4}$ in.   ? $1\frac{7}{8}$ in.

$29\frac{1}{2}$ in.

68.

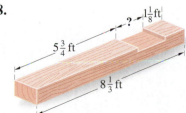

? $1\frac{1}{8}$ ft

$5\frac{3}{4}$ ft

$8\frac{1}{3}$ ft

RELATING CONCEPTS (EXERCISES 69–74)   For Individual or Group Work

Most fraction problems include fractions with different denominators.
Work Exercises 69–74 in order.

69. To add or subtract fractions, we must first rewrite them as like fractions. Rewrite each fraction with the indicated denominator.

(a) $\dfrac{5}{9} = \dfrac{}{54}$

(b) $\dfrac{7}{12} = \dfrac{}{48}$

(c) $\dfrac{5}{8} = \dfrac{}{40}$

(d) $\dfrac{11}{5} = \dfrac{}{120}$

70. When rewriting unlike fractions as like fractions with the least common multiple as a denominator, the new denominator is called the _____ _____ _____ , or LCD.

71. Add or subtract as indicated. Write answers in lowest terms.

(a) $\dfrac{5}{8} + \dfrac{1}{3}$

(b) $\dfrac{19}{20} - \dfrac{5}{12}$

(c) $\begin{array}{r} \dfrac{7}{12} \\ \dfrac{3}{16} \\ + \dfrac{3}{24} \\ \hline \end{array}$

(d) $\begin{array}{r} \dfrac{6}{7} \\ - \dfrac{2}{3} \\ \hline \end{array}$

72. A common method for adding or subtracting mixed numbers is to add or subtract the _____ _____ and then the whole nubmer parts

73. Another method for adding or subtracting mixed numbers is to first change the mixed numbers to _____ fractions. After adding or subtracting, write the answer in lowest terms and as a mixed number when possible.

74. Add or subtract these fractions as indicated. First use the method where you add or subtract fraction parts and then whole number parts. Then use the method where you change each mixed number to an improper fraction before adding or subtracting. Do you get the same answer using both methods? Which method do you prefer?

(a) $\begin{array}{r} 4\dfrac{5}{8} \\ + 3\dfrac{1}{4} \\ \hline \end{array}$

(b) $\begin{array}{r} 12\dfrac{2}{5} \\ - 8\dfrac{7}{8} \\ \hline \end{array}$

3.5 Order Relations and the Order of Operations

There are times when we want to compare the size of two numbers. For example, we might want to know which is the greater amount, the larger size, or the longer distance.

Fractions, like whole numbers, can be located on a number line. For fractions, divide the space between whole numbers into equal parts.

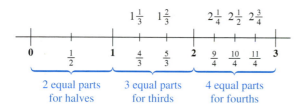

2 equal parts for halves   3 equal parts for thirds   4 equal parts for fourths

OBJECTIVE 1 Identify the greater of two fractions.   To compare the size of two numbers, place the two numbers on a number line and use the following rule.

Comparing the Size of Two Numbers

The number farther to the *left* on the number line is always *less*, and the number farther to the *right* on the number line is always *greater.*

For example, on the number line above, $\frac{1}{2}$ is the *left* of $\frac{4}{3}(1\frac{1}{3})$, so $\frac{1}{2}$ is *less than* $\frac{4}{3}(1\frac{1}{3})$.

> **Work Problem 1 at the Side.** ▶▶▶

Write *order relations* using the symbols shown below.

Symbols Used to Show Order Relations

$<$ is less than $>$ is greater than

EXAMPLE 1 Using Less-Than and Greater-Than Symbols

Rewrite the following using $<$ and $>$ symbols.

(a) $\frac{1}{2}$ is less than $\frac{4}{3}$

$\frac{1}{2}$ **is less than** $\frac{4}{3}$ is written as $\frac{1}{2} < \frac{4}{3}$

(b) $\frac{9}{4}$ is greater than 1

$\frac{9}{4}$ **is greater than** 1 is written as $\frac{9}{4} > 1$

(c) $\frac{5}{3}$ is less than $\frac{11}{4}$

$\frac{5}{3}$ **is less than** $\frac{11}{4}$ is written as $\frac{5}{3} < \frac{11}{4}$

Continued on Next Page

OBJECTIVES

1 Identify the greater of two fractions.

2 Use exponents with fractions.

3 Use the order of operations.

① Locate each fraction on the number line.

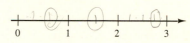

(a) $\frac{2}{3}$

(b) $1\frac{1}{2}$

(c) $2\frac{3}{4}$

2 Use the number line on the previous page to help you write < or > in each blank to make a true statement.

(a) 1 ___$<$___ $\dfrac{5}{4} = 1\frac{1}{4}$

(b) $\dfrac{8}{3}$ ___$>$___ $\dfrac{3}{2}$ $1\frac{1}{2}$

$2\frac{2}{3}$

(c) 0 ___$<$___ 1

(d) $\dfrac{17}{8}$ ___$>$___ $\dfrac{8}{4} = 2$

$2\frac{1}{8}$

3 Write < or > in each blank to make a true statement.

(a) $\dfrac{7}{8}$ _____ $\dfrac{3}{4}$

(b) $\dfrac{13}{8}$ _____ $\dfrac{15}{9}$

(c) $\dfrac{9}{4}$ _____ $\dfrac{7}{3}$

(d) $\dfrac{9}{10}$ _____ $\dfrac{14}{15}$

NOTE
A number line is a very useful tool when working with order relations.

Work Problem 2 at the Side.

The fraction $\frac{7}{8}$ represents 7 of 8 equivalent parts, while $\frac{3}{8}$ means 3 of 8 equivalent parts. Because $\frac{7}{8}$ represents more of the equivalent parts, $\frac{7}{8}$ is greater than $\frac{3}{8}$.

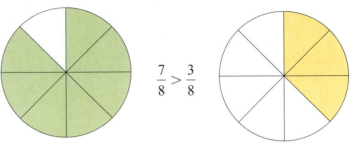

$$\frac{7}{8} > \frac{3}{8}$$

To identify the greater fraction, use the following steps.

Identifying the Greater Fraction

Step 1 Write the fractions as like fractions (same denominators).

Step 2 Compare the numerators. The fraction with the greater numerator is the greater fraction.

EXAMPLE 2 Identifying the Greater Fraction

Determine which fraction in each pair is greater.

(a) $\dfrac{7}{8}, \dfrac{9}{10}$ $\dfrac{35}{40}$ $\dfrac{36}{40}$

First, write the fractions as like fractions. The least common multiple for 8 and 10 is 40.

$$\frac{7}{8} = \frac{7 \cdot \textbf{5}}{8 \cdot \textbf{5}} = \frac{35}{40} \quad \text{and} \quad \frac{9}{10} = \frac{9 \cdot \textbf{4}}{10 \cdot \textbf{4}} = \frac{36}{40}$$

Look at the numerators. Because 36 is greater than 35, $\frac{36}{40}$ is greater than $\frac{35}{40}$. Because $\frac{36}{40}$ is equivalent to $\frac{9}{10}$,

$$\frac{9}{10} > \frac{7}{8} \quad \text{or} \quad \frac{7}{8} < \frac{9}{10}$$

The greater fraction is $\frac{9}{10}$.

(b) $\dfrac{8}{5}, \dfrac{23}{15}$ $\dfrac{40}{15}$ $\dfrac{23}{15}$

The least common multiple of 5 and 15 is 15.

$$\frac{8}{5} = \frac{8 \cdot \textbf{3}}{5 \cdot \textbf{3}} = \frac{24}{15} \quad \text{and} \quad \frac{23}{15} = \frac{23}{15}$$

This shows that $\frac{8}{5}$ is greater than $\frac{23}{15}$, or

$$\frac{8}{5} > \frac{23}{15}$$

Work Problem 3 at the Side.

OBJECTIVE **2** **Use exponents with fractions.** Exponents were used in **Section 1.8** to write repeated multiplication. For example,

$$3^2 = \underbrace{3 \cdot 3}_{\substack{\text{Two} \\ \text{factors of 3}}} = 9 \quad \text{and} \quad 5^3 = \underbrace{5 \cdot 5 \cdot 5}_{\substack{\text{Three} \\ \text{factors of 5}}} = 125$$

Exponent ↗   Exponent ↗

The next example shows exponents used with fractions.

EXAMPLE 3 **Using Exponents with Fractions**

Simplify.

(a) $\left(\dfrac{1}{2}\right)^3$

$$\left(\frac{1}{2}\right)^3 = \overbrace{\frac{1}{2} \cdot \frac{1}{2} \cdot \frac{1}{2}}^{\text{Three factors of } \frac{1}{2}} = \frac{1}{8}$$

(b) $\left(\dfrac{5}{8}\right)^2$

$$\left(\frac{5}{8}\right)^2 = \overbrace{\frac{5}{8} \cdot \frac{5}{8}}^{\text{Two factors of } \frac{5}{8}} = \frac{25}{64}$$

(c) $\left(\dfrac{3}{4}\right)^2 \cdot \left(\dfrac{2}{3}\right)^3$

$$\left(\frac{3}{4}\right)^2 \cdot \left(\frac{2}{3}\right)^3 = \left(\frac{3}{4} \cdot \frac{3}{4}\right) \cdot \left(\frac{2}{3} \cdot \frac{2}{3} \cdot \frac{2}{3}\right)$$

$$= \frac{\overset{1}{\cancel{3}} \cdot \overset{1}{\cancel{3}} \cdot \overset{1}{\cancel{2}} \cdot \overset{1}{\cancel{2}} \cdot \overset{1}{\cancel{2}}}{\underset{2}{\cancel{4}} \cdot \underset{2}{\cancel{4}} \cdot \underset{1}{\cancel{3}} \cdot \underset{1}{\cancel{3}} \cdot 3} \qquad \text{Divide out all the common factors.}$$

$$= \frac{1}{6}$$

Work Problem 4 at the Side. ▶▶▶

OBJECTIVE **3** **Use the order of operations.** Recall the *order of operations* from **Section 1.8.**

Order of Operations
1. Do all operations inside *parentheses or other grouping symbols.*
2. Simplify any expressions with *exponents* and find any *square roots.*
3. *Multiply* or *divide* proceeding from left to right.
4. *Add* or *subtract* proceeding from left to right.

---

4 Simplify.

(a) $\left(\dfrac{1}{2}\right)^4$

(b) $\left(\dfrac{3}{4}\right)^2$

$\dfrac{3}{4} \cdot \dfrac{3}{4} = \dfrac{9}{16}$

(c) $\left(\dfrac{1}{2}\right)^3 \cdot \left(\dfrac{2}{3}\right)^2$

$\dfrac{1}{2} \cdot \dfrac{1}{2} \cdot \dfrac{1}{2} \times \dfrac{2}{3} \cdot \dfrac{2}{3}$

$\dfrac{1}{8} \times \dfrac{4}{9} = \dfrac{4}{72} \quad \dfrac{2}{36} \quad \dfrac{1}{18}$

(d) $\left(\dfrac{1}{5}\right)^2 \cdot \left(\dfrac{5}{3}\right)^2$

ANSWERS
4. (a) $\dfrac{1}{16}$ (b) $\dfrac{9}{16}$ (c) $\dfrac{1}{18}$ (d) $\dfrac{1}{9}$

PEMDAS

5 Simplify by using the order of operations.

(a) $\dfrac{5}{9} - \dfrac{3}{4}\left(\dfrac{2}{3}\right)$

(b) $\dfrac{3}{4}\left(\dfrac{2}{3} \cdot \dfrac{3}{5}\right)$

(c) $\dfrac{7}{8}\left(\dfrac{2}{3}\right) - \left(\dfrac{1}{2}\right)^2$

(d) $\dfrac{\left(\dfrac{5}{6}\right)^2}{\dfrac{4}{3}}$

The next example shows how to apply the order of operations with fractions.

EXAMPLE 4 Using the Order of Operations with Fractions

Simplify by using the order of operations.

(a) $\dfrac{1}{3} + \dfrac{1}{2}\left(\dfrac{4}{5}\right)$

Multiply $\frac{1}{2}\left(\frac{4}{5}\right)$ first.

$$\dfrac{1}{3} + \dfrac{1}{\overset{}{\underset{1}{2}}}\left(\dfrac{\overset{2}{4}}{5}\right) = \dfrac{1}{3} + \dfrac{2}{5}$$

Next, add. The least common denominator of 3 and 5 is 15.

$$\dfrac{1}{3} + \dfrac{2}{5} = \dfrac{5}{15} + \dfrac{6}{15} = \dfrac{11}{15}$$

(b) $\dfrac{3}{8}\left(\dfrac{1}{2} + \dfrac{1}{3}\right)$

$$\dfrac{3}{8}\left(\dfrac{1}{2} + \dfrac{1}{3}\right) = \dfrac{3}{8}\underbrace{\left(\dfrac{3}{6} + \dfrac{2}{6}\right)}$$

Work inside parentheses first.

$$= \dfrac{3}{8}\left(\dfrac{5}{6}\right)$$

$$= \dfrac{\overset{1}{\cancel{3}}}{8}\left(\dfrac{5}{\underset{2}{\cancel{6}}}\right) \quad \text{Divide numerator and denominator by 3.}$$

$$= \dfrac{5}{16} \quad \text{Multiply.}$$

(c) $\left(\dfrac{2}{3}\right)^2 - \dfrac{4}{5}\left(\dfrac{1}{2}\right)$

First simplify the expression with the exponent.
$\frac{2}{3} \cdot \frac{2}{3}$ is $\frac{4}{9}$.

$$\left(\dfrac{2}{3}\right)^2 - \dfrac{4}{5}\left(\dfrac{1}{2}\right) = \dfrac{4}{9} - \dfrac{4}{5}\left(\dfrac{1}{2}\right)$$

$$= \dfrac{4}{9} - \dfrac{\overset{2}{\cancel{4}}}{5}\left(\dfrac{1}{\underset{1}{\cancel{2}}}\right) \quad \text{Next, multiply.}$$

$$= \dfrac{4}{9} - \dfrac{2}{5}$$

$$= \dfrac{20}{45} - \dfrac{18}{45} \quad \text{Subtract last. (Least common denominator is 45.)}$$

$$= \dfrac{2}{45}$$

Handwritten work:
$\dfrac{4}{9} - \dfrac{4}{5}\left(\dfrac{1}{2}\right)$
$\dfrac{4}{9} - \dfrac{2}{5}$
$\dfrac{20}{45}\ \dfrac{18}{45}$
$\dfrac{2}{45}$

Work Problem 5 at the Side.

3.5 Exercises

 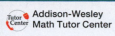

Locate each fraction in Exercises 1–12 on the following number line. See Margin Problem 1.

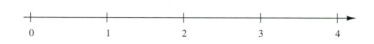

1. $\dfrac{1}{2}$

2. $\dfrac{1}{4}$

3. $\dfrac{3}{2}$

4. $\dfrac{5}{4}$

5. $\dfrac{7}{3}$

6. $\dfrac{11}{4}$

7. $2\dfrac{1}{6}$

8. $3\dfrac{4}{5}$

9. $\dfrac{7}{2}$

10. $\dfrac{7}{8}$

11. $3\dfrac{1}{4}$

12. $1\dfrac{7}{8}$

Write < or > to make a true statement. See Examples 1 and 2.

13. $\dfrac{1}{2}$ ____ $\dfrac{3}{8}$

14. $\dfrac{5}{8}$ ____ $\dfrac{3}{4}$

15. $\dfrac{5}{6}$ ____ $\dfrac{11}{12}$

16. $\dfrac{13}{18}$ ____ $\dfrac{5}{6}$

17. $\dfrac{5}{12}$ ____ $\dfrac{3}{8}$

18. $\dfrac{7}{15}$ ____ $\dfrac{9}{20}$

19. $\dfrac{7}{12}$ ____ $\dfrac{11}{18}$

20. $\dfrac{17}{24}$ ____ $\dfrac{5}{6}$

21. $\dfrac{11}{18}$ ____ $\dfrac{5}{9}$

22. $\dfrac{13}{15}$ ____ $\dfrac{8}{9}$

23. $\dfrac{37}{50}$ ____ $\dfrac{13}{20}$

24. $\dfrac{7}{12}$ ____ $\dfrac{11}{20}$

Simplify. See Example 3.

25. $\left(\dfrac{1}{3}\right)^2$

26. $\left(\dfrac{2}{3}\right)^2$

27. $\left(\dfrac{5}{8}\right)^2$

28. $\left(\dfrac{7}{8}\right)^2$

29. $\left(\dfrac{3}{4}\right)^2$

30. $\left(\dfrac{3}{5}\right)^3$

31. $\left(\dfrac{4}{5}\right)^3$

32. $\left(\dfrac{4}{7}\right)^3$

33. $\left(\dfrac{3}{2}\right)^4$

34. $\left(\dfrac{4}{3}\right)^4$

35. $\left(\dfrac{3}{4}\right)^4$

36. $\left(\dfrac{2}{3}\right)^5$

37. Describe in your own words what a number line is, and draw a picture of one. Be sure to include how it works and how it can be used.

38. You have used the order of operations with whole numbers and again with fractions. List from memory the steps in the order of operations.

Use the order of operations to simplify each expression. See Example 4.

39. $2^4 - 4(3)$

40. $3^2 + 4(1)$

41. $3 \cdot 2^2 - \dfrac{6}{3}$

42. $5 \cdot 2^3 - \dfrac{6}{2}$

43. $\left(\dfrac{1}{2}\right)^2 \cdot 4$

44. $\left(\dfrac{1}{4}\right)^2 \cdot 4$

45. $\left(\dfrac{3}{4}\right)^2 \cdot \left(\dfrac{1}{3}\right)$

46. $\left(\dfrac{2}{3}\right)^3 \cdot \left(\dfrac{1}{2}\right)$

47. $\left(\dfrac{4}{5}\right)^2 \cdot \left(\dfrac{5}{6}\right)^2$

48. $\left(\dfrac{5}{8}\right)^2 \cdot \left(\dfrac{4}{25}\right)^2$

49. $6\left(\dfrac{2}{3}\right)^2\left(\dfrac{1}{2}\right)^3$

50. $9\left(\dfrac{1}{3}\right)^3\left(\dfrac{4}{3}\right)^2$

51. $\dfrac{3}{5}\left(\dfrac{1}{3}\right) + \dfrac{2}{5}\left(\dfrac{3}{4}\right)$

52. $\dfrac{1}{4}\left(\dfrac{3}{4}\right) + \dfrac{3}{8}\left(\dfrac{4}{3}\right)$

53. $\dfrac{1}{2} + \left(\dfrac{1}{2}\right)^2 - \dfrac{3}{8}$

54. $\dfrac{2}{3} + \left(\dfrac{1}{3}\right)^2 - \dfrac{5}{9}$

55. $\left(\dfrac{1}{3} + \dfrac{1}{6}\right) \cdot \dfrac{1}{2}$

56. $\left(\dfrac{3}{5} - \dfrac{3}{20}\right) \cdot \dfrac{4}{3}$

57. $\dfrac{9}{8} \div \left(\dfrac{2}{3} + \dfrac{1}{12}\right)$

58. $\dfrac{6}{5} \div \left(\dfrac{3}{5} - \dfrac{3}{10}\right)$

59. $\left(\dfrac{7}{8} - \dfrac{3}{4}\right) \div \dfrac{3}{2}$

60. $\left(\dfrac{4}{5} - \dfrac{3}{10}\right) \div \dfrac{4}{5}$

61. $\dfrac{3}{8}\left(\dfrac{1}{4} + \dfrac{1}{2}\right) \cdot \dfrac{32}{3}$

62. $\dfrac{1}{3}\left(\dfrac{4}{5} - \dfrac{3}{10}\right) \cdot \dfrac{4}{2}$

63. $\left(\dfrac{3}{4}\right)^2 - \left(\dfrac{1}{2} - \dfrac{1}{6}\right) \div \dfrac{4}{3}$

64. $\left(\dfrac{2}{3}\right)^2 - \left(\dfrac{5}{8} - \dfrac{1}{2}\right) \div \dfrac{3}{2}$

65. $\left(\dfrac{7}{8} - \dfrac{1}{4}\right) - \dfrac{2}{3}\left(\dfrac{3}{4}\right)^2$

66. $\left(\dfrac{5}{6} - \dfrac{7}{12}\right) - \dfrac{3}{4}\left(\dfrac{1}{3}\right)^2$

67. $\left(\dfrac{3}{4}\right)^2\left(\dfrac{2}{3} - \dfrac{5}{9}\right) - \dfrac{1}{4}\left(\dfrac{1}{8}\right)$

68. $\left(\dfrac{2}{3}\right)^2\left(\dfrac{1}{2} - \dfrac{1}{8}\right) - \dfrac{2}{3}\left(\dfrac{1}{8}\right)$

Solve each application problem.

69. In a study of workplace security, $\dfrac{9}{25}$ of the employees thought that more lighting on the grounds and parking lot was most important, while $\dfrac{7}{20}$ of the employees thought that limiting public access was most important. Which group represents the greater number of employees? (*Source:* Society for Human Resource Management.)

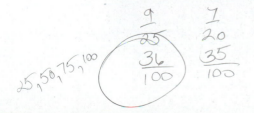

70. When adults were surveyed about conserving and recycling, $\dfrac{3}{4}$ of them said they conserve water while $\dfrac{19}{25}$ of them said they use reusable containers at home. Which group represents the greater number of adults? (*Source:* The NPD Group of the Rechargeable Battery Recycling Corporation.)

You often need to use order relations and the order of operations when solving problems.
Work Exercises 71–80 in order.

71. When comparing the size of two numbers, the symbol _____ means **is less than** and the symbol _____ means **is greater than.**

72. (a) To identify the greater of two or more fractions, we must first write the fractions as _____ fractions and then compare the _____. The fraction with the greater _____ is the greater fraction.

(b) Write four pairs of fractions, all with different denominators. Write the symbol for **less than** or for **greater than** between each pair.

73. Fill in the blanks to complete the order of operations.

1. Do all operations inside _____ or other grouping symbols.

2. Simplify any expressions with _____ and find any _____ roots.

3. _____ or _____ proceeding from left to right.

4. _____ or _____ proceeding from left to right.

74. Use the order of operations to simplify the following.

$$\left(\frac{2}{3}\right)^2 - \left(\frac{4}{5} - \frac{3}{10}\right) \div \frac{5}{4}$$

Simplify, then place the results on the number line.

75. $\left(\dfrac{2}{3}\right)^2$

76. $\left(\dfrac{3}{2}\right)^2$

77. $\left(\dfrac{3}{5}\right)^3$

78. $\left(\dfrac{5}{4}\right)^2$

79. $4 + 2 - 2^2$

80. $\left(\dfrac{5}{8}\right)^2 + \left(\dfrac{7}{8} - \dfrac{1}{4}\right) \div \dfrac{1}{4}$

Summary Exercises on Fractions

Write proper *or* improper *for each fraction.*

1. $\dfrac{3}{4}$

2. $\dfrac{4}{3}$

3. $\dfrac{10}{10}$

4. $\dfrac{11}{12}$

Write each fraction in lowest terms.

5. $\dfrac{30}{36}$

6. $\dfrac{175}{200}$

7. $\dfrac{15}{35}$

8. $\dfrac{115}{235}$

Add, subtract, multiply, or divide as indicated. Simplify all answers.

9. $\dfrac{3}{4} \cdot \dfrac{2}{3}$

10. $\dfrac{7}{12} \cdot \dfrac{9}{14}$

11. $56 \cdot \dfrac{5}{8}$

12. $\dfrac{5}{8} \div \dfrac{3}{4}$

13. $\dfrac{35}{45} \div \dfrac{10}{15}$

14. $21 \div \dfrac{3}{8}$

15. $\dfrac{7}{8} + \dfrac{2}{3}$

16. $\dfrac{5}{8} + \dfrac{3}{4} + \dfrac{7}{16}$

17. $\dfrac{7}{12} + \dfrac{5}{6} + \dfrac{2}{3}$

18. $\dfrac{5}{6} - \dfrac{3}{4}$

19. $\dfrac{7}{8} - \dfrac{5}{12}$

20. $\dfrac{4}{5} - \dfrac{2}{3}$

$\approx$ *First estimate the answer. Then add, subtract, multiply, or divide to find the exact answer.*

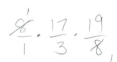

21. *Exact:*

$3\dfrac{1}{2} \cdot 2\dfrac{1}{4}$

Estimate:

___ • ___ = ___

22. *Exact:*

$5\dfrac{3}{8} \cdot 3\dfrac{1}{4}$

Estimate:

___ • ___ = ___

23. *Exact:*

$8 \cdot 5\dfrac{2}{3} \cdot 2\dfrac{3}{8}$

Estimate:

___ • ___ • ___ = ___

24. *Exact:*

$4\dfrac{3}{8} \div 3\dfrac{3}{4}$

Estimate:

___ ÷ ___ = ___

25. *Exact:*

$6\dfrac{7}{8} \div 2$

Estimate:

___ ÷ ___ = ___

26. *Exact:*

$4\dfrac{5}{8} \div \dfrac{3}{4}$

Estimate:

___ ÷ ___ = ___

27. *Estimate:*   *Exact:*

$\xleftarrow{\text{Rounds to}} \left\{ 5\dfrac{2}{3} \right.$

$+ \underline{\quad} \xleftarrow{\text{Rounds to}} \left\{ + 4\dfrac{1}{4} \right.$

28. *Estimate:*   *Exact:*

$18\dfrac{5}{12}$

$+ 9 \dfrac{3}{4}$

29. *Estimate:*   *Exact:*

$14\dfrac{3}{5}$

$+ 10\dfrac{2}{3}$

30. *Estimate:*   *Exact:*

$8\dfrac{3}{4}$

$- 3\dfrac{4}{5}$

31. *Estimate:*   *Exact:*

14

$- 7\dfrac{3}{8}$

32. *Estimate:*   *Exact:*

$31\dfrac{5}{6}$

$- 22\dfrac{7}{12}$

Simplify by using the order of operations.

33. $\dfrac{1}{5}\left(\dfrac{2}{3} - \dfrac{1}{4}\right)$

34. $\dfrac{3}{4} \div \left(\dfrac{1}{2} + \dfrac{1}{3}\right)$

35. $\dfrac{2}{3} + \left(\dfrac{2}{3}\right)^2 - \dfrac{5}{6}$

Find the least common multiple of each set of numbers.

36. 8, 10

37. 9, 18, 24

38. 4, 12, 21

Rewrite each fraction with the indicated denominator.

39. $\dfrac{5}{6} = \dfrac{\quad}{42}$

40. $\dfrac{3}{7} = \dfrac{\quad}{28}$

41. $\dfrac{11}{12} = \dfrac{\quad}{60}$

Write < or > to make a true statement.

42. $\dfrac{7}{8} \underline{\quad} \dfrac{15}{16}$

43. $\dfrac{16}{20} \underline{\quad} \dfrac{23}{30}$

44. $\dfrac{11}{15} \underline{\quad} \dfrac{7}{10}$

3.1	**like fractions**	Fractions with the same denominator are called *like fractions.*
	unlike fractions	Fractions with different denominators are called *unlike fractions.*
3.2	**least common multiple**	Given two or more whole numbers, the least common multiple is the smallest whole number that is divisible by all the numbers.
	LCM	The abbreviation for *least common multiple* is LCM.
3.3	**least common denominator**	When unlike fractions are rewritten as like fractions having the least common multiple as the denominator, the new denominator is the least common denominator.
	LCD	The abbreviation for *least common denominator* is LCD.
3.4	**carrying**	Carrying is the method used in adding mixed numbers when the sum of the fractions is greater than 1. Carry from the fraction to the whole number.
	borrowing	Borrowing is the method used in subtracting mixed numbers when the fraction part of the minuend is less than the fraction part of the subtrahend.

TEST YOUR WORD POWER

See how well you have learned the vocabulary in this chapter. Answers follow the Quick Review.

1. **Like fractions** are
   A. fractions that are equivalent
   B. fractions that are not equivalent
   C. fractions that have the same numerator
   D. fractions that have the same denominator.

2. Two or more fractions are **unlike fractions** if
   A. they are not equivalent
   B. they have different numerators
   C. they have different denominators
   D. they are improper fractions.

3. The **least common multiple** is
   A. the smallest whole number that is divisible by each of two or more numbers
   B. the smallest numerator
   C. the smallest denominator
   D. the smallest whole number that is not divisible by a group of numbers.

4. The abbreviation **LCM** stands for
   A. the largest common multiple
   B. the longest common multiple
   C. the most likely common multiplier
   D. the least common multiple.

5. The **least common denominator** is
   A. needed when multiplying fractions
   B. needed when dividing fractions
   C. the least common multiple of the denominators in a fraction problem
   D. any denominator that is common to a group of fractions.

6. The abbreviation **LCD** stands for
   A. the largest common denominator
   B. the least common denominator
   C. the least common divisor
   D. the most likely common denominator.

Concepts

Examples

3.1 Adding Like Fractions
Add numerators and write in lowest terms.

$$\frac{3}{4} + \frac{1}{4} + \frac{5}{4} = \frac{3 + 1 + 5}{4} = \frac{9}{4} = 2\frac{1}{4}$$

3.1 Subtracting Like Fractions
Subtract numerators and write in lowest terms.

$$\frac{7}{8} - \frac{5}{8} = \frac{7 - 5}{8} = \frac{2 \div 2}{8 \div 2} = \frac{1}{4}$$

3.2 Finding the Least Common Multiple (LCM)
Method of using multiples of the larger number: List the first few multiples of the larger number. Check each one until you find the multiple that is divisible by the smaller number.

$$\frac{1}{3} + \frac{1}{4}$$

4, 8, 12, 16, . . . ← Multiples of 4

First multiple divisible by 3 $(12 \div 3 = 4)$

The least common multiple (LCM) of 3 and 4 is 12.

3.2 Finding the Least Common Multiple (LCM)
Method of prime numbers: First find the prime factorization of each number. Then use the prime factors to build the least common multiple.

Factors of 9

$$9 = 3 \cdot 3$$
$$15 = 3 \cdot 5$$

$$LCM = 3 \cdot 3 \cdot 5 = 45$$

Factors of 15

The least common multiple (LCM) of 9 and 15 is 45.

3.3 Adding Unlike Fractions
Step 1 Find the least common multiple (LCM).
Step 2 Rewrite fractions with the least common multiple as the denominator.
Step 3 Add numerators, placing the sum over the common denominator, and simplify the answer.

$$\frac{1}{3} + \frac{1}{4} + \frac{1}{10} \qquad LCM = 60$$

$$\frac{1}{3} = \frac{20}{60} \quad \frac{1}{4} = \frac{15}{60} \quad \frac{1}{10} = \frac{6}{60}$$

$$\frac{20}{60} + \frac{15}{60} + \frac{6}{60} = \frac{41}{60} \qquad \text{Lowest terms}$$

3.3 Subtracting Unlike Fractions
Step 1 Find the least common multiple (LCM).
Step 2 Rewrite fractions with the least common multiple as the denominator.
Step 3 Subtract numerators, placing the difference over the common denominator, and simplifying the answer.

$$\frac{5}{8} - \frac{1}{3} \qquad LCM = 24$$

$$\frac{5}{8} = \frac{15}{24} \quad \frac{1}{3} = \frac{8}{24}$$

$$\frac{15}{24} - \frac{8}{24} = \frac{7}{24} \qquad \text{Lowest Terms}$$

Concepts	Examples
3.4 *Adding Mixed Numbers*	*Estimate:* *Exact:*

3.4 *Adding Mixed Numbers*

Round the numbers and estimate the answer. Then find the exact answer using these steps.

Step 1 Add fractions using a common denominator.

Step 2 Add whole numbers.

Step 3 Combine the sums of whole numbers and fractions, simplifying the fraction part when necessary.

Compare the exact answer to the estimate to see if it is reasonable.

Estimate:

$$10 \xleftarrow{\text{Rounds to}} \begin{cases} 9\frac{2}{3} = 9\frac{8}{12} \\ \\ + 6\frac{3}{4} = 6\frac{9}{12} \end{cases}$$

$$+ 7 \xleftarrow{\text{Rounds to}}$$

$$17 \qquad\qquad 15\frac{17}{12} = 16\frac{5}{12}$$

The exact answer of $16\frac{5}{12}$ is reasonable because it is close to the estimate of 17.

3.4 *Subtracting Mixed Numbers*

Round the numbers and estimate the answer. Then find the exact answer using these steps.

Step 1 Subtract fractions, using borrowing if necessary.

Step 2 Subtract whole numbers.

Step 3 Combine the differences of whole numbers and fractions, simplifying the fraction part when necessary.

Compare the exact answer to the estimate to see if it is reasonable.

Estimate: *Exact:*

$$9 \xleftarrow{\text{Rounds to}} \begin{cases} 8\frac{5}{8} = 8\frac{15}{24} = 7\frac{39}{24} \\ \\ - 3\frac{11}{12} = 3\frac{22}{24} = 3\frac{22}{24} \end{cases}$$

$$- 4 \xleftarrow{\text{Rounds to}}$$

$$5 \qquad\qquad 4\frac{17}{24}$$

The exact answer of $4\frac{17}{24}$ is reasonable because it is close to the estimate of 5.

3.4 *Adding or Subtracting Mixed Numbers Using an Alternate Method*

Step 1 Change the mixed numbers to improper fractions.

Step 2 Rewrite the unlike fractions as like fractions.

Step 3 Add or subtract the numerators and simplify the answer.

Add.

$$2\frac{2}{3} = \frac{8}{3} = \frac{64}{24}$$

$$+ 1\frac{3}{8} = \frac{11}{8} = + \frac{33}{24}$$ ⟵ Least common denominator

$$\frac{97}{24} = 4\frac{1}{24}$$ Answer as mixed number

↑ Improper fractions

Subtract.

$$8\frac{2}{3} = \frac{26}{3} = \frac{104}{12}$$

$$- 5\frac{3}{4} = \frac{23}{4} = - \frac{69}{12}$$ ⟵ Least common denominator

$$\frac{35}{12} = 2\frac{11}{12}$$ Answer as mixed number

↑ Improper fractions

Concepts	*Examples*

3.5 *Identifying the Larger of Two Fractions*

With unlike fractions, change to like fractions first. The fraction with the greater numerator is the greater fraction. Use these symbols:

$<$ is less than

$>$ is greater than

Identify the greater fraction.

$$\frac{7}{8}, \frac{9}{10}$$

$$\frac{7}{8} = \frac{7 \cdot 5}{8 \cdot 5} = \frac{35}{40}$$

$$\frac{9}{10} = \frac{9 \cdot 4}{10 \cdot 4} = \frac{36}{40}$$

$\frac{35}{40}$ is smaller than $\frac{36}{40}$, so $\frac{7}{8} < \frac{9}{10}$ or $\frac{9}{10} > \frac{7}{8}$.

$\frac{9}{10}$ is greater.

3.5 *Using the Order of Operations with Fractions*

Follow the order of operations.
1. Do all operations inside parentheses or other grouping symbols.
2. Simplify any expressions with exponents and find any square roots.
3. Multiply or divide proceeding from left to right.
4. Add or subtract proceeding from left to right.

Simplify by using the order of operations.

$$\frac{1}{2}\left(\frac{2}{3}\right) - \left(\frac{1}{4}\right)^2 \quad \text{Simplify fraction with exponent.}$$

$$= \frac{1}{\overset{1}{\cancel{2}}}\left(\frac{\overset{1}{\cancel{2}}}{3}\right) - \frac{1}{16} \quad \text{Next, multiply.}$$

$$= \frac{1}{3} - \frac{1}{16}$$

$$= \frac{16}{48} - \frac{3}{48} \quad \text{Change to common denominator and subtract.}$$

$$= \frac{13}{48}$$

ANSWERS TO TEST YOUR WORD POWER

1. D; *Example:* Because the fractions $\frac{3}{8}$ and $\frac{10}{8}$ both have 8 as a denominator, they are like fractions.

2. C; *Example:* The fractions $\frac{2}{3}$ and $\frac{3}{4}$ are unlike fractions because they have different denominators.

3. A; *Example:* The least common multiple of 4 and 5 is 20 because 20 is the smallest number into which 4 and 5 will divide evenly.

4. D; *Example:* LCM is the abbreviation for least common multiple.

5. C; *Example:* The least common denominator of the fractions $\frac{2}{3}$ and $\frac{1}{2}$ is 6 because 6 is the least common multiple of 3 and 2.

   When written using the least common denominator, $\frac{2}{3}$ and $\frac{1}{2}$ become $\frac{4}{6}$ and $\frac{3}{6}$, respectively.

6. B; *Example:* LCD is the abbreviation for least common denominator.

Chapter 3
REVIEW EXERCISES

[3.1] *Add or subtract. Write answers in lowest terms.*

1. $\dfrac{5}{7} + \dfrac{1}{7}$

2. $\dfrac{4}{9} + \dfrac{3}{9}$

3. $\dfrac{1}{8} + \dfrac{3}{8} + \dfrac{2}{8}$

4. $\dfrac{5}{16} - \dfrac{3}{16}$

5. $\dfrac{5}{10} + \dfrac{3}{10}$

6. $\dfrac{5}{12} - \dfrac{3}{12}$

7. $\dfrac{36}{62} - \dfrac{10}{62}$

8. $\dfrac{68}{75} - \dfrac{43}{75}$

Solve each application problem. Write answers in lowest terms.

9. Nurse Suzie Brasher screened $\frac{7}{16}$ of her patients in her first hour on duty and $\frac{5}{16}$ of her patients in the second hour. What fraction of her patients did she screen in the two hours?

10. The Koats for Kids committee members completed $\frac{5}{8}$ of their Web-page design in the morning and $\frac{3}{8}$ in the afternoon. How much less did they complete in the afternoon than in the morning?

[3.2] *Find the least common multiple of each set of numbers.*

11. 5, 2

12. 3, 4

13. 10, 12, 20

14. 3, 8, 4

15. 6, 8, 5, 15

16. 15, 9, 20

Rewrite each fraction using the indicated denominator.

17. $\dfrac{2}{3} = \dfrac{}{12}$

18. $\dfrac{3}{8} = \dfrac{}{56}$

19. $\dfrac{2}{5} = \dfrac{}{25}$

20. $\dfrac{5}{9} = \dfrac{}{81}$

21. $\dfrac{4}{5} = \dfrac{}{40}$

22. $\dfrac{5}{16} = \dfrac{}{64}$

[3.1–3.3] *Add or subtract. Write answers in lowest terms.*

23. $\dfrac{1}{2} + \dfrac{1}{3}$

24. $\dfrac{1}{5} + \dfrac{3}{10} + \dfrac{3}{8}$

25. $\begin{array}{r} \dfrac{5}{12} \\ + \dfrac{5}{24} \\ \hline \end{array}$

26. $\dfrac{2}{3} - \dfrac{1}{4}$

27. $\begin{array}{r} \dfrac{7}{8} \\ - \dfrac{1}{3} \\ \hline \end{array}$

28. $\begin{array}{r} \dfrac{11}{12} \\ - \dfrac{4}{9} \\ \hline \end{array}$

Solve each application problem.

29. The master gardener for a public garden used $\frac{3}{8}$ sack of fertilizer on the lawn, $\frac{1}{4}$ sack of fertilizer on the trees, and $\frac{1}{3}$ sack of fertilizer on the flower beds. How much of the fertilizer did he use?

30. Zallia Todd is planning an 80th birthday party for her husband, Ralph. One-fourth of her budget is for renting a hall, $\frac{1}{3}$ of her budget is for invitations, decorations, and entertainment, and $\frac{2}{5}$ of her budget is for food and beverages. Find the portion of her budget that she will be spending on these three categories.

≈ **[3.4]** *First estimate the answer: Then add or subtract to find the exact answer. Write exact answers as mixed numbers.*

31. *Estimate:*   *Exact:*

⟵ Rounds to $\left\{ 18\dfrac{5}{8} \right.$

$+$ ___ ⟵ Rounds to $\left\{ + 13\dfrac{3}{4} \right.$

___

32. *Estimate:*   *Exact:*

$22\dfrac{2}{3}$

$+$ ___ $+ 15\dfrac{4}{9}$

___

33. *Estimate:*   *Exact:*

$12\dfrac{3}{5}$

$8\dfrac{5}{8}$

$+$ ___ $+ 10\dfrac{5}{16}$

___

34. *Estimate:*   *Exact:*

$31\dfrac{3}{4}$

$-$ ___ $- 14\dfrac{2}{3}$

___

35. *Estimate:*   *Exact:*

34

$-$ ___ $- 15\dfrac{2}{3}$

___

36. *Estimate:*   *Exact:*

$215\dfrac{7}{16}$

$-$ ___ $- 136$

___

Add or subtract by changing mixed numbers to improper fractions. Simplify all answers.

37. $5\dfrac{2}{5}$

$+ 3\dfrac{7}{10}$

___

38. $4\dfrac{3}{4}$

$+ 5\dfrac{2}{3}$

___

39. 5

$- 1\dfrac{3}{4}$

___

40. $6\dfrac{1}{2}$

$- 4\dfrac{5}{6}$

___

41. $8\dfrac{1}{3}$

$- 2\dfrac{5}{6}$

___

42. $5\dfrac{5}{12}$

$- 2\dfrac{5}{8}$

___

≈
First estimate the answer. Then solve each application problem.

43. A medical laboratory had $14\frac{2}{3}$ gallons of distilled water. If $5\frac{1}{2}$ gallons were used in the morning and $6\frac{3}{4}$ gallons were used in the afternoon, find the number of gallons remaining.

Estimate:

Exact:

44. The Boys and Girls Clubs of America collected $28\frac{2}{3}$ tons of newspapers on Saturday and $24\frac{3}{4}$ tons on Sunday. Find the total weight of the newspapers collected.

Estimate:

Exact:

45. In a recent bass-fishing derby, Darrel Holmes caught three largemouth bass weighing $8\frac{7}{8}$ pounds, $9\frac{1}{3}$ pounds, and $6\frac{3}{4}$ pounds. Find their total weight.

Estimate:

Exact:

46. A developer wants to build a shopping center. She bought two parcels of land, one, $1\frac{11}{16}$ acres, and the other, $2\frac{3}{4}$ acres. If she needs a total of $8\frac{1}{2}$ acres for the center, how much additional land does she need to buy?

Estimate:

Exact:

[3.5] *Locate each fraction in Exercises 47–50 on the number line.*

47. $\frac{3}{8}$

48. $\frac{7}{4}$

49. $\frac{8}{3}$

50. $3\frac{1}{5}$

Write < or > in each blank to make a true statement.

51. $\frac{2}{3}$ ___ $\frac{3}{4}$

52. $\frac{3}{4}$ ___ $\frac{7}{8}$

53. $\frac{1}{2}$ ___ $\frac{7}{15}$

54. $\frac{7}{10}$ ___ $\frac{8}{15}$

55. $\frac{9}{16}$ ___ $\frac{5}{8}$

56. $\frac{7}{20}$ ___ $\frac{8}{25}$

57. $\frac{19}{36}$ ___ $\frac{29}{54}$

58. $\frac{19}{132}$ ___ $\frac{7}{55}$

Simplify each expression.

59. $\left(\frac{1}{2}\right)^2$

60. $\left(\frac{2}{3}\right)^2$

61. $\left(\frac{3}{10}\right)^3$

62. $\left(\frac{3}{8}\right)^4$

Simplify by using the order of operations.

63. $8\left(\dfrac{1}{4}\right)^2$

64. $12\left(\dfrac{3}{4}\right)^2$

65. $\left(\dfrac{2}{3}\right)^2 \cdot \left(\dfrac{3}{8}\right)^2$

66. $\dfrac{7}{8} \div \left(\dfrac{1}{8} + \dfrac{3}{4}\right)$

67. $\left(\dfrac{1}{2}\right)^2 \cdot \left(\dfrac{1}{4} + \dfrac{1}{2}\right)$

68. $\left(\dfrac{1}{4}\right)^3 + \left(\dfrac{5}{8} + \dfrac{3}{4}\right)$

MIXED REVIEW EXERCISES

Simplify by using the order of operations as necessary. Write answers in lowest terms and as whole or as mixed numbers when possible.

69. $\dfrac{7}{8} - \dfrac{1}{8}$

70. $\dfrac{7}{10} - \dfrac{3}{10}$

71. $\dfrac{29}{32} - \dfrac{5}{16}$

72. $\dfrac{1}{4} + \dfrac{1}{8} + \dfrac{5}{16}$

73. $\begin{array}{r} 6\frac{2}{3} \\ - 4\frac{1}{2} \\ \hline \end{array}$

74. $\begin{array}{r} 9\frac{1}{2} \\ + 16\frac{3}{4} \\ \hline \end{array}$

75. $\begin{array}{r} 7 \\ - 1\frac{5}{8} \\ \hline \end{array}$

76. $\begin{array}{r} 2\frac{3}{5} \\ 8\frac{5}{8} \\ + \frac{5}{16} \\ \hline \end{array}$

77. $\begin{array}{r} 32\frac{5}{12} \\ - 17 \\ \hline \end{array}$

78. $\dfrac{7}{22} + \dfrac{3}{22} + \dfrac{3}{11}$

79. $\left(\dfrac{1}{4}\right)^2 \cdot \left(\dfrac{2}{5}\right)^3$

80. $\dfrac{3}{8} \div \left(\dfrac{1}{2} + \dfrac{1}{4}\right)$

81. $\left(\dfrac{2}{3}\right)^2 \cdot \left(\dfrac{1}{3} + \dfrac{1}{6}\right)$

82. $\left(\dfrac{2}{3}\right)^3 + \left(\dfrac{2}{3} - \dfrac{5}{9}\right)$

Write $<$ or $>$ in each blank to make a true statement.

83. $\dfrac{2}{3}$ _____ $\dfrac{7}{12}$

84. $\dfrac{8}{9}$ _____ $\dfrac{15}{8}$

85. $\dfrac{17}{30}$ _____ $\dfrac{36}{60}$

86. $\dfrac{5}{8}$ _____ $\dfrac{17}{30}$

Find the least common multiple of each set of numbers.

87. 12, 18

88. 6, 8, 10, 12

89. 9, 14, 21

Rewrite each fraction using the indicated denominator.

90. $\dfrac{2}{3} = \dfrac{}{27}$

91. $\dfrac{9}{12} = \dfrac{}{144}$

92. $\dfrac{4}{5} = \dfrac{}{75}$

≈
First estimate the answer. Then solve each application problem.

93. A cement contractor needs $13\frac{1}{2}$ ft of wire mesh for a concrete walkway and $22\frac{3}{8}$ ft of wire mesh for a driveway. If the contractor starts with a roll of wire that is $92\frac{3}{4}$ ft long, find the number of feet remaining after the two jobs have been completed.

Estimate:

Exact:

94. The Oak Park Garden Group used $10\frac{7}{8}$ cubic yards of mulch for their tomato garden and $25\frac{1}{2}$ cubic yards for their corn and squash garden. If they started with 45 cubic yards of mulch, how many cubic yards remain?

Estimate:

Exact:

Chapter 3

TEST

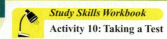
Study Skills Workbook
Activity 10: Taking a Test

Add or subtract. Write answers in lowest terms.

1. $\dfrac{5}{8} + \dfrac{1}{8}$

2. $\dfrac{1}{16} + \dfrac{7}{16}$

3. $\dfrac{7}{10} - \dfrac{3}{10}$

4. $\dfrac{7}{12} - \dfrac{5}{12}$

Find the least common multiple of each set of numbers.

5. 2, 3, 4

6. 6, 3, 5, 15

7. 6, 9, 27, 36

Add or subtract. Write answers in lowest terms.

8. $\dfrac{3}{8} + \dfrac{1}{4}$

9. $\dfrac{2}{9} + \dfrac{5}{12}$

10. $\dfrac{7}{8} - \dfrac{2}{3}$

11. $\dfrac{2}{5} - \dfrac{3}{8}$

$\approx$ *First estimate the answer. Then add or subtract to find the exact answer; simplify exact answers.*

12. $7\dfrac{2}{3} + 4\dfrac{5}{6}$

13. $16\dfrac{2}{5} - 11\dfrac{2}{3}$

14. $18\dfrac{3}{4} + 9\dfrac{2}{5} + 12\dfrac{1}{3}$

15. $24 - 18\dfrac{3}{8}$

1. _____

2. _____

3. _____

4. _____

5. _____

6. _____

7. _____

8. _____

9. _____

10. _____

11. _____

12. Estimate: _____

   Exact: _____

13. Estimate: _____

   Exact: _____

14. Estimate: _____

   Exact: _____

15. Estimate: _____

   Exact: _____

16. _____

16. Most students say that "addition and subtraction of fractions is more difficult than multiplication and division of fractions." Why do you think they say this? Do you agree with these students?

17. _____

17. Devise and explain a method of estimating an answer to addition and subtraction problems involving mixed numbers. Might your estimated answer vary from the exact answer? If it did, what would the estimation accomplish?

≈ *First estimate the answer. Then solve each application problem.*

18. *Estimate:* _____

  Exact: _____

18. A professional football player trains with bodybuilding, jogging, and wind sprints. He trains $5\frac{2}{3}$ hours on Monday, $4\frac{3}{4}$ hours on Tuesday, $3\frac{5}{6}$ hours on Wednesday, $7\frac{1}{3}$ hours on Thursday, and $5\frac{1}{6}$ hours on Friday. Find the total number of hours that he trains.

19. *Estimate:* _____

  Exact: _____

19. A painting contractor arrived at a 6-unit apartment complex with $147\frac{1}{2}$ gallons of exterior paint. If his crew sprayed $68\frac{1}{2}$ gallons on the wood siding, rolled $37\frac{3}{8}$ gallons on the masonry exterior, and brushed $5\frac{3}{4}$ gallons on the trim, find the number of gallons of paint remaining.

Write $<$ or $>$ to make a true statement.

20. _____

20. $\dfrac{3}{4}$ _____ $\dfrac{17}{24}$

21. _____

21. $\dfrac{19}{24}$ _____ $\dfrac{17}{36}$

Simplify. Use the order of operations as needed.

22. _____

22. $\left(\dfrac{1}{3}\right)^3 \cdot 54$

23. _____

23. $\left(\dfrac{3}{4}\right)^2 - \left(\dfrac{7}{8} \cdot \dfrac{1}{3}\right)$

24. _____

24. $4\left(\dfrac{7}{8} - \dfrac{7}{16}\right)$

25. _____

25. $\dfrac{5}{6} + \dfrac{4}{3}\left(\dfrac{3}{8}\right)$

Cumulative Review Exercises

For each number, name the digit that has the given place value.

1. 871

hundreds

ones

2. 5,629,428

millions

thousands

Round each number to the nearest ten, nearest hundred, and nearest thousand.

	Ten	**Hundred**	**Thousand**
3. 1438	_____	_____	_____
4. 59,803	_____	_____	_____

≈ *Use front end rounding to estimate each answer. Then add, subtract, multiply, or divide to find the exact answer.*

5. *Estimate:*   *Exact:*

← Rounds to   2361

← Rounds to   386

← Rounds to   47,304

+ _____ ← Rounds to   + 29,728

6. *Estimate:*   *Exact:*

− _____   24,276

                         − 9 887

7. *Estimate:*   *Exact:*

                                      4468

× _____   ×   280

8. *Estimate:*   *Exact:*

_____)   35)112,385

Add, subtract, multiply, or divide as indicated.

9.   4
         8
         5
       + 9

10.   375,899
          521,742
        + 357,968

11.   1687
        − 1096

12.   3,896,502
        − 1,094,807

13. $3 \times 8 \times 5$

14. $6 \cdot 3 \cdot 7$

15. $5(8)(4)$

16.   57
        × 8

17.   962
        × 384

18.   340
        ×   50

19. 8)1080

20. $13,467 \div 5$

21. 506)16,358

≈ *Use front end rounding to estimate the answer to each application problem. Then find the exact answer.*

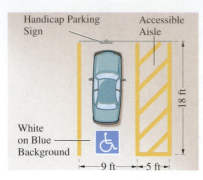

Handicap Parking Sign

Accessible Aisle

White on Blue Background

18 ft

9 ft   5 ft

22. The Americans with Disabilities Act provides the single parking space design shown at the left. Find the perimeter of (distance around) this parking space, including the accessible aisle.

Estimate:

Exact:

23. The single parking space design in Exercise 22 measures 18 ft by 14 ft. Find its area.

Estimate:

Exact:

24. Find the number of 32-ounce cartons of orange juice that can be filled with 40,320 ounces of orange juice.

Estimate:

Exact:

25. A dentist's drill makes 3600 revolutions each minute. How many revolutions would it make in 60 minutes of operation?

Estimate:

Exact:

≈ *Round the mixed numbers in each problem to the nearest whole number and estimate the answer. Then find the exact answer.*

26. The top of a rectangular pool table is $1\frac{3}{4}$ yards by $2\frac{2}{3}$ yards. Find its area.

Estimate:

Exact:

27. A rectangular-shaped off-road vehicle park is $2\frac{5}{8}$ miles wide and $4\frac{2}{3}$ miles long. How many square miles are in the park?

Estimate:

Exact:

$2\frac{2}{3}$ yards

$1\frac{3}{4}$ yards

28. Larry Foxworthy cuts, splits, and delivers firewood. If his truck, when fully loaded, holds $5\frac{1}{4}$ cords of firewood, find the number of cords he could deliver in $3\frac{1}{2}$ loads.

Estimate:

Exact:

29. The Sears Tower in Chicago is 110 stories tall. The total height of the building is $1536\frac{7}{8}$ ft, including a flagpole at the top of the building that is $82\frac{1}{2}$ ft tall. Find the height of the building itself. (*Source: Top 10 of Everything.*)

Estimate:

Exact:

Find the prime factorization of each number. Write the answers using exponents.

30. 100

31. 128

32. 1225

Simplify.

33. $2^4 \cdot 3^2$

34. $4^2 \cdot 2^4$

35. $6^2 \cdot 3^3$

Find each square root.

36. $\sqrt{36}$

37. $\sqrt{81}$

38. $\sqrt{144}$

Simplify by using the order of operations.

39. $5^2 - 2(8)$

$25 - 2(8)$
$25 - 16 = 9$

40. $\sqrt{25} + 5 \cdot 9 - 6$

41. $\dfrac{2}{3}\left(\dfrac{4}{5} - \dfrac{2}{3}\right)$

42. $\dfrac{3}{4} \div \left(\dfrac{1}{3} + \dfrac{1}{2}\right)$

43. $\dfrac{7}{8} + \left(\dfrac{3}{4}\right)^2 - \dfrac{3}{8}$

Write proper or improper for each fraction.

44. $\dfrac{5}{6}$

45. $\dfrac{6}{6}$

46. $\dfrac{11}{10}$

Write each fraction in lowest terms.

47. $\dfrac{25}{60}$

48. $\dfrac{84}{96}$

49. $\dfrac{63}{70}$

Add, subtract, multiply, or divide as indicated. Simplify all answers.

50. $\dfrac{3}{4} \cdot \dfrac{2}{3}$

51. $\dfrac{3}{8} \cdot \dfrac{5}{6}$

52. $42 \cdot \dfrac{7}{8}$

53. $\dfrac{5}{8} \div \dfrac{3}{8}$

54. $\dfrac{25}{40} \div \dfrac{10}{35}$

55. $9 \div \dfrac{2}{3}$

56. $\dfrac{3}{4} + \dfrac{1}{7}$

57. $\dfrac{3}{8} + \dfrac{3}{16} + \dfrac{1}{4}$

58. $\dfrac{11}{18} - \dfrac{5}{12}$

≈ *First estimate the answer. Then add or subtract to find the exact answer. Write exact answers as mixed numbers.*

59. *Estimate:*   *Exact:*

Rounds to $\left\{ 3\dfrac{3}{8} \right.$

$+$ Rounds to $\left\{ + 4\dfrac{1}{2} \right.$

60. *Estimate:*   *Exact:*

$21\dfrac{7}{8}$

$+$ $+ 4\dfrac{5}{12}$

61. *Estimate:*   *Exact:*

5

$-$ $- 2\dfrac{3}{8}$

Find the least common multiple of each set of numbers.

62. 8, 12

63. 3, 8, 15

64. 12, 16, 18

Rewrite each fraction using the indicated denominator.

65. $\dfrac{4}{5} = \dfrac{}{45}$

66. $\dfrac{7}{9} = \dfrac{}{72}$

67. $\dfrac{9}{15} = \dfrac{}{135}$

68. $\dfrac{5}{7} = \dfrac{}{84}$

Locate each fraction in Exercises 69–72 on the number line.

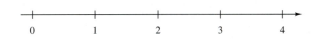

69. $2\dfrac{3}{4}$

70. $\dfrac{1}{9}$

71. $\dfrac{5}{3}$

72. $\dfrac{10}{3}$

Write < or > in each blank to make a true statement.

73. $\dfrac{3}{5}$ ____ $\dfrac{5}{8}$

74. $\dfrac{17}{20}$ ____ $\dfrac{3}{4}$

75. $\dfrac{7}{12}$ ____ $\dfrac{11}{18}$

Decimals

4

Nearly 49 million Americans go fishing at least once a year, making it America's fourth most popular recreational activity. (*Source:* National Sporting Goods Association.) In **Section 4.3,** Exercises 47–50, these friends will use decimal numbers when paying for new equipment. But will decimals help them catch their limit? (See **Section 4.1,** Exercises 59–62, and **Section 4.6,** Exercises 65–68.)

4.1 Reading and Writing Decimals

OBJECTIVES

1. Write parts of a whole using decimals.
2. Identify the place value of a digit.
3. Read and write decimals in words.
4. Write decimals as fractions or mixed numbers.

Fractions are used to represent parts of a whole. In this chapter, **decimals** are used as another way to show parts of a whole. For example, our money system is based on decimals. One dollar is divided into 100 equivalent parts. One cent ($0.01) is one of the parts, and a dime ($0.10) is 10 of the parts. Metric measurement (see **Chapter 7**) is also based on decimals.

OBJECTIVE **1** **Write parts of a whole using decimals.** Decimals are used when a whole is divided into 10 equivalent parts, or into 100 or 1000 or 10,000 equivalent parts. In other words, decimals are fractions with denominators that are a power of 10. For example, the square below is cut into 10 equivalent parts. Written as a fraction, each part is $\frac{1}{10}$ of the whole. Written as a decimal, each part is 0.1. Both $\frac{1}{10}$ and 0.1 are read as "*one tenth.*"

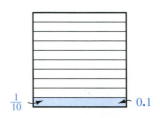

One-tenth of the square is shaded.

The dot in 0.1 is called the **decimal point.**

$$0.1$$
↑
Decimal point

1. There are 10 dimes in one dollar. Each dime is $\frac{1}{10}$ of a dollar. Write the yellow shaded portion of each dollar as a fraction, as a decimal, and in words.

(a)

(b)

(c)

The square at the right has **7** of its 10 parts shaded.

Written as a *fraction,* $\frac{7}{10}$ of the square is shaded.

Written as a *decimal,* 0.**7** of the square is shaded.

Both $\frac{7}{10}$ and 0.7 are read as "*seven tenths.*"

$\frac{7}{10}$

0.7

Seven-tenths of the square is shaded.

◄◄◄ Work Problem 1 at the Side.

The square below is cut into 100 equivalent parts. Written as a *fraction,* each part is $\frac{1}{100}$ of the whole.

Written as a decimal, each part is **0.01** of the whole. Both $\frac{1}{100}$ and 0.01 are read as "one hundredth."

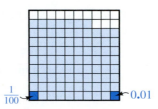

$\frac{1}{100}$ 0.01

The square above has 87 parts shaded.

Written as a fraction, $\frac{87}{100}$ of the total area is shaded.

Written as a decimal, **0.87** of the total area is shaded.
Both $\frac{87}{100}$ and 0.87 are read as "*eighty-seven hundredths.*"

ANSWERS

1. (a) $\frac{1}{10}$; 0.1; one tenth

(b) $\frac{3}{10}$; 0.3; three tenths

(c) $\frac{9}{10}$; 0.9; nine tenths

Work Problem 2 at the Side. ▶▶▶

Example 1 below shows several numbers written as fractions, as decimals, and in words.

EXAMPLE 1 **Using the Decimal Forms of Fractions**

	Fraction	Decimal	Read As
(a)	$\dfrac{4}{10}$	0.4	four tenths
(b)	$\dfrac{9}{100}$	0.09	nine hundredths
(c)	$\dfrac{71}{100}$	0.71	seventy-one hundredths
(d)	$\dfrac{8}{1000}$	0.008	eight thousandths
(e)	$\dfrac{45}{1000}$	0.045	forty-five thousandths
(f)	$\dfrac{832}{1000}$	0.832	eight hundred thirty-two thousandths

Work Problem 3 at the Side. ▶▶▶

OBJECTIVE 2 **Identify the place value of a digit.** The decimal point separates the *whole number part* from the *fractional part* in a decimal number. In the chart below, you see that the **place value** names for fractional parts are similar to those on the whole number side, but end in "*ths*."

Decimal Place Value Chart

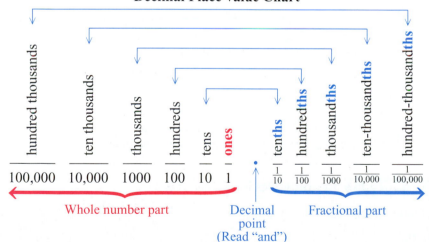

hundred thousands	ten thousands	thousands	hundreds	tens	**ones**	**tenths**	**hundredths**	**thousandths**	**ten-thousandths**	**hundred-thousandths**
100,000	10,000	1000	100	10	1	$\frac{1}{10}$	$\frac{1}{100}$	$\frac{1}{1000}$	$\frac{1}{10,000}$	$\frac{1}{100,000}$

Whole number part   Decimal point (Read "and")   Fractional part

NOTE
Notice that the **ones** place is at the center of the place value chart. There is no "oneths" place.

Also notice that each place is 10 times the value of the place to its right.

Finally, be sure to write a hyphen (dash) in ten-thousand**ths** and hundred-thousand**ths**.

2 Write the portion of each square that is shaded as a fraction, as a decimal, and in words.

(a)

(b)

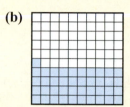

3 Write each decimal as a fraction.

(a) 0.7 $\dfrac{7}{10}$

(b) 0.2 $\dfrac{2}{10}$

(c) 0.03 $\dfrac{3}{100}$

(d) 0.69 $\dfrac{69}{100}$

(e) 0.047 $\dfrac{47}{1000}$

(f) 0.351 $\dfrac{351}{1000}$

ANSWERS

2. **(a)** $\dfrac{3}{10}$; 0.3; three tenths

    (b) $\dfrac{41}{100}$; 0.41; forty-one hundredths

3. **(a)** $\dfrac{7}{10}$ **(b)** $\dfrac{2}{10}$ **(c)** $\dfrac{3}{100}$

    (d) $\dfrac{69}{100}$ **(e)** $\dfrac{47}{1000}$ **(f)** $\dfrac{351}{1000}$

4 Identify the place value of each digit.

(a) 971.54

(b) 0.4

(c) 5.60

(d) 0.0835

5 Tell how to read each decimal in words.

(a) 0.6

(b) 0.46

(c) 0.05

(d) 0.409

(e) 0.0003

(f) 0.0703

(g) 0.088

CAUTION
If a number does *not* have a decimal point, it is a *whole number*. A whole number has no fractional part. If you want to show the decimal point in a whole number, it is just to the *right* of the digit in the ones place. Here are two examples.

$$8 = 8. \qquad\qquad 306 = 306.$$

Decimal point   Decimal point

EXAMPLE 2 Identifying the Place Value of a Digit

Identify the place value of each digit.

(a) 178.36   **(b)** 0.00935

Notice in Example 2(b) that we do *not* use commas on the right side of the decimal point.

◀◀◀ **Work Problem 4 at the Side.**

OBJECTIVE 3 **Read and write decimals in words.** A decimal is read according to its form as a fraction.

ones · tenths
0 . 9

We read 0.9 as "nine tenths" because 0.9 is the same as $\frac{9}{10}$. Notice that 0.9 ends in the tenths place.

ones · tenths hundredths
0 . 0 2

We read 0.02 as "two hundredths" because 0.02 is the same as $\frac{2}{100}$. Notice that 0.02 ends in the hundredths place.

EXAMPLE 3 Reading Decimal Numbers

Tell how to read each decimal in words.

(a) 0.3

Because $0.3 = \frac{3}{10}$, read the decimal as three <u>tenths</u>.

(b) 0.49   Read it as: forty-nine <u>hundredths</u>.

(c) 0.08   Read it as: eight <u>hundredths</u>.

(d) 0.918   Read it as: nine hundred eighteen <u>thousandths</u>.

(e) 0.0106   Read it as: one hundred six <u>ten-thousandths</u>.

◀◀◀ **Work Problem 5 at the Side.**

ANSWERS

4. (a)   hundreds tens ones · tenths hundredths   9 7 1 . 5 4   (b)   ones · tenths   0 . 4

(c)   ones · tenths hundredths   5 . 6 0   (d)   ones · tenths hundredths thousandths ten-thousandths   0 . 0 8 3 5

5. (a) six tenths
   (b) forty-six hundredths
   (c) five hundredths
   (d) four hundred nine thousandths
   (e) three ten-thousandths
   (f) seven hundred three ten-thousandths
   (g) eighty-eight thousandths

Reading Decimal Numbers

Step 1   Read any whole number part to the ***left*** of the decimal point as you normally would.

Step 2   Read the decimal point as "***and.***"

Step 3   Read the part of the number to the ***right*** of the decimal point as if it were an ordinary whole number.

Step 4   Finish with the place value name of the rightmost digit; these names all end in "***ths.***"

> **NOTE**
> If there is *no whole number part*, you will use only Steps 3 and 4.

EXAMPLE 4 **Reading Decimals**

Read each decimal.

(a)

┌──→ 9 is in tenths place. ─┐

16.9

sixteen **and** nine **tenths** ←

16.9 is read "sixteen and nine tenths."

(b)

┌──→ 5 is in hundredths place.─┐

482.35

four hundred eighty-two **and** thirty-five **hundredths** ←

482.35 is read "four hundred eighty-two and thirty-five hundredths."

┌──→ 3 is in thousandths place.

(c) 0.063 is "sixty-three **thousandths**." (No whole number part.)

(d) 11.1085 is "eleven **and** one thousand eighty-five **ten-thousandths**."

> **CAUTION**
> Use "and" *only* when reading a decimal point. A common mistake is to read the whole number 405 as "four hundred *and* five." But there is *no decimal point* shown in 405, so it is read "four hundred five."

Work Problem 6 at the Side. ▶▶▶

OBJECTIVE 4 Write decimals as fractions or mixed numbers. Knowing how to read decimals will help you when writing decimals as fractions.

Writing Decimals as Fractions or Mixed Numbers

Step 1   The digits to the right of the decimal point are the numerator of the fraction.

Step 2   The denominator is 10 for tenths, 100 for hundredths, 1000 for thousandths, 10,000 for ten-thousandths, and so on.

Step 3   If the decimal has a whole number part, the fraction will be a mixed number with the same whole number part.

6 Tell how to read each decimal in words.

(a) 3.8

(b) 15.001

(c) 0.0073

(d) 64.309

7 Write each decimal as a fraction or mixed number.

(a) 0.7

(b) 12.21 $12\frac{21}{100}$

(c) 0.101 $\frac{101}{1000}$

(d) 0.007

(e) 1.3717

8 Write each decimal as a fraction or mixed number in lowest terms.

(a) 0.5

(b) 12.6

(c) 0.85

(d) 3.05

(e) 0.225

(f) 420.0802

EXAMPLE 5 Writing Decimals as Fractions or Mixed Numbers

Write each decimal as a fraction or mixed number.

(a) 0.19

The digits to the right of the decimal point, 19, are the numerator of the fraction. The denominator is 100 for hundredths because the rightmost digit is in the hundredths place.

$$0.1\underset{\uparrow}{9} = \frac{19}{100} \leftarrow 100 \text{ for hundredths}$$

Hundredths place

(b) 0.863

$$0.86\underset{\uparrow}{3} = \frac{863}{1000} \leftarrow 1000 \text{ for thousandths}$$

Thousandths place

(c) 4.0099

The whole number part stays the same.

$$4.009\underset{\uparrow}{9} = 4\frac{99}{10,000} \leftarrow 10,000 \text{ for ten-thousandths}$$

Ten-thousandths place

◀◀◀ Work Problem 7 at the Side.

CAUTION

After you write a decimal as a fraction or a mixed number, make sure the fraction is in lowest terms.

EXAMPLE 6 Writing Decimals as Fractions or Mixed Numbers in Lowest Terms

Write each decimal as a fraction or mixed number in lowest terms.

(a) $0.4 = \dfrac{4}{10} \leftarrow 10$ for tenths

Write $\dfrac{4}{10}$ in lowest terms. $\dfrac{4}{10} = \dfrac{4 \div 2}{10 \div 2} = \dfrac{2}{5} \leftarrow$ Lowest terms

(b) $0.75 = \dfrac{75}{100} = \dfrac{75 \div 25}{100 \div 25} = \dfrac{3}{4} \leftarrow$ Lowest terms

(c) $18.105 = 18\dfrac{105}{1000} = 18\dfrac{105 \div 5}{1000 \div 5} = 18\dfrac{21}{200} \leftarrow$ Lowest terms

(d) $42.8085 = 42\dfrac{8085}{10,000} = 42\dfrac{8085 \div 5}{10,000 \div 5} = 42\dfrac{1617}{2000} \leftarrow$ Lowest terms

◀◀◀ Work Problem 8 at the Side.

ANSWERS

7. (a) $\dfrac{7}{10}$ (b) $12\dfrac{21}{100}$ (c) $\dfrac{101}{1000}$
   (d) $\dfrac{7}{1000}$ (e) $1\dfrac{3717}{10,000}$

8. (a) $\dfrac{1}{2}$ (b) $12\dfrac{3}{5}$ (c) $\dfrac{17}{20}$ (d) $3\dfrac{1}{20}$
   (e) $\dfrac{9}{40}$ (f) $420\dfrac{401}{5000}$

▦ **Calculator Tip** In this book we will write a 0 in the ones place for decimal fractions. We write **0**.45 instead of just .45, to emphasize that there is no whole number. Your *scientific* calculator shows these zeros also. Enter ⊙ ④ ⑤ and notice that the display automatically shows 0.45 even though you did not press 0. For comparison, enter the whole number 45 by pressing ④ ⑤ ⊕ and notice where the decimal point is shown in the display. (It automatically appears to the *right* of the 5.)

4.1 Exercises

FOR EXTRA HELP

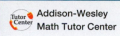 Addison-Wesley Math Tutor Center    MathXL   Digital Video Tutor CD 2 Videotape 7   Student's Solutions Manual   MyMathLab   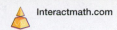 Interactmath.com

Identify the digit that has the given place value. See Example 2.

1. 70.489

tens

ones

tenths

2. 135.296

ones

tenths

tens

3. 0.2518

hundredths

thousandths

ten-thousandths

4. 0.9347

hundredths

thousandths

ten-thousandths

5. 93.01472

thousandths

ten-thousandths

tenths

6. 0.51968

tenths

ten-thousandths

hundredths

7. 314.658

tens

tenths

hundreds

8. 51.325

tens

tenths

hundredths

9. 149.0832

hundreds

hundredths

ones

10. 3458.712

hundreds

hundredths

tenths

11. 6285.7125

thousands

thousandths

hundredths

12. 5417.6832

thousands

thousandths

ones

Write the decimal number that has the specified place values. See Example 2.

13. 0 ones, 5 hundredths, 1 ten, 4 hundreds, 2 tenths   410.25

14. 7 tens, 9 tenths, 3 ones, 6 hundredths, 8 hundreds

15. 3 thousandths, 4 hundredths, 6 ones, 2 ten-thousandths, 5 tenths

16. 8 ten-thousandths, 4 hundredths, 0 ones, 2 tenths, 6 thousandths

17. 4 hundredths, 4 hundreds, 0 tens, 0 tenths, 5 thousandths, 5 thousands, 6 ones

18. 7 tens, 7 tenths, 6 thousands, 6 thousandths, 3 hundreds, 3 hundredths, 2 ones

Write each decimal as a fraction or mixed number in lowest terms. See Examples 1, 5, and 6.

19. 0.7   **20.** 0.1   **21.** 13.4   **22.** 9.8   **23.** 0.25

24. 0.55   **25.** 0.66   **26.** 0.33   **27.** 10.17   **28.** 31.99

29. 0.06   **30.** 0.08   **31.** 0.205   **32.** 0.805

33. 5.002   **34.** 4.008   **35.** 0.686   **36.** 0.492

Tell how to read each decimal in words. See Examples 3 and 4.

37. 0.5   **38.** 0.2

39. 0.78   **40.** 0.55

41. 0.105   **42.** 0.609

43. 12.04   **44.** 86.09

45. 1.075   **46.** 4.025

Write each decimal in numbers. See Examples 3 and 4.

47. six and seven tenths

48. eight and twelve hundredths

49. thirty-two hundredths

50. one hundred eleven thousandths

51. four hundred twenty and eight thousandths

52. two hundred and twenty-four thousandths

53. seven hundred three ten-thousandths

54. eight hundred and six hundredths

55. seventy-five and thirty thousandths

56. sixty and fifty hundredths

57. Anne read the number 4302 as "four thousand three hundred and two." Explain what is wrong with the way Anne read the number.

58. Jerry read the number 9.0106 as "nine and one hundred and six ten-thousandths." Explain the error he made.

The friends on the first page of this chapter need to select the correct fishing line for their reels. Fishing line is sold according to how many pounds of "pull" the line can withstand before breaking. Use the table to answer Exercises 59–62. Write all fractions in lowest terms. (Note: The diameter of the fishing line is its thickness.)

RELATING FISHING LINE DIAMETER TO TEST STRENGTH

Test Strength (pounds)	Average Diameter (inches)
4	0.008
8	0.010
12	0.013
14	0.014
17	0.015
20	0.016

Source: Berkley Outdoor Technologies Group.

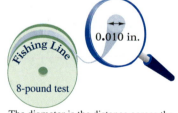

The diameter is the distance across the end of the line (or its thickness).

59. Write the diameter of 8-pound test line in words and as a fraction.

60. Write the diameter of 17-pound test line in words and as a fraction.

61. What is the test strength of the line with a diameter of $\dfrac{13}{1000}$ inch?

62. What is the test strength of the line with a diameter of sixteen thousandths inch?

Suppose your job is to take phone orders for precision parts. Use the table, and in Exercises 63–68, write the correct part number that matches what you hear the customer say over the phone. In Exercises 67–68, write the words you would say to the customer.

Part Number	Size in Centimeters
3-A	0.06
3-B	0.26
3-C	0.6
3-D	0.86
4-A	1.006
4-B	1.026
4-C	1.06
4-D	1.6
4-E	1.602

63. "Please send the six-tenths centimeter bolt."

Part number _____ .

64. "The part missing from our order was the one and six hundredths size."

Part number _____ .

65. "The size we need is one and six thousandths centimeters."

Part number _____ .

66. "Do you still stock the twenty-six hundredths centimeter bolt?"

Part number _____ .

67. "What size is part number 4-E?" Write your answer in words.

68. "What size is part number 4-B?" Write your answer in words.

RELATING CONCEPTS (EXERCISES 69–76) For Individual or Group Work

*Use your knowledge of place value to **work Exercises 69–76 in order.***

69. Look back at the decimal place value chart on page 251. What do you think would be the names of the next four places to the *right* of hundred-thousandths? What information did you use to come up with these names?

70. A common mistake is to think that the first place to the right of the decimal point is "oneths" and the second place is "tenths." Why might someone make that mistake? How would you explain why there is no "oneths" place?

71. Use your answer to Exercise 69 to write 0.72436955 in words.

72. Use your answer to Exercise 69 to write 0.000678554 in words.

73. Write 8006.500001 in words.

74. Write 20,060.000505 in words.

75. Write this decimal in numbers.

three hundred two thousand forty ten-millionths

76. Write this decimal in numbers.

nine billion, eight hundred seventy-six million, five hundred forty-three thousand, two hundred ten and one hundred million two hundred thousand three hundred billionths

4.2 Rounding Decimals

Section 1.7 showed how to round whole numbers. For example, 89 rounded to the nearest ten is 90, and 8512 rounded to the nearest hundred is 8500.

O B J E C T I V E S

1 Learn the rules for rounding decimals.

2 Round decimals to any given place.

3 Round money amounts to the nearest cent or nearest dollar.

OBJECTIVE 1 **Learn the rules for rounding decimals.**   It is also important to be able to **round** decimals. For example, a store is selling 2 candy mints for $0.75 but you want only one mint. The price of each mint is $0.75 ÷ 2, which is $0.375, but you cannot pay part of a cent. Is $0.375 closer to $0.37 or to $0.38? Actually, it's exactly halfway between. When this happens in everyday situations, the rule is to round *up*. The store will charge you $0.38 for the mint.

Rounding Decimals

Step 1   Find the place to which the rounding is being done. Draw a "cut off" line *after* that place to show that you are cutting off and dropping the rest of the digits.

Step 2   Look *only* at the *first* digit you are cutting off.

Step 3(a)   If this digit is *4 or less,* the part of the number you are keeping *stays the same.*

Step 3(b)   If this digit is *5 or more,* you must *round up* the part of the number you are keeping.

Step 4   You can use the ≈ symbol or the ≐ symbol to indicate that the rounded number is now an approximation (close, but not exact). Both symbols mean "is approximately equal to." In this book we will use the ≈ symbol.

CAUTION
Do *not* move the decimal point when rounding.

OBJECTIVE 2 **Round decimals to any given place.**   The following examples show you how to round decimals.

EXAMPLE 1 **Rounding a Decimal Number**

Round 14.39652 to the nearest thousandth.

Step 1   Draw a "cut-off" line after the thousandths place.

$$1\;4\;.\;3\;9\;6\;|\;5\;2$$

You are cutting off the 5 and 2. They will be dropped.

Thousandths

Step 2   Look *only* at the *first* digit you are cutting off. Ignore the other digits you are cutting off.

$$1\;4\;.\;3\;9\;6\;|\;5\;2$$

Look *only* at the 5. Ignore the 2.

Continued on Next Page

① Round to the nearest thousandth.

(a) 0.33492

Step 3 If the first digit you are cutting off is *5 or more,* round up the part of the number you are keeping.

$$
\begin{array}{r}
1\,4\,.\,3\,9\,6\;\;5\;2 \\
+\;\;0\,.\,0\,0\,1 \\
\hline
1\,4\,.\,3\,9\,7
\end{array}
$$

First digit cut is *5 or more,* so round up by adding 1 thousandth to the part you are keeping.

So, 14.39652 rounded to the nearest thousandth is 14.397. We can write 14.39652 ≈ 14.397

CAUTION
When rounding whole numbers in **Section 1.7,** you kept all the digits but changed some to zeros. With decimals, you cut off and *drop the extra digits.* In Example 1 above, 14.39652 rounds to 14.397, *not* 14.39700.

(b) 8.00851

◀◀◀ **Work Problem 1 at the Side.**

In Example 1, the rounded number 14.397 had *three decimal places.* **Decimal places** are the number of digits to the *right* of the decimal point. The first decimal place is tenths, the second is hundredths, the third is thousandths, and so on.

EXAMPLE 2 **Rounding Decimals to Different Places**

Round to the place indicated.

(a) 5.3496 to the nearest tenth

(c) 265.42038

Step 1 Draw a cut-off line after the tenths place.

$$5\,.\,3\;\;4\,9\,6$$

Tenths

You are cutting off the 4, 9, and 6. They will be dropped.

Step 2 $$5\,.\,3\;\;4\,9\,6$$

Look *only* at the 4.

Ignore these digits.

Step 3 $$5\,.\,3\;\;4\,9\,6$$

First digit cut is *4 or less,* so the part you are keeping stays the same.

$$5\,.\,3\;\leftarrow \text{Stays the same}$$

(d) 10.70180

5.3496 rounded to the nearest tenth is 5.3 (*one* decimal place for *tenths*). We can write 5.3496 ≈ 5.3. Notice that 5.3496 does *not* round to 5.3000, which would be ten-thousandths.

(b) 0.69738 to the nearest hundredth

Step 1 $$0\,.\,6\,9\;|\;7\,3\,8$$

Hundredths

Draw a cut-off line after the hundredths place.

Step 2 $$0\,.\,6\,9\;|\;7\,3\,8$$

Look *only* at the 7.

Continued on Next Page

Step 3   0 . 6 9 | 7 3 8   ⌐── First digit cut is *5 or more,* so round up by adding 1 hundredth to the part you are keeping.

```
         1
    0 . 6 9   ← Keep this part.
  + 0 . 0 1   ← To round up, add 1 hundredth.
    0 . 7 0   ← 9 + 1 is 10; write 0 and carry 1 to the tenths place.
```

0.69738 rounded to the nearest hundredth is 0.70. Hundredths is *two* decimal places so you *must* write the 0 in the hundredths place. We can write $0.69738 \approx 0.70$.

> **CAUTION**
> If a *rounded* number has a 0 in the rightmost place, you *must* keep the 0. As shown above, 0.69738 rounded to the nearest hundredth is 0.7**0**. Do **not** write 0.7, which is rounded to tenths instead of hundredths.

(c) 0.01806 to the nearest thousandth

       ⌐── First digit cut is *4 or less,* so the part you are keeping stays the same.

```
   0 . 0 1 8 | 0 6

   0 . 0 1 8
```

0.01806 rounded to the nearest thousandth is 0.018 (*three* decimal places for *thousandths*). We can write $0.01806 \approx 0.018$.

(d) 57.976 to the nearest tenth

       ⌐── First digit cut is *5 or more,* so round up by adding 1 tenth to the part you are keeping.

```
   57.9 | 76

   57.9
 +  0.1
   58.0   ←── 9 + 1 is 10; write the 0 and carry 1
              to the 7 in the ones place.
```

57.976 rounded to the nearest tenth is 58.0. We can write $57.976 \approx 58.0$. You *must* write the 0 in the tenths place to show that the number was rounded to the nearest tenth.

> **CAUTION**
> Check that your rounded answer shows *exactly* the number of decimal places asked for in the problem. Be sure your answer shows *one* decimal place if you rounded to *tenths, two* decimal places for *hundredths, three* decimal places for *thousandths,* and so on.

▶ Work Problem 2 at the Side. ▶▶▶

OBJECTIVE 3 Round money amounts to the nearest cent or nearest dollar. In many everyday situations, such as shopping in a store, money amounts are rounded to the nearest cent. There are 100 cents in a dollar.

$$\text{Each cent is } \frac{1}{100} \text{ of a dollar.}$$

Another way to write $\frac{1}{100}$ is 0.01. So rounding to the *nearest cent* is the same as rounding to the *nearest hundredth of a dollar.*

---

2 Round to the place indicated.

(a) 0.8988 to the nearest hundredth

.90

(b) 5.8903 to the nearest hundredth

5.89

(c) 11.0299 to the nearest thousandth

11.030

(d) 0.545 to the nearest tenth

.5

❸ Round each money amount to the nearest cent.

(a) $14.595

(b) $578.0663

(c) $0.849

(d) $0.0548

EXAMPLE 3 **Rounding to the Nearest Cent**

How much will you pay in each shopping situation? Round each money amount to the nearest cent.

(a) $2.4238 (Is it closer to $2.42 or to $2.43?)

First digit cut is *4 or less,* so the part you are keeping stays the same.

$2.42|**38**

$2.42 ← You pay

You pay $2.42 because $2.4238 is closer to $2.42 than to $2.43.

(b) $0.695 (Is it closer to $0.69 or to $0.70?)

5 or more; round up

$0.69|**5**

$0.69
+ $0.01 ← To round up, add 1 hundredth (1 cent).
——————
$0.70 ← You pay.

Work Problem 3 at the Side.

NOTE
Some stores round *all* money amounts up to the next higher cent, even if the next digit is *4 or less.* In Example 3(a) above, some stores would round $2.4238 *up* to $2.43, even though it is closer to $2.42.

It is also common to round money amounts to the nearest dollar. For example, you can do that on your federal and state income tax returns to make the calculations easier.

EXAMPLE 4 **Rounding to the Nearest Dollar**

Round to the nearest dollar.

(a) $48.69 (Is it closer to $48 or to $49?)

First digit cut is *5 or more,* so round up by adding $1.

$48.|**69**

$48
+ 1
————
$49

$48.69 is closer to $49 than to $48.
So $48.69 rounded to the nearest dollar is $49.

CAUTION
$48.69 rounded to the nearest dollar is $49. Write the answer as $49 to show that the rounding is to the *nearest dollar.* Writing $49.00 would show rounding to the *nearest cent.*

Continued on Next Page

(b) $594.36 (Is it closer to $594 or $595?)

First digit cut is *4 or less,* so the part you keep stays the same.

$594.|36

$594

$594.36 rounded to the nearest dollar is $594.

(c) $349.88 (Is it closer to $349 or to $350?)

5 or more, so round up by adding $1.

$349.|88

$349
+   1
$350

$349.88 rounded to the nearest dollar is $350.

(d) $2689.50 (Is it closer to $2689 or $2690?)

5 or more, so round up by adding $1.

$2689.|50

$2689
+   1
$2690

$2689.50 rounded to the nearest dollar is $2690.

> **NOTE**
> When rounding $2689.50 to the nearest dollar, above, notice that it is exactly halfway between $2689 and $2690. When this happens in every-day situations, the rule is to round *up.* (Scientists working with technical data may use a more complicated rule when rounding numbers that are exactly in the middle.)

(e) $0.61 (Is it closer to $0 or to $1?)

5 or more, so round up.

$0.|61

$0.61 rounded to the nearest dollar is $1.

Calculator Tip Accountants and other people who work with money amounts often set their calculators to automatically round to two decimal places (nearest cent) or to round to zero decimal places (nearest dollar). Your calculator may have this feature.

Work Problem 4 at the Side.

4 Round to the nearest dollar.

(a) $29.10

(b) $136.49

(c) $990.91

(d) $5949.88

(e) $49.60

(f) $0.55

(g) $1.08

Lawn Fertilizer

Gotta Be Green

A lot's being said about personal responsibility these days, and the idea seems to be ending up on the front lawn—literally! Each spring, home-owners across the country gear up to green up their lawns, and the increased use of fertilizer has a lot of environ-mentalists concerned about the potential effects of chemical runoff into nearby rivers and streams.

Each year, according to a study conducted by the University of Minnesota's Department of Agri-culture, each household in the Minneapolis/St. Paul metro area uses an average of 36 pounds of lawn fertilizer. That adds up to 25,529,295 pounds, or 12,765 tons. Add to that another 193,000 pounds of weed killer and you're looking at the total picture for keeping it green in the Twin Cities.

Source: Minneapolis Star Tribune.

1. According to the article,

   (a) How many pounds of lawn fertilizer are used each year in the *entire metro area?*

   (b) Do a division on your calculator to find the number of *households* in the metro area.

   (c) Would it make sense to round your answer to part (b)? If so, how would you round it?

   (d) How many pounds of *weed killer* are used each year in the entire metro area?

   (e) Do a division on your calculator to find the number of pounds of *weed killer* used by each household in the metro area. Round your answer to the nearest hundredth.

2. There are 2000 pounds in one ton.

   (a) Find the number of tons equivalent to 25,529,295 pounds of fertilizer.

   (b) Does your answer match the figure given in the article? If not, what did the author of the article do to get 12,765 tons?

   (c) Is the author's figure accurate? Why or why not?

   (d) Find the number of tons equivalent to 193,000 pounds of weed killer.

   (e) How can you write the division problem 193,000 ÷ 2000 in a simpler way? Explain what you did.

3. According to the article,

   (a) "each household in the Minneapolis/St. Paul metro area uses an average of 36 pounds of lawn fertilizer" each year. What mathematical operation do you do to find an average?

   (b) When the calculations were done to find the average, the answer was probably not *exactly* 36 pounds. List six different values that are *less than* 36 that would round to 36. List two values with one decimal place; two values with two decimal places; and two values with three decimal places.

   (c) List six different values that are *greater than* 36 that would round to 36. List two values each with one, two, and three decimal places.

   (d) What is the *smallest* number that can be rounded to 36? What is the *largest* number?

4.2 Exercises

 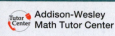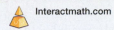
Round each number to the place indicated. See Examples 1 and 2.

1. 16.8974 to the nearest tenth

2. 193.845 to the nearest hundredth

3. 0.95647 to the nearest thousandth

4. 96.81584 to the nearest ten-thousandth

5. 0.799 to the nearest hundredth

6. 0.952 to the nearest tenth

7. 3.66062 to the nearest thousandth

8. 1.5074 to the nearest hundredth

9. 793.988 to the nearest tenth

10. 476.1196 to the nearest thousandth

11. 0.09804 to the nearest ten-thousandth

12. 176.004 to the nearest tenth

13. 48.512 to the nearest one

14. 3.385 to the nearest one

15. 9.0906 to the nearest hundredth

16. 30.1290 to the nearest thousandth

17. 82.000151 to the nearest ten-thousandth

18. 0.400594 to the nearest ten-thousandth

Nardos is grocery shopping. The store will round the amount she pays for each item to the nearest cent. Write the rounded amounts. See Example 3.

19. Soup is three cans for $2.45, so one can is $0.81666. Nardos pays _____.

20. Orange juice is two cartons for $3.89, so one carton is $1.945. Nardos pays _____.

21. Facial tissue is four boxes for $4.89, so one box is $1.2225. Nardos pays _____.

22. Muffin mix is three packages for $1.75, so one package is $0.58333. Nardos pays _____.

23. Candy bars are six for $2.99, so one bar is $0.4983. Nardos pays _____.

24. Spaghetti is four boxes for $4.39, so one box is $1.0975. Nardos pays _____.

As she gets ready to do her income tax return, Ms. Chen rounds each amount to the nearest dollar. Write the rounded amounts. See Example 4.

25. Income from job, $48,649.60

26. Income from interest on bank account, $69.58

27. Union dues, $310.08

28. Federal withholding, $6064.49

29. Donations to charity, $848.91

30. Medical expenses, $609.38

Round each money amount as indicated.

31. $499.98 to the nearest dollar.   **32.** $9899.59 to the nearest dollar   **33.** $0.996 to the nearest cent

34. $0.09929 to the nearest cent.   **35.** $999.73 to the nearest dollar.   **36.** $9999.80 to the nearest dollar.

The table lists speed records for various types of transportation. Use the table to answer Exercises 37–40.

Record	Speed (miles per hour)
Land speed record (specially built car)	763.04
Motorcycle speed record (specially adapted motorcycle)	322.15
Fastest roller coaster	106.9
Fastest military jet	2193.167
Boeing 737-300 airplane (regular passenger service)	495
Indianapolis 500 auto race (fastest average winning speed)	185.981
Daytona 500 auto race (fastest average winning speed)	177.602

Source: The Top 10 of Everything 2003 and The World Almanac 2003.

37. Round these speed records to the nearest whole number.

   (a) Motorcycle

   (b) Roller coaster

38. Round these speed records to the nearest hundredth.

   (a) Daytona 500 average winning speed

   (b) Indianapolis 500 average winning speed

39. Round these speed records to the nearest tenth.

   (a) Indianapolis 500 average winning speed

   (b) Land speed record

40. Round these speed records to the nearest hundred.

   (a) military jet

   (b) Boeing 737-300 airplane

RELATING CONCEPTS (EXERCISES 41–44) For Individual or Group Work

*Use your knowledge about rounding money amounts to **work Exercises 41–44 in order.***

41. Explain what happens when you round $0.499 to the nearest dollar. Why does this happen?

42. Look again at Exercise 41. How else could you round $0.499 that would be more helpful? What kind of guideline does this suggest about rounding to the nearest dollar?

43. Explain what happens when you round $0.0015 to the nearest cent. Why does this happen?

44. Suppose you want to know which of these amounts is less, so you round them both to the nearest cent.

$0.5968 $0.6014

Explain what happens. Describe what you could do instead of rounding to the nearest cent.

4.3 Adding and Subtracting Decimals

OBJECTIVE 1 Add decimals. When adding or subtracting *whole* numbers (**Sections 1.2** and **1.3**), you lined up the numbers in columns so that you were adding ones to ones, tens to tens, and so on. A similar idea applies to adding or subtracting *decimal* numbers. With decimals, you line up the decimal points to make sure you are adding tenths to tenths, hundredths to hundredths, and so on.

Adding and Subtracting Decimals

Step 1 Write the numbers in columns with the decimal points lined up.

Step 2 If necessary, write in zeros so both numbers have the same number of decimal places. Then add or subtract as if they were whole numbers.

Step 3 Line up the decimal point in the answer directly below the decimal points in the problem.

EXAMPLE 1 Adding Decimal Numbers

Add.

(a) 16.92 and 48.34

Step 1 Write the numbers in columns with the decimal points lined up.

$$
\begin{array}{r}
\text{tens ones . tenths hundredths}\\
1\ 6\ .\ 9\ 2\\
+\ 4\ 8\ .\ 3\ 4\\
\end{array}
$$

�; ———— Decimal points are lined up.

Step 2 Add as if these were whole numbers.

$$
\begin{array}{r}
11\\
16\ .\ 92\\
+\ 48\ .\ 34\\
\hline
\end{array}
$$

Step 3
$$65\ .\ 26$$

�; ——— Decimal point in answer is lined up under decimal points in problem.

(b) 5.897 + 4.632 + 12.174

Write the numbers vertically with decimal points lined up. Then add.

$$
\begin{array}{r}
11\ 21\\
5.897\\
4.632\\
+\ 12.174\\
\hline
22.703\\
\end{array}
$$

�; ———— Decimal points are lined up.

> **Work Problem 1 at the Side.** ▶▶▶

In Example 1(a) above, both numbers had *two* decimal places (two digits to the right of the decimal point). In Example 1(b), all the numbers had *three decimal places* (three digits to the right of the decimal point). That made it easy to add tenths to tenths, hundredths to hundredths, and so on.

OBJECTIVES

1 Add decimals.
2 Subtract decimals.
3 Estimate the answer when adding or subtracting decimals.

1 Find each sum.

(a) 2.86 + 7.09

(b) 13.761 + 8.325

(c) 0.319 + 56.007 + 8.252

(d) 39.4 + 0.4 + 177.2

2 Find each sum.

(a) $6.54 + 9.8$

If the number of decimal places does *not* match, you can write in zeros as placeholders to make them match. This is shown in Example 2.

EXAMPLE 2 **Writing Zeros as Placeholders Before Adding**

Add.

(a) $7.3 + 0.85$

There are two decimal places in 0.85 (tenths and hundredths), so write a 0 in the hundredths place in 7.3 so that it has two decimal places also.

$$
\begin{array}{r}
7.3\mathbf{0} \leftarrow \text{One 0 is} \\
+\,0.85 \quad \text{written in.} \\
\hline
8.15
\end{array}
$$

7.30 is equivalent to 7.3 because

$7\dfrac{30}{100}$ in lowest terms is $7\dfrac{3}{10}$

(b) $6.42 + 9 + 2.576$

(b) $6.42 + 9 + 2.576$

Write in zeros so that all the addends have three decimal places. Notice how the whole number 9 is written with the decimal point at the *far right* side. (If you put the decimal point on the *left* side of the 9, you would turn it into the decimal fraction 0.9.)

$$
\begin{array}{r}
6\,.\,4\,2\,\mathbf{0} \leftarrow \text{One 0 is written in.} \\
9\,.\,\mathbf{0\,0\,0} \leftarrow \begin{cases} 9 \text{ is a whole number; decimal point} \\ \text{point and three zeros are written in.} \end{cases} \\
+\,2\,.\,5\,7\,6 \leftarrow \text{No zeros are needed.} \\
\hline
1\,7\,.\,9\,9\,6
\end{array}
$$

Decimal points are lined up.

(c) $8.64 + 39.115 + 3.0076$

> **NOTE**
> Writing zeros to the right of a *decimal* number does *not* change the value of the number, as shown in Example 2(a) above.

◀◀◀ Work Problem 2 at the Side.

(d) $5 + 429.823 + 0.76$

OBJECTIVE **2** **Subtract decimals.**   Subtraction of decimals is done in much the same way as addition of decimals. You can check the answers to subtraction problems using addition, as you did with whole numbers (see **Section 1.3**).

EXAMPLE 3 **Subtracting Decimal Numbers**

Subtract. Check your answers using addition.

(a) 15.82 from 28.93

Step 1
$$
\begin{array}{r}
28\,.\,93 \\
-\,15\,.\,82 \\
\end{array}
$$

Line up decimal points. Then you will be subtracting hundredths from hundredths and tenths from tenths.

Continued on Next Page

Step 2
$$\begin{array}{r} 28.93 \\ -15.82 \\ \hline 13.11 \end{array}$$
Both numbers have two decimal places; no need to write in zeros.

13.11 ← Subtract as if they were whole numbers.

Step 3 └── Decimal point in answer is lined up.

Check the answer by adding 13.11 and 15.82. If the subtraction is done correctly, the sum will be 28.93.

(b) 146.35 minus 58.98
Borrowing is needed here.

$$\begin{array}{r} {}^{0}{\cancel{1}}{}^{13}{\cancel{4}}{}^{15}{\cancel{6}}.{}^{12}{\cancel{3}}{}^{15}{\cancel{5}} \\ -\;\;\;\;5\;8.9\;8 \\ \hline 8\;7.3\;7 \end{array}$$

└── Line up decimal points.

Check the answer by adding 87.37 and 58.98. If you did the subtraction correctly, the sum will be 146.35. (If it *isn't,* you need to rework the problem.)

> **Work Problem 3 at the Side.** ▶▶▶

EXAMPLE 4 **Writing Zeros as Placeholders before Subtracting**

Subtract.

(a) 16.5 from 28.362
Use the same steps as in Example 3 above. Remember to write in zeros so both numbers have three decimal places.

┌── Line up decimal points.

$$\begin{array}{r} 28.362 \\ -16.\mathbf{500} \\ \hline 11.862 \end{array}$$
← Write two zeros.
← Subtract as usual.

Check the answer by adding.
$$\begin{array}{r} 16.500 \\ +11.862 \\ \hline 28.362 \end{array}$$
← Matches minuend in original problem.

(b) 59.7 − 38.914
$$\begin{array}{r} 59.\mathbf{700} \\ -38.914 \\ \hline 20.786 \end{array}$$
← Write two zeros.
← Subtract as usual.

(c) 12 less 5.83
$$\begin{array}{r} 12.\mathbf{00} \\ -\;\;5.83 \\ \hline 6.17 \end{array}$$
← Write a decimal point and two zeros.
← Subtract as usual.

> **Work Problem 4 at the Side.** ▶▶▶

OBJECTIVE 3 Estimate the answer when adding or subtracting decimals.   A common error in working decimal problems by hand is to misplace the decimal point in the answer. Or, when using a calculator, you may accidentally press the wrong key. **Estimating** the answer will help you avoid these mistakes. Start by using *front end rounding* on each number (as you did in **Section 1.7**). Here are several examples. Notice that in the rounded numbers, only the leftmost digit is something other than 0.

3.25	rounds to	3	6.812	rounds to	7
532.6	rounds to	500	26.397	rounds to	30
7094.2	rounds to	7000	351.24	rounds to	400

---

3 Subtract. Check your answers using addition.

(a) 22.7 from 72.9

(b) 6.425 from 11.813

(c) 20.15 − 19.67

4 Subtract. Check your answers using addition.

(a) 18.651 from 25.3

(b) 5.816 − 4.98

(c) 40 less 3.66

(d) 1 − 0.325

$$\begin{array}{r} {}^{0}{\cancel{1}}.{}^{9}{\cancel{0}}{}^{9}{\cancel{0}}{}^{1}{\cancel{0}} \\ -\;0.325 \\ \hline .675 \end{array}$$

ANSWERS
3. (a) 50.2; 50.2 + 22.7 = 72.9
   (b) 5.388; 5.388 + 6.425 = 11.813
   (c) 0.48; 0.48 + 19.67 = 20.15
4. (a) 6.649; 6.649 + 18.651 = 25.3
   (b) 0.836; 0.836 + 4.98 = 5.816
   (c) 36.34; 36.34 + 3.66 = 40
   (d) 0.675; 0.675 + 0.325 = 1

5 First, use front end rounding
≈ and estimate each answer.
Then add or subtract to find
the exact answer.

(a) $2.83 + 5.009 + 76.1$

Estimate:

Exact:

(b) 11.365 from 58

Estimate:

Exact:

(c) $398.81 + 47.658 + 4158.7$

Estimate:

Exact:

(d) Find the difference between
12.837 meters and
46.091 meters.

Estimate:

Exact:

(e) $19.28 plus $1.53

Estimate:

Exact:

EXAMPLE 5 **Estimating Decimal Answers**

Use front end rounding to round each number. Then add or subtract the
rounded numbers to get an estimated answer. Finally, find the exact answer.

(a) Add 194.2 and 6.825.

Estimate:		*Exact:*
200 ← Rounds to		194.200
+ 7 ← Rounds to		+ 6.825
207		201.025

The estimate goes out to the hundreds place (three places to the *left* of the
decimal point), and so does the exact answer. Therefore, the decimal point is
probably in the correct place in the exact answer.

(b) $69.42 + $13.78

Estimate:	*Exact:*
$70 ← Rounds to	$69.42
+ 10 ← Rounds to	+ 13.78
$80	$83.20 ← Exact answer is close to estimate, so it
	is probably correct.

(c) Find the difference between 0.92 ft and 8 ft.
Use subtraction to find the difference between two numbers. The larger
number, 8, is written on top.

Estimate:	*Exact:*
8 ← Rounds to	8.**00** ← Write a decimal point and two zeros.
− 1 ← Rounds to	− 0.92
7	7.08 ft ← Exact answer is close to estimate.

(d) Subtract 1.8614 from 7.3.

Estimate:	*Exact:*
7 ← Rounds to	7.3**000** ← Write three zeros.
− 2 ← Rounds to	− 1.8614
5	5.4386 ← Exact answer is close to estimate.

◀◀◀ Work Problem 5 at the Side.

⊞ **Calculator Tip** If you are *adding* numbers, you can enter them in any
order on your calculator. Try these; jot down the answers.

9.82 ⊕ 1.86 ⊜ _____   1.86 ⊕ 9.82 ⊜ _____

The answers are the same because addition is *commutative*. (See
Section 1.2.) But subtraction is *not* commutative. It *does* matter which
number you enter first. Try these:

9.82 ⊖ 1.86 ⊜ _____   1.86 ⊖ 9.82 ⊜ _____

The second answer has a negative sign (−) next to it. A negative number
is *less* than 0. If it was in your checkbook, you'd be "in the hole" by $7.96.
(Sec **Section 9.1** for more about negative numbers.)

ANSWERS
5. **(a)** *Estimate:* $3 + 5 + 80 = 88$;
      Exact: 83.939
   (b) *Estimate:* $60 − 10 = 50$;
      Exact: 46.635
   (c) *Estimate:* $400 + 50 + 4000 = 4450$;
      Exact: 4605.168
   (d) *Estimate:* $50 − 10 = 40$;
      Exact: 33.254 meters
   (e) *Estimate:* $20 + $2 = $22;
      Exact: $20.81

4.3 Exercises

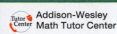

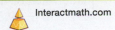

Find each sum or difference. See Examples 1–4.

1. $5.69 + 11.79$

2. $372.1 - 33.7$

3. $24.008 - 0.995$

4. $0.7759 + 9.8883$

5. $8.263 - 0.5$

6. $47.658 - 20.9$

7. $76.5 + 0.506$

8. $1.87 + 9.749$

9. $21 - 0.896$

10. $9 - 1.183$

11. $0.4 - 0.291$

12. $0.35 - 0.088$

13. $39.76005 + 182 + 4.799 + 98.31 + 5.9999$

14. $489.76 + 0.9993 + 38 + 8.55087 + 80.697$

This drawing of a human skeleton shows the average length of the longest bones, in inches. Use the drawing to answer Exercises 15–18.

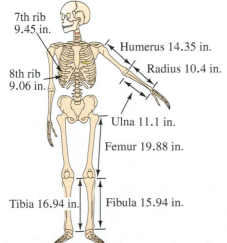

7th rib 9.45 in.

8th rib 9.06 in.

Humerus 14.35 in.

Radius 10.4 in.

Ulna 11.1 in.

Femur 19.88 in.

Tibia 16.94 in.

Fibula 15.94 in.

15. **(a)** What is the combined length of the humerus and radius bones?

   (b) What is the difference in the lengths of these two bones?

16. **(a)** What is the total length of the femur and tibia bones?

   (b) How much longer is the femur than the tibia?

17. **(a)** Find the sum of the lengths of the humerus, ulna, femur, and tibia.

   (b) How much shorter is the 8th rib than the 7th rib?

18. **(a)** What is the difference in the lengths of the two bones in the lower arm?

   (b) What is the difference in the lengths of the two bones in the lower leg?

19. Explain and correct
the error that a student
made when he added
$0.72 + 6 + 39.5$ this way:

$$
\begin{array}{r}
0.72 \\
6 \\
+ \ 39.50 \\
\hline
40.28
\end{array}
$$

20. Explain the difference between saying "subtract 2.9 from 8" and saying "2.9 minus 8."

≈
Use front end rounding to round each number. Then add or subtract the rounded numbers to get an estimated answer. Finally, find the exact answer. See Example 5.

21. *Estimate:*   *Exact:*

$$
\begin{array}{r}
\$19.74 \\
- \quad\quad\quad - \ 6.58 \\
\hline
\end{array}
$$

22. *Estimate:*   *Exact:*

$$
\begin{array}{r}
\$27.96 \\
- \quad\quad\quad - \ 8.39 \\
\hline
\end{array}
$$

23. *Estimate:*   *Exact:*

$$
\begin{array}{r}
392.7 \\
0.865 \\
+ \quad\quad\quad + \ 21.08 \\
\hline
\end{array}
$$

24. *Estimate:*   *Exact:*

$$
\begin{array}{r}
38.55 \\
7.716 \\
+ \quad\quad\quad + \ 0.6 \\
\hline
\end{array}
$$

25. What is 8.6 less 3.751?

   Estimate: *Exact:*

26. What is 31.7 less 4.271?

   Estimate: *Exact:*

27. *Estimate:*   *Exact:*

$$
\begin{array}{r}
62.8173 \\
539.99 \\
+ \quad\quad\quad + \ 5.629 \\
\hline
\end{array}
$$

28. *Estimate:*   *Exact:*

$$
\begin{array}{r}
332.607 \\
12.5 \\
+ \quad\quad\quad + \ 823.3949 \\
\hline
\end{array}
$$

≈
*Use your estimation skills to pick the most reasonable answer for each example.
Do **not** solve the problems. Circle your choice.*

29. $12 - 11.725$

   2.75   0.275   27.5

30. $20 - 1.37$

   0.1863   1.863   18.63

31. $6.5 + 0.007$

   6.507   0.6507   65.07

32. $9.67 + 0.09$

   0.976   9.76   0.00976

33. $456.71 - 454.9$

   18.1   181   1.81

34. $803.25 - 0.6$

   802.65   0.80265   8.0265

35. $6004.003 + 52.7172$

   60.567202   605.67202   6056.7202

36. $128.35 + 97.0093$

   2253.593   225.3593   0.2253593

≈ *First use front end rounding to round each number and estimate the answer. Then find the exact answer. Use the information in the table below for Exercises 37–40.*

INTERNET USERS IN SELECTED COUNTRIES

Country	Number of Users
United States	134.6 million
Japan	33.9 million
China	22.5 million
South Korea	19 million
Canada	15.4 million
France	9 million
World total	413.7 million

Source: Computer Industry Almanac.

37. How many fewer Internet users are there in Canada compared to South Korea?

Estimate:

Exact:

38. How many more users are there in Japan compared to France?

Estimate:

Exact:

39. How many Internet users are there in all the countries listed in the table?

Estimate:

Exact:

40. Using the answer from Exercise 39, calculate the number of worldwide Internet users in countries other than the ones in the table.

Estimate:

Exact:

41. The tallest known land mammal is a prehistoric ancestor of the rhino measuring 6.4 meters. Find the combined heights of these NBA basketball stars: Allen Iverson at 1.83 meters, Shaquille O'Neal at 2.16 meters, and Kevin Garnett at 2.11 meters. Is their combined height greater or less than the prehistoric rhino? By how much? (*Source:* www.NBA.com/players)

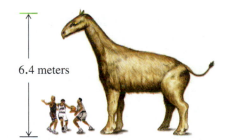

6.4 meters

Estimate:

Exact:

42. At a bakery, Sue Chee bought $7.42 worth of muffins and $10.09 worth of croissants for a staff party and a $0.69 cookie for herself. How much change did she receive from two $10 bills?

Estimate:

Exact:

43. Namiko is comparing two boxes of chicken nuggets. One box weighs 9.85 ounces and the other weighs 10.5 ounces. What is the difference in the weight of the two boxes?

Estimate:

Exact:

44. Sammy works in a veterinarian's office. He weighed two newborn kittens. One was 3.9 ounces and the other was 4.05 ounces. What was the difference in the weight of the two kittens?

Estimate:

Exact:

≈ *Find the perimeter of (distance around) each figure by adding the lengths of the sides.*

45.

19.75 in.

6.3 in.   6.3 in.

19.75 in.

Estimate:

Exact:

46.

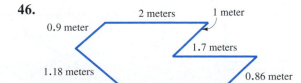

2 meters   1 meter

0.9 meter

1.7 meters

1.18 meters

0.86 meter

2.095 meters

Estimate:

Exact:

≈

The friends buying fishing equipment on the first page of this chapter brought along the store's sale insert from the Sunday paper. Use the information below on sale prices to answer Exercises 47–50. When estimating, round to the nearest whole number.

★ **Fishing Opener Sale** ★
Catch your limit of savings!

Bobbers 3 for 87¢

8-pound test fishing line
regular $4.84
invisible $7.47
fluorescent $5.14   *No-See Line*

Environmentally safe tin split shot
$2.07

Tackle boxes
Two trays $7.96
Three trays $9.96

Leaded split shot
94¢

Spinning reels: $9.88, $12.54, $18.84, $24.96
Spinning rods: $9.97, $18.97, $22.96, $28.94

Source: Wal-Mart.

47. What is the difference in price between the fluorescent and regular fishing line?

Estimate:

Exact:

48. How much more does the least expensive spinning rod cost than the least expensive reel?

Estimate:

Exact:

49. Find the total cost of the second highest priced spinning reel, two packages of tin split shot, and a three-tray tackle box. Sales tax for all the items was $2.31.

Estimate:

Exact:

50. One friend bought a package of bobbers on sale. He also bought some SPF15 sunscreen for $7.53 and a flotation vest for $44.96. Sales tax was $3.74. How much did he spend in all?

Estimate:

Exact:

Olivia Sanchez kept track of her expenses for one month. Use her list to answer Exercises 51–56.

MONTHLY EXPENSES

Rent	$994
Car payment	$190.78
Car repairs, gas	$205
Cable TV	$39.95
Internet access	$19.95
Electricity	$40.80
Telephone	$57.32
Groceries	$186.81
Entertainment	$97.75
Clothing, laundry	$107

51. What were Olivia's total expenses for the month?

52. How much did Olivia pay for telephone, cable TV, and Internet access?

53. What was the difference in the amounts spent for groceries and for the car payment?

54. Compare the amount Olivia spent on entertainment to the amount spent on car repairs and gas. What is the difference?

55. How much more did Olivia spend on rent than on all her car expenses?

56. How much less did Olivia spend on clothing and laundry than on all her car expenses?

Find the length of the dashed line in each rectangle or circle.

57.

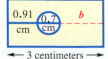

0.91 cm
0.7 cm
b
← 3 centimeters →

58.

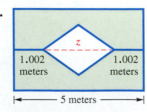

1.002 meters
z
1.002 meters
← 5 meters →

59.

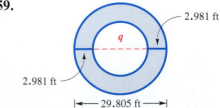

2.981 ft
q
2.981 ft
← 29.805 ft →

4.4 Multiplying Decimals

OBJECTIVE 1 Multiply decimals.   The decimals 0.3 and 0.07 can be multiplied by writing them as fractions.

$$0.3 \quad \times \quad 0.07 \quad = \frac{3}{10} \times \frac{7}{100} = \frac{3 \times 7}{10 \times 100} = \frac{21}{1000} = 0.021$$

1 decimal place + 2 decimal places → 3 decimal places

Can you see a way to multiply decimals without writing them as fractions? Try these steps. Remember that each number in a multiplication problem is called a *factor,* and the answer is called the *product.*

> **Multiplying Decimals**
>
> *Step 1*   Multiply the numbers (the factors) as if they were whole numbers.
>
> *Step 2*   Find the *total* number of decimal places in *both* factors.
>
> *Step 3*   Write the decimal point in the product (the answer) so it has the same number of decimal places as the total from Step 2. You may need to write in extra zeros on the left side of the product to get the correct number of decimal places.

> **NOTE**
> When multiplying decimals, you do ***not*** need to line up decimal points. (You ***do*** need to line up decimal points when adding or subtracting.)

EXAMPLE 1 Multiplying Decimal Numbers

Multiply 8.34 times 4.2.

Step 1   Multiply the numbers as if they were whole numbers.

$$
\begin{array}{r}
8.3\,4 \\
\times \quad 4.2 \\
\hline
1\,6\,6\,8 \\
3\,3\,3\,6 \quad \\
\hline
3\,5\,0\,2\,8
\end{array}
$$

Step 2   Count the total number of decimal places in both factors.

8.3 4 ← 2 decimal places
× 4.2 ← 1 decimal place
1 6 6 8   3 total decimal places
3 3 3 6
3 5 0 2 8

Step 3   Count over 3 places in the product and write the decimal point. Count from *right to left.*

8.3 4 ← 2 decimal places
× 4.2 ← 1 decimal place
1 6 6 8   3 total decimal places
3 3 3 6
3 5.0 2 8 ← 3 decimal places in product

Count over 3 places from right to left to position the decimal point.

Work Problem 1 at the Side.

OBJECTIVES

1. Multiply decimals.
2. Estimate the answer when multiplying decimals.

1 Multiply.

(a) $\begin{array}{r} 2.6 \\ \times\ 0.4 \\ \hline \end{array}$

(b) $\begin{array}{r} 45.2 \\ \times\ 0.25 \\ \hline \end{array}$

(c)   0.104 ← 3 decimal places
       ×   7 ← 0 decimal places
              ← 3 decimal places in the product

(d) $\begin{array}{r} 3.18 \\ \times\ 2.23 \\ \hline \end{array}$

(e) $\begin{array}{r} 611 \\ \times\ 3.7 \\ \hline \end{array}$

② Multiply.

(a) 0.04×0.09

(b) $(0.2)(0.008)$

(c) $(0.003)^2$ *Hint:* Recall that the 2 is an exponent. See **Section 1.8.**

(d) $(0.0081)(0.003)$

(e) $(0.11)(0.0005)$

③ First use front end rounding $\approx$ and estimate the answer. Then find the exact answer.

(a) $(11.62)(4.01)$

(b) $(5.986)(33)$

(c) $8.31(4.2)$

(d) 58.6×17.4

EXAMPLE 2 **Writing Zeros as Placeholders in the Product**

Multiply 0.042 by 0.03.

Start by multiplying, then count decimal places.

$$
\begin{array}{r}
0.0\ 4\ 2 \leftarrow 3 \text{ decimal places} \\
\times\quad 0.0\ 3 \leftarrow 2 \text{ decimal places} \\
\hline
1\ 2\ 6 \leftarrow 5 \text{ decimal places needed in product}
\end{array}
$$

After multiplying, the answer has only three decimal places, but five are needed, so write two zeros on the *left* side of the answer.

$$
\begin{array}{r}
0.0\ 4\ 2 \\
\times\quad 0.0\ 3 \\
\hline
\mathbf{0\ 0}\ 1\ 2\ 6
\end{array}
$$
↑ ↑
Write two zeros on *left* side of answer.

$$
\begin{array}{r}
0.0\ 4\ 2 \leftarrow 3 \text{ decimal places} \\
\times\quad 0.0\ 3 \leftarrow 2 \text{ decimal places} \\
\hline
.0\ 0\ 1\ 2\ 6 \leftarrow 5 \text{ decimal places}
\end{array}
$$
Now count over 5 places and write in the decimal point.

The final product is 0.00126, which has five decimal places.

◀◀◀ Work Problem 2 at the Side

OBJECTIVE **2** **Estimate the answer when multiplying decimals.** If you are doing multiplication problems by hand, estimating the answer helps you check that the decimal point is in the right place. When you are using a calculator, estimating helps you catch an error like pressing the ⊕ key instead of the ⊗ key.

EXAMPLE 3 **Estimating before Multiplying**

First estimate the answer to $(76.34)(12.5)$ using front end rounding. Then find the exact answer.

Estimate:

Rounds to
$$
\begin{array}{r}
80 \leftarrow \\
\times\ 10 \leftarrow \text{Rounds to} \\
\hline
800
\end{array}
$$

Exact:
$$
\begin{array}{r}
7\ 6.3\ 4 \leftarrow 2 \text{ decimal places} \\
\times\quad 1\ 2.5 \leftarrow 1 \text{ decimal place} \\
\hline
3\ 8\ 1\ 7\ 0 \quad 3 \text{ decimal places needed in product} \\
1\ 5\ 2\ 6\ 8 \\
7\ 6\ 3\ 4 \\
\hline
9\ 5\ 4.2\ 5\ 0
\end{array}
$$

Both the estimate and the exact answer go out to the hundreds place, so the decimal point in 954.**2**50 is probably in the correct place.

◀◀◀ Work Problem 3 at the Side.

🖩 **Calculator Tip** When working with money amounts, you may need to write a 0 in your answer. For example, try multiplying 3.54×5 on your calculator. Write down the result.

$$3.54 \ \otimes\ 5\ \ominus \underline{\qquad}$$

Notice that the result is 17.7, which is *not* the way to write a money amount. You have to write the 0 in the hundredths place: $17.7**0** is correct. The calculator does not show the "extra" 0 because:

$$17.70 \text{ or } 17\frac{70}{100} \text{ reduces to } 17\frac{7}{10} \text{ or } 17.7$$

So keep an eye on your calculator—it doesn't know when you're working with money amounts.

4.4 Exercises

FOR EXTRA HELP   Addison-Wesley Math Tutor Center    MathXL   Digital Video Tutor CD 3 Videotape 8   Student's Solutions Manual   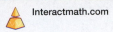 MyMathLab   Interactmath.com

Multiply. See Example 1.

1.   0.042
     × 3.2

2.   0.571
     × 2.9

3.   21.5
     × 7.4

4.   85.4
     × 3.5

5.   23.4
     × 0.666

6.   0.896
     × 0.799

7. $51.88
     ×   665

8. $736.75
     ×   118

Use the fact that $72 \times 6 = 432$ *to solve Exercises 9–16 by simply counting decimal places and writing the decimal point in the correct location. See Examples 1 and 2.*

9. $72 \times 0.6 =$ 4 3 2

10. $7.2 \times 6 =$ 4 3 2

11. $(7.2)(0.06) =$ 4 3 2

12. $(0.72)(0.6) =$ 4 3 2

13. $0.72(0.06) =$ 4 3 2

14. $72(0.0006) =$ 4 3 2

15. $0.0072(0.6) =$ 4 3 2

16. $0.072(0.006) =$ 4 3 2

Multiply. See Example 2.

17. $(0.006)(0.0052)$

18. $(0.0052)(0.009)$

19. $(0.005)^2$

20. $(0.03)^2$

RELATING CONCEPTS (EXERCISES 21–22) For Individual or Group Work

Look for patterns in the multiplications as you **work Exercises 21 and 22 in order.**

21. Do these multiplications:

$(5.96)(10) =$ _____ $(3.2)(10) =$ _____
$(0.476)(10) =$ _____ $(80.35)(10) =$ _____
$(722.6)(10) =$ _____ $(0.9)(10) =$ _____

What pattern do you see? Write a "rule" for multiplying by 10. What do you think the rule is for multiplying by 100? by 1000? Write the rules and try them out on the numbers above.

22. Do these multiplications:

$(59.6)(0.1) =$ _____ $(3.2)(0.1) =$ _____
$(0.476)(0.1) =$ _____ $(80.35)(0.1) =$ _____
$(65)(0.1) =$ _____ $(523)(0.1) =$ _____

What pattern do you see? Write a "rule" for multiplying by 0.1. What do you think the rule is for multiplying by 0.01? by 0.001? Write the rules and try them out on the numbers above.

≈

First use front end rounding to round each number and estimate the answer. Then find the exact answer. See Example 3.

23. *Estimate:*   *Exact:*

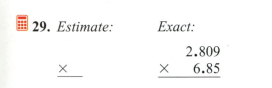

        ← Rounds to   39.6
×   ← Rounds to   ×   4.8

24. *Estimate:*   *Exact:*

         18.7
×   × 2.3

25. *Estimate:*   *Exact:*

         37.1
×   ×   42

26. *Estimate:*   *Exact:*

         5.08
×   ×   71

27. *Estimate:*   *Exact:*

         6.53
×   ×   4.6

28. *Estimate:*   *Exact:*

         7.51
×   ×   8.2

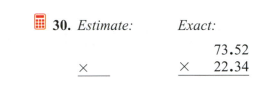

29. *Estimate:*   *Exact:*

         2.809
×   ×   6.85

30. *Estimate:*   *Exact:*

         73.52
×   ×   22.34

≈

Even with most of the problem missing, you can tell whether or not these answers are reasonable. Circle reasonable or unreasonable. If the answer is unreasonable, move the decimal point, or insert a decimal point, to make the answer reasonable.

31. How much was his car payment? $18.90

     reasonable

     unreasonable, should be _____

32. How many hours did she work today? 25 hours

     reasonable

     unreasonable, should be _____

33. How tall is her son? 60.5 in.

     reasonable

     unreasonable, should be _____

34. How much does he pay for rent now? $6.92

     reasonable

     unreasonable, should be _____

35. What is the price of one gallon of milk? $419

     reasonable

     unreasonable, should be _____

36. How long is the living room? 16.8 feet

     reasonable

     unreasonable, should be _____

37. How much did Mrs. Brown's baby weigh?
0.095 pounds

     reasonable

     unreasonable, should be _____

38. What was the sale price of the jacket? $1.49

     reasonable

     unreasonable, should be _____

Solve each application problem. Round money answers to the nearest cent when necessary.

39. LaTasha worked 50.5 hours over the last two weeks. She earns $18.73 per hour. How much did she make?

40. Michael's time card shows 42.2 hours at $10.03 per hour. What are his earnings?

41. Sid needs 0.6 meter of canvas material to make a carry-all bag that fits on his wheelchair. If canvas is $4.09 per meter, how much will Sid spend? (*Note:* $4.09 *per* meter means $4.09 for *one* meter.)

42. How much will Mrs. Nguyen pay for 3.5 yards of lace trim that costs $0.87 per yard?

43. Michelle filled the tank of her pickup truck with regular unleaded gas. Use the information shown on the pump to find how much she paid for gas.

GALLONS	PRICE PER GALLON	GALLONS	PRICE PER GALLON
10.329	$ 1.799	18.605	$ 1.749
SUPRA UNLEADED		UNLEADED REGULAR	
Minimum Octane Rating 90		Minimum Octane Rating 87	

Source: Holiday.

44. Ground beef and chicken legs are on sale. Juma bought 1.7 pounds of legs. Use the information in the ad to find the amount she paid.

45. Ms. Rolack is a real estate broker who helps people sell their homes. Her fee is 0.07 times the price of the home. What was her fee for selling a $229,500 home?

46. Manny Ramirez of the Boston Red Sox had a batting average of 0.349 in the 2002 season. He went to bat 436 times. How many hits did he make? (*Hint:* Multiply the number of times at bat by his batting average.) Round to the nearest whole number. (*Source: World Almanac.*)

Paper money in the United States has not always been the same size. Shown below are the measurements of bills printed before 1929 and the measurements from 1929 on. Use this information to answer Exercises 47–50. Recall that perimeter is the total distance around the edges of a figure (see **Section 1.2***). The area of a rectangle is found by multiplying the length by the width (see* **Section 2.5***) Source: www.moneyfactory.com*

Before 1929

3.125 in.

7.4218 in.

From 1929 on

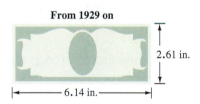

2.61 in.

6.14 in.

47. (a) Find the area of each $20 bill, rounded to the nearest tenth.

(b) What is the difference in the rounded areas?

48. (a) Find the perimeter of each $20 bill, to the nearest hundredth.

(b) How much less is the perimeter of today's bills than the bills printed before 1929?

49. The thickness of one piece of today's paper money is 0.0043 inch.

(a) If you had a pile of 100 bills, how high would the pile be?

(b) How high would a pile of 1000 bills be?

50. (a) Use your answers from Exercise 49 to find the number of bills in a pile that is 43 inches high.

(b) How much money would you have if the pile is all $20 bills?

51. Judy Lewis pays $38.96 per month for basic cable TV. The one-time installation fee was $49. How much will she pay for cable over two years? How much would she pay in two years for the deluxe cable package that costs $89.95 per month?

52. Chuck's SUV payment is $420.27 per month for four years. He also made a down payment of $5000 at the time he bought the SUV. How much will he pay altogether?

53. Paper for the copy machine at the library costs $0.015 per sheet. How much will the library pay for 5100 sheets?

54. A student group collected 2200 pounds of plastic as a fund-raiser. How much will they make if the recycling center pays $0.142 per pound?

55. Barry bought 16.5 meters of rope at $0.47 per meter and three meters of wire at $1.05 per meter. How much change did he get from three $5 bills?

56. Susan bought a 27-inch flat screen TV that cost $549.99. She paid $55 down and $98.83 per month for six months. How much could she have saved by paying cash?

Use the information from the Look Smart mail-order catalog to answer Exercises 57–60.

Knit Shirt Ordering Information		
43-2A	Short sleeved, solid colors	$14.75 each
43-2B	Short sleeved, stripes	$16.75 each
43-3A	Long sleeved, solid colors	$18.95 each
43-3B	Long sleeved, stripes	$21.95 each
XXL size, add $2 per shirt.		
Monogram, $4.95 each. Gift box, $5 each.		

Total Price of All Items (excluding monograms and gift boxes)	Shipping, Packing, and Handling
$0–25.00	$3.50
$25.01–75.00	$5.95
$75.01–125.00	$7.95
$125.01+	$9.95
Shipping to each additional address, add $4.25.	

57. Find the total cost of ordering four long-sleeved, solid-color shirts and two short-sleeved, striped shirts, all in the XXL size, and all shipped to your home.

58. What is the total cost of eight long-sleeved shirts, five in solid colors and three striped? Include the cost of shipping the solid shirts to your home and the striped shirts to your brother's home.

59. (a) What is the total cost, including shipping, of sending three short-sleeved, solid-color shirts, with monograms, in a gift box to your aunt for her birthday?

  (b) How much did the monograms, gift box, and shipping add to the cost of your gift?

60. (a) Suppose you order one of each type of shirt for yourself, adding a monogram to each of the solid-color shirts. At the same time, you order three long-sleeved striped shirts, in the XXL size, shipped to your dad in a gift box. Find the total cost of your order.

  (b) what is the difference in total cost (excluding shipping) between the shirts for yourself and the gift for your dad?

4.5 Dividing Decimals

There are two kinds of decimal division problems: those in which a decimal is divided by a whole number, and those in which a number is divided by a decimal. First recall the parts of a division problem from **Section 1.5.**

$$
\begin{array}{r}
8 \leftarrow \text{Quotient} \\
\text{Divisor} \rightarrow 4)\overline{33} \leftarrow \text{Dividend} \\
\underline{32} \\
1 \leftarrow \text{Remainder}
\end{array}
$$

OBJECTIVE 1 Divide a decimal by a whole number. When the divisor is a whole number, use these steps.

Dividing Decimals by Whole Numbers

Step 1   Write the decimal point in the quotient (answer) directly above the decimal point in the dividend.

Step 2   Divide as if both numbers were whole numbers.

EXAMPLE 1 Dividing Decimals by Whole Numbers

Divide.

(a) 21.93 by 3

Dividend   Divisor

Rewrite the division problem. $3)\overline{21.93}$

Step 1   Write the decimal point in the quotient directly above the decimal point in the dividend.

Decimal points lined up

$3)\overline{21\overset{\textbf{.}}{}93}$

Step 2   Divide as if the numbers were whole numbers.

$3)\overline{21.93}$ quotient 7.31

Check by multiplying the quotient times the divisor.

Check:

$$
\begin{array}{r}
7.31 \\
\times \quad 3 \\
\hline
21.93
\end{array}
$$

Matches, so 7.31 is correct.

The quotient (answer) is 7.31.

(b)

$$9)\overline{470.7}$$ quotient 52.3

Divisor — Dividend

Write the decimal point in the quotient above the decimal point in the dividend. Then divide as if the numbers were whole numbers.

Decimal points lined up

$$
\begin{array}{r}
52.3 \\
9)\overline{470.7} \\
\underline{45} \\
20 \\
\underline{18} \\
27 \\
\underline{27} \\
0
\end{array}
$$

Matches

Check:

$$
\begin{array}{r}
52.3 \\
\times \quad 9 \\
\hline
470.7
\end{array}
$$

The quotient is 52.3.

Work Problem 1 at the Side.

---

1 Divide. Check your answers by multiplying.

(a) $4)\overline{93.6}$

$$
\begin{array}{r}
23.4 \\
4)\overline{93.6} \\
\underline{8} \\
13 \\
\underline{12} \\
1.6
\end{array}
$$

(b) $6)\overline{6.804}$

(c) $11)\overline{278.3}$

(d) $0.51835 \div 5$

(e) $213.45 \div 15$

ANSWERS

1. **(a)** 23.4; (23.4)(4) = 93.6
    (b) 1.134; (1.134)(6) = 6.804
    (c) 25.3; (25.3)(11) = 278.3
    (d) 0.10367; (0.10367)(5) = 0.51835
    (e) 14.23; (14.23)(15) = 213.45

2 Divide. Check your answers by multiplying.

(a) $5\overline{)6.4}$

(b) $30.87 \div 14$

(c) $\dfrac{259.5}{30}$

(d) $0.3 \div 8$

EXAMPLE 2 **Writing Extra Zeros to Complete a Division**

Divide 1.5 by 8.

Keep dividing until the remainder is 0, or until the digits in the quotient begin to repeat in a pattern. In Example 1(b), you ended up with a remainder of 0. But sometimes you run out of digits in the dividend before that happens. If so, write extra zeros on the right side of the dividend so you can continue dividing.

$$\begin{array}{r} 0.1 \\ 8\overline{)1.5} \\ \underline{8} \\ 7 \end{array}$$ ← All digits have been used.

← Remainder is not yet 0.

Write a 0 after the 5 in the dividend so you can continue dividing. Keep writing more zeros in the dividend, if needed. Recall that writing zeros to the *right* of a decimal number does **not** change its value.

$$\begin{array}{r} 0.1\,8\,7\,5 \\ 8\overline{)1.5\,0\,0\,0} \\ \underline{8} \\ 7\,0 \\ \underline{6\,4} \\ 6\,0 \\ \underline{5\,6} \\ 4\,0 \\ \underline{4\,0} \\ 0 \end{array}$$

← Three zeros needed to complete the division

← Stop dividing when the remainder is 0.

Check:

$$\begin{array}{r} 0.1875 \\ \times \qquad 8 \\ \hline 1.5000 \end{array}$$

Matches dividend, so 0.1875 is correct.

CAUTION
When dividing decimals, notice that the dividend may *not* be the larger number, as it was in whole numbers. In Example 2 above, the dividend is 1.5, which is *smaller* than the divisor 8.

◀◀◀ **Work Problem 2 at the Side.**

🖩 **Calculator Tip** When *multiplying* numbers, you can enter them in any order because multiplication is commutative (see **Section 1.4**). But division is *not* commutative. It *does* matter which number you enter first. Try Example 2 both ways; jot down your answers.

$1.5 \ (\div) \ 8 \ (=) \ \underline{0.1875}$ $\qquad$ $8 \ (\div) \ 1.5 \ (=) \ \underline{5.333}$

Notice that the first answer, 0.1875, matches the result from Example 2. But the second answer is much different: 5.333333333. Be careful to enter the dividend first.

The next example shows a quotient (answer) that must be rounded because you will never get a remainder of 0.

EXAMPLE 3 **Rounding a Decimal Quotient**

Divide 4.7 by 3. Round the quotient to the nearest thousandth. Write extra zeros in the dividend so you can continue dividing.

$$
\begin{array}{r}
1.5\,6\,6\,6 \\
3\overline{)4.7\,0\,0\,0} \leftarrow \text{Three zeros added so far} \\
\underline{3} \\
1\,7 \\
\underline{1\,5} \\
2\,0 \\
\underline{1\,8} \\
2\,0 \\
\underline{1\,8} \\
2\,0 \\
\underline{1\,8} \\
2 \leftarrow \text{Remainder is still not 0.}
\end{array}
$$

Notice that the digit 6 in the answer is repeating. It will continue to do so. The remainder will *never be 0*. There are two ways to show that the answer is a **repeating decimal** that goes on forever. You can write three dots after the answer, or you can write a bar above the digits that repeat (in this case, the 6).

$$
1.5\underbrace{666}_{\text{Three dots}} \ldots \quad \text{or} \quad 1.5\underset{\text{Bar above}}{\underline{6}} \leftarrow \text{repeating digit}
$$

When repeating decimals occur, round the quotient according to the directions in the problem. In this example, to round to thousandths, divide out one *more* place, to ten-thousandths.

$$4.7 \div 3 = 1.5666 \ldots \quad \text{rounds to} \quad 1.567$$

Check the answer by multiplying 1.567 by 3. Because 1.567 is a rounded answer, the check will not give exactly 4.7, but it should be very close.

$$(1.567)(3) = 4.701 \leftarrow \begin{array}{l}\text{Does not equal exactly 4.7}\\ \text{because 1.567 was rounded}\end{array}$$

> **CAUTION**
> When checking quotients that you've rounded, the check will *not* match the dividend exactly, but it should be very close.

Work Problem 3 at the Side. ▶▶▶

OBJECTIVE 2 Divide a number by a decimal. To divide by a *decimal* divisor, first change the divisor to a whole number. Then divide as before. To see how this is done, write the problem in fraction form. Here is an example.

$$1.2\overline{)6.36} \quad \text{can be written} \quad \frac{6.36}{1.2}$$

In **Section 3.2** you learned that multiplying the numerator and denominator by the same number gives an equivalent fraction. We want the divisor (1.2) to be a whole number. Multiplying by 10 will accomplish that.

$$\underset{\substack{\text{Decimal} \to \\ \text{divisor}}}{\frac{6.36}{1.2}} = \frac{(6.36)\,(\mathbf{10})}{(1.2)\,(\mathbf{10})} = \underset{\substack{\leftarrow \text{Whole number} \\ \text{divisor}}}{\frac{63.6}{12}}$$

---

3 Divide. Round quotients to the nearest thousandth. If it is a repeating decimal, also write the answer using a bar. Check your answers by multiplying.

(a) $13\overline{)267.01}$

(b) $6\overline{)20.5}$

(c) $\dfrac{10.22}{9}$

(d) $16.15 \div 3$

(e) $116.3 \div 11$

ANSWERS
3. **(a)** 20.539 (rounded); no repeating digits visible on calculator;
         (20.539)(13) = 267.007
    (b) 3.417 (rounded); $3.41\overline{6}$;
         (3.417)(6) = 20.502
    (c) 1.136 (rounded); $1.13\overline{5}$;
         (1.136)(9) = 10.224
    (d) 5.383 (rounded); $5.38\overline{3}$;
         (5.383)(3) = 16.149
    (e) 10.573 (rounded); $10.57\overline{2}$;
         (10.573)(11) = 116.303

4 Divide. If the quotient does not come out even, round to the nearest hundredth.

(a) $0.2\overline{)1.04}$

(b) $0.06\overline{)1.8072}$

(c) $0.005\overline{)32}$

(d) $8.1 \div 0.025$

(e) $\dfrac{7}{1.3}$

(f) $5.3091 \div 6.2$

The short way to multiply by 10 is to move the decimal point *one place* to the *right* in both the divisor and the dividend.

$$1.2\overline{)6.3\,6} \quad \text{is equivalent to} \quad 12\overline{)63.6}$$

> **NOTE**
> Moving the decimal points the **same** number of places in **both** the divisor and dividend will **not** change the answer.

Dividing by Decimals

Step 1   Count the number of decimal places in the divisor and move the decimal point that many places to the *right*. (This changes the divisor to a whole number.)

Step 2   Move the decimal point in the dividend the *same* number of places to the *right*. (Write in extra zeros if needed.)

Step 3   Write the decimal point in the quotient directly above the decimal point in the dividend. Then divide as usual.

EXAMPLE 4 **Dividing by Decimals**

(a) $0.003\overline{)27.69}$

Move the decimal point in the divisor *three* places to the *right* so 0.003 becomes the whole number 3. To move the decimal point in the dividend the same number of places, write in an extra 0.

$$0.003\overline{)27.690.}$$

Move decimal point three places to the right is the same as multiplying by 1000.

Move decimal points in divisor and dividend. Then line up the decimal point in the quotient.

$$\begin{array}{r} 9230. \\ 3\overline{)27690.} \end{array}$$

Divide as usual.

(b) Divide 5 by 4.2. Round to the nearest hundredth.

Move the decimal point in the divisor one place to the right so 4.2 becomes the whole number 42. The decimal point in the dividend starts on the right side of 5 and is also moved one place to the right.

$$\begin{array}{r} 1.1\,9\,0 \\ 4.2\overline{)5.0\,0\,0\,0} \\ \underline{4\,2} \\ 8\,0 \\ \underline{4\,2} \\ 3\,8\,0 \\ \underline{3\,7\,8} \\ 2\,0 \end{array}$$

← In order to round to hundredths, divide out one *more* place, to thousandths.

Round the quotient. It is 1.19 (rounded to the nearest hundredth).

Work Problem 4 at the Side.

ANSWERS
4. (a) 5.2   (b) 30.12   (c) 6400   (d) 324
   (e) 5.38 (rounded)   (f) 0.86 (rounded)

OBJECTIVE 3 Estimate the answer when dividing decimals. Estimating the answer to a division problem helps you catch errors. Compare the estimate to your exact answer. If they are very different, do the division again.

EXAMPLE 5 Estimating before Dividing

First use front end rounding to round each number and estimate the answer. Then divide to find the exact answer.

$$580.44 \div 2.8$$

Here is how one student solved this problem. She rounded 580.44 to 600 and rounded 2.8 to 3 to estimate the answer.

Estimate: *Exact:*

$$\begin{array}{r} \mathbf{200} \\ 3\overline{)600} \end{array} \qquad \begin{array}{r} \mathbf{27.3} \\ 2.8\overline{)580.44} \\ \underline{56} \\ 204 \\ \underline{196} \\ 84 \\ \underline{84} \\ 0 \end{array}$$

Very different; need to rework the problem

Notice that the estimate, which is in the hundreds, is very different from the exact answer, which is only in the tens. This tells the student that she needs to rework the problem. Can you find the error? (The exact answer should be 207.3, which fits with the estimate of 200.)

Work Problem 5 at the Side.

OBJECTIVE 4 Use the order of operations with decimals. Use the order of operations when a decimal problem involves more than one operation, as you did with whole numbers in **Section 1.8.**

Order of Operations
1. Do all operations inside *parentheses* or *other grouping symbols.*
2. Simplify any expressions with *exponents* and find any *square roots.*
3. *Multiply* or *divide*, proceeding from left to right.
4. *Add* or *subtract*, proceeding from left to right.

EXAMPLE 6 Using the Order of Operations

Use the order of operations to simplify each expression.

(a) $2.5 + \mathbf{6.3^2} + 9.62$ Apply the exponent: (6.3)(6.3) is 39.69

$\mathbf{2.5 + 39.69} + 9.62$ Add from left to right.

$42.19 + 9.62$

51.81

(b) $1.82 + \mathbf{(6.7 - 5.2)}(5.8)$ Work inside parentheses.

$1.82 + \mathbf{(1.5)(5.8)}$ Multiply next.

$1.82 + 8.7$ Add last.

10.52

Continued on Next Page

5 Decide whether each answer is reasonable by using front end rounding to estimate the answer. If the exact answer is *not* reasonable, find and correct the error.

(a) $42.75 \div 3.8 = 1.125$

Estimate:

(b) $807.1 \div 1.76 = 458.580$ to nearest thousandth

Estimate:

(c) $48.63 \div 52 = 93.519$ to nearest thousandth

Estimate:

$$\begin{array}{r} .935 \\ 52\overline{)48.6300} \\ \underline{468} \\ 183 \\ \underline{156} \\ 270 \\ \underline{260} \end{array}$$

(d) $9.0584 \div 2.68 = 0.338$

Estimate:

ANSWERS
5. **(a)** Estimate is 40 ÷ 4 = 10; answer is not reasonable; should be 11.25
(b) Estimate is 800 ÷ 2 = 400; answer is reasonable.
(c) Estimate is 50 ÷ 50 = 1; answer is not reasonable, should be 0.935 (rounded)
(d) Estimate is 9 ÷ 3 = 3; answer is not reasonable, should be 3.38

6 Use the order of operations to simplify each expression.

(a) $4.6 - 0.79 + 1.5^2$

(b) $3.64 \div 1.3 \times 3.6$.

(c) $0.08 + 0.6\,(3 - 2.99)$

(d) $10.85 - 2.3\,(5.2) \div 3.2$

(c) $3.7^2 - 1.8 \div 5\,(1.5)$ Apply the exponent.

$13.69 - \underline{1.8 \div 5}\,(1.5)$ Multiply and divide from left to right, so first divide 1.8 by 5.

$13.69 - \underline{0.36\,(1.5)}$ Then multiply 0.36 by 1.5.

$13.69 - 0.54$ Subtract last.

13.15

◀◀◀ **Work Problem 6 at the Side.**

Calculator Tip   Most scientific calculators that have parentheses keys ( ) can handle calculations like those in Example 6 if you just enter the numbers in the order given. For example, the keystrokes for Example 6(b) are:

— Parentheses —

1.82 ⊕ ⦅ 6.7 ⊖ 5.2 ⦆ ⊗ 5.8 ⊜   Answer is 10.52.

Standard, four-function calculators generally do not have parentheses keys and will *not* give the correct answer if you simply enter the numbers in the order given.

Check the instruction manual that came with your calculator for information on "order of calculations" to see if your model has the rules for order of operations built into it. For a quick check, try entering this problem.

2 ⊕ 2 ⊗ 2 ⊜

If the result is 6, the calculator follows the order of operations. If the result is 8, it does *not* have the rules built into it. To see why this test works, do the calculations by hand.

Follow the order of operations.   **Work from left to right.**

$2 + \underline{2 \times 2}$ Multiply before adding. $\underline{2 + 2} \times 2$

$\underline{2 + \quad 4}$ $\underline{4 \quad \times 2}$

6 ◀——— Correct. 8 ◀——— Incorrect

4.5 Exercises

 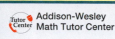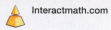
Divide. See Examples 1 and 4.

1. $7\overline{)27.3}$

2. $8\overline{)50.4}$

3. $\dfrac{4.23}{9}$

4. $\dfrac{1.62}{6}$

5. $0.05\overline{)20.01}$

6. $0.08\overline{)16.04}$

7. $1.5\overline{)54}$

8. $2.4\overline{)132}$

Use the fact that $108 \div 18 = 6$ to work Exercises 9–12 simply by moving decimal points.

9. $0.108 \div 1.8$

10. $10.8 \div 18$

11. $0.018\overline{)108}$

12. $0.18\overline{)1.08}$

Divide. Round quotients to the nearest hundredth if necessary. See Examples 3 and 4.

13. $4.6\overline{)116.38}$

14. $2.6\overline{)4.992}$

15. $\dfrac{3.1}{0.006}$

16. $\dfrac{1.7}{0.09}$

Divide. Round quotients to the nearest thousandth.

17. $240 \div 9.88$

18. $7643 \div 5.36$

19. $0.034\overline{)342.81}$

20. $0.043\overline{)1748.4}$

RELATING CONCEPTS (EXERCISES 21–22) For Individual or Group Work

*Look for patterns as you **work Exercises 21 and 22 in order.***

21. Do these division problems:

$3.77 \div 10 =$ _____ $9.1 \div 10 =$ _____

$0.886 \div 10 =$ _____ $30.19 \div 10 =$ _____

$406.5 \div 10 =$ _____ $6625.7 \div 10 =$ _____

(a) What pattern do you see? Write a "rule" for dividing by 10. What do you think the rule is for dividing by 100? by 1000? Write the rules and try them out on the numbers above.

(b) Compare your rules to the ones you wrote in **Section 4.4,** Exercise 21. How are they different?

22. Do these division problems:

$40.2 \div 0.1 =$ _____ $7.1 \div 0.1 =$ _____

$0.339 \div 0.1 =$ _____ $15.77 \div 0.1 =$ _____

$46 \div 0.1 =$ _____ $873 \div 0.1 =$ _____

(a) What pattern do you see? Write a "rule" for dividing by 0.1. What do you think the rule is for dividing by 0.01? by 0.001? Write the rules and try them out on the numbers above.

(b) Compare your rules to the ones you wrote in **Section 4.4,** Exercise 22. How are they different?

≈

Decide whether each answer is reasonable or unreasonable by rounding the numbers and estimating the answer. If the exact answer is not reasonable, find the correct answer. See Example 5.

23. $37.8 \div 8 = 47.25$

Estimate:

24. $345.6 \div 3 = 11.52$

Estimate:

25. $54.6 \div 48.1 = 1.135$

Estimate:

26. $2428.8 \div 4.8 = 50.6$

Estimate:

27. $307.02 \div 5.1 = 6.2$

Estimate:

28. $395.415 \div 5.05 = 78.3$

Estimate:

29. $9.3 \div 1.25 = 0.744$

Estimate:

30. $78 \div 14.2 = 0.182$

Estimate:

Solve each application problem. Round money answers to the nearest cent, if necessary.

31. Rob discovered that his daughter's favorite brand of tights are on sale. He decided to buy one pair as a surprise for her. How much did he pay?

32. The bookstore has a special price on notepads. How much did Randall pay for one notepad?

33. It will take 21 equal monthly payments for Aimee to pay off her credit card balance of $1408.66. How much is she paying each month?

34. Marcella Anderson bought 2.6 meters of suede fabric for $18.19. How much did she pay per meter?

35. Adrian Webb bought 619 bricks to build a barbecue pit, paying $185.70. Find the cost per brick. (*Hint:* Cost *per* brick means the cost for *one* brick.)

36. Lupe Wilson is a newspaper distributor. Last week she paid the newspaper $130.51 for 842 copies. Find the cost per copy.

37. Darren Jackson earned $476.80 for 40 hours of work. Find his earnings per hour.

38. At a CD manufacturing company, 400 CDs cost $289. Find the cost per CD.

39. It took 16.35 gallons of gas to fill the gas tank of Kim's car. She had driven 346.2 miles since her last fill-up. How many miles per gallon did her car get? Round to the nearest tenth.

40. Mr. Rodriquez pays $53.19 each month to Household Finance. How many months will it take him to pay off $1436.13?

Use the table of women's longest long jumps (through the year 2003) to answer Exercises 41–46. To find an average, add up the values you are interested in and then divide the sum by the number of values. Round your answer to the nearest hundredth.

Athlete	Country	Year	Length (meters)
Galina Christyakova	USSR	1988	7.52
Jackie Joyner-Kersee	U.S.	1994	7.49
Heike Drechsler	Germany	1992	7.48
Jackie Joyner-Kersee	U.S.	1987	7.45
Jackie Joyner-Kersee	U.S.	1988	7.40
Jackie Joyner-Kersee	U.S.	1991	7.32
Jackie Joyner-Kersee	U.S.	1996	7.20
Chioma Ajunwa	Nigeria	1996	7.12
Fiona May	Italy	2000	7.09
Ludmila Nova	Australia	1993	7.06

Source: CNNSI.com-Athletics

41. Find the average length of the long jumps made by Jackie Joyner-Kersee.

42. Find the average length of all the long jumps listed in the table.

43. How much longer was the fifth-longest jump than the sixth-longest jump?

44. If the first-place athlete made five jumps of the same length, what would be the total distance jumped?

45. What was the total length jumped by the top three athletes in the table?

46. How much less was the last-place jump than the next-to-last-place jump?

Use the order of operations to simplify each expression. See Example 6

47. $7.2 - 5.2 + 3.5^2$

48. $6.2 + 4.3^2 - 9.72$

49. $38.6 + 11.6 (13.4 - 10.4)$

50. $2.25 - 1.06 (4.85 - 3.95)$

51. $8.68 - 4.6 (10.4) \div 6.4$

52. $25.1 + 11.4 \div 7.5 (3.75)$

53. $33 - 3.2 (0.68 + 9) - 1.3^2$

54. $0.6 + (1.89 + 0.11) \div 0.004 (0.5)$

Solve each application problem.

55. Soup is on sale at six cans for $3.25, or you can purchase individual cans for $0.57. How much will you save per can if you buy six cans? Round to the nearest cent.

56. Nadia's diet says she can eat 3.5 ounces of chicken nuggets. The package weighs 10.5 ounces and contains 15 nuggets. How many nuggets can Nadia eat?

57. The U.S. Treasury prints 37,000,000 pieces of paper money each day. The printing presses run 24 hours a day. How many pieces of money are printed, to the nearest whole number:

(a) each hour?

(b) each minute?

37,000,000 pieces of paper money are printed each day.

(c) each second?

(*Source:* www.moneyfactory.com)

58. Mach 1 is the speed of sound. Dividing a vehicle's speed by the speed of sound gives its speed on the Mach scale. In 1997, a specially built car with two 110,000-horsepower engines broke the world land speed record by traveling 763.035 miles per hour. The speed of sound that day was 748.11 miles per hour. What was the car's Mach speed, to the nearest hundredth? (*Source:* Associated Press.)

General Mills will give a school 10¢ for each box top logo from its cereals and other products. A school can earn up to $10,000 per year. Use this information to answer Exercises 59–62. Round your answers to the nearest whole number when necessary. (Source: General Mills.)

59. How many box tops would a school need to collect in one year to earn the maximum amount?

60. (Complete Exercise 59 first.) If a school has 550 children, how many box tops would each child need to collect in one year to reach the maximum?

61. How many box tops would need to be collected during each of the 38 weeks in the school year to reach the maximum amount?

62. How many box tops would each of the 550 children need to collect during each of the 38 weeks of school to reach the maximum amount?

Summary Exercises on Decimals

Write each decimal as a fraction or mixed number in lowest terms.

1. 0.8

2. 6.004

3. 0.35

Write each decimal in words.

4. 94.5

5. 2.0003

6. 0.706

Write each decimal in numbers.

7. five hundredths

8. three hundred nine ten-thousandths

9. ten and seven tenths

Round to the place indicated.

10. 6.1873 to the nearest hundredth

11. 0.95 to the nearest tenth

12. 0.42025 to the nearest thousandth

13. $0.893 to the nearest cent

14. $3.0017 to the nearest cent

15. $99.64 to the nearest dollar

Simplify.

16. $0.27\,(3.5)$

17. $50 - 0.3801$

18. $0.35 \div 0.007$

19. $\dfrac{90.18}{6}$

20. $(0.004)\,(1.22)$

21. $3.7 - 1.55$

22. $0.95 + 10.005$

23. $3.6 + 0.718 + 9 + 5.0829$

24. $32.305 - 28 + 0.7$

25. $8.9 - 0.4^2 \div 0.02$

26. $18 \div 2.5^2 + 4\,(1.3)$

27. $0.64 \div 16.3$ Round your answer to the nearest hundredth.

28. Find the perimeter and area of this rectangular computer chip. Round your answers to the nearest tenth.

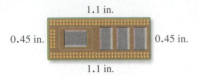

1.1 in.

0.45 in.   0.45 in.

1.1 in.

29. Find the perimeter and area of this rectangular watch face. Round your answers to the nearest hundredth.

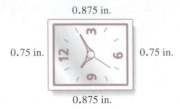

0.875 in.

0.75 in.   0.75 in.

0.875 in.

Use the table on heaviest fruits and vegetables to answer Exercises 30–33.

HEAVIEST FRUITS AND VEGETABLES

Fruit/Vegetable	Weight (in pounds)	Location
Apple	3.5	England
Grapefruit	6.75	Australia
Onion	15.49	England
Peach	1.6	Michigan, USA
Radish	68.563	Japan
Strawberry	0.51	England
Tomato	7.75	Oklahoma, USA
Watermelon	262	Tennessee, USA

Source: Guinness Book of World Records.

30. (a) Identify the heaviest and lightest weight items in the table.

(b) Find the difference in weight between the heaviest and lightest items.

31. If you used the radish, tomato, and onion in a salad, what would be their combined weight?

32. How much would six grapefruits weigh?

33. How many apples would it take to match the weight of the watermelon, to the nearest whole number?

34. The Bell family rented a motor home for $375 per week plus $0.35 per mile. What was the rental cost for their three-week vacation trip of 2650 miles?

35. A craft cooperative paid $40.32 for enough flannel fabric to make eight baby blankets for their store. They sold the blankets for $12.95 each. How much profit was made on the blankets?

4.6 Writing Fractions as Decimals

Writing fractions as equivalent decimals can help you do calculations more easily or compare the size of two numbers.

OBJECTIVE 1 Write fractions as equivalent decimals.   Recall that a fraction is one way to show division (see **Section 1.5**). For example, $\frac{3}{4}$ means $3 \div 4$. If you are doing the division by hand, write it as $4\overline{)3}$. When you do the division, the result is 0.75, the decimal equivalent of $\frac{3}{4}$.

> **Writing a Fraction as an Equivalent Decimal**
>
> *Step 1*   Divide the numerator of the fraction by the denominator.
>
> *Step 2*   If necessary, round the answer to the place indicated.

Work Problem 1 at the Side.

EXAMPLE 1 Writing Fractions or Mixed Numbers as Decimals

(a) Write $\frac{1}{8}$ as a decimal.

$\frac{1}{8}$ means $1 \div 8$. Write it as $8\overline{)1}$. The decimal point in the dividend is on the *right* side of the 1. Write extra zeros in the dividend so you can continue dividing until the remainder is 0.

$$\frac{1}{8} \rightarrow 1 \div 8 \rightarrow 8\overline{)1} \rightarrow \begin{array}{r} 0.125 \\ 8\overline{)1.000} \\ \underline{8} \\ 20 \\ \underline{16} \\ 40 \\ \underline{40} \\ 0 \end{array}$$

Decimal points lined up. Three extra zeros needed. Remainder is 0.

Therefore, $\frac{1}{8} = 0.125$.

To check this, write 0.125 as a fraction, then change it to lowest terms.

$0.125 = \frac{125}{1000}$ In lowest terms $\frac{125 \div 125}{1000 \div 125} = \frac{1}{8}$ ← Original fraction

Calculator Tip   When using your calculator to write fractions as decimals, enter the numbers from the top down. Remember that the order in which you enter the numbers *does* matter in division. Example 1(a) above works like this:

$\frac{1}{8}$ Top down   Enter 1 ÷ 8 =   Answer is 0.125

What happens if you enter 8 ÷ 1 =? Do you see why that cannot possibly be correct? (Answer: $8 \div 1 = 8$. A proper fraction like $\frac{1}{8}$ *cannot* be equivalent to a whole number.)

Continued on Next Page

1 Rewrite each fraction so you could do the division by hand. Do *not* complete the division.

(a) $\frac{1}{9}$ is written $9\overline{)}$

(b) $\frac{2}{3}$ is written $\overline{)}$

(c) $\frac{5}{4}$ is written $\overline{)}$

(d) $\frac{3}{10}$ is written $\overline{)}$

(e) $\frac{21}{16}$ is written $\overline{)}$

(f) $\frac{1}{50}$ is written $\overline{)}$

ANSWERS

1. **(a)** $9\overline{)1}$ **(b)** $3\overline{)2}$ **(c)** $4\overline{)5}$ **(d)** $10\overline{)3}$ **(e)** $16\overline{)21}$ **(f)** $50\overline{)1}$

2 Write each fraction or mixed number as a decimal.

(a) $\dfrac{1}{4}$

(b) $2\dfrac{1}{2}$

(c) $\dfrac{5}{8}$

(d) $4\dfrac{3}{5}$

(e) $\dfrac{7}{8}$

(b) Write $2\dfrac{3}{4}$ as a decimal.

One method is to divide 3 by 4 to get 0.75 for the fraction part. Then add the whole number part to 0.75.

$$
\dfrac{3}{4} \longrightarrow
\begin{array}{r}
0.75 \\
4\overline{)3.00} \\
\underline{2\ 8} \\
20 \\
\underline{20} \\
0
\end{array}
\qquad
\begin{array}{r}
2.00 \leftarrow \text{Whole number part} \\
\text{Fraction part} \longrightarrow +\ 0.75 \\
\hline
2.75
\end{array}
$$

So, $2\dfrac{3}{4} = \mathbf{2.75}$ Check: $2.75 = 2\dfrac{75}{100} = 2\dfrac{3}{4} \leftarrow$ Lowest terms

Whole number parts match.

A second method is to first write $2\dfrac{3}{4}$ as an improper fraction and then divide numerator by denominator.

$$2\dfrac{3}{4} = \dfrac{11}{4}$$

$$
\dfrac{11}{4} \longrightarrow 11 \div 4 \longrightarrow 4\overline{)11} \longrightarrow
\begin{array}{r}
2.75 \\
4\overline{)11.00} \\
\underline{8} \\
3\ 0 \\
\underline{2\ 8} \\
20 \\
\underline{20} \\
0
\end{array}
\leftarrow \text{Two extra zeros needed}
$$

Whole number parts match.

So, $2\dfrac{3}{4} = \mathbf{2.75}$

$\dfrac{3}{4}$ is equivalent to $\dfrac{75}{100}$ or 0.75

▶▶◀ Work Problem 2 at the Side.

EXAMPLE 2 Writing a Fraction as a Decimal with Rounding

Write $\dfrac{2}{3}$ as a decimal and round to the nearest thousandth.

$\dfrac{2}{3}$ means $2 \div 3$. To round to thousandths, divide out one *more* place, to ten-thousandths.

$$
\dfrac{2}{3} \longrightarrow 2 \div 3 \longrightarrow 3\overline{)2} \longrightarrow
\begin{array}{r}
0.6666 \\
3\overline{)2.0000} \\
\underline{1\ 8} \\
20 \\
\underline{18} \\
20 \\
\underline{18} \\
20 \\
\underline{18} \\
2
\end{array}
\begin{array}{l}
\leftarrow \text{Four zeros needed} \\
\quad \text{for ten-thousandths}
\end{array}
$$

$\leftarrow$ Bar above repeating digit

Written as a repeating decimal, $\dfrac{2}{3} = 0.\overline{6}$.

Rounded to the nearest thousandth, $\dfrac{2}{3} \approx 0.667$

Continued on Next Page

🖩 **Calculator Tip**   Try Example 2 on your calculator. Enter 2 ⊝ 3. Which answer do you get?

| 0.666666667 | or | 0.6666666 |

Many scientific calculators will show a 7 as the last digit. Because the sixes keep on repeating forever, the calculator automatically rounds in the last decimal place it has room to show. If you have a 10-digit display space, the calculator is rounding as shown below.

0.6666666666 (11 digits)   rounds to   0.66666666**7**

Next digit is 5 or more, so 6 rounds to 7.

Other calculators, especially standard, four-function ones, may *not* round. They just cut off, or *truncate,* the extra digits. Such a calculator would show 0.6666666 in the display.

Would this difference in calculators show up when changing $\frac{1}{3}$ to a decimal? Why not? (Answer: The repeating digit is 3, which is *4 or less,* so it stays as 3 whether it's rounded or not.)

◀◀◀ **Work Problem 3 at the Side.**

OBJECTIVE 2 Compare the size of fractions and decimals.   You can use a number line to compare fractions and decimals. For example, the number line below shows the space between 0 and 1. The locations of some commonly used fractions are marked, along with their decimal equivalents.

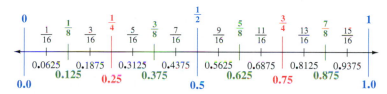

The next number line shows the locations of some commonly used fractions between 0 and 1 that are equivalent to *repeating* decimals. The decimal equivalents use a bar above repeating digits.

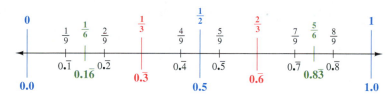

EXAMPLE 3   **Using a Number Line to Compare Numbers**

Use the number lines above to decide whether to write >, <, or = in the blank between each pair of numbers.

(a) 0.6875 _____ 0.625

You learned in **Section 3.5** that the number farther to the right on the number line is the greater number. Look at the first number line above. Because 0.6875 is to the *right* of 0.625, use the > symbol.

0.6875 **is greater than** 0.625   0.6875 > 0.625

—— **Continued on Next Page**

---

3 Write as decimals. Round to 🖩 the nearest thousandth.

(a) $\frac{1}{3}$

(b) $2\frac{7}{9}$

(c) $\frac{10}{11}$

(d) $\frac{3}{7}$

(e) $3\frac{5}{6}$

4 Use the number lines on the previous page to help you decide whether to write $<$, $>$, or $=$ in each blank.

(a) 0.4375 _____ 0.5

(b) 0.75 _____ 0.6875

(c) 0.625 _____ 0.0625

(d) $\dfrac{2}{8}$ _____ 0.375

(e) $0.8\overline{3}$ _____ $\dfrac{5}{6}$

(f) $\dfrac{1}{2}$ _____ $0.\overline{5}$

(g) $0.\overline{1}$ _____ $0.1\overline{6}$

(h) $\dfrac{8}{9}$ _____ $0.\overline{8}$

(i) $0.\overline{7}$ _____ $\dfrac{4}{6}$

(j) $\dfrac{1}{4}$ _____ 0.25

5 Arrange each group in order from smallest to largest.

(a) 0.7, 0.703, 0.7029

(b) 6.39, 6.309, 6.4, 6.401

(c) 1.085, $1\dfrac{3}{4}$, 0.9

(d) $\dfrac{1}{4}, \dfrac{2}{5}, \dfrac{3}{7}, 0.428$

(b) $\dfrac{3}{4}$ _____ 0.75

On the first number line, $\frac{3}{4}$ and 0.75 are at the same point on the number line. They are equivalent.

$$\tfrac{3}{4} = 0.75$$

(c) 0.5 _____ $0.\overline{5}$

On the second number line, 0.5 is to the *left* of $0.\overline{5}$ (which is actually 0.555 . . .) so use the $<$ symbol.

0.5 **is less than** $0.\overline{5}$ $0.5 < 0.\overline{5}$

(d) $\dfrac{2}{6}$ _____ $0.\overline{3}$

Write $\frac{2}{6}$ in lowest terms as $\frac{1}{3}$.
On the second number line you can see that $\frac{1}{3} = 0.\overline{3}$.

 **Work Problem 4 at the Side.**

Fractions can also be compared by first writing each one as a decimal. The decimals can then be compared by writing each one with the same number of decimal places.

EXAMPLE 4 Arranging Numbers in Order

Write each group of numbers in order, from smallest to largest.

(a) 0.49   0.487   0.4903

It is easier to compare decimals if they are all tenths, or all hundredths, and so on. Because 0.4903 has four decimal places (ten-thousandths), write zeros to the right of 0.49 and 0.487 so they also have four decimal places. Writing zeros to the right of a decimal number does *not* change its value (see **Section 4.3**). Then find the smallest and largest number of ten-thousandths.

$0.49 = 0.4900 = $ **4900** ten-thousandths $\leftarrow$ 4900 is in the middle.

$0.487 = 0.4870 = $ **4870** ten-thousandths $\leftarrow$ 4870 is the smallest.

$0.4903 = $ **4903** ten-thousandths $\leftarrow$ 4903 is the largest.

From smallest to largest, the correct order is shown below.

0.487   0.49   0.4903

(b) $2\dfrac{5}{8}$ 2.63   2.6

Write $2\frac{5}{8}$ as $\frac{21}{8}$ and divide $8)\overline{21}$ to get the decimal form, 2.625. Then, because 2.625 has three decimal places, write zeros so all the numbers have three decimal places.

$2\dfrac{5}{8} = 2.625 = $ 2 and **625** thousandths $\leftarrow$ 625 is in the middle.

$2.63 = 2.630 = $ 2 and **630** thousandths $\leftarrow$ 630 is the largest.

$2.6 = 2.600 = $ 2 and **600** thousandths $\leftarrow$ 600 is the smallest.

From smallest to largest, the correct order is shown below.

2.6 $2\dfrac{5}{8}$ 2.63

 **Work Problem 5 at the Side.**

4.6 Exercises

 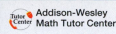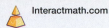
Write each fraction or mixed number as a decimal. Round to the nearest thousandth if necessary. See Examples 1 and 2.

1. $\dfrac{1}{2}$ 2. $\dfrac{1}{4}$ 3. $\dfrac{3}{4}$ 4. $\dfrac{1}{10}$ 5. $\dfrac{3}{10}$

6. $\dfrac{7}{10}$ 7. $\dfrac{9}{10}$ 8. $\dfrac{4}{5}$ 9. $\dfrac{3}{5}$ 10. $\dfrac{2}{5}$

11. $\dfrac{7}{8}$ 12. $\dfrac{3}{8}$ 13. $2\dfrac{1}{4}$ 14. $1\dfrac{1}{2}$ 15. $14\dfrac{7}{10}$

16. $23\dfrac{3}{5}$ 17. $3\dfrac{5}{8}$ 18. $2\dfrac{7}{8}$ ▦ 19. $\dfrac{1}{3}$ ▦ 20. $\dfrac{2}{3}$

▦ 21. $\dfrac{5}{6}$ ▦ 22. $\dfrac{1}{6}$ ▦ 23. $1\dfrac{8}{9}$ ▦ 24. $5\dfrac{4}{7}$

RELATING CONCEPTS (EXERCISES 25–28) For Individual or Group Work

Use your knowledge of fractions and decimals to **work Exercises 25–28 in order.**

25. **(a)** Explain how you can tell that Keith made an error *just by looking at his final answer.* Here is his work.

$$\frac{5}{9} = 5\overline{)9.0}^{\,1.8} \quad \text{so} \quad \frac{5}{9} = 1.8$$

   (b) Show the correct way to change $\frac{5}{9}$ to a decimal. Explain why your answer makes sense.

26. **(a)** How can you prove to Sandra that $2\frac{7}{20}$ is *not* equivalent to 2.035? Here is her work.

$$2\frac{7}{20} = 20\overline{)7.00}^{\,0.35} \quad \text{so} \quad 2\frac{7}{20} = 2.035$$

   (b) What is the correct answer? Show how to prove that it is correct.

27. Ving knows that $\frac{3}{8} = 0.375$. How can he write $1\frac{3}{8}$ as a decimal *without* having to do a division? How can he write $3\frac{3}{8}$ as a decimal? $295\frac{3}{8}$? Explain your answer.

28. Iris has found a shortcut for writing mixed numbers as decimals:

$$2\frac{7}{10} = 2.7 \qquad 1\frac{13}{100} = 1.13.$$

   Does her shortcut work for all mixed numbers? Explain when it works and why it works.

Find each decimal or fraction equivalent. Write fractions in lowest terms.

Fraction	Decimal	Fraction	Decimal
29. _____	0.4	**30.** _____	0.75
31. _____	0.625	**32.** _____	0.111
33. _____	0.35	**34.** _____	0.9
35. $\frac{7}{20}$	_____	**36.** $\frac{1}{40}$	_____
37. _____	0.04	**38.** _____	0.52
39. _____	0.15	**40.** _____	0.85
41. $\frac{1}{5}$	_____	**42.** $\frac{1}{8}$	_____
43. _____	0.09	**44.** _____	0.02

Solve each application problem.

45. The average length of a newborn baby is 20.8 inches. Charlene's baby is 20.08 inches long. Is her baby longer or shorter than the average? By how much?

46. The patient in room 830 is supposed to get 8.3 milligrams of medicine. She was actually given 8.03 milligrams. Did she get too much or too little medicine? What was the difference?

47. The label on the bottle of vitamins says that each capsule contains 0.5 gram of calcium. When checked, each capsule had 0.505 gram of calcium. Was there too much or too little calcium? What was the difference?

48. The glass mirror of the Hubble telescope had to be repaired in space because it would not focus properly. The problem was that the mirror's outer edge had a thickness of 0.6248 centimeter when it was supposed to be 0.625 centimeter. Was the edge too thick or too thin? By how much? (*Source:* NASA.)

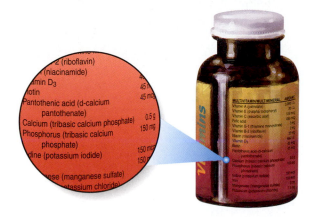

49. Precision Medical Parts makes an artificial heart valve that must measure between 0.998 centimeter and 1.002 centimeters. Circle the lengths that are acceptable:

1.01 cm   0.9991 cm   1.0007 cm   0.99 cm

50. The white rats in a medical experiment must start out weighing between 2.95 ounces and 3.05 ounces. Circle the weights that can be used:

3.0 ounces   2.995 ounces   3.055 ounces

3.005 ounces

51. Ginny Brown hoped her crops would get $3\frac{3}{4}$ inches of rain this month. The newspaper said the area received 3.8 inches of rain. Was that more or less than Ginny had hoped for? By how much?

52. The rats in the experiment in Exercise 50 gained $\frac{3}{8}$ ounce. They were expected to gain 0.3 ounce. Was their actual gain more or less than expected? By how much?

Arrange each group of numbers in order, from smallest to largest. See Example 4.

53. 0.54, 0.5455, 0.5399

54. 0.76, 0.7, 0.7006

55. 5.8, 5.79, 5.0079, 5.804

56. 12.99, 12.5, 13.0001, 12.77

57. 0.628, 0.62812, 0.609, 0.6009

58. 0.27, 0.281, 0.296, 0.3

59. 5.8751, 4.876, 2.8902, 3.88

60. 0.98, 0.89, 0.904, 0.9

61. $0.043, 0.051, 0.006, \dfrac{1}{20}$

62. $0.629, \dfrac{5}{8}, 0.65, \dfrac{7}{10}$

63. $\dfrac{3}{8}, \dfrac{2}{5}, 0.37, 0.4001$

64. $0.1501, 0.25, \dfrac{1}{10}, \dfrac{1}{5}$

Four boxes of fishing line are in the sale bin. The thicker the line, the stronger it is. The diameter of the fishing line is its thickness. Use the information on the boxes to answer Exercises 65–68.

65. Which color box has the strongest line?

66. Which color box has the line with the least strength?

67. What is the difference in line diameter between the weakest and strongest lines?

68. What is the difference in line diameter between the blue and purple boxes?

Some rulers for technical occupations show each inch divided into tenths. Use this scale drawing for Exercises 69–74. Change the measurements on the drawing to decimals and round them to the nearest tenth of an inch.

69. Length **(a)** is _____

70. Length **(b)** is _____

71. Length **(c)** is _____

72. Length **(d)** is _____

73. Length **(e)** is _____

74. Length **(f)** is _____

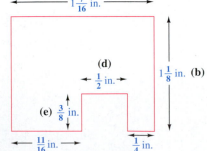

Chapter 4

SUMMARY

KEY TERMS

4.1	**decimals**	Decimals, like fractions, are used to show parts of a whole.
	decimal point	A decimal point is the dot that is used to separate the whole number part from the fractional part of a decimal number.
	place value	A place value is assigned to each place to the right or left of the decimal point. Whole numbers such as ones and tens, are to the *left* of the decimal point. Fractional parts, such as tenths and hundredths, are to the *right* of the decimal point.
4.2	**rounding**	Rounding is "cutting off" a number after a certain place, such as rounding to the nearest hundredth. The rounded number is less accurate than the original number. You can use the symbol "≈" to mean "approximately equal to."
	decimal places	Decimal places are the number of digits to the *right* of the decimal point. For example, 6.37 has two decimal places, and 4.706 has three decimal places.
4.3	**estimating**	Estimating is the process of rounding the numbers in a problem and getting an approximate answer. This helps you check that the decimal point is in the correct place in the exact answer.
4.5	**repeating decimal**	A repeating decimal (like the 6 in 0.1666 . . .) is a decimal number with one or more digits that repeat forever; it never ends. Use three dots to indicate that it is a repeating decimal. Or, write the number with a bar above the repeating digits, as in $0.1\overline{6}$. (Use the dots or the bar, but not both.)

TEST YOUR WORD POWER

See how well you have learned the vocabulary in this chapter. Answers follow the Quick Review.

1. **Decimal numbers** are like fractions in that they both
   A. must be written in lowest terms
   B. need common denominators
   C. have decimal points
   D. represent parts of a whole.

2. **Decimal places** refer to
   A. the digits from 0 to 9
   B. digits to the left of the decimal point
   C. digits to the right of the decimal point
   D. the number of zeros in a decimal number.

3. When a decimal number is **rounded,** it
   A. always ends in 0
   B. has the same number of decimal places as the original number
   C. is less accurate than the original number
   D. is less than one whole.

4. The **decimal point**
   A. separates the whole number part from the fractional part
   B. is always moved when finding a quotient
   C. separates tenths from hundredths
   D. is at the far left side of a whole number.

5. The number $0.\overline{3}$ is an example of
   A. an estimate
   B. a repeating decimal
   C. a rounded number
   D. a truncated number.

6. The **place value** names on the right side of the decimal point are
   A. ones, tens, hundreds, and so on
   B. ones, tenths, hundredths, and so on
   C. zero, one, two, three, four, and so on
   D. tenths, hundredths, thousandths, and so on.

Concepts

Examples

4.1 Reading and Writing Decimals

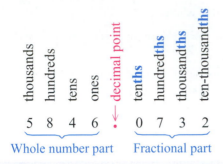

Whole number part   Fractional part

Write each decimal in words.

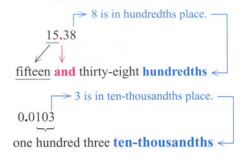

8 is in hundredths place.

15.38

fifteen **and** thirty-eight **hundredths**

3 is in ten-thousandths place.

0.0103

one hundred three **ten-thousandths**

4.1 Writing Decimals as Fractions

The digits to the right of the decimal point are the numerator. The place value of the rightmost digit determines the denominator.

Always write the fractions in lowest terms.

Write 0.45 as a fraction in lowest terms.

The numerator is 45. The rightmost digit, 5, is in the hundredths place, so the denominator is 100. Then write the fraction in lowest terms.

$$\frac{45}{100} = \frac{45 \div 5}{100 \div 5} = \frac{9}{20} \leftarrow \text{Lowest terms}$$

4.2 Rounding Decimals

Find the place to which you are rounding. Draw a cut-off line to the right of that place; the rest of the digits will be dropped. Look *only* at the first digit being cut. If it is *4 or less,* the part you are keeping stays the same. If it is *5 or more,* the part you are keeping rounds up. Do not move the decimal point when rounding. Write "≈" to mean "approximately equal to."

Round 0.17952 to the nearest thousandth.

First digit cut is *5 or more,* so round up.

$$0.179 \mid 52$$

$$0.179 \leftarrow \text{Keep this part.}$$
$$+ 0.001 \leftarrow \text{To round up, add 1 thousandth.}$$
$$\overline{0.180}$$

0.17952 rounds to 0.180. Write 0.17952 ≈ 0.180.

4.3 Adding and Subtracting Decimals

Use front end rounding to round each number and estimate the answer.

To find the exact answer, line up the decimal points. If needed, write in zeros as placeholders. Add or subtract as if they were whole numbers. Line up the decimal point in the answer directly below the decimal points in the problem.

Add 5.68 + 785.3 + 12 + 2.007.

Estimate:	*Exact:*	
6 ⟵	5 . 6**80**	Use zeros as place-holders so that all numbers have three decimal places.
800 ⟵	785 . **300**	
10 ⟵	12 . **000**	
+ 2 ⟵	+ 2 . 007	
818	804 . 987	Line up decimal points.

The estimate and exact answer are both in the hundreds, so the decimal point is probably in the correct place.

4.4 Multiplying Decimals

Step 1 Multiply as you would for whole numbers.
Step 2 Count the total number of decimal places in both factors.
Step 3 Write the decimal point in the product so it has the same number of decimal places as the total from Step 2. You may need to write extra zeros on the left side of the product to get enough decimal places.

Multiply 0.169 × 0.21.

$$
\begin{array}{r}
0.169 \leftarrow \text{3 decimal places} \\
\times \ 0.21 \leftarrow \text{2 decimal places} \\
\hline
169 \quad \leftarrow \text{5 total decimal places} \\
338 \quad \\
\hline
.03549 \leftarrow \text{5 decimal places in product}
\end{array}
$$

Write a 0 in the product so you can count over 5 decimal places. The final answer is 0.03549.

Concepts	Examples

4.5 Dividing by Decimals

Step 1 Change the divisor to a whole number by moving the decimal point to the right.

Step 2 Move the decimal point in the dividend the same number of places to the right.

Step 3 Write the decimal point in the quotient directly above the decimal point in the dividend.

Step 4 Divide as with whole numbers.

Divide 52.8 by 0.75.

$$
\begin{array}{r}
70.4 \\
0.75\overline{)52.800} \\
\underline{525} \\
300 \\
\underline{300} \\
0
\end{array}
$$

Move the decimal point two places to the right in the divisor and dividend. Write zeros in the dividend so you can move the decimal point and continue dividing until the remainder is 0.

To check your answer, multiply 70.4 times 0.75. If the result matches the dividend (52.8), you solved the problem correctly.

4.6 Writing Fractions as Decimals

Divide the numerator by the denominator. If necessary, round to the place indicated.

Write $\frac{1}{8}$ as a decimal.

$\frac{1}{8}$ means $1 \div 8$. Write it as $8\overline{)1}$.

The decimal point is on the right side of 1.

$$
\begin{array}{r}
0.125 \\
8\overline{)1.000} \\
\underline{8} \\
20 \\
\underline{16} \\
40 \\
\underline{40} \\
0
\end{array}
$$

← Write a decimal point and three zeros so you can continue dividing.

Therefore, $\frac{1}{8}$ is equivalent to 0.125.

4.6 Comparing the Size of Fractions and Decimals

Step 1 Write any fractions as decimals.

Step 2 Write zeros so that all the numbers being compared have the same number of decimal places.

Step 3 Use < to mean "is less than," > to mean "is greater than," or list the numbers from smallest to largest.

Arrange in order from smallest to largest.

$$0.505 \qquad \frac{1}{2} \qquad 0.55$$

$0.505 = 505$ thousandths ← 505 is in the middle.

$\frac{1}{2} = 0.5 = 0.500 = 500$ thousandths ← 500 is smallest.

$0.55 = 0.550 = 550$ thousandths ← 550 is largest.

(smallest) $\frac{1}{2}$ 0.505   0.55 (largest)

ANSWERS TO TEST YOUR WORD POWER

1. D; *Example:* For 0.7, the whole is cut into ten parts, and you are interested in 7 of the parts.
2. C; *Examples:* The number 6.87 has two decimal places; 0.309 has three decimal places.
3. C; *Example:* When 0.815 is rounded to 0.8 it is accurate only to the nearest tenth, while the original number was accurate to the nearest thousandth.
4. A; *Example:* In 5.42, the whole number part is 5 ones, and the decimal part is 42 hundredths.
5. B; *Example:* The bar above the 3 in $0.\overline{3}$ indicates that the 3 repeats forever.
6. D; *Example:* In 6.219, the 2 is in the tenths place, the 1 is in the hundredths place, and the 9 is in the thousandths place.

Real-Data Applications

▦ Dollar-Cost Averaging

To gain the most benefit from investing in the stock market, you should purchase shares when stock prices are low and sell stocks when prices are high. Unfortunately, it is very difficult to predict whether prices will rise or fall from one day to the next. Financial advisors recommend using *dollar-cost averaging* instead of trying to guess how the market will change. By investing the same amount of money at regular intervals, such as the beginning of each month, you will be buying more shares when the stock prices are low and fewer shares when stock prices are high. Of course, *dollar-cost averaging* does not guarantee that you will make a profit. However, over time, your shares should cost less than the market average.

Many financial Web sites report current and historical data on stock performance. Suppose you invested $100 at the first of each month in 2003 and purchased shares of Best Buy at its closing price. The table shows data on Best Buy closing prices during 2003.

Best Buy Co. Inc. (BBV)	Amount Invested	Price Per Share	Number of Shares Bought
January	$100	26.09	($100 ÷ 26.09) ≈ 3.8329 shares
February	$100	29.07	
March	$100	26.97	
April	$100	34.58	
May	$100	38.70	
June	$100	43.92	
July	$100	43.65	
August	$100	52.01	
September	$100	47.52	
October	$100	58.31	
November	$100	62.00	
December	$100	60.87	
Total investment			

Source: http://finance.yahoo.com

▦ Use a calculator to help answer these questions.

1. Calculate the total amount invested and enter the value in the table.

2. Calculate the number of shares bought each month. Round the number of shares to the nearest ten-thousandth. The calculation for January is shown as an example.

3. Calculate the average market price per share. (*Hint:* Add the monthly prices per share and divide by 12.)

4. Calculate the average price based on *dollar-cost averaging*. (*Hint:* Divide the total amount invested by the total number of shares.)

5. Rank the following scenarios in order of which was the best investment (most profit or least loss). Show the value of each investment in December 2003 as a basis for your answer.
   (a) $1200 invested in Best Buy in January 2003
   (b) $1200 invested in Best Buy using *dollar-cost averaging*
   (c) $1200 invested in Best Buy in November 2003

Chapter 4
REVIEW EXERCISES

[4.1] *Name the digit that has the given place value.*

1. 243.059
tenths
hundredths

2. 0.6817
ones
tenths

3. $5824.39
hundreds
hundredths

4. 896.503
tenths
tens

5. 20.73861
tenths
ten-thousandths

Write each decimal as a fraction or mixed number in lowest terms.

6. 0.5

7. 0.75

8. 4.05

9. 0.875

10. 0.027

11. 27.8

Write each decimal in words.

12. 0.8

13. 400.29

14. 12.007

15. 0.0306

Write each decimal in numbers.

16. eight and three tenths

17. two hundred five thousandths

18. seventy and sixty-six ten-thousandths

19. thirty hundredths

[4.2] *Round each decimal to the place indicated.*

20. 275.635 to the nearest tenth

21. 72.789 to the nearest hundredth

22. 0.1604 to the nearest thousandth

23. 0.0905 to the nearest thousandth

24. 0.98 to the nearest tenth

Round each money amount to the nearest cent.

25. $15.8333

26. $0.698

27. $17,625.7906

Round each income or expense item to the nearest dollar.

28. Income from pancake breakfast was $350.48.

29. Members paid $129.50 in dues.

30. Refreshments cost $99.61.

31. Bank charges were $29.37.

≈

[4.3] *First use front end rounding to round each number and estimate the answer. Then find the exact answer.*

32. *Estimate:*   *Exact:*

$$
\begin{array}{r}
5.81 \\
423.96 \\
+\ 15.09 \\
\hline
\end{array}
$$

+ ____

33. *Estimate:*   *Exact:*

$$
\begin{array}{r}
75.6 \\
1.29 \\
122.045 \\
0.88 \\
+\ 33.7 \\
\hline
\end{array}
$$

+ ____

34. *Estimate:*   *Exact:*

$$
\begin{array}{r}
308.5 \\
-\ 17.8 \\
\hline
\end{array}
$$

− ____

35. *Estimate:*   *Exact:*

$$
\begin{array}{r}
9.2 \\
-\ 7.9316 \\
\hline
\end{array}
$$

− ____

36. American's favorite recreational activity is exercise walking. About 81.3 million people go walking at least once during the year. Swimming is second with 56.9 million people, and camping is third with 49.9 million people. How many more people go walking than camping? (*Source:* National Sporting Goods Association.)

Estimate:

Exact:

37. Today, Jasmin had $306 in her checking account. She wrote a check to the day care center for $215.53 and a check for $44.67 at the grocery store. What is the new balance in her account?

Estimate:

Exact:

38. Joey spent $1.59 for toothpaste, $5.33 for a gift, and $18.94 for a toaster. He gave the clerk three $10 bills. How much change did he get?

Estimate:

Exact:

39. Roseanne is training for a wheelchair race. She raced 2.3 kilometers on Monday, 4 kilometers on Wednesday, and 5.25 kilometers on Friday. How far did she race altogether?

Estimate:

Exact:

≈

[4.4] *First use front end rounding to round each number and estimate the answer. Then find the exact answer.*

40. *Estimate:*   *Exact:*

$$
\begin{array}{r}
6.138 \\
\times\ \ 3.7 \\
\hline
\end{array}
$$

× ____

41. *Estimate:*   *Exact:*

$$
\begin{array}{r}
42.9 \\
\times\ \ 3.3 \\
\hline
\end{array}
$$

× ____

Multiply.

42. (5.6) (0.002)

43. 0.071 (0.005)

≈ **[4.5]** *Decide whether each answer is reasonable by rounding the numbers and estimating the answer. If the exact answer is not reasonable, find and correct the error.*

44. $706.2 \div 12 = 58.85$

Estimate:

45. $26.6 \div 2.8 = 0.95$

Estimate:

Divide. Round to the nearest thousandth if necessary.

46. $3\overline{)43.4}$

47. $\dfrac{72}{0.06}$

48. $0.00048 \div 0.0012$

≈ **[4.4–4.5]** *Solve each application problem.*

49. Adrienne worked 46.5 hours this week. Her hourly wage is $14.24 for the first 40 hours and 1.5 times that rate over 40 hours. Find her total earnings to the nearest dollar.

50. A book of 12 tickets costs $35.89 at the State Fair midway. What is the cost per ticket, to the nearest cent?

51. Stock in MathTronic sells for $3.75 per share. Kenneth is thinking of investing $500. How many whole shares could he buy?

52. Ground beef is on sale at $0.99 per pound. How much will Ms. Lee pay for a 3.5-pound package, to the nearest cent?

Use the order of operations to simplify each expression.

53. $3.5^2 + 8.7\,(1.95)$

54. $11 - 3.06 \div (3.95 - 0.35)$

[4.6] *Write each fraction or mixed number as a decimal. Round to the nearest thousandth when necessary.*

55. $3\dfrac{4}{5}$

56. $\dfrac{16}{25}$

57. $1\dfrac{7}{8}$

58. $\dfrac{1}{9}$

Arrange each group of numbers in order from smallest to largest.

59. $3.68,\ 3.806,\ 3.6008$

60. $0.215,\ 0.22,\ 0.209,\ 0.2102$

61. $0.17,\ \dfrac{3}{20},\ \dfrac{1}{8},\ 0.159$

MIXED REVIEW EXERCISES

Add, subtract, multiply, or divide as indicated.

62. $89.19 + 0.075 + 310.6 + 5$

63. 72.8×3.5

64. $1648.3 \div 0.46$ Round to the nearest thousandth.

65. $30 - 0.9102$

66. $4.38\,(0.007)$

67. $0.005\overline{)0.047}$

68. $72.105 + 8.2 + 95.37$

69. $81.36 \div 9$

70. $(5.6 - 1.22) + 4.8\,(3.15)$

71. 0.455×18

72. $(1.6)\,(0.58)$

73. $0.218\overline{)7.63}$

74. $21.059 - 20.8$

75. $18.3 - 3^2 \div 0.5$

Use the information in the ad to answer Exercises 76–80. Round money answers to the nearest cent. (Disregard any sales tax.)

Grand Opening Sale!
Save on Clothing for the Entire Family

Jeans for Teens
only $19.95 each
women's sizes $24.99

Athletic Shoes
regularly priced
$89.99 to $149.50
NOW just $71 to $119.60

Men's socks NOW 3 pairs for $8.99
Children's socks 6 pairs for $5

Hurry in — TWO DAYS ONLY

76. How much would one pair of men's socks cost?

77. How much more would one pair of men's socks cost than one pair of children's socks?

78. How much would Fernando pay for a dozen pair of men's socks?

79. How much would Akiko pay for five pairs of teen jeans and four pairs of women's jeans?

80. What is the difference between the cheapest sale price for athletic shoes and the highest regular price?

To decrease your risk of clogged arteries, it is recommended that you get at least 2 milligrams of vitamin B-6 each day. Use the information in the table to answer Exercises 81–82.

SOURCES OF VITAMIN B-6 (AMOUNTS IN MILLIGRAMS)	
1 cup orange juice	0.11
1 banana	0.66
1 baked potato with skin	0.7
$\frac{1}{2}$ cup strawberries	0.45
3 ounces skinless chicken	0.5
3 ounces water-packed tuna	0.3
$\frac{1}{2}$ cup chickpeas	0.57

Source: Harvard Health Letter.

81. (a) Which food item has the highest amount of vitamin B-6?

(b) Which food item has the lowest amount?

(c) What is the difference in the amount of vitamin B-6 between the food items with the highest and lowest amounts?

82. (a) Suppose you ate a banana, 3 ounces of skinless chicken, and 1 cup of strawberries. How many milligrams of vitamin B-6 would you get?

(b) Did you get more or less than the recommended amount? By how much?

Chapter **4**

T E S T

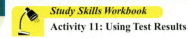

Study Skills Workbook
Activity 11: Using Test Results

Write each decimal as a fraction or mixed number in lowest terms.

1. 18.4   **2.** 0.075

Write each decimal in words.

3. 60.007   **4.** 0.0208

≈ *Round each decimal to the place indicated.*

5. 725.6089 to the nearest tenth

6. 0.62951 to the nearest thousandth

7. $1.4945 to the nearest cent

8. $7859.51 to the nearest dollar

≈ *First use front end rounding to round each number and estimate the answer. Then find the exact answer.*

9. 7.6 + 82.0128 + 39.59   **10.** 79.1 − 3.602

11. 5.79 (1.2)   **12.** 20.04 ÷ 4.8

Add, subtract, multiply, or divide as indicated.

13. 53.1 + 4.631 + 782 + 0.031   **14.** 670 − 0.996

15. (0.0069) (0.007)   **16.** $0.15\overline{)72}$

1. _____

2. _____

3. _____

4. _____

5. _____

6. _____

7. _____

8. _____

9. *Estimate:* _____
   Exact: _____

10. *Estimate:* _____
    Exact: _____

11. *Estimate:* _____
    Exact: _____

12. *Estimate:* _____
    Exact: _____

13. _____

14. _____

15. _____

16. _____

17. _____

17. Write $2\frac{5}{8}$ as a decimal. Round to the nearest thousandth, if necessary.

Arrange in order from smallest to largest.

18. _____

18. 0.44, 0.451, $\frac{9}{20}$, 0.4506

Use the order of operations to simplify this expression.

19. _____

19. $6.3^2 - 5.9 + 3.4\,(0.5)$

Solve each application problem.

20. _____

20. Jennifer had $71.15 in her checking account. Yesterday her account earned $0.95 interest for the month, and she deposited a paycheck for $390.77. The bank charged her $16 for new checks. What is the new balance in her account?

21. _____

21. Three types of ducks that are hunted in the United States are gadwalls, wigeons, and pintails. In 2003, the estimated populations of these ducks were: 2.5 million gadwalls, 2.551 million wigeons, and 2.56 million pintails. List the ducks in order from the highest number to the lowest. (*Source:* U.S. Fish and Wildlife Service.)

22. _____

22. Mr. Yamamoto bought 1.85 pounds of cheese at $2.89 per pound. How much did he pay for the cheese, to the nearest cent?

23. _____

23. Loren's baby had a temperature of 102.7 degrees. Later in the day it was 99.9 degrees. How much had the baby's temperature dropped?

24. _____

24. Pat bought 3.4 meters of fabric. She paid $15.47. What was the cost per meter?

25. _____

25. Write your own application problem using decimals. Make it different from problems 20–24. Then show how to solve your problem.

Cumulative Review Exercises

CHAPTERS 1–4

Name the digit that has the given place value.

1. 19,076,542
   hundreds
   millions
   ones

2. 83.0754
   tenths
   thousandths
   tens

Round each number as indicated.

3. 499,501 to the nearest thousand

4. 602.4937 to the nearest hundredth

5. $709.60 to the nearest dollar

6. $0.0528 to the nearest cent

≈ *First use front end rounding to round each number and estimate the answer. Then find the exact answer.*

7. *Estimate:*   *Exact:*

$$\begin{array}{r} 3672 \\ 589 \\ + \underline{9078} \end{array}$$

$+ \underline{}$

8. *Estimate:*   *Exact:*

$$\begin{array}{r} 4.06 \\ 15.7 \\ + \underline{0.923} \end{array}$$

$+ \underline{}$

9. *Estimate:*   *Exact:*

$$\begin{array}{r} 5018 \\ - \underline{1809} \end{array}$$

$- \underline{}$

10. *Estimate:*   *Exact:*

$$\begin{array}{r} 51.6 \\ - \underline{7.094} \end{array}$$

$- \underline{}$

11. *Estimate:*   *Exact:*

$$\begin{array}{r} 3317 \\ \times \underline{166} \end{array}$$

$\times \underline{}$

12. *Estimate:*   *Exact:*

$$\begin{array}{r} 6.82 \\ \times \underline{7.3} \end{array}$$

$\times \underline{}$

13. *Estimate:*   *Exact:*

$\underline{})\overline{}$

$46)\overline{123{,}740}$

14. *Estimate:*   *Exact:*

$\underline{})\overline{}$

$8.4)\overline{37.8}$

15. *Estimate:*

$\underline{} \cdot \underline{} = \underline{}$

Exact:

$1\frac{9}{10} \cdot 3\frac{3}{4}$

16. *Estimate:*

$\underline{} \div \underline{} = \underline{}$

Exact:

$2\frac{1}{3} \div \frac{5}{6}$

17. *Estimate:*

$\underline{} + \underline{} = \underline{}$

Exact:

$1\frac{4}{5} + 1\frac{2}{3}$

18. *Estimate:*

$\underline{} - \underline{} = \underline{}$

Exact:

$4\frac{1}{2} - 1\frac{7}{8}$

Add, subtract, multiply, or divide as indicated.

19. $10 - 0.329$

20. $2\dfrac{3}{5} \cdot \dfrac{5}{9}$

21. $9 + 72{,}417 + 799$

22. $11\dfrac{1}{5} \div 8$

23. $5006 - 92$

24. $0.7 + 85 + 7.903$

25. Write your answer using R for the remainder.

$7\overline{)2831}$

26. $\dfrac{5}{6} + \dfrac{7}{8}$

27. 332×704

28. $(0.006)(5.44)$

29. $3.2(2.5)$

30. $25.2 \div 0.56$

31. $\dfrac{2}{3} \div 5\dfrac{1}{6}$

32. $5\dfrac{1}{4} - 4\dfrac{7}{12}$

33. $4.7 \div 9.3$
   Round to the nearest hundredth.

Use the order of operations to simplify each expression.

34. $10 - 4 \div 2 \cdot 3$

35. $\sqrt{36} + 3(8) - 4^2$

36. $\dfrac{2}{3}\left(\dfrac{7}{8} - \dfrac{1}{2}\right)$

37. $0.9^2 + 10.6 \div 0.53$

38. $4^3 \cdot 3^2$

39. $\sqrt{196}$

40. Find the prime factorization of 200. Write your answer using exponents.

41. Write 40.035 in words.

42. Write three hundred six ten-thousandths in numbers.

Write each decimal as a fraction or mixed number in lowest terms.

43. 0.125

44. 3.08

Write each fraction or mixed number as a decimal. Round to the nearest thousandth if necessary.

45. $2\dfrac{3}{5}$

46. $\dfrac{7}{11}$

47. Write **<** or **>** in the blank to make a true statement: $\dfrac{5}{8}$ ___ $\dfrac{4}{9}$

Arrange each group of numbers in order, from smallest to largest.

48. 7.005, 7.5005, 7.5, 7.505

49. $\dfrac{7}{8}$, 0.8, $\dfrac{21}{25}$, 0.8015

Use the information in the table to answer Exercises 50–53. All measurements are in inches.

Children's Hats	Head Size	Order Size:
Measure around head above eyebrow ridges.	16½"–18"	XXS
	18¼"–19"	XS
	19¼"–20"	S
	20¼"–21⅛"	M
	21½"–22¼"	L

Source: Lands' End.

50. If the distance around your child's head is $21\frac{1}{16}$ in., which hat size should you order?

51. Find the difference between the smaller and larger measurements for each of the hat sizes.

52. Change the measurements for the medium (M) size to decimals. Then find the difference in the measurements. Show why this answer is equivalent to your answer from Exercise 51.

53. Did you prefer doing the subtraction using fractions (as in Exercise 51) or using decimals (as in Exercise 52)? Explain your reasoning.

First round the numbers and estimate the answer to each application problem. Then find the exact answer.

54. Lameck had two $20 bills. He spent $17.96 on gasoline and $0.87 for a candy bar at the convenience store. How much money does he have left?

Estimate:

Exact:

55. Manuela's daughter is 50 inches tall. Last year she was $46\frac{5}{8}$ inches tall. How much has she grown? (When estimating, round to the nearest whole number.)

Estimate:

Exact:

56. Sharon records textbooks on tape for students who are blind. Her hourly wage is $11.63. How much did she earn working 16.5 hours last week, to the nearest cent?

Estimate:

Exact:

57. The Farnsworth Elementary School has eight classrooms with 22 students in each one and 12 classrooms with 26 students in each one. How many students attend the school?

Estimate:

Exact:

58. Toshihiro bought $2\frac{1}{3}$ yd of cotton fabric and $3\frac{7}{8}$ yd of wool fabric. How many yards did he buy in all?

Estimate:

Exact:

59. Kimberly had $29.44 in her checking account. She wrote a check for $40 and deposited a $220.06 paycheck into her account, but not in time to prevent an $18 overdraft charge. What is the new balance in her account?

Estimate:

Exact:

60. Paulette bought 2.7 pounds of grapes for $6.18. What was the cost per pound, to the nearest cent?

Estimate:

Exact:

61. Carter Community College received a $78,000 grant from a local hospital to help students pay tuition for nursing classes. How much money could be given to each of 107 students? Round to the nearest dollar.

Estimate:

Exact:

Use the information in the table to answer Exercises 62–64.

Animal	Average Weight of Animal (ounces)	Average Weight of Food Eaten Each Day (ounces)
Hamster	3.5	0.4
Queen bee	0.004	?
Hummingbird	?	0.07

Source: NCTM News Bulletin

62. (a) In how many days will a hamster eat enough food to equal its body weight? Round to the nearest whole number of days.

(b) If a 140-pound woman ate her body weight of food in the same number of days as the hamster, how much would she eat each day? Round to the nearest tenth.

63. While laying eggs, a queen bee eats eighty times her weight each day. Use this information to fill in one of the missing values in the table.

64. A hummingbird's body weight is about $1\frac{3}{5}$ times the weight of its daily food intake. Find its body weight using decimal numbers and using fractions. Then prove that the two answers are equivalent.

Ratio and Proportion

5

Two out of every three American households have cellular phones. Worldwide, about 1 out of 6 people use a cell phone! (*Source:* eBrain Market Research.)

You can use unit rates to find the best deal on cell phone service plans. But most plans do not cover long-distance calls to certain U.S. locations or to other countries. Then you can use unit rates to find the cheapest long-distance calling card. (See **Section 5.2,** Exercises 29–32 for calling cards and Exercises 37–40 for cell phone plans; see Cumulative Review, Exercises 51–54 for international calling cards.)

5.1 Ratios

A **ratio** compares two quantities. You can compare two numbers, such as 8 and 4, or two measurements that have the *same* type of units, such as 3 *days* and 12 *days*. (*Rates* compare measurements with different types of units and are covered in the next section.)

Ratios can help you see important relationships. For example, if the ratio of your monthly expenses to your monthly income is 10 to 9, then you are spending $10 for every $9 you earn and going deeper into debt.

OBJECTIVE 1 Write ratios as fractions. A ratio can be written in three ways.

> **Writing a Ratio**
>
> The ratio of $7 **to** $3 can be written:
>
> $$7 \text{ to } 3 \quad \text{ or } \quad 7{:}3 \quad \text{ or } \quad \frac{7}{3} \leftarrow \text{Fraction bar indicates "to."}$$
>
> ":" indicates "**to**"

Writing a ratio as a fraction is the most common method, and the one we will use here. All three ways are read, "the ratio of 7 **to** 3." The word **to** separates the quantities being compared.

> **Writing a Ratio as a Fraction**
>
> Order is important when writing a ratio. The quantity mentioned **first** is the **numerator**. The quantity mentioned **second** is the **denominator**. For example:
>
> $$\text{The ratio of } 5 \text{ to } 12 \text{ is written } \frac{5}{12}$$

EXAMPLE 1 Writing Ratios

Ancestors of the Pueblo Indians built multistory apartment towns in New Mexico about 1100 years ago. A room might measure 14 ft long, 11 ft wide, and 15 ft high.

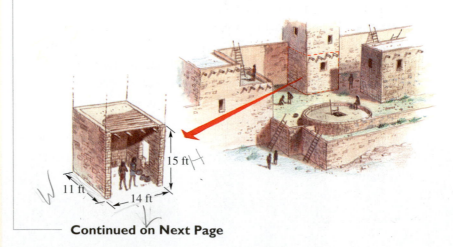

15 ft

11 ft   14 ft

Continued on Next Page

Write each ratio as a fraction, using the room measurements.

(a) Ratio of length to width

The ratio of **length** to **width** is $\dfrac{14 \text{ ft}}{11 \text{ ft}} = \dfrac{14}{11}$

Numerator   Denominator
(mentioned first)   (mentioned second)

You can divide out common *units* just like you divided out common *factors* when writing fractions in lowest terms. (See **Section 2.4.**) However, do *not* rewrite the fraction as a mixed number. Keep it as the ratio of 14 to 11.

(b) Ratio of width to height

The ratio of width **to** height is $\dfrac{11 \text{ ft}}{15 \text{ ft}} = \dfrac{11}{15}$

> **CAUTION**
> Remember, the order of the numbers is important in a ratio. Look for words "ratio of **a** to **b**." Write the ratio as $\dfrac{a}{b}$, *not* $\dfrac{b}{a}$. The quantity mentioned first is the numerator.

Work Problem 1 at the Side. ▶▶▶

Any ratio can be written as a fraction. Therefore, you can write a ratio in *lowest terms,* just as you do with any fraction.

EXAMPLE 2 **Writing Ratios in Lowest Terms**

Write each ratio in lowest terms.

(a) 60 days to 20 days

The ratio is $\frac{60}{20}$. Write this ratio in lowest terms by dividing the numerator and denominator by 20.

$$\frac{60}{20} = \frac{60 \div 20}{20 \div 20} = \frac{3}{1} \leftarrow \left\{ \begin{array}{l} \text{Ratio in} \\ \text{lowest terms} \end{array} \right.$$

So, the ratio of 60 days to 20 days is 3 to 1 or, written as a fraction, $\frac{3}{1}$.

> **CAUTION**
> In the fractions chapters you would have rewritten $\frac{3}{1}$ as 3. But a *ratio* compares *two* quantities, so you need to keep both parts of the ratio and write it as $\frac{3}{1}$.

(b) 50 ounces of medicine to 120 ounces of medicine

The ratio is $\frac{50}{120}$. Divide the numerator and denominator by 10.

$$\frac{50}{120} = \frac{50 \div 10}{120 \div 10} = \frac{5}{12} \leftarrow \left\{ \begin{array}{l} \text{Ratio in} \\ \text{lowest terms} \end{array} \right.$$

So, the ratio of 50 ounces to 120 ounces is $\frac{5}{12}$.

Continued on Next Page

---

1 Shane spent $14 on meat, $5 on milk, and $7 on fresh fruit. Write each ratio as a fraction.

(a) The ratio of amount spent on fruit to amount spent on milk.

(b) The ratio of amount spent on milk to amount spent on meat.

(c) The ratio of amount spent on meat to amount spent on milk.

ANSWERS

1. **(a)** $\dfrac{7}{5}$ **(b)** $\dfrac{5}{14}$ **(c)** $\dfrac{14}{5}$

2 Write each ratio as a fraction in lowest terms.

(a) 9 hours to 12 hours

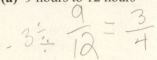

(b) 100 meters to 50 meters

(c) Write the ratio of width to length for this rectangle.

Length
48 ft

Width
24 ft

3 Write each ratio as a ratio of whole numbers in lowest terms.

(a) The price of Tamar's favorite brand of lipstick increased from $5.50 to $7.00. Find the ratio of the increase in price to the original price.

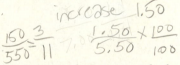

(b) Last week, Lance worked 4.5 hours each day. This week he cut back to 3 hours each day. Find the ratio of the decrease in hours to the original number of hours.

$$\frac{1.5}{4.5} = \frac{15}{45} = \frac{3}{9} = \frac{1}{3}$$

(c) 15 people in a large van to 6 people in a small van

$$\text{The ratio is } \frac{15}{6} = \frac{15 \div 3}{6 \div 3} = \frac{5}{2} \leftarrow \left\{ \begin{array}{l} \text{Ratio in} \\ \text{lowest terms} \end{array} \right.$$

> **NOTE**
> Although $\frac{5}{2} = 2\frac{1}{2}$, ratios are *not* written as mixed numbers. Nevertheless, in Example 2(c) above, the ratio $\frac{5}{2}$ does mean the large van holds $2\frac{1}{2}$ times as many people as the small van.

◀◀◀ Work Problem 2 at the Side.

OBJECTIVE 2 Solve ratio problems involving decimals or mixed numbers. Sometimes a ratio compares two decimal numbers or two fractions. It is easier to understand if we rewrite the ratio as a ratio of two whole numbers.

EXAMPLE 3 Using Decimal Numbers in a Ratio

The price of a Sunday newspaper increased from $1.50 to $1.75. Find the ratio of the increase in price **to** the original price.

The words increase in price are mentioned first, so the increase will be the numerator. How much did the price to go up? Use subtraction.

$$\text{new price} - \text{original price} = \text{increase}$$
$$\$1.75 \quad - \quad \$1.50 \quad = \quad \$0.25$$

The words the original price are mentioned second, so the original price of $1.50 is the denominator.

The ratio of increase in price **to** original price is shown below.

$$\frac{0.25}{1.50} \begin{array}{l} \leftarrow \text{increase} \\ \leftarrow \text{original price} \end{array}$$

Now rewrite the ratio as a ratio of whole numbers. Recall that if you multiply both the numerator and denominator of a fraction by the same number, you get an equivalent fraction. The decimals in this example are hundredths, so multiply by 100 to get whole numbers. (If the decimals are tenths, multiply by 10. If thousandths, multiply by 1000.) Then write the ratio in lowest terms.

$$\frac{0.25}{1.50} = \frac{0.25 \times 100}{1.50 \times 100} = \frac{25}{150} = \frac{25 \div 25}{150 \div 25} = \frac{1}{6} \leftarrow \left\{ \begin{array}{l} \text{Ratio in} \\ \text{lowest terms} \end{array} \right.$$

Ratio as two whole numbers

◀◀◀ Work Problem 3 at the Side.

EXAMPLE 4 Using Mixed Numbers in Ratios

Write each ratio as a comparison of whole numbers in lowest terms.

(a) 2 days to $2\frac{1}{4}$ days

Write the ratio as follows. Divide out the common units.

$$\frac{2 \text{ days}}{2\frac{1}{4} \text{ days}} = \frac{2}{2\frac{1}{4}}$$

Continued on Next Page

Next, write 2 as $\frac{2}{1}$ and $2\frac{1}{4}$ as the improper fraction $\frac{9}{4}$.

$$\frac{2}{2\frac{1}{4}} = \frac{\frac{2}{1}}{\frac{9}{4}}$$

Now rewrite the problem in horizontal format, using the "÷" symbol for division. Finally, multiply by the reciprocal of the divisor, as you did in **Section 2.7.**

$$\frac{\frac{2}{1}}{\frac{9}{4}} = \frac{2}{1} \div \frac{9}{4} = \frac{2}{1} \cdot \frac{4}{9} = \frac{8}{9}$$

Reciprocals

The ratio, in lowest terms, is $\frac{8}{9}$.

(b) $3\frac{1}{4}$ to $1\frac{1}{2}$

   Write the ratio as $\dfrac{3\frac{1}{4}}{1\frac{1}{2}}$. Then write $3\frac{1}{4}$ and $1\frac{1}{2}$ as improper fractions.

$$3\frac{1}{4} = \frac{13}{4} \quad \text{and} \quad 1\frac{1}{2} = \frac{3}{2}$$

The ratio is shown below.

$$\frac{3\frac{1}{4}}{1\frac{1}{2}} = \frac{\frac{13}{4}}{\frac{3}{2}}$$

Rewrite as a division problem in horizontal format, using the "÷" symbol. Then multiply by the reciprocal of the divisor.

$$\frac{13}{4} \div \frac{3}{2} = \frac{13}{\overset{2}{\cancel{4}}} \cdot \frac{\overset{1}{\cancel{2}}}{3} = \frac{13}{6} \leftarrow \begin{cases} \text{Ratio in} \\ \text{lowest terms} \end{cases}$$

> **Work Problem 4 at the Side.**

OBJECTIVE 3 **Solve ratio problems after converting units.** When a ratio compares measurements, both measurements must be in the *same* units. For example, *feet* must be compared to *feet, hours* to *hours, pints* to *pints,* and *inches* to *inches*.

EXAMPLE 5 **Ratio Applications Using Measurement**

(a) Write the ratio of the length of the shorter board on the left to the length of the longer board on the right. Compare in inches.

24 in.
2 ft   30 in.

$\dfrac{24}{30}$

First, express 2 ft in inches. Because 1 ft has 12 in., 2 ft is

$$2 \cdot \mathbf{12 \text{ in.}} = 24 \text{ in.}$$

── **Continued on Next Page**

---

4 Write each ratio as a ratio of whole numbers in lowest terms.

(a) $3\frac{1}{2}$ to 4

(b) $5\frac{5}{8}$ pounds to $3\frac{3}{4}$ pounds

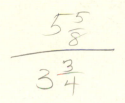

$$\frac{5\frac{5}{8}}{3\frac{3}{4}}$$

(c) $3\frac{1}{2}$ in. to $\frac{7}{8}$ in.

5 Write each ratio as a fraction in lowest terms. (*Hint:* Recall that it is usually easier to write the ratio using the smaller measurement unit.)

(a) 9 in. to 6 ft

(b) 2 days to 8 hours

(c) 7 yd to 14 ft

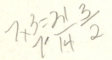

(d) 3 quarts to 3 gallons

(e) 25 minutes to 2 hours

(f) 4 pounds to 12 ounces

On the previous page, the length of the board on the left is 24 in., so the ratio of the lengths is

$$\frac{2 \text{ ft}}{30 \text{ in.}} = \frac{24 \text{ in.}}{30 \text{ in.}} = \frac{24}{30}$$

Write the ratio in lowest terms.

$$\frac{24}{30} = \frac{24 \div 6}{30 \div 6} = \frac{4}{5} \quad \leftarrow \begin{cases} \text{Ratio in} \\ \text{lowest terms} \end{cases}$$

The shorter board is $\frac{4}{5}$ the length of the longer board.

> **NOTE**
>
> Notice in the example above that we wrote the ratio using the smaller unit (inches are smaller than feet). Using the smaller unit will help you avoid working with fractions. If we wrote the ratio using feet, then
>
> $$30 \text{ in.} = 2\frac{1}{2} \text{ ft.}$$
>
> So the ratio in feet is shown below.
>
> $$\frac{2 \text{ ft}}{2\frac{1}{2} \text{ ft}} = \frac{2}{1} \div \frac{5}{2} = \frac{2}{1} \cdot \frac{2}{5} = \frac{4}{5} \quad \leftarrow \text{Same result}$$
>
> The ratio is the same, but it takes more steps to get the answer. Using the smaller unit is usually easier.

(b) Write the ratio of 28 days to 3 weeks.

Since it is easier to write the ratio using the smaller measurement unit, compare in *days* because days are shorter than weeks.

First express 3 weeks in days. Because 1 weeks has 7 days, 3 weeks is

$$3 \cdot 7 \text{ days} = 21 \text{ days}.$$

So the ratio in days is shown below.

$$\frac{28 \text{ days}}{3 \text{ weeks}} = \frac{28 \text{ days}}{21 \text{ days}} = \frac{28}{21} = \frac{28 \div 7}{21 \div 7} = \frac{4}{3} \quad \leftarrow \begin{cases} \text{Ratio in} \\ \text{lowest terms} \end{cases}$$

The following table will help you set up ratios that compare measurements. You will work with these measurements again in **Chapter 7.**

Measurement Comparisons

Length	Capacity (Volume)
1 foot = 12 inches	1 pint = 2 cups
1 yard = 3 feet	1 quart = 2 pints
1 mile = 5280 feet	1 gallon = 4 quarts
Weight	**Time**
1 pound = 16 ounces	1 minute = 60 seconds
1 ton = 2000 pounds	1 hour = 60 minutes
	1 day = 24 hours
	1 week = 7 days

◀◀◀ Work Problem 5 at the Side.

5.1   Exercises

 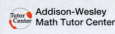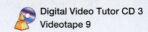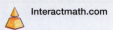
Write each ratio as a fraction in lowest terms. See Examples 1 and 2.

1. 8 to 9

2. 11 to 15

3. $100 to $50

4. 35¢ to 7¢

5. 30 minutes to 90 minutes

6. 9 pounds to 36 pounds

7. 80 miles to 50 miles

8. 300 people to 450 people

9. 6 hours to 16 hours

10. 45 books to 35 books

Write each ratio as a ratio of whole numbers in lowest terms. See Examples 3 and 4.

11. $4.50 to $3.50

12. $0.08 to $0.06

13. 15 to $2\frac{1}{2}$

14. 5 to $1\frac{1}{4}$

15. $1\frac{1}{4}$ to $1\frac{1}{2}$

16. $2\frac{1}{3}$ to $2\frac{2}{3}$

Write each ratio as a fraction in lowest terms. For help, use the table of measurement relationships on the previous page. See Example 5.

17. 4 ft to 30 in.

18. 8 ft to 4 yd

19. 5 minutes to 1 hour

20. 8 quarts to 5 pints

21. 15 hours to 2 days

22. 3 pounds to 6 ounces

23. 5 gallons to 5 quarts

24. 3 cups to 3 pints

The table shows the number of greeting cards that Americans buy for various occasions.
Use the information to answer Exercises 25–30. Write each ratio as a fraction in lowest terms.

Holiday/Event	Cards Sold
Valentine's Day	900 million
Mother's Day	150 million
Father's Day	95 million
Graduation	60 million
Thanksgiving	30 million
Halloween	25 million

Source: Hallmark Cards.

25. Find the ratio of Thanksgiving cards to graduation cards.

26. Find the ratio of Halloween cards to Mother's Day cards.

27. Find the ratio of Valentine's Day cards to Halloween cards.

28. Find the ratio of Mother's Day cards to Father's Day cards.

29. Explain how you might use the information in the table if you owned a shop selling gifts and greeting cards.

30. Why is the ratio of Valentine's Day cards to graduation cards $\frac{15}{1}$? Give two possible reasons.

$$\frac{900}{60} \qquad \frac{15}{1}$$

The bar graph shows worldwide sales of the most popular songs of all time. Use the graph to complete Exercises 31–34. Write each ratio as a fraction in lowest terms.

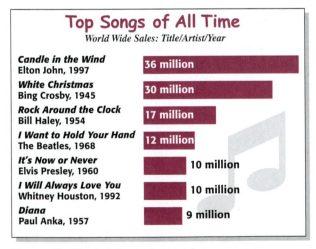

Top Songs of All Time
World Wide Sales: Title/Artist/Year

Candle in the Wind
Elton John, 1997 — 36 million

White Christmas
Bing Crosby, 1945 — 30 million

Rock Around the Clock
Bill Haley, 1954 — 17 million

I Want to Hold Your Hand
The Beatles, 1968 — 12 million

It's Now or Never
Elvis Presley, 1960 — 10 million

I Will Always Love You
Whitney Houston, 1992 — 10 million

Diana
Paul Anka, 1957 — 9 million

Source: The Music Information Database.

31. Write two ratios that compare the top-selling song to the second best seller and the top-selling song to the third best seller.

32. Write two ratios that compare sales of Elvis Presley's song with sales of the songs just ahead and just behind it in the graph.

33. Sales of which two songs give a ratio of $\frac{3}{1}$? There may be more than one correct answer.

34. Sales of which two songs give a ratio of $\frac{5}{6}$? There may be more than one correct answer.

Use the circle graph of one person's monthly budget to complete Exercises 35–36.
Write each ratio as a fraction in lowest terms.

35. Find the ratio of

(a) rent to utilities $\dfrac{750}{125}$

(b) rent to food

(c) rent and utilities to total expenses.

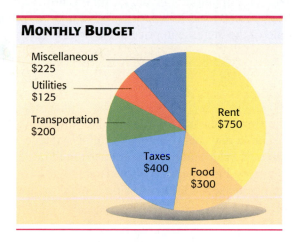

MONTHLY BUDGET

Miscellaneous $225

Utilities $125

Transportation $200

Taxes $400

Food $300

Rent $750

36. Find the ratio of

(a) taxes to rent

(b) food to transportation

(c) taxes and food to rent and utilities.

For each figure, find the ratio of the length of the longest side to the length of the shortest side. Write each ratio as a fraction in lowest terms.

37.

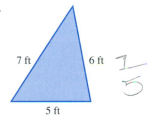

7 ft   6 ft $\dfrac{7}{5}$

5 ft

38.

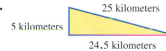

25 kilometers

5 kilometers

24.5 kilometers

39.

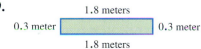

1.8 meters

0.3 meter   0.3 meter

1.8 meters

40.

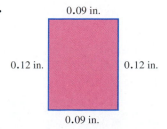

0.09 in.

0.12 in.   0.12 in.

0.09 in.

41.

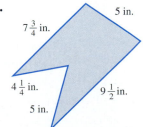

5 in.

$7\frac{3}{4}$ in.

$4\frac{1}{4}$ in. $9\frac{1}{2}$ in.

5 in.

42.

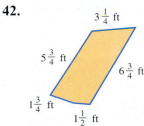

$3\frac{1}{4}$ ft

$5\frac{3}{4}$ ft $6\frac{3}{4}$ ft

$1\frac{3}{4}$ ft $1\frac{1}{2}$ ft

Write each ratio as a fraction in lowest terms.

43. The price of automobile engine oil recently went from $10.80 to $13.50 per case of 12 quarts. Find the ratio of the increase in price to the original price.

44. The price that a pharmacy pays for an antibiotic decreased from $8.80 to $5.60 for 10 tablets. Find the ratio of the decrease in price to the original price.

45. The first time a movie was made in Minnesota, the cast and crew spent $59\frac{1}{2}$ days filming winter scenes. The next year, another movie was filmed in $8\frac{3}{4}$ weeks. Find the ratio of the first movie's filming time to the second movie's time. Compare in weeks.

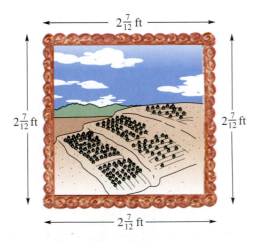

46. The percheron, a large draft horse, measures about $5\frac{3}{4}$ ft at the shoulder. The prehistoric ancestor of the horse measured only $15\frac{3}{4}$ in. at the shoulder. Find the ratio of the percheron's height to its prehistoric ancestor's height. Compare in inches. (*Source: Eyewitness Books: Horse.*)

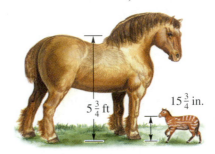

RELATING CONCEPTS (EXERCISES 47–50) For Individual or Group Work

Use your knowledge of ratios to **work Exercises 47–50 in order.**

47. In this painting, what is the ratio of the length of the longest side to the length of the shortest side? What other measurements could the painting have and still maintain the same ratio?

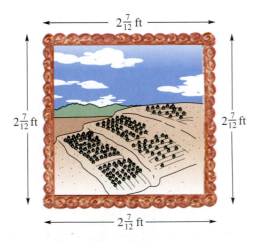

48. The ratio of my son's age to my daughter's age is 4 to 5. One possibility is that my son is 4 years old and my daughter is 5 years old. Find six other possibilities that fit the 4 to 5 ratio.

49. Amelia said that the ratio of her age to her mother's age is 5 to 3. Is this possible? Explain your answer.

50. Would you prefer that the ratio of your income to your friend's income be 1 to 3 or 3 to 1? Explain your answers.

5.2 Rates

A *ratio* compares two measurements with the same type of units, such as 9 feet to 12 feet (both length measurements). But many of the comparisons we make use measurements with different types of units, such as the following.

$$\begin{array}{ll} \text{160 dollars } \textbf{for } \text{8 hours} & \textbf{(money to time)} \\ \text{450 miles } \textbf{on } \text{15 gallons} & \textbf{(distance to capacity)} \end{array}$$

This type of comparison is called a **rate**.

OBJECTIVE 1 **Write rates as fractions.** Suppose you hiked 18 miles in 4 hours. The *rate* at which you hiked can be written as a fraction in lowest terms.

$$\frac{18 \text{ miles}}{4 \text{ hours}} = \frac{18 \text{ miles} \div 2}{4 \text{ hours} \div 2} = \frac{9 \text{ miles}}{2 \text{ hours}} \leftarrow \left\{ \begin{array}{l} \text{Rate in} \\ \text{lowest terms} \end{array} \right.$$

In a rate, you often find these words separating the quantities you are comparing.

in   for   on   per   from

> **CAUTION**
> When writing a rate, always include the units, such as miles, hours, dollars, and so on. Because the units in a rate are different, the units do *not* divide out.

EXAMPLE 1 **Writing Rates in Lowest Terms**

Write each rate as a fraction in lowest terms.

(a) 5 gallons of chemical **for** $60.

$$\frac{5 \text{ gallons} \div 5}{60 \text{ dollars} \div 5} = \frac{1 \text{ gallon}}{12 \text{ dollars}} \leftarrow \left\{ \begin{array}{l} \text{Write the units:} \\ \text{gallons and dollars.} \end{array} \right.$$

(b) $1500 wages **in** 10 weeks

$$\frac{1500 \text{ dollars} \div 10}{10 \text{ weeks} \div 10} = \frac{150 \text{ dollars}}{1 \text{ week}}$$

(c) 2225 miles **on** 75 gallons of gas

$$\frac{2225 \text{ miles} \div 25}{75 \text{ gallons} \div 25} = \frac{89 \text{ miles}}{3 \text{ gallons}}$$

> **Work Problem 1 at the Side.** ▶▶▶

OBJECTIVE 2 **Find unit rates.** When the *denominator* of a rate is 1, it is called a **unit rate.** We use unit rates frequently. For example, you earn $12.75 for *1 hour* of work. This unit rate is written:

$$\$12.75 \textbf{ per } \text{hour} \quad \text{or} \quad \$12.75/\text{hour}.$$

Or, you drive 28 miles on *1 gallon* of gas. This unit rate is written

$$28 \text{ miles } \textbf{per } \text{gallon} \quad \text{or} \quad 28 \text{ miles/gallon}.$$

Use **per** or a slash mark (/) when writing unit rates.

OBJECTIVES

1 Write rates as fractions.
2 Find unit rates.
3 Find the best buy based on cost per unit.

1 Write each rate as a fraction in lowest terms.

(a) $6 for 30 packages

$$\frac{6}{30} = \frac{3}{15} \quad \frac{1}{5}$$

(b) 500 miles in 10 hours

$$\frac{500}{10} \qquad \frac{50}{1}$$

(c) 4 teachers for 90 students

(d) 1270 bushels from 30 acres

ANSWERS
1. **(a)** $\dfrac{\$1}{5 \text{ packages}}$ **(b)** $\dfrac{50 \text{ miles}}{1 \text{ hour}}$
   (c) $\dfrac{2 \text{ teachers}}{45 \text{ students}}$ **(d)** $\dfrac{127 \text{ bushels}}{3 \text{ acres}}$

2 Find each unit rate.

(a) $4.35 for 3 pounds of cheese

(b) 304 miles on 9.5 gallons of gas

(c) $850 in 5 days

(d) 24-pound turkey for 15 people

EXAMPLE 2 Finding Unit Rates

Find each unit rate.

(a) 337.5 miles on 13.5 gallons of gas
Write the rate as a fraction.

$$\frac{337.5 \text{ miles}}{13.5 \text{ gallons}} \leftarrow \text{The fraction bar indicates division.}$$

Divide 337.5 by 13.5 to find the unit rate.

$$13.5\overline{)337.5} = \frac{2\,5.}{}$$

$$\frac{337.5 \text{ miles} \div \mathbf{13.5}}{13.5 \text{ gallons} \div \mathbf{13.5}} = \frac{25 \text{ miles}}{1 \text{ gallon}}$$

The unit rate is 25 miles **per** gallon, or 25 miles/gallon.

(b) 549 miles in 18 hours.

$$\frac{549 \text{ miles}}{18 \text{ hours}} \qquad \text{Divide: } 18\overline{)549.0} = 30.5$$

The unit rate is 30.5 miles/hour

(c) $810 in 6 days

$$\frac{810 \text{ dollars}}{6 \text{ days}} \qquad \text{Divide: } 6\overline{)810} = 135$$

The unit rate is $135/day.

◀◀◀ Work Problem 2 at the Side.

OBJECTIVE 3 Find the best buy based on cost per unit. When shopping for groceries, household supplies, and health and beauty items, you will find many different brands and package sizes. You can save money by finding the lowest *cost per unit*.

> **Cost per Unit**
>
> **Cost per unit** is a rate that tells how much you pay for *one* item or *one* unit. Examples are $1.95 per gallon, $47 per shirt, and $2.98 per pound.

EXAMPLE 3 Determining the Best Buy

The local store charges the following prices for pancake syrup. Find the best buy.

Continued on Next Page

The best buy is the container with the *lowest* cost per unit. All the containers are measured in *ounces* (oz), so you first need to find the *cost per ounce* for each one. Divide the price of the container by the number of ounces in it. Round to the nearest thousandth, if necessary.

Let the *order* of the *words* help you set up the rate.

cost ⟶ **$1.28**
per (means divide) ⟶ ─────────
ounce ⟶ **12 ounces**

Size	Cost per Unit (rounded)
12 ounces	$\dfrac{\$1.28}{12 \text{ ounces}} \approx \$0.107$ per ounce (highest)
24 ounces	$\dfrac{\$1.81}{24 \text{ ounces}} \approx \$0.075$ per ounce (lowest)
36 ounces	$\dfrac{\$2.73}{36 \text{ ounces}} \approx \$0.076$ per ounce

The lowest cost per ounce is $0.075, so the 24-ounce container is the best buy.

> **NOTE**
> Earlier we rounded money amounts to the nearest hundredth (nearest cent). But when comparing unit costs, rounding to the nearest thousandth will help you see the difference between very similar unit costs. Notice that the 24-ounce and 36-ounce syrup containers above would both have rounded to $0.08 per ounce if we had rounded to hundredths.

Work Problem 3 at the Side ▶▶▶

🖩 **Calculator Tip**   When using a calculator to find unit prices, remember that division is *not* commutative. In Example 3 you wanted to find cost per ounce. Let the *order* of the *words* help you enter the numbers in the correct order.

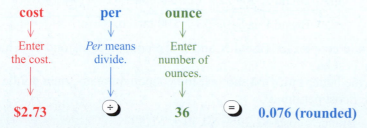

cost	**per**	**ounce**
↓	↓	↓
Enter the cost.	*Per* means divide.	Enter number of ounces.
↓	↓	↓
$2.73	÷	**36** = **0.076 (rounded)**

If you entered 36 ÷ 2.73 =, you'd get the number of *ounces* per *dollar.* How could you use that information to find the best buy? (*Answer:* The best buy would be to get the greatest number of ounces per dollar.)

Finding the best buy is sometimes a complicated process. Things that affect the cost per unit can include "cents off" coupons and differences in how much use you'll get out of each unit.

③ Find the best buy (lowest cost per unit) for each purchase.

   (a) 2 quarts for $3.25
   3 quarts for $4.95
   4 quarts for $6.48

   (b) 6 cans of cola for $1.99
   12 cans of cola for $3.49
   24 cans of cola for $7

ANSWERS
3. **(a)** 4 quarts, at $1.62 per quart
    (b) 12 cans, at $0.291 per can (rounded)

4 Solve each problem.

(a) Some batteries claim to last longer than others. If you believe these claims, which brand is the best buy?

Four-pack of AA-size batteries for $2.79

One AA-size battery for $1.19; lasts twice as long

(b) Which tube of toothpaste is the better buy? You have a coupon for 85¢ off Brand C and a coupon for 20¢ off Brand D.

Brand C is $3.89 for 6 ounces.

Brand D is $1.59 for 2.5 ounces.

EXAMPLE 4 Solving Best Buy Applications

Solve each application problem.

(a) There are many brands of liquid laundry detergent. If you feel they all do a good job of cleaning your clothes, you can base your purchase on cost per unit. But some brands are "concentrated" so you can use less detergent for each load of clothes. Which of the choices shown below is the best buy?

try **SUDZY** to clean your clothes!
50 fluid ounces for $3.99
Does same number of washloads as the old 64-ounce bottle!

WHITE-O gets out ALL the stains!
One gallon (128 ounces) for $9.89
Does twice the washloads of the old gallon bottle!

To find Sudzy's unit cost, divide $3.99 by 64 ounces, not 50 ounces. You're getting as many clothes washed as if you bought 64 ounces. Similarly, to find White-O's unit cost, divide $9.89 by 256 ounces (twice 128 ounces, or 2 • 128 ounces = 256 ounces).

$$\text{Sudzy} \quad \frac{\$3.99}{64 \text{ ounces}} \approx \$0.062 \text{ per ounce}$$

$$\text{White-O} \quad \frac{\$9.89}{256 \text{ ounces}} \approx \$0.039 \text{ per ounce}$$

White-O has the lower cost per ounce and is the better buy. (However, if you try it and it really doesn't get out all the stains, Sudzy may be worth the extra cost.)

(b) "Cents-off" coupons also affect the best buy. Suppose you are looking at these choices for "extra-strength" pain reliever. Both brands have the same amount of pain reliever in each tablet.

Brand X is $2.29 for 50 tablets.

Brand Y is $10.75 for 200 tablets.

You have a 40¢ coupon for Brand X and a 75¢ coupon for Brand Y. Which choice is the best buy?

To find the better buy, first subtract the coupon amounts, then divide to find the lower cost per ounce.

Brand X costs $2.29 − $0.40 = $1.89

$$\frac{\$1.89}{50 \text{ tablets}} \approx \$0.038 \text{ per tablet}$$

Brand Y costs $10.75 − $0.75 = $10.00

$$\frac{\$10.00}{200 \text{ tablets}} = \$0.05 \text{ per tablet}$$

Brand X has the lower cost per tablet and is the better buy.

Work Problem 4 at the Side.

ANSWERS

4. (a) One battery that lasts twice as long (like getting two) is the better buy. The cost per unit is $0.595 per battery. The four-pack is $0.698 per battery (rounded).

(b) Brand C with the 85¢ coupon is the better buy at $0.507 per ounce (rounded). Brand D with the 20¢ coupon is $0.556 per ounce.

5.2 Exercises

FOR
EXTRA
HELP
 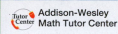 Addison-Wesley
Math Tutor Center
 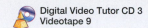 MathXL
 Digital Video Tutor CD 3
Videotape 9
 Student's
Solutions
Manual
MyMathLab
MyMathLab
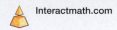 Interactmath.com

Write each rate as a fraction in lowest terms. See Example 1.

1. 10 cups for 6 people

$\frac{10}{6}$ $\frac{5}{3}$

2. $12 for 30 pens

3. 15 feet in 35 seconds

4. 100 miles in 30 hours

5. 72 miles on 4 gallons

6. 132 miles on 8 gallons

Find each unit rate. See Example 2.

7. $60 in 5 hours

8. $2500 in 20 days

$\frac{2500}{20}$

9. 7.5 pounds for 6 people

$\frac{7.5}{6}$ $\frac{2.5}{2} = \frac{1.25}{1}$

10. 44 bushels from 8 trees

11. $413.20 for 4 days

12. $74.25 for 9 hours

Earl kept the following record of the gas he bought for his car. For each entry, find the number of miles he traveled and the unit rate. Round your answers to the nearest tenth.

	Date	Odometer at Start	Odometer at End	Miles Traveled	Gallons Purchased	Miles per Gallon
13.	2/4	27,432.3	27,758.2		15.5	
14.	2/9	27,758.2	28,058.1		13.4	
15.	2/16	28,058.1	28,396.7		16.2	
16.	2/20	28,396.7	28,704.5		13.3	

Source: Author's car records.

Find the best buy (based on the cost per unit) for each item. See Example 3. (Source: Cub Foods.)

17. Black pepper

18. Shampoo

19. Cereal
13 ounces for $2.80
15 ounces for $3.15
18 ounces for $3.98

20. Soup
2 cans for $1.90
3 cans for $2.89
5 cans for $4.58

$$\frac{1.90}{2} \quad \frac{2.89}{3} \quad \frac{4.58}{5}$$

21. Chunky peanut butter
12 ounces for $1.29
18 ounces for $1.79
28 ounces for $3.39
40 ounces for $4.39

$$\frac{1.29}{12} \quad \frac{1.79}{18} \quad \frac{3.39}{28} \quad \frac{4.39}{40}$$

.108 .100 .121 .109

22. Baked beans
8 ounces for $0.59
16 ounces for $0.77
21 ounces for $0.99
28 ounces for $1.36

23. Suppose you are choosing between two brands of chicken noodle soup. Brand A is $0.88 per can and Brand B is $0.98 per can. But Brand B has more chunks of chicken in it. Which soup is the better buy? Explain your choice.

24. A small bag of potatoes costs $0.19 per pound. A large bag costs $0.15 per pound. But there are only two people in your family, so half the large bag would probably rot before you used it up. Which bag is the better buy? Explain.

Solve each application problem. See Examples 2–4.

25. Makesha lost 10.5 pounds in six weeks. What was her rate of loss in pounds per week?

$$\frac{10.5}{6} = 1.75 \text{ pounds/week}$$

26. Enrique's taco recipe uses three pounds of meat to feed 10 people. Give the rate in pounds per person.

27. Russ works 7 hours to earn $85.82. What is his pay rate per hour?

$$\frac{85.82}{7} \quad 12.26 \text{ hr}$$

28. Find the cost of 1 gallon of gas if 18 gallons cost $33.48.

The table lists information about three long-distance calling cards. The connection fee is charged each time you make a call, no matter how long the call lasts. Use the table to answer Exercises 29–32. Round answers to the nearest thousandth when necessary.

LONG-DISTANCE CALLING CARDS (U.S.)

Card Name	Cost per Minute	Connection Fee
Radiant Penny	$0.01	$0.39
IDT Special	0.022	0.14
Access America	0.047	0.00

Source: www.1callcard.com

29. (a) Find the *actual* total cost, including the connection charge, for a five-minute call using each card.
(b) Find the cost per minute for this call using each card and select the best buy.

30. (a) Find the *actual* total cost, including the connection charge, for a 30-minute call using each card.
(b) Find the cost per minute for this call using each card and select the best buy.

31. Find the *actual* total cost per minute for a 15-minute call and a 25-minute call using each card. Then select the best buy for each call.

32. All the cards round calls up to the next full minute.
(a) Suppose you call the wrong number. How much would you pay for this 40-*second* call on each card?
(b) How much would you save on this call by using Access America instead of Radiant Penny?

33. If you believe the claims that some batteries last longer, which is the better buy?

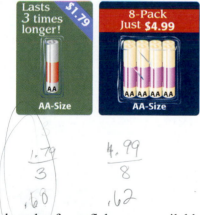

34. Which is the better buy, assuming these laundry detergents both clean equally well?

35. Three brands of cornflakes are available. Brand G is priced at $2.39 for 10 ounces. Brand K is $3.99 for 20.3 ounces and Brand P is $3.39 for 16.5 ounces. You have a coupon for 50¢ off Brand P and a coupon for 60¢ off Brand G. Which cereal is the best buy based on cost per unit?

36. Two brands of facial tissue are available. Brand K is specially priced at three boxes of 175 tissues each for $5. Brand S is priced at $1.29 per box of 125 tissues. You have a coupon for 20¢ off one box of Brand S and a coupon for 45¢ off one box of Brand K. How can you get the best buy on one box of tissue?

RELATING CONCEPTS (EXERCISES 37–40) For Individual or Group Work

On the first page of this chapter, we said that unit rates can help you get the best deal on cell phone service. Use the information in the table to work Exercises 37–40 in order.

Company	Anytime Minutes	One-time Activation Fee	Monthly Charge	Termination Fee
Verizon	400	$35	$59.99	$175
T-Mobile	600	$35	$39.99	$200
Nextel	500	$35	$45.99	$200
Sprint	500	$36	$55	$150

Source: Advertisements appearing in *Minneapolis Star Tribune.*

Notes:

1. All companies require that you sign a contract for one year of service and charge a one-time termination fee if you quit early.
2. Unused minutes cannot be carried over to the next month.

37. How much will each company's activation fee cost you on a *monthly* basis during the one-year contract?

38. All the plans allow unlimited calls on nights and weekends, so you would be using the "anytime minutes" on weekdays. Figure out the average number of weekdays per month. Then, for each plan, how many minutes could you use per weekday? Round to the nearest whole minute.

39. Find the actual average cost per "anytime minute" during the one-year contract for each company, including the activation fee. Assume you use all the minutes and no more. Decide how to round your answers so you can find the best buy.

40. Suppose that after two months you canceled your service because you found that you only used 100 "anytime minutes" per month. Under those conditions, find the actual cost per "anytime minute" for each company, to the nearest cent.

5.3 Proportions

OBJECTIVES

1 Write proportions.
2 Determine whether proportions are true or false.
3 Find cross products.

OBJECTIVE 1 Write proportions. A **proportion** states that two ratios (or rates) are equivalent. For example,

$$\frac{\$20}{4 \text{ hours}} = \frac{\$40}{8 \text{ hours}}$$

is a proportion that says the rate $\frac{\$20}{4 \text{ hours}}$ is equivalent to the rate $\frac{\$40}{8 \text{ hours}}$.
As the amount of money doubles, the number of hours also doubles. This proportion is read:

20 dollars **is to** 4 hours **as** 40 dollars **is to** 8 hours.

EXAMPLE 1 Writing Proportions

Write each proportion.

(a) 6 ft is to 11 ft **as** 18 ft is to 33 ft.

$$\frac{6 \text{ ft}}{11 \text{ ft}} = \frac{18 \text{ ft}}{33 \text{ ft}} \quad \text{so} \quad \frac{6}{11} = \frac{18}{33} \qquad \text{The common units (ft) divide out and are not written.}$$

(b) \$9 is to 6 liters **as** \$3 is to 2 liters.

$$\frac{\$9}{6 \text{ liters}} = \frac{\$3}{2 \text{ liters}} \qquad \text{Units must be written.}$$

Work Problem 1 at the Side.

OBJECTIVE 2 Determine whether proportions are true or false.
There are two ways to see whether a proportion is true. One way is to *write both of the ratios in lowest terms.*

EXAMPLE 2 Writing Both Ratios in Lowest Terms

Are the following proportions true?

(a) $\dfrac{5}{9} = \dfrac{18}{27}$

Write each ratio in lowest terms.

$$\frac{5}{9} \leftarrow \begin{array}{l}\text{Already in} \\ \text{lowest terms}\end{array} \qquad \frac{18 \div 9}{27 \div 9} = \frac{2}{3} \leftarrow \begin{array}{l}\text{Lowest} \\ \text{terms}\end{array}$$

Because $\frac{5}{9}$ is *not* equivalent to $\frac{2}{3}$, the proportion is *false.*

(b) $\dfrac{16}{12} = \dfrac{28}{21}$.

Write each ratio in lowest terms.

$$\frac{16 \div 4}{12 \div 4} = \frac{4}{3} \quad \text{and} \quad \frac{28 \div 7}{21 \div 7} = \frac{4}{3}$$

Both ratios are equivalent to $\frac{4}{3}$, so the proportion is *true.*

Work Problem 2 at the Side.

---

1 Write each proportion.

(a) \$7 is to 3 cans as \$28 is to 12 cans

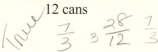

(b) 9 meters is to 16 meters as 18 meters is to 32 meters

(c) 5 is to 7 as 35 is to 49

(d) 10 is to 30 as 60 is to 180

2 Determine whether each proportion is true or false by writing both ratios in lowest terms.

(a) $\dfrac{6}{12} = \dfrac{15}{30}$

(b) $\dfrac{20}{24} = \dfrac{3}{4}$

(c) $\dfrac{25}{40} = \dfrac{30}{48}$

(d) $\dfrac{35}{45} = \dfrac{12}{18}$

OBJECTIVE **3** **Find cross products.** Another way to test whether the ratios in a proportion are equivalent is to compare *cross products*.

Using Cross Products to Determine Whether a Proportion Is True

To see whether a proportion is true, first multiply along one diagonal, then multiply along the other diagonal, as shown here.

$$5 \cdot 4 = \mathbf{20}$$

$$\frac{2}{5} = \frac{4}{10}$$

Cross products are equal.

$$2 \cdot 10 = \mathbf{20}$$

In this case the **cross products** are both 20. When cross products are *equal*, the proportion is *true*. If the cross products are *unequal*, the proportion is *false*.

> **NOTE**
> The cross products test is based on rewriting both fractions with a common denominator of $5 \cdot 10$, or 50.
>
> $$\frac{2 \cdot \mathbf{10}}{5 \cdot \mathbf{10}} = \frac{20}{50} \quad \text{and} \quad \frac{4 \cdot \mathbf{5}}{10 \cdot \mathbf{5}} = \frac{20}{50}$$
>
> We see that $\frac{2}{5}$ and $\frac{4}{10}$ are equivalent because both can be rewritten as $\frac{20}{50}$. The cross product test takes a shortcut by comparing only the two numerators ($20 = 20$).

EXAMPLE 3 **Using Cross Products**

Use cross products to see whether each proportion is true or false.

(a) $\dfrac{3}{5} = \dfrac{12}{20}$

Multiply along one diagonal and then along the other diagonal.

$$5 \cdot 12 = \mathbf{60}$$

$$\frac{3}{5} = \frac{12}{20}$$

Equal

$$3 \cdot 20 = \mathbf{60}$$

The cross products are *equal,* so the proportion is *true*.

> **CAUTION**
> Use cross products *only* when working with *proportions*. Do **not** use cross products when multiplying fractions, adding fractions, or writing fractions in lowest terms.

Continued on Next Page

(b) $\dfrac{2\frac{1}{3}}{3\frac{1}{3}} = \dfrac{9}{16}$

Find the cross products.

Changed to improper fractions

$$3\frac{1}{3} \cdot 9 = \frac{10}{\cancel{3}} \cdot \frac{\cancel{9}^{3}}{1} = \frac{30}{1} = \mathbf{30}$$

$$\dfrac{2\frac{1}{3}}{3\frac{1}{3}} = \dfrac{9}{16}$$

Unequal

$$2\frac{1}{3} \cdot 16 = \frac{7}{3} \cdot \frac{16}{1} = \frac{112}{3} = \mathbf{37\frac{1}{3}}$$

The cross products are *unequal,* so the proportion is *false.*

NOTE

The numbers in a proportion do *not* have to be whole numbers. They can be fractions, mixed numbers, decimal numbers, and so on.

Work Problem 3 at the Side. ▶▶▶

---

③ Find the cross products to see whether each proportion is true or false.

(a) $\dfrac{5}{9} = \dfrac{10}{18}$ ⁹⁰   ⁹⁰

(b) $\dfrac{32}{15} = \dfrac{16}{8}$ ²⁴⁰   ²⁵⁶

(c) $\dfrac{10}{17} = \dfrac{20}{34}$

(d) $\dfrac{2.4}{6} = \dfrac{5}{12}$ $(6)(5) =$ $(2.4)(12) =$

(e) $\dfrac{3}{4.25} = \dfrac{24}{34}$

(f) $\dfrac{1\frac{1}{6}}{2\frac{1}{3}} = \dfrac{4}{8}$

Focus on Real-Data Applications

Tour of the West

Highlights of a tour for British visitors to the western United States are visits to Mount Rushmore, Devil's Tower, Yellowstone National Park, the Grand Tetons, Bryce Canyon, Zion National Park, the Painted Desert, and the Grand Canyon. The itinerary is shown on the map to the right.

The travel distances and times between the daily stopping points are estimates, based on information from the Web site www.mapquest.com. The table below gives the daily route, the travel distances, and the travel times between the locations.

 1. Calculate the average speed in miles per hour (mph) for each segment of the trip, rounded to the nearest whole number. Notice that you are working with rates that compare distance to time (miles to hours). *Hint:* Divide the distance traveled by the time elapsed. You must first rewrite the time in decimal form. For example, on Day 4, 1 hr 40 min is $1 + \frac{40}{60}$ hr or 1.6666666 hr, which rounds to 1.67 hr. The average speed is 56 mi $\div 1.67$ hr ≈ 33.6 or 34 mph.

Day	Location	Distance	Time	Average Speed
1	London, England to Denver, Colorado (CO)	4693 mi	14 hr flight	
2	Denver, CO			
3	Denver, CO to Custer, South Dakota (SD)	365 mi	8 hr	
4	Custer, SD to Lead, SD	56 mi	1 hr, 40 min	34 mph
5	Lead, SD to Cody, Wyoming (WY)	361 mi	7 hr, 50 min	
6	Cody, WY to Yellowstone National Park, WY	110 mi	3 hr, 10 min	
7	Yellowstone, WY to Jackson, WY	131 mi	3 hr, 45 min	
8	Jackson, WY to Salt Lake City, Utah (UT)	269 mi	7 hr	
9	Salt Lake City, UT			
10	Salt Lake City, UT to Cedar City, UT	251 mi	4 hr, 20 min	
11	Cedar City, UT to Page, Arizona (AZ)	155 mi	4 hr, 30 min	
12	Page, AZ to Grand Canyon, AZ	138 mi	4 hr	
13	Grand Canyon, AZ to Las Vegas, Nevada (NV)	279 mi	6 hr, 30 min	
14	Las Vegas, NV			
15	Las Vegas, NV to San Francisco, CA	420 mi	1 hr 40 min flight	
	San Francisco, CA to London, England	5376 mi	17 hr flight	
	Totals (driving portion of the tour)			

2. Calculate the total driving distance and total driving time during this tour. Find the overall average speed.

3. Why do you think the average driving speeds for this trip are not closer to 55 mph?

4. If a friend from Scotland asked your opinion about how interesting and feasible these tourist attractions would be, what advice would you give?

5.3 Exercises

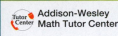

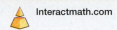
Write each proportion. See Example 1.

1. $9 is to 12 cans as $18 is to 24 cans.

2. 28 people is to 7 cars as 16 people is to 4 cars.

3. 200 adults is to 450 children as 4 adults is to 9 children.

4. 150 trees is to 1 acre as 1500 trees is to 10 acres.

5. 120 ft is to 150 ft as 8 ft is to 10 ft.

$$\frac{120}{150} \quad \frac{8}{10}$$

6. $6 is to $9 as $10 is to $15.

Determine whether each proportion is true or false by writing the ratios in lowest terms. Show the reduced ratios and then write true *or* false*. See Example 2.*

7. $\dfrac{6}{10} = \dfrac{3}{5}$

8. $\dfrac{1}{4} = \dfrac{9}{36}$

9. $\dfrac{5}{8} = \dfrac{25}{40}$

$$\frac{5}{8}$$

10. $\dfrac{2}{3} = \dfrac{20}{27}$

11. $\dfrac{150}{200} = \dfrac{200}{300}$

12. $\dfrac{100}{120} = \dfrac{75}{100}$

13. $\dfrac{42}{15} = \dfrac{28}{10}$

14. $\dfrac{18}{16} = \dfrac{36}{32}$

15. $\dfrac{32}{18} = \dfrac{48}{27}$

16. $\dfrac{15}{48} = \dfrac{10}{24}$

17. $\dfrac{7}{6} = \dfrac{54}{48}$

18. $\dfrac{28}{21} = \dfrac{44}{33}$

Use cross products to determine whether each proportion is true or false. Show the cross products and circle true *or* false*. See Example 3.*

19. $\dfrac{2}{9} = \dfrac{6}{27}$

$\quad$ 54

$\quad$ 54

$\quad$ (True)   False

20. $\dfrac{20}{25} = \dfrac{4}{5}$

$\quad$ True   False

21. $\dfrac{20}{28} = \dfrac{12}{16}$

$\quad$ True   False

22. $\dfrac{16}{40} = \dfrac{22}{55}$

 True   False

23. $\dfrac{110}{18} = \dfrac{160}{27}$

 True   False

24. $\dfrac{600}{420} = \dfrac{20}{14}$

 True   False

25. $\dfrac{3.5}{4} = \dfrac{7}{8}$

 True   False

26. $\dfrac{36}{23} = \dfrac{9}{5.75}$

 True   False

27. $\dfrac{18}{16} = \dfrac{2.8}{2.5}$

 True   False

28. $\dfrac{0.26}{0.39} = \dfrac{1.3}{1.9}$

 True   False

29. $\dfrac{6}{3\frac{2}{3}} = \dfrac{18}{11}$

 True   False

30. $\dfrac{16}{13} = \dfrac{2}{1\frac{5}{8}}$

 True   False

31. $\dfrac{2\frac{5}{8}}{3\frac{1}{4}} = \dfrac{21}{26}$

 True   False

32. $\dfrac{28}{17} = \dfrac{9\frac{1}{3}}{5\frac{2}{3}}$

 True   False

33. $\dfrac{\frac{2}{3}}{2} = \dfrac{2.7}{8}$

 True   False

34. $\dfrac{3.75}{1\frac{1}{4}} = \dfrac{7.5}{2\frac{1}{2}}$

 True   False

35. $\dfrac{2\frac{3}{10}}{8.05} = \dfrac{\frac{1}{4}}{0.9}$

 True   False

36. $\dfrac{3}{\frac{5}{6}} = \dfrac{1.5}{\frac{7}{12}}$

 True   False

37. Suppose Ichiro Suzuki of the Seattle Mariners had 16 hits in 50 times at bat and Sammy Sosa of the Chicago Cubs was at bat 400 times and got 128 hits. Paul is trying to convince Jamie that the two men hit equally well. Show how you could use a proportion and cross products to see whether Paul is correct.

38. Jay worked 3.5 hours and packed 91 cartons. Craig packed 126 cartons in 5.25 hours. To see whether the men worked equally fast, Barry set up this proportion:

$$\dfrac{3.5}{91} = \dfrac{126}{5.25}$$

Explain what is wrong with Barry's proportion and write a correct one. Is the correct proportion true or false?

5.4 Solving Proportions

OBJECTIVES

1 Find the unknown number in a proportion.

2 Find the unknown number in a proportion with mixed numbers or decimals.

OBJECTIVE 1 Find the unknown number in a proportion. Four numbers are used in a proportion. If any three of these numbers are known, the fourth can be found. For example, find the unknown number that will make this proportion true.

$$\frac{3}{5} = \frac{x}{40}$$

The x represents the unknown number. Start by finding the cross products.

$$\frac{3}{5} = \frac{x}{40} \quad \begin{array}{l} 5 \cdot x \\ 3 \cdot 40 \end{array} \right\} \text{Cross products.}$$

To make the proportion true, the cross products must be equal.

$$5 \cdot x = \underbrace{3 \cdot 40}$$
$$5 \cdot x = 120$$

The equal sign says that $5 \cdot x$ and 120 are equivalent. If $5 \cdot x$ and 120 are *both* divided by 5, the results will still be equivalent.

$$\frac{5 \cdot x}{5} = \frac{120}{5} \leftarrow \text{Divide both sides by 5.}$$

Divide out the common factor of 5.
$$\frac{\overset{1}{\cancel{5}} \cdot x}{\underset{1}{\cancel{5}}} = 24 \qquad \text{On the right side, divide 120 by 5 to get 24.}$$

Multiplying by 1 does *not* change a number, so in the numerator on the left side, $1 \cdot x$ is the same as x.

$$\frac{x}{1} = 24$$

Dividing by 1 does *not* change a number, so on the left side, $\frac{x}{1}$ is the same as x.

$$x = 24$$

The unknown number in the proportion is 24.

The complete proportion is shown below.

$$\frac{3}{5} = \frac{24}{40} \leftarrow x \text{ is 24.}$$

Check by finding the cross products. If they are equal, you solved the problem correctly. If they are unequal, rework the problem.

$$\frac{3}{5} = \frac{24}{40} \quad \begin{array}{l} 5 \cdot 24 = \textbf{120} \\ 3 \cdot 40 = \textbf{120} \end{array} \right\} \text{Equal; proportion is true.}$$

The cross products are equal, so the solution, $x = 24$, is correct.

CAUTION
The solution is 24, which is the unknown number in the proportion. 120 is *not* the solution; it is the cross product you get when *checking* the solution.

Solve a proportion for an unknown number by using the following steps.

Finding an Unknown Number in a Proportion

Step 1 Find the cross products.

Step 2 Show that the cross products are equivalent.

Step 3 Divide both products by the number multiplied by *x* (the number next to *x*).

Step 4 Check by writing the solution in the *original* proportion and finding the cross products.

EXAMPLE 1 Solving Proportions for Unknown Numbers

Find the unknown number in each proportion. Round answers to the nearest hundredth when necessary.

(a) $\dfrac{16}{x} = \dfrac{32}{20}$

Recall that ratios can be rewritten in lowest terms. If desired, you can do that *before* finding the cross products. In this example, write $\frac{32}{20}$ in lowest terms as $\frac{8}{5}$, which gives the proportion $\dfrac{16}{x} = \dfrac{8}{5}$.

Step 1
$$\dfrac{16}{x} \quad = \quad \dfrac{8}{5}$$
$x \cdot 8 \leftarrow$
$16 \cdot 5 \leftarrow$ Find the cross products.

Step 2 $x \cdot 8 = \underline{16 \cdot 5} \leftarrow$ Show that cross products are equivalent.
$$x \cdot 8 = \quad 80$$

Step 3 $\dfrac{x \cdot \overset{1}{\cancel{8}}}{\cancel{8}_1} = \dfrac{80}{8} \leftarrow$ Divide both sides by 8.

$$x = 10 \leftarrow \text{Find } x. \text{ (No rounding necessary)}$$

The unknown number in the proportion is 10.

Step 4   Write the solution in the *original* proportion and check by finding cross products.

$10 \cdot 32 = \mathbf{320} \leftarrow$

$x \text{ is } 10 \rightarrow \dfrac{16}{10} = \dfrac{32}{20}$ Equal; proportion is true.

$16 \cdot 20 = \mathbf{320} \leftarrow$

The cross products are equal, so 10 is the correct solution.

NOTE
It is not necessary to write the ratios in lowest terms before solving. However, if you do, you will have smaller numbers to work with.

Continued on Next Page

(b) $\dfrac{7}{12} = \dfrac{15}{x}$

Not =

Step 1 $\dfrac{7}{12} \,\times\, \dfrac{15}{x}$ ← Find the cross products.

$12 \cdot 15 = 180$

Step 2 $7 \cdot x = 180$ ← Show that cross products are equal.

Step 3 $\dfrac{\overset{1}{\cancel{7}} \cdot x}{\cancel{7}_{1}} = \dfrac{180}{7}$ ← Divide both sides by 7.

$x \approx 25.71$ ← Rounded to nearest hundredth

When the division does not come out even, check for directions on how to round your answer. Divide out one more place, then round.

$$\begin{array}{r} 25.714 \\ 7\overline{)180.000} \end{array}$$ ← Divide out to thousandths so you can round to hundredths.

The unknown number in the proportion is 25.71 (rounded).

Step 4 Write the solution in the original proportion and check by finding the cross products.

$12 \cdot 15 = \mathbf{180}$

$\dfrac{7}{12} \,\times\, \dfrac{15}{25.71}$

$7 \cdot 25.71 = \mathbf{179.97}$

Very close, but not equal

The cross products are slightly different because you rounded the value of x. However, they are close enough to see that the problem was done correctly and that 25.71 is the approximate solution.

> **Work Problem 1 at the Side.** ▶▶▶

OBJECTIVE 2 **Find the unknown number in a proportion with mixed numbers or decimals.** The next example shows how to work with mixed numbers or decimals in a proportion.

EXAMPLE 2 **Solving Proportions with Mixed Numbers and Decimals**

Find the unknown number in each proportion.

(a) $\dfrac{2\frac{1}{5}}{6} = \dfrac{x}{10}$

$\dfrac{2\frac{1}{5}}{6} \,\times\, \dfrac{x}{10}$ ← Find the cross products.

$6 \cdot x$

$2\frac{1}{5} \cdot 10$

Find $2\frac{1}{5} \cdot 10$.

$$2\frac{1}{5} \cdot 10 = \frac{11}{5} \cdot \frac{10}{1} = \frac{11}{\cancel{5}} \cdot \frac{\overset{2}{\cancel{10}}}{1} = \frac{22}{1} = 22$$

Changed to improper fraction

Continued on Next Page

---

❶ Find the unknown numbers. Round to hundredths when necessary. Check your answers by finding the cross products.

(a) $\dfrac{1}{2} = \dfrac{x}{12}$

(b) $\dfrac{6}{10} = \dfrac{15}{x}$

(c) $\dfrac{28}{x} = \dfrac{21}{9}$

(d) $\dfrac{x}{8} = \dfrac{3}{5}$

(e) $\dfrac{14}{11} = \dfrac{x}{3}$

2 Find the unknown numbers. Round to hundredths on the decimal problems, if necessary. Check your answers by finding the cross products.

(a) $\dfrac{3\frac{1}{4}}{2} = \dfrac{x}{8}$

(b) $\dfrac{x}{3} = \dfrac{1\frac{2}{3}}{5}$

(c) $\dfrac{0.06}{x} = \dfrac{0.3}{0.4}$

(d) $\dfrac{2.2}{5} = \dfrac{13}{x}$

(e) $\dfrac{x}{6} = \dfrac{0.5}{1.2}$

(f) $\dfrac{0}{2} = \dfrac{x}{7.092}$

2. (a) $x = 13$ **(b)** $x = 1$ **(c)** $x = 0.08$
(d) $x \approx 29.55$ (rounded to nearest hundredth)
(e) $x = 2.5$ **(f)** $x = 0$

Show that the cross products are equivalent.

$$6 \cdot x = 22$$

Divide both sides by 6.

$$\dfrac{\overset{1}{\cancel{6}} \cdot x}{\underset{1}{\cancel{6}}} = \dfrac{22}{6}$$

Write the answer as a mixed number in lowest terms.

$$x = \dfrac{22 \div 2}{6 \div 2} = \dfrac{11}{3} = 3\dfrac{2}{3}$$

The unknown number is $3\frac{2}{3}$.

Write the solution in the proportion and check by finding the cross products.

$$\dfrac{2\frac{1}{5}}{6} = \dfrac{3\frac{2}{3}}{10}$$

$$6 \cdot 3\dfrac{2}{3} = \dfrac{\overset{2}{\cancel{6}}}{1} \cdot \dfrac{11}{\underset{1}{\cancel{3}}} = \dfrac{22}{1} = 22$$

$$2\dfrac{1}{5} \cdot 10 = \dfrac{11}{\underset{1}{\cancel{5}}} \cdot \dfrac{\overset{2}{\cancel{10}}}{1} = \dfrac{22}{1} = 22$$

Equal

The cross products are equal, so $3\frac{2}{3}$ is the correct solution.

(b) $\dfrac{1.5}{0.6} = \dfrac{2}{x}$

Show that cross products are equivalent.

$$(1.5)(x) = (0.6)(2)$$
$$(1.5)(x) = 1.2$$

Divide both sides by 1.5.

$$\dfrac{\overset{1}{\cancel{(1.5)}}(x)}{\underset{1}{\cancel{1.5}}} = \dfrac{1.2}{1.5}$$

$$x = \dfrac{1.2}{1.5}$$

Complete the division.

$$x = 0.8 \qquad 1.5\overline{)1.20}\ \overset{.8}{}$$

So the unknown number is 0.8.
Check the solution by finding the cross products.

$$\dfrac{1.5}{0.6} = \dfrac{2}{0.8}$$

$$(0.6)(2) = 1.2$$
$$(1.5)(0.8) = 1.2$$

Equal

The cross products are equal, so 0.8 is the correct solution.

Work Problem 2 at the Side.

5.4 Exercises

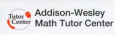
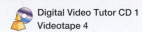

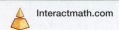

Find the unknown number in each proportion. Round your answers to hundredths, if
necessary. Check your answers by finding the cross products. See Examples 1 and 2.

1. $\dfrac{1}{3} = \dfrac{x}{12}$

$$\dfrac{3 \cdot x}{3} = \dfrac{12}{3} = 4$$

2. $\dfrac{x}{6} = \dfrac{15}{18}$

3. $\dfrac{15}{10} = \dfrac{3}{x}$

4. $\dfrac{5}{x} = \dfrac{20}{8}$

5. $\dfrac{x}{11} = \dfrac{32}{4}$

6. $\dfrac{12}{9} = \dfrac{8}{x}$

7. $\dfrac{42}{x} = \dfrac{18}{39}$

8. $\dfrac{49}{x} = \dfrac{14}{18}$

9. $\dfrac{x}{25} = \dfrac{4}{20}$

10. $\dfrac{6}{x} = \dfrac{4}{8}$

11. $\dfrac{8}{x} = \dfrac{24}{30}$

$$24 \cdot x = 8 \cdot 30$$
$$\dfrac{24 \cdot x}{24} \quad \dfrac{240}{24} = 10$$
$$x = 10$$

12. $\dfrac{32}{5} = \dfrac{x}{10}$

13. $\dfrac{99}{55} = \dfrac{44}{x}$

14. $\dfrac{x}{12} = \dfrac{101}{147}$

15. $\dfrac{0.7}{9.8} = \dfrac{3.6}{x}$

16. $\dfrac{x}{3.6} = \dfrac{4.5}{6}$

17. $\dfrac{250}{24.8} = \dfrac{x}{1.75}$

$$24.8 \cdot x = 250 \cdot 1.75$$
$$\dfrac{24.8 \cdot x}{24.8} = \dfrac{437.50}{24.8} = 17.641$$
$$17.64$$

18. $\dfrac{4.75}{17} = \dfrac{43}{x}$

$$4.75 \cdot x = 4$$

Find the unknown number in each proportion. Write your answers as whole or mixed numbers when possible. See Example 2.

19. $\dfrac{15}{1\frac{2}{3}} = \dfrac{9}{x}$ x=1

$15 \cdot x = \dfrac{9}{1} \cdot \dfrac{5}{3} = \dfrac{45}{3} = \dfrac{15}{15}$

$\dfrac{15 \cdot x}{15}$ x=1

20. $\dfrac{x}{\frac{3}{10}} = \dfrac{2\frac{2}{9}}{1}$

21. $\dfrac{2\frac{1}{3}}{1\frac{1}{2}} = \dfrac{x}{2\frac{1}{4}}$

22. $\dfrac{1\frac{5}{6}}{x} = \dfrac{\frac{3}{14}}{\frac{6}{7}}$

Solve each proportion two different ways. First change all the numbers to decimal form and solve. Then change all the numbers to fraction form and solve; write your answers in lowest terms.

23. $\dfrac{\frac{1}{2}}{x} = \dfrac{2}{0.8}$ = .4

$\dfrac{x \cdot 2}{2} = \dfrac{1}{2} \cdot \dfrac{0.8}{1} = \dfrac{.4}{2} = .2$

24. $\dfrac{\frac{3}{20}}{0.1} = \dfrac{0.03}{x}$

25. $\dfrac{x}{\frac{3}{50}} = \dfrac{0.15}{1\frac{4}{5}}$ .005   .009   = .009

$\dfrac{x \cdot 1\frac{4}{5}}{1\frac{4}{5}} = \dfrac{0.15}{1} \cdot \dfrac{3}{50} = \dfrac{.45}{50} = \dfrac{.009}{1\frac{4}{5}} = \dfrac{.009}{1.8} = .005 = .036$

$1\frac{4}{5}$

26. $\dfrac{8\frac{4}{5}}{1\frac{1}{10}} = \dfrac{x}{0.4}$

RELATING CONCEPTS (EXERCISES 27–28)   For Individual or Group Work

Work Exercises 27–28 in order. *First prove that the proportions are **not** true. Then create four true proportions for each exercise by changing one number at a time.*

27. $\dfrac{10}{4} = \dfrac{5}{3}$

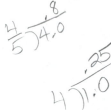

28. $\dfrac{6}{8} = \dfrac{24}{30}$

Summary Exercises on Ratios, Rates, and Proportions

Use the circle graph of one college's enrollment to complete Exercises 1–4. Write each ratio as a fraction in lowest terms.

1. Write the ratio of freshmen to juniors.

2. What is the ratio of freshmen to the total college enrollment?

3. Find the ratio of seniors and sophomores to juniors.

4. Write the ratio of freshmen and sophomores to juniors and seniors.

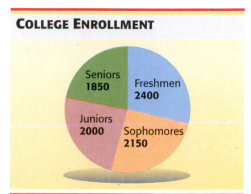

COLLEGE ENROLLMENT

Seniors 1850
Freshmen 2400
Juniors 2000
Sophomores 2150

The bar graph shows the number of Americans who play various instruments. Use the graph to complete Exercises 5 and 6. Write each ratio as a fraction in lowest terms.

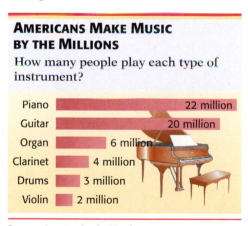

AMERICANS MAKE MUSIC BY THE MILLIONS

How many people play each type of instrument?

Piano	22 million
Guitar	20 million
Organ	6 million
Clarinet	4 million
Drums	3 million
Violin	2 million

Source: America by the Numbers.

5. Write six ratios that compare the least popular instrument to each of the other instruments.

6. Which two instruments give each of these ratios:
(a) $\frac{5}{1}$; **(b)** $\frac{2}{1}$? There may be more than one correct answer.

The table lists data on the top three individual scoring NBA basketball games of all time. Use the data to answer Exercises 7 and 8. Round answers to the nearest tenth.

7. What was Wilt Chamberlain's scoring rate in points per minute and in minutes per point?

Player	Date	Points	Min.
Wilt Chamberlain	3/2/62	100	48
David Thompson	4/9/78	73	43
David Robinson	4/24/94	71	44

Source: www.NBA.com

8. Find David Robinson's scoring rate in points per minute and minutes per point.

9. Lucinda's paycheck showed gross pay of $652.80 for 40 hours of work and $195.84 for 8 hours of overtime work. Find her regular hourly pay rate and her overtime rate.

10. Satellite TV is being offered to new subscribers at $25 per month for 50 channels, $39 per month for 100 channels, or $48 per month for 150 channels. What is the monthly cost per channel under each plan? (*Source:* Dish1Up Satellites.)

11. Find the best buy on coffee.

    26 ounces for $4.78

    34 ounces for $5.44

    39 ounces for $7.49

    (*Source:* Cub Foods.)

12. Which brand of cat food is the best buy? You have a coupon for $2 off Brand P, and another for $1 off Brand N.

    Brand N is $3.75 for 3.5 pounds.

    Brand P is $5.99 for 7 pounds.

    Brand R is $10.79 for 18 pounds.

Use either the method of writing in lowest terms or the method of finding cross products to decide whether each proportion is true or false. Show your work and then write true *or* false.

13. $\dfrac{28}{21} = \dfrac{44}{33}$

14. $\dfrac{2.3}{8.05} = \dfrac{0.25}{0.9}$

15. $\dfrac{2\frac{5}{8}}{3\frac{1}{4}} = \dfrac{21}{26}$

Solve each proportion to find the unknown number. Round your answers to hundredths when necessary.

16. $\dfrac{7}{x} = \dfrac{25}{100}$

17. $\dfrac{15}{8} = \dfrac{6}{x}$

18. $\dfrac{x}{84} = \dfrac{78}{36}$

19. $\dfrac{10}{11} = \dfrac{x}{4}$

20. $\dfrac{x}{17} = \dfrac{3}{55}$

21. $\dfrac{2.6}{x} = \dfrac{13}{7.8}$

22. $\dfrac{0.14}{1.8} = \dfrac{x}{0.63}$

23. $\dfrac{\frac{1}{3}}{8} = \dfrac{x}{24}$

24. $\dfrac{6\frac{2}{3}}{4\frac{1}{6}} = \dfrac{\frac{6}{5}}{x}$

5.5 Solving Application Problems with Proportions

O B J E C T I V E 1 Use proportions to solve application problems.
Proportions can be used to solve a wide variety of problems. Watch for problems in which you are given a ratio or rate and then are asked to find part of a corresponding ratio or rate. Remember that a ratio or rate compares two quantities and often includes one of the following indicator words.

in   for   on   per   from   to

Use the six problem-solving steps you learned in **Section 1.10.**

Step 1 **Read** the problem.

Step 2 **Work out a plan.**

Step 3 **Estimate** a reasonable answer.

Step 4 **Solve** the problem.

Step 5 **State the answer.**

Step 6 **Check** your work.

EXAMPLE 1 Solving a Proportion Application

Mike's car can travel 163 **miles on** 6.4 **gallons** of gas. How far can it travel on a full tank of 14 **gallons** of gas? Round to the nearest whole mile.

Step 1 **Read** the problem. The problem asks for the number of miles the car can travel on 14 gallons of gas.

Step 2 **Work out a plan.** Decide what is being compared. This example compares **miles** to **gallons**. Write a proportion using the two rates. Be sure that *both* rates compare miles to gallons in the same order. In other words, miles is in both numerators and gallons is in both denominators. Use a letter to represent the unknown number.

Matching units

This rate compares **miles** to **gallons** → $\dfrac{163 \text{ miles}}{6.4 \text{ gallons}} = \dfrac{x \text{ miles}}{14 \text{ gallons}}$ ← This rate compares **miles** to **gallons**.

Matching units

Step 3 **Estimate** a reasonable answer. To estimate the answer, notice that 14 gallons is a little more than *twice as much* as 6.4 gallons, so the car should travel a little more than *twice as far.* So use 2 • 163 miles = 326 miles as our estimate.

Step 4 **Solve** the problem. Ignore the units while solving for x.

$$\frac{163 \text{ miles}}{6.4 \text{ gallons}} = \frac{x \text{ miles}}{14 \text{ gallons}}$$

$(6.4)(x) = (163)(14)$ Show that cross products are equivalent.

$(6.4)(x) = 2282$

$$\frac{(6.4)(x)}{6.4} = \frac{2282}{6.4}$$ Divide both sides by 6.4.

$x = 356.5625$ Round to 357.

Continued on Next Page

1 Set up and solve a proportion for each problem.

(a) If 2 pounds of fertilizer will cover 50 square feet of garden, how many pounds are needed for 225 square feet?

(b) A U.S. map has a scale of 1 inch to 75 miles. Lake Superior is 4.75 inches long on the map. What is the lake's actual length in miles?

(c) Cough syrup is to be given at the rate of 30 milliliters for each 100 pounds of body weight. How much should be given to a 34-pound child? Round to the nearest whole milliliter.

Step 5   **State the answer.** Rounded to the nearest mile, the car can travel about 357 miles on a full tank of gas.

Step 6   **Check** your work. The answer, 357 miles, is a little more than the estimate of 326 miles, so it is reasonable.

CAUTION
When setting up a proportion do *not* mix up the units in the rates.

$$\text{compares \textbf{miles} to \textbf{gallons}} \left\{ \frac{163 \text{ \textbf{miles}}}{6.4 \text{ \textbf{gallons}}} = \frac{14 \text{ \textbf{gallons}}}{x \text{ \textbf{miles}}} \right\} \text{compares \textbf{gallons} to \textbf{miles}}$$

These rates do *not* compare things in the same order and *cannot* be set up as a proportion.

◄◄◄ Work Problem 1 at the Side.

EXAMPLE 2   **Solving a Proportion Application**

A newspaper report says that 7 out of 10 people surveyed watch the news on TV. At that rate, how many of the 3200 people in town would you expect to watch the news?

Step 1   **Read** the problem. The problem asks how many of the 3200 people in town would be expected to watch TV news.

Step 2   **Work out a plan.** You are comparing people who watch the news to people surveyed. Set up a proportion using the two rates described in the example. Be sure that both rates make the same comparison. "People who watch the news" is mentioned first, so it should be in the numerator of *both* rates.

$$\text{People who watch news} \rightarrow \frac{7}{10} = \frac{x}{3200} \leftarrow \text{People who watch news}$$
Total group → (people surveyed)   ← Total group (people in town)

Step 3   **Estimate** a reasonable answer. To estimate the answer, notice that 7 out of 10 people is more than half the people, but less than all the people. Half of 3200 people is $3200 \div 2 = 1600$, so our estimate is between 1600 and 3200 people.

Step 4   **Solve** the problem. Solve for the unknown number in the proportion.

$$\frac{7}{10} = \frac{x}{3200}$$

$$10 \cdot x = 7 \cdot 3200 \qquad \text{Show that cross products are equivalent.}$$

$$10 \cdot x = 22{,}400$$

$$\frac{\overset{1}{\cancel{10}} \cdot x}{\underset{1}{\cancel{10}}} = \frac{22{,}400}{10} \qquad \text{Divide both sides by 10.}$$

$$x = 2240$$

Continued on Next Page

Step 5 **State the answer.** You would expect 2240 people in town to watch the news on TV.

Step 6 **Check** your work. The answer, 2240 people, is between 1600 and 3200, as called for in the estimate.

CAUTION

Always check that your answer is reasonable. If it is not, look at the way your proportion is set up. Be sure you have matching units in the numerators and matching units in the denominators.

For example, suppose you had set up the last proportion *incorrectly* as shown here.

$$\frac{7}{10} = \frac{3200}{x} \quad \leftarrow \text{Incorrect setup}$$

$$7 \cdot x = 10 \cdot 3200$$

$$\frac{\overset{1}{\cancel{7}} \cdot x}{\underset{1}{\cancel{7}}} = \frac{32,000}{7}$$

$$x \approx 4571 \text{ people} \quad \leftarrow \text{Unreasonable answer}$$

This answer is *unreasonable* because there are only 3200 people in the town; it is **not** possible for 4571 people to watch the news.

Work Problem 2 at the Side. ▶▶▶

---

2 Solve each problem to find a reasonable answer. Then flip one side of your proportion to see what answer you get with an *incorrect* setup. Explain why the second answer is *unreasonable*.

(a) A survey showed that 2 out of 3 people would like to lose weight. At this rate, how many people in a group of 150 want to lose weight?

$$\frac{2}{3} \quad \frac{x}{150}$$

$$3 \cdot x = 2 \cdot 150$$

$$\frac{3 \cdot x}{3} = \frac{300}{3} = 100$$

(b) In one state, 3 out of 5 college students receive financial aid. At this rate, how many of the 4500 students at Central Community College receive financial aid?

(c) An advertisement says that 9 out of 10 dentists recommend sugarless gum. If the ad is true, how many of the 60 dentists in our city would recommend sugarless gum?

ANSWERS

2. **(a)** 100 people (reasonable); incorrect setup gives 225 people (only 150 people in the group).
 (b) 2700 students (reasonable); incorrect setup gives 7500 students (only 4500 students at the college).
 (c) 54 dentists (reasonable); incorrect setup gives about 67 dentists (only 60 dentists in the city).

Feeding Hummingbirds

A recipe can be used to make as much of a mixture as you need as long as the ingredients are kept proportional. Use the recipe for a homemade mixture of sugar water for hummingbird feeders to answer these problems.

1. What is the ratio of sugar to water in the recipe?

   What is the ratio of water to sugar in the recipe?

2. Complete each table.

Sugar	Water
1 cup	4 cups
	5 cups
	6 cups
	7 cups
2 cups	8 cups

Sugar	Water
1 cup	4 cups
	3 cups
	2 cups
	1 cup

3. How much water would you need
   (a) if you wanted to use 3 cups of sugar?
   (b) if you wanted to use 4 cups of sugar?
   (c) if you wanted to use $\frac{1}{3}$ cup of sugar?

4. One cup of sugar weighs about 200 grams and one cup of water weighs about 235 grams. If you mix 1 cup of sugar and 4 cups of water, what is the approximate weight of the resulting mixture?

5. The article says that the nectar from wildflowers visited by hummingbirds has an average sugar concentration of 21 percent. That represents a ratio of 21 to 100. If you wanted to mix a sugar solution in the same proportion as the wildflower nectar, how much water should you use for 1 cup of sugar? What proportion did you set up to solve this problem?

Feeding Hummingbirds

After getting a hummingbird feeder, the next step is to fill it! You have two choices at this point: you can either buy one of the commercial mixtures or you can make your own solution:

> **Recipe for Homemade Mixture:**
> **1 part sugar (not honey)**
> **4 parts water**
> **Boil for 1 to 2 minutes. Cool.**
> **Store extra in refrigerator.**

The concentration of the sugar is important. A 1 to 4 ratio of sugar to water is recommended because it approximates the ratio of sugar to water found in the nectar of many hummingbird flowers. A recent study of native California wildflowers visited by hummingbirds showed that their nectar had an average sugar concentration of 21 percent. This is sweet enough to attract the hummers without being too sweet. If you increase the concentration of sugar, it may be harder for the birds to digest; if you decrease the concentration, they may lose interest.

Boiling the solution helps retard fermentation. Sugar-and-water solutions are subject to rapid spoiling, especially in hot weather.

Source: The Hummingbird Book.

6. As you change the amounts of water and sugar, should you change the length of time that you boil the mixture? Explain your answer.

7. Will the length of time it takes to get the water hot enough to start boiling change? Explain your answer.

5.5 Exercises

 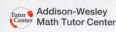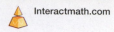
Set up and solve a proportion for each application problem. See Example 1.

1. Caroline can sketch four cartoon strips in five hours. How long will it take her to sketch 18 strips?

   $\dfrac{4}{5} = \dfrac{18}{x}$

   $4 \cdot x = 90 = 22.5$

2. The Cosmic Toads recorded eight songs on their first CD in 26 hours. How long will it take them to record 14 songs for their second CD?

3. Sixty newspapers cost $27. Find the cost of 16 newspapers.

4. Twenty-two guitar lessons cost $396. Find the cost of 12 lessons.

5. If three pounds of fescue grass seed cover about 350 square feet of ground, how many pounds are needed for 4900 square feet?

6. Anna earns $1242.08 in 14 days. How much does she earn in 260 days?

7. Tom makes $455.75 in 5 days. How much does he make in 3 days?

8. If 5 ounces of a medicine must be mixed with 8 ounces of water, how many ounces of medicine would be mixed with 20 ounces of water?

9. The bag of rice noodles shown below makes 7 servings. At that rate, how many ounces of noodles do you need for 12 servings, to the nearest ounce?

10. This can of sweet potatoes is enough for 4 servings. How many ounces are needed for 9 servings, to the nearest ounce?

11. Three quarts of a latex enamel paint will cover about 270 square feet of wall surface. How many quarts will you need to cover 350 square feet of wall surface in your kitchen and 100 square feet of wall surface in your bathroom?

12. One gallon of clear gloss wood finish covers about 550 square feet of surface. If you need to apply three coats of finish to 400 square feet of surface, how many gallons do you need, to the nearest tenth?

Use the floor plan shown to complete Exercises 13–16. On the plan, one inch represents four feet.

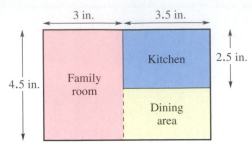

13. What is the actual length and width of the kitchen?

14. What is the actual length and width of the family room?

15. What is the actual length and width of the dining area?

16. What is the actual length and width of the entire floor plan?

The table below lists recommended amounts of food to order for 25 party guests. Use the table to answer Exercises 17 and 18. (Source: CubFoods.)

17. How much of each food item should Nathan and Amanda order for a graduation party with 60 guests?

FOOD FOR 25 GUESTS

Item	Amount
Fried chicken	40 pieces
Lasagna	14 pounds
Deli meats	4.5 pounds
Sliced cheese	$2\frac{1}{3}$ pounds
Bakery buns	3 dozen
Potato salad	6 pounds

18. Taisha is having 20 neighbors over for a Fourth of July picnic. How much food should she buy?

Set up a proportion to solve each problem. Check to see whether your answer is reasonable. Then flip one side of your proportion to see what answer you get with an incorrect setup. Explain why the second answer is unreasonable. See Example 2.

19. About 7 out of 10 people entering a community college need to take a refresher math course. If there are 2950 entering students, how many will probably need refresher math? (*Source:* Minneapolis Community and Technical College.)

20. In a survey, only 3 out of 100 people like their eggs poached. At that rate, how many of the 60 customers who ordered eggs at Soon-Won's restaurant this morning asked to have them poached? Round to the nearest whole person.

21. About 1 out of 8 people choose vanilla as their favorite ice cream flavor. If 238 people attend an ice cream social, how many would you expect to choose vanilla? Round to the nearest whole person.

22. In a test of 200 sewing machines, only one had a defect. At that rate, how many of the 5600 machines shipped from the factory have defects?

23. About 98 out of 100 U.S. households have at least one TV set. There were 107,500,000 U.S. households in 2002. How many households have one or more TVs? (*Source:* Nielsen Media Research.)

24. In a survey, 3 out of 100 dog owners wash their pets by having the dogs go into the shower with them. If the survey is accurate, how many of the 31,200,000 dog owners in the United States use this method? (*Source:* Teledyne Water Pik, American Veterinary Medical Association.)

Set up and solve a proportion for each problem.

25. The stock market report says that 5 stocks went up for every 6 stocks that went down. If 750 stocks went down yesterday, how many went up?

26. The human body contains 90 pounds of water for every 100 pounds of body weight. How many pounds of water are in a child who weighs 80 pounds?

27. The ratio of the length of an airplane wing to its width is 8 to 1. If the length of a wing is 32.5 meters, how wide must it be? Round to the nearest hundredth.

28. The Rosebud School District wants a student-to-teacher ratio of 19 to 1. How many teachers are needed for 1850 students? Round to the nearest whole number.

29. The number of calories you burn is proportional to your weight. A 150-pound person burns 222 calories during 30 minutes of tennis. How many calories would a 210-pound person burn, to the nearest whole number? (*Source: Wellness Encyclopedia.*)

30. (Complete Exercise 29 first.) A 150-pound person burns 189 calories during 45 minutes of grocery shopping. How many calories would a 115-pound person burn, to the nearest whole number? (*Source: Wellness Encyclopedia.*)

31. At 3 P.M., Coretta's shadow is 1.05 meters long. Her height is 1.68 meters. At the same time, a tree's shadow is 6.58 meters long. How tall is the tree? Round to the nearest hundredth.

32. Refer to Exercise 31. Later in the day, Coretta's shadow was 2.95 meters long. How long a shadow did the tree have at that time? Round to the nearest hundredth.

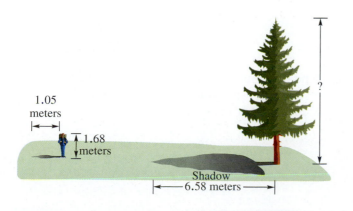

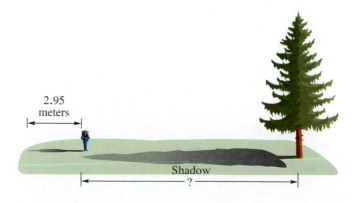

33. Can you set up a proportion to solve this problem? Explain why or why not. Jim is 25 years old and weighs 180 pounds. How much will he weigh when he is 50 years old?

34. Write your own application problem that can be solved by setting up a proportion. Also show the proportion and the steps needed to solve your problem.

35. A survey of college students shows that 4 out of 5 drink coffee. Of the students who drink coffee, 1 out of 8 adds cream to it. How many of the 48,000 students at Ohio State University would be expected to use cream in their coffee?

36. About 9 out of 10 adults think it is a good idea to exercise regularly. But of the ones who think it is a good idea, only 1 in 6 actually exercises at least three times a week. At this rate, how many of the 300 employees in our company exercise regularly?

37. The nutrition information on a bran cereal box says that a $\frac{1}{3}$-cup serving provides 80 calories and 8 grams of dietary fiber. At that rate, how many calories and grams of fiber are in a $\frac{1}{2}$-cup serving? (*Source:* Kraft Foods, Inc.)

38. A $\frac{2}{3}$-cup serving of penne pasta has 210 calories and 2 grams of dietary fiber. How many calories and grams of fiber would be in a 1-cup serving? (*Source:* Borden Foods.)

RELATING CONCEPTS (EXERCISES 39–42) For Individual or Group Work

A box of instant mashed potatoes has the list of ingredients shown in the table. Use this information to **work Exercises 39–42 in order.**

Ingredient	For 12 Servings
Water	$3\frac{1}{2}$ cups
Margarine	6 Tbsp
Milk	$1\frac{1}{2}$ cups
Potato flakes	4 cups

Source: General Mills.

39. Find the amount of each ingredient needed for 6 servings. Show *two* different methods for finding the amounts. One method should use proportions.

40. Find the amount of each ingredient needed for 18 servings. Show *two* different methods for finding the amounts, one using proportions and one using your answers from Exercise 39.

41. Find the amount of each ingredient needed for 3 servings, using your answers from either Exercise 39 or Exercise 40.

42. Find the amount of each ingredient needed for 9 servings, using your answers from either Exercise 40 or Exercise 41.

Chapter 5

SUMMARY

KEY TERMS

5.1	**ratio**	A ratio compares two quantities having the same type of units. For example, the ratio of 6 apples to 11 apples is written in fraction form as $\frac{6}{11}$. The common units (apples) divide out.
5.2	**rate**	A rate compares two measurements with different types of units. Examples are 96 dollars for 8 hours, or 450 miles on 18 gallons.
	unit rate	A unit rate has 1 in the denominator.
	cost per unit	Cost per unit is a rate that tells how much you pay for one item or one unit. The lowest cost per unit is the best buy.
5.3	**proportion**	A proportion states that two ratios or rates are equivalent.
	cross products	Multiply along one diagonal and then along the other diagonal to find the cross products of a proportion. If the cross products are equal, the proportion is true.

TEST YOUR WORD POWER

See how well you have learned the vocabulary in this chapter. Answers follow the Quick Review.

1. A ratio
  A. can be written only as a fraction
  B. compares two quantities that have the same type of units
  C. compares two quantities that have different types of units
  D. is the reciprocal of a rate.

2. A rate
  A. can be written only as a decimal
  B. compares two quantities that have the same type of units
  C. compares two quantities that have different types of units
  D. is the reciprocal of a ratio.

3. A unit rate
  A. has a numerator of 1
  B. has a denominator of 1
  C. is found by cross multiplying
  D. is usually written in fraction form.

4. Cost per unit is
  A. the best buy
  B. a ratio written in lowest terms
  C. found by comparing cross products
  D. the price of one item or one unit.

5. A proportion
  A. shows that two ratios or rates are equivalent
  B. contains only whole numbers or decimals
  C. always has one unknown number
  D. states that two improper fractions are equivalent.

6. Cross products are
  A. used only with ratios, not with rates
  B. equal when a proportion is false.
  C. used to find the best buy
  D. equal when a proportion is true.

QUICK REVIEW

Concepts

5.1 *Writing a Ratio*
A ratio compares two quantities that have the same type of units. A ratio is usually written as a fraction with the number that is mentioned first in the numerator. The common units divide out and are not written in the answer. Check that the fraction is in lowest terms.

Examples

Write this ratio as a fraction in lowest terms.

60 ounces of medicine **to** 160 ounces of medicine

$$\frac{60 \text{ ounces}}{160 \text{ ounces}} = \frac{60 \div 20}{160 \div 20} = \frac{3}{8} \quad \left\{ \begin{array}{l} \text{Ratio in lowest} \\ \text{terms} \end{array} \right.$$

↑
Divide out common units.

355

Concepts	Examples

5.1 Using Mixed Numbers in a Ratio

If a ratio has mixed numbers, change the mixed numbers to improper fractions. Rewrite the problem in horizontal format using the "÷" symbol for division. Finally, multiply by the reciprocal of the divisor.

Write as a ratio of whole numbers in lowest terms.

$$2\tfrac{1}{2} \quad \text{to} \quad 3\tfrac{3}{4}$$

$$\dfrac{2\tfrac{1}{2}}{3\tfrac{3}{4}} = \dfrac{\tfrac{5}{2}}{\tfrac{15}{4}}$$

Reciprocal

$$= \dfrac{5}{2} \div \dfrac{15}{4} = \dfrac{5}{2} \cdot \dfrac{4}{15}$$

$$= \dfrac{\overset{1}{\cancel{5}}}{2} \cdot \dfrac{\overset{2}{\cancel{4}}}{\underset{3}{\cancel{15}}} = \dfrac{2}{3} \quad \leftarrow \text{Ratio in lowest terms}$$

5.1 Using Measurements in Ratios

When a ratio compares measurements, both measurements must be in the *same* units. It is usually easier to compare the measurements using the smaller unit, for example, inches instead of feet.

Write as a ratio in lowest terms.

$$8 \text{ in. to } 6 \text{ ft}$$

Compare using the smaller unit, inches. Because 1 ft has 12 in., 6 ft is

$$6 \cdot \mathbf{12 \text{ in.}} = 72 \text{ in.}$$

The ratio is shown below.

$$\dfrac{8 \cancel{\text{ in.}}}{72 \cancel{\text{ in.}}} = \dfrac{8 \div 8}{72 \div 8} = \dfrac{1}{9}$$

Divide out common units.

5.2 Writing Rates

A rate compares two measurements with different types of units. The units do *not* divide out, so you must write them as part of the rate.

Write the rate as a fraction in lowest terms.

$$475 \text{ miles in 10 hours}$$

$$\dfrac{475 \text{ miles} \div 5}{10 \text{ hours} \div 5} = \dfrac{95 \text{ miles}}{2 \text{ hours}} \quad \leftarrow \text{Must write units: miles and hours}$$

5.2 Finding a Unit Rate

A unit rate has 1 in the denominator. To find the unit rate, divide the numerator by the denominator. Write unit rates using the word **per** or a / mark.

Write as a unit rate: $1278 in 9 days.

$$\dfrac{\$1278}{9 \text{ days}} \quad \leftarrow \text{The fraction bar indicates division.}$$

$$9\overline{)1278}^{\,142} \quad \text{so} \quad \dfrac{\$1278 \div 9}{9 \text{ days} \div 9} = \dfrac{\$142}{1 \text{ day}}$$

Write the answer as $142 **per** day or $142/day.

Concepts	**Examples**

5.2 *Finding the Best Buy*

The best buy is the item with the lowest cost per unit. Divide the price by the number of units. Round to thousandths, if necessary. Then compare to find the lowest cost per unit.

Find the best buy on cheese. You have a coupon for 50¢ off on 2 pounds or 75¢ off on 3 pounds.

$$2 \text{ pounds for } \$2.75$$

$$3 \text{ pounds for } \$4.15$$

Find the cost per unit (cost per pound) after subtracting the coupon.

$$2 \text{ pounds cost } \$2.75 - \$0.50 = \$2.25$$

$$\frac{\$2.25}{2} = \$1.125 \text{ per pound}$$

$$3 \text{ pounds cost } \$4.15 - \$0.75 = \$3.40$$

$$\frac{\$3.40}{3} \approx \$1.133 \text{ per pound}$$

The lowest cost per pound is $1.125, so 2 pounds of cheese is the best buy.

5.3 *Writing Proportions*

A proportion states that two ratios or rates are equivalent. The proportion "5 is to 6 as 25 is to 30" is written as shown below.

$$\frac{5}{6} = \frac{25}{30}$$

To see whether a proportion is true or false, multiply along one diagonal, then multiply along the other diagonal. If the two cross products are equal, the proportion is true. If the two cross products are unequal, the proportion is false.

Write as a proportion: 8 is to 40 as 32 is to 160

$$\frac{8}{40} = \frac{32}{160}$$

Is this proportion true or false?

$$\frac{6}{8\frac{1}{2}} = \frac{24}{34}$$

Find the cross products.

$$\frac{6}{8\frac{1}{2}} = \frac{24}{34}$$

$$8\frac{1}{2} \cdot 24 = \frac{17}{2} \cdot \frac{\overset{12}{24}}{1} = \mathbf{204}$$

$$6 \cdot 34 = \mathbf{204} \quad \text{Equal}$$

The cross products are equal, so the proportion is true.

5.4 *Solving Proportions*

Solve for an unknown number in a proportion by using the steps shown on the next page.

Find the unknown number.

$$\frac{12}{x} = \frac{6}{8}$$

$$\frac{12}{x} = \frac{3}{4} \quad \longleftarrow \text{ Lowest terms}$$

(*continued*)

Concepts	*Examples*

Step 1 Find the cross products. (If desired, you can rewrite the ratios in lowest terms before finding the cross products.)

Step 1

$$\frac{12}{x} \times \frac{3}{4}$$

$x \cdot 3$ ← $\Big]$ Find cross products
$12 \cdot 4$ ←

Step 2 Show that the cross products are equivalent.

Step 2 $x \cdot 3 = \underbrace{12 \cdot 4}$ Show that cross products are equivalent.

$x \cdot 3 = \quad 48$

Step 3 Divide both products by the number multiplied by *x* (the number next to *x*).

Step 3 $\dfrac{x \cdot \overset{1}{\cancel{3}}}{\underset{1}{\cancel{3}}} = \dfrac{48}{3}$ Divide both sides by 3.

$x = 16$

Step 4 Check by writing the solution in the original proportion and finding the cross products.

Step 4

x is 16. →
$$\frac{12}{16} \times \frac{6}{8}$$

$16 \cdot 6 = \textbf{\color{red}96}$ ←
$\Big]$ Equal
$12 \cdot 8 = \textbf{\color{red}96}$ ←

The cross products are equal, so 16 is the correct solution.

5.5 *Solving Application Problems with Proportions*
Decide what is being compared. Set up and solve a proportion using the two rates described in the problem. Be sure that *both* rates compare things in the *same order*. Use a letter, like *x,* to represent the unknown number.

Use the six problem-solving steps.

Step 1 **Read** the problem carefully.

Step 2 **Work out a plan.**

Step 3 **Estimate** a reasonable answer.

Step 4 **Solve** the problem.

If 3 pounds of grass seed cover 450 square feet of lawn, how much seed is needed for 1500 square feet of lawn?

Step 1 The problem asks for the pounds of grass seed needed for 1500 square feet of lawn.

Step 2 Pounds of seed is compared to square feet of lawn. Set up and solve a proportion using the two given rates. Be sure that pounds of seed is in both numerators and square feet of lawn is in both denominators.

Step 3 Because 1500 square feet is about three times as much lawn as 450 square feet, about three times as much seed is needed. So, 3 • 3 pounds = 9 pounds as our estimate.

Step 4 With the proportion set up correctly, solve for the unknown number.

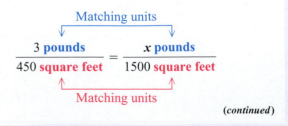

(continued)

Concepts	Examples
	Both sides compare pounds to square feet. Ignore the units while finding the cross products and solving for x.

$$450 \cdot x = \underbrace{3 \cdot 1500}$$ Show that cross products are equivalent.

$$450 \cdot x = \quad 4500$$

$$\frac{\overset{1}{\cancel{450}} \cdot x}{\underset{1}{\cancel{450}}} = \frac{4500}{450}$$ Divide both sides by 450.

$$x = 10$$

Step 5 **State the answer.**	*Step 5* 10 pounds of grass seed are needed.
Step 6 **Check** your work.	*Step 6* The exact answer, 10 pounds of seed, is close to our estimate of 9 pounds, so it is reasonable.

ANSWERS TO TEST YOUR WORD POWER

1. B; *Example:* The ratio of 3 miles to 4 miles is $\frac{3}{4}$; the common units (miles) divide out.

2. C; *Example:* $4.50 for 3 pounds is a rate comparing dollars to pounds.

3. B; *Example:* $\dfrac{\$1.79}{1 \text{ pound}}$ is a unit rate. We write it as $1.79 per pound or $1.79/pound.

4. D; *Example:* $1.95 per gallon tells the price of one gallon (one unit).

5. A; *Example:* $\dfrac{5}{6} = \dfrac{25}{30}$ is a proportion because $\dfrac{5}{6}$ is equivalent to $\dfrac{25}{30}$.

6. D; *Example:* The cross products for $\dfrac{5}{6} = \dfrac{25}{30}$ are $6 \cdot 25 = 150$ and $5 \cdot 30 = 150$.

Real-Data Applications

Currency Exchange

When you travel between countries, you will exchange U.S. dollars for the local currency. The exchange rate between currencies changes daily, and you can easily find the updated rates using the Internet or any major newspaper. The table shown below has been extracted from the Oanda Web page, www.oanda.com. It shows how much of each country's currency was equivalent to 1 U.S. dollar on February 14, 2004.

NORTH AMERICA/CARIBBEAN CURRENCY RATES
(FEBRUARY 14, 2004)

Currency	Symbol	Value
Canadian dollar	CAD	1.3194
Cayman Islands dollar	KYD	0.82
Jamaican dollar	JMD	65.1
Mexican peso	MXN	10.99
United States dollar	USD	1.00

From the table we can see that the currency exchange rate from U.S. dollars to Mexican pesos was given as follows:

$1.00 U.S. was equivalent to 10.99 Mexican pesos

You can set up a proportion to convert dollars to pesos. For example, suppose you want to determine the number of pesos that is equivalent to $50.00.

$$\frac{\$1}{10.99 \text{ pesos}} = \frac{\$50}{x \text{ pesos}} \quad \text{or} \quad \frac{1}{10.99} = \frac{50}{x}$$

$$(1)(x) = (10.99)(50)$$

$$x = 549.50 \text{ pesos}$$

So $50 buys 549 pesos and 50 centavos.

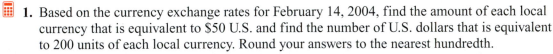

1. Based on the currency exchange rates for February 14, 2004, find the amount of each local currency that is equivalent to $50 U.S. and find the number of U.S. dollars that is equivalent to 200 units of each local currency. Round your answers to the nearest hundredth.

   (a) $50 = _____ Canadian dollars, and 200 Canadian dollars = _____ U.S. dollars.

   (b) $50 = _____ Cayman Island dollars, and
   200 Cayman Island dollars = _____ U.S. dollars.

   (c) $50 = _____ Jamaican dollars, and 200 Jamaican dollars = _____ U.S. dollars.

2. Set up a proportion to find the number of U.S. dollars that was equivalent to 1 Mexican Peso. Round your answer to the nearest cent. 1 Mexican peso was equivalent to $_____ (U.S.).

3. From Problem 2, you should recognize the conversion rate based on 1 Mexican peso as the expression $\frac{1}{10.99}$. What is the mathematical word that describes the relationship between the conversion rates 10.99 and $\frac{1}{10.99}$?

Chapter 5
REVIEW EXERCISES

[5.1] *Write each ratio as a fraction in lowest terms. Change to the same units when necessary, using the table of measurement comparisons in* **Section 5.1.** *Use the information in the graph to answer Exercises 1–3.*

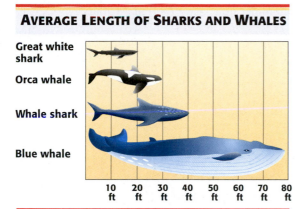

AVERAGE LENGTH OF SHARKS AND WHALES

Great white shark

Orca whale

Whale shark

Blue whale

| 10 ft | 20 ft | 30 ft | 40 ft | 50 ft | 60 ft | 70 ft | 80 ft |

Source: Grolier Multimedia Encyclopedia.

1. Ratio of orca whale's length to whale shark's length

2. Ratio of blue whale's length to great white shark's length

3. Which two animals' lengths give a ratio of $\frac{1}{2}$? There are several answers.

4. $2.50 to $1.25

5. $0.30 to $0.45

6. $1\frac{2}{3}$ cups to $\frac{2}{3}$ cup

7. $2\frac{3}{4}$ miles to $16\frac{1}{2}$ miles

8. 5 hours to 100 minutes

9. 9 in. to 2 ft

10. 1 ton to 1500 pounds

11. 8 hours to 3 days

12. Jake sold $350 worth of his kachina figures. Ramona sold $500 worth of her pottery. What is the ratio of her sales to his?

13. Ms. Wei's new car gets 35 miles per gallon. Her old car got 25 miles per gallon. Find the ratio of the new car's mileage to the old car's mileage.

14. This fall, 6000 students are taking math courses and 7200 students are taking English courses. Find the ratio of math students to English students.

[5.2] *Write each rate as a fraction in lowest terms.*

15. $88 for 8 dozen

16. 96 children in 40 families

17. In his keyboarding class, Patrick can type four pages in 20 minutes. Give his rate in pages per minute and minutes per page.

18. Elena made $24 in three hours. Give her earnings in dollars per hour and hours per dollar.

Find the best buy.

19. Minced onion

   13 ounces for $2.29

   8 ounces for $1.45

   3 ounces for $0.95

20. Dog food; you have a coupon for $1 off on 25 pounds or more.

   50 pounds for $19.95

   25 pounds for $10.40

   8 pounds for $3.40

[5.3] *Use either the method of writing in lowest terms or the method of finding cross products to decide whether each proportion is true or false. Show your work and then write* true *or* false.

21. $\dfrac{6}{10} = \dfrac{9}{15}$

22. $\dfrac{6}{48} = \dfrac{9}{36}$

23. $\dfrac{47}{10} = \dfrac{98}{20}$

24. $\dfrac{64}{36} = \dfrac{96}{54}$

25. $\dfrac{1.5}{2.4} = \dfrac{2}{3.2}$

26. $\dfrac{3\frac{1}{2}}{2\frac{1}{3}} = \dfrac{6}{4}$

[5.4] *Find the unknown number in each proportion. Round answers to the nearest hundredth, if necessary.*

27. $\dfrac{4}{42} = \dfrac{150}{x}$

28. $\dfrac{16}{x} = \dfrac{12}{15}$

29. $\dfrac{100}{14} = \dfrac{x}{56}$

30. $\dfrac{5}{8} = \dfrac{x}{20}$

31. $\dfrac{x}{24} = \dfrac{11}{18}$

32. $\dfrac{7}{x} = \dfrac{18}{21}$

33. $\dfrac{x}{3.6} = \dfrac{9.8}{0.7}$

34. $\dfrac{13.5}{1.7} = \dfrac{4.5}{x}$

35. $\dfrac{0.82}{1.89} = \dfrac{x}{5.7}$

[5.5] *Set up and solve a proportion for each application problem.*

36. The ratio of cats to dogs at the animal shelter is 3 to 5. If there are 45 dogs, how many cats are there?

37. Danielle had 8 hits in 28 times at bat during last week's games. If she continues to hit at the same rate, how many hits will she gets in 161 times at bat?

38. If 3.5 pounds of ground beef cost $9.77, what will 5.6 pounds cost? Round to the nearest cent.

39. About 4 out of 10 students are expected to vote in campus elections. There are 8247 students. How many are expected to vote? Round to the nearest whole number.

40. The scale on Brian's model railroad is 1 in. to 16 ft. One of the scale model boxcars is 4.25 in. long. What is the length of a real boxcar in feet?

41. Marvette makes necklaces to sell at a local gift shop. She made 2 dozen necklaces in $16\frac{1}{2}$ hours. How long will it take her to make 40 necklaces?

42. A 180-pound person burns 284 calories playing basketball for 25 minutes. At this rate, how many calories would the person burn in 45 minutes, to the nearest whole number? (*Source: Wellness Encyclopedia.*)

43. In the hospital pharmacy, Michiko sees that a medicine is to be given at the rate of 3.5 milligrams for every 50 pounds of body weight. How much medicine should be given to a patient who weighs 210 pounds?

MIXED REVIEW EXERCISES

Find the unknown number in each proportion. Round answers to the nearest hundredth, if necessary.

44. $\dfrac{x}{45} = \dfrac{70}{30}$

45. $\dfrac{x}{52} = \dfrac{0}{20}$

46. $\dfrac{64}{10} = \dfrac{x}{20}$

47. $\dfrac{15}{x} = \dfrac{65}{100}$

48. $\dfrac{7.8}{3.9} = \dfrac{13}{x}$

49. $\dfrac{34.1}{x} = \dfrac{0.77}{2.65}$

Find cross products to decide whether each proportion is true or false. Show the cross products and then circle true *or false.*

50. $\dfrac{55}{18} = \dfrac{80}{27}$

51. $\dfrac{5.6}{0.6} = \dfrac{18}{1.94}$

52. $\dfrac{\frac{1}{5}}{2} = \dfrac{1\frac{1}{6}}{11\frac{2}{3}}$

    True   False

    True   False

    True   False

Write each ratio as a fraction in lowest terms. Change to the same units when necessary.

53. 4 dollars to 10 quarters

54. $4\frac{1}{8}$ in. to 10 in.

55. 10 yd to 8 ft

56. $3.60 to $0.90

57. 12 eggs to 15 eggs

58. 37 meters to 7 meters

59. 3 pints to 4 quarts

60. 15 minutes to 3 hours

61. $4\frac{1}{2}$ miles to $1\frac{3}{10}$ miles

62. Nearly 7 out of 8 fans buy something to drink at rock concerts. How many of the 28,500 fans at today's concert would be expected to buy a beverage? Round to the nearest hundred fans.

63. Emily spent $150 on car repairs and $400 on car insurance. What is the ratio of the amount spent on insurance to the amount spent on repairs?

64. Antonio is choosing among three packages of plastic wrap. Is the best buy 25 ft for $0.78; 75 ft for $1.99; or 100 ft for $2.59? He has a coupon for 50¢ off either of the larger two packages.

65. On this scale drawing of a backyard patio, 0.5 in. represents 6 ft. If the patio measures 1.75 in. long and 1.25 in. wide on the drawing, what will be the actual length and width of the patio when it is built?

0.5 in. = 6 ft

66. A lawn mower uses 0.8 gallon of gas every 3 hours. The gas tank holds 2 gallons. How long can the mower run on a full tank?

67. An antibiotic is to be given at the rate of $1\frac{1}{2}$ teaspoons for every 24 pounds of body weight. How much should be given to an infant who weighs 8 pounds?

68. Charles made 251 points during 169 minutes of playing time last year. If he plays 14 minutes in tonight's game, how many points would you expect him to make? Round to the nearest whole number.

69. Refer to Exercise 67. Explain each step you took in solving the problem. Be sure to tell how you decided which way to set up the proportion and how you checked your answer.

70. A vitamin supplement for cats is to be given at the rate of 1000 milligrams for a 5-pound cat. (*Source:* St. Jon Pet Care Products.)

(a) How much should be given to a 7-pound cat?

(b) How much should be given to an 8-ounce kitten?

Chapter **5**

TEST

Write each rate or ratio as a fraction in lowest terms. Change to the same units when necessary.

1. 16 fish to 20 fish

2. 300 miles on 15 gallons

3. $15 for 75 minutes

4. 3 hours to 40 minutes

5. The college's theater has 320 seats. The auditorium has 1200 seats. Find the ratio of auditorium seats to theater seats.

6. Use the information in the table about Quiznos chicken with bacon sub sandwich to find the best buy.

Size	Length of Sub	Price
Small	4.5 inches	$4.29
Regular	8 inches	$5.49
Large	12 inches	$8.29

7. Find the best buy on spaghetti sauce. You have a coupon for 75¢ off Brand X and a coupon for 25¢ off Brand Y.

28 ounces of Brand X for $3.89

18 ounces of Brand Y for $1.89

13 ounces of Brand Z for $1.29

8. Suppose the ratio of your income last year to your income this year is 3 to 2. Explain what this means. Give an example of the dollars earned last year and this year that fits the 3 to 2 ratio.

Determine whether each proportion is true or false. Show your work and then write true *or* false.

9. $\dfrac{6}{14} = \dfrac{18}{45}$

10. $\dfrac{8.4}{2.8} = \dfrac{2.1}{0.7}$

1. _____

2. _____

3. _____

4. _____

5. _____

6. _____

7. _____

8. _____

9. _____

10. _____

Find the unknown number in each proportion. Round the answers to the nearest hundredth, if necessary.

11. _____

12. _____

13. _____

14. _____

15. _____

16. _____

17. _____

18. _____

19. _____

20. _____

11. $\dfrac{5}{9} = \dfrac{x}{45}$

12. $\dfrac{3}{1} = \dfrac{8}{x}$

13. $\dfrac{x}{20} = \dfrac{6.5}{0.4}$

14. $\dfrac{2\frac{1}{3}}{x} = \dfrac{\frac{8}{9}}{4}$

Set up and solve a proportion for each application problem.

15. Pedro typed 240 words in five minutes in his keyboarding class. At that rate, how many words could he type in 12 minutes?

16. Just 0.8 ounce of wildflower seeds is enough for 50 square feet of ground. What weight of seeds is needed for a garden with 225 square feet? (*Source:* White Swan Ltd.)

17. About 2 out of every 15 people are left-handed. How many of the 650 students in our school would you expect to be left-handed? Round to the nearest whole number.

18. A student set up the proportion for Exercise 17 this way and arrived at an answer of 4875.

$$\dfrac{2}{15} = \dfrac{650}{x} \qquad Check: \qquad \dfrac{2}{15} \bowtie \dfrac{650}{4875}$$

$$15 \cdot 650 = 9750$$

$$2 \cdot 4875 = 9750$$

Because the cross products are equal, the student said the answer is correct. Is the student right? Explain why or why not.

19. A medication is given at the rate of 8.2 grams for every 50 pounds of body weight. How much should be given to a 145-pound person? Round to the nearest tenth.

20. On a scale model, 1 in. represents 8 ft. If a building in the model is 7.5 in. tall, what is the actual height of the building in feet?

Cumulative Review Exercises

CHAPTERS 1–5

Name the digit that has the given place value.

1. 216,475,038
thousands
tens
millions
hundred thousands

2. 340.6915
hundredths
ones
ten-thousandths
hundreds

Round each number as indicated.

3. 9903 to the nearest hundred

4. 617.0519 to the nearest tenth

5. $99.81 to the nearest dollar

6. $3.0555 to the nearest cent

≈ *First use front end rounding to round each number and estimate the answer. Then find the exact answer.*

7. *Estimate:*   *Exact:*

$$\begin{array}{r} 28 \\ 5206 \\ +\ \ 351 \\ \hline \end{array}$$

$+\ \rule{1cm}{0.4pt}$

8. *Estimate:*   *Exact:*

$$\begin{array}{r} 63.1 \\ -\ 5.692 \\ \hline \end{array}$$

$-\ \rule{1cm}{0.4pt}$

9. *Estimate:*   *Exact:*

$$\begin{array}{r} 4716 \\ \times\ \ \ 804 \\ \hline \end{array}$$

$\times\ \rule{1cm}{0.4pt}$

10. *Estimate:*   *Exact:*

$$\begin{array}{r} 0.982 \\ \times\ \ 17.8 \\ \hline \end{array}$$

$\times\ \rule{1cm}{0.4pt}$

11. *Estimate:*   *Exact:*

$53\overline{)48{,}071}$

12. *Estimate:*   *Exact:*

$4.5\overline{)1638}$

13. *Estimate:*   *Exact:*

$\rule{1cm}{0.4pt} \cdot \rule{1cm}{0.4pt} = \rule{1cm}{0.4pt}$ $1\dfrac{5}{6} \cdot 3\dfrac{3}{5}$

14. *Estimate:*   *Exact:*

$\rule{1cm}{0.4pt} \div \rule{1cm}{0.4pt} = \rule{1cm}{0.4pt}$ $5\dfrac{1}{4} \div \dfrac{7}{8}$

15. *Estimate:*   *Exact:*

$\rule{1cm}{0.4pt} - \rule{1cm}{0.4pt} = \rule{1cm}{0.4pt}$ $2\dfrac{4}{5} - 1\dfrac{5}{6}$

16. *Estimate:*   *Exact:*

$\rule{1cm}{0.4pt} + \rule{1cm}{0.4pt} = \rule{1cm}{0.4pt}$ $2\dfrac{9}{10} + 10\dfrac{1}{2}$

Add, subtract, multiply, or divide as indicated.

17. $988 + 373{,}422 + 6$

18. $30 - 0.66$

19. Write your answer using R for the remainder.

$33\overline{)20{,}157}$

20. $(1.9)(0.004)$

21. $3020 - 708$

22. $0.401 + 62.98 + 5$

23. $1.39 \div 0.025$

24. $(6392)(5609)$

Use the order of operations to simplify each expression.

25. $36 + 18 \div 6$

26. $8 \div 4 + (10 - 3^2) \cdot 4^2$

27. $88 \div \sqrt{121} \cdot 2^3$

28. $(16.2 - 5.85) - 2.35(4)$

29. Write 0.0105 in words.

30. Write sixty and seventy-one thousandths in numbers.

Nest boxes for birds are made with different sizes of entry holes. If the opening is the right size, it allows only certain types of birds to use the nest box and helps prevent entry by predators. Use the information in the table to answer Exercises 31–34. The diameter is the distance across the opening at its widest point.

Eastern bluebird nest box with 1.5-inch opening

Bird	Diameter of Opening
Eastern bluebird	1.5 inches
Western bluebird	$1\frac{9}{16}$ inches
Chickadee	1.25 inches
Swallow	$1\frac{3}{8}$ inches
Wren	$1\frac{1}{8}$ inches

Source: Duncraft.

31. Write each mixed number measurement in the table as a decimal. Round to the nearest thousandth, if necessary.

32. List the nest box openings in order from smallest to largest.

33. Find the difference in diameter between the largest and smallest openings. Write your answer as a fraction or mixed number in lowest terms.

34. What is the difference in diameter between the openings for an eastern bluebird and a western bluebird? Write your answer as a fraction and as a decimal.

Write each rate or ratio as a fraction in lowest terms. Change to the same units when necessary. Use the circle graph for Exercises 35–37.

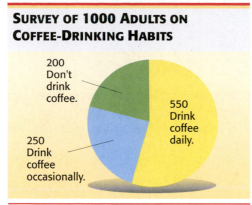

SURVEY OF 1000 ADULTS ON COFFEE-DRINKING HABITS

200 Don't drink coffee.

550 Drink coffee daily.

250 Drink coffee occasionally.

Source: National Coffee Association of USA Inc.

35. Write the ratio of those who drink coffee daily to those who drink it occasionally.

36. What is the ratio of those who do not drink coffee to the total number of people in the survey?

37. Find the ratio of all the adults who drink coffee to those who never drink it.

38. Write the ratio of 20 minutes to 4 hours.

39. Find the ratio of 2 ft to 8 in.

40. Find the best buy on instant mashed potatoes. You have a coupon for 50¢ off either the 36-serving or 48-serving box.

A box that makes 20 servings for $1.59

A box that makes 36 servings for $3.24

A box that makes 48 servings for $4.99

Find the unknown number in each proportion. Round your answers to the nearest hundredth, if necessary.

41. $\dfrac{9}{12} = \dfrac{x}{28}$

42. $\dfrac{7}{12} = \dfrac{10}{x}$

43. $\dfrac{x}{\frac{3}{4}} = \dfrac{2\frac{1}{2}}{\frac{1}{6}}$

44. $\dfrac{6.7}{x} = \dfrac{62.8}{9.15}$

Solve each application problem.

45. The college honor society has a goal of collecting 1500 pounds of food to fill Thanksgiving baskets. So far they've collected $\frac{5}{6}$ of their goal. How many more pounds do they need?

46. Tara has a photo that is 10 centimeters wide by 15 centimeters long. If the photo is enlarged to a length of 40 centimeters, find the new width, to the nearest tenth.

47. The distance around Dunning Pond is $1\frac{1}{10}$ miles. Norma ran around the pond four times in the morning and $2\frac{1}{2}$ times in the afternoon. How far did she run in all?

48. Rodney used 49.8 gallons of gas for his SUV while driving 896.5 miles on a vacation. How many miles per gallon did he get, rounded to the nearest tenth?

49. In a survey, 5 out of 8 apartment residents said they are sometimes bothered by noise from their neighbors. How many of the 224 residents at Harris Towers would you expect to be bothered by noise?

50. The directions on a bottle of plant food call for $\frac{1}{2}$ teaspoon in two quarts of water. How much plant food is needed for five quarts? (*Source:* Schultz Company.)

Use the information in the table on international long-distance calling card rates to answer Exercise 51–54. Round answers to the nearest whole minute when necessary.

$10 International Phone Card *No Connnection Fee!*	
Place a call to	**Cost per minute**
Mexico City	$0.10
Philippines	$0.333
Japan	$0.14
Hong Kong	$0.111
Canada	$0.08
Ghana	$0.172
South Korea	$0.104

Source: Walgreens, IDT

51. List the per-minute rates from least to greatest.

52. If you buy a $10 calling card, how long a call could you make to Japan?

53. How many $10 cards would you have to buy to make two hours' worth of calls to the Philippines?

54. What is the ratio of minutes you can call Canada for $10 to minutes you can call Mexico City for $10?

If you start a small business, you may choose to run an ad in a neighborhood advertising circular. One company that prints and mails circulars in Chicago, Detroit, and Minneapolis has advertising rates shown below. Use this information to answer Exercises 55 and 56. Round answers to the nearest thousandth when necessary.

Why are you still paying so much for direct mail?

Full page, in full color, to 25,000 homes!

Plus, get $\frac{1}{2}$ off on your second page!

$595

Half page $375

Source: MarketShare.

55. How much would your business spend to reach each home with a

   (a) full-page ad?

   (b) half-page ad?

56. **(a)** How much would it cost for a two-page ad sent to 25,000 homes?

   (b) What is the cost per home for a two-page ad?

Percent

Percents are a large part of our everyday lives. For example, percents are used for interest rates on automobile loans, home loans, and other installment loans. Sales tax, commission rates, and discounts on sale items are other everyday applications that show the importance of understanding percent. For example, knowing how to calculate the sales tax on any item you purchase allows you to know the true cost of anything you buy. (See Exercise 55 in **Section 6.5** and Examples 1 and 2, Exercises 27–30, 47, and 48 in **Section 6.6**.)

6.1 Basics of Percent

OBJECTIVES

1. Learn the meaning of percent.
2. Write percents as decimals.
3. Write decimals as percents.
4. Understand 100%, 200%, and 300%.
5. Use 50%, 10%, and 1%.

1. Write as percents.

(a) In a group of 100 people, 46 are homeowners. What percent are homeowners?

(b) The sales tax is $6 per $100. What percent is this?

(c) Out of 100 students, 68 are working full- or part-time. What percent are working?

Notice that the figure below has one hundred squares of equal size. Eleven of the squares are shaded. The shaded portion is $\frac{11}{100}$, or 0.11, of the total figure.

Shaded portion is 11 out of 100 parts, or $\frac{11}{100}$, or 0.11 or 11%.

$.11 = 11\%$

The shaded portion is also 11% of the total, or "eleven parts out of 100 parts." Read **11%** as "eleven percent."

OBJECTIVE 1 Learn the meaning of percent. As we just saw, a percent is a ratio with a denominator of 100.

> **The Meaning of Percent**
> **Percent** means *per one hundred*. The "%" sign is used to show the number of parts out of one hundred parts.

EXAMPLE 1 Understanding Percent

(a) If 43 out of 100 students are men, then 43 per (out of) 100 or $\frac{43}{100}$ or **43%** of the students are men.

(b) If a person pays a tax of $7 on every $100 of purchases, then the tax rate is $7 per $100. The ratio is $\frac{7}{100}$ and the percent of tax is **7%**.

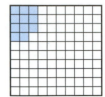 Work Problem 1 at the Side.

OBJECTIVE 2 Write percents as decimals. If 8% means 8 parts out of 100 parts or $\frac{8}{100}$, then p% means p parts out of 100 parts or $\frac{p}{100}$. Because $\frac{p}{100}$ is another way to write the division $p \div 100$, we have

$$p\% = \frac{p}{100} = p \div 100 \qquad 100\overline{)p}$$

> **Writing a Percent as a Decimal**
> $$p\% = \underbrace{\frac{p}{100}}_{\text{As a fraction}} \qquad \text{or} \qquad p\% = \underbrace{p \div 100}_{\text{As a decimal}}$$

EXAMPLE 2 Writing Percents as Decimals

Write each percent as a decimal.

(a) 47%

$$p\% = p \div 100$$

$$47\% = 47 \div 100 = 0.47 \qquad \text{Decimal form}$$

$\frac{47}{100} = .47$

Continued on Next Page

(b) 76%   76% = 76 ÷ 100 = 0.76   Decimal form

(c) 28.2%   28.2% = 28.2 ÷ 100 = 0.282   Decimal form

(d) 100%   100% = 100 ÷ 100 = 1.00   Decimal form

> **CAUTION**
> In Example 2(d) above, notice that 100% is 1.00, or 1, which is a whole number. Whenever you have a percent that is 100% or greater, the equivalent decimal number will be *1 or greater than 1*.

Work Problem 2 at the Side. ▶▶▶

The resulting answers in Example 2 above suggest the following procedure for writing a percent as a decimal.

> **Writing a Percent as a Decimal**
> *Step 1*   Drop the percent sign.
>
> *Step 2*   Divide by 100.

> **NOTE**
> Recall from **Section 4.6** that a quick way to divide a number by 100 is to move the decimal point **two places to the left.**

EXAMPLE 3   **Writing Percents as Decimals by Moving the Decimal Point**

Write each percent as a decimal by moving the decimal point two places to the left.

(a) 17%

17% = 17.%   Decimal point starts at far right side.

0.17 ← Percent sign is dropped. (Step 1)

—— Decimal point is moved two places to the left. (Step 2)

17% = 0.17

(b) 160%

160% = 160.% = 1.60 or 1.6   Decimal starts at far right side.
                                1.60 is equivalent to 1.6.

(c) 4.9%

.049%   0 is attached so the decimal point can be moved
            two places to the left.

4.9% = 0.049

(d) 0.6%

0.6% = 0.006   Two zeros are attached so the decimal
                  point can be moved two places to the left.

—— **Continued on Next Page**

---

2 Write each percent as a decimal.

(a) 68%

$68 \div 100 = .68$

(b) 34% $34 \div 100$

$.34$

(c) 58.5%

$58.5 \div 100 =$

$.585$

(d) 175%

(e) 200%

3 Write each percent as a decimal.

(a) 96%

(b) 6%

(c) 24.8%

(d) 0.9%

$9 \div 100$
$.009$

CAUTION
Look at Example 3(d) on the previous page, where 0.6% is less than 1%. Because 0.6% is $\frac{6}{10}$ of 1%, it is *less than 1%*. Any fraction of a percent is *less than 1%*.

Work Problem 3 at the Side.

OBJECTIVE 3 Write decimals as percents. You can write any decimal as a percent. For example, the decimal 0.78 is the same as the fraction

$$\frac{78}{100}$$

This fraction means 78 out of 100 parts, or 78%. The following steps give the same result.

Writing a Decimal as a Percent

Step 1 Multiply by 100.

Step 2 Attach a percent sign.

NOTE
A quick way to divide or multiply a number by 100 is to move the decimal point two places to the left or two places to the right, respectively.

Multiply decimal by 100; move decimal point two places to the **right**.

Decimal   **Percent**

Divide percent by 100; move decimal point two places to the **left**.

EXAMPLE 4 Writing Decimals as Percents by Moving the Decimal Point

Write each decimal as a percent by moving the decimal point two places to the right.

(a) 0.21

0.21% ← Percent sign is attached. (Step 1)

↑ Decimal point is moved two places to the right. (Step 2)

0.21 = 21%

↑ Decimal point is not written with whole number percents.

(b) 0.529 = 52.9%

(c) 1.92 = 192%

Continued on Next Page

(d) 2.5

2.5**0**%   0 is attached so the decimal point can be moved two places to the right.

2.5 = 250%

(e) 3

3. = 3.**00**%   Two zeros are attached so the decimal point can be moved two places to the right.

so 3 = 300%

> **CAUTION**
> Look at Examples 4(c), 4(d), and 4(e) above, where 1.92, 2.5, and 3 are greater than 1. Because the number 1 is equivalent to 100%, all numbers greater than 1 will be *greater than 100%*.

Work Problem 4 at the Side. ▶▶▶

OBJECTIVE 4 **Understand 100%, 200%, and 300%.**   When working with percents, it is helpful to have several reference points. 100%, 200%, and 300% are three such helpful reference points.

100% means 100 parts out of 100 parts. That's **all** of the parts. If 100% of the 18 people attending last week's meeting attended this week's meeting, then 18 people (**all** of them) attended this week.

If attendance at the meeting this week is 200% of last week's attendance of 18 people, then this week's attendance is 36 people, or *two* times as many people (2 • 18 = 36). Likewise, if attendance is 300% of last week's attendance, then *three* times as many people, or 54 people, attended (3 • 18 = 54).

EXAMPLE 5   **Finding 100%, 200%, and 300% of a Number**

Fill in the blanks.

(a) 100% of 82 people is _____.
100% is **all** of the people. So, 100% of 82 people is <u>82 people</u>.

(b) 200% of $63 is _____.
200% is twice (2 times) as much money. So, 200% of $63 is 2 • $63 = <u>$126</u>.

(c) 300% of 32 employees is _____.
300% is 3 times as many employees. So, 300% of 32 employees is 3 • 32 = <u>96 employees</u>.

Work Problem 5 at the Side. ▶▶▶

OBJECTIVE 5 **Use 50%, 10%, and 1%.**   50% means 50 parts out of 100 parts, which is **half** of the parts ($\frac{50}{100} = \frac{1}{2}$). 50% of $18 is $9 (**half** of the money).

When using 10%, we have 10 parts out of 100 parts, which is $\frac{1}{10}$ of the parts ($\frac{10}{100} = \frac{1}{10}$). To find 10% or $\frac{1}{10}$ of a number, we move the decimal point **one** place to the left. 10% of $285 is $28.50 (because $28.5. = $28.50).

To find 1% of a number ($\frac{1}{100}$), we move the decimal point **two** places to the left. 1% of $198 is $1.98 (because $1.98. = $1.98).

---

4 Write each decimal as a percent.

(a) 0.74   **(b)** 0.15

(c) 0.09   **(d)** 0.617

(e) 0.834   **(f)** 5.34

(g) 2.8   **(h)** 4

5 Fill in the blanks.

(a) 100% of $7.80 is _____.

(b) 100% of 1850 workers is _____.

(c) 200% of 24 photographs is _____.

(d) 300% of 8 miles is _____.

6 Fill in the blanks.

(a) 50% of 200 patients is _____.

(b) 50% of 64 e-mails is _____.

(c) 10% of 3850 elm trees is _____.

(d) 1% of 240 ft is _____.

EXAMPLE 6 Finding 50%, 10%, and 1% of a Number

Fill in the blanks.

(a) 50% of 24 hours is _____.

50% is half of the hours. So, 50% of 24 hours is 12 hours .

(b) 10% of 280 pages is _____.

10% is $\frac{1}{10}$ of the pages. Move the decimal point *one* place to the left. So, 10% of 280. pages is 28 pages .

(c) 1% of $540 is _____.

1% is $\frac{1}{100}$ of the money. Move the decimal point *two* places to the left. So, 1% of $540. is $5.40 .

Work Problem 6 at the Side.

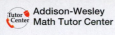 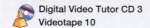 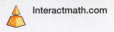

6.1 Exercises

| FOR EXTRA HELP | Addison-Wesley Math Tutor Center | MathXL | Digital Video Tutor CD 3 Videotape 10 | Student's Solutions Manual | MyMathLab | Interactmath.com |

Write each percent as a decimal. See Examples 2 and 3.

1. 12%

$12 \div 100 =$
$.12$

2. 57%

3. 70%

4. 40%

5. 25%

6. 35%

7. 140%

8. 250%

9. 5.5%

10. 6.7%

11. 100%

12. 600%

13. 0.5%

14. 0.25%

15. 0.35%

16. 0.75%

Write each decimal as a percent. See Example 4.

17. 0.6

18. 0.9

$.9 \times 100 =$
90

19. 0.58

20. 0.25

21. 0.01

22. 0.07

23. 0.125

24. 0.875

25. 0.375

26. 0.625

27. 2

28. 5

29. 3.7

30. 2.2

31. 0.0312

32. 0.0625

33. 4.162

34. 8.715

35. 0.0028

36. 0.0064

37. Fractions, decimals, and percents are all used to describe a part of something. The use of percents is much more common than fractions and decimals. Why do you suppose this is true?

38. List five uses of percent that are or will be part of your life. Consider the activities of working, shopping, saving, and planning for the future.

Write each percent as a decimal and each decimal as a percent. See Examples 2–4.

39. In Lincoln, 45% of the refuse is recycled.

.45

40. At College of DuPage, 82% of the students work part-time.

41. At Bett's Boutique, 18% of the items sold are returned.

42. There was a 43.2% voter turnout at the election.

43. The property tax rate in Alpine County is 0.035.

44. A church building fund has 0.89 of the money needed.

45. The number of people successfully completing CPR training this session is 2 times that of the last session.

46. The number of newspaper subscribers was 4 times as great as last quarter.

47. Only 0.005 of the total population has this genetic defect.

48. The return rate of defective keyboards is 0.0075 of total output.

49. The patient's blood pressure was 153.6% of normal.

50. Success with the diet was 248.7% greater than anticipated.

Fill in the blanks. Remember that 100% is all of something, 200% is two times as many, and 300% is three times as many. See Example 5.

51. There are 20 children in the preschool class. 100% of the children are served breakfast and lunch. How many children are served both meals?

_____

52. The company owns 345 vans. 100% of the vans are painted white with blue lettering. How many vans are painted white with blue lettering?

_____

53. Last year we had 210 employees. This year we have 200% of that number. How many employees do we have this year? _____

54. This week's expenses are 200% of last week's $380. This week's expenses are _____.

55. This week we'll need 300% of the 90 chairs that were used for last week's meeting. We'll need _____.

56. Jim's new car gets 300% of the 12 miles per gallon that his old car got. His new car gets _____.

Fill in the blanks. Remember that 50% is half of something, 10% is found by moving the decimal point one place to the left, and 1% is found by moving the decimal point two places to the left. See Example 6.

57. Jacob owes $755 for tuition. Financial aid will pay 50% of the cost. Financial aid will pay _____.

58. Elizabeth gets 3200 "off-peak" minutes of calling time on her cell phone plan. Last month she used 50% of her allowed minutes. The number of minutes used was _____.

59. Only 10% of 8200 commuters are carpooling to work. How many commuters carpool? _____

60. Sarah expects that 10% of the 240 dozen plants in her greenhouse will not be sold. The expected number of unsold plants is _____.

61. The naturalist said that 1% of the 2600 plants in the park are poisonous. How many plants are poisonous? _____

62. Of the 4800 accidents, only 1% was caused by mechanical failure. How many accidents were caused by mechanical failure? _____.

63. (a) Describe a shortcut method of finding 100% of a number.

(b) Show an example using your shortcut.

64. (a) Describe a shortcut method of finding 50% of a number.

(b) Show an example using your shortcut.

65. (a) Describe a shortcut method of finding 200% of a number.

    (b) Show an example using your shortcut.

66. (a) Describe a shortcut method of finding 300% of a number.

    (b) Show an example using your shortcut.

67. (a) Describe a shortcut method of finding 10% of a number.

$$\frac{1}{10} \qquad \frac{10}{100} \times \frac{x}{1}$$

    (b) Show an example using your shortcut.

68. (a) Describe a shortcut method of finding 1% of a number.

$$\frac{1}{100} = \frac{x}{1}$$

    (b) Show an example using your shortcut.

Homeowners were asked what home improvement projects they would take on themselves. The bar graph shows the ranking of the top projects and the percent of homeowners selecting each project. Use this graph to answer Exercises 69–72. Write each answer as a percent and as a decimal.

69. What portion of the homeowners selected painting and wallpapering as a project to do themselves?

70. What portion of the homeowners selected carpentry projects as a project to do themselves?

71. (a) What was the second-most-selected project?

    (b) Write the portion of the homeowners who selected that project.

72. (a) What was the third-most-selected project?

    (b) Write the portion of the homeowners who selected that project.

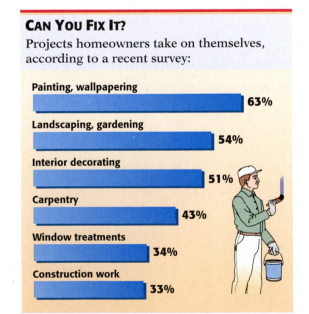

CAN YOU FIX IT?

Projects homeowners take on themselves, according to a recent survey:

Painting, wallpapering — 63%
Landscaping, gardening — 54%
Interior decorating — 51%
Carpentry — 43%
Window treatments — 34%
Construction work — 33%

Note: Multiple responses allowed.
Source: American Express Home Improvement Index telephone survey.

There are more than 90,000 women serving in the U.S. military. The circle graph shows the percent of these women serving in each branch of the armed forces. Use this graph to answer Exercises 73–76. Write each answer as a percent and as a decimal.

73. What portion of the women in the U.S. military are in the Air Force?

74. What portion of the women in the U.S. military are in the Navy?

75. (a) Which branch of the U.S. military has the lowest portion of women?

   (b) What portion is this?

$$\text{Marine Corp.} \quad 5\%$$
$$\frac{5}{100} \quad .05$$

76. (a) Which branch of the U.S. military has the highest portion of women?

   (b) What portion is this?

Army
$$\frac{45}{100}$$

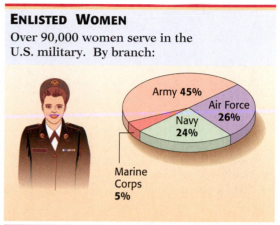

ENLISTED WOMEN

Over 90,000 women serve in the U.S. military. By branch:

Army 45%
Air Force 26%
Navy 24%
Marine Corps 5%

Source: U.S. Department of Defense

In the United States 12.7% of the population is 65 years of age or older. The bar graph shows the countries with the highest percent of people 65 years of age or older. Use this graph to answer Exercises 77–80. Write each answer as a percent and as a decimal.

77. What portion of the population of Spain is 65 or older?

78. What portion of the population of Greece is 65 or older?

79. (a) Which two countries in the graph have the lowest portion of the population 65 or older?

   (b) What portion is this?

80. (a) Which country has the highest portion of the population 65 or older?

   (b) What portion is this?

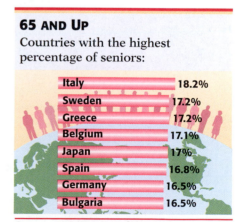

65 AND UP

Countries with the highest percentage of seniors:

Country	Percent
Italy	18.2%
Sweden	17.2%
Greece	17.2%
Belgium	17.1%
Japan	17%
Spain	16.8%
Germany	16.5%
Bulgaria	16.5%

Source: U.S. Bureau of the Census International Programs Center.

Write a percent for both the shaded and unshaded parts of each figure.

81.

10 × 10 = 100

$\frac{95}{100}$

82.

83.

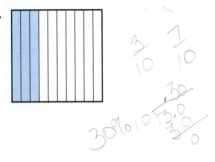

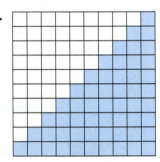

$\frac{3}{10}$ $\frac{7}{10}$

30% 10⟌3.0

84.

85.

86.

6.2 Percents and Fractions

OBJECTIVES

1. Write percents as fractions.
2. Write fractions as percents.
3. Use the table of percent equivalents.

OBJECTIVE 1 Write percents as fractions. Percents can be written as fractions by using what we learned in the previous section.

Writing a Percent as a Fraction

$$p\% = \frac{p}{100}, \quad \text{as a fraction.}$$

EXAMPLE 1 Writing Percents as Fractions

Write each percent as a fraction or mixed number in lowest terms.

(a) 25%

As we saw in the last section, 25% can be written as a decimal.

$$25\% = 25 \div 100 = 0.25 \quad \text{Percent sign dropped}$$

Because 0.25 means 25 hundredths,

$$0.25 = \frac{25}{100} = \frac{25 \div 25}{100 \div 25} = \frac{1}{4} \quad \text{Lowest terms}$$

It is not necessary, however, to write 25% as a decimal first. Just write

$$25\% = \frac{25}{100} \quad \text{25 per 100}$$

$$= \frac{1}{4} \quad \text{Lowest terms}$$

(b) 76%

The percent becomes the numerator.

Write 76% as $\frac{76}{100}$

The *denominator* is always 100 because percent means *parts per 100.*

Write $\frac{76}{100}$ in lowest terms.

$$\frac{76 \div 4}{100 \div 4} = \frac{19}{25} \quad \text{Lowest terms}$$

(c) 150%

$$150\% = \frac{150}{100} = \frac{150 \div 50}{100 \div 50} = \frac{3}{2} = 1\frac{1}{2} \quad \text{Mixed number}$$

NOTE
Remember that percent means *per 100.*

Work Problem 1 at the Side ▶▶▶

1 Write each percent as a fraction or mixed number in lowest terms.

(a) 50%

$50 \div 100 =$

$\frac{5}{10} = \frac{1}{2}$ .5

(b) 75%

(c) 48%

(d) 23%

(e) 125%

$125 \div 100 = 1.25$

$1. \frac{25}{100} \frac{5}{20} \frac{1}{4}$

(f) 250%

2 Write each percent as a fraction in lowest terms.

(a) 37.5%

[handwritten: 375/1000, 75/200, 15/40, 3/8, 37.5/100, 37.5)37.5, 300, 75]

(b) 62.5%

(c) 4.5%

(d) $66\frac{2}{3}\%$

[handwritten: $66\frac{2}{3}$ over 100, 200/3 over 100, 200 ÷ 100, 200/3, 200/3 × 1/100 = 200/300]

(e) $10\frac{1}{3}\%$

(f) $87\frac{1}{2}\%$

The next example shows how to write decimal and fraction percents as fractions.

EXAMPLE 2 **Writing Decimal or Fraction Percents as Fractions**

Write each percent as a fraction in lowest terms.

(a) 15.5%

Write 15.5 over 100

$$15.5\% = \frac{15.5}{100}$$

[handwritten: .155, 155/1000]

To get a whole number in the numerator, multiply the numerator and denominator by 10. (Recall that multiplying by $\frac{10}{10}$ is the same as multiplying by 1.)

$$\frac{15.5}{100} = \frac{15.5(10)}{100(10)} = \frac{155}{1000}$$

Now write the fraction in lowest terms.

$$\frac{155 \div 5}{1000 \div 5} = \frac{31}{200}$$

(b) $33\frac{1}{3}\%$

Write $33\frac{1}{3}$ over 100

$$33\frac{1}{3}\% = \frac{33\frac{1}{3}}{100}$$

[handwritten: 33.3, 3)1.00, 9, 10, 33.3, 33/100, 33 3/10]

When we have a mixed number in the numerator, we must write the mixed number as an improper fraction.

Here, we write $33\frac{1}{3}$ as the improper fraction $\frac{100}{3}$

$$\frac{33\frac{1}{3}}{100} = \frac{\frac{100}{3}}{100}$$

Now rewrite the division problem in a horizontal form. Then we multiply by the reciprocal of the divisor.

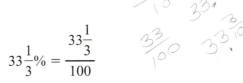

Multiply by the reciprocal.

$$\frac{\frac{100}{3}}{100} = \frac{100}{3} \div 100 = \frac{100}{3} \div \frac{100}{1} = \frac{\overset{1}{\cancel{100}}}{3} \cdot \frac{1}{\underset{1}{\cancel{100}}} = \frac{1}{3}$$

NOTE
In Example 2(a) at the top of the page we could have changed 15.5% to $15\frac{1}{2}\%$ and then written it as the improper fraction $\frac{31}{2}$ over 100, as was done in part (b). It is usually easier not to change decimals to fractions but to leave decimal percents as they are.

◀◀◀ Work Problem 2 at the Side.

OBJECTIVE 2 Write fractions as percents. We will use the formula given at the beginning of this section to write fractions as percents.

$$p\% = \frac{p}{100}$$

EXAMPLE 3 Writing Fractions as Percents

Write each fraction as a percent. Round to the nearest tenth if necessary.

(a) $\dfrac{3}{5}$

$$5\overline{)3.00}\quad.6$$

Write $\frac{3}{5}$ as a percent by solving for p in the proportion below.

$$\frac{3}{5} = \frac{p}{100}$$

Find cross products and show that they are equivalent.

$$5 \cdot p = 3 \cdot 100$$
$$5 \cdot p = 300$$

Divide both sides by 5.

$$\frac{\overset{1}{\cancel{5}} \cdot p}{\underset{1}{\cancel{5}}} = \frac{300}{5}$$

$$p = 60$$

This result means that $\frac{3}{5} = \frac{60}{100}$ or 60%.

> **NOTE**
> Solving proportions can be reviewed in **Section 5.4.**

(b) $\dfrac{7}{8}$

Write a proportion.

$$\frac{7}{8} = \frac{p}{100}$$

$$8 \cdot p = 7 \cdot 100 \qquad \text{Show that cross products are equivalent.}$$

$$8 \cdot p = 700$$

$$\frac{\overset{1}{\cancel{8}} \cdot p}{\underset{1}{\cancel{8}}} = \frac{700}{8} \qquad \text{Divide both sides by 8.}$$

$$p = 87.5$$

Finally, $\frac{7}{8} = 87.5\%$.

> **NOTE**
> If you think of $\frac{700}{8}$ as an improper fraction, changing it to a mixed number gives an answer of $87\frac{1}{2}$. So $\frac{7}{8} = 87.5\%$ or $87\frac{1}{2}\%$.

Continued on Next Page

3 Write as percents. Round to the nearest tenth if necessary.

(a) $\frac{1}{4}$

$$\frac{1}{4} = \frac{p}{100}$$

$$1 \cdot 100 = 4 \cdot p$$

$$\frac{100}{4} = \frac{4 \cdot p}{4} = 25\%$$

(b) $\frac{3}{10}$

(c) $\frac{6}{25}$

(d) $\frac{5}{8}$

$$\frac{5}{8} = \frac{p}{100}$$

$$p \cdot 8 = \frac{500}{8} = 62.5\%$$

(e) $\frac{1}{6}$

(f) $\frac{2}{9}$

$$\frac{2}{9} = \frac{p}{100}$$

$$9 \cdot p = \frac{2 \cdot 100}{9} = 22.2\%$$

(c) $\frac{5}{6}$

Start with a proportion.

$$\frac{5}{6} = \frac{p}{100}$$

$$6 \cdot p = 5 \cdot 100 \quad \text{Show that cross products are equivalent.}$$

$$6 \cdot p = 500$$

$$\frac{\overset{1}{\cancel{6}} \cdot p}{\underset{1}{\cancel{6}}} = \frac{500}{6} \quad \text{Divide both sides by 6.}$$

$$p = 83.\overline{3} \quad \text{The answer is a repeating decimal.}$$

$$p \approx 83.3 \quad \text{Round to the nearest tenth.}$$

Solving this proportion shows

$$\frac{5}{6} = 83.\overline{3}\% \approx 83.3\% \quad \text{(rounded)}.$$

NOTE
You can treat $\frac{500}{6}$ as an improper fraction to get an exact answer of $83\frac{1}{3}\%$.

Work Problem 3 at the Side.

OBJECTIVE 3 **Use the table of percent equivalents.** The more you work with fractions and mixed numbers and their decimal and percent equivalents, the more familiar you will become with them.

The table on the next two pages shows common and not so common fractions and mixed numbers and their decimal and percent equivalents.

EXAMPLE 4 **Using the Table of Percent Equivalents**

Read the following from the table.

(a) $\frac{1}{12}$ as a percent
Find $\frac{1}{12}$ in the "fraction" column. The equivalent percent is 8.3% (rounded) or $8\frac{1}{3}\%$ (exact).

(b) 0.375 as a fraction
Look in the "decimal" column for 0.375. The equivalent fraction is $\frac{3}{8}$.

(c) $\frac{13}{16}$ as a percent
Find $\frac{13}{16}$ in the "fraction" column. The equivalent percent is 81.25% or $81\frac{1}{4}\%$.

Calculator Tip Example 4 (c) above could be solved on the calculators as shown below.

13 ÷ 16 = 0.8125 × 100 = **81.25**

Multiplying by 100 changes the decimal to a percent.
Or, if your calculator has a percent key, follow these steps.

13 ÷ 16 % **81.25**

Press % key instead of = key.

On scientific calculators, you may need to press the **2nd** key to access the % function.

> **NOTE**
> When a fraction like $\frac{1}{12}$ is changed to a decimal, it is a repeating decimal that goes on forever, 0.083333. . . . In the table below these decimals are rounded to the nearest thousandth. When the decimal is then changed to a percent, it will be to the nearest tenth of a percent. Decimals that do not repeat are usually not rounded.

Work Problem 4 at the Side. ▶▶▶

Percent, Decimal, and Fraction Equivalents

Percent (rounded to tenths when necessary)	Decimal	Fraction
1%	0.01	$\frac{1}{100}$
2%	0.02	$\frac{1}{50}$
4%	0.04	$\frac{1}{25}$
5%	0.05	$\frac{1}{20}$
6.25% or $6\frac{1}{4}$%	0.0625	$\frac{1}{16}$
8.3% (rounded) or $8\frac{1}{3}$% (exact)	$0.08\overline{3}$ rounds to 0.083	$\frac{1}{12}$
10%	0.1	$\frac{1}{10}$
12.5% or $12\frac{1}{2}$%	0.125	$\frac{1}{8}$
16.7% (rounded) or $16\frac{2}{3}$% (exact)	$0.1\overline{6}$ rounds to 0.167	$\frac{1}{6}$
18.75% or $18\frac{3}{4}$%	0.1875	$\frac{3}{16}$
20%	0.2	$\frac{1}{5}$
25%	0.25	$\frac{1}{4}$
30%	0.3	$\frac{3}{10}$
31.25% or $31\frac{1}{4}$%	0.3125	$\frac{5}{16}$
33.3% (rounded) or $33\frac{1}{3}$% (exact)	$0.\overline{3}$ rounds to 0.333	$\frac{1}{3}$
37.5% or $37\frac{1}{2}$%	0.375	$\frac{3}{8}$
40%	0.4	$\frac{2}{5}$

(continued)

4 Read the following fractions, mixed numbers, decimals, and percents from the table on this or the next page. If you already know the answer or can solve the answer quickly, don't use the table.

(a) $\frac{3}{4}$ as a percent

(b) 10% as a fraction

(c) $0.\overline{6}$ as a fraction

(d) $37\frac{1}{2}$% as a fraction

(e) $\frac{7}{8}$ as a percent

(f) $\frac{1}{2}$ as a percent

(g) $33\frac{1}{3}$% as a fraction

(h) $1\frac{3}{4}$ as a percent

ANSWERS

4. (a) 75% (b) $\frac{1}{10}$ (c) $\frac{2}{3}$ (d) $\frac{3}{8}$
   (e) 87.5% (f) 50% (g) $\frac{1}{3}$ (h) 175%

Percent, Decimal, and Fraction Equivalents (*continued*)

Percent (rounded to tenths when necessary)	Decimal	Fraction
43.75% or $43\frac{3}{4}$%	0.4375	$\frac{7}{16}$
50%	0.5	$\frac{1}{2}$
56.25% or $56\frac{1}{4}$%	0.5625	$\frac{9}{16}$
60%	0.6	$\frac{3}{5}$
62.5% or $62\frac{1}{2}$%	0.625	$\frac{5}{8}$
66.7% (rounded) or $66\frac{2}{3}$% (exact)	$0.\overline{6}$ rounds to 0.667	$\frac{2}{3}$
68.75% or $68\frac{3}{4}$%	0.6875	$\frac{11}{16}$
70%	0.7	$\frac{7}{10}$
75%	0.75	$\frac{3}{4}$
80%	0.8	$\frac{4}{5}$
81.25% or $81\frac{1}{4}$%	0.8125	$\frac{13}{16}$
83.3% (rounded) or $83\frac{1}{3}$% (exact)	$0.8\overline{3}$ rounds to 0.833	$\frac{5}{6}$
87.5% or $87\frac{1}{2}$%	0.875	$\frac{7}{8}$
90%	0.9	$\frac{9}{10}$
93.75% or $93\frac{3}{4}$%	0.9375	$\frac{15}{16}$
100%	1.0	1
110%	1.1	$1\frac{1}{10}$
125%	1.25	$1\frac{1}{4}$
133.3% (rounded) or $133\frac{1}{3}$% (exact)	$1.\overline{3}$ rounds to 1.333	$1\frac{1}{3}$
150%	1.5	$1\frac{1}{2}$
166.7% (rounded) or $166\frac{2}{3}$% (exact)	$1.\overline{6}$ rounds to 1.667	$1\frac{2}{3}$
175%	1.75	$1\frac{3}{4}$
200%	2.0	2

(handwritten annotation near 0.8125 row: $\frac{8125}{10000}$ $\frac{325}{400}$ $\frac{13}{16}$)

6.2 **Exercises**

FOR EXTRA HELP   Addison-Wesley Math Tutor Center   MathXL   Digital Video Tutor CD 4 Videotape 10   Student's Solutions Manual   MyMathLab   Interactmath.com

Write each percent as a fraction or mixed number in lowest terms. See Examples 1 and 2.

1. 25%   **2.** 30%   **3.** 75%   **4.** 80%

5. 85%   **6.** 45%   **7.** 62.5%   **8.** 87.5%

9. 6.25%   **10.** 43.75%   **11.** $16\frac{2}{3}\%$ **12.** $66\frac{2}{3}\%$

13. $6\frac{2}{3}\%$ **14.** $46\frac{2}{3}\%$ **15.** 0.5%   **16.** 0.8%

17. 180%   **18.** 140%   **19.** 375%   **20.** 225%

Write each fraction as a percent. Round percents to the nearest tenth if necessary.
See Example 3.

21. $\frac{1}{2}$ **22.** $\frac{4}{10}$ **23.** $\frac{4}{5}$ **24.** $\frac{3}{10}$

25. $\frac{7}{10}$ **26.** $\frac{3}{4}$ **27.** $\frac{37}{100}$ **28.** $\frac{63}{100}$

29. $\frac{5}{8}$ **30.** $\frac{1}{8}$ **31.** $\frac{7}{8}$ **32.** $\frac{3}{8}$

33. $\frac{12}{25}$ **34.** $\frac{15}{25}$ **35.** $\frac{23}{50}$ **36.** $\frac{18}{50}$

37. $\dfrac{7}{20}$ **38.** $\dfrac{9}{20}$ **39.** $\dfrac{5}{6}$ **40.** $\dfrac{1}{6}$

41. $\dfrac{5}{9}$ **42.** $\dfrac{7}{9}$ **43.** $\dfrac{1}{7}$ **44.** $\dfrac{5}{7}$

Complete the chart. Round decimals to the nearest thousandth and percents to the nearest tenth if necessary. See Examples 3 and 4.

	Fraction	Decimal	Percent
45.	_____	0.5	_____
46.	$\dfrac{1}{4}$	_____	_____
47.	_____	_____	87.5%
48.	$\dfrac{3}{4}$	_____	_____
49.	_____	0.8	_____
50.	_____	_____	60%
51.	$\dfrac{1}{6}$	_____	_____
52.	$\dfrac{1}{3}$	_____	_____
53.	_____	0.7	_____

	Fraction	**Decimal**	**Percent**
54.	_____	_____	37.5%
55.	_____	_____	12.5%
56.	_____	0.625	_____
57.	$\dfrac{2}{3}$	_____	_____
58.	$\dfrac{5}{6}$	_____	_____
59.	$\dfrac{3}{50}$	_____	_____
60.	$\dfrac{3}{10}$	_____	_____
61.	$\dfrac{8}{100}$	_____	_____
62.	_____	_____	100%
63.	$\dfrac{1}{200}$	_____	_____
64.	$\dfrac{1}{400}$	_____	_____

Fraction	Decimal	Percent
65. _____	2.5	_____
66. _____	1.7	_____
67. $3\frac{1}{4}$	_____	_____
68. $2\frac{4}{5}$	_____	_____

69. Select a decimal percent and write it as a fraction. Select a different fraction and write it as a percent. Write an explanation of each step of your work.

70. Prepare a table showing fraction, decimal, and percent equivalents for five fractions and mixed numbers of your choice.

In the following application problems, write the answer as a fraction in lowest terms, as a decimal, and as a percent.

71. Many pet owners say they have used the Internet to find pet information. Of 500 people who used the Internet for this purpose, 90 said they used it when buying a pet. What portion used the Internet when buying a pet? (*Source:* American Animal Hospital Association.)

72. About $\frac{1}{3}$ of all books purchased last year were for children. Of these children's books, 27 of every 100 purchased included a coloring activity. What portion of the children's books included a coloring activity? (*Source:* Consumer Research Study on Book Publishing, the American Booksellers, and the Book Industry Study Group.)

73. Only 13 out of every 100 adults consume the recommended 1000 milligrams of calcium daily. What portion consumes the recommended daily amount? (*Source:* Market Facts for Milk Mustache.)

74. In a survey on how people learn to parent, 360 parents out of 800 said they were most influenced by relatives, friends, and spouses. What portion learned to parent this way? (*Source:* Bama Research.)

75. Of the 15 people at the office, 9 are single parents. What portion are single parents?

76. A pizza shop has 25 employees. Of these, 14 are students. What portion are students?

77. An insurance office has 80 employees. If 64 of the employees have cellular phones, what portion of the employees do *not* have cellular phones?

78. A zoo has 125 types of animals, including 25 that are endangered. What portion is *not* endangered?

79. An antibiotic is used to treat 380 people. If 342 people do not have side effects from the antibiotic, find the portion that do have side effects.

80. An apple grower's co-operative has 250 members. If 100 of the growers use a certain insecticide, find the portion that do *not* use the insecticide.

The circle graph shows how much the 10,800 students at a college expect to earn this summer. Use this graph to answer Exercises 81–84, giving the answer as a fraction, as a decimal, and as a percent.

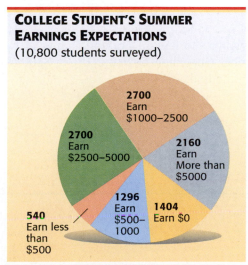

COLLEGE STUDENT'S SUMMER EARNINGS EXPECTATIONS
(10,800 students surveyed)

2700 Earn $1000–2500
2700 Earn $2500–5000
2160 Earn More than $5000
540 Earn less than $500
1296 Earn $500–1000
1404 Earn $0

Source: U.S. Collegeclub.com

81. What portion of the students expect to earn $2500 to $5000?

82. What portion of the students expect to earn $500 to $1000?

83. Find the portion who expect to earn less than $500.

84. Find the portion who expect to earn more than $5000.

Understanding the basics of percent is an important topic in mathematics. **Work Exercises 85–94 in order.**

85. 100% of a number means all of the parts or

_____ parts out of _____ parts. 200%

means two times as many parts and 300% means three times as many parts.

86. Fill in the blanks.

    (a) 100% of 765 workers is _____.

    (b) 200% of 48 letters is _____.

    (c) 300% of 7 videos is _____.

87. 50% of a number is _____ parts out of _____

parts, which is _____ of the parts.

88. 10% of a number is _____ parts out of _____

parts and is found by moving the decimal point

_____ place to the _____.

89. 1% of a number is _____ part out of _____

parts and is found by moving the decimal point

_____ places to the _____.

90. Fill in the blanks.

    (a) 50% of 1050 homes is _____.

    (b) 10% of 370 printers is _____.

    (c) 1% of $8 sales is _____.

Using the shortcut methods for finding 1%, 10%, 50%, 100%, 200%, *and* 300%, *answer Exercises 91–94.*

91. Devise a shortcut method for finding 15% of a number. Use your method to find 15% of 160.

92. Devise a shortcut method for finding 150% of a number. Use your method to find 150% of 160.

93. Explain a shortcut method for finding 90% of a number. Show how to find 90% of $450 using your method.

94. Explain a shortcut method for finding 210% of a number. Show how to find 210% of $800 using your method.

6.3 Using the Percent Proportion and Identifying the Components in a Percent Problem

There are two ways to solve percent problems. One method uses proportions and is discussed in this and the next section. The other method uses the percent equation and is explained in **Section 6.5.**

OBJECTIVE 1 Learn the percent proportion. We have seen that a statement of two equivalent ratios is called a proportion.

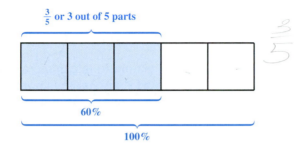

$\frac{3}{5}$ or 3 out of 5 parts

60%

100%

For example, the fraction $\frac{3}{5}$ is the same as the ratio 3 to 5, and 60% is the same as the ratio 60 to 100. As the figure above shows, these two ratios are equivalent and make a proportion.

Work Problem 1 at the Side. ▶▶▶

The percent proportion can be used to solve percent problems.

Percent Proportion
Part is to *whole* as *percent* is to *100*.

$$\frac{\textbf{part}}{\textbf{whole}} = \frac{\textbf{percent}}{\textbf{100}} \quad \leftarrow \text{Always 100 because percent means per 100}$$

In the figure at the top of the page, the **whole** is 5 (the entire quantity), the **part** is 3 (the part of the whole), and the **percent** is 60. Write the percent proportion as follows.

$$\frac{\textbf{part} \rightarrow 3}{\textbf{whole} \rightarrow 5} = \frac{60 \leftarrow \textbf{percent}}{100 \leftarrow \textbf{100}}$$

OBJECTIVE 2 Solve for an unknown value in a percent proportion. As shown in **Section 5.4**, if any three of the four values in a proportion are known, the fourth can be found by solving the proportion.

EXAMPLE 1 Using the Percent Proportion

Use the percent proportion and solve for the unknown value. Let x represent the unknown value.

(a) part = 12, percent = 25; find the whole.

$$\frac{\text{part}}{\text{whole}} = \frac{\text{percent}}{100} \quad \textcolor{blue}{\text{Percent proportion}}$$

Percent

$$\text{Part} \rightarrow \frac{12}{\text{Whole (unknown)} \rightarrow x} = \frac{25}{100} \quad \text{or} \quad \frac{12}{x} = \frac{1}{4} \quad \tfrac{25}{100} \text{ is } \tfrac{1}{4} \text{ in lowest terms.}$$

Continued on Next Page

OBJECTIVES

1 Learn the percent proportion.

2 Solve for an unknown value in a percent proportion.

3 Identify the percent.

4 Identify the whole.

5 Identify the part.

1 As a review of proportions, use the method of comparing cross products to decide whether each proportion is *true* or *false.*

(a) $\dfrac{1}{2} = \dfrac{25}{50}$

(b) $\dfrac{3}{4} = \dfrac{150}{200}$

(c) $\dfrac{7}{8} = \dfrac{180}{200}$

(d) $\dfrac{32}{53} = \dfrac{160}{265}$

(e) $\dfrac{112}{41} = \dfrac{332}{123}$

2 Use the percent proportion $\left(\dfrac{\text{part}}{\text{whole}} = \dfrac{\text{percent}}{100}\right)$ and solve for the unknown value.

(a) part = 12, percent = 16

(b) part = 30, whole = 120

(c) whole = 210, percent = 20

(d) whole = 4000, percent = 32

(e) part = 74, whole = 185

Find the cross products to solve this proportion.

$$\frac{12}{x} = \frac{1}{4}$$

with cross products $x \cdot 1$ and $12 \cdot 4$

Show that the cross products are equivalent.

$$x \cdot 1 = 12 \cdot 4$$
$$x = 48$$

The whole is 48.

> **CAUTION**
> You cannot divide out common factors from a numerator and denominator in a proportion. This can be done *only* when the fractions are being multiplied.

(b) part = 30, whole = 50; find the percent.
Use the percent proportion.

$$\underset{\text{Whole} \to}{\overset{\text{Part} \to}{}}\ \frac{30}{50} = \frac{x}{100} \underset{\text{(unknown)}}{\overset{\text{Percent}}{\downarrow}} \qquad \text{Percent proportion}$$

$$\frac{3}{5} = \frac{x}{100} \qquad \tfrac{30}{50} \text{ is } \tfrac{3}{5} \text{ in lowest terms.}$$

$$5 \cdot x = 3 \cdot 100 \qquad \text{Find cross products.}$$

$$5 \cdot x = 300$$

$$\frac{\overset{1}{\cancel{5}} \cdot x}{\underset{1}{\cancel{5}}} = \frac{300}{5} \qquad \text{Divide both sides by 5.}$$

$$x = 60$$

The percent is 60, written as 60%.

(c) whole = 150, percent = 18; find the part.

$$\underset{\text{Whole} \to}{\overset{\text{Part (unknown)} \to}{}}\ \frac{x}{150} = \frac{18}{100} \overset{\text{Percent}}{\downarrow} \quad \text{or} \quad \frac{x}{150} = \frac{9}{50} \qquad \tfrac{18}{100} \text{ is } \tfrac{9}{50} \text{ in lowest terms.}$$

$$x \cdot 50 = 150 \cdot 9 \qquad \text{Find cross products.}$$

$$x \cdot 50 = 1350$$

$$\frac{x \cdot \overset{1}{\cancel{50}}}{\underset{1}{\cancel{50}}} = \frac{1350}{50} \qquad \text{Divide both sides by 50.}$$

$$x = 27$$

The part is 27.

◀◀◀ Work Problem 2 at the Side.

As a help in solving percent problems, keep in mind this basic idea.

> ### Percent Problems
>
> All percent problems involve a comparison between a part of something and the whole.

Solving these problems requires identifying the three components of a percent proportion: part, whole, and percent.

OBJECTIVE 3 Identify the percent.   Look for the percent first. It is the easiest to identify.

> ### Percent
>
> The **percent** is the ratio of a part to a whole, with 100 as the denominator. In a problem, the percent appears with the word **percent** or with the symbol **"%"** after it.

EXAMPLE 2 Finding the Percent in Percent Problems

Find the percent in the following.

(a) 32% of the 900 men were retired.

   Percent

The percent is 32. The number 32 appears with the symbol %.

(b) $150 is 25 percent of what number?

   Percent

The percent is 25 because 25 appears with the word *percent*.

(c) What percent of the 350 women will go?

   Percent (unknown)

The word *percent* has no number with it, so the percent is the unknown part of the problem.

Work Problem 3 at the Side. ▷▷▷

OBJECTIVE 4 Identify the whole.   Next, look for the whole.

> ### Whole
>
> The **whole** is the entire quantity. In a percent problem, the whole often appears after the word **of**.

---

3 Identify the percent.

(a) Of the 900 blood glucose tests, 25% will be completed by Adrian.

Whole = 900
perc = 25%

$$\frac{P}{W} = \frac{P\%}{100} \qquad \frac{x}{900} \qquad \frac{25\%}{100}$$

$$x \cdot 100 = 25 \times 900$$

(b) Of the 620 preschool students, 65% will be served breakfast and lunch.

(c) Find the amount of sales tax by multiplying $590 and $6\frac{1}{2}$ percent.

6.5%

(d) 105 is 3% of what number?

(e) What percent of the 380 guests will return this year?

4 Identify the whole.

(a) Of the 900 blood glucose tests, 25% will be completed by Adrian.

(b) Of the 620 preschool students, 65% will be served breakfast and lunch.

(c) Find the amount of sales tax by multiplying sales of $590 and $6\frac{1}{2}$ percent.

(d) $105 is 3% of what number?

(e) What percent of the 380 guests will return this year?

5 Identify the part, then set up the percent proportion.

(a) Of the 900 blood glucose tests, 25% or 225 will be completed by Adrian.

(b) Of the 620 preschool students, 65% or 403 will be served breakfast and lunch.

(c) Find the sales tax by multiplying $590 and $6\frac{1}{2}$ percent.

(d) $105 is 3% of what number?

(e) 80% of the 380 guests will return this year.

ANSWERS

4. **(a)** 900 **(b)** 620 **(c)** $590
   (d) what number (an unknown) **(e)** 380

5. **(a)** 225; $\dfrac{\text{Part} \to 225}{\text{Whole} \to 900} = \dfrac{25 \leftarrow \text{Percent}}{100 \leftarrow \text{Always 100}}$

   (b) 403; $\dfrac{\text{Part} \to 403}{\text{Whole} \to 620} = \dfrac{65 \leftarrow \text{Percent}}{100 \leftarrow \text{Always 100}}$

   (c) Unknown;

   $\dfrac{\text{Part} \to \text{unknown}}{\text{Whole} \to 590} = \dfrac{6\frac{1}{2} \leftarrow \text{Percent}}{100 \leftarrow \text{Always 100}}$

   (d) 105;

   $\dfrac{\text{Part} \to 105}{\text{Whole} \to \text{unknown}} = \dfrac{3 \leftarrow \text{Percent}}{100 \leftarrow \text{Always 100}}$

   (e) unknown;

   $\dfrac{\text{Part} \to \text{unknown}}{\text{Whole} \to 380} = \dfrac{80 \leftarrow \text{Percent}}{100 \leftarrow \text{Always 100}}$

EXAMPLE 3 Finding the Whole in Percent Problems

Identify the whole in the following.

(a) 32% **of** the 900 men were too large for the imported car.

Whole

The whole is 900. The number 900 appears after the word *of*.

(b) $150 is 25 percent **of** what number?

Whole   The whole is the unknown part of the problem.

(c) 85% **of** 7000 is what number?

Whole

Work Problem 4 at the Side.

OBJECTIVE 5 Identify the part.   Finally, look for the part.

Part
The **part** is the portion being compared with the whole.

NOTE
If you have trouble identifying the part, find the percent and whole first. The remaining number is the part.

EXAMPLE 4 Finding the Part in Percent Problems

Identify the part in the following. Then set up the percent proportion.

(a) 54% **of** 700 students is 378 students.
   First find the percent and the whole.

   54% **of** 700 students is 378 students.

   Percent; with % sign   Whole; follows "of"

   The remaining number, 378, is the part.

   54% **of** 700 students is 378 students. $\dfrac{\text{Part} \to 378}{\text{Whole} \to 700} = \dfrac{54 \leftarrow \text{Percent}}{100 \leftarrow \text{Always 100}}$

   Percent   Whole   Part

(b) $150 is 25% **of** what number? $\dfrac{\text{Part} \to 150}{\text{Whole} \to \text{unknown}} = \dfrac{25 \leftarrow \text{Percent}}{100 \leftarrow \text{Always 100}}$

   Percent   Whole (unknown)

   $150 is the remaining number, so the part is $150.

(c) 85% **of** 7000 is what number? $\dfrac{\text{Part} \to \text{unknown}}{\text{Whole} \to 7000} = \dfrac{85 \leftarrow \text{Percent}}{100 \leftarrow \text{Always 100}}$

   Percent   Whole   Part (unknown)

Work Problem 5 at the Side.

6.3 Exercises

FOR EXTRA HELP
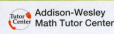 Tutor Center Addison-Wesley Math Tutor Center
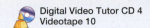 MathXL
 Digital Video Tutor CD 4 Videotape 10
Student's Solutions Manual
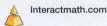 MyMathLab
Interactmath.com

Find the unknown value in the percent proportion $\dfrac{part}{whole} = \dfrac{percent}{100}$. Round to the nearest tenth if necessary. If the answer is a percent, be sure to include a percent sign (%). See Example 1.

1. part = 5, percent = 10

2. part = 20, percent = 25

3. part = 30, percent = 20

4. part = 25, percent = 25

5. part = 28, percent = 40

6. part = 11, percent = 5

7. part = 15, whole = 60

8. part = 105, whole = 35

9. part = 36, whole = 24

10. part = 1.5, whole = 4.5

11. part = 9.25, whole = 27.75

12. part = 12.8, whole = 9.6

13. whole = 52, percent = 50

14. whole = 160, percent = 35

15. whole = 72, percent = 30

16. whole = 115, percent = 38

17. whole = 94.4, part = 25

18. whole = 89.6, part = 50

$$\frac{part}{whole} = \frac{P}{100} \qquad \frac{25}{94.4} = \frac{x}{100}$$

$$\frac{94.4 \cdot x}{94.4} = \frac{2500}{94.4} = \frac{26.48}{26.5}$$

Solve each problem. If the answer is a percent, be sure to include a percent sign (%). See Examples 2–4.

19. Find the whole if the part is 46 and the percent is 40.

20. The percent is 45 and the whole is 160. Find the part.

21. The whole is 5000 and the part is 20. Find the percent.

22. Suppose the part is 15 and the whole is 2500. Find the percent.

23. Find the percent if the whole is 4300 and the part is $107\frac{1}{2}$.

24. What is the part, if the percent is $12\frac{3}{4}$ and the whole is 5600?

25. The whole is 6480 and the part is 19.44. Find the percent.

26. Suppose the part is 281.25 and the percent is $1\frac{1}{4}$. Find the whole.

*In Exercises 27–42, set up the percent proportion, and write "unknown" for the value that is not given. Recall that the percent proportion is $\dfrac{part}{whole} = \dfrac{percent}{100}$. Do **not** try to solve for any unknowns. See Examples 2–4.*

27. 10% of how many bicycles is 60 bicycles?

28. 58% of how many preschoolers is 203 preschoolers?

29. 75% of $800 is $600.

30. 93% of $1500 is $1395.

31. What is 25% of $970?

32. What is 61% of 830 homes?

33. 12 injections is 20% of what number of injections?

34. 92 servings is 26% of what number of servings?

35. 34 trophies is 50% of 68 trophies.

36. 410 pallets is $33\frac{1}{3}$% of 1230 pallets.

37. What percent of $296 is $177?

38. What percent of $121 is $30?

39. 54.34 is 3.25% of what number?

40. 16.74 is 11.9% of what number?

41. 0.68% of $487 is what amount?

42. What amount is 6.21% of $704.35?

43. Identify the three components in a percent problem. In your own words, write one sentence telling how you will identify each of these three components.

44. Write one short sentence or statement using numbers and words. The statement should include a percent, a whole, and a part. Identify each of these three components.

*Set up the percent proportion for each application problem. Do **not** try to solve for any unknowns.*

45. In a tree-planting project, 640 of the 810 trees planted were still living one year later. What percent of the trees planted were still living?

$$\frac{640}{810} = \frac{x}{100} \qquad x \cdot 810 = 64000 \quad 79.01$$

46. Ivory Soap is $99\frac{44}{100}\%$ pure. If a bar of Ivory Soap weighs 9 ounces, how many ounces are pure? (*Source:* Procter & Gamble.)

47. Of the 142 people attending a movie theater, 86 bought buttered popcorn. What percent bought buttered popcorn?

48. On her first check from the Pizza Hut Restaurant, 15% is withheld from Maria's total earnings of $225. What amount is withheld?

49. Of the customers buying a salad at McDonald's, 23% prefer Newman's Light Salad Dressing of salad customers is 610, find the number who prefer Newman's Light Dressing.

50. Of the total candy bars contained in a vending machine, 240 bars have been sold. If 25% of the bars have been sold, find the total number of candy bars that were in the machine.

51. There are 680 computer chips in a secured storage area designed to hold 2000 computer chips. What percent of the storage area is filled?

52. There have been 36 cups of coffee served from a banquet-sized coffee pot. If this is 30% of the capacity of the pot, find the capacity of the pot.

53. In a recent survey of 480 adults, 55% said that they would prefer to have their wedding at a religious site. How many said they would prefer the religious site? (*Source:* National Family Opinion Research.)

54. Sue Ann needs 64 credits to graduate. If she has completed 48 of the credits needed, what percent of the credits has she already completed?

55. In a poll of 822 people, 49.5% said that they get their news from television. Find the number of people who said they get their news from television. (*Source: Brills Content.*)

56. The sales tax on a new car is $820. If the sales tax rate is 5%, find the price of the car before the sales tax is added.

57. A medical clinic found that 16.8% of the patients were late for their appointments. The number of patients who were late was 504. Find the total number of patients.

58. The state troopers tested 924 cars for safety. There were 231 cars that failed the safety test for one or more reasons. Find the percent of cars that failed the test.

59. Jerry Azzaro has listed 680 antique toys on eBay. If 45% of these were antique toy trains, find the number that were toy trains.

60. There are 1470 students at American River College who have purchased textbooks from amazon.com. If the number of students at the college is 24,650, what percent have made textbook purchases from amazon.com?

6.4 Using Proportions to Solve Percent Problems

This is the percent proportion that you learned about in the previous section.

$$\frac{\text{part}}{\text{whole}} = \frac{\text{percent}}{100}$$

Recall that if one of the vaues is unknown, you can find it by solving the percent proportion.

OBJECTIVES

1 Use the percent proportion to find the part.

2 Find the whole using the percent proportion.

3 Find the percent using the percent proportion.

OBJECTIVE **1** **Use the percent proportion to find the part.** The first example shows how to use the percent proportion to find the part.

EXAMPLE 1 **Finding the Part with the Percent Proportion**

Find 15% of $160.

$\frac{P}{W} = \frac{P}{100\%}$ $\frac{x}{160}$ $\frac{15}{100}$

Here the percent is 15 and the whole is 160. (Recall that the whole often comes after the word *of.*) Now find the part. Let x represent the unknown part.

$$\frac{\text{part}}{\text{whole}} = \frac{\text{percent}}{100} \quad \text{so} \quad \frac{x}{160} = \frac{15}{100} \quad \text{or} \quad \frac{x}{160} = \frac{3}{20} \quad \substack{\frac{15}{100} \text{ is } \frac{3}{20} \text{ in lowest} \\ \text{terms.}}$$

Find the cross products in the proportion and show that they are equivalent.

$$x \cdot 20 = 160 \cdot 3 \quad \text{Cross products}$$

$$x \cdot 20 = 480$$

$$\frac{x \cdot \overset{1}{\cancel{20}}}{\underset{1}{\cancel{20}}} = \frac{480}{20} \quad \text{Divide both sides by 20.}$$

$$x = 24 \quad \text{The unknown part is 24.}$$

15% of $160 is **$24**.

Work Problem 1 at the Side.

Just as with the fraction application problems in **Section 2.6** the word *of* is an indicator word meaning *multiply.* Here is an example.

$$15\% \text{ of } 160$$
$$\downarrow$$
$$15\% \cdot 160$$

Because of this, there is another way to find the part.

Finding the Part Using Multiplication

To find the part:

Step 1 Identify the percent. Write the percent as a decimal. $.15$

Step 2 Multiply this decimal by the whole. $.15 \times 160$

1 Use the percent proportion to find the part.

(a) 10% of 1250 allergy tests

$\frac{x}{1250}$ $\frac{10}{100}$

$x \cdot 100 = 10 \cdot 1250$

125 $\frac{100}{100}$ $\frac{12500}{100}$

(b) 15% of $3220

(c) 7% of 2700 miles

(d) 39% of 1220 meters

2 Use multiplication to find the part.

(a) 55% of 10,000 injections

(b) 16% of 120 miles

(c) 135% of 60 dosages

(d) 0.5% of $238

EXAMPLE 2 Finding the Part Using Multiplication

Use multiplication to find the part.

(a) Find 42% of 830 yards.

> *Step 1*   Here, the percent is 42. Write 42% as the decimal 0.42.
>
> *Step 2*   Multiply 0.42 and the whole, which is 830.
>
> $$\text{part} = (0.42)(830)$$
> $$= 348.6 \text{ yd}$$

≈ It is a good idea to estimate the answer, to make sure no mistakes were made with decimal points. Round 42% to 40% or 0.4, and round 830 as 800. Next, 40% of 800 is

$$(0.4)(800) = 320 \leftarrow \text{Estimate}$$

so the exact answer of 348.6 is reasonable.

(b) Find 25% of 1680 cars.

Identify the percent as 25. Write 25% in decimal form as 0.25. Now, multiply 0.25 and 1680.

$$\text{part} = (0.25)(1680) = 420 \text{ cars} \quad \text{Multiply.}$$

You can also use a shortcut to find the answer. Since 25% means 25 parts out of 100 parts, this is the same as $\frac{1}{4}$ of the whole ($\frac{25}{100} = \frac{1}{4}$). Do you see a shortcut here? You can find $\frac{1}{4}$ of a number by dividing the number by 4. So, this shortcut gives us the exact answer, $1680 \div 4 = 420$.

(c) Find 140% of 60 miles.

In this problem, the percent is 140. Write 140% as the decimal 1.40. Next, multiply 1.40 and 60.

$$\text{part} = (1.40)(60) = 84 \text{ miles} \quad \text{Multiply.}$$

≈ You can estimate the answer by realizing that 140% is close to 150% (which is $1\frac{1}{2}$) and $1\frac{1}{2}$ times 60 is 90. So, 84 miles is a reasonable answer.

(d) Find 0.4% of 50 kilometers.

$$\text{part} = (0.004)(50) = 0.2 \text{ kilometer} \quad \text{Multiply.}$$
$$\uparrow$$
Write 0.4% as a decimal.

≈ Estimate the answer 0.4% is less than 1%.

$$1\% \text{ of } 50 \text{ kilometers} = \underset{\sim}{50.} = 0.5 \text{ kilometer}$$

So our exact answer should be *less than* 0.5 kilometer, and 0.2 kilometer fits this requirement.

◀◀◀ **Work Problem 2 at the Side.**

EXAMPLE 3 Solving for the Part in an Application Problem

Raley's Markets has 850 employees. Of these employees, 28% are students. How many of the employees are students?

Continued on Next Page

Step 1   **Read** the problem. The problem asks us to find the number of employees who are students.

Step 2   **Work out a plan.** Look for the word *of* as an indicator word for multiplication.

$$28\,\% \text{ of the employees are students.}$$

      — Indicator word

The total number of employees is 850, so the whole is 850. The percent is 28. To find the number of students, find the part.

Step 3   **Estimate** a reasonable answer. You can estimate the answer by rounding 28% to 25% and 850 to 900. Remember 25% is 25 parts out of 100, which is equivalent to $\frac{1}{4}$. So divide 900 by 4.

$$900 \div 4 = 225 \text{ students} \leftarrow \text{Estimate}$$

Step 4   **Solve** the problem.

$$\text{part} = (0.28)(850) = 238 \qquad \text{Multiply.}$$
          ↑
        — Write 28% as a decimal.

Step 5   **State the answer.** Raley's Markets has 238 student employees.

Step 6   **Check.** The exact answer, 238 students, is close to our estimate of 225 students.

Work Problem 3 at the Side. ▶▶▶

▦ **Calculator Tip**   If you are using a calculator, you could solve Example 3 above like this.

$$0.28 \;\otimes\; 850 \;\ominus\; \mathbf{238}$$

Or, you can use this alternate approach on calculators with a % key.

$$850 \;\otimes\; 28 \;\%\; \ominus\; \mathbf{238}$$

OBJECTIVE 2   Find the whole using the percent proportion.   The next example shows how to use the percent proportion to find the whole.

> **NOTE**
> Remember, the whole is the entire quantity.

EXAMPLE 4   Finding the Whole with the Percent Proportion

(a) 8 tables is 4% of what number of tables?
    Here the percent is 4, the whole is unknown, and the part is 8. Use the percent proportion to find the whole. Let x represent the unknown whole.

$$\frac{\text{part}}{\text{whole}} = \frac{\text{percent}}{100} \quad \text{so} \quad \frac{8}{x} = \frac{4}{100} \quad \text{or} \quad \frac{8}{x} = \frac{1}{25} \qquad \text{$\frac{4}{100}$ is $\frac{1}{25}$ in lowest terms.}$$

$$x \cdot 1 = 8 \cdot 25 \qquad \text{Cross products}$$

$$x = 200$$

8 tables is 4% of **200 tables**.

    Continued on Next Page

---

3 Use the six problem-solving steps to solve each problem.

(a) One day on Jacob's mail route there were 2920 pieces of mail. If 45% of those were advertising pieces, find the number of advertising pieces.

(b) There were 14,100 customers at the Java City Coffee Shop this month. If 23% of the customers ordered iced mochas, find the number of iced mocha customers.

ANSWERS
3. (a) 1314 advertising pieces
   (b) 3243 iced mocha customers

4 Use the percent proportion to find the unknown whole.

(a) 750 Super Lotto Tickets is 25% of what number of tickets?

(b) 28 antiques is 35% of what number of antiques?

(c) 387 customers is 36% of what number of customers?

▦ (d) 292.5 miles is 37.5% of what number of miles?

(b) 135 tourists is 15% of what number of tourists?
The percent is 15 and the part is 135, so

$$\text{Part} \rightarrow \frac{135}{x} = \frac{15}{100} \leftarrow \text{Percent}$$
$$\text{Whole (unknown)} \rightarrow \quad\quad \leftarrow \text{Always 100}$$

$$\frac{135}{x} = \frac{3}{20} \quad\quad \text{Lowest terms}$$

$$x \cdot 3 = 135 \cdot 20 \quad\quad \text{Cross products}$$

$$x \cdot 3 = 2700$$

$$\frac{x \cdot \overset{1}{\cancel{3}}}{\cancel{3}_1} = \frac{2700}{3} \quad\quad \text{Divide both sides by 3.}$$

$$x = 900.$$

135 tourists is 15% of **900 tourists**.

◀◀◀ **Work Problem 4 at the Side.**

EXAMPLE 5 **Applying the Percent Proportion**

At Newark Salt Works, 78 employees are absent because of illness. If this is 5% of the total number of employees, how many employees does the company have?

Step 1 **Read** the problem. The problem asks for the total number of employees.

Step 2 **Work out a plan.** From the information in the problem, the percent is 5 and the part of the total number of employees is 78. The total number of employees or entire quantity, which is the whole, is the unknown.

Step 3 **Estimate** a reasonable answer. Round the number of employees from 78 to 80. Then, 5% is equivalent to the fraction $\frac{1}{20}$. Since 80 is $\frac{1}{20}$ of the total number of employees.

$$80 \cdot 20 = 1600 \text{ employees} \leftarrow \text{Estimate}$$

Step 4 **Solve** the problem. Use the percent proportion to find the whole (the total number of employees).

$$\text{Part} \rightarrow \frac{78}{x} = \frac{5}{100} \leftarrow \text{Percent}$$
$$\text{Whole (unknown)} \rightarrow \quad\quad \leftarrow \text{Always 100}$$

$$\frac{78}{x} = \frac{1}{20} \quad\quad \text{Lowest terms}$$

Find the cross products.

$$x \cdot 1 = 78 \cdot 20$$

$$x = 1560$$

Step 5 **State the answer.** The company has **1560 employees**.

Step 6 **Check.** The exact answer, 1560 employees, is close to our estimate of 1600 employees.

Continued on Next Page

NOTE
To estimate the answer to Example 5 on the previous page, the 5% was changed to its fraction equivalent, $\frac{1}{20}$. Because 80 (rounded) is $\frac{1}{20}$ of the total employees, 80 was multiplied by 20 to get 1600, the estimated answer.

Work Problem 5 at the Side.

OBJECTIVE **3** **Find the percent using the percent proportion.**   Finally, if the part and the whole are known, the percent proportion can be used to find the percent.

EXAMPLE 6 **Using the Percent Proportion to Find the Percent**

(a) 13 coupons is what percent of 52 coupons?
    The whole is 52 (follows *of*) and the part is 13. Next, find the percent.

$$\frac{\text{part}}{\text{whole}} = \frac{\text{percent}}{100}$$

$$\text{Part} \rightarrow \frac{13}{52} = \frac{x}{100} \quad \begin{array}{l} \leftarrow \text{Percent (unknown)} \\ \leftarrow \text{Always 100} \end{array}$$

Lowest terms $\quad \dfrac{1}{4} = \dfrac{x}{100}$

Find the cross products.

$$4 \cdot x = 1 \cdot 100$$

$$\frac{\overset{1}{\cancel{4}} \cdot x}{\underset{1}{\cancel{4}}} = \frac{100}{4} \qquad \text{Divide both sides by 4.}$$

$$x = 25$$

13 coupons is **25%** of 52 coupons.

(b) What percent of $500 is $100?
    The whole is 500 (follows *of*) and the part is 100, so

$$\frac{100}{500} = \frac{x}{100} \quad \longleftarrow \text{Percent (unknown)}$$

Lowest terms $\qquad \dfrac{1}{5} = \dfrac{x}{100}$

$$5 \cdot x = 1 \cdot 100 \qquad \text{Cross products}$$

$$5 \cdot x = 100$$

$$\frac{\overset{1}{\cancel{5}} \cdot x}{\underset{1}{\cancel{5}}} = \frac{100}{5} \qquad \text{Divide both sides by 5.}$$

$$x = 20$$

20% of $500 is $100.

—— **Continued on Next Page**

5 Use the six problem-solving steps and the percent proportion to solve each problem.

(a) A freeze resulted in a loss of 52% of an avocado crop. If the loss was 182 tons, find the total number of tons in the crop.

(b) A metal alloy contains 450 pounds of zinc, which is 8% of the alloy. Find the total weight of the alloy.

ANSWERS
5. (a) 350 tons **(b)** 5625 pounds

6 Use the percent proportion to solve each problem.

(a) $21 is what percent of $105?

> **CAUTION**
> When finding the percent, be sure to label your answer with the percent symbol (%).

Work Problem 6 at the Side.

(b) What percent of 320 Internet companies is 48 Internet companies?

EXAMPLE 7 Applying the Percent Proportion

A roof is expected to last 20 years before needing replacement. If the roof is now 15 years old, what percent of the roof's life has been used?

Step 1 **Read** the problem. The problem asks for the percent of the roof's life that is already used.

Step 2 **Work out a plan.** The expected life of the roof is the entire quantity or *whole*, which is 20. The part of the roof's life that is already used is 15. Use the percent proportion to find the percent of the roof's life used.

(c) What percent of 2280 court trials is 1026 trials?

Step 3 **Estimate** a reasonable answer. Since the roof is 15 years old, it is $\frac{15}{20}$ or $\frac{3}{4}$ used. Remember that $\frac{3}{4}$ is equivalent to 75%, our estimate.

Step 4 **Solve** the problem. Let x represent the unknown percent.

(d) 432 snowboarders is what percent of 108 snowboarders?

$$\text{Part} \to \frac{15}{20} = \frac{x}{100} \quad \text{or} \quad \frac{3}{4} = \frac{x}{100} \qquad \text{$\frac{15}{20}$ is $\frac{3}{4}$ in lowest terms.}$$
$$\text{Whole} \to$$

Find the cross products.

$$\frac{432}{108} = \frac{x}{100}$$
$$\frac{x \cdot 108}{108} = \frac{43200}{1082} = 400$$

$$4 \cdot x = 3 \cdot 100$$
$$4 \cdot x = 300$$

$$\frac{\overset{1}{\cancel{4}} \cdot x}{\underset{1}{\cancel{4}}} = \frac{300}{4} \qquad \text{Divide both sides by 4.}$$

$$x = 75$$

7 Solve each problem.

(a) The bid price on an auction item is $289 while the minimum acceptable price is $425. The bid price is what percent of the minimum?

$$\frac{289}{425} \qquad \frac{x}{100}$$
$$68$$

Step 5 **State the answer.** **75%** of the roof's life has been used.

Step 6 **Check.** The exact answer, 75% matches our estimate of 75%.

Work Problem 7 at the Side.

(b) A late-model domestic car gets 38 miles per gallon on the highway and 32.3 miles per gallon around town. What percent of the highway mileage does the car get around town?

Whole

$$\frac{32.3}{38} \qquad \frac{x}{100} = 85$$

EXAMPLE 8 Applying the Percent Proportion

whole

Rainfall this year was 33 inches, while normal rainfall is only 30 inches. What percent of normal rainfall is this year's rainfall?

Step 1 **Read** the problem. The problem asks us to find what percent this year's rainfall is of normal rainfall.

Step 2 **Work out a plan.** The normal rainfall is the *whole*, which is 30. This year's rainfall is all of normal rainfall and more, or 33 (part = 33). You need to find the percent that this year's rainfall is of normal rainfall.

ANSWERS
6. (a) 20% (b) 15% (c) 45% (d) 400%
7. (a) 68% (b) 85%

Continued on Next Page

Step 3   **Estimate** a reasonable answer. The increase in rainfall is 3 inches and the whole is 30 inches. The increase is $\frac{3}{30} = \frac{1}{10}$ or 10%. The whole is 100%, so 100% + 10% = 110%, our estimate.

Step 4   **Solve** the problem. Let x represent the unknown percent.

$$\frac{33}{30} = \frac{x}{100} \quad \text{or} \quad \frac{11}{10} = \frac{x}{100} \qquad \text{$\frac{33}{30}$ is $\frac{11}{10}$ in lowest terms.}$$

Find the cross products.

$$10 \cdot x = 11 \cdot 100$$

$$10 \cdot x = 1100$$

$$\frac{\overset{1}{\cancel{10}} \cdot x}{\underset{1}{\cancel{10}}} = \frac{1100}{10} \qquad \text{Divide both sides by 10.}$$

$$x = 110$$

Step 5   **State the answer.** This year's rainfall is **110%** of normal rainfall.

Step 6   **Check.** The exact answer, 110%, matches our estimate of 110%.

> **Work Problem 8 at the Side.** ▶▶▶

---

8 Solve each problem.

(a) The number of students who usually take the ←*whole* accounting class is 300. If 450 students are taking the class this semester, what percent of the usual number are now taking the accounting class?

$$\frac{450}{300} \qquad \frac{x}{100}$$

$$150\%$$

(b) The service department set a goal of 360 service calls this week. If they made 504 service calls, find the percent of their goal that they completed.

$$\frac{504}{360} \qquad \frac{x}{100}$$

$$140\%$$

Educational Tax Incentives

The government sponsors tax incentive programs to make education more affordable. To qualify for the programs, you have to have an adjusted gross income below a certain level (typically $40,000). You can find specific information at the Internal Revenue Service Web site: www.irs.ustreas.gov.

- The Hope Scholarship offers 100% of the first $1000 spent for certain expenses, such as tuition and books, during the first year of college, plus 50% of the next $1000 incurred during the second year of college. The scholarship money is payable as a tax refund. The student cannot have completed the first two years of post-secondary education and must meet certain educational goals and workload criteria.

- Lifetime Learning Credits are based on qualified expenses, including tuition and books, and equal 20% of the first $5000 in expenses. It is not based on a student's workload and is not limited to only two years.

- Only one of the credits can be claimed for each student.

Suppose you are paying your own educational costs, and your adjusted gross income meets the guidelines to qualify for the Hope Scholarship or Lifetime Learning Credits. Your goals are to earn an Associate of Arts degree from a community college and then transfer to a state university to complete a Bachelor's degree. Tuition costs for resident students at North Harris Montgomery Community College District (NHMCCD) in Texas are used as an example of educational expenses. (Expenses vary among schools, and you can easily find that information in the college's catalog or Internet site.)

Residents of NHMCCD pay a $12 registration fee for each semester enrolled plus $30 per semester hour tuition and fees. Assume that you must study a total of 15 semester hours in developmental work in mathematics, reading, and writing, and to complete an Associate of Arts degree you must study 60 additional semester hours. You decide to limit your course load to 15 credit hours each semester. Assume that one course is 3 semester hours, and you will have to purchase books at an approximate cost of $75 per course.

1. How many semesters and how many courses will it take you to finish the requirements for an Associate of Arts degree?

2. What is the total cost to complete the Associate of Arts degree for **(a)** books and **(b)** tuition and fees?

3. Calculate the total cost for tuition, fees, and books during the first two years (four semesters). What is the maximum tax incentive payable under the Hope Scholarship during **(a)** the first year and **(b)** the second year?

4. Assume that you are returning to school and do not qualify for the Hope Scholarship. What is the maximum tax incentive payable under the Lifetime Learning Credits during **(a)** the first year and **(b)** the second year?

5. How much additional tax incentive would be payable under the Lifetime Learning Credits for the remaining coursework to complete the Associate of Arts degree? How much additional tax incentive would be payable under the Lifetime Learning Credits to complete a Bachelor's degree?

$\frac{P}{whole} \qquad \frac{x\ P(\%)}{100}$

6.4 Exercises

FOR EXTRA HELP

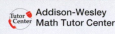 Addison-Wesley Math Tutor Center

 MathXL

 Digital Video Tutor CD 4 Videotape 10

📖 Student's Solutions Manual

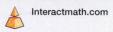 MyMathLab MyMathLab

▲ Interactmath.com

Find the part using the multiplication shortcut. See Example 2.

1. 35% of 120 test tubes

$\frac{x}{120} \quad \frac{35}{100} \quad x \cdot 100 = 35 \cdot 120$

$\frac{100}{100} \quad \frac{4200}{100} = 42$

2. 20% of 1800 rentals

3. 45% of 4080 military personnel

4. 12% of 3650 Web sites

5. 4% of 120 ft

6. 9% of $150

7. 150% of 210 files

8. 130% of 60 trees

9. 52.5% of 1560 trucks

10. 38.2% of 4250 loads

11. 2% of $164

12. 6% of $434

13. 225% of 680 tables

14. 135% of 800 commuters

15. 17.5% of 1040 homes

16. 46.1% of 843 kilograms

17. 0.9% of $2400

$\frac{x}{2400} \quad \frac{.9}{100} \quad 21.60$

$x \cdot 100 = 2160 0 \frac{}{100} = 216$

18. 0.3% of $1400

$\frac{x}{1400} \quad \frac{3}{100}$

Find the whole using the percent proportion. See Example 4.

19. 80 e-mails is 25% of what number of e-mails?

20. 32 medical exams is 5% of what number of medical exams?

21. 30% of what number of hay bales is 48 hay bales?

22. 55% of what number of experiments is 209 experiments?

23. 495 successful students is 90% of what number of students?

24. 84 letters is 28% of what number of letters?

25. 462 mountain bikes is 140% of what number of mountain bikes?

26. 1496 graduates is 110% of what number of graduates?

27. $12\frac{1}{2}$% of what number is 350?

$\left(\textit{Hint:}\text{ Write }12\frac{1}{2}\%\text{ as }12.5\%.\right)$

28. $5\frac{1}{2}$% of what number is 176?

$\left(\textit{Hint:}\text{ Write }5\frac{1}{2}\%\text{ as }5.5\%.\right)$

Find the percent using the percent proportion. Round your answers to the nearest tenth if necessary. See Example 6.

29. 18 bean burritos is what percent of 36 bean burritos?

30. 62 hospital rooms is what percent of 248 hospital rooms?

31. 390 SUVs is what percent of 750 SUVs?

32. 650 liters is what percent of 1000 liters?

33. 32 patients is what percent of 400 patients?

34. 7 bridges is what percent of 350 bridges?

35. 54 CDs is what percent of 3600 CDs?

36. 60 cartons is what percent of 2400 cartons?

37. What percent of $344 is $64?

38. What percent of $398 is $14?

39. What percent of 250 tires is 23 tires?

40. What percent of 105 employees is 54 employees?

41. A student turned in the following answers on a test. You can tell that two of the answers are incorrect without even working the problems. Find the incorrect answers and explain how you identified them (without actually solving the problems).

50% of $84 is $ _42_ .

150% of $30 is $ _20_ .

25% of $16 is $ _32_ .

100% of $217 is $ _217_ .

42. Write a percent problem on any topic you choose. Be sure to include only two of the three components so that you can solve for the third component. Identify each component of the problem and then solve it.

Solve each application problem. Round percent answers to the nearest tenth if necessary. See Examples 3, 5, 7, and 8.

43. Aimee Toit, who works part-time, earns $240 per week and has 22% of this amount withheld for taxes, Social Security, and Medicare. Find the amount withheld.

44. Kirsten Speed needs 124 credits to graduate. If she has already completed 75% of the necessary credits, find the number of credits completed.

45. The guided-missile destroyer USS *Sullivans* has a 335-person crew, of which 13% are female. Find the number of female crew members. Round to the nearest whole number. (*Source:* U.S. Navy.)

46. In a survey on where to hold their weddings, 45% of 480 adults preferred a nonreligious site. How many adults said they would prefer a religious site? (*Source:* National Family Opinion Research.)

47. This year, there are 550 scholarship applicants. If 40% of the applicants will receive a scholarship, find the number of students who will receive a scholarship.

48. A U.S. Food and Drug Administration (FDA) biologist found that canned tuna is "relatively clean." Extraneous matter was found in 5% of the 1600 cans of tuna tested. How many cans of tuna contained extraneous matter?

49. On average, a family of two adults and two children pays $95 per night for lodging and $104 a day for food while on vacation this year. If this increases by 3% over the next year, find the total cost per day for lodging and meals for such a family after the increase.

50. The sales of Tropicana pure orange juice in China were only $100 million this year. Market researchers estimate that sales will increase by 35% next year. Find the amount of orange juice sales estimated for next year. (*Source:* Seagram's Tropicana Beverage Group.)

The graph (pictograph) shows the percent of children 6–11 years of age who are neglecting dental hygiene. Use the information to answer Exercises 51–54.

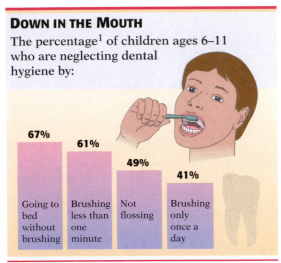

DOWN IN THE MOUTH

The percentage[1] of children ages 6–11 who are neglecting dental hygiene by:

- 67% — Going to bed without brushing
- 61% — Brushing less than one minute
- 49% — Not flossing
- 41% — Brushing only once a day

[1]Respondents allowed to choose multiple answers.
Source: Services for Crest.

51. What percent of the children brush their teeth before going to bed?

52. What percent of the children floss their teeth?

53. If 3400 children answered the questions for this survey, how many of the children brush less than one minute?

54. How many of the 3400 children in the survey brush only once a day?

55. A recent study examined 48,000 military jobs, such as Army attack helicopter pilot or Navy gunner's mate. It was found that only 960 of these jobs are filled by women. What percent of these jobs are filled by women? (*Source:* Rand's National Defense Research Institute.)

56. There are more than 55,000 words in *Webster's Dictionary,* but most educated people can identify only 20,000 of these words. What percent of the words in the dictionary can these people identify?

57. Ebony Durrant has 7.5% of her earnings deposited into her retirement plan. If $240 per month is deposited in the plan, find her monthly and yearly earnings.

58. About 61% of the 43,000,000 people who receive Social Security benefits are paid with a direct deposit to their bank. How many of the people receiving benefits are paid with a direct deposit?

The graph (pictograph) shows the percent of chicken noodle soup sold during the cold-and-flu season. Use this information to answer Exercises 59–62.

SOUP'S ON
60% of the 350 million cans of chicken noodle soup sold each year are purchased during cold-and-flu season, with January being the number 1 month.

10%   9%   8%   15%   11%   7%

October November December January February March

Source: USA Today.

59. Which of the cold-and-flu season months had the lowest sales of chicken noodle soup?

60. What percent of the chicken noodle soup sales take place in the *non-flu* season months?

61. Find the number of cans of soup sold in the highest sales month.

62. How many more cans of soup were sold in October than in November?

The circle graph shows the sales at fast-food hamburger chains as a percent of total fast-food hamburger sales. Use this information to answer Exercises 63–66.

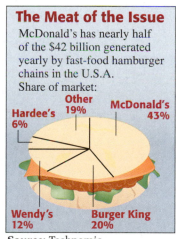

The Meat of the Issue
McDonald's has nearly half of the $42 billion generated yearly by fast-food hamburger chains in the U.S.A.
Share of market:

Other 19%   McDonald's 43%
Hardee's 6%
Wendy's 12%   Burger King 20%

Source: Technomic.

63. Which of the hamburger chains had the lowest sales?

64. What percent of the total hamburger sales were made by the two companies having the lowest sales?

65. Find the total annual sales for McDonald's.

66. Find the total annual sales for Burger King.

67. A collection agency, specializing in collecting past-due child support, charges $25 as an application fee plus 20% of the amount collected. What is the total charge for collecting $3100 in past-due child support?

68. Raw steel production by the nation's steel mills decreased by 2.5% from last week. The decrease amounted to 50,475 tons. Find the steel production last week.

69. Marketing Intelligence Service says that there were 15,401 new products introduced last year. If 86% of the products introduced last year failed to reach their business objectives, find the number of products that were successful. (Round to the nearest whole number.)

70. A family of four with a monthly income of $2900 spends 90% of its earnings and saves the balance. Find **(a)** the monthly savings and **(b)** the annual savings of this family.

Knowing and using the percent proportion is useful when solving percent problems.
Work Exercises 71–77 in order.

71. In the percent proportion, part is to _____ as

percent is to _____.

72. All percent problems involve a comparison between

a part of something and the _____.

*Use this Ramen Noodles label of nutrition facts to answer Exercises 73–77. Read the label
very carefully. Round to the nearest tenth of a percent.*

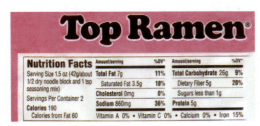

Source: Ramen Noodles package.

73. How many calories per serving are from total
carbohydrates?

74. The package label shows that the 27 grams (g) of
carbohydrates in one serving are 9% of the recommended Daily Value (DV). What is the recommended daily value of carbohydrates?

75. Find the number of grams of total fat that are
needed to meet the recommended Daily Value (DV)
of fat. Round to the nearest gram.

76. Will a person eating two packages of Ramen Noodles in one day exceed their recommended daily
value of sodium? Explain your answer.

77. How many packages of Ramen Noodles must be
eaten in a day to meet the recommended daily value
of fiber? Would this be possible? Would this result
in good nutrition? (Round to the nearest whole
package.)

6.5 Using the Percent Equation

In the last section you were shown how to use a proportion to solve percent problems. In this section we show another way to solve these problems by using the **percent equation.** The percent equation is just a rearrangement of the percent proportion.

O B J E C T I V E S

1 Use the percent equation to find the part.

2 Find the whole using the percent equation.

3 Find the percent using the percent equation.

> **Percent Equation**
>
> $$\text{part} = \text{percent} \cdot \text{whole}$$
>
> *Be sure to write the percent as a decimal before using the equation.*

When using the percent proportion, we did *not* have to write the percent as a decimal because of the 100 in the denominator of the proportion. However, because there is no 100 in the percent *equation, we must* first write the percent as a decimal by dividing by 100.

Some of the examples solved earlier will be reworked by using the percent equation. If you want to, you can look back at **Section 6.4** to see how some of these same problems were solved using proportions. This will give you a comparison of the two methods.

O B J E C T I V E 1 Use the percent equation to find the part.   The first example shows how to find the part.

EXAMPLE 1 **Finding the Part**

(a) Find 15% of $160.
  Write 15% as the decimal 0.15. The whole, which comes after the word *of,* is 160. Next, use the percent equation. Let x represent the unknown part.

$$\text{part} = \text{percent} \cdot \text{whole}$$
$$x = (0.15)(160)$$

Multiply 0.15 and 160.

$$x = 24$$

15% of $160 is **$24**.

(b) Find 110% of 80 cases.
  Write 110% as the decimal 1.10. The whole is 80. Let x represent the unknown part.

$$\text{part} = \text{percent} \cdot \text{whole}$$
$$x = (1.10)(80)$$
$$x = 88$$

110% of 80 cases is **88 cases**.

(c) Find 0.4% of 250 patients.
  Write 0.4% as the decimal 0.004. The whole is 250. Let x represent the unknown part.

$$\text{part} = \text{percent} \cdot \text{whole}$$
$$x = (0.004)(250)$$
$$x = 1$$

0.4% of 250 patients is **1 patient**.

———— **Continued on Next Page**

1 Use the percent equation to find the part.

(a) 15% of 880 policyholders

(b) 23% of 840 gallons

(c) 120% of $220

(d) 135% of $1080

(e) 0.5% of 1200 test tubes

(f) 0.25% of 1600 lab tests

≈ Estimate the answer 0.4% is approximately 0.5 or $\frac{1}{2}$ of 1%. Because 1% is $\frac{1}{100}$, 1% of 250 is

$$250 \div 100 = 2.5$$

Since 2.5 is equal to 1%, half of 2.5 is equal to 1.25 (because $2.5 \div 2 = 1.25$). So, the exact answer of 1 patient is reasonable.

Work Problem 1 at the Side.

CAUTION
When using the percent equation, the percent must always be *changed to a decimal* before multiplying.

OBJECTIVE 2 Find the whole using the percent equation. The next example shows how to use the percent equation to find the whole.

NOTE
Remember that the word *of* is an indicator word for *multiply*.

EXAMPLE 2 Solving for the Whole

(a) 8 tables is 4% of what number of tables?
The part is 8 and the percent is 4% or the decimal 0.04. The whole is unknown.

8   is   4% **of** what number?
└─ Indicator word

Next, use the percent equation

part = percent • whole

$$8 = (0.04)(x) \quad \text{Let } x \text{ represent the unknown whole.}$$

$$\frac{8}{0.04} = \frac{(0.04)(x)}{0.04} \quad \text{Divide both sides by 0.04.}$$

$$200 = x \longleftarrow \text{Whole}$$

8 tables is 4% of **200 tables**.

(b) 135 tourists is 15% of what number of tourists?
Write 15% as 0.15. The part is 135. Next, use the percent equation to find the whole.

part = percent • whole

$$135 = (0.15)(x) \quad \text{Let } x \text{ represent the unknown whole.}$$

$$\frac{135}{0.15} = \frac{(0.15)(x)}{0.15} \quad \text{Divide both sides by 0.15.}$$

$$900 = x \longleftarrow \text{Whole}$$

135 tourists is 15% of **900 tourists**.

Continued on Next Page

(c) $8\frac{1}{2}\%$ of what number is 102?

Write $8\frac{1}{2}\%$ as 8.5%, or the decimal 0.085. The part is 102. Use the percent equation.

$$\textbf{part = percent} \cdot \textbf{whole}$$

$$102 = (0.085)(x) \qquad \text{\color{blue}Let } x \text{ represent the unknown whole.}$$

$$\frac{102}{0.085} = \frac{(0.\overset{1}{\cancel{085}})(x)}{\underset{1}{\cancel{0.085}}} \qquad \text{\color{blue}Divide both sides by 0.085.}$$

$$1200 = x \longleftarrow \text{\color{blue}Whole}$$

102 is $8\frac{1}{2}\%$ of **1200.**

$\approx$ Estimate the answer. Notice that $8\frac{1}{2}\%$ is close to 10%. If 102 is 10% of a number, then the number is 10 times 102, or 1020. So the exact answer, 1200, is reasonable.

> **CAUTION**
>
> In Example 2(c) above, the $8\frac{1}{2}\%$ was changed to 8.5%, which is the decimal form of $8\frac{1}{2}\%$ ($8\frac{1}{2} = 8.5$). The percent sign still remained in 8.5%. Then 8.5% was changed to the decimal 0.085 before dividing.

Work Problem 2 at the Side. ▶▶▶

OBJECTIVE 3 **Find the percent using the percent equation.** The final example shows how to use the percent equation to find the percent.

EXAMPLE 3 Finding the Percent

(a) 13 auto mechanics is what percent of 52 auto mechanics?

Because 52 follows *of,* the whole is 52. The part is 13, and the percent is unknown. Use the percent equation.

$$\textbf{part = percent} \cdot \textbf{whole}$$

$$13 = x \cdot 52 \qquad \text{\color{blue}Let } x \text{ represent the unknown percent.}$$

$$\frac{13}{52} = \frac{x \cdot \overset{1}{\cancel{52}}}{\underset{1}{\cancel{52}}} \qquad \text{\color{blue}Divide both sides by 52.}$$

$$0.25 = x$$

$$0.25 \text{ is } 25\% \qquad \text{\color{blue}Write the decimal as a percent.}$$

13 auto mechanics is **25%** of 52 auto mechanics.

The equation can also be set up using *of* as an indicator word for multiplication and *is* as an indicator word for "is equal to."

$$\begin{array}{ccccc} 13 & \text{is} & \text{what percent} & \text{of} & 52? \\ \downarrow & \downarrow & \downarrow & \downarrow & \downarrow \\ 13 & = & x & & \cdot 52 \end{array}$$

$$13 = x \cdot 52 \qquad \text{\color{blue}Same equation as above}$$

Continued on Next Page

---

2 Find the whole using the percent equation.

(a) 18 supervisors is 45% of what number of supervisors?

whole = X
part = 18
p% = 45

$$\frac{18}{45} = \frac{45 \cdot X}{45}$$

(b) 67.5 containers is 27% of what number of containers?

(c) 666 inoculations is 45% of what number of inoculations?

(d) $5\frac{1}{2}\%$ of what number of policies is 66 policies?

3 Find the percent using the percent equation.

(a) What percent of 35 monitors is 7 monitors?

(b) 34 post office boxes is what percent of 85 post office boxes?

(c) What percent of 920 invitations is 1288 invitations?

(d) 9 world-class runners is what percent of 1125 runners?

(b) What percent of $500 is $100?

The whole is 500 and the part is 100. Let x represent the unknown percent.

$$\textbf{part} = \textbf{percent} \cdot \textbf{whole}$$

$$100 = x \cdot 500 \qquad \text{Let } x \text{ represent the unknown percent.}$$

$$\frac{100}{500} = \frac{x \cdot \overset{1}{\cancel{500}}}{\underset{1}{\cancel{500}}} \qquad \text{Divide both sides by 500.}$$

$$0.20 = x$$

$$0.20 \text{ is } 20\% \longleftarrow \text{Write the decimal as a percent.}$$

20% of $500 is $100.

(c) What percent of $300 is $390?

The whole is 300 and the part is 390. Let x represent the unknown percent.

$$\textbf{part} = \textbf{percent} \cdot \textbf{whole}$$

$$390 = x \cdot 300 \qquad \text{Let } x \text{ represent the unknown percent.}$$

$$\frac{390}{300} = \frac{x \cdot \overset{1}{\cancel{300}}}{\underset{1}{\cancel{300}}} \qquad \text{Divide both sides by 300.}$$

$$1.3 = x$$

$$1.3 \text{ is } 130\% \longleftarrow \text{Write the decimal as a percent.}$$

130% of $300 is $390.

(d) 6 ladders is what percent of 1200 ladders?

Since 1200 follows *of,* the whole is 1200. The part is 6. Let x represent the unknown percent.

$$\textbf{part} = \textbf{percent} \cdot \textbf{whole}$$

$$6 = x \cdot 1200 \qquad \text{Let } x \text{ represent the unknown percent.}$$

$$\frac{6}{1200} = \frac{x \cdot \overset{1}{\cancel{1200}}}{\underset{1}{\cancel{1200}}} \qquad \text{Divide both sides by 1200.}$$

$$0.005 = x$$

$$0.005 \text{ is } 0.5\% \longleftarrow \text{Write the decimal as a percent.}$$

6 ladders is **0.5%** of 1200 ladders.

$\approx$ You can estimate the answer because 1% of 1200 ladders is found by moving the decimal point two places to the left in 1200, resulting in 12. Since 6 ladders is half of 12 ladders, our answer should be $\frac{1}{2}$ of 1% or 0.5%. Our exact answer matches the estimate.

> **CAUTION**
> When you use the percent equation to solve for an unknown percent, the answer will always be in decimal form. Notice that in Example 3(a), (b), (c), and (d) above, **the decimal answer had to be changed to a percent** by multiplying by 100 and attaching the percent sign. The answers became: (a) $0.25 = 25\%$; (b) $0.20 = 20\%$; (c) $1.3 = 130\%$; and (d) $0.005 = 0.5\%$.

ANSWERS

3. **(a)** 20% **(b)** 40% **(c)** 140% **(d)** 0.8%

◀◀◀ Work Problem 3 at the Side.

6.5 Exercises

 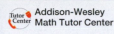

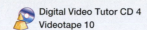
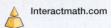
Find the part using the percent equation. See Example 1.

1. 25% of 1080 blood donors

2. 19% of 700 shirts

3. 45% of 3000 bath towels

4. 75% of 360 dosages

5. 32% of 260 quarts

6. 44% of 430 liters

7. 140% of 2500 air bags

8. 145% of 580 hamburgers

9. 12.4% of 8300 meters

10. 26.4% of 4700 miles

11. 0.8% of $520

12. 0.3% of $480

Find the whole using the percent equation. See Example 2.

13. 24 patients is 15% of what number of patients?

14. 32 classrooms is 20% of what number of classrooms?

15. 40% of what number of salads is 130 salads?

16. 75% of what number of wrenches is 675 wrenches?

17. 476 circuits is 70% of what number of circuits?

18. 270 lab tests is 45% of what number of lab tests?

19. $12\frac{1}{2}$% of what number of people is 135 people?

20. $18\frac{1}{2}$% of what number of circuit breakers is 370 circuit breakers?

21. $1\frac{1}{4}$% of what number of gallons is 3.75 gallons?

22. $2\frac{1}{4}$% of what number of files is 9 files?

Find the percent using the percent equation. See Example 3.

23. 70 shipments is what percent of 140 shipments?

24. 180 telemarketers is what percent of 450 telemarketers?

25. 114 tuxedos is what percent of 150 tuxedos?

26. 75 offices is what percent of 125 offices?

27. What percent of $264 is $330?

28. What percent of $480 is $696?

29. What percent of 160 liters is 2.4 liters?

30. What percent of 600 meters is 7.5 meters?

31. 170 cartons is what percent of 68 cartons?

32. 612 orders is what percent of 425 orders?

33. When using the percent equation, the percent must always be changed to a decimal before doing any calculations. Show and explain how to change a fraction percent to a decimal. Use $2\frac{1}{2}\%$ in your explanation.

34. Suppose a problem on your homework assignment was, "Find $\frac{1}{2}\%$ of $1300." Your classmates got answers of $0.65, $6.50, $65, and $650. Which answer is correct? How and why are they getting all of these answers? Explain.

Solve each application problem.

35. A study of office workers found that 27% would like more storage space. If there are 14 million office workers, how many want more storage space? (*Source:* Steelcase Workplace Index.)

36. Most shampoos contain 75% to 90% water. If a 16-ounce bottle of shampoo contains 78% water, find the number of ounces of water in the bottle. Round to the nearest tenth of an ounce.

37. The household lubricant WD-40 is used in 79% of U.S. homes. If there are 104.2 million U.S. homes, find the number of homes in which WD-40 is used. Round to the nearest tenth of a million.

38. Spam and Spam Lite together accounted for 62.2% of the $148 million canned lunch meat sales over a 52-week period (the entire year). Find the total annual sales of these products. Round to the nearest hundredth of a million. (*Source:* Hormel Foods Corporation.)

39. In a recent survey of 1100 employers, it was found that 84% offer only one health plan to employees. How many of these employers offer only one health plan?

40. For a tour of the eastern United States, a travel agent promised a trip of 3300 miles. Exactly 35% of the trip was by air. How many miles were traveled by air?

41. In the United States, 5 of the 50 states have no sales tax. The remaining states do have a sales tax. (*Source: Mathematics for Business,* 7e, Salzman, Miller, Clendenen.)

   (a) What percent of the states do not have a sales tax?

   (b) What percent have a sales tax?

42. Among the 50 companies receiving the greatest number of U.S. patents last year, 18 were Japanese companies. (*Source: Wall Street Journal.*)

   (a) What percent of the companies were Japanese companies?

   (b) What percent of the companies were not Japanese companies?

43. In a survey of 1250 Americans, 461 rated their health as excellent. What percent of these Americans rate their health as excellent? Round to the nearest tenth of a percent. (*Source:* National Health Interview Survey.)

44. General Nutrition Center now has 3200 stores and plans to add 450 more stores. Find the percent of additional stores that they have planned.

The graph shows the average time spent preparing weekday dinners. Assume that 5400 people were surveyed to gather this data. Use this graph to answer Exercises 45–48.

DINNER TIME

Average time required to prepare a weekday dinner.

16–30 minutes 39%

31–45 minutes 35%

6% 15 minutes or less

4% 61+ minutes

0.5% No response

0.5% Other

15% 46–60 minutes

Source: The NPD Group.

45. Find the number of people who said they spend "16–30 minutes" preparing weekday dinners.

46. What total number of people answered "No response" and "Other"?

47. How many people said they spend over 30 minutes preparing weekday dinners?

48. Find the number of people who said they spend 30 minutes or less preparing weekday dinners.

49. Chemical Banking Corporation made $338 million worth of mortgage loans to minorities last year. If this represented 18.6% of all their mortgages, find the total value of all mortgages that they made last year. Round to the nearest tenth of a million.

50. Rachel Williams has 8.5% of her earnings deposited into the credit union. If this amounts to $131.75 per month, find her annual earnings.

51. The Chevy Camaro was introduced in 1967. Sales that year were 220,917, which was 46.2% of the number of Ford Mustangs sold in the same year. Find the number of Mustangs sold in 1967. Round to the nearest whole number.

52. Chris Goodwin is a waiter and has sales of $822.25 on Saturday. If this is 28.6% of his sales for the week, find his weekly sales.

53. J & K Mustang has increased the sale of auto parts by $32\frac{1}{2}$% over last year. If the sale of parts last year amounted to $385,200, find the volume of sales this year.

54. An ad for steel-belted radial tires promises 15% better mileage. If mileage had been 25.6 miles per gallon in the past, what mileage could be expected after these tires are installed? Round to the nearest tenth of a mile.

55. A Polaris Vac-Sweep is priced at $524 with an allowed trade-in of $125 on an old unit. If sales tax of $7\frac{3}{4}$% is charged on the price of the new Polaris unit before the trade-in, find the total cost to the customer after receiving the trade-in. (*Hint:* Trade-in is subtracted last.)

56. A Sony Personal Entertainment Organizer originally priced at $248 is marked down 25%. Find the price of the organizer after the markdown.

6.6 Solving Application Problems with Percent

Percent has many applications in our daily lives. This section discusses percent as it applies to sales tax, commissions, discounts, and the percent of change (increase and decrease).

OBJECTIVE 1 Find sales tax. States, countries, and cities often collect taxes on sales to customers. The **sales tax** is a percent of the total sale. The following formula for finding sales tax is based on the percent equation.

Sales Tax Formula

$$\text{part} = \text{percent} \cdot \text{whole}$$

amount of sales tax = rate of tax $\cdot$ cost of item

EXAMPLE 1 Solving for Sales Tax

Office Max sells a flat panel display for $299. If the sales tax rate is 5%, how much tax was paid? What was the total cost of the flat panel display?

Step 1 **Read** the problem. The problem asks for the total cost of the flat panel display including the sales tax.

Step 2 **Work out a plan.** Use the sales tax formula to find the amount of sales tax. Write the tax rate (5%) as a decimal (0.05). The cost of the item is $299. Use the letter *a* to represent the unknown *amount* of tax. Add the sales tax to the cost of the item.

Step 3 **Estimate** a reasonable answer. Round $299 to $300. Recall that 5% is equivalent to $\frac{1}{20}$, so divide $300 by 20.

$$\$300 \div 20 = \$15 \text{ tax}$$

The total estimated cost is $300 + $15 = $315. ← Estimate

Step 4 **Solve** the problem.

$$\text{part} = \text{percent} \cdot \text{whole}$$

amount of sales tax = rate of tax $\cdot$ cost of item

$$a = (5\%)(\$299)$$
$$a = (0.05)(\$299)$$
$$a = \$14.95 \quad \text{Sales tax}$$

The tax paid on the flat panel display was $14.95. The customer would pay a total cost of $299 + $14.95 = $313.95.

Step 5 **State the answer.** The total cost of the flat panel display is $313.95.

Step 6 **Check.** The exact answer, $313.95, is close to our estimate of $315.

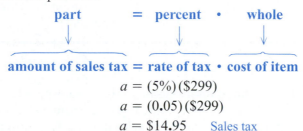
Work Problem 1 at the Side.

---

1 Suppose the sales tax rate in your state is 6%. Find the amount of the tax and the total you would pay for each item.

(a) $29 Little League bat

(b) $119 Game Boy game console

(c) $1287 leather chair and ottoman

(d) $19,300 pickup truck

(a) The tax on a $320 patio set is $25.60.

(b) The tax on a $24 sweatshirt is $1.56.

(c) The tax on a $13,520 Chrysler PT Cruiser is $540.80.

EXAMPLE 2 Finding the Sales Tax Rate

The sales tax on a $14,800 Honda Civic is $962. Find the rate of the sales tax.

Step 1 **Read** the problem. This problem asks us to find the sales tax rate.

Step 2 **Work out a plan.** Use the sales tax formula.

sales tax = rate of tax • cost of item

Solve for the rate of tax, which is the percent. The cost of the Honda Civic (the whole) is $14,800, and the amount of sales tax (the part) is $962. Use r to represent the unknown *rate* of tax (the percent).

Step 3 **Estimate** a reasonable answer. Round $14,800 to $15,000 and round $962 to $1000. The sales tax is $\frac{1000}{15,000}$ or $\frac{1}{15}$ of the cost of the car. So divide 1 by 15 to find the percent (rate) of sales tax.

$$\frac{1}{15} = 0.06\overline{6} \approx 7\% \quad \leftarrow \text{Rounded estimate}$$

Step 4 **Solve** the problem.

$$\text{sales tax} = \text{rate of tax} \cdot \text{cost of item}$$
$$\$962 = r \cdot \$14,800$$

$$\frac{962}{14,800} = \frac{r \cdot 14,800}{14,800} \quad \text{Divide both sides by 14,800.}$$

$$0.065 = r$$
$$0.065 \text{ is } 6.5\% \quad \text{Write the decimal as a percent.}$$

Step 5 **State the answer.** The sales tax rate is 6.5% or $6\frac{1}{2}\%$.

Step 6 **Check.** The exact answer, $6\frac{1}{2}\%$, is close to our estimate of 7%.

Work Problem 2 at the Side.

> **NOTE**
> You can use the sales tax formula to find the amount of sales tax, the cost of an item, or the rate of sales tax (the percent).

OBJECTIVE 2 Find commissions. Many salespeople are paid by *commission* rather than an hourly wage. If you are paid by **commission,** you are paid a certain percent of your total sales dollars. The formula below for finding the commission is based on the percent equation.

Commission Formula

part = percent • whole

amount of commission = rate of commission • amount of sales

EXAMPLE 3 Determining the Amount of Commission

Scott Samuels had pharmaceutical sales of $42,500 last month. If his commission rate is 9%, find the amount of his commission.

Step 1 **Read** the problem. The problem asks for the amount of commission that Samuels earned.

Step 2 **Work out a plan.** Use the commission formula. Write the rate of commission (9%) as a decimal (0.09). The amount of Samuels' sales ($42,500) is the whole. Use c to represent the unknown *amount* of commission.

Step 3 **Estimate** a reasonable answer. Round the commission rate of 9% to 10%. Round the amount of sales from $42,500 to $40,000. Since 10% is equivalent to $\frac{1}{10}$, divide $40,000 by 10 to find the amount of commission.

$$\$40,000 \div 10 = \$4000 \quad \leftarrow \text{Estimate}$$

Step 4 **Solve** the problem.

amount of commission = rate of commission • amount of sales

$$c = (9\%)(\$42,500)$$
$$c = (0.09)(\$42,500)$$
$$c = \$3825 \quad \text{Amount of commission}$$

Step 5 **State the answer.** Samuels earned a commission of $3825 for selling the pharmaceuticals.

Step 6 **Check.** The exact answer, $3825, is close to our estimate of $4000.

Work Problem 3 at the Side. ▶▶▶

EXAMPLE 4 Finding the Rate of Commission

Chris Knudson earned a commission of $510 for selling $17,000 worth of shipping supplies. Find the rate of commission.

Step 1 **Read** the problem. In this problem, we must find the rate (percent) of commission.

Step 2 **Work out a plan.** You could use the commission formula. Another approach is to use the percent proportion. The *whole* is $17,000, the *part* is $510, and the *percent* is unknown. (The rate of commission is the percent.)

Step 3 **Estimate** a reasonable answer. Round the commission, $510, to $500. The commission in fraction form is $\frac{\$500}{\$20,000}$, which divides out to $\frac{1}{40}$. Changing $\frac{1}{40}$ to a percent gives $2\frac{1}{2}\%$ (rounded), our estimate.

Step 4 **Solve** the problem.

$$\frac{\text{part}}{\text{whole}} = \frac{x}{100} \quad \leftarrow \text{Percent (unknown)}$$

$$\frac{510}{17,000} = \frac{x}{100}$$

$$17,000 \cdot x = 510 \cdot 100 \quad \text{Cross multiply.}$$

$$\frac{\overset{1}{\cancel{17,000}} \cdot x}{\underset{1}{\cancel{17,000}}} = \frac{51,000}{17,000} \quad \text{Divide both sides by 17,000.}$$

$$x = 3$$

Continued on Next Page

3 Find the amount of commission.

(a) Jill Buteo sells dental equipment at a commission rate of 12% and has sales for the month of $28,750.

(b) Last month Pam Prentice sold a home for $175,500 for a client and earned a commission of 6%.

4 Find the rate of commission.

(a) A commission of $450 is earned on one sale of computer products worth $22,500.

(b) Jamal Story earns $2898 for selling office furniture worth $32,200.

5 Find the amount of the discount and the sale price.

(a) An Easy-Boy leather recliner originally priced at $950 is offered at a 42% discount.

(b) Wal-Mart has women's sweater sets on sale at 35% off. One sweater set was originally priced at $30.

Step 5 **State the answer.** The rate of commission is 3%.

Step 6 **Check.** The exact answer of 3% is close to our estimate of $2\frac{1}{2}$%.

◀◀◀ Work Problem 4 at the Side.

OBJECTIVE 3 Find the discount and sale price. Most of us prefer buying things when they are on sale. A store will reduce prices, or **discount,** to attract additional customers. Use the following formula to find the discount and the sale price.

Discount Formula and Sale Price Formula

amount of discount = rate (or percent) of discount • original price

sale price = original price − amount of discount

EXAMPLE 5 Finding a Sale Price

The Oak Mill Furniture Store has a home entertainment center with an original price of $840 on sale at 15% off. Find the sale price of the entertainment center.

Step 1 **Read** the problem. This problem asks for the price of an entertainment center after a discount of 15%.

Step 2 **Work out a plan.** The problem is solved in two steps. First, find the amount of the discount, that is, the amount that will be "taken off" (subtracted), by multiplying the original price ($840) by the rate of the discount (15%). The second step is to subtract the amount of discount from the original price. This gives you the sale price, which is what you will actually pay for the entertainment center.

Step 3 **Estimate** a reasonable answer. Round the original price from $840 to $800, and the rate of discount from 15% to 20%. Since 20% is equivalent to $\frac{1}{5}$, the discount is $800 ÷ 5 = $160. The sale price is $800 − $160 = $640, our estimate.

Step 4 **Solve** the problem. First find the amount of the discount.

amount of discount = rate of discount • original price

$$a = (0.15)(\$840) \quad \text{Write 15\% as a decimal.}$$
$$a = \$126 \quad \text{Amount of discount}$$

Now find the sale price of the entertainment center by subtracting the amount of the discount ($126) from the original price.

sale price = original price − amount of discount

$$= \$840 - \$126$$
$$= \$714 \quad \text{Sale price}$$

Step 5 **State the answer.** The sale price of the entertainment center is $714.

Step 6 **Check.** The exact answer, $714, is close to our estimate of $640.

◀◀◀ Work Problem 5 at the Side.

ANSWERS
4. (a) 2% (b) 9%
5. (a) $399; $551 (b) $10.50; $19.50

🖩 **Calculator Tip**   In Example 5 on the previous page, you can use a calculator to find the amount of discount and subtract the discount from the original price all in one step.

$$840 \ominus .15 \otimes 840 \ominus 714$$

↑	↑	↑
Original price	Amount of discount	Sale price

A scientific calculator observes the order of operations, so it will automatically do the multiplication before the subtraction.

OBJECTIVE **4** **Find the percent of change.**   We are often interested in looking at increases or decreases in sales, production, population, and many other areas. This type of problem involves finding the *percent of change.* Use the following steps to find the **percent of increase.**

> **Finding the Percent of Increase**
>
> *Step 1*   Use subtraction to find the amount of increase.
>
> *Step 2*   Use the percent proportion to find the percent of increase.
>
> $$\frac{\textbf{amount of increase (part)}}{\textbf{original value (whole)}} = \frac{\textbf{percent}}{\textbf{100}}$$

EXAMPLE 6 **Finding the Percent of Increase**

Attendance at county parks climbed from 18,300 last month to 56,730 this month. Find the percent of increase.

Step 1   **Read** the problem. The problem asks for the percent of increase.

Step 2   **Work out a plan.** Subtract the attendance last month (18,300) from the attendance this month (56,730) to find the amount of increase in attendance. Next, use the percent proportion. The whole is 18,300 (last month's original attendance), the part is 38,430 (amount of increase in attendance), and the percent is unknown.

Step 3   **Estimate** a reasonable answer. Round 18,300 to 20,000 and 56,730 to 60,000. The amount of increase is 60,000 − 20,000 = 40,000. Since 40,000 (the increase) is *twice* as large as the original amount, the percent of increase is 200%, our estimate.

Step 4   **Solve** the problem.

$$56,730 - 18,300 = 38,430 \quad \text{Amount of increase in attendance}$$

Amount of increase → $\dfrac{38,430}{18,300} = \dfrac{x}{100}$ Percent proportion
Original value →

Solve this proportion to find that $x = 210$.

Step 5   **State the answer.** The percent of increase is 210%.

Step 6   **Check.** The exact answer, 210%, is close to our estimate of 200%.

Work Problem 6 at the Side. ▶▶▶

6 Find the percent of increase.

(a) A manufacturer of snowboards increased production from 14,100 units last year to 19,035 this year.

(b) The number of flu cases rose from 496 cases last week to 620 this week.

Use the following steps to find the **percent of decrease.**

7 Find the percent of decrease.

(a) The number of service calls fell from 380 last month to 285 this month.

> **Finding the Percent of Decrease**
>
> *Step 1*   Use subtraction to find the amount of decrease.
>
> *Step 2*   Use the percent proportion to find the percent of decrease.
>
> $$\frac{\textbf{amount of decrease (part)}}{\textbf{original value (whole)}} = \frac{\textbf{percent}}{\textbf{100}}$$

EXAMPLE 7   **Finding the Percent of Decrease**

The number of production employees this week fell to 1406 people from 1480 people last week. Find the percent of decrease.

Step 1   **Read** the problem. The problem asks for the percent of decrease.

Step 2   **Work out a plan.** Subtract the number of employees this week (1406) from the number of employees last week (1480) to find the amount of decrease. Next, use the percent proportion. The whole is 1480 (last week's original number of employees), the part is 74 (decrease in employees), and the percent is unknown.

(b) The number of workers applying for unemployment fell from 4850 last month to 3977 this month.

Step 3   **Estimate** a reasonable answer. Estimate the answer by rounding 1406 to 1400 and 1480 to 1500. The decrease is $1500 - 1400 = 100$. Since 100 is $\frac{1}{15}$ of 1500, our estimate is $1 \div 15 \approx 0.07$ or 7%.

Step 4   **Solve** the problem.

$$1480 - 1406 = 74 \qquad \text{Decrease in number of employees}$$

$$\text{Amount of decrease} \rightarrow \frac{74}{1480} = \frac{x}{100} \qquad \text{Percent proportion}$$
$$\text{Original value} \rightarrow$$

Solve this proportion to find that $x = 5$.

Step 5   **State the answer.** The percent of decrease is 5%.

Step 6   **Check.** The exact answer, 5%, is close to our estimate of 7%.

> **CAUTION**
> When solving for percent of increase or decrease, the *whole is always the original value* or *value before the change occurred.* The part is the change in values, that is, how much something went up or went down.

◀◀◀ **Work Problem 7 at the Side.**

6.6 **Exercises**

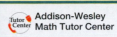

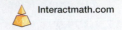
Find the amount of sales tax or the tax rate and the total cost (amount of sale + amount of tax = total cost). Round money answers to the nearest cent if necessary. See Examples 1 and 2.

	Amount of Sale	Tax Rate	Amount of Tax	Total Cost
1.	$6	4%	_____	_____
2.	$45	5%	_____	_____
3.	$425	_____	$12.75	_____
4.	$322	_____	$19.32	_____
5.	$284	_____	$14.20	_____
6.	$84	_____	$5.88	_____
7.	$12,229	$5\frac{1}{2}\%$	_____	_____
8.	$11,789	$7\frac{1}{2}\%$	_____	_____

Find the commission earned or the rate of commission. Round money answers to the nearest cent if necessary. See Examples 3 and 4.

	Sales	Rate of Commission	Commission
9.	$280	8%	_____
10.	$660	10%	_____
11.	$3000	_____	$600
12.	$7800	_____	$1170
13.	$6183.50	3%	_____
14.	$4416.70	7%	_____
15.	$73,500	9%	_____
16.	$55,800	6%	_____

Find the amount or rate of discount and the sale price after the discount. Round money answers to the nearest cent if necessary. See Example 5.

	Original Price	Rate of Discount	Amount of Discount	Sale Price
17.	$199.99	10%	_____	_____
18.	$29.95	15%	_____	_____
19.	$180	_____	$54	_____
20.	$38	_____	$9.50	_____
21.	$17.50	25%	_____	_____
22.	$76	60%	_____	_____
23.	$58.40	15%	_____	_____
24.	$99.80	30%	_____	_____

25. You are trying to decide between Company A paying a 10% commission and Company B paying an 8% commission. For which company would you prefer to work? What considerations other than commission rate would be important to you?

26. Give four examples of where you might use the percent of increase or the percent of decrease in your own personal activities. Think in terms of work, school, home, hobbies, and sports.

Solve each application problem. Round money answers to the nearest cent and rates to the nearest tenth of a percent if necessary. See Examples 1–7.

Country Store has a unique selection of merchandise that it sells by mail and over the Internet. Use the shipping and insurance delivery chart below and a sales tax rate of 5% to solve Exercises 27–30. There is no sales tax on shipping and insurance. (*Source:* Country Store Catalog.)

SHIPPING AND INSURANCE DELIVERY CHART	
Up to $15.00	add $3.95
$15.01 to $25.00	add $5.95
$25.01 to $35.00	add $6.95
$35.01 to $50.00	add $7.95
$50.01 to $70.00	add $8.95
$70.01 to $99.99	add $9.95
$100.00 or more	add $10.95

27. Find the total cost of six Small Fry Handi-Pan electric skillets priced at $29.99 each.

28. A customer ordered five sets of flour-sack towels priced at $12.99 per set. What is the total cost?

29. Find the total cost of three pop-up hampers at $9.99 each and four nonstick mini doughnut pans at $10.99 each.

30. What is the total cost of five coach lamp bird feeders at $19.99 each and six garden weather centers at $14.99 each?

31. An Anderson wood-frame French door is priced at $1980 with a sales tax of $99. Find the rate of sales tax.

32. Textbooks for three classes cost $245 plus tax of $17.15. Find the sales tax rate.

33. Harley-Davidson, the only major U.S.-based motorcycle manufacturer, says that it expects to sell 267,285 motorcycles this year, up from 234,461 last year. Find the percent of increase in production. (*Source:* Harley-Davidson, Inc.)

34. Americans are eating more fish. This year the average American will eat 15.5 pounds compared to only 12.5 pounds per year a decade ago. Find the percent of increase. (*Source: Consumer Reports.*)

35. The number of industrial accidents this month fell to 989 accidents from 1276 accidents last month. Find the percent of decrease.

36. The average number of hours worked in manufacturing jobs last week fell from 41.1 to 40.9. Find the percent of decrease.

37. A "60% off sale" begins today at Wanda's Women's Wear. What is the sale price of women's wool coats normally priced at $335?

38. What is the sale price of a $769 Kenmore washer/dryer set with a discount of 25%?

The weekly sales record of the top four salespeople at the Family Shoe Store is shown in the table. Use this information to answer Exercises 39–42.

FAMILY SHOE STORE

Employee	Sales	Rate of Commission	Commission
McKee, J.	$9500	3%	_____
Brown, D.	$10,200	3%	_____
Poznick, M.	$8875	___	$355.00
Washington, R.	$11,560	___	$578.00

39. Find the commission for McKee.

40. Find the commission for Brown.

41. What is the rate of commission for Poznick?

42. What is the rate of commission for Washington?

43. An 8-millimeter camcorder normally priced at $590 is on sale at 18% off. Find the discount and the sale price.

44. A Honda Pilot is offered at 12% off the manufacturer's suggested retail price. Find the discount and the sale price of this SUV, originally priced at $32,500.

45. The price per share of Toys Я Us stock fell from $35.50 to $33.50. Find the percent of decrease in price.

46. In the past five years, the cost of generating electricity from the sun has been brought down from 24 cents per kilowatt hour to 8 cents (less than the newest nuclear power plants.) Find the percent of decrease.

47. College students are offered a 6% discount on a dictionary that sells for $18.50. If the sales tax is 6%, find the cost of the dictionary including the sales tax.

48. A personal computer and monitor priced at $698 are marked down 8%. If the sales tax is also 8%, find the cost of the computer and monitor including sales tax.

49. A real estate agent sells a condominium for $129,605. A sales commission of 6% is charged. The agent gets 55% of this commission. How much money does the agent get?

50. The local real estate agents' association collects a fee of 2% on all money received by its members. The members charge 6% of the selling price of a property as their fee. How much does the association get, if its members sell property worth a total of $8,680,000?

51. What is the total price of a ski boat with an original price of $10,214, if it is sold at a 15% discount? The sales tax rate is $3\frac{3}{4}$%.

52. A commercial security alarm system originally priced at $10,800 is discounted 22%. Find the total price of the system if the sales tax rate is $7\frac{1}{4}$%.

RELATING CONCEPTS (EXERCISES 53–58) For Individual or Group Work

Knowing how to use the percent equation is important when solving application problems involving sales tax. **Work Exercises 53–58 in order.**

53. The percent equation is

part = _____ _____ .

54. The formula used to find sales tax is an application of the percent equation. The sales tax formula is

sales tax = _____ _____ .

In the United States there are certain items on which an excise tax is charged in addition to a sales tax. A table of federal excise taxes is shown here. Use this table to answer Exercises 55–58. (Excise tax is calculated on the amount of the sale before sales tax is added.) Round answers to the nearest cent.

FEDERAL EXCISE TAXES

Product or Service	Rate	Product or Service	Rate
Telephone service	3%	Tires (by weight)	
Teletypewriter service	3%	Under 40 pounds	No tax
Air transportation	7.5%	40–69 pounds	15¢/pound over 40 pounds
International air travel	$13.40/person	70–89 pounds	$4.50 plus 30¢/pound
Air freight	6.25%		over 70 pounds
		90 pounds and more	$10.50 plus 50¢/pound
			over 90 pounds
		Truck and trailer, chassis and bodies	12%
Fishing rods	10%	Inland waterways fuel	24.4¢/gallon
Bows and arrows	12.4%	Ship passenger tax	$3/passenger
Gasoline	18.4¢ gallon	Vaccines	75¢/dose
Diesel fuel	24.4¢ gallon		
Aviation fuel	21.9¢ gallon		

Source: Publication 510, I.R.S., Excise Taxes.

55. Some archery equipment (bows and arrows) is priced at $123. Use the federal excise tax table and a sales tax rate of $6\frac{1}{2}$% to find the cost of the equipment, including both taxes. (Round to the nearest cent.)

56. Refer to Exercise 55. Calculate the two taxes separately and then add them together. Now, add the two tax rates together and then find the tax. Are your answers the same? Why or why not? (*Hint:* Recall the commutative and associative properties of multiplication.)

57. The price of an international airline ticket is $1248. Use the federal excise tax table and a sales tax rate of $7\frac{3}{4}$% to find the total cost of one ticket. (Sales tax is not charged on the $13.40 federal excise tax.)

58. Refer to Exercise 57. Can the federal excise tax be added to the sales tax rate to find the total tax? Why or why not?

6.7 Simple Interest

When we open a savings account, we are actually lending money to the bank or credit union. It will in turn lend this money to individuals and businesses. These people then become borrowers. The bank or credit union pays a fee to the savings account holders and charges a higher fee to its borrowers. These fees are called *interest*.

Interest is a fee paid or a charge made for lending or borrowing money. The amount of money borrowed is called the **principal.** The charge for interest is often given as a percent, called the interest rate or **rate of interest.** The rate of interest is assumed to be *per year,* unless stated otherwise. Time is always expressed in years or fractions of a year.

OBJECTIVE 1 Find the simple interest on a loan. In most cases, interest on a loan is computed on the *original principal* and is called **simple interest.** We use the following **interest formula** to find simple interest.

> **Formula for Simple Interest**
>
> $$\text{Interest} = \text{principal} \cdot \text{rate} \cdot \text{time}$$
>
> The formula is usually written in letters.
>
> $$I = p \cdot r \cdot t$$

> **NOTE**
> Simple interest is used for most short-term business loans, most real estate loans, and many automobile and consumer loans.

EXAMPLE 1 Finding Interest for a Year

Find the interest on $5000 at 6% for 1 year.

The amount borrowed, or principal (p), is $5000. The interest rate (r) is 6%, which is 0.06 as a decimal, and the time of the loan (t) is 1 year. Use the formula.

$$I = p \cdot r \cdot t$$
$$I = (5000)(0.06)(1)$$
$$I = 300$$

The interest is $300.

> **Work Problem 1 at the Side.** ▶▶▶

EXAMPLE 2 Finding Interest for More Than a Year

Find the interest on $4200 at 8% for three and a half years.

The principal (p) is $4200. The rate (r) is 8%, or 0.08 as a decimal, and the time (t) is $3\frac{1}{2}$ or 3.5 years. Use the formula.

$$I = p \cdot r \cdot t$$
$$I = (4200)(0.08)(3.5)$$
$$I = 1176$$

The interest is $1176.

> **Work Problem 2 at the Side.** ▶▶▶

OBJECTIVES

1 Find the simple interest on a loan.

2 Find the total amount due on a loan.

① Find the interest.

(a) $1000 at 5% for 1 year

(b) $3650 at 2% for 1 year

② Find the interest.

(a) $820 at 6% for $3\frac{1}{2}$ years

(b) $4850 at 8% for $2\frac{1}{2}$ years

(c) $16,800 at 3% for $2\frac{3}{4}$ years

ANSWERS
1. (a) $50 (b) $73
2. (a) $172.20 (b) $970 (c) $1386

3 Find the interest.

(a) $1800 at 3% for 4 months

(b) $28,000 at $9\frac{1}{2}$% for 3 months

4 Find the total amount due on each loan.

(a) $3800 at $6\frac{1}{2}$% for 6 months

(b) $12,400 at 5% for 5 years

(c) $2400 at 11% for $2\frac{3}{4}$ years

CAUTION
It is best to change fractions of percents or fractions of years into their decimal form. In Example 2 on the previous page, $3\frac{1}{2}$ years becomes 3.5 years.

Interest rates are given **per year.** For loan periods of less than one year, be careful to express time as a fraction of a year.

If time is given in months, for example, use a denominator of 12, because there are 12 months in a year. A loan for 9 months would be for $\frac{9}{12}$ of a year.

EXAMPLE 3 Finding Interest for Less Than 1 Year

Find the interest on $840 at $8\frac{1}{2}$% for 9 months.

The principal is $840. The rate is $8\frac{1}{2}$% or 0.085, and the time is $\frac{9}{12}$ of a year. Use the formula $I = p \cdot r \cdot t$.

$$I = \underbrace{(840)(0.085)}\left(\frac{9}{12}\right) \quad \text{9 months} = \tfrac{9}{12} \text{ of a year.}$$

$$= 71.4 \quad \left(\frac{3}{4}\right) \quad \tfrac{9}{12} \text{ in lowest terms is } \tfrac{3}{4}.$$

$$= \frac{(71.4)(3)}{4}$$

$$= \frac{214.2}{4} = 53.55$$

The interest is $53.55.

Calculator Tip The calculator solution to Example 3 above uses chain calculations.

840 ✕ 0.085 ✕ 9 ÷ 12 = **53.55**

Work Problem 3 at the Side.

OBJECTIVE 2 Find the total amount due on a loan. When a loan is repaid, the interest is added to the original principal to find the total amount due.

Formula for Amount Due

amount due = principal + interest

EXAMPLE 4 Calculating the Total Amount Due

A loan of $3240 has been made at 12% for 3 months. Find the total amount due.

First find the interest, then add the principal and the interest to find the total amount due.

$$I = (3240)(0.12)\left(\frac{3}{12}\right) \quad \text{3 months} = \tfrac{3}{12} \text{ of a year.}$$

$$I = \$97.20$$

The interest is $97.20.

amount due = principal + interest
= $3240 + $97.20 = $3337.20

The total amount due is $3337.20

Work Problem 4 at the Side.

6.7 **Exercises**

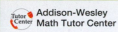

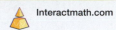

Find the interest. See Examples 1 and 2.

	Principal	Rate	Time in Years	Interest
1.	$100	6%	1	_____
2.	$200	3%	1	_____
3.	$700	5%	3	_____
4.	$900	2%	4	_____
5.	$240	4%	3	_____
6.	$190	3%	2	_____
7.	$2300	$8\frac{1}{2}\%$	$2\frac{1}{2}$	_____
8.	$4700	$5\frac{1}{2}\%$	$1\frac{1}{2}$	_____
9.	$10,800	$7\frac{1}{2}\%$	$2\frac{3}{4}$	_____
10.	$12,400	$6\frac{1}{2}\%$	$3\frac{3}{4}$	_____

Find the interest. Round to the nearest cent if necessary. See Example 3.

	Principal	Rate	Time in Months	Interest
11.	$400	5%	6	_____
12.	$600	2%	5	_____
13.	$820	6%	12	_____

Principal	Rate	Time in Months	Interest
14. $780	8%	24	_____
15. $940	3%	18	_____
16. $178	4%	12	_____
17. $1225	$5\frac{1}{2}\%$	3	_____
18. $2660	$7\frac{1}{2}\%$	3	_____
19. $15,300	$7\frac{1}{4}\%$	7	_____
20. $13,700	$3\frac{3}{4}\%$	11	_____

Find the total amount due on the following loans. Round to the nearest cent if necessary. See Example 4.

Principal	Rate	Time	Total Amount Due
21. $200	5%	1 year	_____
22. $400	2%	6 months	_____
23. $740	6%	9 months	_____
24. $1180	3%	2 years	_____

	Principal	Rate	Time	Total Amount Due
25.	$1800	9%	18 months	_____
26.	$9000	6%	7 months	_____
27.	$3250	10%	6 months	_____
28.	$7600	5%	1 year	_____
29.	$16,850	$7\frac{1}{2}\%$	9 months	_____
30.	$19,450	$5\frac{1}{2}\%$	6 months	_____

31. The amount of interest paid on savings accounts and charged on loans can vary from one institution to another. However, when the amount of interest is calculated, three factors are used in the calculation. Name these three factors and describe them in your own words.

32. Interest rates are usually given as a rate per year (annual rate). Explain what must be done when time is given in months. Write your own problem where time is given in months and then show how to solve it.

Solve each application problem. Round to the nearest cent if necessary.

33. Reann Chang deposits $825 at 5% for 1 year. How much interest will she earn?

34. The Jidobu family invests $18,000 at 9% for 6 months. What amount of interest will the family earn?

35. The Bank of America loans $380,000 to a business at 6% for 18 months. How much interest will the bank earn?

36. Helen Dale, a professional golfer, deposits $80,000 of her tournament winnings at 7% for 4 years. How much interest will she earn?

37. A student borrows $1200 at 8% for 5 months to pay for books and tuition. Find the total amount due.

38. Jill borrows $2750 from her dad for a used car. The loan will be paid back with 8% interest at the end of 9 months. Find the total amount due.

39. Marie Perino deposits $14,800 in her school credit union account for 10 months. If the credit union pays $2\frac{1}{4}\%$ interest, find the amount of interest she will earn.

40. Silvo DiLoreto, owner of Sunset Realtors, borrows $27,000 to update his office computer system. If the loan is for 24 months at $7\frac{1}{4}\%$, find the amount of interest he will owe.

41. An investment fund pays $7\frac{1}{4}\%$ interest. If Beverly Habecker deposits $8800 in her account for $\frac{1}{4}$ year, find the amount of interest she will earn.

42. Ms. Henderson owes $1900 in taxes. She is charged a penalty of $12\frac{1}{4}\%$ annual interest and pays the taxes and penalty after 6 months. Find the total amount she must pay.

43. A gift shop owner invests his profits of $11,500 at $8\frac{3}{4}\%$ interest for $\frac{3}{4}$ year. Find the total amount in his account at the end of this time.

44. A pawn shop owner lends $35,400 to another business for $\frac{1}{2}$ year at an interest rate of 14.9%. How much interest will be earned on the loan?

45. The owners of Clear Lake Marina bought six canoes to be used as rentals at a cost of $550 per canoe. If they borrowed 60% of the total cost for 9 months at $12\frac{1}{2}\%$ interest, find the amount due.

46. The owners of Baily and Daughters Excavating purchased four earth movers at a cost of $485,000 each. If they borrowed 80% of the total purchase price for $2\frac{1}{2}$ years at $10\frac{1}{2}\%$ interest, find the total amount due.

Summary Exercises on Percent

Write each percent as a decimal and each decimal as a percent.

1. 6.25%

2. 380%

3. 0.375

4. 0.006

Write each percent as a fraction or mixed number in lowest terms and each fraction as a percent.

5. 87.5%

6. 160%

7. $\dfrac{5}{8}$

8. $\dfrac{1}{125}$

Find the part, whole, or percent as indicated. Round percent answers to the nearest tenth if necessary.

9. 6% of $780 is what amount?

10. 70 rolls of film is 14% of how many rolls of film?

11. What percent of 320 policies is 176 policies?

12. 0.8% of 3500 screening exams is how many exams?

13. 1016.4 acres is 280% of what number of acres?

14. What percent of 658 circuits is 18 circuits?

Find the amount of sales tax or the tax rate and the total cost. Round to the nearest cent.

	Amount of Sale	Tax Rate	Amount of Tax	Total Cost
15.	$176	$5\dfrac{1}{2}\%$	_____	_____
16.	$926	_____	$64.82	_____

Find the commission earned or the rate of commission.

	Sales	Rate of Commission	Commission
17.	$3765	8%	_____
18.	$22,482	_____	$1461.33

Find the amount or rate of discount and the amount paid after the discount. Round to the nearest cent.

Original Price	Rate of Discount	Amount of Discount	Sale Price
19. $49.95	18%	_____	_____
20. $684	_____	$239.40	_____

Find the interest and the total amount due on the following simple interest loans.

Principal	Rate	Time	Interest	Total Amount Due
21. $1000	3%	1 year	_____	_____
22. $2380	$7\frac{1}{2}\%$	3 years	_____	_____
23. $1470	4%	9 months	_____	_____
24. $6820	$6\frac{1}{2}\%$	18 months	_____	_____

Solve each application problem. Round percent answers to the nearest tenth of a percent if necessary.

25. The number of Harley-Davidson motorcycles sold last year was 267,285, up from 111,800 motorcycles five years ago. Find the percent of increase. (*Source:* Harley-Davidson, Inc.)

26. The number of ski lift tickets sold this week was 3419, down from last week's sales of 3820 tickets. Find the percent of decrease.

6.8 Compound Interest

The interest we studied in **Section 6.7** was *simple interest* (interest only on the original principal). A common type of interest used with savings accounts and most investments is **compound interest** or interest paid on past interest as well as on the principal.

OBJECTIVE 1 Understand compound interest. Suppose that you make a single deposit of $1000 in a savings account that earns 5% per year. What will happen to your savings over 3 years? At the end of the first year, 1 year's interest on the original deposit is found. Use the simple interest formula.

$$\text{Interest} = \text{principal} \cdot \text{rate} \cdot \text{time}$$

Year 1   ($1000)(0.05)(1) = **$50**

Add the interest to the $1000 to find the amount in your account at the end of the first year. $1000 + **$50** = **$1050**
The interest for the second year is found on $1050; that is, the interest is **compounded.**

Year 2   (**$1050**)(0.05)(1) = **$52.50**

Add this interest to the $1050 to find the amount in your account at the end of the second year. **$1050** + **$52.50** = **$1102.50**.
The interest for the third year is found on $1102.50.

Year 3   (**$1102.50**)(0.05)(1) ≈ **$55.13**

Add this interest to the **$1102.50.**   **$1102.50** + **$55.13** = **$1157.63**

At the end of 3 years, you will have **$1157.63** in your savings account. The $1157.63 that you have in your account is called the **compound amount.**

If you had earned only *simple* interest for 3 years, your interest would be as follows.

$$I = (\$1000)(0.05)(3)$$

$$= \$150$$

At the end of 3 years, you would have $1000 + $150 = $1150 in your account. Compounding the interest increased your earnings by $7.63 because $1157.63 − $1150 = $7.63.

With *compound* interest, the interest earned during the second year is greater than that earned during the first year, and the interest earned during the third year is greater than that earned during the second year. This happens because the interest earned each year is *added* to the principal, and the new total is used to find the amount of interest in the next year.

Compound Interest

Interest paid on principal plus past interest is called **compound interest.**

1 Find the compound amount given the following deposits. Round to the nearest cent if necessary.

(a) $500 at 4% for 2 years

(b) $2000 at 7% for 3 years

OBJECTIVE 2 **Understand compound amount.** Find the compound amount as shown below.

EXAMPLE 1 **Finding the Compound Amount**

Nancy Wegener deposits $3400 in an account that pays 6% interest compounded annually for 4 years. Find the compound amount. Round to the nearest cent when necessary.

Year	Interest	Compound Amount
1	($3400)(0.06)(1) = **$204**	
	$3400 + **$204** =	**$3604**
2	($3604)(0.06)(1) = **$216.24**	
	$3604 + **$216.24** =	**$3820.24**
3	($3820.24)(0.06)(1) ≈ **$229.21**	
	$3820.24 + **$229.21** =	**$4049.45**
4	($4049.45)(0.06)(1) ≈ **$242.97**	
	$4049.45 + **$242.97** =	**$4292.42**

The compound amount is $4292.42.

◀◀◀ Work Problem 1 at the Side.

---

2 Find the compound amount by multiplying the original deposit by 100% plus the compound interest rate. Round to the nearest cent if necessary.

(a) $1800 at 2% for 3 years

$(1800)(1.02)(1.02)(1.02) =$

_____

(b) $900 at 3% for 2 years

(c) $2500 at 5% for 4 years

OBJECTIVE 3 **Find the compound amount.** A more efficient way of finding the compound amount is to add the interest rate to 100% and then multiply by the original deposit. Notice that in Example 1 above, at the end of the first year, you will have $3400 (100% of the original deposit) plus 6% (of the original deposit) or 106% (100% + 6% = 106%).

EXAMPLE 2 **Finding the Compound Amount**

Find the compound amount in Example 1 using multiplication.

Year 1 Year 2 Year 3 Year 4

($3400)(1.06)(1.06)(1.06)(1.06) ≈ $4292.42

Original deposit   100% + 6% = 106% = 1.06   Compound amount

Our answer, $4292.42, is the same as in Example 1 above.

◀◀◀ Work Problem 2 at the Side.

NOTE
By adding the compound interest rate to 100%, we can then multiply by the original deposit. This will give us the compound amount at the end of each compound interest period.

Calculator Tip If you use a calculator for Example 2 above, you can use the y^x key (exponent key).

3400 ⊗ 1.06 y^x 4 ⊜ **4292.42 (rounded)**

The 4 following the y^x key represents the number of compound interest periods.

OBJECTIVE **4** **Use a compound interest table.** The calculation of compound interest can be quite tedious. For this reason, compound interest tables have been developed.

Suppose you deposit $1 in a savings account today that earns 4% compounded annually and you allow the deposit to remain for 3 years. The diagram below shows the compound amount at the end of each of the 3 years.

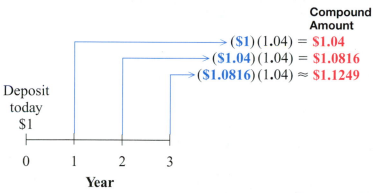

Compound Amount

→ ($1)(1.04) = **$1.04**
→ ($1.04)(1.04) = **$1.0816**
→ ($1.0816)(1.04) ≈ **$1.1249**

Deposit today
$1

0   1   2   3
Year

Using the compound amounts for $1, a table can be formed. Look at the table below and find the column headed 4%. The first three numbers for years 1, 2, and 3 are the same as those we have calculated for $1 at 4% for 3 years. This table, giving the compound amounts on a $1 deposit for given lengths of time and interest rates, can be used for finding the compound amount on any amount of deposit.

COMPOUND INTEREST

Time Periods	3.00%	3.50%	4.00%	4.50%	5.00%	5.50%	6.00%	8.00%	Time Periods
1	1.0300	1.0350	1.0400	1.0450	1.0500	1.0550	1.0600	1.0800	1
2	1.0609	1.0712	1.0816	1.0920	1.1025	1.1130	1.1236	1.1664	2
3	1.0927	1.1087	1.1249	1.1412	1.1576	1.1742	1.1910	1.2597	3
4	1.1255	1.1475	1.1699	1.1925	1.2155	1.2388	1.2625	1.3605	4
5	1.1593	1.1877	1.2167	1.2462	1.2763	1.3070	1.3382	1.4693	5
6	1.1941	1.2293	1.2653	1.3023	1.3401	1.3788	1.4185	1.5869	6
7	1.2299	1.2723	1.3159	1.3609	1.4071	1.4547	1.5036	1.7138	7
8	1.2668	1.3168	1.3686	1.4221	1.4775	1.5347	1.5938	1.8509	8
9	1.3048	1.3629	1.4233	1.4861	1.5513	1.6191	1.6895	1.9990	9
10	1.3439	1.4106	1.4802	1.5530	1.6289	1.7081	1.7908	2.1589	10
11	1.3842	1.4600	1.5395	1.6229	1.7103	1.8021	1.8983	2.3316	11
12	1.4258	1.5111	1.6010	1.6959	1.7959	1.9012	2.0122	2.5182	12

EXAMPLE 3 **Using a Compound Interest Table**

Find each compound amount using the compound interest table. Round answers to the nearest cent.

(a) $1 is deposited at a 5% interest rate for 10 years.

Look down the column headed 5%, and across to row 10 (because 10 years = 10 time periods). At the intersection of the column and row, read the compound amount, **1.6289**, which can be rounded to $1.63.

(b) $1 is deposited at $3\frac{1}{2}$% for 6 years.

The intersection of the $3\frac{1}{2}$% (3.50%) column and row 6 shows 1.2293 as the compound amount. Round this to $1.23.

Work Problem 3 at the Side. ▶▶▶

---

3 Find the compound amount using the compound interest table. Round to the nearest cent.

(a) $1 at 3% for 6 years

(b) $1 at 6% for 8 years

(c) $1 at $4\frac{1}{2}$% for 12 years

4 Find the compound amount and the interest.

(a) $4000 at 3% for 10 years

(b) $12,600 at $5\frac{1}{2}$% for 8 years

(c) $32,700 at 6% for 12 years

OBJECTIVE 5 Find the compound amount and the amount of compound interest. Find the compound amount and interest as follows.

Finding the Compound Amount and the Interest

Compound Amount

Find the compound amount for any amount of principal by multiplying the principal by the compound amount for $1.

Interest

Find the amount of interest earned on a deposit by subtracting the amount originally deposited from the compound amount.

EXAMPLE 4 Finding Compound Amount and Interest

Find the compound amount and the interest.

(a) $1000 at $5\frac{1}{2}$% interest for 12 years.

Look in the table on page 445 for $5\frac{1}{2}$% (5.50%) and 12 periods; find the number 1.9012 but do *not* round it. Multiply this number and the principal of $1000.

$$(\$1000)\,(\textbf{1.9012}) = \textbf{\$1901.20}$$

The account will contain $1901.20 after 12 years.

Find the amount of interest by subtracting the original deposit from the compound amount.

$$\textbf{\$1901.20} - \$1000 = \textbf{\$901.20}$$

(b) $6400 at 8% for 7 years

Look in the table for 8% and 7 periods, finding 1.7138. Multiply.

$$(\$6400)\,(\textbf{1.7138}) = \textbf{\$10,968.32} \qquad \text{Compound amount}$$

Subtract the original deposit from the compound amount.

$$\textbf{\$10,968.32} - \$6400 = \textbf{\$4568.32} \qquad \text{Interest}$$

A total of $4568.32 in interest was earned.

◀◀◀ Work Problem 4 at the Side.

6.8 Exercises

FOR EXTRA HELP

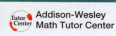 Addison-Wesley Math Tutor Center

 MathXL

 Digital Video Tutor CD 4 Videotape 11

 Student's Solutions Manual

MyMathLab

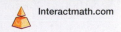 Interactmath.com

Find the compound amount given the following deposits. Calculate the interest each year, then add it to the previous year's amount. See Example 1.

1. $500 at 4% for 2 years

2. $1000 at 5% for 3 years

3. $1800 at 3% for 3 years

4. $2000 at 8% for 3 years

5. $3500 at 7% for 4 years

6. $5500 at 6% for 4 years

Find each compound amount by multiplying the original amount deposited by 100% plus the compound rate given in the following. See Example 2. Round to the nearest cent if necessary.

7. $1000 at 5% for 2 years

8. $500 at 4% for 3 years

9. $1400 at 6% for 5 years

10. $2500 at 3% for 4 years

11. $1180 at 7% for 8 years

12. $12,800 at 6% for 7 years

13. $10,940 at 8% for 6 years

14. $15,710 at 10% for 8 years

Use the table on page 445 to find the compound amount and the interest. Interest is compounded annually. Round to the nearest cent if necessary. See Example 3.

15. $1000 at 4% for 5 years

16. $10,000 at 3% for 4 years

17. $8000 at 6% for 10 years

18. $7800 at 5% for 8 years

19. $8428.17 at $4\frac{1}{2}$% for 6 years

20. $10,472.88 at $5\frac{1}{2}$% for 12 years

21. Write a definition for compound interest. Describe in your own words what compound interest means to you.

22. What is the difference between the compound amount and compound interest?

Use the table on page 445 to solve each application problem. Round to the nearest cent if necessary. See Examples 3 and 4.

23. Jane Chavez deposited $8450 in an account that pays 6% interest compounded annually. Find the amount she will have (compound amount) at the end of 8 years.

24. Rob Diamond borrowed $22,500 from his uncle to open Bookkeeping and More. He will repay the loan at the end of 5 years at 8% interest compounded annually. Find the amount he will repay.

25. Al Granard lends $76,000 to the owner of Rick's Limousine Service. He will be repaid at the end of 9 years at 6% interest compounded annually. Find **(a)** the total amount that he should be repaid and **(b)** the amount of interest earned.

26. Sadie Simms has $44,500 in an individual retirement account (IRA) that pays 5% interest compounded annually. Find **(a)** the total amount she will have at the end of 8 years and **(b)** the amount of interest earned.

27. Jennifer Barrister deposits $30,000 at 6% interest compounded annually. Two years after she makes the first deposit, she deposits another $40,000, also at 6% compounded annually.

   (a) What total amount will she have five years after her first deposit?

   (b) What amount of interest will she have earned?

28. Kara Ivee invests $25,000 at 8% interest compounded annually. Three years after she makes the first deposit, she deposits another $25,000, also at 8% compounded annually.

   (a) What total amount will she have five years after her first deposit?

   (b) What amount of interest will she have earned?

Knowing how to solve interest problems is important to businesspeople and consumers alike. **Work Exercises 29–34 in order.**

29. Simple interest calculation is used for most short-term business loans, most real estate loans, and many automobile and consumer loans. The formula for simple interest is

Interest = _____ • _____ • _____

or $I =$ _____ • _____ • _____ .

30. When a loan is repaid, the interest is added to the original principal. The formula for amount due is

amount due = _____ + _____ .

31. Compound interest is paid on most savings accounts and many other types of investments. Compound interest is interest calculated on _____ plus past _____ .

32. The compound amount is the total amount in an account at the end of a period of time. Compound amount is the original _____ + compound _____ .

33. Donna Gonsalves has two choices. She can invest $4350 at 6% simple interest for 6 years, or she can invest the same amount at 6% compounded annually for 6 years.

   (a) Find the difference in the amount of interest earned in these two accounts.

   (b) If the length of time is doubled from 6 to 12 years, will the difference in the amount of interest earned also double?

   (c) Use your own examples to determine that what you found in part (b) is true with other interest rates and other lengths of time.

34. One account is opened with $20,000 at 8% simple interest for 12 years. Another account is opened with $16,000 at 8% compounded annually for 12 years.

   (a) At the end of the 12 years, which account has the higher balance and by how much?

   (b) What does this tell you about compound interest? Why?

Chapter 6
SUMMARY

NEW FORMULAS

To write percents as decimals: $p\% = p \div 100$

To write percents as fractions: $p\% = \dfrac{p}{100}$

Percent proportion: $\dfrac{\text{part}}{\text{whole}} = \dfrac{\text{percent}}{100}$

Percent equation: part = percent • whole

Amount of sales tax: amount of sales tax = rate of tax • cost of item

Amount of commission: amount of commission = rate of commission • amount of sales

Amount of discount: amount of discount = rate of discount • original price

Sale price: sale price = original price − amount of discount

Percent of increase: $\dfrac{\text{amount of increase}}{\text{original value}} = \dfrac{\text{percent}}{100}$

Percent of decrease: $\dfrac{\text{amount of decrease}}{\text{original value}} = \dfrac{\text{percent}}{100}$

Interest: $I = p \cdot r \cdot t$ or   Interest = principal • rate • time

Amount due: amount due = principal + interest

TEST YOUR WORD POWER

See how well you have learned the vocabulary in this chapter. Answers follow the Quick Review.

1. To write a **percent as a decimal,** you drop the percent sign
   A. after finding the decimal point
   B. after removing the decimal point
   C. after moving the decimal point two places to the right
   D. after moving the decimal point two places to the left.

2. To write a **decimal as a percent,** you add the percent sign
   A. after finding the decimal point
   B. after removing the decimal point
   C. after moving the decimal point two places to the right
   D. after moving the decimal point two places to the left.

3. **Percent** means
   A. the same as interest
   B. per every ten
   C. per one thousand
   D. per one hundred.

4. When you use $\dfrac{\text{part}}{\text{whole}} = \dfrac{\text{percent}}{100}$ to solve percent problems, you are using the
   A. simple interest formula
   B. percent proportion
   C. percent equation
   D. percent sales tax formula.

5. The **percent equation** is
   A. part = percent • whole
   B. $I = p \cdot r \cdot t$
   C. $p\% = \dfrac{p}{100}$
   D. amount due = principal + interest.

6. In a **percent of increase or decrease** problem, the increase or decrease is a percent of
   A. the largest amount
   B. the original amount
   C. the new or most recent amount
   D. all the amounts.

7. In the formula $I = p \cdot r \cdot t,$ the p stands for
   A. proportion
   B. product
   C. principal
   D. percent.

8. The term **rate** in an interest problem represents the
   A. whole
   B. percent
   C. part
   D. amount of interest.

QUICK REVIEW

Concepts	Examples

6.1 *Basics of Percent*

Writing a Percent as a Decimal
To write a percent as a decimal, move the decimal point two places to the left and drop the % sign.

$50\% \ (.50\%) = 0.50$ or just 0.5

$3\% \ (.03\%) = 0.03$

Writing a Decimal as a Percent
To write a decimal as a percent, move the decimal point two places to the right and attach a % sign.

$0.75 \ (0.75) = 75\%$

$3.6 \ (3.60) = 360\%$

6.2 *Writing a Fraction as a Percent*

Use a proportion and solve for p to change a fraction to percent.

$\dfrac{2}{5} = \dfrac{p}{100}$ Proportion

$5 \cdot p = 2 \cdot 100$ Cross products

$5 \cdot p = 200$

$\dfrac{\overset{1}{\cancel{5}} \cdot p}{\underset{1}{\cancel{5}}} = \dfrac{200}{5}$ Divide both sides by 5.

$p = 40$

$\dfrac{2}{5} = 40\%$ Attach % sign.

Concepts	Examples

6.3 *Learning the Percent Proportion*

Part is to whole as percent is to 100.

$$\frac{\textbf{part}}{\textbf{whole}} = \frac{\textbf{percent}}{\textbf{100}} \leftarrow \text{Always } 100$$

Use the percent proportion to solve for the unknown value.
part = 30, whole = 50; find the percent.

$$\begin{array}{cc} & \overset{\displaystyle\text{Percent (unknown)}}{\downarrow} \\ \text{Part} \rightarrow \dfrac{30}{50} = \dfrac{x}{100} & \leftarrow \text{Always } 100 \\ \text{Whole} \rightarrow & \end{array}$$

$$\frac{3}{5} = \frac{x}{100} \qquad \tfrac{30}{50} \text{ is } \tfrac{3}{5} \text{ in lowest terms.}$$

$$5 \cdot x = 3 \cdot 100 \qquad \text{Cross products}$$

$$5 \cdot x = 300$$

$$\frac{\overset{1}{\cancel{5}} \cdot x}{\cancel{5}} = \frac{300}{5} \qquad \text{Divide both sides by 5.}$$

$$x = 60$$

The percent is 60, which is written as 60%.

6.3 *Identifying Percent, Whole, and Part in a Percent Problem*

The percent appears with the word **percent** or with the symbol **%**.

Find the percent, whole, and part in the following.

10% **of** the 500 pies is how many pies?

Percent · Whole · Part (unknown)

The whole often appears after the word **of**. The whole is the entire quantity or total.

20 cats is 5% **of** what number of cats?

Part · Percent · Whole (unknown)

The part is the portion of the total. If the percent and the whole are found first, the remaining number is the part.

What percent **of** $220 is $33?

Percent (unknown) · Whole · Part

6.4 *Applying the Percent Proportion*

Read the problem and identify the percent, whole, and part. Use the percent proportion to solve for the unknown quantity.

A liquid mixture in a tank contains 35% distilled water. If 28 gallons of distilled water are in the tank when it is full, find the capacity of the tank.

$$\text{percent} = 35 \quad \text{and} \quad \text{part} = 28$$

Use the percent proportion to find the whole.

$$\text{Whole (unknown)} \rightarrow \frac{\text{part}}{x} = \frac{\text{percent}}{100}$$

$$\frac{28}{x} = \frac{35}{100}$$

$$\frac{28}{x} = \frac{7}{20} \qquad \text{Lowest terms}$$

$$x \cdot 7 = 560 \qquad \text{Cross products}$$

$$\frac{x \cdot \overset{1}{\cancel{7}}}{\cancel{7}} = \frac{560}{7} \qquad \text{Divide both sides by 7.}$$

$$x = 80$$

The capacity of the tank is 80 gallons.

Concepts	**Examples**

6.5 *Using the Percent Equation*

The percent equation is part = percent • whole. Identify the percent, whole, and part and solve for the unknown quantity. Always write the percent as a decimal before using the equation.

Solve each problem.

(a) Find 20% of 220 applicants.

$$\textbf{part (unknown)} = \text{percent} \cdot \text{whole}$$

$$x = (0.2)(220)$$

$$x = 44$$

20% of 220 applicants is 44 applicants.

(b) 8 balls is 4% of what number of balls?

$$\text{part} = \text{percent} \cdot \textbf{whole (unknown)}$$

$$8 = (0.04)(x)$$

$$\frac{8}{0.04} = \frac{\overset{1}{\cancel{(0.04)}}(x)}{\underset{1}{\cancel{0.04}}}$$

$$x = 200$$

8 balls is 4% of 200 balls.

(c) $13 is what percent of $52?

$$\text{part} = \textbf{percent (unknown)} \cdot \text{whole}$$

$$13 = x \cdot 52$$

$$\frac{13}{52} = \frac{x \cdot \overset{1}{\cancel{52}}}{\underset{1}{\cancel{52}}}$$

$$x = 0.25 = 25\%$$

$13 is 25% of $52.

6.6 *Solving Application Problems with Proportions*

To solve for **sales tax,** use this formula.

amount of sales tax = rate of tax • cost of item

To find **commissions,** use this formula.

amount of commission = rate of commission • amount of sales

To find the **discount** and the **sale price,** use these formulas.

amount of discount = rate of discount • original price

sale price = original price − amount of discount

The price of a 35-inch television is $699, and the sales tax is 5%. Find the sales tax.

amount of sales tax = (5%) ($699)

= (0.05) ($699) = $34.95

The sales are $92,000 with a commission rate of 3%. Find the commission.

amount of commission = (3%) ($92,000)

= (0.03) ($92,000)

= $2760

A gas oven originally priced at $480 is offered at a 25% discount. Find the amount of the discount and the sale price.

discount = (0.25) ($480) = **$120**

sale price = $480 − **$120** = **$360**

Concepts	Examples

6.6 Solving Application Problems with Proportions (continued)

To find the **percent of change,** subtract to find the amount of change (increase or decrease), which is the part. The whole is the original value or value before the change.

The number of parking violations rose from 1980 violations to 2277. Find the percent of increase.

$$2277 - 1980 = 297 \quad \text{Increase}$$

$$\frac{297}{1980} = \frac{\text{percent}}{100}$$

Solve the proportion to find that the percent = 15, so the percent of increase is 15%.

6.7 Finding Simple Interest

Use the formula $I = p \cdot r \cdot t$

$$\text{Interest} = \text{principal} \cdot \text{rate} \cdot \text{time}$$

Time (t) is in years. When the time is given in months, use a fraction with 12 in the denominator because there are 12 months in a year.

$2800 is deposited at 8% for 3 months. Find the amount of interest.

$$I = p \cdot r \cdot t$$
$$= (2800)(0.08)\left(\frac{3}{12}\right)$$
$$= (224)\left(\frac{1}{4}\right) = \frac{(224)(1)}{4} = \$56$$

6.8 Finding Compound Amount and Compound Interest

There are three methods for finding the compound amount.
1. Calculate the interest for each compound interest period, then add it back to the principal.

Find the compound amount and interest if $1500 is deposited at 5% interest for 3 years.

1.

	Interest	Compound Amount
Year 1	($1500)(0.05)(1) = **$75**	$1500 + **$75** = **$1575**
Year 2	($1575)(0.05)(1) = **$78.75**	$1575 + **$78.75** = **$1653.75**
Year 3	($1653.75)(0.05) ≈ **$82.69**	$1653.75 + **$82.69** = **$1736.44**

2. Multiply the original deposit by 100% plus the compound interest rate.

2. ($1500)(1.05)(1.05)(1.05) ≈ $1736.44

Original deposit   Compound amount

100% + 5% = 105% = 1.05

3. Use the table on page 445 to find the interest on $1. Then, multiply the table value by the principal

The compound interest is found with the formula.

$$\text{compound interest} = \text{compound amount} - \text{original deposit}$$

3. Locate 5% across the top of the table and 3 periods at the left. The table value is 1.1576.

compound amount = ($1500)(1.1576) = $1736.40*

interest = $1736.40 - $1500 = $236.40

* The difference in the compound amount results from rounding in the table.

ANSWERS TO TEST YOUR WORD POWER

1. D; *Example:* 50% written as a decimal is 0.50 or 0.5.

2. C; *Example:* 0.25 written as a percent is 0.25% or 25%.

3. D; *Example:* 8% means 8 per 100.

4. B; *Example:* Part = 4, and whole = 25. To find the percent, $\dfrac{4}{25} = \dfrac{x}{100}$; $x = 16$ or 16%.

5. A; *Example:* Percent = 25, and whole = 300. To find the part, $0.25 \times 300 = 75$.

6. B; *Example:* Original value = \$200, and amount of increase or decrease = \$40. To find the percent of increase or decrease,

$\dfrac{40}{200} = 0.2 = 20\%$ of increase or decrease.

7. C; *Example:* Principal (p) = \$800, rate (r) = 5%, and time (t) = 1 year. Then, $I = p \cdot r \cdot t$ so $I = (\$800)(0.05)(1) = \$40$.

8. B; *Example:* Principal (p) = \$1650, rate (r) = 4%, and time (t) = $\dfrac{1}{2}$ year. To find the interest (I),

$I = (\$1650)(0.04)\left(\dfrac{1}{2}\right) = \$33$.

Chapter 6
REVIEW EXERCISES

[6.1] *Write each percent as a decimal and each decimal as a percent.*

$35 \div 100 = .35$

1. 35% *.35*

2. 150% *1.5*

3. 99.44% *.9944*

4. 0.085% *.00085*

5. 3.15 *315%* 3.15×100

6. 0.02 *2%*

7. 0.875 *87.5%*

8. 0.002 *.2%*

[6.2] *Write each percent as a fraction or mixed number in lowest terms and each fraction as a percent.*

9. 15% *.15 $\frac{15}{100}$ $\frac{3}{20}$*

10. 37.5% *.375 $\frac{375}{1000}$ $\frac{15}{40}$ $\frac{3}{8}$*

11. 175% *1.75 $\frac{175}{100}$ $\frac{3}{4}$*

12. 0.25% *.0025 $\frac{25}{10000}$ $\frac{5}{2000}$ $\frac{1}{400}$*

13. $\frac{3}{4}$ *.75 $\frac{75}{100}$ $\frac{3}{4}$*

14. $\frac{5}{8}$ *.625 $\frac{625}{1000}$ $\frac{25}{40}$ $\frac{5}{8}$ 62.5 or $62\frac{1}{2}$%*

15. $3\frac{1}{4}$ *3.25 325%*

16. $\frac{1}{200}$ *.005 .5%*

Complete this chart.

Fraction	Decimal	Percent
$\frac{1}{8}$	**17.** ____	**18.** ____
19. $\frac{1}{4}$	0.25	**20.** *25%* $\frac{25}{100}$
21. $1\frac{8}{10}$ $\frac{4}{5}$	**22.** *1.8*	180%

[6.3] *Find the unknown value in the percent proportion* $\dfrac{part}{whole} = \dfrac{percent}{100}$.

23. part = 25, percent = 10 $\dfrac{25}{x} = \dfrac{10}{100}$ $x \cdot 10 = 2500$ 250

24. whole = 480, percent = 5

Identify each component and then set up each problem using the percent proportion, $\dfrac{part}{whole} = \dfrac{percent}{100}$. *Do not try to solve for the unknown value.*

25. 35% of 820 mailboxes is 287 mailboxes.

$\dfrac{287}{820} = \dfrac{35}{100}$

26. 73 brooms is what percent of 90 brooms?

27. Find 14% of 160 bicycles.

28. 418 curtains is 16% of what number of curtains?

29. A golfer lost three of his eight golf balls. What percent were lost?

30. Only 88% of the door keys cut will operate properly. If there are 1280 keys cut, find the number of keys that will operate properly.

[6.4] *Find the part using the percent proportion or the multiplication shortcut.*

31. 18% of 950 programs

$.18 \times 950 = 171$

32. 60% of 1450 reference books

33. 0.6% of 5200 acres

$.006 \times 5200 = 31.2$

34. 0.2% of 1400 kilograms

Find the whole using the percent proportion.

$\dfrac{part}{whole} = \dfrac{\%}{100}$

35. 105 crates is 14% of what number of crates?

$\dfrac{105}{x} \qquad \dfrac{14}{100}$

$\dfrac{x \cdot 14}{14} = \dfrac{10500}{14} \qquad 750$

36. 348 test tubes is 15% of what number of test tubes?

37. 677.6 miles is 140% of what number of miles?

whole

38. 2.5% of what number of cases is 425 cases?

part   whole

*Find the percent using the **percent proportion**. Round percent answers to the nearest tenth if necessary.*

39. 649 tulip bulbs is what percent of 1180 tulip bulbs?

$\dfrac{649}{1180} \qquad \dfrac{x}{100}$

$x \cdot 1180 = 64900 = 55\%$

40. What percent of 1620 dinner rolls is 85 dinner rolls?

41. What percent of 380 pairs of socks is 36 pairs?

95

1.2

1.2 95/100   95/100   1.14/x   95=114

42. What percent of 650 soup cans is 200 soup cans?

[6.1–6.4] *Solve each application problem. Round percent answers to the nearest tenth if necessary.*

43. The average cost of 30 seconds of advertising during the Super Bowl five years ago was $1.2 million. If the increase in cost over the last five years has been 95%, find the average cost of 30 seconds of advertising during the Super Bowl this year. Round to the nearest tenth of a million. (*Source:* NFL Research.)

$\dfrac{x}{1,200,000} \qquad \dfrac{95}{100}$

$\dfrac{x \cdot 100}{100} = \dfrac{114,000,000}{100}$

$1,140,000$

$1,200,000$

$1,200,000 \qquad \dfrac{95}{100}$

$\dfrac{.95}{100}$

$1.2 \times .95$ $1.2 = .95 \times 1.14$

44. Scientists tell us that there are 9600 species of birds and that 1000 of these species are in danger of extinction. What percent of the bird species are in danger of extinction?

$x \cdot 95 = 129,000,000$

[6.5] *Use the percent equation to answer each question.*

45. 32% of $454 is what amount?

$P = 32 \cdot 454$

$145.28

46. 155% of 120 trucks is how many trucks?

47. 0.128 ounce is what percent of 32 ounces?

48. 304.5 meters is what percent of 174 meters?

49. 33.6 miles is 28% of what number of miles?

50. $92 is 16% of what number?

[6.6] *Find the amount of sales tax or the tax rate and the total cost. Round to the nearest cent if necessary.*

	Amount of Sale	Tax Rate	Amount of Tax	Total Cost
51.	$630	5%	_____	_____
52.	$780	_____	$58.50	_____

Find the commission earned or the rate of commission.

	Sales	Rate of Commission	Commission
53.	$3450	8%	_____
54.	$65,300	_____	$3265

Find the amount or rate of discount and the amount paid after the discount. Round to the nearest cent if necessary.

	Original Price	Rate of Discount	Amount of Discount	Sale Price
55.	$112.50	30%	_____	_____
56.	$252	_____	$63	_____

[6.7] *Find the single interest due on each loan.*

Principal	Rate	Time in Years	Interest
57. $200	4%	1	_____
58. $1080	5%	$1\frac{1}{4}$	_____

Find the simple interest paid on each investment.

Principal	Rate	Time in Months	Interest
59. $400	7%	3	_____
60. $1560	$6\frac{1}{2}\%$	18	_____

Find the total amount due on each simple interest loan.

Principal	Rate	Time	Total Amount Due
61. $750	$5\frac{1}{2}\%$	2 years	_____
62. $1530	6%	9 months	_____

[6.8] *Find the compound amount and compound interest in the following. Interest is compounded annually. You may use the table on page 445. Round to the nearest cent if necessary.*

Principal	Rate	Time in Years	Compound Amount	Compound Interest
63. $4000	3%	10	_____	_____
64. $1870	4%	4	_____	_____
65. $3600	8%	3	_____	_____
66. $12,500	$5\frac{1}{2}\%$	5	_____	_____

MIXED REVIEW EXERCISES

Find the unknown value in the percent proportion $\dfrac{part}{whole} = \dfrac{percent}{100}$.

67. whole = 80, percent = 15

68. part = 738, percent = 45

Use the percent proportion or percent equation to answer each question.

69. 12% of 194 meters is how many meters?

70. 327 cars is what percent of 218 cars?

71. 0.6% of $85 is what amount?

72. 396 employees is 20% of what number of employees?

73. 76 chickens is what percent of 190 chickens?

74. 214.484 liters is 43% of what number of liters?

Write each percent as a decimal and each decimal as a percent.

75. 55%

76. 300%

77. 5

78. 4.71

79. 8.6%

80. 0.621

81. 0.375%

82. 0.0006

Write each percent as a fraction in lowest terms and each fraction as a percent.

83. $\dfrac{3}{4}$

84. 42%

85. 87.5%

86. $\dfrac{3}{8}$

87. $32\dfrac{1}{2}\%$

88. $\dfrac{3}{5}$

89. 0.25%

90. $3\dfrac{3}{4}$

Solve each application problem. Round percent answers to the nearest tenth and money amounts to the nearest cent if necessary.

91. Jim Bralley invests $16,850 at $10\frac{1}{2}\%$ for 9 months. Find the amount of interest earned.

92. Jim Havey borrows $14,750 at 8% for 18 months to buy a small Lionel train collection. Find the total amount due.

93. A Hotpoint refrigerator has a capacity of 11.5 cubic feet in the refrigerator and 5.5 cubic feet in the freezer. What percent of the total capacity is the capacity of the freezer?

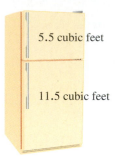

5.5 cubic feet

11.5 cubic feet

94. Tommy Downs invests the money he inherited from his aunt at 6% compounded annually for 4 years. If the amount of money invested is $12,500, find

(a) the compound amount at the end of 4 years and

(b) the amount of interest that he earned. Do not use the table.

95. Tom Dugally, a real estate agent, sold two properties, one for $125,000 and the other for $290,000. After all of his expenses, he receives a commission of $1\frac{1}{2}$% of total sales. Find the commission that he earned.

96. Our mail carrier, Norm, saw his route expand from 481 residential stops to 520 residential stops. Find the percent of increase.

97. A Sears Kenmore washer/dryer set priced at $958 is marked down 18%. If the sales tax is 8%, find the cost of the washer/dryer set including the sales tax.

98. In a telephone survey, 749 people said diabetes is a serious problem in the United States. If this was 71% of the survey group, find the total number of people in the telephone survey. Round to the nearest whole number. (*Source:* Hoffman-LaRouche Company.)

99. Stephen and Heather Hall established a budget allowing 25% for rent, 30% for food, 8% for clothing, 20% for travel and recreation, and the remainder for savings. Stephen takes home $2075 per month, and Heather takes home $32,500 per year. How much money will the couple save in a year?

100. The mileage on a car dropped from 32.8 miles per gallon to 28.5 miles per gallon. Find the percent of decrease.

Chapter **6**

T E S T

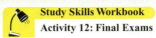
Study Skills Workbook
Activity 12: Final Exams

Write each percent as a decimal and each decimal as a percent.

1. 65%

2. 0.8

3. 1.75

4. 0.875

5. 300%

6. 0.05%

Write each percent as a fraction in lowest terms.

7. 12.5%

8. 0.25%

Write each fraction or mixed number as a percent.

9. $\dfrac{3}{5}$

10. $\dfrac{5}{8}$

11. $2\dfrac{1}{2}$

Solve each problem.

12. 32 sacks is 4% of what number of sacks?

13. $680 is what percent of $3400?

14. Ann Barnes has saved 65% of the amount needed for a down payment on a home. If she has saved $12,025, find the total down payment needed.

15. The price of a diamond engagement ring is $3240 plus sales tax of $6\frac{1}{2}$%. Find the total cost of the engagement ring including sales tax.

16. An insurance company pays its salespeople on commission. If a commission of $628 is earned on insurance sales of $7850, find the rate of commission paid.

1. _____

2. _____

3. _____

4. _____

5. _____

6. _____

7. _____

8. _____

9. _____

10. _____

11. _____

12. _____

13. _____

14. _____

15. _____

16. _____

17. _____

17. Attendance at the homecoming game decreased from 5760 fans last year to 4320 fans this year. Find the percent of decrease.

18. _____

18. A problem includes last year's salary, this year's salary, and asks for the percent of increase. Explain how you would identify the part, the whole, and the percent in the problem. Show the percent proportion that you would use.

19. _____

19. Write the formula used to find interest. Explain the difference in what to do if the time is expressed in months or in years. Write a problem that involves finding interest for 9 months and another problem that involves finding interest for $2\frac{1}{2}$ years. Use your own numbers for the principal and the rate. Show how to solve your problems.

Find the amount of discount and the sale price.

	Original Price	**Rate of Discount**
20.	$96	12%
21.	$280	32.5%

20. _____

21. _____

Find the simple interest on each loan.

	Principal	**Rate**	**Time**
22.	$4200	6%	$1\frac{1}{2}$ years
23.	$6400	5%	9 months

22. _____

23. _____

24. _____

24. A parent borrows $5300 to help her son start college. The loan is for 9 months at 9% interest. Find the total amount due on the loan.

25. (a) _____

(b) _____

25. The River City School PTA Emergency Fund deposited $4000 at 6% compounded annually. Two years after the first deposit, they deposit another $5000, also at 6% compounded annually. Use the compound interest table on page 445.

(a) What total amount will they have 4 years after their first deposit? Round to the nearest dollar.

(b) What amount of interest will they have earned?

Cumulative Review Exercises

CHAPTERS 1-6

≈ *First use front end rounding to round each number and estimate the answer. Then find the exact answer.*

1. *Estimate:*   *Exact:*

         5608
           94
  +___   + 739

2. *Estimate:*   *Exact:*

          0.56
         49.614
  +___   + 8.4

3. *Estimate:*   *Exact:*

          75,078
  ___   − 46,090

4. *Estimate:*   *Exact:*

          7.8
  ___   − 3.5029

5. *Estimate:*   *Exact:*

          6538
  ×___   × 708

6. *Estimate:*   *Exact:*

          65.3
  ×___   − 8.7

7. *Estimate:*   *Exact:*

)___   43)38,786

8. *Estimate:*   *Exact:*

)___   7.6)2432

9. *Estimate:*   *Exact:*

)___   0.8)6.76

Use the order of operations to simplify each expression.

10. $6^2 - 3(6)$

11. $\sqrt{49} + 5 \cdot 4 - 8$

12. $9 + 6 \div 3 + 7(4)$

Round each number to the place shown.

13. 4677 to the nearest ten

14. 7,583,281 to the nearest hundred thousand

15. $513.499 to the nearest dollar

16. $362.735 to the nearest cent

Add, subtract, multiply, or divide as indicated. Write answers in lowest terms and as whole or mixed numbers when possible.

17. $\dfrac{3}{4} + \dfrac{5}{8}$

18. $\dfrac{1}{3} + \dfrac{3}{4}$

19. $\begin{aligned} & 5\dfrac{3}{4} \\ +\, & 7\dfrac{5}{8} \\ \hline \end{aligned}$

20. $\dfrac{7}{8} - \dfrac{3}{4}$

21. $\begin{aligned} & 8\dfrac{3}{8} \\ -\, & 4\dfrac{1}{2} \\ \hline \end{aligned}$

22. $\begin{aligned} & 26\dfrac{1}{3} \\ -\, & 17\dfrac{4}{5} \\ \hline \end{aligned}$

23. $\left(\dfrac{7}{8}\right)\left(\dfrac{2}{3}\right)$

24. $7\dfrac{3}{4} \cdot 3\dfrac{3}{8}$

25. $36 \cdot \dfrac{4}{5}$

26. $\dfrac{5}{8} \div \dfrac{5}{7}$

27. $12 \div \dfrac{3}{4}$

28. $2\dfrac{3}{4} \div 7\dfrac{1}{2}$

Write < or > to make a true statement.

29. $\dfrac{5}{8}$ _____ $\dfrac{2}{3}$

30. $\dfrac{8}{15}$ _____ $\dfrac{11}{20}$

31. $\dfrac{2}{3}$ _____ $\dfrac{7}{12}$

Simplify each expression. Use the order of operations as needed.

32. $\dfrac{2}{3}\left(\dfrac{7}{8} - \dfrac{1}{2}\right)$

33. $\dfrac{7}{8} \div \left(\dfrac{3}{4} + \dfrac{1}{8}\right)$

34. $\left(\dfrac{5}{6} - \dfrac{5}{12}\right) - \left(\dfrac{1}{2}\right)^2 \cdot \dfrac{2}{3}$

Write each fraction as a decimal. Round to the nearest thousandth if necessary.

35. $\dfrac{3}{4}$

36. $\dfrac{3}{8}$

37. $\dfrac{7}{12}$

38. $\dfrac{11}{20}$

Write each ratio as a fraction in lowest terms. Be sure to make all necessary conversions.

39. 2 hours to 40 minutes

40. If there are 27 people and 36 life jackets, what is the ratio of life jackets to people?

41. $1\dfrac{5}{8}$ to 13

Use cross products to decide whether each proportion is true *or* false. *Show the cross products and then circle* True *or* False.

42. $\dfrac{5}{8} = \dfrac{45}{72}$

    True   False

43. $\dfrac{64}{144} = \dfrac{48}{108}$

    True   False

Find the unknown value in each proportion.

44. $\dfrac{1}{5} = \dfrac{x}{30}$

45. $\dfrac{224}{32} = \dfrac{28}{x}$

46. $\dfrac{8}{x} = \dfrac{72}{144}$

47. $\dfrac{x}{120} = \dfrac{7.5}{30}$

Write each percent as a decimal. Write each decimal as a percent.

48. 78%

49. 3%

50. 200%

51. 0.5%

52. 0.87

53. 3.8

54. 0.023

Write each percent as a fraction or mixed number in lowest terms. Write each fraction as a percent.

55. 8%

56. 62.5%

57. 175%

58. $\dfrac{7}{8}$

59. $\dfrac{3}{10}$

60. $4\dfrac{1}{5}$

Solve each percent problem.

61. 65% of 3280 DVDs is how many DVDs?

62. $5\dfrac{1}{2}$% of $720 is how much?

63. 72 tires is 40% of what number of tires?

64. $4\dfrac{1}{2}$% of what number of miles is 76.5 miles?

65. What percent of 656 books is 328 books?

66. 72 hours is what percent of 180 hours?

Find the amount of sales tax or the tax rate and the total cost. Round to the nearest cent if necessary.

	Amount of Sale	Tax Rate	Amount of Tax	Total Cost
67.	$8.50	4%	_____	_____
68.	$460	_____	$29.90	_____

Find the commission earned or the rate of commission.

	Sales	Rate of Commission	Commission
69.	$12,538	8%	_____
70.	$225,300	_____	$5632.50

Find the amount or rate of discount and the amount paid after the discount. Round to the nearest cent if necessary.

	Original Price	Rate of Discount	Amount of Discount	Sale Price
71.	$456	45%	_____	_____
72.	$1085	_____	$162.75	_____

Find the total amount due on each simple interest loan. Round to the nearest cent if necessary.

	Principal	Rate	Time	Total Amount to be Repaid
73.	$970	10%	$1\frac{1}{2}$ years	_____
74.	$22,250	7%	6 months	_____

Set up and solve a proportion for each problem.

75. Carol can test the hearing of 11 patients in 4 hours. Find the number of hearing tests that she can do in 12 hours.

76. If 12.5 ounces of Roundup weed and grass killer is needed to make 5 gallons of spray, how much Roundup is needed for 102 gallons of spray?

The number of existing single-family homes sold in four regions of the country in the same month of two separate years are shown in the figure below. Use this information to answer Exercises 77–80. Round answers to the nearest tenth of a percent if necessary.

AT HOME
Existing home sales by region.

West 54,000 / 49,600
Midwest 65,000 / 66,300
Northeast 32,000 / 36,000
South 82,000 / 77,500

■ = Last year
■ = This year

77. Find the percent of increase in sales in the northeastern region.

78. Find the percent of increase in sales in the midwestern region.

79. What is the percent of decrease in sales in the southern region?

80. What is the percent of decrease in sales in the western region?

81. Joan Ong has $26,880 invested in her home. If this amount is 25% of her total assets, find her total assets.

82. Pat Ueda deposits $15,000 in a savings account that pays 5.5% interest compounded annually. Find

    (a) the amount that he will have in the account at the end of 5 years, and

    (b) the amount of interest earned. Do not use the table.

Measurement

7

NASA's Hubble space telescope took this picture of Mars when it was approximately 43 million miles from Earth.
(*Source:* NASA.)

A greatly enlarged photo of a computer microchip.

We are constantly measuring things, from very large to very small.

For example, on August 27, 2003, the planet Mars was closer to Earth than it had been at any time in the last 60,000 years. Mars was "only" 34,646,437 miles away! (See **Section 7.1,** Exercises 63–64.)

On the other hand, computer microchips often measure a tiny 5 mm long by 1 mm wide. (Learn about millimeters in **Section 7.2** and then try Exercise 39.)

7.1 Problem Solving with English Measurement

OBJECTIVES

1 Learn the basic measurement units in the English system.

2 Convert among measurement units using multiplication or division.

3 Convert among measurement units using unit fractions.

4 Solve application problems using English measurement.

We measure things all the time: the distance traveled on vacation, the floor area we want to cover with carpet, the amount of milk in a recipe, the weight of the bananas we buy at the store, the number of hours we work, and many more.

In the United States, we still use the **English system** of measurement for many everyday activities. Examples of English units are inches, feet, quarts, ounces, and pounds. However, the fields of science, medicine, sports, and manufacturing use the **metric system** (meters, liters, and grams). And, because the rest of the world uses only the metric system, U.S. businesses have been changing to the metric system in order to compete internationally.

OBJECTIVE 1 Learn the basic measurement units in the English system. Until the switch to the metric system is complete, we still need to know how to use the English system of measurement. The table below lists the relationships you should memorize. The time relationships are used in both the English and metric systems.

1 After memorizing the measurement conversions, answer these questions.

(a) 1 c = _____ fl oz

(b) _____ qt = 1 gal

(c) 1 wk = _____ days

(d) _____ ft = 1 yd

(e) 1 ft = _____ in.

(f) _____ oz = 1 lb

(g) 1 T = _____ lb

(h) _____ min = 1 hr

(i) 1 pt = _____ c

(j) _____ hr = 1 day

(k) 1 min = _____ sec

(l) 1 qt = _____ pt

(m) _____ ft = 1 mi

English Measurement Relationships

Length	Weight
1 foot (ft) = 12 inches (in.)	1 pound (lb) = 16 ounces (oz)
1 yard (yd) = 3 feet (ft)	1 ton (T) = 2000 pounds (lb)
1 mile (mi) = 5280 feet (ft)	

Capacity	Time
1 cup (c) = 8 fluid ounces (fl oz)	1 minute (min) = 60 seconds (sec)
1 pint (pt) = 2 cups (c)	1 hour (hr) = 60 minutes (min)
1 quart (qt) = 2 pints (pt)	1 day = 24 hours (hr)
1 gallon (gal) = 4 quarts (qt)	1 week (wk) = 7 days

As you can see, there is no simple way to convert among these various measures. The units evolved over hundreds of years and were based on a variety of "standards." For example, one yard was the distance from the tip of a king's nose to his thumb when his arm was outstretched. An inch was three dried barleycorns laid end to end.

EXAMPLE 1 Knowing English Measurement Units

Memorize the English measurement conversions shown above. Then answer these questions.

(a) 24 hr = _____ day   Answer: 1 day

(b) 1 yd = _____ ft   Answer: 3 ft

◀◀◀ Work Problem 1 at the Side.

OBJECTIVE 2 Convert among measurement units using multiplication or division. You often need to convert from one unit of measure to another. Two methods of converting measurements are shown here. Study each way and use the method you prefer. The first method involves deciding whether to multiply or divide.

Converting among Measurement Units

1. **Multiply** when converting from a larger unit to a smaller unit.
2. **Divide** when converting from a smaller unit to a larger unit.

ANSWERS

1. **(a)** 8   **(b)** 4   **(c)** 7   **(d)** 3   **(e)** 12
   (f) 16   **(g)** 2000   **(h)** 60   **(i)** 2   **(j)** 24
   (k) 60   **(l)** 2   **(m)** 5280

> **EXAMPLE 2** **Converting from One Unit of Measure to Another**

Convert each measurement.

(a) 7 ft to inches

You are converting from a *larger* unit to a *smaller* unit (a *foot* is longer than an *inch*), so multiply.

Because *1 ft = **12** in.,* multiply by 12.

$$7 \text{ ft} = 7 \cdot \mathbf{12} = 84 \text{ in.}$$

(b) $3\frac{1}{2}$ lb to ounces

You are converting from a *larger* unit to a *smaller* unit (a *pound* is heavier than an *ounce*), so multiply.

Because *1 lb = **16** oz,* multiply by 16.

$$3\frac{1}{2} \text{ lb} = 3\frac{1}{2} \cdot \mathbf{16} = \frac{7}{\overset{}{\underset{1}{2}}} \cdot \frac{\overset{8}{16}}{1} = \frac{56}{1} = 56 \text{ oz}$$

(c) 20 qt to gallons

You are converting from a *smaller* unit to a *larger* unit (a *quart* is smaller than a *gallon*) so divide.

Because *4 qt = 1 gal,* divide by 4.

$$20 \text{ qt} = \frac{20}{4} = 5 \text{ gal}$$

Divide by 4. ⬑

(d) 45 min to hours

You are converting from a *smaller* unit to a *larger* unit (a *minute* is less than an *hour*), so divide.

Because *60 min = 1 hr,* divide by 60 and write the fraction in lowest terms.

$$45 \text{ min} = \frac{45}{\mathbf{60}} = \frac{45 \div \mathbf{15}}{60 \div \mathbf{15}} = \frac{3}{4} \text{ hr} \leftarrow \text{Lowest terms}$$

Divide by 60. ⬑

> **Work Problem 2 at the Side.** ▶▶▶

OBJECTIVE 3 Convert among measurement units using unit fractions. If you have trouble deciding whether to multiply or divide when converting measurements, use *unit fractions* to solve the problem. You'll also find this method useful in science classes. A **unit fraction** is equivalent to 1. Here is an example.

$$\frac{12 \text{ in.}}{12 \text{ in.}} = \frac{\overset{1}{\cancel{12} \text{ in.}}}{\underset{1}{\cancel{12} \text{ in.}}} = 1$$

Use the table of measurement relationships on the previous page to find that 12 in. is the same as 1 ft. So you can substitute 1 ft for 12 in. in the numerator, or you can substitute 1 ft for 12 in. in the denominator. This makes two useful unit fractions.

$$\frac{\mathbf{1 \text{ ft}}}{12 \text{ in.}} = 1 \qquad \text{or} \qquad \frac{12 \text{ in}}{\mathbf{1 \text{ ft}}} = 1$$

To convert from one measurement unit to another, just multiply by the appropriate unit fraction. Remember, a unit fraction is equivalent to 1. Multiplying something by 1 does *not* change its value.

---

2 Convert each measurement using multiplication or division.

(a) $5\frac{1}{2}$ ft to inches

(b) 64 oz to pounds

(c) 6 yd to feet

(d) 2 T to pounds

(e) 35 pt to quarts

(f) 20 min to hours

(g) 4 wk to days

3 First write the unit fraction needed to make each conversion. Then complete the conversion.

(a) 36 in. to feet

unit fraction } $\dfrac{1\ ft}{12\ in.}$

(b) 14 ft to inches

unit fraction } $\dfrac{in.}{ft}$

(c) 60 in. to feet

unit fraction } _____

(d) 4 yd to feet

unit fraction } _____

(e) 39 ft to yards

unit fraction } _____

(f) 2 mi to feet

unit fraction } _____

Use these guidelines to choose the correct unit fraction.

> **Choosing a Unit Fraction**
>
> The *numerator* should use the measurement unit you want in the *answer*.
>
> The *denominator* should use the measurement unit you want to *change*.

EXAMPLE 3 Using Unit Fractions with Length Measurements

(a) Convert 60 in. to feet.

Use a unit fraction with feet (the unit for your answer) in the numerator, and inches (the unit being changed) in the denominator. Because *1 ft = 12 in.,* the necessary unit fraction is

$$\dfrac{1\ ft}{12\ in.}\quad \leftarrow \text{Unit for your answer is feet.}$$
$$\leftarrow \text{Unit being changed is inches.}$$

Next, multiply 60 in. times this unit fraction. Write 60 in. as the fraction $\dfrac{60\ in.}{1}$. Then divide out common units and factors wherever possible.

$$60\ \textbf{in.} \cdot \dfrac{1\ ft}{12\ \textbf{in.}} = \dfrac{\overset{5}{\cancel{60}}\ \textbf{in.}}{1} \cdot \dfrac{1\ ft}{\underset{1}{\cancel{12}}\ \textbf{in.}} = \dfrac{5 \cdot 1\ ft}{1} = 5\ ft$$

These units should match.
Divide out inches.
Divide 60 and 12 by 12.

(b) Convert 9 ft to inches.

Select the correct unit fraction to change 9 ft to inches.

$$\dfrac{12\ in.}{1\ ft}\quad \leftarrow \text{Unit for your answer is inches.}$$
$$\leftarrow \text{Unit being changed is feet.}$$

Multiply 9 ft times the unit fraction.

$$9\ \textbf{ft} \cdot \dfrac{12\ in.}{1\ \textbf{ft}} = \dfrac{9\ \cancel{ft}}{1} \cdot \dfrac{12\ in.}{1\ \cancel{ft}} = \dfrac{9 \cdot 12\ in.}{1} = 108\ in.$$

These units should match.
Divide out feet.

> **CAUTION**
> If no units will divide out, you made a mistake in choosing the unit fraction.

◄◄◄ Work Problem 3 at the Side.

EXAMPLE 4 Using Unit Fractions with Capacity and Weight Measurements

(a) Convert 9 pt to quarts.

First select the correct unit fraction.

$$\dfrac{1\ qt}{2\ pt}\quad \leftarrow \text{Unit for your answer is quarts.}$$
$$\leftarrow \text{Unit being changed is pints.}$$

— **Continued on Next Page**

ANSWERS

3. **(a)** 3 ft **(b)** $\dfrac{12\ in.}{1\ ft}$; 168 in.

(c) $\dfrac{1\ ft}{12\ in.}$; 5 ft **(d)** $\dfrac{3\ ft}{1\ yd}$; 12 ft

(e) $\dfrac{1\ yd}{3\ ft}$; 13 yd **(f)** $\dfrac{5280\ ft}{1\ mi}$; 10,560 ft

Next multiply.

Write as mixed number.

$$9 \text{ pt} \cdot \frac{1 \text{ qt}}{2 \text{ pt}} = \frac{9 \text{ pt}}{1} \cdot \frac{1 \text{ qt}}{2 \text{ pt}} = \frac{9}{2} \text{ qt} = 4\frac{1}{2} \text{ qt}$$

These units should match.

Divide out pints.

(b) Convert $7\frac{1}{2}$ gal to quarts.

Write as an improper fraction.

$$\frac{7\frac{1}{2} \text{ gal}}{1} \cdot \frac{4 \text{ qt}}{1 \text{ gal}} = \frac{15}{2} \cdot \frac{4}{1} \text{ qt}$$

Divide out gallons.

$$= \frac{15}{\underset{1}{2}} \cdot \frac{\overset{2}{4}}{1} \text{ qt}$$

$$= 30 \text{ qt}$$

(c) Convert 36 oz to pounds.

$$\frac{\overset{9}{36} \text{ oz}}{1} \cdot \frac{1 \text{ lb}}{\underset{4}{16} \text{ oz}} = \frac{9}{4} \text{ lb} = 2\frac{1}{4} \text{ lb}$$

> **NOTE**
> In Example 4(c) above you get $\frac{9}{4}$ lb. Recall that $\frac{9}{4}$ means $9 \div 4$. If you do $9 \div 4$ on your calculator, you get **2.25 lb**. English measurements usually use fractions or mixed numbers, like $2\frac{1}{4}$ lb. However, 2.25 lb is also correct and is the way grocery stores often show weights of produce, meat, and cheese.

Work Problem 4 at the Side.

EXAMPLE 5   **Using Several Unit Fractions**

Sometimes you may need to use two or three unit fractions to complete a conversion.

(a) Convert 63 in. to yards.

Use the unit fraction $\frac{1 \text{ ft}}{12 \text{ in.}}$ to change inches to feet and the unit fraction $\frac{1 \text{ yd}}{3 \text{ ft}}$ to change feet to yards. Notice how all the units divide out except yards, which is the unit you want in the answer.

$$\frac{63 \text{ in.}}{1} \cdot \frac{1 \text{ ft}}{12 \text{ in.}} \cdot \frac{1 \text{ yd}}{3 \text{ ft}} = \frac{63}{36} \text{ yd} = \frac{63 \div 9}{36 \div 9} \text{ yd} = \frac{7}{4} \text{ yd} = 1\frac{3}{4} \text{ yd}$$

Continued on Next Page

---

4 Convert using unit fractions.

(a) 16 qt to gallons

(b) 3 c to pints

(c) $3\frac{1}{2}$ T to pounds

(d) $1\frac{3}{4}$ lb to ounces

(e) 4 oz to pounds

⑤ Convert using two or three unit fractions.

(a) 4 T to ounces

You can also divide out common factors in the numbers.

$$\frac{\overset{7}{\cancel{\underset{1}{\cancel{63}}}}}{1} \cdot \frac{1}{\underset{4}{\cancel{12}}} \cdot \frac{1}{\underset{1}{\cancel{3}}} = \frac{7}{4} = 1\frac{3}{4}\,\text{yd}$$

Instead of changing $\frac{7}{4}$ to $1\frac{3}{4}$, you can enter $7 \div 4$ on your calculator to get 1.75 yd. Both answers are correct because 1.75 is equivalent to $1\frac{3}{4}$.

(b) Convert 2 days to seconds.

Use three unit fractions. The first one changes days to hours, the next one changes hours to minutes, and the last one changes minutes to seconds. All the units divide out except seconds, which is what you want in your answer.

$$\frac{2\ \cancel{\text{days}}}{1} \cdot \frac{24\ \cancel{\text{hr}}}{1\ \cancel{\text{day}}} \cdot \frac{60\ \cancel{\text{min}}}{1\ \cancel{\text{hr}}} \cdot \frac{60\ \text{seconds}}{1\ \cancel{\text{min}}} = 172{,}800\ \text{seconds}$$

Divide out **days**.

Divide out **hr**.

Divide out **min**.

(b) 3 mi to inches

◀◀◀ **Work Problem 5 at the Side.**

OBJECTIVE 4 Solve application problems using English measurement. To solve application problems, we will use the steps you learned in **Section 1.10.** Those steps are summarized here.

Step 1 **Read** the problem carefully.

Step 2 **Work out a plan.**

Step 3 **Estimate** a reasonable answer.

Step 4 **Solve** the problem.

Step 5 **State the answer.**

Step 6 **Check** your work.

(c) 36 pt to gallons

(d) 2 wk to minutes

EXAMPLE 6 Solving English Measurement Applications

(a) A 36 oz can of coffee is on sale at Jerry's Foods for $7.89. What is the cost per pound, to the nearest cent? (*Source:* Jerry's Foods.)

Step 1 **Read** the problem. The problem asks for the cost per *pound* of coffee.

Step 2 **Work out a plan.** The weight of the coffee is given in *ounces* but the answer must be cost *per pound*. Convert ounces to pounds. The word *per* indicates division. You need to divide the cost by the number of pounds.

Step 3 **Estimate** a reasonable answer. To estimate, round $7.89 to $8. Then, there are 16 oz in a pound, so 36 oz are a little more than 2 pounds. So, $8 ÷ 2 = $4 per pound as our estimate.

ANSWERS
5. **(a)** 128,000 oz **(b)** 190,080 in.
(c) $4\frac{1}{2}$ gal or 4.5 gal **(d)** 20,160 min

— **Continued on Next Page**

Step 4 **Solve** the problem. Use a unit fraction to convert 36 oz to pounds.

$$\frac{\overset{9}{\cancel{36}\ \cancel{oz}}}{1} \cdot \frac{1\ lb}{\underset{4}{\cancel{16}\ \cancel{oz}}} = \frac{9}{4}\ lb = 2.25\ lb$$

Then divide to find the *cost* per *pound*.

$$\begin{array}{c}\text{Cost} \rightarrow\\ \text{per} \rightarrow\\ \text{pound} \rightarrow\end{array}\ \frac{\$7.89}{2.25\ lb} = 3.50\overline{6} \approx 3.51 \quad \text{(rounded)}$$

Step 5 **State the answer.** The coffee costs $3.51 per pound (to the nearest cent).

Step 6 **Check** your work. The exact answer of $3.51 is close to our estimate of $4.

(b) Bilal's favorite cake recipe uses $1\frac{2}{3}$ cups of milk. If he makes six cakes for a bake sale at his son's school, how many quarts of milk will he need?

Step 1 **Read** the problem. The problem asks for the number of *quarts* of milk needed for six cakes.

Step 2 **Work out a plan.** Multiply to find the number of *cups* of milk for six cakes. Then convert *cups* to *quarts* (the unit required in the answer).

Step 3 **Estimate** a reasonable answer. To estimate, round $1\frac{2}{3}$ cups to 2 cups. Then, 2 cups times 6 = 12 cups. There are 4 cups in a quart, so 12 cups ÷ 4 = 3 quarts as our estimate.

Step 4 **Solve** the problem. First multiply. Then use unit fractions to convert.

$$1\frac{2}{3} \cdot 6 = \frac{5}{\underset{1}{\cancel{3}}} \cdot \frac{\overset{2}{\cancel{6}}}{1} = \frac{10}{1} = 10\ cups \quad \left\{\begin{array}{l}\text{Milk needed for}\\ \text{six cakes}\end{array}\right.$$

$$\frac{\overset{5}{\cancel{10}}\ cups}{1} \cdot \frac{1\ \cancel{pt}}{\underset{1}{\cancel{2}}\ \cancel{cups}} \cdot \frac{1\ qt}{2\ \cancel{pt}} = \frac{5}{2}\ qt = 2\frac{1}{2}\ qt$$

Step 5 **State the answer.** Bilal needs $2\frac{1}{2}$ qt (or 2.5 qt) of milk.

Step 6 **Check** your work. The exact answer of $2\frac{1}{2}$ qt is close to our estimate of 3 qt.

NOTE
In Step 2 above, we *first multiplied* $1\frac{2}{3}$ cups by 6 to find the number of cups needed, then *converted* 10 cups to $2\frac{1}{2}$ quarts. It would also work to *first convert* $1\frac{2}{3}$ cups to $\frac{5}{12}$ qt, then *multiply* $\frac{5}{12}$ qt by 6 to get $2\frac{1}{2}$ qt.

Work Problem 6 at the Side. ▶▶▶

6 Solve each application problem using the six problem-solving steps.

(a) Kristin paid $3.29 for 12 oz of extra sharp cheddar cheese. What is the price per pound, to the nearest cent?

(b) A moving company estimates 11,000 lb of furnishings for an average 3-bedroom house. If the company made five such moves last week, how many tons of furnishings did they move? (*Source:* North American Van Lines.)

ANSWERS
6. **(a)** $4.39 per pound (rounded)
(b) 27.5 or $27\frac{1}{2}$ T

Real-Data Applications

Growing Sunflowers

The front and back of a seed packet for sunflowers are shown at the left. Look at the top of the packet first.

Sunflower, Mammoth Grey Stripe

Tall Plants, Huge Flowers

Net Wt. 3.5 g $1.19

Sunflower, Mammoth Grey Stripe
The stalk of this sunflower will grow to 12'(4 m). Flowers will range from 6"(15cm) to 15"(38cm) in diameter. Sunflowers can thrive in poor soil with little moisture.

Type	Height	Planting Depth	Seed Spacing	Thinning Height	Spacing After Thinning	Days to Germination
Annual	8-10' 2.4-3 m	1/2" 13 mm	6" 15 cm	3" 8 cm	2' 61 cm	10-20

Select a sunny or lightly shaded location and plant outdoors, where plants are to remain, after all danger of frost is past. For tallest plants, sow in good soil with moderate moisture.

Stock #1185   7 18964 98119 7

Source: Olds Seed Solutions.

1. There were 42 seeds in the packet. If 40 of the seeds sprouted, what was the cost per sprout, to the nearest cent?

2. If vegetable and flower seeds were on sale at 30% off, what was the cost per sprout, to the nearest cent?

3. What percent of the seeds sprouted, to the nearest whole percent?

4. How many seeds would weigh 1 gram?

5. The table at the bottom of the packet uses the symbol (') for feet, and the symbol (") for inches.

   (a) How tall will the plants grow, in feet?

   (b) How tall will they grow in inches?

   (c) How tall will they grow in yards?

6. If you plant all 42 seeds in one long row, using the spacing given on the package, how long will your row be in feet?

7. How many inches tall should the plants be when you thin them (remove less vigorous plants to give others room to grow)? How tall is that in feet?

8. What is the range in the diameter of the flowers, in inches, and in feet? Diameter is the distance across the circular flower.

9. (a) Using the information in the article at the right, how many gum wrappers are needed to make 1 foot of chain?

   (b) To make 1 inch of chain?

   (c) How many inches of chain, to the nearest hundredth, are made from one wrapper?

10. Is the article correct in saying that 125 miles is 34 million gum wrappers?

A Work of Art?

Sometimes a person's life's work makes it into a museum. So it figures that Michael Knutson's 128-foot chain of gum wrappers now resides in the Yellow Medicine County Museum in Granite Falls, Minnesota.

According to the *Redwood Gazette,* the wrapper chain started in 1974 when Knutson, now a 43-year-old woodworker, became bored during study hall. "I've never been much of a gum chewer, but I found most of the wrappers on the streets of the city," he said. (FYI: It takes 6602 wrappers to make a 128-foot chain.)

Granite Falls is about 125 miles—or 34 million gum wrappers—west of Minneapolis/St. Paul.

Source: Minneapolis Star Tribune.

7.1 Exercises

FOR
EXTRA
HELP

Tutor Center Addison-Wesley
Math Tutor Center

Math XL
MathXL

Digital Video Tutor CD 4
Videotape 12

Student's
Solutions
Manual

MyMathLab
MyMathLab

Interactmath.com

Fill in the blanks with the measurement relationships you have memorized. See Example 1.

1. 1 yd = _____ ft

2. 1 ft = _____ in.

3. _____ fl oz = 1 c

4. _____ qt = 1 gal

5. 1 mi = _____ ft

6. 1 wk = _____ days

7. _____ lb = 1 T

8. _____ oz = 1 lb

9. 1 min = _____ sec

10. 1 day = _____ hr

Convert each measurement using unit fractions. See Examples 3 and 4.

11. 120 sec = _____ min

12. 180 min = _____ hr

13. 8 qt = _____ gal

14. 6 gal = _____ qt

15. An adult African elephant could weigh 7 to 8 tons. How many pounds could it weigh? (*Source: The Top 10 of Everything.*)

16. A reticulated python snake is the world's longest snake. It grows to a length of 18 to 33 feet. How many yards long can the snake be? (*Source: The Top 10 of Everything.*)

17. 9 yd = _____ ft

18. 20,000 lb = _____ T

19. 7 lb = _____ oz

20. 96 oz = _____ lb

21. 5 qt = _____ pt

22. 26 pt = _____ qt

23. 90 min = _____ hr

24. 45 sec = _____ min

25. 3 in. = _____ ft

26. 30 in. = _____ ft

27. 24 oz = _____ lb

28. 36 oz = _____ lb

29. 5 c = _____ pt

30. 15 qt = _____ gal

Use the information in the bar graph below to answer Exercises 31–32.

Thickness of Lake Ice Needed for Safe Walking/Driving

15 in.	Pickup truck
12 in.	Car
5 in.	Snowmobile or ATV
4 in.	Person walking

Source: Wisconsin DNR.

31. If the ice on a lake is $\frac{1}{2}$ ft thick, what will it safely support?

32. How many feet of ice are needed to safely drive a pickup truck on a lake?

33. $2\frac{1}{2}$ T = _____ lb

34. $4\frac{1}{2}$ pt = _____ c

35. $4\frac{1}{4}$ gal = _____ qt

36. $2\frac{1}{4}$ hr = _____ min

37. After 15 years, a saguaro cactus is still only one-third to two-thirds of a foot tall, depending upon rainfall. How tall could the cactus be in inches? (*Source: Ecology of the Saguaro III.*)

38. Yao Ming, an NBA basketball player from China, is $7\frac{1}{2}$ ft tall. What is his height in inches? (*Source: www.NBA.com*)

Use two or three unit fractions to make each conversion. See Example 5.

39. 6 yd = _____ in.

40. 2 T = _____ oz

41. 112 c = _____ qt

42. 336 hr = _____ wk

43. 6 days = _____ sec

44. 5 gal = _____ c

45. $1\frac{1}{2}$ T = _____ oz

46. $3\frac{1}{3}$ yd = _____ in.

47. The statement 8 = 2 is *not* true. But with appropriate measurement units, it *is* true.

$$8 \; quarts = 2 \; gallons$$

Attach measurement units to these numbers to make the statement true.

(a) 1 _____ = 16 _____

(b) 10 _____ = 20 _____

(c) 120 _____ = 2 _____

(d) 2 _____ = 24 _____

(e) 6000 _____ = 3 _____

(f) 35 _____ = 5 _____

48. Explain in your own words why you can add 2 feet + 12 inches to get 3 feet, but you cannot add 2 feet + 12 pounds.

Convert each measurement. See Example 5.

49. $2\frac{3}{4}$ mi = _____ in.

50. $5\frac{3}{4}$ tons = _____ oz

51. $6\frac{1}{4}$ gal = _____ fl oz

52. $3\frac{1}{2}$ days = _____ sec

53. 24,000 oz = _____ T

54. 57,024 in. = _____ mi

Solve each application problem. Show your work. See Example 6.

55. Geralyn bought 20 oz of strawberries for $2.29. What was the price per pound for the strawberries, to the nearest cent?

56. Zach paid $0.79 for a 1.6 oz candy bar. What was the cost per pound?

57. Dan orders supplies for the science labs. Each of the 24 stations in the chemistry lab needs 2 ft of rubber tubing. If rubber tubing sells for $8.75 per yard, how much will it cost to equip all the stations?

58. In 2001, Marquette, Michigan, had 219 inches of snowfall, while Detroit, Michigan, had 18 inches. What was the difference in snowfall between the two cities, in feet? Round to the nearest tenth. (*Source: World Almanac.*)

59. Tropical cockroaches are the fastest land insects. They can run about 5 feet per second. At this rate, how long would it take the cockroach to travel one mile? (*Source: Guinness Book of World Records.*)

Give your answer

(a) in seconds;

(b) in minutes, to the nearest tenth.

60. A snail moves at an average speed of 2 feet every 3 minutes. At that rate, how long would it take the snail to travel one mile? (*Source: Beakman and Jax.*)

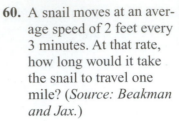

Give your answer

(a) in hours;

(b) in days.

61. At the day care center, each of the 15 toddlers drinks about $\frac{2}{3}$ cup of milk with lunch. The center is open 5 days a week.

(a) How many quarts of milk will the center need for one week of lunches?

(b) If the center buys milk in gallon jugs, how many jugs should be ordered for one week?

62. Bob's Candies in Albany, Georgia, makes 135,000 pounds of candy canes each day. (*Source:* Bob's Candies, Inc.)

(a) How many tons of candy canes are produced during a 5-day workweek?

(b) The plant operates 24 hours per day. How many tons of candy canes are produced each hour, to the nearest tenth?

RELATING CONCEPTS (EXERCISES 63–64) For Individual or Group Work

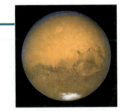

*On the first page of this chapter, we said that the planet Mars was only 34,646,437 miles from Earth in August 2003. Use this information and learn more about Mars as you **work Exercises 63–64 in order.** (Source: NASA and Associated Press.)*

63. In order to better understand how far away Mars was in August 2003, we need to compare the distance to something more familiar.

(a) The distance around Earth is about 24,907 miles. How many times would you travel around Earth before going 34,646,437 miles? Round your answer to the nearest whole number.

(b) The driving distance from Los Angeles to New York City is 2786 miles. How many times would you have to make that drive to match the distance from Earth to Mars? Round to the nearest whole number

64. The divisions you did in Exercise 63 were long and messy, so a calculator was very helpful. Now let's see how we can do it *without* a calculator.

(a) Round the distance from Mars to Earth to the nearest million. Round the distance around Earth to the nearest thousand. Now do the division without a calculator, using the shortcut for division by multiples of 10 in **Section 1.6.**

(b) Round the driving distance from Los Angeles to New York City to the nearest hundred. Now do the division by hand using the shortcut.

(c) Compare your answers above to those from Exercise 63. Are they just as useful as the answers in Exercise 63?

7.2 The Metric System—Length

Around 1790, a group of French scientists developed the metric system of measurement. It is an organized system based on multiples of 10, like our number system and our money. After you are familiar with metric units, you will see that they are easier to use than the hodgepodge of English measurement relationships you used in **Section 7.1**.

> **NOTE**
> The metric system information in this text is consistent with usage guidelines from the National Institute of Standards and Technology, www.nist.gov/metric.

OBJECTIVE 1 Learn the basic metric units of length.   The basic unit of length in the metric system is the **meter** (also spelled *metre*). Use the symbol **m** for meter; do not put a period after it. If you put five of the pages from this textbook side by side, they would measure about 1 meter. Or, look at a yardstick—a meter is just a little longer. A yard is 36 inches long; a meter is about 39 inches long.

|— 1 meter (about 39 in.) —|
|— 1 yard (36 in.) —|

In the metric system, you use meters for things like buying fabric for sewing projects, measuring the length of your living room, talking about heights of buildings, or describing track and field athletic events.

Buy 2 m
of fabric
(about 2 yd)

6 m
(about 20 ft)

15 m
(about
49 ft)

Work Problem 1 at the Side.

To make longer or shorter length units in the metric system, **prefixes** are written in front of the word *meter*. For example, the prefix *kilo* means **1000**, so a *kilo*meter is **1000** meters. The table below shows how to use the prefixes for length measurements. It is helpful to memorize the prefixes because they are also used with weight and capacity measurements. The blue boxes are the units you will use most often in daily life.

Prefix	kilo-meter	hecto-meter	deka-meter	meter	deci-meter	centi-meter	milli-meter
Meaning	1000 meters	100 meters	10 meters	1 meter	$\frac{1}{10}$ of a meter	$\frac{1}{100}$ of a meter	$\frac{1}{1000}$ of a meter
Symbol	**km**	**hm**	**dam**	m	**dm**	**cm**	**mm**

Length units that are used most often

OBJECTIVES

1 Learn the basic metric units of length.

2 Use unit fractions to convert among units.

3 Move the decimal point to convert among units.

1 Circle the items that measure about 1 meter.

Length of a pencil

Length of a baseball bat

Height of doorknob from the floor

Height of a house

Basketball player's arm length

Length of a paper clip

ANSWERS
1. baseball bat, height of doorknob, basketball player's arm length

❷ Write the most reasonable metric unit in each blank. Choose from km, m, cm, and mm.

(a) The woman's height is

168 _____.

(b) The man's waist is

90 _____ around.

(c) Louise ran the 100 _____

dash in the track meet.

(d) A postage stamp is

22 _____ wide.

(e) Michael paddled his

canoe 2 _____ down

the river.

(f) The pencil lead is

1 _____ thick.

(g) A stick of gum is

7 _____ long.

(h) The highway speed limit

is 90 _____ per hour.

(i) The classroom was

12 _____ long.

(j) A penny is about

18 _____ across.

Here are some comparisons to help you get acquainted with the commonly used length units: km, m, cm, mm.

*Kilo*meters are used instead of miles. A kilometer is **1000** meters. It is about 0.6 mile (a little more than half a mile) or about 5 to 6 city blocks. If you participate in a 10 km run, you'll run about 6 miles.

A meter is divided into 100 smaller pieces called *centi*meters. Each centimeter is $\frac{1}{100}$ of a meter. Centimeters are used instead of inches. A centimeter is a little shorter than $\frac{1}{2}$ inch. The cover of this textbook is about 21 cm wide. A nickel is about 2 cm across. Measure the width and length of your little finger on this centimeter ruler. The width of your little finger is probably about 1 cm.

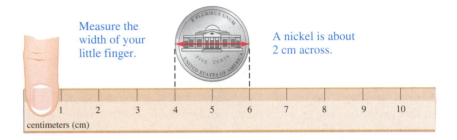

A meter is divided into 1000 smaller pieces called *milli*meters. Each millimeter is $\frac{1}{1000}$ of a meter. It takes 10 mm to equal 1 cm, so it is a very small length. The thickness of a dime is about 1 mm. Measure the width of your pen or pencil and the width of your little finger on this millimeter ruler.

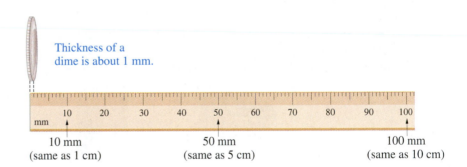

EXAMPLE 1 Using Metric Length Units

Write the most reasonable metric unit in each blank. Choose from km, m, cm, and mm.

(a) The distance from home to work is 20 _____.

20 **km** because kilometers are used instead of miles.

20 km is about 12 miles.

(b) My wedding ring is 4 _____ wide.

4 **mm** because the width of a ring is very small.

(c) The newborn baby is 50 _____ long.

50 **cm**; which is half of a meter; a meter is about 39 inches so half a meter is around 20 inches.

◀◀◀ **Work Problem 2 at the Side.**

OBJECTIVE 2 Use unit fractions to convert among units. You can convert among metric length units using unit fractions. Keep these relationships in mind when setting up the unit fractions.

Metric Length Relationships

1 km = 1000 m so the unit fractions are:	**1 m = 1000 mm** so the unit fractions are:
$\dfrac{1 \text{ km}}{1000 \text{ m}}$ or $\dfrac{1000 \text{ m}}{1 \text{ km}}$	$\dfrac{1 \text{ m}}{1000 \text{ mm}}$ or $\dfrac{1000 \text{ mm}}{1 \text{ m}}$
1 m = 100 cm so the unit fractions are:	**1 cm = 10 mm** so the unit fractions are:
$\dfrac{1 \text{ m}}{100 \text{ cm}}$ or $\dfrac{100 \text{ cm}}{1 \text{ m}}$	$\dfrac{1 \text{ cm}}{10 \text{ mm}}$ or $\dfrac{10 \text{ mm}}{1 \text{ cm}}$

EXAMPLE 2 **Using Unit Fractions to Convert Length Measurements**

Convert each measurement using unit fractions.

(a) 5 km to m

Put the unit for the answer (meters) in the numerator of the unit fraction; put the unit you want to change (km) in the denominator.

Unit fraction equivalent to 1 $\left\{\dfrac{1000 \text{ m}}{1 \text{ km}}\right.$ ← Unit for answer
← Unit being changed

Multiply. Divide out common units where possible.

$$5 \text{ km} \cdot \dfrac{1000 \text{ m}}{1 \text{ km}} = \dfrac{5 \text{ km}}{1} \cdot \dfrac{1000 \text{ m}}{1 \text{ km}} = \dfrac{5 \cdot 1000 \text{ m}}{1} = 5000 \text{ m}$$

These units should match.

5 km = 5000 m

The answer makes sense because a kilometer is much longer than a meter, so 5 km will contain many meters.

(b) 18.6 cm to m

Multiply by a unit fraction that allows you to divide out centimeters.

Unit fraction

$$\dfrac{18.6 \text{ cm}}{1} \cdot \dfrac{1 \text{ m}}{100 \text{ cm}} = \dfrac{18.6}{100} \text{ m} = 0.186 \text{ m}$$

18.6 cm = 0.186 m

There are 100 cm in a meter, so 18.6 cm will be a small part of a meter. The answer makes sense.

Work Problem 3 at the Side. ▶▶▶

---

3 First write the unit fraction needed to make each conversion. Then complete the conversion.

(a) 3.67 m to cm

unit fraction $\left\{\dfrac{100 \text{ cm}}{1 \text{ m}}\right.$

(b) 92 cm to m

unit fraction $\left\{\dfrac{\text{m}}{\text{cm}}\right.$

(c) 432.7 cm to m

unit fraction $\left\{\underline{\quad}\right.$

(d) 65 mm to cm

unit fraction $\left\{\underline{\quad}\right.$

(e) 0.9 m to mm

unit fraction $\left\{\underline{\quad}\right.$

(f) 2.5 cm to mm

unit fraction $\left\{\underline{\quad}\right.$

ANSWERS

3. **(a)** 367 cm **(b)** $\dfrac{1 \text{ m}}{100 \text{ cm}}$; 0.92 m

(c) $\dfrac{1 \text{ m}}{100 \text{ cm}}$; 4.327 m

(d) $\dfrac{1 \text{ cm}}{10 \text{ mm}}$; 6.5 cm

(e) $\dfrac{1000 \text{ m}}{1 \text{ m}}$; 900 mm

(f) $\dfrac{10 \text{ mm}}{1 \text{ cm}}$; 25 mm

4 Do each multiplication or division by hand or on a calculator. Compare your answer to the one you get by moving the decimal point.

(a) $(43.5)(10) =$ _____

43.5 gives 435.

(b) $43.5 \div 10 =$ _____

43.5 gives _____

(c) $(28)(100) =$ _____

28.00 gives _____

(d) $28 \div 100 =$ _____

28. gives _____

(e) $(0.7)(1000) =$ _____

0.700 gives _____

(f) $0.7 \div 1000 =$ _____

000.7 gives _____

OBJECTIVE 3 Move the decimal point to convert among units. By now you have probably noticed that conversions among metric units are made by multiplying or dividing by 10, by 100, or by 1000. A quick way to *multiply* by 10 is to move the decimal point one place to the *right*. Move it two places to the right to multiply by 100, three places to multiply by 1000. *Dividing* is done by moving the decimal point to the *left* in the same manner.

◀◀◀ **Work Problem 4 at the Side.**

An alternate conversion method to unit fractions is moving the decimal point using this **metric conversion line.**

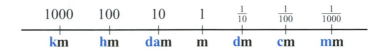

Here are the steps for using the conversion line.

Using the Metric Conversion Line

Step 1 Find the unit you are given on the metric conversion line.

Step 2 Count the number of places to get from the unit you are given to the unit you want in the answer.

Step 3 Move the decimal point the **same number of places** and in the **same direction** as you did on the conversion line.

EXAMPLE 3 Using the Metric Conversion Line

Use the metric conversion line to make the following conversions.

(a) 5.702 km to m

Find **km** on the metric conversion line. To get to **m**, you move *three places* to the *right*. So move the decimal point in 5.702 *three places* to the *right*.

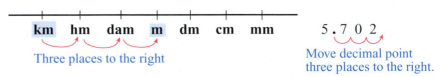

5.702 km = 5702 m

(b) 69.5 cm to m

Find **cm** on the conversion line. To get to **m**, move *two places* to the *left*.

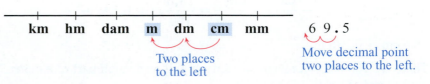

69.5 cm = 0.695 m

Continued on Next Page

(c) 8.1 cm to mm

From **cm** to **mm** is *one place* to the *right*.

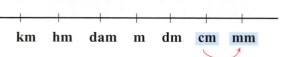

km   hm   dam   m   dm   **cm   mm**   8 . 1

One place   Move decimal point
to the right   one place to the right.

8.1 cm = 81 mm

Work Problem 5 at the Side. ▶▶▶

EXAMPLE 4 **Practicing Length Conversions**

Convert using the metric conversion line.

(a) 1.28 m to mm

Moving from **m** to **mm** is going *three places to the right*. In order to move the decimal point in 1.28 three places to the right, you must write a 0 as a placeholder.

1.28**0**   Zero is written in as a placeholder.

Move decimal point
three places to the right.

1.28 m = 1280 mm

(b) 60 cm to m

From **cm** to **m** is two places to the left. The decimal point in 60 starts at the *far right side* because 60 is a whole number. Then move it two places to the left.

60.   60.
↑
Decimal point starts here.   Move decimal point
                             two places to the left.

60 cm = 0.60 m, which is equivalent to 0.6 m.

(c) 8 m to km

From **m** to **km** is three places to the left. The decimal point in 8 starts at the far right side. In order to move it three places to the left, you must write two zeros as placeholders.

                         Two zeros are written
                         in as placeholders.
8.   **008.**
↑
Decimal point starts here.   Move decimal point
                             three places to the left.

8 m = 0.008 km.

Work Problem 6 at the Side. ▶▶▶

---

5 Convert using the metric conversion line.

(a) 12.008 km to m

(b) 561.4 m to km

(c) 20.7 cm to m

(d) 20.7 cm to mm

(e) 4.66 m to cm

(f) 85.6 mm to cm

6 Convert using the metric conversion line.

(a) 9 m to mm

(b) 3 cm to m

(c) 14.6 km to m

(d) 5 mm to cm

(e) 70 m to km

(f) 0.8 m to cm

ANSWERS
5. (a) 12,008 m **(b)** 0.5614 km
   (c) 0.207 m **(d)** 207 mm
   (e) 466 cm **(f)** 8.56 cm
6. (a) 9000 mm **(b)** 0.03 m
   (c) 14,600 m **(d)** 0.5 cm
   (e) 0.07 km **(f)** 80 cm

Measuring Up

1. How much do nails grow in one week? one month? one year?

2. How much does scalp hair grow in one week? one month? one year? (Use metric units.)

3. When you have finished **Section 7.5**, come back to this article. Is the statement about hair growing 6 inches a year accurate? Explain your answer.

Hair and Nail Growth

Q How fast do hair and nails grow? Do they grow faster in the summer?

A Fingernails grow on average, about one-tenth of a millimeter per day, although there is considerable variation among individuals. Fingernails grow faster than toenails, and nails on the longest fingers appear to grow the fastest.

Fingernails, as well as hair and skin, grow faster in the summer, presumably under the influence of sunlight, which expands blood vessels, bringing more oxygen and nutrients to the area and allowing for faster growth.

The rate the scalp hair grows is 0.3 to 0.4 millimeter per day, or about 6 inches a year.

Source: Minneapolis Star Tribune

New device measures distances within billionths of an inch

U.S. officials have unveiled "the ultimate ruler," a measuring device that can gauge distances to within billionths of an inch—the length of five individual atoms—and may help revolutionize high-tech manufacturing.

Developed at a cost of $8 million by the National Institute of Standards and Technology with other agencies, the Molecular Measuring Machine can measure distances to within 40 nanometers, or billionths of a meter. After further refinement it is expected to measure within 1 nanometer.

By way of comparison, the period at the end of this sentence is about 300,000 nanometers wide.

The machine is expected to prove a boon to U.S. manufacturers of computer chips and other tiny high-tech items that must meet exacting specifications. It can also help manufacturers better calibrate their own super-accurate equipment so that, for example, even more devices could be put onto silicon chips to increase their computing power.

"This is a key project where the United States is a long way in front of any other country," said Trevor Howe, a University of Connecticut professor of metallurgy and director of its Precision Manufacturing Center.

Source: Boston Globe.

4. The article on the left states, "the Molecular Measuring Machine can measure distances to within 40 nanometers, or billionths of a meter." This suggests that the prefix *nano* means what?

5. Write the two unit fractions you would use to convert between meters and nanometers. Use your unit fractions to change 40 nanometers to meters, and to change 300,000 nanometers to meters.

6. Why do you suppose the headline and first paragraph talk about "billionths of an inch" when the rest of the article specifies nanometers, which are billionths of a meter?

7. Use the information in the article to find the length of one atom in inches. Write your answer as a decimal number and in words.

7.2 Exercises

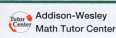
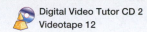

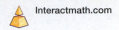
Use your knowledge of the meaning of metric prefixes to fill in the blanks.

1. *kilo* means _____ so

1 km = _____ m

2. *deka* means _____ so

1 dam = _____ m

3. *milli* means _____ so

1 mm = _____ m

4. *deci* means _____ so

1 dm = _____ m

5. *centi* means _____ so

1 cm = _____ m

6. *hecto* means _____ so

1 hm = _____ m

Use this ruler to measure the width of your thumb and hand for Exercises 7–10.

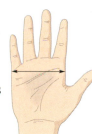

7. The width of your hand in centimeters

8. The width of your hand in millimeters

9. The width of your thumb in millimeters

10. The width of your thumb in centimeters

Write the most reasonable metric length unit in each blank. Choose from km, m, cm, and mm.
See Example 1.

11. The child was 91 _____ tall.

12. The cardboard was 3 _____ thick.

13. Ming-Na swam in the 200 _____ backstroke race.

14. The bookcase is 75 _____ wide.

15. Adriana drove 400 _____ on her vacation.

16. The door is 2 _____ high.

17. An aspirin tablet is 10 _____ across.

18. Lamard jogs 4 _____ every morning.

19. A paper clip is about 3 _____ long.

20. My pen is 145 _____ long.

21. Dave's truck is 5 _____ long.

22. Wheelchairs need doorways that are at least 80 _____ wide.

23. Describe at least three examples of metric length units that you have come across in your daily life.

24. Explain one reason the metric system would be easier for a child to learn than the English system.

Convert each measurement. Use unit fractions or the metric conversion line. See Examples 2–4.

25. 7 m to cm

26. 18 m to cm

27. 40 mm to m

28. 6 mm to m

29. 9.4 km to m

30. 0.7 km to m

31. 509 cm to m

32. 30 cm to m

33. 400 mm to cm

34. 25 mm to cm

35. 0.91 m to mm

36. 4 m to mm

37. Is 82 cm greater than or less than 1 m? What is the difference in the lengths?

38. Is 1022 m greater than or less than 1 km? What is the difference in the lengths?

39. On the first page of this chapter, we said that computer microchips may be only 5 mm long and 1 mm wide. Using the ruler on the previous page, draw a rectangle that measures 5 mm by 1 mm. Then convert each measurement to centimeters.

A greatly enlarged photo of a computer microchip.

40. Many cameras use film that is 35 mm wide. Film for movie theaters may be 70 mm wide. Using the ruler on the previous page, draw a line that is 35 mm long and a line 70 mm long. Then convert each measurement to centimeters.

41. The Roe River near Great Falls, Montana, is the shortest river in the world, with a north fork that is just under 18 m long. How many kilometers long is the north fork of the river? (*Source: Guinness Book of Amazing Nature.*)

42. There are 60,000 km of blood vessels in the human body. How many meters of blood vessels are in the body? (*Source: Big Book of Knowledge.*)

43. The median height for U.S. females who are 20 to 29 years old is about 1.64 m. Convert this height to centimeters and to millimeters. (*Source: U.S. National Center for Health Statistics.*)

44. The median height for 20- to 29-year-old males in the United States is about 177 cm. Convert this height to meters and to millimeters. (*Source: U.S. National Center for Health Statistics.*)

45. Use two unit fractions to convert 5.6 mm to km.

46. Use two unit fractions to convert 16.5 km to mm.

7.3 The Metric System—Capacity and Weight (Mass)

We use capacity units to measure liquids, such as the amount of milk in a recipe, the gasoline in our car tank, and the water in an aquarium. (The English capacity units used in the United States are cups, pints, quarts, and gallons.) The basic metric unit for capacity is the **liter** (also spelled *litre*). The capital letter **L** is the symbol for liter, to avoid confusion with the numeral 1.

OBJECTIVE 1 Learn the basic metric units of capacity. The liter is related to metric length in this way: a box that measures 10 cm on every side holds exactly one liter. (The volume of the box is 10 cm • 10 cm • 10 cm = 1000 cubic centimeters. Volume is discussed in **Section 8.7**.) A liter is just a little more than 1 quart.

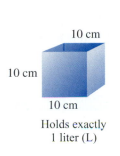

10 cm

10 cm

10 cm

Holds exactly 1 liter (L)

About 1 liter of milk

About 1 liter of oil for your car

A liter is a little more than one quart (just $\frac{1}{4}$ cup more).

In the metric system you use liters for things like buying milk and soda at the store, filling a pail with water, and describing the size of your home aquarium.

Buy a 2 L bottle of soda

Use a 12 L pail to wash floors

Watch the fish in your 40 L aquarium

Work Problem 1 at the Side.

To make larger or smaller capacity units, we use the same **prefixes** as we did with length units. For example, *kilo* means 1000 so a *kilo*meter is 1000 meters. In the same way, a *kilo*liter is 1000 liters.

Prefix	kilo-liter	hecto-liter	deka-liter	liter	deci-liter	centi-liter	milli-liter
Meaning	1000 liters	100 liters	10 liters	1 liter	$\frac{1}{10}$ of a liter	$\frac{1}{100}$ of a liter	$\frac{1}{1000}$ of a liter
Symbol	kL	hL	daL	L	dL	cL	mL

Capacity units used most often

1 Which things can be measured in liters?

Amount of water in the bathtub

Length of the bathtub

Width of your car

Amount of gasoline you buy for your car

Weight of your car

Height of a pail

Amount of water in a pail

2 Write the most reasonable metric unit in each blank. Choose from L and mL.

(a) I bought 8 _____ of milk at the store.

(b) The nurse gave me 10 _____ of cough syrup.

(c) This is a 100 _____ garbage can.

(d) It took 10 _____ of paint to cover the bedroom walls.

(e) My car's gas tank holds 50 _____.

(f) I added 15 _____ of oil to the pancake mix.

(g) The can of orange soda holds 350 _____.

(h) My friend gave me a 30 _____ bottle of expensive perfume.

The capacity units you will use most often in daily life are liters (L) and *milli*liters (mL). A tiny box that measures 1 cm on every side holds exactly one milliliter. (In medicine, this small amount is also called 1 cubic centimeter, or 1 cc for short.) It takes 1000 ml to make 1 L. Here are some useful comparisons.

Holds exactly 1 milliliter (mL)	Teaspoon holds 5 mL	One cup holds about 250 mL

EXAMPLE 1 **Using Metric Capacity Units**

Write the most reasonable metric unit in each blank. Choose from L and mL.

(a) The bottle of shampoo held 500 _____.
500 **mL** because 500 L would be about 500 quarts, which is too much.

(b) I bought a 2 _____ carton of orange juice.
2 **L** because 2 mL would be less than a teaspoon.

◀◀◀ **Work Problem 2 at the Side.**

OBJECTIVE 2 Convert among metric capacity units. Just as with length units, you can convert between milliliters and liters using unit fractions.

Metric Capacity Relationships

1 L = 1000 mL, so the unit fractions are:

$$\frac{1\text{ L}}{1000\text{ mL}} \quad \text{or} \quad \frac{1000\text{ mL}}{1\text{ L}}$$

Or you can use a metric conversion line to decide how to move the decimal point.

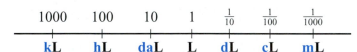

EXAMPLE 2 **Converting among Metric Capacity Units**

Convert using the metric conversion line or unit fractions.

(a) 2.5 L to mL
Using the metric conversion line:
From **L** to **mL** is *three places* to the *right*.

2.**500**   Write two zeros as placeholders.

2.5 L = 2500 mL

Using unit fractions:

Multiply by a unit fraction that allows you to divide out liters.

$$\frac{2.5\text{ }\cancel{L}}{1} \cdot \frac{1000\text{ mL}}{1\text{ }\cancel{L}} = 2500\text{ mL}$$

Continued on Next Page

ANSWERS
2. **(a)** L **(b)** mL **(c)** L
   (d) L **(e)** L **(f)** mL
   (g) mL **(h)** mL

(b) 80 mL to L

Using the metric conversion line:

From **mL** to **L** is *three places* to the *left*.

80.   080.

Decimal point starts here.   Move decimal point three places to the left

80 mL = 0.080 L or 0.08 L

Using unit fractions:

Multiply by a unit fraction that allows you to divide out mL.

$$\frac{80 \text{ mL}}{1} \cdot \frac{1 \text{ L}}{1000 \text{ mL}}$$

$$= \frac{80}{1000} \text{ L} = 0.08 \text{ L}$$

Work Problem 3 at the Side. ▶▶▶

OBJECTIVE 3 Learn the basic metric units of weight (mass). The **gram** is the basic metric unit for *mass*. Although we often call it "weight," there is a difference. Weight is a measure of the pull of gravity; the farther you are from the center of Earth, the less you weigh. In outer space you become weightless, but your mass, the amount of matter in your body, stays the same regardless of where you are. In science courses, it will be important to distinguish between the weight of an object and its mass. But for everyday purposes, we will use the word *weight*.

The gram is related to metric length in this way: The weight of the water in a box measuring 1 cm on every side is 1 gram. This is a very tiny amount of water (1 mL) and a very small weight. One gram is also the weight of a dollar bill or a single raisin. A nickel weighs 5 grams. A regular-sized hamburger and bun weighs from 175 to 200 grams.

The 1 mL of water in this tiny box weighs 1 gram.

A nickel weighs 5 grams.

A dollar bill weighs 1 gram.

A hamburger weighs 175 to 200 grams.

Work Problem 4 at the Side. ▶▶▶

3 Convert.

(a) 9 L to mL

(b) 0.75 L to mL

(c) 500 mL to L

(d) 5 mL to L

(e) 2.07 L to mL

(f) 3275 mL to L

4 Which things would weigh about 1 gram?

A small paper clip

A pair of scissors

One playing card from a deck of cards

A calculator

An average-sized apple

The check you wrote at the grocery store

ANSWERS

3. **(a)** 9000 mL **(b)** 750 mL **(c)** 0.5 L
    (d) 0.005 L **(e)** 2070 mL **(f)** 3.275 L

4. paper clip, playing card, check

5 Write the most reasonable metric unit in each blank. Choose from kg, g, and mg.

(a) A thumbtack weighs

800 _____.

(b) A teenager weighs

50 _____.

(c) This large cast-iron

frying pan weighs

1 _____.

(d) Jerry's basketball

weighed 600 _____.

(e) Tamlyn takes a

500 _____ calcium

tablet every morning.

(f) On his diet, Greg can eat

90 _____ of meat for

lunch.

(g) One strand of hair weighs

2 _____.

(h) One banana might weigh

150 _____.

To make larger or smaller weight units, we use the same **prefixes** as we did with length and capacity units. For example, *kilo* means 1000 so a *kilo*meter is 1000 meters, a *kilo*liter is 1000 liters, and a *kilo*gram is 1000 grams.

Prefix	kilo-gram	hecto-gram	deka-gram	gram	deci-gram	centi-gram	milli-gram
Meaning	1000 grams	100 grams	10 grams	1 gram	$\frac{1}{10}$ of a gram	$\frac{1}{100}$ of a gram	$\frac{1}{1000}$ of a gram
Symbol	kg	hg	dag	g	dg	cg	mg

Weight (mass) units that are used most often

The units you will use most often in daily life are kilograms (kg), grams (g), and milligrams (mg). *Kilo*grams are used instead of pounds. A kilogram is 1000 grams. It is about 2.2 pounds. Two packages of butter plus one stick of butter weigh about 1 kg. An average newborn baby weighs 3 to 4 kg; a college football player might weigh 100 to 130 kg.

1 kilogram is about 2.2 pounds   100 to 130 kg   3 to 4 kg

Extremely small weights are measured in *milli*grams. It takes 1000 mg to make 1 g. Recall that a dollar bill weighs about 1 g. Imagine cutting it into 1000 pieces; the weight of one tiny piece would be 1 mg. Dosages of medicine and vitamins are given in milligrams. You will also use milligrams in science classes.

Cut a dollar bill into 1000 pieces.
One tiny piece weighs 1 milligram.

EXAMPLE 3 Using Metric Weight Units

Write the most reasonable metric unit in each blank. Choose from kg, g, and mg.

(a) Ramon's suitcase weighed 20 _____.
    20 **kg** because kilograms are used instead of pounds.
    20 kg is about 44 pounds.

(b) LeTia took a 350 _____ aspirin tablet.
    350 **mg** because 350 g would be more than the weight of a hamburger, which is too much.

(c) Jenny mailed a letter that weighed 30 _____.
    30 **g** because 30 kg would be much too heavy and 30 mg is less than the weight of a dollar bill.

◀◀◀ Work Problem 5 at the Side.

OBJECTIVE 4 Convert among metric weight (mass) units. As with length and capacity, you can convert among metric weight units by using unit fractions. The unit fractions you need are shown here.

Metric Weight (Mass) Relationships

1 kg = 1000 g so the unit fractions are:

$$\frac{1 \text{ kg}}{1000 \text{ g}} \quad \text{or} \quad \frac{1000 \text{ g}}{1 \text{ kg}}$$

1 g = 1000 mg so the unit fractions are:

$$\frac{1 \text{ g}}{1000 \text{ mg}} \quad \text{or} \quad \frac{1000 \text{ mg}}{1 \text{ g}}$$

Or you can use a metric conversion line to decide how to move the decimal point.

1000	100	10	1	$\frac{1}{10}$	$\frac{1}{100}$	$\frac{1}{1000}$
kg	**hg**	**dag**	**g**	**dg**	**cg**	**mg**

EXAMPLE 4 Converting among Metric Weight Units

Convert using the metric conversion line or unit fractions.

(a) 7 mg to g

Using the metric conversion line:
From **mg** to **g** is *three places* to the *left*.

7.   007.

↑
Decimal point starts here.

Move decimal point three places to the left.

7 mg = 0.007 g

Using unit fractions:

Multiply by a unit fraction that allows you to divide out mg.

$$\frac{7 \text{ mg}}{1} \cdot \frac{1 \text{ g}}{1000 \text{ mg}} = \frac{7}{1000} \text{ g}$$

$$= 0.007 \text{ g}$$

(b) 13.72 kg to g

Using the metric conversion line:
From **kg** to **g** is *three* places to the *right*.

13.72**0**

Decimal point moves three places to the right.

13.72 kg = 13,720 g

A comma
(not a decimal point)

Using unit fractions:

Multiply by a unit fraction that allows you to divide out kg.

$$\frac{13.72 \text{ kg}}{1} \cdot \frac{1000 \text{ g}}{1 \text{ kg}} = 13,720 \text{ g}$$

A comma
(not a decimal point)

Work Problem 6 at the Side.

---

6 Convert.

(a) 10 kg to g

(b) 45 mg to g

(c) 6.3 kg to g

(d) 0.077 g to mg

(e) 5630 g to kg

(f) 90 g to kg

ANSWERS
6. (a) 10,000 g **(b)** 0.045 g **(c)** 6300 g
(d) 77 mg **(e)** 5.63 kg **(f)** 0.09 kg

7 First decide which type of units are needed: length, capacity, or weight. Then write the most appropriate unit in the blank. Choose from km, m, cm, mm, L, mL, kg, g, and mg.

(a) Gail bought a 4 _____ can of paint.

Use _____ units.

(b) The bag of chips weighed 450 _____.

Use _____ units.

(c) Give the child 5 _____ of cough syrup.

Use _____ units.

(d) The width of the window is 55 _____.

Use _____ units.

(e) Akbar drives 18 _____ to work.

Use _____ units.

(f) Each computer weighs 5 _____.

Use _____ units.

(g) A credit card is 55 _____ wide.

Use _____ units.

OBJECTIVE 5 Distinguish among basic metric units of length, capacity, and weight (mass). As you encounter things to be measured at home, on the job, or in your classes at school, be careful to use the correct type of measurement unit.

Use *length units* (kilometers, meters, centimeters, millimeters) to measure:

how long	how high	how far away
how wide	how tall	how far around (perimeter)
how deep	distance	

Use *capacity units* (liters, milliliters) to measure liquids (things that can be poured) such as:

water	shampoo	gasoline
milk	perfume	oil
soft drinks	cough syrup	paint

Also use liters and milliliters to describe how much liquid something can hold, such as an eyedropper, measuring cup, pail, or bathtub.

Use *weight units* (kilograms, grams, milligrams) to measure:

the weight of something   how heavy something is

In **Chapter 8** you will use square units (such as square meters) to measure area, and cubic units (such as cubic centimeters) to measure volume.

EXAMPLE 5 Using a Variety of Metric Units

First decide which type of units are needed: length, capacity, or weight. Then write the most appropriate metric unit in the blank. Choose from km, m, cm, mm, L, mL, kg, g, and mg.

(a) The letter needs another stamp because it weighs 40 _____.

Use _____ units.

The letter weighs 40 **g** because 40 mg is less than the weight of a dollar bill and 40 kg would be about 88 pounds.

Use **weight** units because of the word "weighs."

(b) The swimming pool is 3 _____ deep at the deep end.

Use _____ units.

The pool is 3 **m** deep because 3 cm is only about an inch and 3 km is more than a mile.

Use **length** units because of the word "deep."

(c) This is a 340 _____ can of juice.

Use _____ units.

It is a 340 **mL** can because 340 liters would be more than 340 quarts.

Use **capacity** units because juice is a liquid.

Work Problem 7 at the Side.

7.3 Exercises

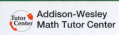

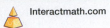
Write the most reasonable metric unit in each blank. Choose from L, mL, kg, g, and mg. See Examples 1 and 3.

1. The glass held 250 _____ of water.

2. Hiromi used 12 _____ of water to wash the kitchen floor.

3. Dolores can make 10 _____ of soup in that pot.

4. Jay gave 2 _____ of vitamin drops to the baby.

5. Our yellow Labrador dog grew up to weigh 40 _____

6. A small safety pin weighs 750 _____.

7. Lori caught a small sunfish weighing 150 _____.

8. One dime weighs 2 _____.

9. Andre donated 500 _____ of blood today.

10. Barbara bought the 2 _____ bottle of cola.

11. The patient received a 250 _____ tablet of medication each hour.

12. The 8 people on the elevator weighed a total of 500 _____.

13. The gas can for the lawn mower holds 4 _____.

14. Kevin poured 10 _____ of vanilla into the mixing bowl.

15. Pam's backpack weighs 5 _____ when it is full of books.

16. One grain of salt weighs 2 _____.

Today, medical measurements are usually given in the metric system. Since we convert among metric units of measure by moving the decimal point, it is possible that mistakes can be made. Examine the following dosages and indicate whether they are reasonable or unreasonable. If a dose is unreasonable, indicate whether it is too much or too little.

17. Drink 4.1 L of Kaopectate after each meal.

18. Drop 1 mL of solution into the eye twice a day.

19. Soak your feet in 5 kg of Epsom salts per liter of water.

20. Inject 0.5 L of insulin each morning.

21. Take 15 mL of cough syrup every four hours.

22. Take 200 mg of vitamin C each day.

23. Take 350 mg of aspirin three times a day.

24. Buy a tube of ointment weighing 0.002 g.

25. Describe at least two examples of metric capacity units and two examples of metric weight units that you have come across in your daily life.

26. Explain in your own words how the meter, liter, and gram are related.

27. Describe how you decide which unit fraction to use when converting 6.5 kg to grams.

28. Write an explanation of each step you would use to convert 20 mg to grams using the metric conversion line.

Convert each measurement. Use unit fractions or the metric conversion line. See Examples 2 and 4.

29. 15 L to mL

30. 6 L to mL

31. 3000 mL to L

32. 18,000 mL to L

33. 925 mL to L

34. 200 mL to L

35. 8 mL to L

36. 25 mL to L

37. 4.15 L to mL

38. 11.7 L to mL

39. 8000 g to kg

40. 25,000 g to kg

41. 5.2 kg to g

42. 12.42 kg to g

43. 0.85 g to mg

44. 0.2 g to mg

45. 30,000 mg to g

46. 7500 mg to g

47. 598 mg to g

48. 900 mg to g

49. 60 mL to L

50. 6.007 kg to g

51. 3 g to kg

52. 12 mg to g

53. 0.99 L to mL

54. 13,700 mL to L

Write the most appropriate metric unit in each blank. Choose from km, m, cm, mm, L, mL, kg, g, and mg. See Example 5.

55. The masking tape is 19 _____ wide.

56. The roll has 55 _____ of tape on it.

57. Buy a 60 _____ jar of acrylic paint for art class.

58. One onion weighs 200 _____.

59. My waist measurement is 65 _____.

60. Add 2 _____ of windshield washer fluid to your car.

61. A single postage stamp weighs 90 _____.

62. The hallway is 10 _____ long.

Solve each application problem. Show your work. (Source for Exercises 63–68: Top 10 of Everything.)

63. Human skin has about 3 million sweat glands, which release an average of 300 mL of sweat per day. How many liters of sweat are released each day?

64. In hot climates, the sweat glands in a person's skin may release up to 3.5 L of sweat in one day. How many milliliters is that?

65. The average weight of a human brain is 1.34 kg. How many grams is that?

66. A healthy human heart pumps about 70 mL of blood per beat. How many liters of blood does it pump per beat?

67. On average, we breathe in and out roughly 900 mL of air every 10 seconds. How many liters of air is that?

68. In the Victorian era, people believed that heavier brains meant greater intelligence. They were impressed that Otto von Bismarck's brain weighed 1907 g, which is how many kilograms?

69. A small adult cat weighs from 3000 g to 4000 g. How many kilograms is that? (*Source:* Lyndale Animal Hospital.)

70. If the letter you are mailing weighs 29 g, you must put additional postage on it. How many kilograms does the letter weigh? (*Source:* U.S. Postal Service.)

71. Is 1005 mg greater than or less than 1 g? What is the difference in the weights?

72. Is 990 mL greater than or less than 1 L? What is the difference in the amounts?

73. One nickel weighs 5 g. How many nickels are in 1 kg of nickels?

74. The ratio of the total length of all the fish to the amount of water in an aquarium can be 3 cm of fish for every 4 L of water. What is the total length of all fish you can put in a 40 L aquarium? (*Source: Tropical Aquarium Fish.*)

RELATING CONCEPTS (EXERCISES 75–78) For Individual or Group Work

*Recall that the prefix **kilo** means 1000, so a **kilo**meter is 1000 meters. You'll learn about other prefixes for numbers greater than 1000 as you **work Exercises 75–78 in order.***

75. (a) The prefix *mega* (abbreviated with a capital M) means one million. So a *mega*meter (Mm) is how many meters?

1 Mm = _____ m

(b) Figure out a unit fraction that you can use to convert megameters to meters. Then use it to convert 3.5 Mm to meters.

76. (a) The prefix *giga* (abbreviated with a capital G) means one billion. So a *giga*meter (Gm) is how many meters?

1 Gm = _____ m

(b) Figure out a unit fraction you can use to convert meters to gigameters. Then use it to convert 2500 m to gigameters.

77. (a) The prefix *tera* (abbreviated with a capital T) means one trillion.

So 1 Tm = _____ m.

(b) Think carefully before you fill in the blanks:

1 Tm = _____ Gm

1 Tm = _____ Mm

78. A computer's memory is measured in *bytes*. A byte can represent a single letter, a digit, or a punctuation mark. The memory for a desktop computer may be measured in megabytes (abbreviated MB) or gigabytes (abbreviated GB). Using the meanings of *mega* and *giga,* it would seem that

1 MB = _____ bytes and

1 GB = _____ bytes.

However, because computers use a base 2 or binary system, 1 MB is actually 2^{20} and 1 GB is 2^{30}. Use your calculator to find the actual values.

2^{20} = _____ 2^{30} = _____

7.4 Problem Solving with Metric Measurement

OBJECTIVE **1** **Solve application problems involving metric measurements.** One advantage of the metric system is the ease of comparing measurements in application situations. Just be sure that you are comparing similar units: mg to mg, km to km, and so on.

Once again, we will use the six problem-solving steps you learned in **Section 1.10.**

OBJECTIVE

1 Solve application problems involving metric measurements.

EXAMPLE 1 **Solving a Metric Application**

Cheddar cheese is on sale at $8.99 per kilogram. Jake bought 350 g of the cheese. How much did he pay, to the nearest cent?

Step 1   **Read** the problem. The problem asks for the cost of 350 g of cheese.

Step 2   **Work out a plan.** The price is $8.99 per *kilogram*, but the amount Jake bought is given in *grams*. Convert grams to kilograms (the unit in the price). Then multiply the weight by the cost per kilogram.

Step 3   **Estimate** a reasonable answer. Round the cost of 1 kg from $8.99 to $9. There are 1000 g in a kilogram, so 350 g is about $\frac{1}{3}$ of a kilogram. Jake is buying about $\frac{1}{3}$ of a kilogram, so $\frac{1}{3}$ of $9 = $3 as our estimate.

Step 4   **Solve** the problem. Use a unit fraction to convert 350 g to kilograms.

$$\frac{350 \text{ g}}{1} \cdot \frac{1 \text{ kg}}{1000 \text{ g}} = \frac{350}{1000} \text{ kg} = 0.35 \text{ kg}$$

Now multiply 0.35 kg times the cost per kilogram.

$$\frac{\$8.99}{1 \text{ kg}} \cdot \frac{0.35 \text{ kg}}{1} = \$3.1465 \approx \$3.15 \quad \text{(rounded)}$$

Step 5   **State the answer.** Jake paid $3.15, rounded to the nearest cent.

Step 6   **Check** your work. The exact answer of $3.15 is close to our estimate of $3.

Work Problem 1 at the Side. ▶▶▶

1 Solve this problem using the six problem-solving steps.

Satin ribbon is on sale at $0.89 per meter. How much will 75 cm cost, to the nearest cent?

EXAMPLE 2 **Solving a Metric Application**

Olivia has 2.6 m of lace. How many centimeters of lace can she use to trim each of six hair ornaments? Round to the nearest tenth of a centimeter.

Step 1   **Read** the problem. The problem asks for the number of centimeters of lace for each of six hair ornaments.

Step 2   **Work out a plan.** The given amount of lace is in *meters*, but the answer must be in *centimeters*. Convert meters to centimeters, then divide by 6 (the number of hair ornaments).

Continued on Next Page

ANSWER
1. $0.67 (rounded)

2 Lucinda's doctor wants her to take 1.2 g of medication each day in three equal doses. How many milligrams should be in each dose? Use the six problem-solving steps.

Step 3 **Estimate** a reasonable answer. To estimate, round 2.6 m of lace to 3 m. Then, 3 m = 300 cm, and 300 cm ÷ 6 = 50 cm as our estimate.

Step 4 **Solve** the problem. On the metric conversion line, moving from **m** to **cm** is two places to the right, so move the decimal point in 2.6 m two places to the right. Then divide by 6.

$$2.60 \text{ m} = 260 \text{ cm} \qquad \frac{260 \text{ cm}}{6 \text{ ornaments}} \approx 43.3 \text{ cm per ornament}$$

Step 5 **State the answer.** Olivia can use about 43.3 cm of lace on each ornament.

Step 6 **Check** your work. The exact answer of 43.3 cm is close to our estimate of 50 cm.

◀◀◀ **Work Problem 2 at the Side.**

> **NOTE**
> In Example 1 we used a unit fraction to convert the measurement, and in Example 2 we moved the decimal point. Use whichever method you prefer. Also, there is more than one way to solve an application problem. Another way to solve Example 2 is to divide 2.6 m by 6 to get 0.4333 m of lace for each ornament. Then convert 0.4333 m to 43.3 cm (rounded to the nearest tenth).

3 Andrea has two pieces of fabric. One measures 2 m 35 cm and the other measures 1 m 85 cm. How many meters of fabric does she have in all? Use the six problem-solving steps.

EXAMPLE 3 **Solving a Metric Application**

Rubin measured a board and found that the length was 3 m plus an additional 5 cm. He cut off a piece measuring 1 m 40 cm for a shelf. Find the length in meters of the remaining piece of board.

Step 1 **Read** the problem. Part of a board is cut off. The problem asks what length of board, in meters, is left over. It may help to make a drawing of the board and label the lengths given in the problem.

Step 2 **Work out a plan.** The lengths involve two units, m and cm. Rewrite both lengths in meters (the unit called for in the answer), and then subtract.

Step 3 **Estimate** a reasonable answer. To estimate, 3 m 5 cm can be rounded to 3 m, because 5 cm is less than half of a meter (less than 50 cm). Round 1 m 40 cm down to 1 m. Then, 3 m − 1 m = 2 m as our estimate.

Step 4 **Solve** the problem. Rewrite the lengths in meters. Then subtract.

$$
\begin{array}{ll}
3 \text{ m} \longrightarrow & 3.00 \text{ m} \\
\text{plus 5 cm} \longrightarrow & \underline{+\ 0.05 \text{ m}} \\
& 3.05 \text{ m}
\end{array}
\qquad
\begin{array}{ll}
1 \text{ m} \longrightarrow & 1.0 \text{ m} \\
\text{plus 40 cm} \longrightarrow & \underline{+\ 0.4 \text{ m}} \\
& 1.4 \text{ m}
\end{array}
$$

Subtract to find leftover length.

$$
\begin{array}{ll}
3.05 \text{ m} & \longleftarrow \text{Board} \\
\underline{-\ 1.40 \text{ m}} & \longleftarrow \text{Shelf} \\
1.65 \text{ m} & \longleftarrow \text{Leftover piece}
\end{array}
$$

Step 5 **State the answer.** The length of the remaining piece of board is 1.65 m.

Step 6 **Check** your work. The exact answer of 1.65 m is close to our estimate of 2 m.

◀◀◀ **Work Problem 3 at the Side.**

7.4 Exercises

FOR
EXTRA
HELP

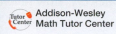 Addison-Wesley
Math Tutor Center

 MathXL
MathXL

 Digital Video Tutor CD 4
Videotape 12

Student's
Solutions
Manual

 MyMathLab
MyMathLab

Interactmath.com

Solve each application problem. Show your work. Round money answers to the nearest cent. See Examples 1–3.

1. Bulk rice is on special at $0.65 per kilogram. Pam scooped some rice into a bag and put it on the scale. How much will she pay for 2 kg 50 g of rice?

2. Lanh is buying a piece of plastic tubing measuring 3 m 15 cm for the science lab. The price is $4.75 per meter. How much will Lanh pay?

3. A miniature Yorkshire terrier, one of the smallest dogs, may weigh only 500 g. But a St. Bernard, the heaviest dog, could easily weigh 90 kg. What is the difference in the weights of the two dogs, in kilograms? (*Source: Big Book of Knowledge*)

4. The world's longest insect is the giant stick insect of Indonesia, measuring 33 cm. The fairy fly, the smallest insect, is just 0.2 mm long. How much longer is the giant stick insect, in millimeters? (*Source: Big Book of Knowledge.*)

5. An adult human body contains about 5 L of blood. If each beat of the heart pumps 70 mL of blood, how many times must the heart beat to pass all the blood through the heart? Round to the nearest whole number of beats. (*Source: Harper's Index.*)

6. A floor tile measures 30 cm by 30 cm and weighs 185 g. How many kilograms would a stack of 24 tiles weigh? How much would five stacks of tiles weigh? (*Source:* The Tile Shop.)

7. Each piece of lead for a mechanical pencil has a thickness of 0.5 mm and is 60 mm long. Find the total length in centimeters of the lead in a package with 30 pieces. If the price of the package is $3.29, find the cost per centimeter for the lead. (*Source:* Pentel.)

60 mm

8. The apartment building caretaker puts 750 mL of chlorine into the swimming pool every day. How many liters should he order to have a one-month (30-day) supply on hand? If chlorine is sold in containers that hold 4 L, how many containers should be ordered for one month? How much chlorine will be left over at the end of the month?

9. Rosa is building a bookcase. She has one board that is 2 m 8 cm long and another that is 2 m 95 cm long. How long are the two boards together, in meters?

10. Janet has 10 m 30 cm of fabric. She wants to make curtains for three windows that are all the same size. How much fabric is available for each window, to the nearest tenth of a meter?

11. In a chemistry lab, each of the 45 students needs 85 mL of acid. How many 1 L bottles of acid need to be ordered?

12. James needs two 1.3 m pieces and two 85 cm pieces of wood molding to frame a picture. The price is $5.89 per meter plus 7% sales tax. How much will James pay?

Use the bar graph below to answer Exercises 13 and 14.

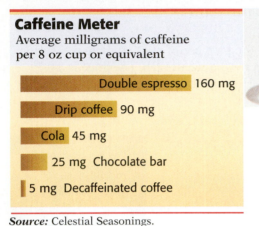

Caffeine Meter
Average milligrams of caffeine per 8 oz cup or equivalent

Double espresso 160 mg
Drip coffee 90 mg
Cola 45 mg
25 mg Chocolate bar
5 mg Decaffeinated coffee

Source: Celestial Seasonings.

13. If Agnete usually drinks three 8 oz cups of drip coffee each day, how many grams of caffeine will she consume in one week?

14. Lorenzo's doctor has suggested that he cut down on caffeine. So Lorenzo switched from drinking four 8 oz cups of cola every day to drinking two 8 oz cups of decaffeinated coffee. How many fewer grams of caffeine is he consuming each week?

15. During August 2003, Mars moved closer to Earth at a rate of about 10,000 m per second. How much closer, in kilometers, did Mars get to Earth:

   (a) in one second,

   (b) in one minute,

   (c) in one hour?

   (*Source:* NASA.)

16. Some of the newest football stadiums have Field Turf instead of grass. Use the drawing below to find the total thickness in centimeters of the top two layers of Field Turf.

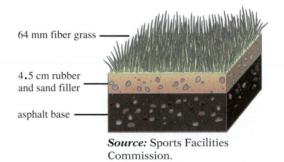

64 mm fiber grass

4.5 cm rubber and sand filler

asphalt base

Source: Sports Facilities Commission.

RELATING CONCEPTS (EXERCISES 17–20) For Individual or Group Work

It is difficult to weigh very light objects, such as a single sheet of paper or a single staple (unless you have a very expensive scientific scale). One way around this problem is to weigh a large number of the items and then divide to find the weight of one item. Of course, before dividing, you must subtract the weight of the box or wrapper that the items are packaged in to find the net weight. **Work Exercises 17–20 in order,** *to complete the table.*

	Item	Total Weight	Weight of Packaging	Net Weight	Weight of One Item in Grams	Weight of One Item in Milligrams
17.	Box of 50 envelopes	255 g	40 g	_____	_____	_____
18.	Box of 1000 staples	350 g	20 g	_____	_____	_____
19.	Ream of paper (500 sheets)	_____	50 g	_____	_____	3000 mg
20.	Box of 100 small paper clips	_____	5 g	_____	_____	500 mg

7.5 Metric–English Conversions and Temperature

OBJECTIVE 1 Use unit fractions to convert between metric and English units. Until the United States has switched completely from the English system to the metric system, it will be necessary to make conversions from one system to the other. *Approximate* conversions can be made with the help of the table below, in which the values have been rounded to the nearest hundredth or thousandth. (The only value that is exact, not rounded, is 1 inch = 2.54 cm.)

OBJECTIVES

1 Use unit fractions to convert between metric and English units.

2 Learn common temperatures on the Celsius scale.

3 Convert temperatures using formulas and following the order of operations.

Metric to English		English to Metric	
1 kilometer	≈ 0.62 mile	1 mile	≈ 1.61 kilometers
1 meter	≈ 1.09 yards	1 yard	≈ 0.91 meter
1 meter	≈ 3.28 feet	1 foot	≈ 0.30 meter
1 centimeter	≈ 0.39 inch	1 inch	= 2.54 centimeters
1 liter	≈ 0.26 gallon	1 gallon	≈ 3.78 liters
1 liter	≈ 1.06 quarts	1 quart	≈ 0.95 liter
1 kilogram	≈ 2.20 pounds	1 pound	≈ 0.45 kilogram
1 gram	≈ 0.035 ounce	1 ounce	≈ 28.35 grams

EXAMPLE 1 Converting between Metric and English Length Units

Convert 10 m to yards using unit fractions. Round your answer to the nearest tenth if necessary.

We're changing from a *metric* unit to an *English* unit. In the "Metric to English" side of the table above, you see that 1 meter ≈ 1.09 yards. Two unit fractions can be written using that information.

$$\frac{1 \text{ m}}{1.09 \text{ yd}} \quad \text{or} \quad \frac{1.09 \text{ yd}}{1 \text{ m}}$$

Multiply by the unit fraction that allows you to divide out meters (that is, meters is in the denominator).

$$10 \text{ **m**} \cdot \frac{1.09 \text{ yd}}{1 \text{ **m**}} = \frac{10 \text{ m}}{1} \cdot \frac{1.09 \text{ yd}}{1 \text{ m}} = \frac{(10)(1.09 \text{ yd})}{1} = 10.9 \text{ yd}$$

These units should match.

10 m ≈ 10.9 yd

NOTE
In Example 1 above, you could also use the other numbers from the table involving meters and yards: 1 yard ≈ 0.91 meter.

$$\frac{10 \text{ m}}{1} \cdot \frac{1 \text{ yd}}{0.91 \text{ m}} = \frac{10}{0.91} \text{ yd} \approx 10.99 \text{ yd}$$

The answer is slightly different because the values in the table are rounded. Also, you have to divide instead of multiply, which is usually more difficult to do without a calculator. We will use the first method in this chapter.

1 Convert using unit fractions. Round your answers to the nearest tenth.

(a) 23 m to yards

(b) 40 cm to inches

(c) 5 mi to kilometers (Look at the "English to Metric" side of the table.)

(d) 12 in. to centimeters

Work Problem 1 at the Side.

ANSWERS
1. (a) 23 m ≈ 25.1 yd **(b)** 40 cm ≈ 15.6 in.
   (c) 5 mi ≈ 8.1 km **(d)** 12 in. ≈ 30.5 cm

2 Convert. Use the values from the table on the previous page to make unit fractions. Round answers to the nearest tenth.

(a) 17 kg to pounds

(b) 5 L to quarts

(c) 90 g to ounces

(d) 3.5 gal to liters

(e) 145 lb to kilograms

(f) 8 oz to grams

EXAMPLE 2 **Converting between Metric and English Weight and Capacity Units**

Convert using unit fractions. Round your answers to the nearest tenth.

(a) 3.5 kg to pounds

Look in the "Metric to English" side of the table on the previous page to see that 1 kilogram ≈ 2.20 pounds. Use this information to write a unit fraction that allows you to divide out kilograms.

$$\frac{3.5 \text{ kg}}{1} \cdot \frac{2.20 \text{ lb}}{1 \text{ kg}} = \frac{(3.5)(2.20 \text{ lb})}{1} = 7.7 \text{ lb}$$

3.5 kg ≈ 7.7 lb

(b) 18 gal to liters

Look in the "English to Metric" side of the table to see that 1 gallon ≈ 3.78 liters. Write a unit fraction that will allow you to divide out gallons.

$$\frac{18 \text{ gal}}{1} \cdot \frac{3.78 \text{ L}}{1 \text{ gal}} = \frac{(18)(3.78 \text{ L})}{1} = 68.04 \text{ L}$$

68.04 rounded to the nearest tenth is 68.0.
18 gal ≈ 68.0 L

(c) 300 g to ounces

In the "Metric to English" side of the table, 1 gram ≈ 0.035 ounce.

$$\frac{300 \text{ g}}{1} \cdot \frac{0.035 \text{ oz}}{1 \text{ g}} = \frac{(300)(0.035 \text{ oz})}{1} = 10.5 \text{ oz}$$

300 g ≈ 10.5 oz

CAUTION
Because the metric and English systems were developed independently, almost all comparisons are approximate. Your answers should be written with the "≈" symbol to show they are approximate.

Work Problem 2 at the Side.

OBJECTIVE 2 **Learn common temperatures on the Celsius scale.** In the metric system, temperature is measured on the **Celsius** scale. On the Celsius scale, water freezes at 0 °C and boils at 100 °C. The small raised circle stands for "degrees" and the capital **C** is for Celsius. Read the temperatures like this:

Water freezes at 0 degrees Celsius (0 °C).

Water boils at 100 degrees Celsius (100 °C).

The English temperature system, used only in the United States, is measured on the **Fahrenheit** scale. On this scale:

Water freezes at 32 degrees Fahrenheit (32 °F).

Water boils at 212 degrees Fahrenheit (212 °F).

The thermometer below shows some typical temperatures in both Celsius and Fahrenheit. For example, comfortable room temperature is about 20 °C or 68 °F, and normal body temperature is about 37 °C or 98.6 °F.

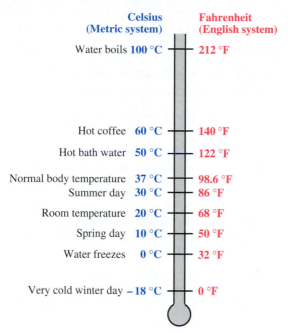

NOTE
The freezing and boiling temperatures are exact. The other temperatures are approximate. Even normal body temperature varies slightly from person to person.

EXAMPLE 3 Using Celsius Temperatures

Circle the Celsius temperature that is most reasonable for each situation.

(a) Warm summer day   29 °C   64 °C   90 °C

29 °C is reasonable. 64 °C and 90 °C are too hot; they're both above the temperature of hot bath water (above 122 °F).

(b) Inside a freezer   −10 °C   3 °C   25 °C

−10 °C is the reasonable temperature because it is the only one below the freezing point of water (0 °C). Your frozen foods would start thawing at 3 °C or 25 °C.

Work Problem 3 at the Side.

OBJECTIVE **3** **Convert temperatures using formulas and following the order of operations.** You can use these formulas to convert between Celsius and Fahrenheit temperatures.

Celsius–Fahrenheit Conversion Formulas

Converting from Fahrenheit (F) to Celsius (C)

$$C = \frac{5(F - 32)}{9}$$

Converting from Celsius (C) to Fahrenheit (F)

$$F = \frac{9 \cdot C}{5} + 32$$

3 Circle the Celsius temperature that is *most* reasonable for each situation.

(a) Set the living room thermostat at:
11 °C   21 °C   71 °C

(b) The baby has a fever of:
29 °C   39 °C   49 °C

(c) Wear a sweater outside because it's:
15 °C   25 °C   50 °C

(d) My iced tea is:
−5 °C   5 °C   30 °C

(e) Time to go swimming! It's:
95 °C   65 °C   35 °C

(f) Inside a refrigerator (not the freezer) it's:
−15 °C   0 °C   3 °C

(g) There's a blizzard outside. It's:
10 °C   0 °C   −20 °C

(h) I need hot water to get these clothes clean. It should be:
55 °C   105 °C   200 °C

4 Convert to Celsius.

(a) 59 °F

(b) 41 °F

(c) 212 °F

(d) 98.6 °F

5 Convert to Fahrenheit.

(a) 100 °C

(b) 25 °C

(c) 80 °C

(d) 5 °C

As you use these formulas, be sure to follow the order of operations from **Section 1.8.**

> **Order of Operations**
> 1. Do all operations inside *parentheses* or *other grouping symbols.*
> 2. Simplify any expressions with *exponents* and find any *square roots.*
> 3. *Multiply* or *divide,* proceeding from left to right.
> 4. *Add* or *subtract,* proceeding from left to right.

EXAMPLE 4 **Converting Fahrenheit to Celsius**

Convert 68 °F to Celsius.

Use the formula and follow the order of operations.

$$C = \frac{5(F - 32)}{9}$$

$$= \frac{5(68 - 32)}{9} \qquad \text{Work inside parentheses first.}$$

$$= \frac{5(36)}{9}$$

$$= \frac{5(\overset{4}{\cancel{36}})}{\underset{1}{\cancel{9}}} \qquad \begin{array}{l}\text{Divide out common factors.}\\ \text{Multiply.}\end{array}$$

$$= 20$$

Thus, 68 °F = 20 °C.

◀◀◀ **Work Problem 4 at the Side.**

EXAMPLE 5 **Converting Celsius to Fahrenheit**

Convert 15 °C to Fahrenheit.

Use the formula and follow the order of operations.

$$F = \frac{9 \cdot C}{5} + 32$$

$$= \frac{9 \cdot 15}{5} + 32$$

$$= \frac{9 \cdot \overset{3}{\cancel{15}}}{\underset{1}{\cancel{5}}} + 32 \qquad \begin{array}{l}\text{Divide out common factors.}\\ \text{Multiply.}\end{array}$$

$$= 27 + 32 \qquad \text{Add.}$$

$$= 59$$

Thus, 15 °C = 59 °F.

◀◀◀ **Work Problem 5 at the Side.**

ANSWERS
4. (a) 15 °C **(b)** 5 °C
   (c) 100 °C **(d)** 37 °C
5. (a) 212 °F **(b)** 77 °F
   (c) 176 °F **(d)** 41 °F

7.5   Exercises

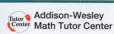

Use the table on the first page of this section and unit fractions to make approximate conversions from metric to English or English to metric. Round your answers to the nearest tenth. See Examples 1 and 2.

1. 20 m to yards

2. 8 km to miles

3. 80 m to feet

4. 85 cm to inches

5. 16 ft to meters

6. 3.2 yd to meters

7. 150 g to ounces

8. 2.5 oz to grams

9. 248 lb to kilograms

10. 7.68 kg to pounds

11. 28.6 L to quarts

12. 15.75 L to gallons

13. For the 2000 Olympics, the 3M Company used 5 g of pure gold to coat Michael Johnson's track shoes. (*Source:* 3M Company.)

   (a) How many ounces of gold were used, to the nearest tenth?

   (b) Was this enough extra weight to slow him down?

14. The label on a Van Ness auto feeder for cats and dogs says it holds 1.4 kg of dry food. How many pounds of food does it hold, to the nearest tenth? (*Source:* Van Ness Plastics.)

15. The heavy-duty wash cycle in a dishwater uses 8.4 gal of water. How many liters does it use, to the nearest tenth? (*Source:* Frigidaire.)

16. The rinse-and-hold cycle in a dishwasher uses only 4.5 L of water. How many gallons does it use, to the nearest tenth? (*Source:* Frigidaire.)

17. The smallest pet fish are dwarf gobies, which are half an inch long. How many centimeters long is a dwarf gobie, to the nearest tenth? (*Hint:* Write half an inch in decimal form.)

18. The fastest nerve signals in the human body travel 120 m per second. How many feet per second do the signals travel? (*Source: Big Book of Knowledge.*)

Circle the more reasonable temperature for each situation. See Example 3.

19. A snowy day

28 °C   28 °F

20. Brewing coffee

80 °C   80 °F

21. A high fever

40 °C   40 °F

22. Swimming pool water

78 °C   78 °F

23. Oven temperature

150 °C   150 °F

24. Light jacket weather

10 °C   10 °F

25. Would a drop in temperature of 20 Celsius degrees be more or less than a drop of 20 Fahrenheit degrees? Explain your answer.

26. Describe one advantage of switching from the Fahrenheit temperature scale to the Celsius scale. Describe one disadvantage.

Use the conversion formulas from this section and the order of operations to convert Fahrenheit temperatures to Celsius and Celsius temperatures to Fahrenheit. Round your answers to the nearest degree if necessary. See Examples 4 and 5.

27. 60 °F

28. 80 °F

29. 104 °F

30. 36 °F

31. 8 °C

32. 18 °C

33. 35 °C

34. 0 °C

Solve each application problem. Round your answers to the nearest degree if necessary.

35. The highest temperature ever recorded on Earth was 136 °F at El Aziza, Libya, in 1922. Convert this temperature to Celsius. (*Source: The World Almanac.*)

36. Hummingbirds have a normal body temperature of 107 °F. But on cold nights they go into a state of torpor where their body temperature drops to 39 °F. What are these temperatures in Celsius? (*Source: Wildbird.*)

37. (a) Here is the tag on a pair of Sorel boots. In what kind of weather would you wear these boots?

**Comfort range
24 °C to 4 °C**

Source: Sorel.

(b) For what Fahrenheit temperatures are the boots designed?

(c) What range of metric temperatures do you have in January where you live? Would you be comfortable in these boots?

38. Sleeping bags made by Eddie Bauer are sold around the world. Each type of sleeping bag is designed for outdoor camping in certain temperatures.

Junior bag	5 °C or warmer
Removable liner bag	0 °C to 15 °C
Conversion bag	−7 °C to 0 °C

(*Source:* Eddie Bauer.)

(a) At what Fahrenheit temperatures should you use the Junior bag?

(b) What Fahrenheit temperatures is the removable liner bag designed for?

The article below appeared in American newspapers. However, both Newfoundland (part of Canada) and Ireland use the metric system. Their newspapers would have reported all the measurements in metric units. Complete the conversions to metric, rounding answers to the nearest tenth.

Q: A recent news brief reported on some men who flew a model airplane from Newfoundland to Ireland. Can you provide some details of the flight?

A: The model plane is 6 feet long and weighs 11 pounds. Made of balsa wood and mylar, it crossed the Atlantic—the flight path took it 1,888.3 miles—in 38 hours, 23 minutes. It soared at a cruising altitude of 1,000 feet. The plane used a souped-up piston engine and carried less than a gallon of fuel, as mandated by rules of the Federation Aeronautique Internationale, the governing body of model airplane building. When it landed in County Galway, Ireland, it had less than 2 fluid ounces of fuel left. The plane was built by Maynard Hill of Silver Spring, Md.

(Source: New York Times.)

39. Length of model plane

40. Weight of plane

41. Length of flight path

42. Time of flight

43. Cruising altitude

44. Fuel at the start, in milliliters

45. Fuel left after landing, in milliliters
(*Hint:* First convert 2 fl oz to quarts.)

46. What *percent* of the fuel was left at the end of the flight?

Chapter 7

7.1	**English system**	The English system of measurement (U.S. system of units) is used for many daily activities only in the United States. Commonly used units in this system include quarts, pounds, feet, miles, and degrees Fahrenheit.
	metric system	The metric system of measurement is an international system used in manufacturing, science, medicine, sports, and other fields. Commonly used units in this system include meters, liters, grams, and degrees Celsius.
	unit fraction	A unit fraction involves measurement units and is equivalent to 1. Unit fractions are used to convert among different measurements.
7.2	**meter**	The meter is the basic unit of length in the metric system. The symbol **m** is used for meter. One meter is a little longer than a yard.
	prefixes	Attaching a prefix to meter, liter, or gram produces larger or smaller units. For example, the prefix *kilo* means 1000 so a *kilo*meter is 1000 meters.
	metric conversion line	The metric conversion line is a line showing the various metric measurement prefixes and their size relationship to each other.
7.3	**liter**	The liter is the basic unit of capacity in the metric system. The symbol **L** is used for liter. One liter is a little more than one quart.
	gram	The gram is the basic unit of weight (mass) in the metric system. The symbol **g** is used for gram. One gram is the weight of 1 milliliter of water or one dollar bill.
7.5	**Celsius**	The Celsius scale is used to measure temperature in the metric system. Water boils at 100 °C and freezes at 0 °C.
	Fahrenheit	The Fahrenheit scale is used to measure temperature in the English system. Water boils at 212 °F and freezes at 32 °F.

NEW FORMULAS

Converting from Celsius to Fahrenheit:

$$F = \frac{9 \cdot C}{5} + 32$$

Converting from Fahrenheit to Celsius:

$$C = \frac{5(F - 32)}{9}$$

TEST YOUR WORD POWER

See how well you have learned the vocabulary in this chapter. Answers follow the Quick Review.

1. The **metric system**
 - A. uses meters, liters, and degrees Fahrenheit
 - B. is based on multiples of 10
 - C. is used only in the United States
 - D. has evolved over centuries.

2. The **English system** of measurement
 - A. is used throughout the world
 - B. is based on multiples of 12
 - C. uses feet, inches, quarts, and pounds
 - D. was developed by a group of scientists in 1790.

3. A **unit fraction**
 - A. has the unit you want to change in the numerator
 - B. has a denominator of 1
 - C. must be written in lowest terms
 - D. is equivalent to 1.

4. A **gram** is
 - A. the weight of 1 mL of water
 - B. abbreviated gm
 - C. equivalent to 1000 kg
 - D. approximately equal to 2.2 pounds.

5. A **meter** is
 - A. equivalent to 1000 cm
 - B. approximately equal to $\frac{1}{2}$ inch
 - C. abbreviated m with no period after it
 - D. the basic unit of capacity in the metric system.

6. The **Celsius** temperature scale
 - A. shows water freezing at 32°
 - B. is used in the English system of measurement
 - C. shows water boiling at 100°
 - D. cannot be converted to the Fahrenheit temperature scale.

QUICK REVIEW

Concepts	Examples

7.1 *The English System of Measurement*

Memorize the basic measurement relationships. Then, to convert units, multiply when changing from a larger unit to a smaller unit; divide when changing from a smaller unit to a larger unit.

Convert each measurement.

(a) 5 ft to inches

$$5 \text{ ft} = 5 \cdot \mathbf{12} = 60 \text{ in.}$$

(b) 3 lb to ounces

$$3 \text{ lb} = 3 \cdot \mathbf{16} = 48 \text{ oz}$$

(c) 15 qt to gallons

$$15 \text{ qt} = \frac{15}{\mathbf{4}} = 3\frac{3}{4} \text{ gal}$$

7.1 *Using Unit Fractions*

Another, more useful, conversion method is multiplying by a unit fraction. The unit you want in the answer should be in the numerator. The unit you want to change should be in the denominator.

Convert 32 oz to pounds.

$$32 \text{ oz} \cdot \frac{1 \text{ lb}}{16 \text{ oz}} \left.\right\} \text{ Unit fraction}$$

These units should match.

$$= \frac{\overset{2}{\cancel{32} \text{ oz}}}{1} \cdot \frac{1 \text{ lb}}{\underset{1}{\cancel{16} \text{ oz}}} \quad \begin{array}{l} \text{Divide out ounces.} \\ \text{Divide out common factors.} \end{array}$$

$$= 2 \text{ lb}$$

7.1 *Solving English Measurement Application Problems*

To solve application problems, use the six problem-solving steps.

Use the six steps to solve this problem.

Mr. Green has 10 yd of rope. He is cutting it into eight pieces so his sailing class can practice knot tying. How many feet of rope will each of his eight students get?

Step 1   **Read** the problem carefully.

Step 1   The problem asks how many feet of rope can be given to each of eight students.

Step 2   **Work out a plan.**

Step 2   Convert 10 yd to feet (the unit required in the answer). Then divide by eight students.

Step 3   **Estimate** a reasonable answer.

Step 3   There are 3 ft in one yard, so there are 30 ft in 10 yd. Then 30 ft ÷ 8 ≈ 4 ft as our estimate.

Step 4   **Solve** the problem.

Step 4   Use a unit fraction to convert 10 yd to feet, then divide.

$$\frac{10 \text{ yd}}{1} \cdot \frac{3 \text{ ft}}{1 \text{ yd}} = 30 \text{ ft}$$

$$\frac{30 \text{ ft}}{8 \text{ students}} = 3\frac{3}{4} \text{ ft or } 3.75 \text{ ft per student}$$

Step 5   **State the answer.**

Step 6   **Check** your work.

Step 5   Each student gets $3\frac{3}{4}$ ft or 3.75 ft of rope.

Step 6   The exact answer of 3.75 ft is close to our estimate of 4 ft.

Concepts	Examples

7.2 Basic Metric Length Units

Use approximate comparisons to judge which length units are appropriate:

1 mm is the thickness of a dime.

1 cm is about $\frac{1}{2}$ inch.

1 m is a little more than 1 yard.

1 km is about 0.6 mile.

Write the most reasonable metric unit in each blank. Choose from km, m, cm, and mm.

The room is 6 __m__ long.

A paper clip is 30 __mm__ long.

He drove 20 __km__ to work.

7.2 and 7.3 Converting within the Metric System

Using Unit Fractions

One conversion method is to multiply by a unit fraction. Use a fraction with the unit you want in the answer in the numerator and the unit you want to change in the denominator.

Convert.

(a) 9 g to kg

$$\frac{9 \not{g}}{1} \cdot \frac{1 \text{ kg}}{1000 \not{g}} = \frac{9}{1000} \text{ kg} = 0.009 \text{ kg}$$

(b) 3.6 m to cm

$$\frac{3.6 \not{m}}{1} \cdot \frac{100 \text{ cm}}{1 \not{m}} = 360 \text{ cm}$$

Using the Metric Conversion Line

Another conversion method is to find the unit you are given on the metric conversion line. Count the number of places to get from the unit you are given to the unit you want. Move the decimal point the same number of places and in the same direction.

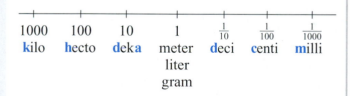

1000	100	10	1	$\frac{1}{10}$	$\frac{1}{100}$	$\frac{1}{1000}$
kilo	hecto	dek**a**	meter liter gram	deci	centi	milli

Convert.

(a) 68.2 kg to g

From **kg** to **g** is three places to the right.

6 8.2 0 0   Decimal point is moved three places to the right.

68.2 kg = 68,200 g

(b) 300 mL to L

From **mL** to **L** is three places to the left.

3 0 0.   Decimal point is moved three places to the left.

300 mL = 0.3 L

(c) 825 cm to m

From **cm** to **m** is two places to the left.

8 2 5.   Decimal point is moved two places to the left.

825 cm = 8.25 m

Concepts	*Examples*

7.3 *Basic Metric Capacity Units*
Use approximate comparisons to judge which capacity units are appropriate:

> 1 L is a little more than 1 quart.
>
> 1 mL is the amount of water in a cube 1 cm on each side.
>
> 5 mL is about one teaspoon.
>
> 250 mL is about one cup.

Write the most appropriate metric unit in each blank. Choose from L and mL.

The pail holds 12 __L__ .

The milk carton from the vending machine holds 250 __mL__ .

7.3 *Basic Metric Weight (Mass) Units*
Use approximate comparisons to judge which weight units are appropriate:

> 1 kg is about 2.2 pounds.
>
> 1 g is the weight of 1 mL of water or one dollar bill.
>
> 1 mg is $\frac{1}{1000}$ of a gram; very tiny!

Write the most appropriate metric unit in each blank. Choose from kg, g, and mg.

The wrestler weighed 95 __kg__ .

She took a 500 __mg__ aspirin tablet.

One banana weighs 150 __g__ .

7.4 *Solving Metric Application Problems*
Convert units so you are comparing kg to kg, cm to cm, and so on. When a measurement involves two units, such as 6 m 20 cm, write it in terms of the unit called for in the answer (6.2 m or 620 cm).

Use the six problem-solving steps.
Step 1 **Read** the problem carefully.

Step 2 **Work out a plan.**

Step 3 **Estimate** a reasonable answer.

Step 4 **Solve** the problem.

Step 5 **State the answer.**

Step 6 **Check** your work.

Use the six steps to solve this problem.

George cut 1 m 35 cm off of a 3 m board. How long was the leftover piece, in meters?

Step 1 The problem asks for the length of the leftover piece in meters.

Step 2 Convert the cut-off measurement to meters (the unit required in the answer), and then subtract to find the "leftover."

Step 3 To estimate, round 1 m 35 cm to 1 m, because 35 cm is less than half a meter. Then, 3 m − 1 m = 2 m as our estimate.

Step 4 Convert the cut-off measurement to meters, then subtract.

$$
\begin{array}{ll}
1\text{ m} \rightarrow \quad 1.00\text{ m} & 3.00\text{ m} \leftarrow \text{Board} \\
\text{plus 35 cm} \rightarrow \underline{+\ 0.35\text{ m}} & \underline{-\ 1.35\text{ m}} \leftarrow \text{Cut off} \\
\quad\quad\quad\quad 1.35\text{ m} & 1.65\text{ m} \leftarrow \text{Left}
\end{array}
$$

Step 5 The leftover piece is 1.65 m long.

Step 6 The exact answer of 1.65 m is close to our estimate of 2 m.

Concepts

7.5 *Converting between Metric and English Units*
Write a unit fraction using the values in the table of conversion factors on the first page of **Section 7.5.** Because the values in the table are rounded, your answers will be approximate.

7.5 *Common Celsius Temperatures*
Use approximate and exact comparisons to judge which temperatures are appropriate:

Exact comparisons:
0 °C is the freezing point of water (32 °F).

100 °C is the boiling point of water (212 °F).

Approximate comparisons:
10 °C for a spring day (50 °F)

20 °C for room temperature (68 °F)

30 °C for summer day (86 °F)

37 °C for normal body temperature (98.6 °F)

7.5 *Converting between Fahrenheit and Celsius Temperatures*
Use this formula to convert from Fahrenheit (F) to Celsius (C).

$$C = \frac{5(F - 32)}{9}$$

Examples

Convert. Round answers to the nearest tenth.

(a) 23 m to yards
From the table, 1 meter ≈ 1.09 yards.

$$\frac{23 \ \cancel{m}}{1} \cdot \frac{1.09 \ yd}{1 \ \cancel{m}} = 25.07 \ yd$$

25.07 rounds to 25.1, so 23 m ≈ 25.1 yd.

(b) 4 oz to grams
From the table, 1 ounce ≈ 28.35 grams.

$$\frac{4 \ \cancel{oz}}{1} \cdot \frac{28.35 \ g}{1 \ \cancel{oz}} = 113.4 \ g$$

So 4 oz ≈ 113.4 g.

Circle the Celsius temperature that is most reasonable.

(a) Hot summer day:
  (35 °C)   90 °C   110 °C

(b) The first snowy day in winter:
  −20 °C   (0 °C)   15 °C

Convert 176 °F to Celsius.

$$C = \frac{5(\mathbf{176} - 32)}{9} \qquad \text{Replace F with 176.}$$

$$= \frac{5(\overset{16}{\cancel{144}})}{\underset{1}{\cancel{9}}} \qquad \begin{array}{l}\text{Divide out common factors.} \\ \text{Then multiply.}\end{array}$$

$$= 80$$

176 °F = 80 °C

Continued

Concepts	Examples

7.5 Converting between Fahrenheit and Celsius Temperatures (continued)

Use this formula to convert from Celsius (C) to Fahrenheit (F).

$$F = \frac{9 \cdot C}{5} + 32$$

Convert 80 °C to Fahrenheit.

$$F = \frac{9 \cdot \mathbf{80}}{5} + 32 \qquad \text{Replace C with 80.}$$

$$= \frac{9 \cdot \overset{16}{\cancel{80}}}{\underset{1}{\cancel{5}}} + 32 \qquad \text{Divide out common factors.}$$
$$\text{Then multiply.}$$

$$= 144 + 32 \qquad \text{Add.}$$

$$= 176$$

$$80\ °C = 176\ °F$$

ANSWERS TO TEST YOUR WORD POWER

1. B; *Examples:* 10 meters = 1 dekameter; 100 meters = 1 hectometer; 1000 meters = 1 kilometer.
2. C; *Examples:* feet and inches are used to measure length, quarts to measure capacity, pounds to measure weight.
3. D; *Example:* Because 12 in. = 1 ft, the unit fraction $\frac{12 \text{ in.}}{1 \text{ ft}}$ is equivalent to $\frac{12 \text{ in.}}{12 \text{ in.}} = 1$.
4. A; *Example:* A small box measuring 1 cm on every edge holds exactly 1 mL of water, and the water weighs 1 g.
5. C; *Example:* A measurement of 16 meters is written 16 m (without a period).
6. C; *Example:* In the metric system, water freezes at 0 °C and boils at 100 °C. The English system uses the Fahrenheit temperature scale where water freezes at 32 °F and boils at 212 °F.

Chapter 7

REVIEW EXERCISES

[7.1] *Fill in the blanks with the measurement relationships you have memorized.*

1. 1 lb = _____ oz

2. _____ ft = 1 yd

3. 1 T = _____ lb

4. _____ qt = 1 gal

5. 1 hr = _____ min

6. 1 c = _____ fl oz

7. _____ sec = 1 min

8. _____ ft = 1 mi

9. _____ in. = 1 ft

Convert using unit fractions.

10. 4 ft = _____ in.

11. 6000 lb = _____ T

12. 64 oz = _____ lb

13. 18 hr = _____ day

14. 150 min = _____ hr

15. $1\frac{3}{4}$ lb = _____ oz

16. $6\frac{1}{2}$ ft = _____ in.

17. 7 gal = _____ c

18. 4 days = _____ sec

19. The average depth of the world's oceans is 12,460 ft. (*Source: Handy Ocean Answer Book.*)

   (a) What is the average depth in yards?

   (b) What is the average depth in miles, to the nearest tenth?

20. During the first year of a program to recycle office paper, a company recycled 123,260 pounds of paper. The company received $40 per ton for the paper. How much money did the company make? Use the six problem-solving steps. (*Source:* I. C. System.)

[7.2] *Write the most reasonable metric length unit in each blank. Choose from km, m, cm, and mm.*

21. My thumb is 20 _____ wide.

22. Her waist measurement is 66 _____.

23. The two towns are 40 _____ apart.

24. A basketball court is 30 _____ long.

25. The height of the picnic bench is 45 _____.

Height

26. The eraser on the end of my pencil is 5 _____ long.

Convert using unit fractions or the metric conversion line.

27. 5 m to cm

28. 8.5 km to m

29. 85 mm to cm

30. 370 cm to m   **31.** 70 m to km   **32.** 0.93 m to mm

[7.3] *Write the most reasonable metric unit in each blank. Choose from L, mL, kg, g, and mg.*

33. The eyedropper
holds 1 _____.

34. I can heat 3 _____
of water in this
saucepan.

35. Loretta's hammer weighed 650 _____.

36. Yongshu's suitcase weighed 20 _____ when it
was packed.

37. My fish tank holds
80 _____ of water.

38. I'll buy the 500 _____
bottle of mouthwash.

39. Mara took a 200 _____ antibiotic pill.

40. This piece of chicken weighs 100 _____

Convert using unit fractions or the metric conversion line.

41. 5000 mL to L   **42.** 8 L to mL   **43.** 4.58 g to mg

44. 0.7 kg to g   **45.** 6 mg to g   **46.** 35 mL to L

[7.4] *Solve each application problem. Show your work. Use the six problem-solving steps.*

47. Each serving of punch at the wedding reception will
be 180 mL. How many liters of punch are needed
for 175 servings?

48. Jason is serving a 10 kg turkey to 28 people. How
many grams of meat is he allowing for each person?
Round to the nearest whole gram.

49. Yerald weighed 92 kg.
Then he lost 4 kg 750 g.
What is his weight now,
in kilograms?

50. Young-Mi bought 2 kg 20 g of onions. The price
was $1.49 per kilogram. How much did she pay, to
the nearest cent?

[7.5] *Use the table on the first page of **Section 7.5** and unit fractions to make approximate
conversions. Round your answers to the nearest tenth if necessary.*

51. 6 m to yards   **52.** 30 cm to inches

53. 108 km to miles

54. 800 mi to kilometers

55. 23 qt to liters

56. 41.5 L to quarts

*Write the appropriate **metric** temperature in each blank.*

57. Water freezes at _____.

58. Water boils at _____.

59. Normal body temperature is about _____.

60. Comfortable room temperature is about _____.

*Use the conversion formulas in **Section 7.5** to convert each temperature to Fahrenheit or to Celsius. Round to the nearest degree if necessary.*

61. 77 °F

62. 92 °F

63. 6 °C

64. Water coming into a dishwasher should be at least 49 °C to clean the dishes properly. What Fahrenheit temperature is that? (*Source:* Frigidaire.)

MIXED REVIEW EXERCISES

Write the most reasonable metric unit in each blank. Choose from km, m, cm, mm, L, mL, kg, g, and mg.

65. I added 1 _____ of oil to my car.

66. The box of books weighed 15 _____.

67. Larry's shoe is 30 _____ long.

68. Jan used 15 _____ of shampoo on her hair.

69. My fingernail is 10 _____ wide.

70. I walked 2 _____ to school.

71. The tiny bird weighed 15 _____.

72. The new library building is 18 _____ wide.

73. The cookie recipe uses 250 _____ of milk.

74. Renee's pet mouse weighs 30 _____.

75. One postage stamp weighs 90 _____.

76. I bought 30 _____ of gas for my car.

Convert the following using unit fractions, the metric conversion line, or the temperature conversion formulas.

77. 10.5 cm to millimeters

78. 45 min to hours

79. 90 in. to feet

80. 1.3 m to centimeters

81. 25 °C to Fahrenheit

82. $3\frac{1}{2}$ gal to quarts

83. 700 mg to grams

84. 0.81 L to milliliters

85. 5 lb to ounces

86. 60 kg to grams

87. 1.8 L to milliliters

88. 86 °F to Celsius

89. 0.36 m to centimeters

90. 55 mL to liters

Solve each application problem. Use the six problem-solving steps in Exercises 91–92.

91. Peggy had a board measuring 2 m 4 cm. She cut off 78 cm. How long is the board now, in meters?

92. During the 12-day Minnesota State Fair, one of the biggest in the United States, Sweet Martha's booth sells an average of 3000 pounds of cookies per day. How many tons of cookies are sold in all? (*Source: Minneapolis Star Tribune.*)

93. Olivia is sending a recipe to her mother in Mexico. Among other things, the recipe calls for 4 oz of rice and a baking temperature of 350 °F. Convert these measurements to metric, rounding to the nearest gram and nearest degree.

94. While on vacation in Canada, Jalo became ill and went to a health clinic. They said he weighed 80.9 kg and was 1.83 m tall. Find his weight in pounds and height in feet. Round to the nearest tenth.

The largest two-axle trucks in the world are mining trucks used to haul huge loads of rock and iron ore. Some information about these $1.5 million trucks is given below. Fill in the blank spaces in the table. Then use the information to answer Exercises 101–102. Round answers to the nearest tenth.

WORLD'S LARGEST TWO-AXLE TRUCK

	Mining Truck	Measurements	
		Metric Units	**English Units**
95.	*Length of truck*	_____ m	44 ft
96.	*Height of truck*	6.7 m	_____ ft
97.	*Height of tire*	_____ m	4 yd
98.	*Width of tire*	102 cm	_____ in.
99.	*Weight of load truck can carry*	220,000 kg	_____ lb
100.	*Fuel needed to travel 1 mile*	_____ to _____ L	5 to 6 gal

Source: Hull Rust Mahoning Mine.

101. Using English measurements:

    (a) The truck can carry a load weighing how many tons?

    (b) How many inches high is each tire?

102. Using metric measurements:

    (a) What is the width of each tire in meters?

    (b) How tall is the truck in centimeters?

Chapter **7**

T E S T

Convert each measurement.

1. 9 gal = _____ qt

2. 45 ft = _____ yd

3. 135 min = _____ hr

4. 9 in. = _____ ft

5. $3\frac{1}{2}$ lb = _____ oz

6. 5 days = _____ min

Write the most reasonable metric unit in each blank. Choose from km, m, cm, mm, L, mL, kg, g, and mg.

7. My husband weighs 75 _____.

8. I hiked 5 _____ this morning.

9. She bought 125 _____ of cough syrup.

10. This apple weighs 180 _____.

11. This page is about 21 _____ wide.

12. My watch band is 10 _____ wide.

13. I bought 10 _____ of soda for the picnic.

14. The bracelet is 16 _____ long.

Convert the following measurements. Show your work.

15. 250 cm to meters

16. 4.6 km to meters

17. 5 mm to centimeters

18. 325 mg to grams

19. 16 L to milliliters

20. 0.4 kg to grams

21. 10.55 m to centimeters

22. 95 mL to liters

1. _____

2. _____

3. _____

4. _____

5. _____

6. _____

7. _____

8. _____

9. _____

10. _____

11. _____

12. _____

13. _____

14. _____

15. _____

16. _____

17. _____

18. _____

19. _____

20. _____

21. _____

22. _____

23. _____

23. The rainiest place in the world is Mount Waialeale in Hawaii, which receives 460 inches of rain each year. What is the average rainfall per month, in feet, to the nearest tenth? (_Source:_ National Geographic Society.)

24. (a) _____

(b) _____

24. A 6-inch Subway Vegie Delite sandwich has 590 mg of sodium. A 6-inch Super Subway Melt sandwich has 2.9 g of sodium. (_Source:_ Subway.)

(a) How much more sodium is in the Super Melt sandwich, in milligrams, than in the Vegie Delite?

(b) The recommended amount of sodium is less than 2400 mg daily. How much more or less sodium does the Super Melt have than the recommended daily amount?

Pick the metric temperature that is most appropriate in each situation.

25. _____

25. The water is almost boiling.

210 °C   155 °C   95 °C

26. _____

26. The tomato plants may freeze tonight.

30 °C   20 °C   0 °C

Use the table from **Section 7.5** and unit fractions to convert each measurement. Round your answers to the nearest tenth if necessary.

27. _____

27. 6 ft to meters

28. 125 lb to kilograms

28. _____

29. _____

29. 50 L to gallons

30. 8.1 km to miles

30. _____

Use the conversion formulas to convert each temperature. Round your answers to the nearest degree if necessary.

31. _____

31. 74 °F to Celsius

32. 2 °C to Fahrenheit

32. _____

Solve this application problem.

33. _____

33. Denise is making five matching pillows. She needs 1 m 20 cm of braid to trim each pillow. If the braid costs $3.98 per yard, how much will she spend to trim the pillows, to the nearest cent? (First find the number of meters of braid Denise needs.)

34. _____

34. Describe two benefits the United States would achieve by switching entirely to the metric system.

Cumulative Review Exercises

CHAPTERS 1–7

$\approx$ *First use front end rounding to round each number and estimate the answer. Then find the exact answer.*

1. *Estimate:*   *Exact:*

$$\begin{array}{r} 107.5 \\ 2.548 \\ +\quad\quad 68.79 \\ \hline \end{array}$$

2. *Estimate:*   *Exact:*

$$\begin{array}{r} 31{,}007 \\ -\quad\quad 829 \\ \hline \end{array}$$

3. *Estimate:*   *Exact:*

$$\begin{array}{r} 92{,}075 \\ \times\quad\quad 183 \\ \hline \end{array}$$

4. *Estimate:*   *Exact:*

$$\begin{array}{r} 56.52 \\ \times\quad 4.7 \\ \hline \end{array}$$

5. *Estimate:*   *Exact:*

$$37\overline{)19{,}610}$$

6. *Estimate:*   *Exact:*

$$8.3\overline{)38.18}$$

7. *Estimate:*   *Exact:*

$$\begin{array}{r} 1\frac{7}{10} \\ +\ 3\frac{4}{5} \\ \hline \end{array}$$

8. *Estimate:*   *Exact:*

$$\begin{array}{r} 5\frac{1}{2} \\ -\ 1\frac{2}{7} \\ \hline \end{array}$$

9. *Exact:*

$$3\frac{1}{6} \cdot 4\frac{2}{3} =$$

Estimate:

$$\underline{\quad} \cdot \underline{\quad} = \underline{\quad}$$

10. *Exact:*

$$2\frac{1}{4} \div \frac{9}{10} =$$

Estimate:

$$\underline{\quad} \div \underline{\quad} = \underline{\quad}$$

Add, subtract, multiply, or divide as indicated. Write answers to fraction problems in lowest terms and as whole or mixed numbers when possible.

11. $3 - 2\frac{5}{16}$

12. $7 + 484{,}099 + 3939$

13. $12 \cdot 2\frac{2}{9}$

14. $0.86 \div 0.066$
Round your answer to the nearest tenth.

15. $\frac{3}{8} + \frac{5}{6}$

16. $8 - 0.9207$

17. $3\frac{3}{4} \div 6$

18. Write your answer using R for the remainder.

$$47\overline{)14{,}467}$$

19. $(2.54)(0.003)$

Use the order of operations to simplify each expression.

20. $24 - 12 \div 6(8) + (25 - 25)$

21. $3^2 + 2^5 \cdot \sqrt{64}$

22. Write 307.19 in words.

23. Write eighty-two ten-thousandths in numbers.

24. Arrange in order from smallest to largest.

0.67   0.067   0.6   0.6007

25. Find the best buy on extra large disposable diapers.

Package of 16 for $3.87
Package of 22 for $5.96
Package of 36 for $11.69

Complete the table.

Fraction or Mixed Number	Decimal	Percent
$\frac{7}{8}$	**26.** _____	**27.** _____
28. _____	0.05	**29.** _____
30. _____	**31.** _____	350%

The bar graph shows the total grams of fat in 3-ounce servings of various meats and fish. Use the graph to answer Exercises 32–34. Write each ratio as a fraction in lowest terms.

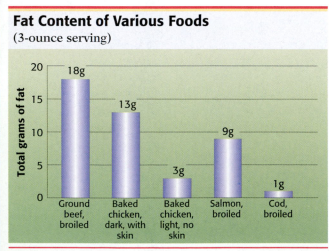

Fat Content of Various Foods
(3-ounce serving)

Source: Consumers Union of U.S.

32. What is the ratio of fat grams in dark chicken with skin compared to ground beef?

33. Find the ratio of fat grams in ground beef compared to light chicken with no skin.

34. Which two food items give a ratio of $\frac{3}{1}$? There may be more than one answer.

Find the unknown number in each proportion. Round your answers to hundredths if necessary.

35. $\dfrac{x}{16} = \dfrac{3}{4}$

36. $\dfrac{0.9}{0.75} = \dfrac{2}{x}$

Solve each percent problem.

37. $4 is what percent of $80?

38. 36 hours is 120% of what number of hours?

*Convert the following measurements. Use the table in **Section 7.5** and the temperature conversion formulas when necessary.*

39. $2\frac{1}{2}$ ft to inches

40. 105 sec to minutes

41. 2.8 m to centimeters

42. 65 mg to grams

43. 198 km to miles

44. 50 °F to Celsius

Write the most reasonable metric unit in each blank. Choose from km, m, cm, mm, L, mL, kg, g, and mg.

45. Ron bought the tube of toothpaste weighing 100 _____.

46. The teacher's desk is 140 _____ long.

47. The hallway is 3 _____ wide.

48. Joe's hammer weighed 1 _____.

49. Anne took a 500 _____ tablet of vitamin C.

50. Tia added 125 _____ of milk to her cereal.

Circle the metric temperature that is most appropriate in each situation.

51. John has a slight fever.

    38 °C   70 °C   99 °C

52. You'll need a light jacket outside.

    0 °C   12 °C   45 °C

Solve each application problem.

53. Calbert works at a Wal-Mart store and used his employee discount to buy a $189.94 digital camera at 10% off. Find the amount of his discount, to the nearest cent, and the sale price. (*Source:* Wal-Mart.)

54. Danielle had a roll of 35 mm film developed. She received 24 prints for $10.35 plus $6\frac{1}{2}$% sales tax. What was the cost per print, to the nearest cent?

55. Bags of slivered almonds weigh 4 oz each. They are packed in a carton that weighs 12 oz. How many pounds would a carton containing 48 bags weigh?

56. On the Illinois map, one centimeter represents 12 km. The center of Springfield is 7.8 cm from the center of Bloomington on the map.

(a) What is the actual distance in kilometers?

(b) What is the actual distance in miles, to the nearest tenth?

57. Dimitri took out a $3\frac{1}{2}$ year car loan for $8750 at 9% simple interest. Find the interest and the total amount due on the loan.

58. On a 35-problem math test, Juana solved 31 problems correctly. What percent of the problems were correct? Round to the nearest tenth of a percent.

59. Mark bought 650 g of maple sugar candy on his vacation in Montreal. The candy is priced at $14.98 per kilogram. How much did Mark pay, to the nearest cent?

60. The Jackson family is making three kinds of holiday cookies that require brown sugar. The recipes call for $2\frac{1}{4}$ cups, $1\frac{1}{2}$ cups, and $\frac{3}{4}$ cup, respectively. They bought two packages of brown sugar, each holding $2\frac{1}{3}$ cups. The amount bought is how much more or less than the amount needed?

61. Akuba is knitting a scarf. Six rows of knitting result in 5 cm of scarf. At that rate, how many rows will she have to knit to make a 100 cm scarf?

62. A survey of the 5600 students on our campus found that $\frac{3}{8}$ of the students work 20 hours or more per week. How many students work 20 hours or more?

63. A spray-on bandage applies a transparent film over a wound that protects it from bacteria. The cost is $6 for about 40 applications. What is the cost per application? (*Source:* Curad.)

64. The average hospital stay is now about 5 days, compared to about 8 days in 1970. What is the percent of decrease in the length of hospital stays? (*Source:* National Center for Health Statistics.)

Use the information in the table below to answer Exercises 65–66. Round answers to the nearest tenth of a percent when necessary.

WHAT'S IN YOUR MAILBOX EVERY WEEK?

Received in the average U.S. mailbox
13 advertisements
3 bills
1 financial statement
Received by the average U.S. e-mail account
12 spam (junk) messages
12 legitimate messages

(*Source:* U.S. Postal Rate Commission and Ferris Research.)

65. What percent of the U.S. mailbox items are:

(a) advertisements,

(b) financial statements.

66. What part of the U.S. e-mail messages are spam (junk)? Write your answer as a

(a) fraction,

(b) decimal,

(c) percent

Geometry

8

An F-1 tornado (wind speeds of 73 to 112 miles per hour) blew down many large trees in Shoreview, Minnesota. As part of an internship, forestry major Bill Masterson made an inventory of Shoreview's remaining trees to help the city plan replacements. He measured the circumference of each tree at chest height. Then, using the circle formulas in **Section 8.6,** he calculated the diameter. This information helped him analyze growth patterns and tree age. (See Exercise 31 in **Section 8.6.**) (*Source: Shoreview Press.*)

8.1 Basic Geometric Terms

OBJECTIVES

1. Identify lines, line segments, and rays.
2. Identify parallel and intersecting lines.
3. Identify and name angles.
4. Classify angles as right, acute, straight, or obtuse.
5. Identify perpendicular lines.

Geometry was developed centuries ago when people needed a way to measure land. The name *geometry* comes from the Greek words *ge*, meaning earth, and *metron*, meaning measure. Today we still use geometry to measure land. It is also important in architecture, construction, navigation, art and design, physics, chemistry, and astronomy. You can use it at home when you buy carpet or wallpaper, hang a picture, or do home repairs. This chapter discusses the basic terms of geometry and the common geometric shapes that are all around us.

Geometry starts with the idea of a point. A **point** is a location in space. It has no length or width. A point is represented by a dot and is named by writing a capital letter next to the dot.

Point *P*

OBJECTIVE **1** **Identify lines, line segments, and rays.** A **line** is a straight row of points that goes on forever in both directions. A line is drawn by using arrowheads to show that it never ends. The line is named using the letters of any two points on the line.

Line *AB*, written $\overleftrightarrow{AB}$

A piece of a line that has two endpoints is called a **line segment.** A line segment is named using its endpoints. The segment with endpoints *P* and *Q* is shown below. It can be named $\overline{PQ}$ or $\overline{QP}$.

Line segment *PQ*, written $\overline{PQ}$

A **ray** is a part of a line that has only one endpoint and goes on forever in one direction. A ray is named by using the endpoint and some other point on the ray. The endpoint is always mentioned first.

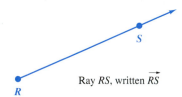

Ray *RS*, written $\overrightarrow{RS}$

1 Identify each figure as a line, line segment, or ray.

(a)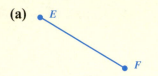

(b)

(c)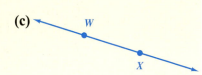

(d)

EXAMPLE 1 **Identifying Lines, Rays, and Line Segments**

Identify each figure below as a line, line segment, or ray.

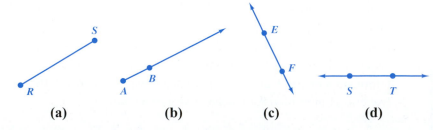

Figure **(a)** has two endpoints, so it is a *line segment.*

Figure **(b)** starts at point *A* and goes on forever in one direction, so it is a *ray.*

Figures **(c)** and **(d)** go on forever in both directions, so they are *lines.*

ANSWERS

1. **(a)** line segment **(b)** ray **(c)** line
   (d) line segment

◀◀◀ **Work Problem 1 at the Side.**

OBJECTIVE 2 Identify parallel and intersecting lines. A *plane* is an infinitely large flat surface. A floor or a wall is a part of a plane. Lines that are in the *same plane,* but that never intersect (never cross), are called **parallel lines,** while lines that cross are called **intersecting lines.** (Think of an intersection, where two streets cross each other.)

EXAMPLE 2 **Identifying Parallel and Intersecting Lines**

Label each pair of the lines as appearing to be parallel or as intersecting.

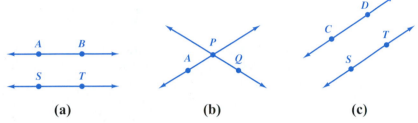

(a)   (b)   (c)

The lines in Figures **(a)** and **(c)** do not intersect; they appear to be *parallel lines.* The lines in Figure **(b)** cross at *P,* so they are *intersecting lines.*

CAUTION

Appearances may be deceiving! Do not assume that lines are parallel unless it is stated that they are parallel.

Work Problem 2 at the Side. ▶▶▶

OBJECTIVE 3 Identify and name angles. An **angle** is made up of two rays that start at a common endpoint. This common endpoint is called the *vertex.*

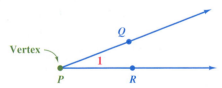

$\overrightarrow{PQ}$ and $\overrightarrow{PR}$ are called the *sides* of the angle. The angle can be named in four different ways, as shown below.

Naming an Angle

When naming an angle, the vertex is written alone or it is written in the middle of two other points. If two or more angles have the **same vertex,** as in Example 3 on the next page, do **not** use the vertex alone to name an angle.

2 Label each pair of lines as appearing to be parallel or as intersecting.

(a)

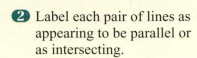

(b)

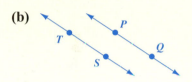

(c)

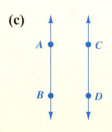

3 **(a)** Name the highlighted angle in three different ways.

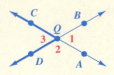

(b) Darken the rays that make up ∠ZTW.

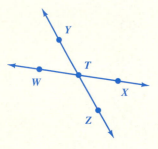

(c) Name this angle in four different ways.

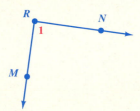

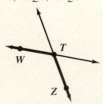
EXAMPLE 3 **Identifying and Naming an Angle**

Name the highlighted angle.

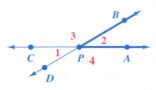

The angle can be named ∠*BPA*, ∠*APB*, or ∠2. It cannot be named ∠*P*, using the vertex alone, because four different angles have *P* as their vertex.

▶▶◀ Work Problem 3 at the Side.

OBJECTIVE **4** **Classify angles as right, acute, straight, or obtuse.** Angles can be measured in **degrees**. The symbol for degrees is a small, raised circle °. Think of the minute hand on a clock as a ray of an angle. Suppose it is at 12:00. During one hour of time, the minute hand moves around in a complete circle. It moves 360 *degrees*, or 360°. In half an hour, at 12:30, the minute hand has moved halfway around the circle or 180°. An angle of 180° is called a **straight angle**. When two rays go in opposite directions and form a straight line, then the rays form a straight angle.

Complete circle
360°

Straight angle
(half a circle)
180°

In a quarter of an hour, at 12:15, the minute hand has moved $\frac{1}{4}$ of the way around the circle, or 90°. An angle of 90° is called a **right angle**. Sometimes you hear it called a *square angle*. The minute hands at 12:00 and 12:15 form one corner of a square. So, to show that an angle is a **right angle**, we draw a **small square** at the vertex.

Right angle
($\frac{1}{4}$ of a circle)
90°

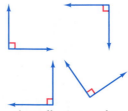

A small square at the vertex identifies right angles.

An angle that measures 1° is shown below. You can see that an angle of 1° is very small.

1° angle

Some other terms used to describe angles are shown below.

Acute angles measure less than 90°.

Examples of acute angles

Obtuse angles measure more than 90° but less than 180°.

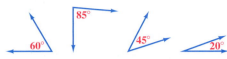

Examples of obtuse angles

Section 10.1 shows you how to use a tool called a *protractor* to measure the number of degrees in an angle.

> **Classifying Angles**
> **Acute angles** measure less than 90°.
> **Right angles** measure *exactly* 90°.
> **Obtuse angles** measure more than 90° but less than 180°.
> **Straight angles** measure *exactly* 180°.

> **NOTE**
> Angles can also be measured in radians, which you will learn about in a later math course.

EXAMPLE 4 **Classifying Angles**

Label each angle as acute, right, obtuse, or straight.

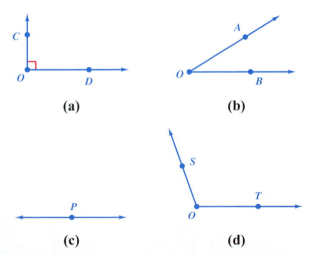

(a)

(b)

(c)

(d)

Figure **(a)** shows a *right angle* (exactly 90° and identified by a small square at the vertex).
Figure **(b)** shows an *acute angle* (less than 90°).
Figure **(c)** shows a *straight angle* (exactly 180°).
Figure **(d)** shows an *obtuse angle* (more than 90° but less than 180°).

Work Problem 4 at the Side. ▶▶▶

4 Label each figure as an acute, right, obtuse, or straight angle.

(a)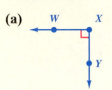

(b)

(c)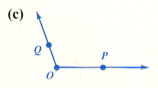

(d)

5 Which pair of lines is perpendicular? How can you describe the other pair of lines?

(a)

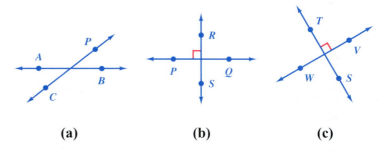

OBJECTIVE 5 Identify perpendicular lines. Two lines are called **perpendicular lines** if they intersect to form a right angle.

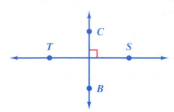

$\overleftrightarrow{CB}$ and $\overleftrightarrow{ST}$ are **perpendicular lines** because they intersect at right angles. This can be written in the following way: $\overleftrightarrow{CB} \perp \overleftrightarrow{ST}$.

EXAMPLE 5 Identifying Perpendicular Lines

Which pairs of lines are perpendicular?

(a) **(b)** **(c)**

The lines in Figures **(b)** and **(c)** are *perpendicular* to each other because they intersect at right angles.

The lines in Figure **(a)** are *intersecting lines,* but they are *not* perpendicular because they do not form a right angle.

Work Problem 5 at the Side.

(b)

8.1   Exercises

Identify each figure as a line, line segment, or ray and name it using the appropriate symbol. See Example 1.

1.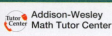

2.

3.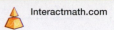

4.

5.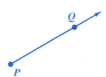

6.

Label each pair of lines as appearing to be parallel, as perpendicular, or as intersecting.
See Examples 2 and 5.

7.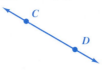

8.

9.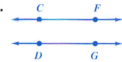

10.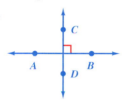

11.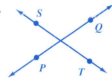

12.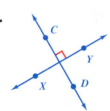

Name each highlighted angle by using the three-letter form of identification. See Example 3.

13.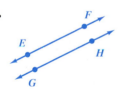

14.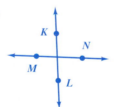

15.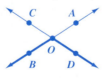

16.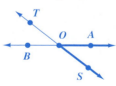

17.

18.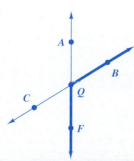

Label each angle as acute, right, obtuse, or straight. For right and straight angles, indicate the number of degrees in the angle. See Example 4.

19.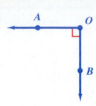

20.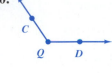

21.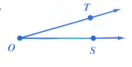

22.

23.

24.

25. Explain what is happening in each sentence.

   (a) The road was so slippery that my car did a 360.

   (b) After the election, the governor's view on taxes took a 180° turn.

26. Find at least four examples of right angles in your home, at work, or on the street. Make a sketch of each example and label the right angle.

RELATING CONCEPTS (EXERCISES 27–32)   For Individual or Group Work

*Use the figure below to **work Exercises 27–32 in order.** Decide whether each statement is **true** or **false.** If it is true, explain why. If it is false, rewrite it to make it a true statement.*

27. ∠*UST* is 90°.

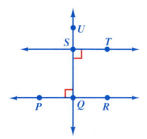

28. $\overleftrightarrow{SQ}$ and $\overleftrightarrow{PQ}$ are perpendicular.

29. The measure of ∠*USQ* is less than the measure of ∠*PQR*.

30. $\overleftrightarrow{ST}$ and $\overleftrightarrow{PR}$ are intersecting.

31. $\overleftrightarrow{QU}$ and $\overleftrightarrow{TS}$ are parallel.

32. ∠*UST* and ∠*UQR* measure the same number of degrees.

8.2 Angles and Their Relationships

OBJECTIVE **1** **Identify complementary angles and supplementary angles and find the measure of a complement or supplement of a given angle.** Two angles are called **complementary angles** if the sum of their measures is 90°. If two angles are complementary, each angle is the *complement* of the other.

EXAMPLE 1 **Identifying Complementary Angles**

Identify each pair of complementary angles.

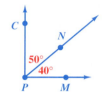

∠*MPN* (40°) and ∠*NPC* (50°) are complementary angles because

$$40° + 50° = \mathbf{90°}$$

∠*CAB* (30°) and ∠*FHG* (60°) are complementary angles because

$$30° + 60° = \mathbf{90°}$$

Work Problem 1 at the Side.

EXAMPLE 2 **Finding the Complement of Angles**

Find the complement of each angle.

(a) 30°
Find the complement of 30° by subtracting. $\mathbf{90°} - 30° = \mathbf{60°}$ ←Complement

(b) 40°
Find the complement of 40° by subtracting. $\mathbf{90°} - 40° = \mathbf{50°}$ ←Complement

Work Problem 2 at the Side.

Two angles are called **supplementary angles** if the sum of their measures is 180°. If two angles are supplementary, each angle is the *supplement* of the other.

EXAMPLE 3 **Identifying Supplementary Angles**

Identify each pair of supplementary angles.

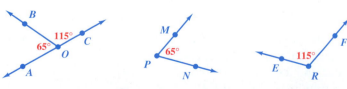

∠*BOA* and ∠*BOC*, because 65° + 115° = **180°**
∠*BOA* and ∠*ERF*, because 65° + 115° = **180°**
∠*BOC* and ∠*MPN*, because 115° + 65° = **180°**
∠*MPN* and ∠*ERF*, because 65° + 115° = **180°**

Work Problem 3 at the Side.

1 Identify each pair of complementary angles.

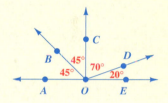

2 Find the complement of each angle.

(a) 35°

(b) 80°

3 Identify each pair of supplementary angles.

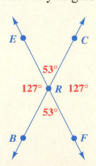

ANSWERS
1. ∠*AOB* and ∠*BOC*; ∠*COD* and ∠*DOE*
2. (a) 55° **(b)** 10°
3. ∠*CRF* and ∠*BRF*; ∠*CRE* and ∠*ERB*;
     ∠*BRF* and ∠*BRE*; ∠*CRE* and ∠*CRF*

4 Find the supplement of each angle.

(a) 175°

(b) 30°

5 Identify the angles that are congruent.

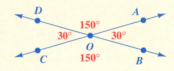

6 Identify the vertical angles.

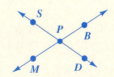

EXAMPLE 4 **Finding the Supplement of Angles**

Find the supplement of each angle.

(a) 70°
Find the supplement of 70° by subtracting.   **180° − 70° = 110°** ← Supplement

(b) 140°
Find the supplement of 140° by subtracting.   **180° − 140° = 40°** ← Supplement

Work Problem 4 at the Side.

O B J E C T I V E **2** **Identify congruent angles and vertical angles and use this knowledge to find the measures of angles.**   Two angles are called **congruent angles** if they measure the same number of degrees. If two angles are congruent, this is written as ∠A ≅ ∠B and read as, "angle A **is congruent to** angle B." Here is an example.

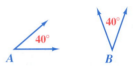

Congruent angles: ∠A ≅ ∠B

EXAMPLE 5 **Identifying Congruent Angles**

Identify the angles that are congruent.

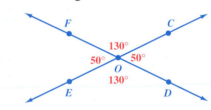

∠FOC ≅ ∠EOD   and   ∠COD ≅ ∠EOF

Work Problem 5 at the Side.

Angles that share a common side and a common vertex are called *adjacent* angles, such as ∠FOC and ∠COD in Example 5 above. Angles that do *not* share a common side are called *nonadjacent* angles. Two nonadjacent angles formed by intersecting lines are called **vertical angles.**

EXAMPLE 6 **Identifying Vertical Angles**

Identify the vertical angles in this figure.

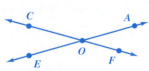

∠AOF and ∠COE are vertical angles because they do not share a common side and they are formed by two intersecting lines ($\overleftrightarrow{CF}$ and $\overleftrightarrow{EA}$).

∠COA and ∠EOF are also vertical angles.

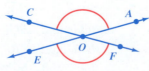

Work Problem 6 at the Side.

Look back at Example 5 on the previous page. Notice that the two *congruent* angles that measure 130° are also *vertical* angles. Also, the two congruent angles that measure 50° are vertical angles. This illustrates the following property.

> **Vertical Angles Are Congruent**
>
> If two angles are *vertical* angles, they are *congruent,* that is, they measure the same number of degrees.

EXAMPLE 7 Finding the Measures of Vertical Angles

In the figure below, find the measure of each unlabeled angle. Then write the measure of each angle on the figure.

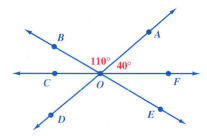

(a) ∠COD
  ∠COD and ∠AOF are vertical angles, so they are congruent. This means they measure the same number of degrees.

  The measure of ∠AOF is 40°   so   the measure of ∠COD is 40° also.

(b) ∠DOE
  ∠DOE and ∠BOA are vertical angles, so they are congruent.

  The measure of ∠BOA is 110°   so   the measure of ∠DOE is 110° also.

(c) ∠COB
  Look at ∠COB, ∠BOA, and ∠AOF. Notice that $\overrightarrow{OC}$ and $\overrightarrow{OF}$ go in opposite directions. Therefore, ∠COF is a straight angle and measures 180°. To find the measure of ∠COB, subtract the sum of the other two angles from 180°.

$$180° - (110° + 40°) = 180° - (150°) = 30°$$

The measure of ∠COB is 30°.

(d) ∠EOF
  ∠EOF and ∠COB are vertical angles, so they are congruent. We know from part (c) above that the measure of ∠COB is 30° so the measure of ∠EOF is 30° also.

The figure with all the angle measures is shown below.

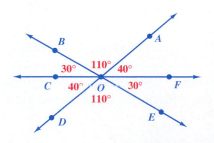

Work Problem 7 at the Side.

---

7 In the figure below, find the number of degrees in each unlabeled angle. Then write the angle measures on the figure.

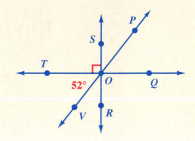

(a) ∠TOS

(b) ∠QOR

(c) ∠POQ

(d) ∠VOR

(e) ∠POS

ANSWERS
7. **(a)** 90° **(b)** 90° **(c)** 52°
   (d) 38° **(e)** 38°

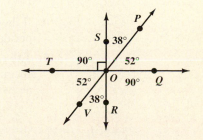

The London Eye

The British Airways London Eye was built as a symbol of the turn of the century and represents the cycle of life. It was designed by architects David Marks and Julia Barfield to suspend over the River Thames in Jubilee Gardens, on the bank opposite the Houses of Parliament and Big Ben. It first opened on December 31, 1999.

The London Eye is the largest observation wheel ever designed. It is a unique form of a Ferris wheel that features 32 evenly-spaced, oval-shaped capsules. Each capsule holds 25 people. The wheel structure was assembled on platforms erected in the riverbed, and once assembled was lifted into place by a floating crane. The London Eye is the fourth tallest structure in London (and the only one of the four that is open to the public), and the capsules were designed so that passengers have an unobstructed view throughout the ride. On a clear day, passengers can see 25 miles to the 900-year-old Windsor Castle, the world's largest royal residence, used by Queen Elizabeth II.

Another unique feature of the London Eye is that it is designed to rotate continuously at walking speed. At one end of the loading platform, passengers walk out of the capsules, and at the other end of the platform, passengers walk into the capsules. Because the wheel rotates so slowly, the passengers can stand and walk around the capsules during their ride, and enjoy a 360° view of the London skyline. The wheel does stop to allow people in wheelchairs to board. (*Source: The Bankside Press,* 2001.)

Specifications for the London Eye	
Diameter	135 meters
Speed	0.26 meters per second
Time to revolve	30 minutes

 1. **(a)** What is the diameter of the wheel in feet? (*Hint:* Use 1 m ≈ 3.28 ft.)
   (b) The circumference of the wheel is the distance around the outside rim. Find the circumference by multiplying the diameter times 3.1416 and rounding the answer to the nearest foot. (See **Section 8.6** for more information about circumference.)

2. Find the distance, to the nearest tenth of a foot, along the rim arc between two adjacent capsules. Recall that there are 32 capsules.

3. If the London Eye is operating at full capacity, how many people will be riding in a 90° section of the wheel? in a 45° section? in a 180° section? in a 270° section? in a 360° section?

4. If the London Eye operates from 8:00 A.M. until 6:00 P.M. at full capacity, how many people per day can ride?

5. The speed is given as 0.26 meter per second. Find the speed of the London Eye in feet per second. Round the answer to the nearest hundredth. Do you believe that the wheel revolves at "walking speed"? (Try it and see!)

8.2 Exercises

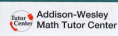
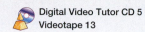

 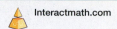
Identify each pair of complementary angles. See Example 1.

1.

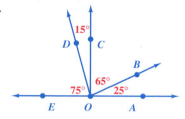

2.

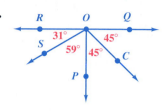

Identify each pair of supplementary angles. See Example 3.

3.

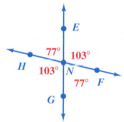

4.

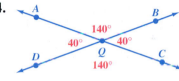

Find the complement of each angle. See Example 2.

5. 40°   **6.** 35°   **7.** 86°   **8.** 59°

Find the supplement of each angle. See Example 4.

9. 130°   **10.** 75°   **11.** 90°   **12.** 5°

In each figure, identify the angles that are congruent. See Example 5.

13.

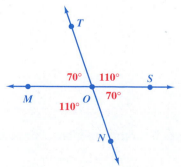

14.

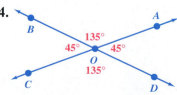

Use your knowledge of vertical angles to answer Exercises 15 and 16. See Examples 6 and 7.

15. In the figure below, ∠*AOH* measures 37° and ∠*COE* measures 63°. Find the measure of each of the other angles.

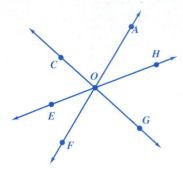

16. In the figure below, ∠*POU* measures 105° and ∠*UOT* measures 40°. Find the measure of each of the other angles.

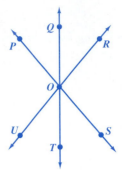

17. In your own words, write a definition of complementary angles and a definition of supplementary angles. Draw a picture to illustrate each definition.

18. Make up a test problem in which a student has to use knowledge of vertical angles. Include a drawing with some angles labeled and ask the student to find the size of the remaining angles. Give the correct answer for your problem.

In each figure, $\overrightarrow{BA}$ is parallel to $\overrightarrow{CD}$. Identify two pairs of congruent angles and the number of degrees in each congruent angle.

19.

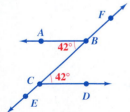

20.

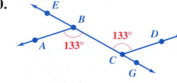

21. Can two obtuse angles be supplementary? Explain why or why not.

22. Can two acute angles be complementary? Explain why or why not.

8.3 Rectangles and Squares

A **rectangle** is a figure with four sides that meet to form 90° angles. Each set of opposite sides is *parallel* and *congruent* (has the same length).

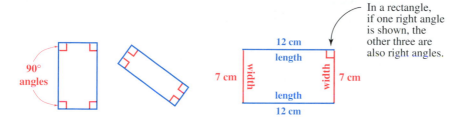

In a rectangle, if one right angle is shown, the other three are also right angles.

Each longer side of a rectangle is called the length (*l*) and each shorter side is called the width (*w*).

> **Work Problem 1 at the Side.** ▶▶▶

OBJECTIVE 1 Find the perimeter and area of a rectangle. The distance around the outside edges of a figure is the **perimeter** of the figure. Think of how much fence you would need to put around the sides of a garden plot, or how far you would walk if you walked around the outside edges of your living room. In either case you would add up the lengths of the sides. Look at the rectangle above that has the lengths of the sides labeled. To find its perimeter, you add the lengths of the sides.

Perimeter = **12 cm** + **12 cm** + **7 cm** + **7 cm** = 38 cm

Because the two long sides are both 12 cm, and the two short sides are both 7 cm, you can also use this formula.

> **Finding the Perimeter of a Rectangle**
>
> Perimeter of a rectangle = length + length + width + width
> $$P = (2 \cdot \text{length}) + (2 \cdot \text{width})$$
> $$P = 2 \cdot l + 2 \cdot w$$

EXAMPLE 1 Finding the Perimeter of Rectangles

Find the perimeter of each rectangle.

(a)

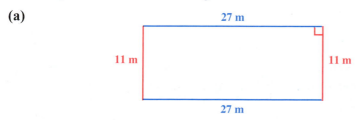

The length of this rectangle is **27 m** and the width is **11 m**.
Use the formula $P = 2 \cdot l + 2 \cdot w$

$P = 2 \cdot $ **l** $ + 2 \cdot $ **w**   Replace *l* with 27 m and *w* with 11 m.
$P = $ 2 · **27 m** $ + $ 2 · **11 m**   Do the multiplications first.
$P = $ 54 m $ + $ 22 m   Add last.
$P = 76$ m

The perimeter of the rectangle (the distance you would walk around the outside edges of the rectangle) is 76 m.

— **Continued on Next Page**

OBJECTIVES

1. Find the perimeter and area of a rectangle.
2. Find the perimeter and area of a square.
3. Find the perimeter and area of a composite shape.

---

1 Identify all the rectangles.

(a)

(b)

(c)

(d)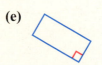

(e)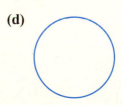

(f)

(g)

ANSWERS
1. **(a), (b),** and **(e)** are rectangles; **(c), (d), (f),** and **(g)** are not.

② Find the perimeter of each rectangle by using the formula or by adding the lengths of the sides.

(a)

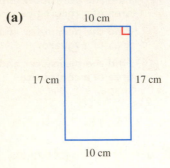

10 cm

17 cm   17 cm

10 cm

As a check, you can add up the lengths of the four sides.

$$P = \textbf{27 m} + \textbf{27 m} + \textbf{11 m} + \textbf{11 m}$$

$$P = 76 \text{ m} \leftarrow \text{Same result as using the formula}$$

(b) A rectangle 8.9 ft by 12.3 ft
You can use the formula, as shown below.

$$P = 2 \cdot \quad l \quad + 2 \cdot \quad w$$

$$P = 2 \cdot \textbf{12.3 ft} + 2 \cdot \textbf{8.9 ft}$$

$$P = \quad 24.6 \text{ ft} \quad + \quad 17.8 \text{ ft}$$

$$P = 42.4 \text{ ft}$$

Or, you can add up the lengths of the four sides.

$$P = \textbf{12.3 ft} + \textbf{12.3 ft} + \textbf{8.9 ft} + \textbf{8.9 ft}$$

$$P = 42.4 \text{ ft} \leftarrow \text{Same result as using the formula}$$

Either method will give you the correct result.

◀◀◀ Work Problem 2 at the Side.

(b)

10.5 ft

7 ft

The *perimeter* of a rectangle is the distance around the *outside edges*. The **area** of a rectangle is the amount of surface *inside* the rectangle. We measure area by seeing how many squares of a certain size are needed to cover the surface inside the rectangle. Think of covering the floor of a rectangular living room with carpet. Carpet is measured in square yards, that is, square pieces that measure 1 yard along each side. Here is a drawing of a living room floor.

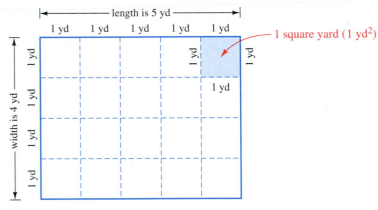

(c) 6 m wide and 11 m long

You can see from the drawing that it takes 20 squares to cover the floor. We say that the area of the floor is 20 *square yards*. A shorter way to write square yards is yd².

20 **square yards**   can be written as   20 **yd²**

To find the number of squares, you can count them, or you can multiply the number of squares in the length (5) times the number of squares in the width (4) to get 20. The formula is given below.

(d) 0.9 km by 2.8 km

Finding the Area of a Rectangle

Area of a rectangle = length • width

$$A = l \cdot w$$

Remember to use *square units* when measuring area.

Squares of many sizes can be used to measure area. For smaller areas, you might use the ones shown below.

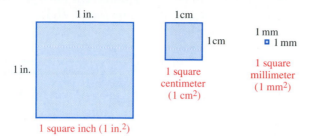

1 square inch (1 in.²)

Actual-size drawings

Other sizes of squares that are often used to measure area are listed here, but they are too large to draw on this page.

1 square meter (1 m²)   1 square foot (1 ft²)

1 square kilometer (1 km²)   1 square yard (1 yd²)

                                     1 square mile (1 mi²)

> **CAUTION**
> The raised 2 in 4^2 means that you multiply $4 \cdot 4$ to get 16. The raised 2 in cm² or yd² is a short way to write the word *square*. When you see 5 cm², say "five square centimeters." Do *not* multiply $5 \cdot 5$. The exponent applies to cm, *not* to the number.

EXAMPLE 2 **Finding the Area of Rectangles**

Find the area of each rectangle.

(a)

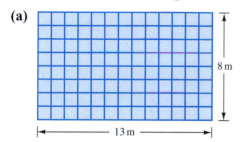

8 m

13 m

The length of this rectangle is 13 m and the width is 8 m. Use the formula $A = l \cdot w$.

$A = \quad l \quad \cdot \quad w$ Replace *l* with 13 m and *w* with 8 m.

$A = \mathbf{13\ m} \cdot \mathbf{8\ m}$ Multiply.

$A = 104$ square meters

"Square meters" can be written as m², so the area is 104 m².

(b) A rectangle measuring 7 cm by 21 cm

First make a sketch of the rectangle. The length is 21 cm (the longer measurement) and the width is 7 cm. Then use the formula for the area of a rectangle, $A = l \cdot w$.

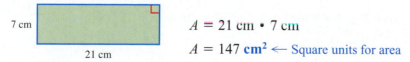

7 cm

21 cm

$A = 21\ \text{cm} \cdot 7\ \text{cm}$

$A = 147\ \mathbf{cm^2} \leftarrow$ Square units for area

The area of the rectangle is 147 cm².

Continued on Next Page

3 Find the area of each rectangle.

(a)

9 ft

4 ft — 4 ft

9 ft

(b) A rectangle is 6 m long and 0.5 m wide. (First make a sketch of the rectangle and label the lengths of the sides.)

(c) A rectangular patio measures 3.5 yd by 2.5 yd. (First make a sketch of the patio and label the lengths of the sides.)

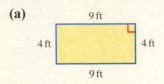
CAUTION
The units for *area* will always be *square* units (cm^2, m^2, yd^2, mi^2, and so on). The units for *perimeter* will always be *linear* units (cm, m, yd, mi, and so on) *not* square units.

◀◀◀ **Work Problem 3 at the Side.**

O B J E C T I V E **2** **Find the perimeter and area of a square.** A **square** is a rectangle with all sides the same length. Two squares are shown below. Notice the 90° angles.

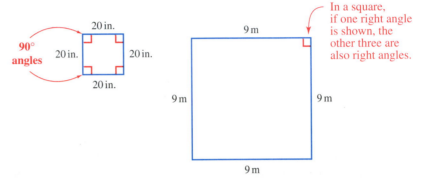

90° angles   20 in.   20 in.   20 in.   20 in.

9 m   9 m   9 m   9 m

In a square, if one right angle is shown, the other three are also right angles.

To find the *perimeter* of (distance around) the square on the right, you could add 9 m + 9 m + 9 m + 9 m to get 36 m. A shorter way is to multiply the length of one side times 4, because all four sides are the same length.

> **Finding the Perimeter of a Square**
> Perimeter of a square = side + side + side + side
> or, $P = 4 \cdot$ side
> $P = 4 \cdot s$

As with a rectangle, you can multiply length times width to find the *area* of (surface inside) a square. Because the length and the width are the same in a square, the formula is written as shown below.

> **Finding the Area of a Square**
> Area of a square = side $\cdot$ side
> $A = s \cdot s$
> $A = s^2$
>
> Remember to use **square units** when measuring area.

EXAMPLE 3 **Finding the Perimeter and Area of a Square**

(a) Find the perimeter of the square shown above where each side measures 9 m.

Use the formula.

$P = 4 \cdot s$

$P = 4 \cdot 9$ m

$P = 36$ m

Or add up the four sides.

$P = 9$ m $+ 9$ m $+ 9$ m $+ 9$ m

$P = 36$ m

Same answer

Continued on Next Page

(b) Find the area of the same square where each side measures 9 m.

$$A = s^2$$
$$A = \ s \ \bullet \ s$$
$$A = 9\text{ m} \bullet 9\text{ m}$$
$$a = 81\text{ m}^2 \leftarrow \text{Square units for area}$$

> **CAUTION**
> Be careful! s^2 means $s \bullet s$. It does **not** mean $2 \bullet s$. In Example 3(b) above, s is 9 m, so s^2 is 9 m $\bullet$ 9 m $= 81$ m². It is **not** $2 \bullet 9$ m $= 18$ m.

Work Problem 4 at the Side. ▶▶▶

OBJECTIVE 3 Find the perimeter and area of a composite shape. As with any other shape, you can find the perimeter of (distance around) an irregular shape by adding up the lengths of the sides. To find the area (surface inside the shape), try to break it up into pieces that are squares or rectangles. Find the area of each piece and then add them together.

> **CAUTION**
> **Perimeter** is the **distance around the outside edges** of a flat shape. It is always measured in *linear units* such as cm, m, yd, and so on.
>
> **Area** is the amount of **surface inside** a flat shape. It is always measured in *square units* such as cm², m², yd², and so on.

EXAMPLE 4 Finding the Perimeter and Area of a Composite Figure

The floor of a room has the shape shown below.

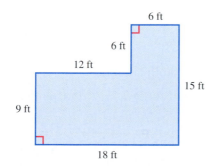

(a) Suppose you want to put a new wallpaper border along the top of all the walls. How much material do you need?

Find the perimeter of the room by adding up the lengths of the sides.

$$P = 9\text{ ft} + 12\text{ ft} + 6\text{ ft} + 6\text{ ft} + 15\text{ ft} + 18\text{ ft}$$
$$P = 66\text{ ft}$$

You need 66 ft of wallpaper border.

Continued on Next Page

---

4 Find the perimeter and area of each square.

(a)

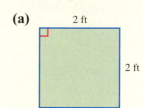

(b) 10.5 cm on each side (Make a sketch of the square.)

(c) 2.1 mi on a side (Make a sketch of the square.)

ANSWERS
4. **(a)** $P = 8$ ft; $A = 4$ ft²
   (b) $P = 42$ cm;
   $A = 110.25$ cm²

   (c) $P = 8.4$ mi;
   $A = 4.41$ mi²

5 Carpet costs $19.95 per square yard. Find the cost of carpeting each room. Round your answers to the nearest cent if necessary.

(a)

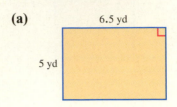
6.5 yd
5 yd

(b)

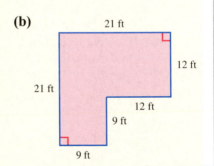
21 ft
21 ft
12 ft
12 ft
9 ft
9 ft

(c) A rectangular classroom is 24 ft long and 18 ft wide. (Make a sketch of the classroom.)

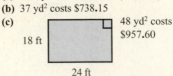
(b) The carpet you like costs $20.50 per square yard. How much will it cost to carpet the room?

First change the measurements from feet to yards, because the carpet is sold in square yards. There are 3 ft in 1 yd, so multiply by the unit fraction that allows you to divide out feet. Let's start with 9 ft.

$$\frac{\overset{3}{\cancel{9}} \cancel{ft}}{1} \cdot \frac{1 \text{ yd}}{\underset{1}{\cancel{3}} \cancel{ft}} = 3 \text{ yd}$$

— Divide out ft.
— Divide 9 and 3 by 3.

Use the same unit fraction to change the other measurements from feet to yards.

$$\frac{\overset{4}{\cancel{12}} \cancel{ft}}{1} \cdot \frac{1 \text{ yd}}{\underset{1}{\cancel{3}} \cancel{ft}} = 4 \text{ yd} \qquad \frac{\overset{2}{\cancel{6}} \cancel{ft}}{1} \cdot \frac{1 \text{ yd}}{\underset{1}{\cancel{3}} \cancel{ft}} = 2 \text{ yd}$$

$$\frac{\overset{5}{\cancel{15}} \cancel{ft}}{1} \cdot \frac{1 \text{ yd}}{\underset{1}{\cancel{3}} \cancel{ft}} = 5 \text{ yd} \qquad \frac{\overset{6}{\cancel{18}} \cancel{ft}}{1} \cdot \frac{1 \text{ yd}}{\underset{1}{\cancel{3}} \cancel{ft}} = 6 \text{ yd}$$

Next, break up the room into two pieces. Use just the measurements for the length and width of each piece.

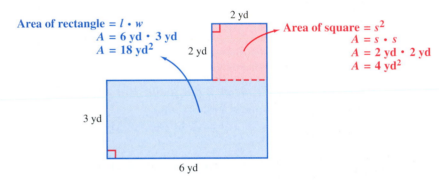

Area of rectangle $= l \cdot w$
$A = 6 \text{ yd} \cdot 3 \text{ yd}$
$A = 18 \text{ yd}^2$

Area of square $= s^2$
$A = s \cdot s$
$A = 2 \text{ yd} \cdot 2 \text{ yd}$
$A = 4 \text{ yd}^2$

2 yd
2 yd
3 yd
6 yd

Total area $= \mathbf{18 \text{ yd}^2} + \mathbf{4 \text{ yd}^2} = 22 \text{ yd}^2$

Multiply to find the cost of the carpet.

$$\frac{22 \text{ yd}^2}{1} \cdot \frac{\$20.50}{1 \text{ yd}^2} = \$451.00$$

It will cost $451.00 to carpet the room.

You could have cut the room into two rectangles as shown below. The total area is the same

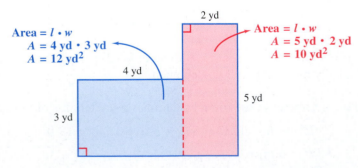

Area $= l \cdot w$
$A = 4 \text{ yd} \cdot 3 \text{ yd}$
$A = 12 \text{ yd}^2$

Area $= l \cdot w$
$A = 5 \text{ yd} \cdot 2 \text{ yd}$
$A = 10 \text{ yd}^2$

2 yd
4 yd
5 yd
3 yd

Total area $= \mathbf{12 \text{ yd}^2} + \mathbf{10 \text{ yd}^2} = 22 \text{ yd}^2$ ← Same answer as above

◄◄◄ Work Problem 5 at the Side.

8.3 Exercises

FOR
EXTRA
HELP

 Addison-Wesley
Math Tutor Center

 MathXL

 Digital Video Tutor CD 5
Videotape 13

 Student's
Solutions
Manual

MyMathLab
MyMathLab

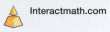 Interactmath.com

Study Skills Workbook
Activity 9: Mind Maps

Find the perimeter and area of each rectangle or square. See Examples 1–3.

1.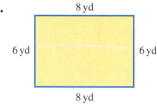
8 yd
6 yd   6 yd
8 yd

2.
7 in.
18 in.   18 in.
7 in.

3.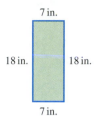
0.9 km   0.9 km
0.9 km   0.9 km

4.
7.5 m
7.5 m

Draw a sketch of each square or rectangle and label the lengths of the sides. Then find the perimeter and the area. (Sketches may vary, show your sketches to your instructor.)

5. 10 ft by 10 ft

6. 8 cm by 17 cm

7. 14 m by 0.5 m

8. 2.35 km by 8.4 km

9. A storage building that is 76.1 ft by 22 ft

10. A science lab measuring 12 m by 12 m

11. A square nature preserve 3 mi wide

12. A square of cardboard 20.3 cm on a side

Find the perimeter and area of each figure. See Example 4.

13.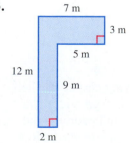
7 m
3 m
5 m
12 m
9 m
2 m

14.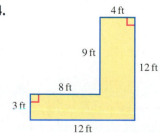
4 ft
9 ft
12 ft
8 ft
3 ft
12 ft

15.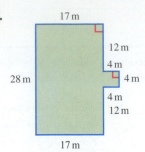

16.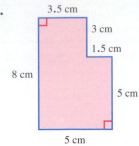

First find the length of the unlabeled side in each figure. Then find the perimeter and area of each figure.

17.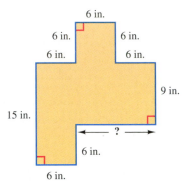

18.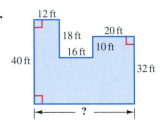

Solve each application problem. In Exercises 19–24, draw a sketch for each problem and label it with the appropriate measurements. (Sketches may vary; show your sketches to your instructor.)

19. Gymnastic floor exercises are performed on a square mat that is 12 meters on a side. Find the perimeter and area of a mat. (*Source:* www.nist.gov)

20. A regulation volleyball court is 18 meters by 9 meters. Find the perimeter and area of a regulation court. (*Source:* www.nist.gov)

21. The Wang's family room measures 20 ft by 25 ft. They are covering the floor with square tiles that measure 1 ft on a side and cost $0.92 each. How much will they spend on tile?

22. A page in this book measures 27.5 cm from top to bottom and 20.5 cm from side to side. Find the perimeter and the area of the page.

23. Tyra's kitchen is 4.4 m wide and 5.1 m long. She is pasting a decorative border strip that costs $4.99 per meter around the top edge of all the walls. How much will she spend?

24. Mr. and Mrs. Gomez are buying carpet for their square-shaped bedroom that is 5 yd wide. The carpet is $23 per square yard and padding and installation is another $6 per square yard. How much will they spend in all?

25. Advanced Photo System (APS) cameras allow you to choose from three different print sizes each time you snap a photo. The choices are shown below. Find the perimeter and area of each size print. (*Source:* Kodak.)

Panoramic
4 in. × 14 in.

4 in. × 6 in.

4 in. × 7 in.

26. The table below shows information on two tents for camping.

Tents	Coleman Family Dome	Eddie Bauer Dome Tent
Dimensions	13 ft × 13 ft	12 ft × 12 ft
Sleeps	8 campers	6 campers
Sale price	$127	$99

(*Source:* target.com)

(a) For the Coleman tent, find the perimeter, area, and number of square feet of floor space for each camper. Round to the nearest whole number.

(b) Find the same information for the Eddie Bauer tent.

27. A regulation football field is rectangular, 100 yd long (excluding end zones), and has an area of 5300 yd². Find the width of the field. (*Source:* National Football League.)

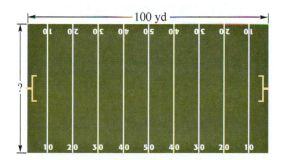

100 yd

28. There are 14,790 ft² of ice in the rectangular playing area for a major league hockey game (excluding the area behind the goal lines). If the playing area is 85 ft wide, how long is it? (*Source:* National Hockey League.)

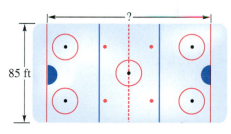

85 ft

?

29. A rectangular lot is 124 ft by 172 ft. County rules require that nothing be built on land within 12 ft of any edge of the lot. First, add labels to the sketch of the lot, showing the land that cannot be build on. Then find the area of the land that cannot be built on.

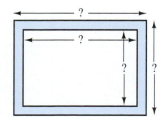

30. Find the cost of fencing needed for this rectangular field. Fencing along the country roads costs $4.25 per foot. Fencing for the other two sides costs $2.75 per foot.

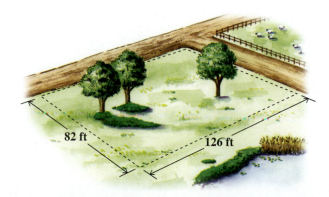

82 ft   126 ft

RELATING CONCEPTS (EXERCISES 31–36)   For Individual or Group Work

Use your knowledge of perimeter and area to **work Exercises 31–36 in order.**

31. Suppose you have 12 ft of fencing to make a square or rectangular garden plot. Draw sketches of *all* the possible plots that use exactly 12 ft of fencing and label the lengths of the sides. Use only *whole number* lengths. (*Hint:* There are three possibilities.)

32. (a) Find the area of each plot in Exercise 31.

   (b) Which plot has the greatest area?

33. Repeat Exercise 31 using 16 ft of fencing. Be sure to draw *all* possible plots that have whole number lengths for the sides.

34. (a) Find the area of each plot in Exercise 33.

   (b) Compare your results to those from Exercise 32. What do you notice about the plots with the greatest area?

35. (a) Draw a sketch of a rectangular plot 3 ft by 2 ft. Find the perimeter and area.

   (b) Suppose you *double* the length of the plot and *double* the width. Draw a sketch of the enlarged plot and find the perimeter and area.

   (c) The *perimeter* of the enlarged plot is how many times greater than the perimeter of the original plot? The *area* of the enlarged plot is how many times greater than the original area?

36. (a) Refer to part (a) of Exercise 35. Suppose you *triple* the length and width of the original plot. Draw a sketch of the enlarged plot and find the perimeter and area.

   (b) How many times greater is the *perimeter* of the enlarged plot? How many times greater is the *area* of the enlarged plot?

   (c) Suppose you make the length and width *four times greater* in the enlarged plot. What would you predict will happen to the perimeter and area, compared to the original plot?

8.4 Parallelograms and Trapezoids

A **parallelogram** is a four-sided figure with opposite sides parallel, such as the ones shown below. Notice that opposite sides have the same length.

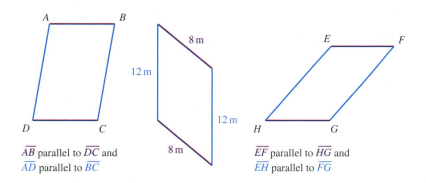

$\overline{AB}$ parallel to $\overline{DC}$ and $\overline{AD}$ parallel to $\overline{BC}$

$\overline{EF}$ parallel to $\overline{HG}$ and $\overline{EH}$ parallel to $\overline{FG}$

OBJECTIVES

1 Find the perimeter and area of a parallelogram.

2 Find the perimeter and area of a trapezoid.

OBJECTIVE 1 Find the perimeter and area of a parallelogram. Perimeter is the distance around a flat shape, so the easiest way to find the perimeter of a parallelogram is to add the lengths of the four sides.

EXAMPLE 1 Finding the Perimeter of a Parallelogram

Find the perimeter of the middle parallelogram above.

$$P = \textbf{12 m} + \textbf{12 m} + \textbf{8 m} + \textbf{8 m} = 40 \text{ m}$$

Work Problem 1 at the Side. ⟩⟩⟩

To find the area of a parallelogram, first draw a dashed line inside the figure as shown here.

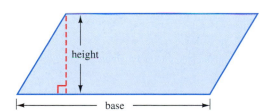

Try this yourself by tracing this parallelogram onto a piece of paper.

The length of the dashed line is the *height* of the parallelogram. It forms a *right angle* with the base. The height is the shortest distance between the base and the opposite side.

Now cut off the triangle created on the left side of the parallelogram above and move it to the right side, as shown below.

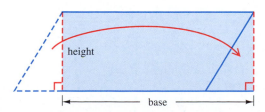

The parallelogram has been made into a rectangle. You can see that the area of the parallelogram and the rectangle are the same.

Equal areas
→ Area of the   rectangle   = length • width
↓   ↓   ↓
→ Area of the parallelogram = base • height

1 Find the perimeter of each parallelogram

(a)

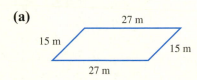

(b)

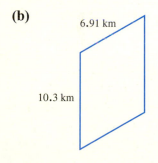

ANSWERS
1. (a) $P = 84$ m   **(b)** $P = 34.42$ km

2 Find the area of each parallelogram.

(a)

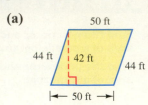

(b)

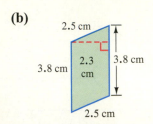

(c) A parallelogram with base $12\frac{1}{2}$ mi and height $4\frac{3}{4}$ mi (*Hint:* Write $12\frac{1}{2}$ as 12.5 and $4\frac{3}{4}$ as 4.75.)

3 Find the perimeter of each trapezoid.

(a)

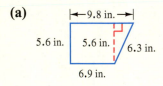

(b)

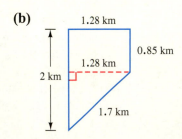

(c) A trapezoid with sides 39.7 cm, 29.2 cm, 74.9 cm, and 16.4 cm

---

Finding the Area of a Parallelogram

Area of a parallelogram = base • height

$$A = b \bullet h$$

Remember to use *square units* when measuring area.

EXAMPLE 2 Finding the Area of Parallelograms

Find the area of each parallelogram.

(a)

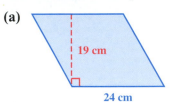

The base is 24 cm and the height is 19 cm. Use the formula $A = b \bullet h$.

$$A = \quad b \quad \bullet \quad h$$

$$A = 24 \text{ cm} \bullet 19 \text{ cm}$$

$$A = 456 \text{ cm}^2 \quad \text{Square units for area}$$

(b)

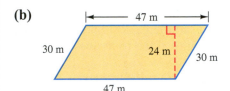

$$A = 47 \text{ m} \bullet 24 \text{ m}$$

$$A = 1128 \text{ m}^2 \quad \text{Square units for area}$$

Notice that you do *not* use the 30 m sides when finding the area. But you would use them when finding the *perimeter* of the parallelogram.

Work Problem 2 at the Side.

OBJECTIVE 2 Find the perimeter and area of a trapezoid.
A **trapezoid** is a four-sided figure with one pair of parallel sides, such as the figures shown below. Unlike parallelograms, opposite sides of a trapezoid might *not* have the same length.

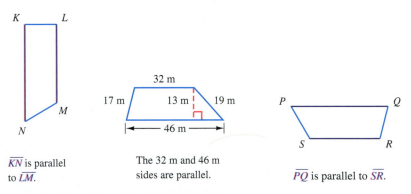

$\overline{KN}$ is parallel to $\overline{LM}$.

The 32 m and 46 m sides are parallel.

$\overline{PQ}$ is parallel to $\overline{SR}$.

EXAMPLE 3 Finding the Perimeter of a Trapezoid

Find the perimeter of the middle trapezoid above.
You can find the perimeter of any flat shape by adding the lengths of the sides.

$$P = 17 \text{ m} + 32 \text{ m} + 19 \text{ m} + 46 \text{ m}$$

$$P = 114 \text{ m}$$

Notice that the height (13 m) is *not* part of the perimeter, because the height is *not* one of the *outside edges* of the shape.

Work Problem 3 at the Side.

Use this formula to find the *area* of a trapezoid.

Finding the Area of a Trapezoid

$$\text{Area} = \frac{1}{2} \cdot \text{heigh} \cdot (\text{short base} + \text{long base})$$

$$A = \frac{1}{2} \cdot h \cdot (b + B)$$

or $\quad A = 0.5 \cdot h \cdot (b + B)$

Remember to use **square units** when measuring area.

EXAMPLE 4 **Finding the Area of a Trapezoid**

Find the area of this trapezoid. The short base and long base are the *parallel* sides.

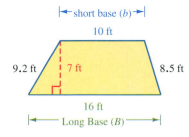

The height (**h**) is **7 ft**, the short base (**b**) is **10 ft**, and the long base (**B**) is **16 ft**. You do *not* need the 9.2 ft or 8.5 ft sides to find the area.

$$A = \frac{1}{2} \cdot h \cdot (b + B)$$

$$A = \frac{1}{2} \cdot 7 \text{ ft} \cdot (10 \text{ ft} + 16 \text{ ft}) \quad \text{Work inside parentheses first.}$$

$$A = \frac{1}{\overset{\cancel{2}}{1}} \cdot 7 \text{ ft} \cdot (\overset{13}{\cancel{26}} \text{ ft})$$

$$A = 91 \text{ ft}^2 \quad \text{Square units for area}$$

You can also find the area by using 0.5, the decimal equivalent for $\frac{1}{2}$, in the formula.

$$A = 0.5 \cdot h \cdot (b + B)$$

$$A = 0.5 \cdot 7 \cdot (10 + 16)$$

$$A = 0.5 \cdot 7 \cdot \quad 26$$

$$A = 91 \text{ ft}^2 \quad \text{Same answer as above}$$

⊞ Calculator Tip Use the parentheses keys on your scientific calculator to work Example 4 above.

0.5 ⊗ 7 ⊗ ⦅ 10 ⊕ 16 ⦆ ⊜ 91

What happens if you do *not* use the parentheses keys? What order of operations will the calculator follow then? (Answer: The calculator will multiply 0.5 times 7 times 10, and then add 16, giving an *incorrect* answer of 51.)

Work Problem 4 at the Side. ▶▶▶

4 Find the area of each trapezoid.

(a)

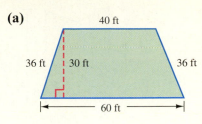

(b)

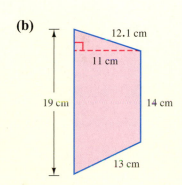

(c) A trapezoid with height 4.7 m, short base 9 m, and long base 10.5 m (First draw a sketch and label the bases and height).

ANSWERS
4. **(a)** $A = 1500 \text{ ft}^2$ **(b)** $A = 181.5 \text{ cm}^2$

  (c) $A = 45.825 \text{ m}^2$

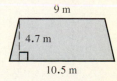

5 Find the area of each floor.

(a)

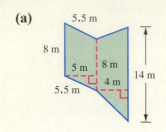

(b)

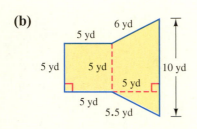

6 Find the cost of carpeting the floors in Problem 5 above. The cost of carpet is as follows:

(a) Floor (a), $18.50 per square meter.

(b) Floor (b), $28 per square yard.

EXAMPLE 5 Finding the Area of a Composite Figure

Find the area of this figure.

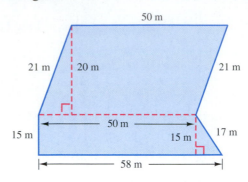

Break the figure into two pieces, a parallelogram and a trapezoid. Find the area of each piece, and then add the areas.

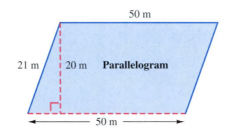

Area of parallelogram

$A = b \cdot h$

$A = 50 \text{ cm} \cdot 20 \text{ cm}$

$A = \mathbf{1000 \text{ m}^2}$

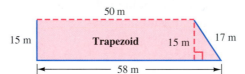

Area of trapezoid

$A = \dfrac{1}{2} \cdot h \cdot (b + B)$

$A = 0.5 \cdot 15 \text{ m} \cdot (50 \text{ m} + 58 \text{ m})$

$A = \mathbf{810 \text{ m}^2}$

Total area = **1000 m² + 810 m² = 1810 m²**

The area of the figure is 1810 m².

◀◀◀ Work Problem 5 at the Side.

EXAMPLE 6 Applying Knowledge of Area

Suppose the figure in Example 5 above represents the floor plan of a hotel lobby. What is the cost of labor to install tile on the floor if the labor charge is $35.11 per square meter?

From Example 5, the floor area is 1810 m². To find the labor cost, multiply the number of square meters times the cost of labor per square meter.

$$\text{cost} = \frac{1810 \text{ m}^2}{1} \cdot \frac{\$35.11}{1 \text{ m}^2}$$

$$\text{cost} = \$63,549.10$$

The cost of the labor is $63,549.10.

◀◀◀ Work Problem 6 at the Side.

ANSWERS
5. **(a)** $A = 40 \text{ m}^2 + 44 \text{ m}^2 = 84 \text{ m}^2$
   (b) $A = 25 \text{ yd}^2 + 37.5 \text{ yd}^2 = 62.5 \text{ yd}^2$
6. **(a)** $1554   **(b)** $1750

8.4 Exercises

Find the perimeter of each figure. See Examples 1 and 3.

1.

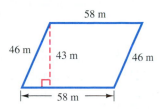

58 m

46 m   43 m   46 m

58 m

2.

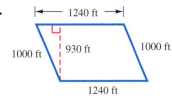

|← 1240 ft →|

1000 ft   930 ft   1000 ft

1240 ft

3.

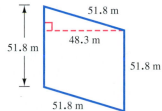

51.8 m

51.8 m   48.3 m

51.8 m

51.8 m

4.

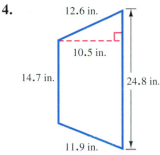

12.6 in.

10.5 in.

14.7 in.   24.8 in.

11.9 in.

5.

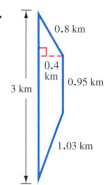

0.8 km

0.4 km   0.95 km

3 km

1.03 km

6.

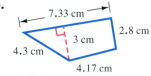

7.33 cm

2.8 cm

4.3 cm   3 cm

4.17 cm

Find the area of each figure. See Examples 2 and 4.

7.

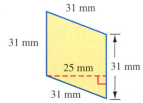

31 mm

31 mm   31 mm

25 mm   31 mm

31 mm

8.

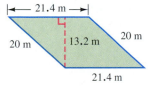

|← 21.4 m →|

20 m   13.2 m   20 m

21.4 m

9.

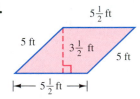

$5\frac{1}{2}$ ft

5 ft $3\frac{1}{2}$ ft   5 ft

$5\frac{1}{2}$ ft

10.

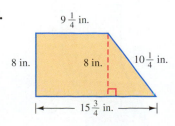

$9\frac{1}{4}$ in.

8 in.   8 in. $10\frac{1}{4}$ in.

$15\frac{3}{4}$ in.

11.

42 cm

61.4 cm   86.2 cm

42 cm

48.8 cm

12.

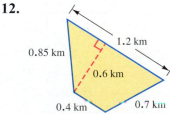

1.2 km

0.85 km

0.6 km

0.4 km   0.7 km

Solve each application problem. First label the bases and heights on the sketches. See Example 6.

13. The backyard of a new home is shaped like a trapezoid with a height of 45 ft and bases of 80 ft and 110 ft. What is the cost of putting sod on the yard if the landscaper charges $0.33 per square foot for sod?

14. A swimming pool is in the shape of a parallelogram with a height of 9.6 m and base of 12.4 m. Find the labor cost to make a custom solar cover for the pool at a cost of $4.92 per square meter.

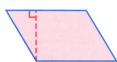

15. A piece of fabric for a quilt design is in the shape of a parallelogram. The base is 5 in. and the height is 3.5 in. What is the total area of the 25 parallelogram pieces needed for the quilt?

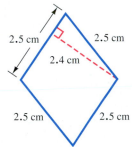

16. An accountant is paying $832 per month to rent an office in an old building. Her office is shaped like a trapezoid, with bases of 32 ft and 20 ft and a height of 20 ft. How much rent is she paying per square foot?

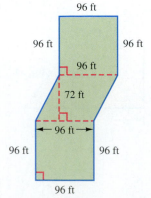

*Find **two** errors in each student's solution below. Write a sentence explaining each error. Then show how to work the problem correctly.*

17.

2.5 cm 2.5 cm
2.4 cm
2.5 cm 2.5 cm

$P = 2.5 \text{ cm} + 2.4 \text{ cm} + 2.5 \text{ cm} + 2.5 \text{ cm} + 2.5 \text{ cm}$

$P = 12.4 \text{ cm}^2$

18.

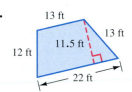

13 ft
13 ft
12 ft
11.5 ft
22 ft

$A = (0.5)(11.5 \text{ ft}) \cdot (12 \text{ ft} + 13 \text{ ft})$

$A = 143.75 \text{ ft}$

Find the area of each figure. See Example 5.

19.

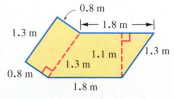

0.8 m
1.3 m
1.8 m
1.1 m
1.3 m
1.3 m
0.8 m
1.8 m

20.

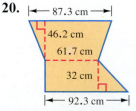

87.3 cm
46.2 cm
61.7 cm
32 cm
92.3 cm

21.

96 ft
96 ft   96 ft
96 ft
72 ft
96 ft
96 ft   96 ft
96 ft

8.5 Triangles

A **triangle** is a figure with exactly three sides. Some examples are shown below.

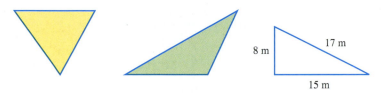

8 m   17 m
15 m

OBJECTIVE 1 Find the perimeter of a triangle. To find the perimeter of a triangle (the distance around the edges), add the lengths of the three sides.

EXAMPLE 1 Finding the Perimeter of a Triangle

Find the perimeter of the triangle above on the right.

$$P = 8 \text{ m} + 15 \text{ m} + 17 \text{ m}$$

$$P = 40 \text{ m}$$

Work Problem 1 at the Side. ▷▷▷

As with parallelograms, you can find the *height* of a triangle by measuring the distance from one vertex of the triangle to the opposite side (the base). The height line must be *perpendicular* to the base; that is, it must form a right angle with the base. Sometimes you have to extend the base in order to draw the height perpendicular to it, as shown on the right in the figures below.

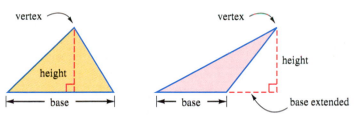

vertex   vertex
height   height
base   base   base extended

If you cut out two identical triangles and turn one upside down, you can fit them together to form a parallelogram.

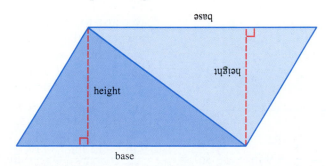

base
height
height
height
base

Recall from **Section 8.4** that the area of the parallelogram is *base* times *height*. Because each triangle is *half* of the parallelogram, the area of one triangle is

$$\frac{1}{2} \text{ of base times height.}$$

1 Find the perimeter of each triangle.

(a)   31 mm
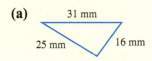
25 mm   16 mm

(b)   25.9 m
11.7 m
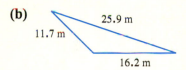
16.2 m

(c) A triangle with sides $6\frac{1}{2}$ yd, $9\frac{3}{4}$ yd, and $11\frac{1}{4}$ yd

2 Find the area of each triangle.

(a)

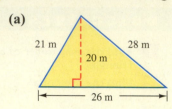

(b)

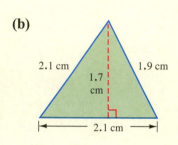

(c)

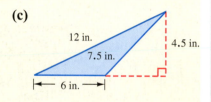

(d)

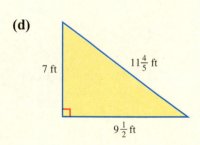

ANSWERS

2. (a) $A = 260 \text{ m}^2$ (b) $A = 1.785 \text{ cm}^2$

(c) $A = 13.5 \text{ in.}^2$

(d) $A = 33.25 \text{ ft}^2$ or $33\frac{1}{4} \text{ ft}^2$

OBJECTIVE 2   Find the area of a triangle.   Use the following formula to find the *area* of a triangle.

Finding the Area of a Triangle

$$\text{Area of a triangle} = \frac{1}{2} \cdot \text{base} \cdot \text{height}$$

$$A = \frac{1}{2} \cdot b \cdot h$$

$$\text{or} \quad A = 0.5 \cdot b \cdot h$$

Remember to use **square units** when measuring area.

EXAMPLE 2   Finding the Area of Triangles

Find the area of each triangle.

(a)

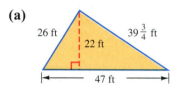

The base is 47 ft and the height is 22 ft. You do *not* need the 26 ft or $39\frac{3}{4}$ ft sides to find the area.

$$A = \frac{1}{2} \cdot \boldsymbol{b} \cdot \boldsymbol{h}$$

$$A = \frac{1}{\underset{1}{\cancel{2}}} \cdot \mathbf{47 \text{ ft}} \cdot \overset{11}{\cancel{22} \text{ ft}} \qquad \begin{array}{l}\text{Divide out}\\\text{common}\\\text{factor of 2.}\end{array}$$

$$A = 517 \text{ ft}^2 \qquad \text{Square units for area}$$

(b)

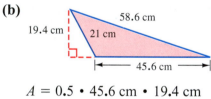

$$A = 0.5 \cdot 45.6 \text{ cm} \cdot 19.4 \text{ cm}$$

$$A = 442.32 \text{ cm}^2$$

The base must be extended to draw the height. However, still use 45.6 cm for *b* in the formula. Because the measurements are decimal numbers, it is easier to use 0.5 (the decimal equivalent of $\frac{1}{2}$) in the formula.

(c)

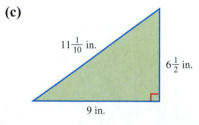

Because two sides of the triangle are perpendicular to each other, use those sides as the base and the height. (Recall that the height must be perpendicular to the base.) You can use fractions or use the equivalent decimal numbers.

Using frac-tions $A = \dfrac{\mathbf{1}}{\mathbf{2}} \cdot 9 \text{ in.} \cdot \mathbf{6}\dfrac{\mathbf{1}}{\mathbf{2}} \text{ in.} = \dfrac{1}{2} \cdot \dfrac{9 \text{ in.}}{1} \cdot \dfrac{13 \text{ in.}}{2} = \dfrac{117}{4} \text{ in.}^2 = 29\dfrac{1}{4} \text{ in.}^2$

Using deci-mals $A = \mathbf{0.5} \cdot 9 \text{ in.} \cdot \mathbf{6.5} \text{ in.} = 29.25 \text{ in.}^2 \leftarrow \text{Equivalent}$

◀◀◀ Work Problem 2 at the Side.

EXAMPLE 3 **Using the Concept of Area**

Find the area of the shaded part in this figure.

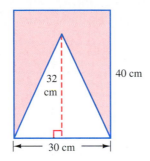

The *entire* figure is a rectangle. Find the area of the rectangle.

$$A = l \cdot w$$
$$A = 30 \text{ cm} \cdot 40 \text{ cm}$$
$$A = 1200 \text{ cm}^2$$

The *un*shaded part is a triangle. Find the area of the triangle.

$$A = \frac{1}{2} \cdot \overset{15}{\cancel{30}} \text{ cm} \cdot 32 \text{ cm}$$

$$A = 480 \text{ cm}^2$$

Subtract to find the area of the shaded part.

$$A = \underbrace{1200 \text{ cm}^2}_{\text{Entire area}} - \underbrace{480 \text{ cm}^2}_{\text{Unshaded part}} = \underbrace{720 \text{ cm}^2}_{\text{Shaded part}}$$

The area of the shaded part of the figure is 720 cm^2.

Work Problem 3 at the Side. ▶▶▶

EXAMPLE 4 **Applying the Concept of Area**

The Department of Transportation cuts triangular signs out of rectangular pieces of metal using the measurements shown above in Example 3. If the metal costs $0.02 per square centimeter, how much does the metal cost for the sign? What is the cost of the metal that is *not* used?

From Example 3 above, the area of the triangle (the sign) is 480 cm^2. Multiply the area of the sign times the cost per square centimeter.

$$\text{cost of sign} = \frac{480 \text{ cm}^2}{1} \cdot \frac{\$0.02}{1 \text{ cm}^2} = \$9.60$$

The metal that is *not* used is the *shaded* part from Example 3. The unused area is 720 cm^2.

$$\text{cost of unused metal} = \frac{720 \text{ cm}^2}{1} \cdot \frac{\$0.02}{1 \text{ cm}^2} = \$14.40$$

The cost of the metal for the triangular sign is $9.60. The unused metal costs $14.40.

Work Problem 4 at the Side. ▶▶▶

3 Find the area of the shaded part in this figure.

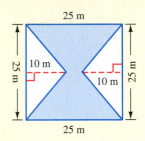

4 Suppose the figure in Problem 3 above is an auditorium floor plan. The shaded part will be covered with carpet costing $27 per square meter. The rest will be covered with vinyl floor covering costing $18 per square meter. What is the total cost of covering the floor?

ANSWERS
3. $A = 625 \text{ m}^2 - 125 \text{ m}^2 - 125 \text{ m}^2 = 375 \text{ m}^2$
4. $10,125 + $2250 + $2250 = $14,625

5 Find the number of degrees in the third angle of each triangle.

(a)
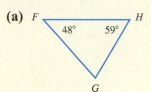

(b)
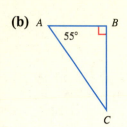

(c)
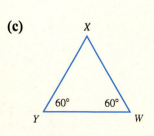

The *tri* in *tri*angle means *three*. So the name tells you that a triangle has three angles. The sum of the measures of the three angles in any triangle is *always* 180°. You can see it by drawing a triangle, cutting off the three angles, and rearranging them to make a straight angle (180°).

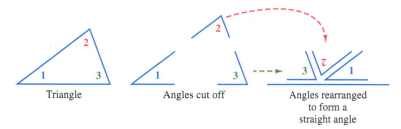

Triangle   Angles cut off   Angles rearranged to form a straight angle

---

Finding the Unknown Angle Measurement in a Triangle

Step 1   Add the number of degrees in the measures of the two given angles.

Step 2   Subtract the sum from 180°.

EXAMPLE 5 **Finding an Angle Measurement in Triangles**

Find the number of degrees in the indicated angle.

(a) Angle *R*

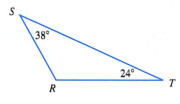

Step 1   Add the two angle measurements you are given.

$$38° + 24° = 62°$$

Step 2   Subtract the sum from 180°.

$$180° - 62° = 118°$$

∠*R* measures 118°.

(b) Angle *F*

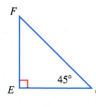

∠*E* is a right angle, so it measures 90°.

Step 1 $90° + 45° = 135°$

Step 2 $180° - 135° = 45°$

∠*F* measures 45°.

◀◀◀ Work Problem 5 at the Side.

8.5 **Exercises**

FOR EXTRA HELP

Tutor Center Addison-Wesley Math Tutor Center

Math XL MathXL

Digital Video Tutor CD 5 Videotape 14

Student's Solutions Manual

MyMathLab MyMathLab

Interactmath.com

Find the perimeter and area of each triangle. See Examples 1 and 2.

1.
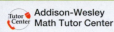

58 m
66 m
72 m   72 m

2.

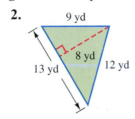

9 yd
8 yd
13 yd   12 yd

3.

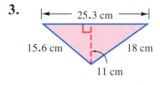

25.3 cm
15.6 cm   18 cm
11 cm

4.
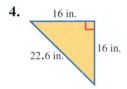

16 in.
16 in.
22.6 in.

5.
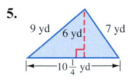

9 yd   6 yd   7 yd
$10\frac{1}{4}$ yd

6.
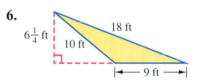

$6\frac{1}{4}$ ft   18 ft
10 ft
9 ft

7.
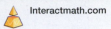

35.5 cm
21.3 cm
28.4 cm

8.

7.2 ft   7.2 ft
6.2 ft
7.2 ft

Find the shaded area in each figure. See Example 3.

9.
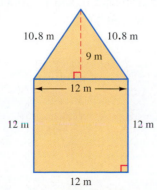

10.8 m   10.8 m
9 m
12 m
12 m   12 m
12 m

10.
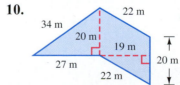
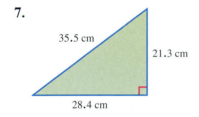
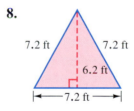

34 m   22 m
20 m
19 m
27 m   20 m
22 m

11.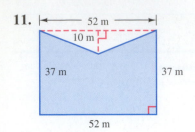

12.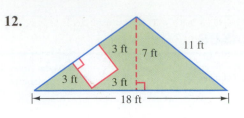

Find the number of degrees in the third angle of each triangle. See Example 5.

13.
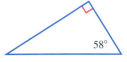

14.

15.
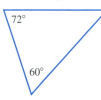

16.

17. Can a triangle have two right angles? Explain your answer.

18. In your own words, explain where the $\frac{1}{2}$ comes from in the formula for area of a triangle. Draw a sketch to illustrate your explanation.

Solve each application problem. See Example 4.

19. A triangular tent flap measures $3\frac{1}{2}$ ft along the base and has a height of $4\frac{1}{2}$ ft. How much canvas is needed to make the flap?

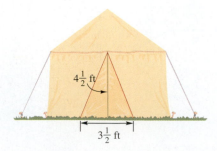

20. A wooden sign in the shape of a right triangle has perpendicular sides measuring 1.5 m and 1.2 m. How much surface area does the sign have?

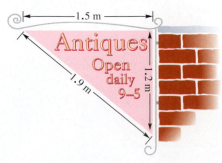

21. A triangular space between three streets has the measurements shown below.

   (a) How much new curbing will be needed to go around the space?

   (b) How much sod will be needed to cover the space?

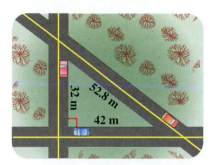

22. Each gable end of a new house has a span of 36 ft and a rise of 9.5 ft. What is the total area of both gable ends of the house?

23. All sides of the house below are congruent and all roof sections are congruent.

   (a) Find the area of one side of the house.

   (b) Find the area of one roof section.

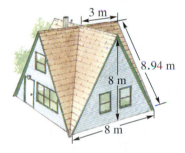

24. The sketch shows the plan for an office building. The shaded area will be a parking lot. What is the cost of building the parking lot if the contractor charges $28.00 per square yard for materials and labor?

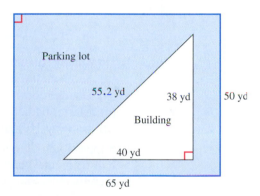

25. A city lot with an unusual shape is shown below.

   (a) How much frontage (distance along streets) does the lot have?

   (b) What is the area of the lot?

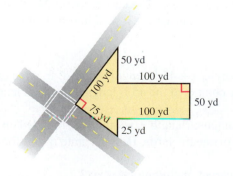

26. A car dealership wants three nylon pennants to hang in its front window. The height of the two smaller pennants is 3.5 ft, and the larger pennant has a height of 6.5 ft. How much nylon fabric is needed for all the pennants?

Real-Data Applications

Interior Design

Suppose that you have just bought a small waterfront house and you plan to remodel the interior by installing tile and carpet on the floors. A sketch of the floor plan is shown below. All measurements are in feet and represent the measurements rounded up to the nearest foot.

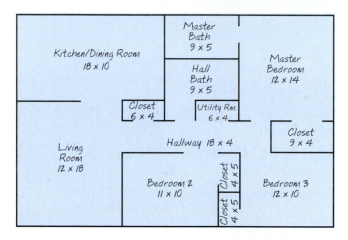

1. How many square feet of floor space are in each of the following rooms?
   (a) Kitchen/dining room
   (b) Living room
   (c) Master bath, and hall bath (including entry) combined
   (d) Hallway, hall closet, and utility room combined

2. Suppose you plan to install floor tiles in the kitchen/dining room, living room, both baths, hall bath entry, hallway, hall closet, and utility room. The flooring company salesperson recommends that you increase your square footage requirement by 5% to compensate for waste. How many square feet of tile are needed?

3. Tiles are sold in boxes of twelve 1-foot-square tiles, and partial boxes are not sold. The tile that you selected costs $5.75 per square foot, installed, based on the total number of tiles purchased. The sales tax rate is 8.25%.
   (a) How many boxes of tile must you purchase?
   (b) How much will it cost to install the tile, including sales tax?

4. How many square feet of floor is in all the bedrooms and bedroom closets, combined?

5. Suppose you plan to carpet the bedrooms and closets. The carpet costs $24.95 per square *yard,* installed, and the sales tax rate is 8.25%. The carpet is sold in rolls that are 12 ft wide, so a 3 ft length equals 4 yd². Why?

   The salesperson recommends that you purchase a 45 ft length of the 12 ft. wide carpet.
   (a) Write how the salesperson may have computed the 45 ft length. The 45 ft length is how many square *yards* of carpet?
   (b) A common industry formula is to compute (length • width) ÷ 8, rounded up to the next square yard, to estimate the number of square yards. How accurate is that formula for this job?
   (c) How much will it cost to install the carpet, including sales tax?

8.6 Circles

OBJECTIVES

1 Find the radius and diameter of a circle.

2 Find the circumference of a circle.

3 Find the area of a circle.

4 Become familiar with Latin and Greek prefixes used in math terminology.

OBJECTIVE 1 Find the radius and diameter of a circle.   Suppose you start with one dot on a piece of paper. Then you draw many dots that are each 2 cm away from the first dot. If you draw enough dots (points) you'll end up with a *circle*. Each point on the circle is exactly 2 cm away from the *center* of the circle. The 2 cm distance is called the *radius, r,* of the circle. The distance across the circle (passing through the center) is called the *diameter, d,* of the circle.

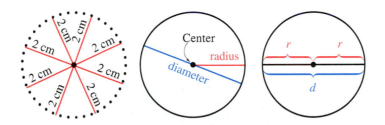

Circle, Radius, and Diameter

A **circle** is a two-dimensional (flat) figure with all points the same distance from a fixed center point.

The **radius** (r) is the distance from the center of the circle to any point on the circle.

The **diameter** (d) is the distance across the circle passing through the center.

Using the circle above on the right as a model, you can see some relationships between the radius and diameter.

Finding the Diameter and Radius of a Circle

$$\text{diameter} = 2 \cdot \text{radius}$$

$$d = 2 \cdot r$$

$$\text{and} \quad r = \frac{d}{2}$$

EXAMPLE 1 Finding the Diameter and Radius of Circles

Find the unknown length of the diameter or radius in each circle.

(a)

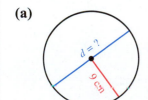

Because the radius is 9 cm, the diameter is twice as long.

$$d = 2 \cdot \ \boldsymbol{r}$$

$$d = 2 \cdot \boldsymbol{9\ cm}$$

$$d = 18 \text{ cm}$$

Continued on Next Page

1 Find the unknown length of the diameter or radius in each circle.

(a)

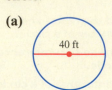

40 ft

(b)

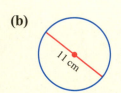

11 cm

(c)

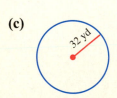

32 yd

(d)

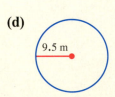

9.5 m

(b)

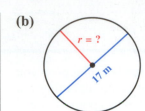

$r = ?$

17 m

The radius is half the diameter.

$$r = \frac{d}{2}$$

$$r = \frac{17 \text{ m}}{2}$$

$$r = 8.5 \text{ m} \quad \text{or} \quad 8\frac{1}{2} \text{ m}$$

◀◀◀ **Work Problem 1 at the Side.**

OBJECTIVE **2** **Find the circumference of a circle.** The perimeter of a circle is called its **circumference.** Circumference is the distance around the edge of a circle.

The diameter of the can in the drawing is about 10.6 cm, and the circumference of the can is about 33.3 cm. Dividing the circumference of the circle by the diameter gives an interesting result.

$$\frac{\text{circumference}}{\text{diameter}} = \frac{33.3}{10.6} \approx 3.14 \qquad \text{Rounded to the nearest hundredth}$$

Dividing the circumference of *any* circle by its diameter *always* gives an answer close to 3.14. This means that going around the edge of any circle is a littler more than 3 times as far as going straight across the circle.

This ratio of circumference to diameter is called π (the Greek letter **pi**, pronounced PIE). There is no decimal that is exactly equal to π, but here is the *approximate* value.

$$\pi \approx 3.14159265359$$

> **Rounding the Value of Pi (π)**
>
> We usually round π to 3.14. Therefore, calculations involving π will give approximate answers and should be written using the $\approx$ symbol.

Use the following formulas to find the *circumference* of a circle.

> **Finding the Circumference (Distance around a Circle)**
>
> $$\text{Circumference} = \pi \cdot \text{diameter}$$
> $$C = \pi \cdot d$$
>
> or, because $d = 2 \cdot r$, then $C = \pi \cdot 2 \cdot r$ usually written $C = 2 \cdot \pi \cdot r$
>
> Remember to use linear units such as ft, yd, m, and cm when measuring circumference (**not** square units).

EXAMPLE 2 Finding the Circumference of Circles

Find the circumference of each circle. Use 3.14 as the approximate value for π. Round answers to the nearest tenth.

(a)

38 m

The *diameter* is 38 m, so use the formula with d in it.

$$C = \pi \cdot d$$
$$C \approx 3.14 \cdot 38 \text{ m}$$
$$C \approx 119.3 \text{ m} \qquad \text{Rounded}$$

(b)

11.5 cm

In this example, the *radius* is labeled, so it is easier to use the formula with r in it.

$$C = 2 \cdot \pi \cdot r$$
$$C \approx 2 \cdot 3.14 \cdot 11.5 \text{ cm}$$
$$C \approx 72.2 \text{ cm} \qquad \text{Rounded}$$

> **Calculator Tip**   Many *scientific* calculators have a π key. Try pressing it. With a 10-digit display, you'll see the value of π to the nearest billionth.
>
> $$3.141592654$$
>
> But this is still an approximate value, although it is more precise than rounding π to 3.14. Try finding the circumference in Example 2(a) above using the π key.
>
> π $\times$ 38 $=$ Answer is 119.3805208; rounds to 119.4
>
> When you used 3.14 as the approximate value of π, the result rounded to 119.3, so the answers are slightly different. In this book, we will use 3.14 instead of the π key. Our measurements of radius and diameter are given as whole numbers or with tenths, so it is acceptable to round π to hundredths. And, some students may be using standard calculators without a π key or doing the calculations by hand.

Work Problem 2 at the Side. ▶▶▶

OBJECTIVE 3 Find the area of a circle.   To find the formula for the area of a circle, start by cutting two circles into many pie-shaped pieces.

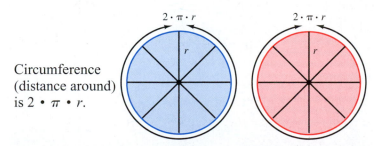

Circumference (distance around) is $2 \cdot \pi \cdot r$.

Unfold the circles, much as you might "unfold" a peeled orange, and put them together as shown here.

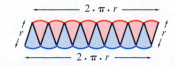

---

2 Find the circumference of each circle. Use 3.14 as the approximate value for π. Round answers to the nearest tenth.

(a)

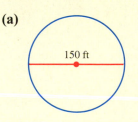

150 ft

(b)

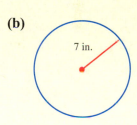

7 in.

(c) diameter 0.9 km

(d) radius 4.6 m

3 Find the area of each circle. Use 3.14 for π. Round your answers to the nearest tenth.

(a)

1 cm

(b)

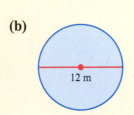

12 m

(*Hint:* The diameter is 12 m so $r =$ _____ m)

(c)

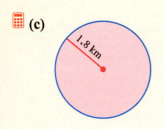

1.8 km

(d)

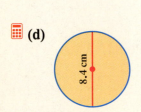

8.4 cm

The figure is approximately a rectangle with width r (the radius of the original circle) and length $2 \cdot \pi \cdot r$ (the circumference of the original circle). The area of the "rectangle" is length times width.

$$\text{Area} = \qquad l \qquad \cdot \; w$$
$$\text{Area} = \overbrace{2 \cdot \pi \cdot r} \cdot \underbrace{r}$$
$$\text{Area} = 2 \cdot \pi \cdot \quad r^2 \quad \leftarrow \text{Recall that } r \cdot r \text{ is } r^2$$

Because the "rectangle" was formed from *two* circles, the area of *one* circle is half as much.

$$\frac{1}{\overset{1}{\underset{1}{2}}} \cdot \overset{1}{2} \cdot \pi \cdot r^2 = 1 \cdot \pi \cdot r^2 \quad \text{or simply} \quad \pi \cdot r^2$$

> **Finding the Area of a Circle**
> $$\text{Area of a circle} = \pi \cdot \text{radius} \cdot \text{radius}$$
> $$A = \pi \cdot r^2$$
> Remember to use *square units* when measuring area.

EXAMPLE 3 Finding the Area of Circles

Find the area of each circle. Use 3.14 for π. Round your answers to the nearest tenth.

(a) A circle with a radius of 8.2 cm
Use the formula $A = \pi \cdot r^2$, which means $\pi \cdot r \cdot r$.

$$A = \quad \pi \quad \cdot \quad r \quad \cdot \quad r$$
$$A \approx 3.14 \cdot 8.2 \text{ cm} \cdot 8.2 \text{ cm}$$
$$A \approx 211.1 \text{ cm}^2 \qquad \text{Rounded; square units for area}$$

(b)

10 ft

To use the area formula, you need to know the *radius* (r). In this circle, the *diameter* is 10 ft. First find the radius.

$$r = \frac{d}{2}$$
$$r = \frac{10 \text{ ft}}{10} = 5 \text{ ft}$$

Now find the area.

$$A \approx 3.14 \cdot 5 \text{ ft} \cdot 5 \text{ ft}$$
$$A \approx 78.5 \text{ ft}^2 \quad \text{Square units for area}$$

> **CAUTION**
> When finding *circumference,* you can start with either the radius or the diameter. When finding *area,* you must use the *radius.* If you are given the diameter, divide it by 2 to find the radius. Then find the area.

ANSWERS

3. **(a)** $A \approx 3.1 \text{ cm}^2$ **(b)** $A \approx 113.0 \text{ m}^2$
   (c) $A \approx 10.2 \text{ km}^2$ **(d)** $A \approx 55.4 \text{ cm}^2$

◀◀◀ Work Problem 3 at the Side.

⊞ **Calculator Tip** You can use your calculator to find the area of the circle in Example 3(a) on the previous page. The first method works on both scientific and standard calculators:

$$3.14 \;\textcircled{×}\; 8.2 \;\textcircled{×}\; 8.2 \;\textcircled{=}\;$$ Answer is 211.1336

You round the answer to 211.1 (nearest tenth).

On a *scientific* calculator you can also use the $\textcircled{x^2}$ key, which automatically squares the number you enter (that is, multiplies the number times itself):

$$3.14 \;\textcircled{×}\; 8.2 \;\textcircled{x^2}\; \underbrace{67.24}\; \textcircled{=}\;$$ Answer is 211.1336

Appears automatically;
8.2 × 8.2 is 67.24

In the next example, we find the area of a *semicircle,* which is half the area of a circle.

EXAMPLE 4 Finding the Area of a Semicircle

Find the area of the semicircle. Use 3.14 for π. Round your answer to the nearest tenth.

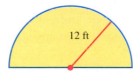

First, find the area of a whole circle with a radius of 12 ft.

$$A = \pi \cdot r \cdot r$$

$$A \approx 3.14 \cdot 12 \text{ ft} \cdot 12 \text{ ft}$$

$$A \approx 452.16 \text{ ft}^2 \; \leftarrow \text{ Do not round yet.}$$

Divide the area of the whole circle by 2 to find the area of the semicircle.

$$\frac{452.16 \text{ ft}^2}{2} = 226.08 \text{ ft}^2$$

The *last* step is rounding 226.08 to the nearest tenth.

Area of semicircle ≈ 226.1 ft² ← Rounded

▶ **Work Problem 4 at the Side.** ▶▶▶

EXAMPLE 5 Applying the Concept of Circumference

A circular rug is 8 ft in diameter. The cost of fringe for the edge is $2.25 per foot. What will it cost to add fringe to the rug? Use 3.14 for π.

$$\text{Circumference} = \pi \cdot d$$

$$C \approx 3.14 \cdot 8 \text{ ft}$$

$$C \approx 25.12 \text{ ft}$$

$$\text{cost} = \text{cost per foot} \cdot \text{circumference}$$

$$\text{cost} = \frac{\$2.25}{1 \text{ ft}} \cdot \frac{25.12 \text{ ft}}{1}$$

$$\text{cost} = \$56.52$$

The cost of adding fringe to the rug is $56.25.

▶ **Work Problem 5 at the Side.** ▶▶▶

④ Find the area of each ⊞ semicircle. Use 3.14 for π. Round your answers to the nearest tenth.

(a)

24 m

(b)

35.4 ft

(c)
9.8 m

⑤ Find the cost of binding ⊞ around the edge of a circular rug that is 3 m in diameter. The binder charges $4.50 per meter. Use 3.14 for π.

6 Find the cost of covering the underside of the rug in margin problem 5 with a nonslip rubber backing. The rubber backing costs $2 per square meter.

EXAMPLE 6 Applying the Concept of Area

Find the cost of covering the rug in Example 5 (on the previous page) with a plastic cover. The material for the cover costs $1.50 per square foot. Use 3.14 for π.

First find the radius. $\qquad r = \dfrac{d}{2} = \dfrac{8 \text{ ft}}{2} = 4 \text{ ft}$

Then find the area. $\qquad A = \pi \cdot r^2$

$$A \approx 3.14 \cdot 4 \text{ ft} \cdot 4 \text{ ft}$$

$$A \approx 50.24 \text{ ft}^2$$

$$\text{cost} = \dfrac{\$1.50}{1 \text{ ft}^2} \cdot \dfrac{50.24 \text{ ft}^2}{1} = \$75.36$$

The cost of the plastic cover is $75.36.

◀◀◀ Work Problem 6 at the Side.

---

7 **(a)** Here are some more prefixes you have seen in this textbook. List at least one math term and one nonmathematical word that use each prefix.

dia- (through):

fract- (break):

par- (beside):

per- (divide):

peri- (around);

rad- (ray):

rect- (right):

sub- (below):

OBJECTIVE **4** **Become familiar with Latin and Greek prefixes used in math terminology.** Many English words are built from Latin or Greek root words and prefixes. Knowing the meaning of the more common ones can help you figure out the meaning of terms in many subject areas, including math.

EXAMPLE 7 Using Prefixes to Understand Math Terms

(a) Listed below are some Latin and Greek root words and prefixes with their meanings in parentheses. You've already seen math terms in this textbook that use these prefixes. List at least one math term and one nonmathematical word that use each prefix or root word.

cent- (100): ***cent*imeter; *cent*ury**

circum- (around): ***circum*ference; *circum*vent**

de- (down): ***de*nominator; *de*cline**

dec- (10): ***dec*imal; *Dec*ember** (originally the 10th month in the old calendar)

> There are many answers. These are some of the possibilities.

(b) Suppose you have trouble remembering which part of a fraction is the denominator. How could your knowledge of prefixes help in this situation?

The *de-* prefix in ***de*nominator** means "down" so the denominator is the number *down* below the fraction bar.

◀◀◀ Work Problem 7 at the Side.

(b) How could you use your knowledge of prefixes to remember the difference between perimeter and area?

ANSWERS

6. $14.13
7. **(a)** Some possibilities are:
   *dia*meter; *dia*gonal
   *fract*ion; *fract*ure
   *par*allel; *par*amedic
   *per*cent; *per* capita
   *peri*meter; *peri*scope
   *rad*ius; *rad*iate
   *rect*angle; *rect*ify
   *sub*tract; *sub*marine
   (b) *Peri-* in perimeter means "around," so perimeter is the distance *around* the edges of a shape.

NOTE

Here are some additional prefixes and root words and their meanings that you will see in the rest of **Chapter 8,** in **Chapter 9,** and in other math classes. An example of a math term and a nonmathematical word are shown for each one.

equ- (equal): ***equ*ation; *equ*inox**

lateral (side): **quadri*lateral*; bi*lateral***

hemi- (half): ***hemi*sphere; *hemi*trope**

re- (back or again); ***re*ciprocal; *re*duce**

8.6 Exercises

Find the unknown length in each circle. See Example 1.

1. 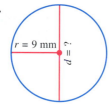   **2.** 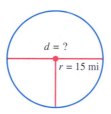   **3.**    **4.**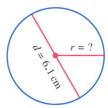

Find the circumference and area of each circle. Use 3.14 as the approximate value for π. Round your answers to the nearest tenth. See Examples 2 and 3.

5. 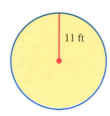   **6.** 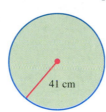   **7.** 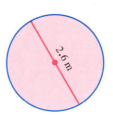   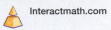 **8.**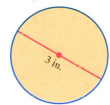

Find the circumference and area of circles having the following diameters. Use 3.14 for π. Round your answers to the nearest tenth. See Examples 2 and 3.

9. $d = 15$ cm   **10.** $d = 39$ ft   **11.** $d = 7\frac{1}{2}$ ft

12. $d = 4\frac{1}{2}$ yd   **13.** $d = 8.65$ km   **14.** $d = 19.5$ mm

Find each shaded area. Note that Exercises 15 and 18 contain semicircles. Use 3.14 as the approximate value of π. Round your answers to the nearest tenth if necessary. See Example 4.

15.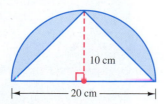

16.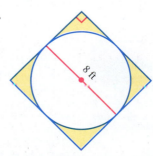

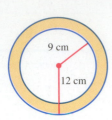

17.

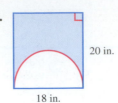

18.

19. How would you explain π to a friend who is not in your math class? Write an explanation. Then make up a test question that requires the use of π, and show how to solve it.

20. Explain how circumference and perimeter are alike. How are they different? Make up two problems, one involving perimeter, the other circumference. Show how to solve your problems.

Solve each application problem. Use 3.14 as the approximate value of π. Round your answers to the nearest tenth. See Examples 5 and 6.

21. An irrigation system moves around a center point to water a circular area for crops. If the irrigation system is 50 yd long, how large is the watered area?

22. If you swing a ball held at the end of a string 20 cm long, how far will the ball travel on each turn?

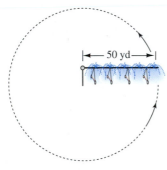

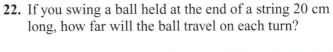

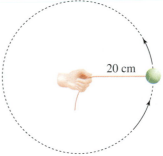

23. A Michelin Cross Terrain SUV tire has an overall diameter of 29.10 inches. How far will a point on the tire tread move in one complete turn? (*Source:* Michelin.) *Bonus question:* How many revolutions does the tire make per mile? Round to the nearest whole number.

24. In September 2003, hurricane Isabel slammed into the East Coast of the United States. The huge circular storm had a diameter of 400 miles. What area did the storm cover?

🖩 *For Exercises 25–30, first draw a circle and label the radius or diameter. Then solve the problem. Use 3.14 for π and round answers to the nearest tenth.*

25. A radio station can be heard 150 miles in all directions during evening hours. How many square miles are in the station's broadcast area?

26. An earthquake was felt by people 900 km away in all directions from the epicenter (the source of the earthquake). How much area was affected by the quake?

27. The diameter of Diana Hestwood's wristwatch is 1 in. and the radius of the clock face on her kitchen wall is 3 in. Find the circumference and the area of each clock face.

28. The diameter of the largest known ball of twine is 12 ft 9 in. The sign posted near the ball says it has a circumference of 40 ft. Is the sign correct? *Hint:* First change 9 in. to feet and add it to 12 ft. (*Source: Guinness Book of World Records.*)

29. Blaine Fenstad wants to buy a pair of two-way radios. Some models have a range of 2 miles under ideal conditions. More expensive models have a range of 5 miles. What is the difference in the area covered by the 2-mile and 5-mile models? (*Source:* Best Buy.)

30. The National Audubon Society holds an end-of-year bird count. Volunteers count all the birds they see in a circular area during a 24-hour period. Each circle has a diameter of 15 miles. About 1700 circular areas are counted across the United States each December. What is the total area covered by the count?

31. On the first page of this chapter, you read about a forester measuring the circumferences of trees.

    (a) If the circumference of one tree is 144 cm, what is the diameter?

    (b) Explain how you solved part (a).

32. In Atlanta, Interstate 285 circles the city and is known as the "perimeter." If the circumference of the circle made by the highway is 62.8 miles, find:

    (a) the diameter of the circle.

    (b) the area inside the circle.

    (*Source: Greater Atlanta Newcomer's Guide.*)

33. Find the cost of sod, at $1.76 per square foot, for this playing field that has a semicircle on each end.

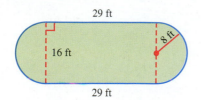

34. Find the area of this skating rink.

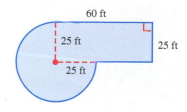

Use the information about prefixes in Example 7 to answer Exercises 35 and 36.

35. Explain how you could use the information about prefixes to remember the difference between radius, diameter, and circumference.

36. Explain how you could use the information about prefixes to avoid confusion between parallel and perpendicular lines.

RELATING CONCEPTS (EXERCISES 37–42)   For Individual or Group Work

Use the table below to **work Exercises 37–42 in order.**

Find the best buy for each type of pizza. The best buy is the lowest cost per square inch of pizza. All the pizzas are circular in shape, and the measurement given on the menu board is the diameter of the pizza in inches. Use 3.14 *as the approximate value of* π. *Round the area to the nearest tenth. Round cost per square inch to the nearest thousandth.*

Pizza Menu	Small $7\frac{1}{2}''$	Medium 13″	Large 16″
Cheese only	$2.80	$ 6.50	$ 9.30
"The Works"	$3.70	$ 8.95	$14.30
Deep-dish combo	$4.35	$10.95	$15.65

37. Find the area of a small pizza.

38. Find the area of a medium pizza.

39. Find the area of a large pizza.

40. What is the cost per square inch for each size of cheese pizza? Which size is the best buy?

41. What is the cost per square inch for each size of "The Works" pizza? Which size is the best buy?

42. You have a coupon for 95¢ off any small pizza. What is the cost per square inch for each size of deep-dish combo pizza? Which size is the best buy?

Summary Exercises on Perimeter, Circumference, and Area

1. Draw a sketch of each of these shapes: **(a)** square, **(b)** rectangle, **(c)** parallelogram. On each sketch, indicate 90° angles and show which sides are the same length.

2. (a) Draw a sketch of a circle and show the radius.

(b) Draw another circle and show the diameter.

(c) Describe the relationship between the radius and diameter of a circle.

3. Describe how you can find the perimeter of any flat shape with straight sides.

4. In your own words, describe the difference between finding the perimeter of a shape and finding the area of the shape.

5. Match each shape to its corresponding area formula.

Shapes   **Area Formulas**

parallelogram _____   **(a)** $A = \dfrac{1}{2} \cdot b \cdot h$

square _____   **(b)** $A = \pi \cdot r^2$

trapezoid _____   **(c)** $A = l \cdot w$

circle _____   **(d)** $A = \dfrac{1}{2} \cdot h \cdot (b + B)$

rectangle _____   **(e)** $A = s^2$

triangle _____   **(f)** $A = b \cdot h$

6. (a) If you know the *radius* of a circle, which formula do you use to find its circumference?

(b) If you know the *diameter* of a circle, which formula do you use to find its circumference?

(c) If you know the *diameter* of a circle, what must you do *before* using the formula for finding the area?

7. Find the perimeter and area of each triangle, to the nearest tenth.

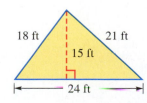

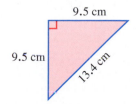

8. Some answers from a student's test paper are listed below. The *number* part of each answer is correct, but the *units* are not. Rewrite each answer with the correct units.

(a) $A = 12$ cm   **(b)** $P = 6\dfrac{1}{2}$ ft^2

(c) $C \approx 28.5$ m^2 **(d)** $A = 307$ in.

Name each figure and find its perimeter and area. Round answers to the nearest tenth.

9.

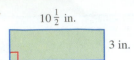

10 ½ in.

3 in.

10.

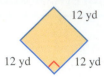

12 yd

12 yd 12 yd

11.

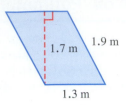

1.7 m 1.9 m

1.3 m

12.

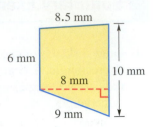

8.5 mm

6 mm

8 mm

9 mm

10 mm

For each circle, find **(a)** *the diameter or radius,* **(b)** *the circumference, and* **(c)** *the area.*
Use 3.14 as the approximate value of π. *Round answers to the nearest tenth.*

13.

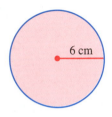

6 cm

14.

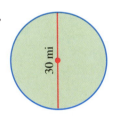

30 mi

▦ **15.**

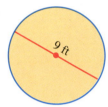

9 ft

Find the area of each shaded region. Note that Exercise 18 contains semicircles. Use 3.14
as the approximate value of π. *Round answers to the nearest tenth.*

16.

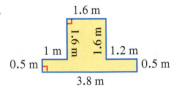

1.6 m

1.6 m 1.6 m

1 m 1.2 m

0.5 m 0.5 m

3.8 m

17.

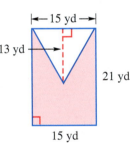

|← 15 yd →|

13 yd

21 yd

15 yd

18.

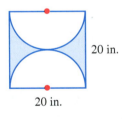

20 in.

20 in.

Solve each application problem. Use 3.14 as the approximate value of π. *Round answers*
to the nearest tenth.

▦ **19.** The Mormons traveled west to Utah by covered
wagon in 1847. They tied a rag to a wagon wheel to
keep track of the distance they traveled. The radius
of the wheel was 2.33 ft. How far did the rag travel
each time the wheel made a complete revolution?
(*Source: Trail of Hope.*) *Bonus question:* How many
wheel revolutions equaled one mile?

20. The rectangular front door for a new log home is
0.9 m wide and 2 m high. How much will it cost for
weather strip material to go around all edges of the
door, if weather strip costs $0.77 per meter?

2 m

|← 0.9 m →|

8.7 Volume

OBJECTIVE 1 Find the volume of a rectangular solid. A shoe box and a cereal box are examples of three-dimensional (or solid) figures. The three dimensions are length, width, and height. (A rectangle or square is a two-dimensional figure. The two dimensions are length and width.)

  If you want to know how much a shoe box will hold, you find its *volume*. We measure volume by seeing how many cubes of a certain size will fill the space inside the box. Three sizes of *cubic units* are shown below. Notice that all the edges of a cube have the same length and all the sides meet at right angles.

OBJECTIVES

Find the volume of a
1 rectangular solid;
2 sphere;
3 cylinder;
4 cone and pyramid.

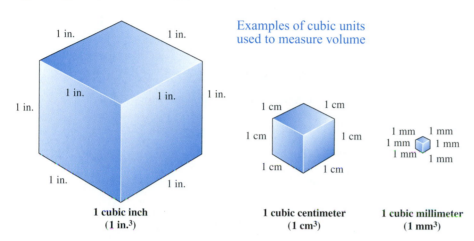

Examples of cubic units used to measure volume

1 cubic inch (1 in.³)   1 cubic centimeter (1 cm³)   1 cubic millimeter (1 mm³)

Some other sizes of cubes that are used to measure volume are 1 cubic foot (1 ft³), 1 cubic yard (1 yd³), and 1 cubic meter (1 m³).

CAUTION

The raised 3 in 4^3 means that you multiply $4 \cdot 4 \cdot 4$ to get 64. The raised 3 in cm³ or ft³ is a short way to write the word *cubic*. When you see 5 cm³, say "five *cubic* centimeters." Do *not* multiply $5 \cdot 5 \cdot 5$. The exponent applies to cm, *not* to the number.

Volume

Volume is a measure of the space inside a solid shape. The volume of a solid is how many cubic units it takes to fill the solid.

  Use the formula below to find the *volume* of *rectangular solids* (box-like shapes).

Finding the Volume of Rectangular Solids

Volume of a rectangular solid = length • width • height

$$V = l \cdot w \cdot h$$

Remember to use **cubic units** when measuring volume.

1 Find the volume of each box. Round your answers to the nearest tenth if necessary.

(a)

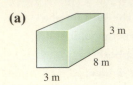

3 m
8 m
3 m

(b)

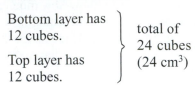

23.4 cm
52.3 cm
15.2 cm

(c) Length $6\frac{1}{4}$ ft, width $3\frac{1}{2}$ ft, height 2 ft

---

EXAMPLE 1 Finding the Volume of Rectangular Solids

Find the volume of each box.

(a)

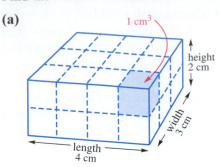

1 cm³
height 2 cm
width 3 cm
length 4 cm

Each cube that fits in the box is 1 cubic centimeter (1 cm³). To find the volume, you can count the number of cubes.

Bottom layer has 12 cubes.
Top layer has 12 cubes.
} total of 24 cubes (24 cm³)

Or, you can use the formula for rectangular solids.

$$V = l \cdot w \cdot h$$
$$V = 4 \text{ cm} \cdot 3 \text{ cm} \cdot 2 \text{ cm}$$
$$V = 24 \text{ cm}^3 \quad \text{Cubic units for volume}$$

(b)

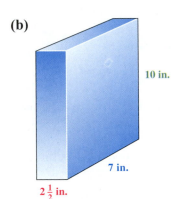

10 in.
7 in.
$2\frac{1}{2}$ in.

Use the formula $V = l \cdot w \cdot h$.

$$V = 7 \text{ in.} \cdot 2\frac{1}{2} \text{ in.} \cdot 10 \text{ in.}$$

$$V = \frac{7 \text{ in.}}{1} \cdot \frac{5 \text{ in.}}{\overset{2}{\cancel{2}}_1} \cdot \frac{\overset{5}{\cancel{10}} \text{ in.}}{1} = 175 \text{ in.}^3$$

If you like, use 2.5 instead of $2\frac{1}{2}$; they are equivalent.

$$V = 7 \text{ in.} \cdot 2.5 \text{ in.} \cdot 10 \text{ in.} = 175 \text{ in.}^3$$

◀◀◀ Work Problem 1 at the Side.

OBJECTIVE 2 Find the volume of a sphere. A *sphere* is shown below. Examples of spheres include baseballs, oranges, and Earth. (They aren't perfect spheres, but they're close.)

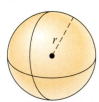

r

As with circles, the *radius* of a sphere is the distance from the center to the edge of the sphere. Use the following formula to find the *volume* of a *sphere*.

Finding the Volume of a Sphere

$$\text{Volume of a sphere} = \frac{4}{3} \cdot \pi \cdot r \cdot r \cdot r$$

$$V = \frac{4}{3} \cdot \pi \cdot r^3 \quad \text{or} \quad \frac{4 \cdot \pi \cdot r^3}{3}$$

Remember to use **cubic units** when measuring volume.

ANSWERS
1. **(a)** $V = 72 \text{ m}^3$ **(b)** $V \approx 18{,}602.1 \text{ cm}^3$
   (c) $V = 43\frac{3}{4} \text{ ft}^3$ or $V \approx 43.8 \text{ ft}^3$

EXAMPLE 2 Finding the Volume of Spheres

Find the volume of each sphere with the help of a calculator. Use 3.14 as the approximate value of π. Round your answers to the nearest tenth.

(a)

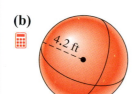

$$V = \frac{4}{3} \cdot \pi \cdot r^3$$

$$V \approx \frac{4 \cdot 3.14 \cdot 9 \text{ m} \cdot 9 \text{ m} \cdot 9 \text{ m}}{3}$$

$V \approx 3052.08$ Now round to tenths.

$V \approx 3052.1 \text{ m}^3$ Cubic units for volume

(b)

$$V \approx \frac{4 \cdot 3.14 \cdot 4.2 \text{ ft} \cdot 4.2 \text{ ft} \cdot 4.2 \text{ ft}}{3}$$

$V \approx 310.18176$ Now round to tenths.

$V \approx 310.2 \text{ ft}^3$ Cubic units for volume

Calculator Tip You can find the volume of the sphere in Example 2(b) above on your calculator. The first method works on both scientific and standard calculators:

4 ⊗ 3.14 ⊗ 4.2 ⊗ 4.2 ⊗ 4.2 ⊘ 3 ⊜   Answer is 310.18176

Round the answer to 310.2 ft³.

On a *scientific* calculator you can use the ⓨˣ key to calculate r^3 (to multiply the radius times itself three times).

4 ⊗ 3.14 ⊗ 4.2 ⓨˣ 3 ⊘ 3 ⊜   Answer is 310.18176
                    r^3

Recall that we are using 3.14 as the approximate value for π instead of using the ⓟ key.

You can also use the ⓨˣ key with other exponents. For example:

To find 2^5, press 2 ⓨˣ 5 ⊜   Answer is 32

To find 6^4, press 6 ⓨˣ 4 ⊜   Answer is 1296

Work Problem 2 at the Side. ▶▶▶

Half a sphere is called a *hemisphere*. The volume of a hemisphere is *half* the volume of a sphere. Use the following formula to find the *volume* of a hemisphere.

Finding the Volume of a Hemisphere

$$\text{Volume of a hemisphere} = \frac{1}{\cancel{2}_1} \cdot \frac{\cancel{4}^2}{3} \cdot \pi \cdot r^3$$

$$V = \frac{2}{3} \cdot \pi \cdot r^3 \quad \text{or} \quad \frac{2 \cdot \pi \cdot r^3}{3}$$

Remember to use **cubic units** when measuring volume.

2 Find the volume of each sphere. Use 3.14 for π. Round your answers to the nearest tenth.

(a)

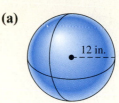

(b)

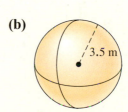

(c) Sphere with a radius of 2.7 cm

3 Find the volume of each hemisphere. Use 3.14 for π. Round your answers to the nearest tenth.

(a)

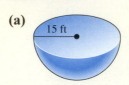

15 ft

(b)

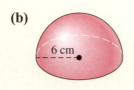

6 cm

4 Find the volume of each cylinder. Use 3.14 for π. Round your answers to the nearest tenth.

(a)

12 ft — 4 ft

(b)

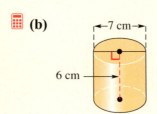

7 cm
6 cm

EXAMPLE 3 Finding the Volume of a Hemisphere

Find the volume of the hemisphere with the help of a calculator. Use 3.14 for π. Round your answer to the nearest tenth.

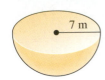

7 m

$$V = \frac{2 \cdot \pi \cdot r^3}{3}$$

$$V \approx \frac{2 \cdot 3.14 \cdot 7\text{ m} \cdot 7\text{ m} \cdot 7\text{ m}}{3}$$

$V \approx 718.0$ m³   Rounded to nearest tenth

Work Problem 3 at the Side.

OBJECTIVE 3 **Find the volume of a cylinder.** Several *cylinders* are shown below.

The height must be perpendicular to the circular top and bottom of the cyclinder.

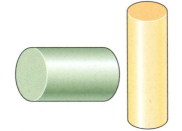

radius

height

These are called *right circular cylinders* because the top and bottom are circles, and the side makes a right angle with the top and bottom. Examples of cylinders are a soup can, a home water heater, and a piece of pipe.

Use the formula below to find the *volume* of a *cylinder*. Notice that the first part of the formula, $\pi \cdot r^2$, is the area of the circular base.

Finding the Volume of a Cylinder

Volume of a cylinder $= \pi \cdot r \cdot r \cdot h$

$$V = \pi \cdot r^2 \cdot h$$

Remember to use **cubic units** when measuring volume.

EXAMPLE 4 Finding the Volume of Cylinders

Find the volume of each cylinder. Use 3.14 as the approximate value of π. Round your answers to the nearest tenth if necessary.

(a)

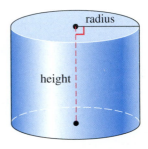

20 m
9 m

The diameter is 20 m so the radius is $\frac{20\text{ m}}{2} = 10$ m. The height is 9 m. Use the formula to find the volume.

$$V = \pi \cdot r^2 \cdot h$$

$$V \approx 3.14 \cdot 10\text{ m} \cdot 10\text{ m} \cdot 9\text{ m}$$

$V \approx 2826$ m³   Cubic units for volume.

(b)

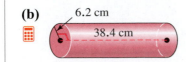

6.2 cm
38.4 cm

$$V \approx 3.14 \cdot 6.2\text{ cm} \cdot 6.2\text{ cm} \cdot 38.4\text{ cm}$$

$V \approx 4634.94144$ Now round to tenths.

$V \approx 4634.9$ cm³   Cubic units for volume

Work Problem 4 at the Side.

ANSWERS
3. **(a)** $V \approx 7065$ ft³   **(b)** $V \approx 452.2$ cm³
4. **(a)** $V \approx 602.9$ ft³   **(b)** $V \approx 230.8$ cm³

OBJECTIVE 4 Find the volume of a cone and a pyramid. A cone and a pyramid are shown below. Notice that the height line is perpendicular to the base in both solids.

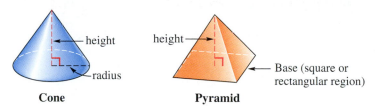

Cone   **Pyramid**

Use the formula below to find the *volume* of a cone.

Finding the Volume of a Cone

$$\text{Volume of a cone} = \frac{1}{3} \cdot B \cdot h$$

$$\text{or} \quad V = \frac{B \cdot h}{3}$$

where B is the area of the circular base of the cone and h is the height of the cone.

Remember to use *cubic units* when measuring volume.

EXAMPLE 5 Finding the Volume of a Cone

Find the volume of the cone. Use 3.14 for π. Round your answer to the nearest tenth.

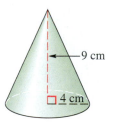

—9 cm

☐ 4 cm

First find the value of B in the formula, which is the *area of the circular base*. Recall that the formula for the area of a circle is πr^2.

$$B = \pi \cdot r \cdot r$$

$$B \approx 3.14 \cdot 4 \text{ cm} \cdot 4 \text{ cm}$$

$B \approx 50.24 \text{ cm}^2$ ← Do not round to tenths yet.

Next, find the volume. The height is 9 cm.

$$V = \frac{B \cdot h}{3}$$

$$V \approx \frac{50.24 \text{ cm}^2 \cdot 9 \text{ cm}}{3}$$

$V \approx 150.72 \text{ cm}^3$ Now round to tenths.

$V \approx 150.7 \text{ cm}^3$ Cubic units for volume

Work Problem 5 at the Side. ▶▶▶

---

5 Find the volume of a cone with base radius 2 ft and height 11 ft. Use 3.14 for π. Round your answer to the nearest tenth.

ANSWERS
5. $V \approx 46.1 \text{ ft}^3$

6 Find the volume of a pyramid with a square base 10 m by 10 m and a height of 8 m. Round your answer to the nearest tenth.

Use the same formula to find the *volume* of a *pyramid* as you did to find the *volume* of a *cone*.

> ### Finding the Volume of a Pyramid
>
> $$\text{Volume of a pyramid} = \frac{1}{3} \cdot B \cdot h$$
>
> $$\text{or} \quad V = \frac{B \cdot h}{3}$$
>
> where B is the area of the square or rectangular base of the pyramid and h is the height of the pyramid.
>
> Remember to use **cubic units** when measuring volume.

> **NOTE**
> In this book, we will work only with pyramids that have a base with four sides (square or rectangle). In later math courses you may work with pyramids that have a base with three sides (triangle), five sides (pentagon), six sides (hexagon), and so on.

EXAMPLE 6 Finding the Volume of a Pyramid

Find the volume of this pyramid with a rectangular base. Round your answer to the nearest tenth.

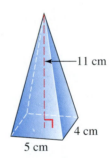
11 cm
4 cm
5 cm

First find the value of B in the formula, which is the *area of the rectangular base.* Recall that the area of a rectangle is found by multiplying length times width.

$$B = 5 \text{ cm} \cdot 4 \text{ cm}$$

$$\mathbf{B = 20 \text{ cm}^2}$$

Next, find the volume.

$$V = \frac{\textcolor{red}{B \cdot h}}{3}$$

$$V = \frac{\textcolor{blue}{20 \text{ cm}^2 \cdot 11 \text{ cm}}}{3}$$

$$V \approx 73.3 \text{ cm}^3 \qquad \text{Rounded to nearest tenth}$$

Work Problem 6 at the Side.

8.7 Exercises

Name each solid and find its volume. Use 3.14 as the approximate value of π. Round your answers to the nearest tenth if necessary. See Examples 1–6.

1.

12.5 cm

11 cm

4 cm

2.

$4\frac{1}{2}$ ft

$4\frac{1}{2}$ ft

$4\frac{1}{2}$ ft

3.

22 m

4.

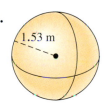

1.53 m

5.

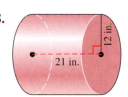

12 in.

6.

7.4 in.

7.

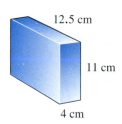

5 ft

6 ft

8.

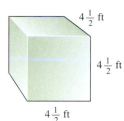

12 in

21 in.

9.

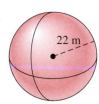

16 m

5 m

10.

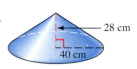

28 cm

40 cm

11.

20 cm

15 cm

8 cm

12.

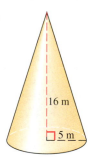

15 m

5 m

7 m

Solve each application problem. Use 3.14 as the approximate value of π. Round your final answers to the nearest tenth if necessary.

13. A pencil box measures 3 in. by 8 in. by $\frac{3}{4}$ in. high. Find the volume of the box. (*Source:* Faber Castell.)

14. A train is being loaded with shipping crates. Each one is 6 m long, 3.4 m wide, and 2 m high. How much space will each crate take?

15. An oil candle globe made of hand-blown glass has a diameter of 16.8 cm. What is the volume of the globe?

16. A metal sphere used as part of a fountain has a diameter of $6\frac{1}{2}$ ft. Find its volume.

17. One of the ancient stone pyramids in Egypt has a square base that measures 145 m on each side. The height is 93 m. What is the volume of the pyramid? (*Source: The Columbia Encyclopedia.*)

18. A cylindrical woven basket made by a Northwest Coast tribe is 8 cm high and has a diameter of 11 cm. What is the volume of the basket?

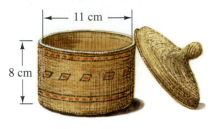

19. A city sewer pipe has a diameter of 5 ft and a length of 200 ft. Find the volume of the pipe.

20. An ice cream cone has a diameter of 2 in. and a height of 4 in. Find its volume.

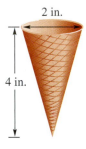

21. Explain the *two* errors made by a student in finding the volume of a cylinder with a diameter of 7 cm and a height of 5 cm. Find the correct answer.

$$V \approx 3.14 \cdot 7 \cdot 7 \cdot 5$$
$$V \approx 769.3 \text{ cm}^2$$

22. Compare the steps in finding the volume of a cylinder and a cone. How are they similar? Suppose you know the volume of a cylinder. How can you find the volume of a cone with the same radius and height by doing just a one-step calculation?

23. Find the volume.

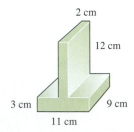

24. Find the volume. (*Hint:* Notice the square hole that goes through the center of the shape.)

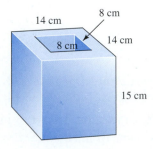

8.8 Pythagorean Theorem

In **Section 8.3** you used this formula for the area of a square, $A = s^2$. The blue square below has an area of 25 cm^2 because 5 cm • 5 cm = 25 cm^2.

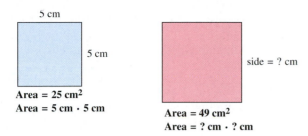

5 cm
5 cm
Area = 25 cm^2
Area = 5 cm · 5 cm

side = ? cm
Area = 49 cm^2
Area = ? cm · ? cm

OBJECTIVES

1 Find square roots using the square root key on a calculator.

2 Find the unknown length in a right triangle.

3 Solve application problems involving right triangles.

The red square above has an area of 49 cm^2. To find the length of a side, ask yourself, "What number can be multiplied by itself to give 49?" Because 7 • 7 = 49, the length of each side is 7 cm.

Also, because 7 • 7 = 49, we say that 7 is the *square root* of 49, or $\sqrt{49} = 7$. Also, $\sqrt{81} = 9$, because 9 • 9 = 81. (See **Section 1.8** for further review.)

Work Problem 1 at the Side. ▶▶▶

A number that has a whole number as its square root is called a *perfect square*. For example, 9 is a perfect square because $\sqrt{9} = 3$, and 3 is a whole number.

The first few perfect squares are listed below.

① Find each square root.

(a) $\sqrt{36}$

(b) $\sqrt{25}$

The First Twelve Perfect Squares

$\sqrt{1} = 1$	$\sqrt{16} = 4$	$\sqrt{49} = 7$	$\sqrt{100} = 10$
$\sqrt{4} = 2$	$\sqrt{25} = 5$	$\sqrt{64} = 8$	$\sqrt{121} = 11$
$\sqrt{9} = 3$	$\sqrt{36} = 6$	$\sqrt{81} = 9$	$\sqrt{144} = 12$

(c) $\sqrt{9}$

OBJECTIVE 1 **Find square roots using the square root key on a calculator.** If a number is *not* a perfect square, then you can find its *approximate* square root by using a calculator with a square root key.

(d) $\sqrt{100}$

⊞ **Calculator Tip** To find a square root, use the ⟨√⟩ key on a standard calculator or the ⟨√x⟩ key on a scientific calculator. In either case, you do *not* need to use the ⟨=⟩ key. Try these. Jot down your answers.

To find $\sqrt{16}$ press: 16 ⟨√x⟩   Answer is 4

To find $\sqrt{7}$ press: 7 ⟨√x⟩   Answer is 2.645751311

(e) $\sqrt{121}$

For $\sqrt{7}$, your calculator shows 2.645751311, which is an *approximate* answer. We will be rounding to the nearest thousandth, so $\sqrt{7} \approx 2.646$. To check, multiply 2.646 times 2.646. Do you get 7 as the result? No, you get 7.001316, which is very close to 7. The difference is due to rounding.

2 Use a calculator with a ⌨ square root key to find each square root. Round to the nearest thousandth if necessary.

(a) $\sqrt{11}$

(b) $\sqrt{40}$

(c) $\sqrt{56}$

(d) $\sqrt{196}$

(e) $\sqrt{147}$

EXAMPLE 1   **Finding the Square Root of Numbers**

⌨ Use a calculator to find each square root. Round your answers to the nearest thousandth.

(a) $\sqrt{35}$ Calculator shows 5.916079783; round to 5.916

(b) $\sqrt{124}$ Calculator shows 11.13552873; round to 11.136

(c) $\sqrt{200}$ Calculator shows 14.14213562; round to 14.142

◀◀◀ **Work Problem 2 at the Side.**

OBJECTIVE 2   Find the unknown length in a right triangle.   One place you will use square roots is when working with the *Pythagorean Theorem.* This theorem applies only to *right* triangles (triangles with a 90° angle). The longest side of a right triangle is called the **hypotenuse.** It is opposite the right angle. The other two sides are called *legs*. The legs form the right angle.

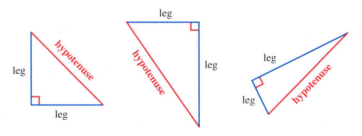

Examples of right triangles

Pythagorean Theorem

$$(\text{hypotenuse})^2 = (\text{leg})^2 + (\text{leg})^2$$

In other words, square the length of each side. After you have squared all the sides, the sum of the squares of the two legs will equal the square of the hypotenuse. An example is shown below.

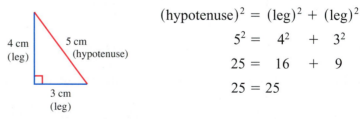

$$(\text{hypotenuse})^2 = (\text{leg})^2 + (\text{leg})^2$$
$$5^2 = 4^2 + 3^2$$
$$25 = 16 + 9$$
$$25 = 25$$

The theorem is named after Pythagoras, a Greek mathematician who lived about 2500 years ago. He and his followers may have used floor tiles to prove the theorem, as shown below.

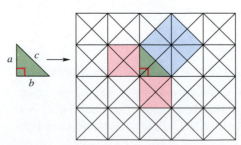

The green right triangle in the center of the floor tiles has sides *a, b,* and *c.* The pink square drawn on side *a* contains four triangular tiles. The pink square on side *b* contains four tiles. The blue square on side *c* contains eight tiles. The number of tiles in the square on side *c* equals the sum of the number of tiles in the squares on sides *a* and *b,* that is, 8 tiles = 4 tiles + 4 tiles. As a result, you often see the Pythagorean Theorem written as $c^2 = a^2 + b^2$.

ANSWERS

2. (a) $\sqrt{11} \approx 3.317$ **(b)** $\sqrt{40} \approx 6.325$
   (c) $\sqrt{56} \approx 7.483$ **(d)** $\sqrt{196} = 14$
   (e) $\sqrt{147} \approx 12.124$

If you know the lengths of any two sides in a right triangle, you can use the Pythagorean Theorem to find the length of the third side.

> **Formulas Based on the Pythagorean Theorem**
> To find the hypotenuse, use this formula:
> $$\text{hypotenuse} = \sqrt{(\text{leg})^2 + (\text{leg})^2}$$
> To find a leg, use this formula:
> $$\text{leg} = \sqrt{(\text{hypotenuse})^2 - (\text{leg})^2}$$

> **CAUTION**
> *Remember:* A small square drawn in one angle of a triangle indicates a right angle. You can use the Pythagorean Theorem *only* on triangles that have a right angle.

EXAMPLE 2 Finding the Unknown Length in Right Triangles

Find the unknown length in each right triangle. Round answers to the nearest tenth if necessary.

(a)

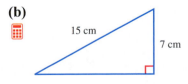

3 ft
4 ft

The unknown length is the side opposite the right angle, which is the hypotenuse. Use the formula for finding the hypotenuse.

$$\text{hypotenuse} = \sqrt{(\text{leg})^2 + (\text{leg})^2} \qquad \textcolor{blue}{\text{Find the hypotenuse.}}$$

$$\text{hypotenuse} = \sqrt{(3)^2 + (4)^2} \qquad \textcolor{blue}{\text{Legs are 3 and 4}}$$

$$= \sqrt{9 + 16} \qquad \textcolor{blue}{3 \cdot 3 \text{ is } 9 \quad \text{and} \quad 4 \cdot 4 \text{ is } 16}$$

$$= \sqrt{25}$$

$$= 5$$

The hypotenuse is 5 ft long.

(b)

15 cm
7 cm

We *do* know the length of the hypotenuse (15 cm), so it is the length of one of the legs that is unknown. Use the formula for finding a leg.

$$\text{leg} = \sqrt{(\text{hypotenuse})^2 - (\text{leg})^2} \qquad \textcolor{blue}{\text{Find a leg.}}$$

$$\text{leg} = \sqrt{(15)^2 - (7)^2} \qquad \textcolor{blue}{\text{Hypotenuse is 15, one leg is 7}}$$

$$= \sqrt{225 - 49} \qquad \textcolor{blue}{15 \cdot 15 \text{ is } 225 \quad \text{and} \quad 7 \cdot 7 \text{ is } 49}$$

$$= \sqrt{176} \qquad \textcolor{blue}{\text{Use calculator to find } \sqrt{176}}$$

$$\approx 13.3 \qquad \textcolor{blue}{\text{Round } 13.26649916 \text{ to } 13.3}$$

The length of the leg is approximately 13.3 cm.

Work Problem 3 at the Side.

3 Find the unknown length in each right triangle. Round your answers to the nearest tenth if necessary.

(a)

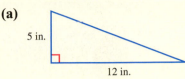

5 in.
12 in.

(b)

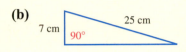

7 cm
90°
25 cm

(c)

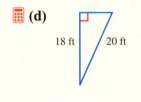

13 m
17 m

(d)

18 ft
20 ft

(e)

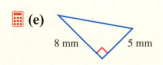

8 mm
5 mm

ANSWERS

3. **(a)** $\sqrt{169} = 13$ in.   **(b)** $\sqrt{576} = 24$ cm
   (c) $\sqrt{458} \approx 21.4$ m   **(d)** $\sqrt{76} \approx 8.7$ ft
   (e) $\sqrt{89} \approx 9.4$ mm

4 These problems show ladders leaning against buildings. Find the unknown lengths. Round to the nearest tenth of a foot if necessary.

(a)

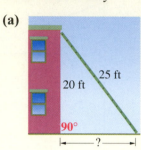

How far away from the building is the bottom of the ladder?

(b)

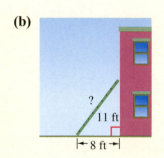

How long is the ladder?

(c) A 17 ft ladder is leaning against a building. The bottom of the ladder is 10 ft from the building. How high up on the building will the ladder reach? (*Hint:* Start by drawing the building and the ladder.)

OBJECTIVE 3 **Solve application problems involving right triangles.**
The next example shows an application of the Pythagorean Theorem.

EXAMPLE 3 **Using the Pythagorean Theorem**

A television antenna is on the roof of a house, as shown below. Find the length of the support wire. Round your answer to the nearest tenth of a meter if necessary.

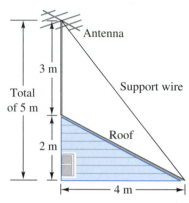

A right triangle is formed. The total length of the leg on the left is 3 m + 2 m = 5 m.

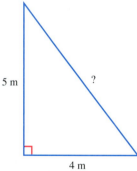

Notice that the support wire is opposite the right angle, so it is the hypotenuse of the right triangle.

$$\text{hypotenuse} = \sqrt{(\text{leg})^2 + (\text{leg})^2} \qquad \color{blue}{\text{Find the hypotenuse.}}$$

$$\text{hypotenuse} = \sqrt{(5)^2 + (4)^2} \qquad \color{blue}{\text{Legs are 5 and 4}}$$

$$= \sqrt{25 + 16} \qquad \color{blue}{5^2 \text{ is 25 and } 4^2 \text{ is 16}}$$

$$= \sqrt{41} \qquad \color{blue}{\text{Use } \sqrt{x} \text{ key on a calculator.}}$$

$$\approx 6.4 \qquad \color{blue}{\text{Round 6.403124237 to 6.4}}$$

The length of the support wire is approximately **6.4 m.**

CAUTION
You use the Pythagorean Theorem to find the **length** of one side, *not* the area of the triangle. Your answer will be in linear units, such as ft, yd, cm, m, and so on (*not* ft², yd², cm², m²).

◀◀◀ **Work Problem 4 at the Side.**

ANSWERS
4. **(a)** $\sqrt{225} = 15$ ft   **(b)** $\sqrt{185} \approx 13.6$ ft
   (c) $\sqrt{189} \approx 13.7$ ft

8.8 Exercises

Find each square root. Starting with Exercise 5, use the square root key on a calculator. Round your answers to the nearest thousandth if necessary. See Example 1.

1. $\sqrt{16}$ **2.** $\sqrt{4}$ **3.** $\sqrt{64}$ **4.** $\sqrt{81}$

5. $\sqrt{11}$ **6.** $\sqrt{23}$ **7.** $\sqrt{5}$ **8.** $\sqrt{2}$

9. $\sqrt{73}$ **10.** $\sqrt{80}$ **11.** $\sqrt{101}$ **12.** $\sqrt{125}$

13. $\sqrt{190}$ **14.** $\sqrt{160}$ **15.** $\sqrt{1000}$ **16.** $\sqrt{2000}$

17. You know that $\sqrt{25} = 5$ and $\sqrt{36} = 6$. Using just that information (no calculator), describe how you could *estimate* $\sqrt{30}$. How would you estimate $\sqrt{26}$ or $\sqrt{35}$? Now check your estimates using a calculator.

18. Explain the relationship between *squaring* a number and finding the *square root* of a number. Include two examples to illustrate your explanation.

Find the unknown length in each right triangle. Use a calculator to find square roots. Round your answers to the nearest tenth if necessary. See Example 2.

19.
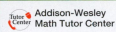
15 ft   90°   36 ft

20.
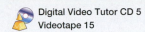
9 cm   12 cm

21.

8 in.   90°   15 in.

22.
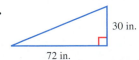
30 in.   72 in.

23.

16 mm   20 mm

24.
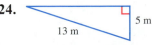
5 m   13 m

25.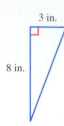
3 in.
8 in.

26.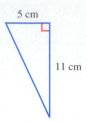
5 cm
11 cm

27.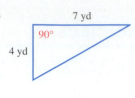
7 yd
90°
4 yd

28.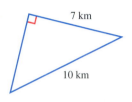
7 km
10 km

29.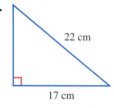
22 cm
17 cm

30.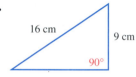
16 cm
9 cm
90°

31.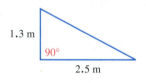
1.3 m
90°
2.5 m

32.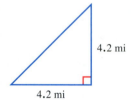
4.2 mi
4.2 mi

33.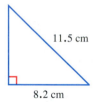
11.5 cm
8.2 cm

34.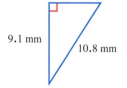
9.1 mm
10.8 mm

35.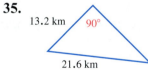
13.2 km   90°
21.6 km

36.
26.5 ft
37.4 ft

*Solve each application problem. Round your answers to the nearest tenth if necessary.
See Example 3.*

37. Find the length of this loading ramp.

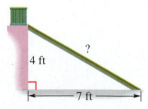
4 ft
?
7 ft

38. Find the unknown length in this roof plan.

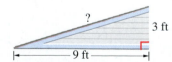
?
3 ft
9 ft

39. How high is the airplane above the ground?

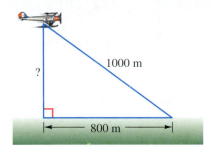

40. Find the height of this farm silo.

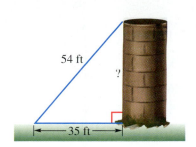

41. How long is the diagonal brace on this rectangular gate?

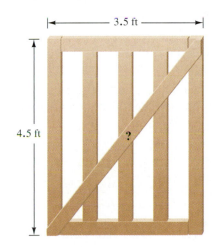

42. Find the height of this rectangular television screen. (*Source:* Sears.)

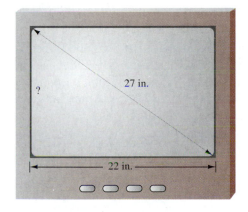

43. To reach his ladylove, a knight placed a 12 ft ladder against the castle wall. If the base of the ladder is 3 ft from the building, how high on the castle will the top of the ladder reach? Draw a sketch of the castle and ladder and solve the problem.

44. William drove his car 15 miles north, then made a right turn and drove 7 miles east. How far is he, in a straight line, from his starting point? Draw a sketch to illustrate the problem and solve it.

45. Explain the *two* errors made by a student in solving this problem. Also find the correct answer. Round to the nearest tenth.

$$? = \sqrt{(9)^2 + (7)^2}$$
$$= \sqrt{18 + 14}$$
$$= \sqrt{32} \approx 5.657 \text{ in.}$$

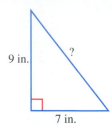

9 in.   ?   7 in.

46. Explain the *two* errors made by a student in solving this problem. Also find the correct answer. Round to the nearest tenth.

$$? = \sqrt{(13)^2 + (20)^2}$$
$$= \sqrt{169 + 400}$$
$$= \sqrt{569} \approx 23.9 \text{ m}^2$$

?   13 m   20 m

READING CONCEPTS (EXERCISES 47–50) For Individual or Group Work

*Use your knowledge of the Pythagorean Theorem to **work Exercises 47–50 in order.** Round answers to the nearest tenth.*

47. A major league baseball diamond is a square shape measuring 90 ft on each side. If the catcher throws a ball from home plate to second base, how far is he throwing the ball? (*Source:* American League of Professional Baseball Clubs.)

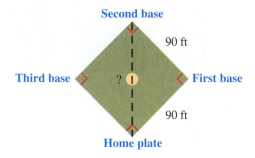

Second base

90 ft

Third base   ?   First base

90 ft

Home plate

48. A softball diamond is only 60 ft on each side. (*Source:* Amateur Softball Association.)

(a) Draw a sketch of the softball diamond and label the bases and the lengths of the sides.

(b) How far is it to throw a ball from home plate to second base?

49. Look back at your answer to Exercise 47. Explain how you can tell the distance from third base to first base without doing any further calculations.

50. Show how you could set up a proportion to answer Exercise 48 instead of using the Pythagorean Theorem. (You'll need your answer from Exercise 47.)

8.9 Similar Triangles

Two triangles with the same *shape* (but not necessarily the same size) are called **similar triangles.** Three pairs of similar triangles are shown below.

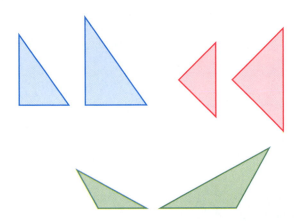

OBJECTIVE ⬛**1** **Identify corresponding parts in similar triangles.**   The two triangles shown below are different sizes but have the same shape, so they are *similar triangles.* Angles *A* and *P* measure the same number of degrees and are called *corresponding angles.* Angles *B* and *Q* are corresponding angles, as are angles *C* and *R.* The triangles have the same shape because the corresponding angles have the same measure.

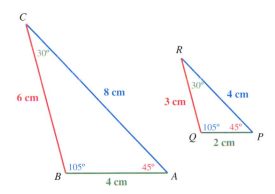

$\overline{PR}$ and $\overline{AC}$ are called *corresponding sides* because they are *opposite* corresponding angles. Also, $\overline{QR}$ and $\overline{BC}$ are corresponding sides, as are $\overline{PQ}$ and $\overline{AB}$. Although corresponding angles measure the same number of degrees, corresponding sides do *not* need to be the same length. In the triangles here, each side in the smaller triangle is *half* the length of the corresponding side in the larger triangle.

Work Problem 1 at the Side. ▶▶▶

OBJECTIVE ⬛**2** **Find the unknown lengths of sides in similar triangles.** Similar triangles are useful because of the following definition.

Similar Triangles

If two triangles are **similar,** then
1. Corresponding angles have the same measure, and
2. The ratios of the lengths of corresponding sides are equal.

OBJECTIVES

⬛**1** Identify corresponding parts in similar triangles.

⬛**2** Find the unknown lengths of sides in similar triangles.

⬛**3** Solve application problems involving similar triangles.

1 Identify corresponding angles and sides in these similar triangles.

(a)

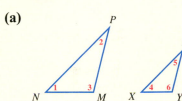

Angles:
1 and _____
2 and _____
3 and _____

Sides:
$\overline{PN}$ and _____
$\overline{PM}$ and _____
$\overline{NM}$ and _____

(b)

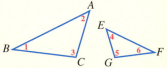

Angles:
1 and _____
2 and _____
3 and _____

Sides:
$\overline{AB}$ and _____
$\overline{BC}$ and _____
$\overline{AC}$ and _____

ANSWERS
1. **(a)** 4; 5; 6; $\overline{ZX}$; $\overline{ZY}$; $\overline{XY}$
   (b) 6; 4; 5; $\overline{EF}$; $\overline{FG}$; $\overline{EG}$

2 Find the length of $\overline{EF}$ in Example 1 by setting up and solving a proportion. Let x represent the unknown length.

EXAMPLE 1   Finding the Unknown Lengths of Sides in Similar Triangles

Find the length of y in the smaller triangle. Assume the triangles are similar.

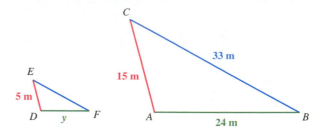

    $\overline{DF}$, the length you want to find in the smaller triangle, corresponds to $\overline{AB}$ in the larger triangle. Then, notice that $\overline{ED}$ in the smaller triangle corresponds to $\overline{CA}$ in the larger triangle, and you know both of their lengths. Because the *ratios* of the lengths of corresponding sides are equal, you can set up a proportion. (Recall that a proportion states that two ratios are equal.)

$$\text{Corresponding sides} \begin{cases} DF \rightarrow \\ AB \rightarrow \end{cases} \frac{y}{24} = \frac{5}{15} \begin{cases} \leftarrow ED \\ \leftarrow CA \end{cases} \text{Corresponding sides}$$

$$\frac{y}{24} = \frac{1}{3} \qquad \text{Write } \tfrac{5}{15} \text{ in lowest terms as } \tfrac{1}{3}.$$

Find the cross products.

$$\frac{y}{24} = \frac{1}{3} \qquad \begin{array}{l} 24 \cdot 1 = 24 \\ y \cdot 3 \end{array}$$

Show that the cross products are equivalent.

$$y \cdot 3 = 24$$

Divide both sides by 3.

$$\frac{y \cdot \overset{1}{\cancel{3}}}{\underset{1}{\cancel{3}}} = \frac{24}{3}$$

$$y = 8$$

$\overline{DF}$ has a length of 8 m.

◀◀◀ Work Problem 2 at the Side.

EXAMPLE 2   **Finding an Unknown Length and the Perimeter**

Find the perimeter of the smaller triangle. Assume the triangles are similar.

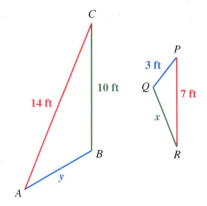

First find x, the length of $\overline{RQ}$, then add the lengths of all three sides to find the perimeter.

The smaller triangle is turned "upside down" compared to the larger triangle, so be careful when identifying corresponding sides. $\overline{AC}$ is the longest side in the larger triangle, and $\overline{PR}$ is the longest side in the smaller triangle. So $\overline{PR}$ and $\overline{AC}$ are corresponding sides and you know both of their lengths. $\overline{QR}$, the length you want to find in the smaller triangle, corresponds to $\overline{BC}$ in the larger triangle. The ratios of the lengths of corresponding sides are equal, so you can set up a proportion.

$$\begin{matrix} QR \rightarrow \\ BC \rightarrow \end{matrix} \frac{x}{10} = \frac{7}{14} \begin{matrix} \leftarrow PR \\ \leftarrow AC \end{matrix}$$

$$\frac{x}{10} = \frac{1}{2} \qquad \text{Write } \tfrac{7}{14} \text{ in lowest terms as } \tfrac{1}{2}.$$

Find the cross products.

$$\frac{x}{10} \bowtie \frac{1}{2} \qquad \begin{array}{l} 10 \cdot 1 = 10 \\ \\ x \cdot 2 \end{array}$$

Show that the cross products are equivalent.

$$x \cdot 2 = 10$$

Divide both sides by 2.

$$\frac{x \cdot \overset{1}{\cancel{2}}}{\underset{1}{\cancel{2}}} = \frac{10}{2}$$

$$x = 5$$

$\overline{RQ}$ has a length of 5 ft.

Now add the lengths of all three sides to find the perimeter of the smaller triangle.

$$\text{Perimeter} = 5 \text{ ft} + 3 \text{ ft} + 7 \text{ ft} = 15 \text{ ft}$$

Work Problem 3 at the Side. ▶▶▶

3 **(a)** Find the perimeter of triangle ABC in Example 2 at the left.

(b) Find the perimeter of each triangle. Assume the triangles are similar.

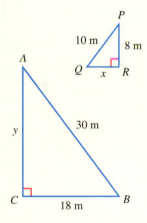

4 Find the height of each flagpole.

(a)

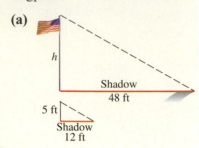

(b)

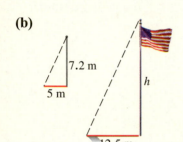

OBJECTIVE 3 Solve application problems involving similar triangles.
The next example shows an application of similar triangles.

EXAMPLE 3 Using Similar Triangles in an Application

A flagpole casts a shadow 99 m long at the same time that a pole 10 m tall casts a shadow 18 m long. Find the height of the flagpole.

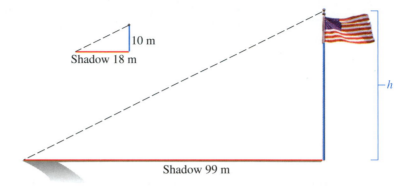

The triangles shown are similar, so write a proportion to find h.

Height in larger triangle $\rightarrow \dfrac{h}{10} = \dfrac{99}{18} \leftarrow$ **Shadow** in larger triangle
Height in smaller triangle $\rightarrow \phantom{\dfrac{h}{10}} \phantom{\dfrac{99}{18}} \leftarrow$ **Shadow** in smaller triangle

Find the cross products and show that they are equivalent.

$$h \cdot 18 = 10 \cdot 99$$
$$h \cdot 18 = 990$$

Divide both sides by 18.

$$\frac{h \cdot \overset{1}{\cancel{18}}}{\underset{1}{\cancel{18}}} = \frac{990}{18}$$

$$h = 55$$

The flagpole is 55 m high.

NOTE
There are several other correct ways to set up the proportion in Example 3 above. One way is to simply flip the ratios on *both* sides of the equal sign.

$$\frac{10}{h} = \frac{18}{99}$$

But there is another option, shown below.

Height in **larger** triangle $\rightarrow \dfrac{h}{99} = \dfrac{10}{18} \leftarrow$ Height in **smaller** triangle
Shadow in **larger** triangle $\rightarrow \phantom{\dfrac{h}{99}} \phantom{\dfrac{10}{18}} \leftarrow$ Shadow in **smaller** triangle

Notice that both ratios compare *height* to *shadow* in the same order. The ratio on the left describes the larger triangle, and the ratio on the right describes the smaller triangle.

◀◀◀ **Work Problem 4 at the Side.**

ANSWERS
4. (a) $h = 20$ ft **(b)** $h = 18$ m

8.9 Exercises

FOR EXTRA HELP

Tutor Center Addison-Wesley Math Tutor Center

MathXL MathXL

Digital Video Tutor CD 5 Videotape 15

Student's Solutions Manual

MyMathLab MyMathLab

Interactmath.com

Which pairs of triangles appear to be similar? Write similar *or* not similar *for each pair.*

1.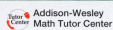

2.

3.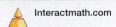

4.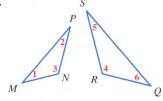

5.

6.

Name the corresponding angles and the corresponding sides in each pair of similar triangles. See Margin Problem 1.

7.

8.

9.

10.

Write the ratio for each pair of corresponding sides in the similar triangles shown below.
Write the ratios as fractions in lowest terms. See Examples 1 and 2.

11. $\dfrac{AB}{PQ}; \dfrac{AC}{PR}; \dfrac{BC}{QR}$

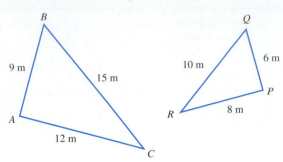

12. $\dfrac{AB}{PQ}; \dfrac{AC}{PR}; \dfrac{BC}{QR}$

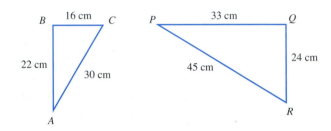

Find the unknown lengths in each pair of similar triangles. See Example 1.

13.

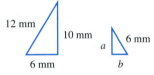

14.

15.

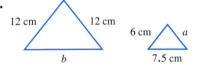

16.

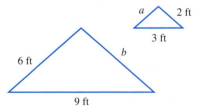

Find the perimeter of each triangle. Assume the triangles are similar. See Example 2.

17.

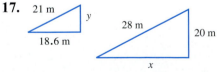

18.

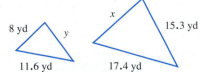

19. Triangles *CDE* and *FGH* are similar. Find the perimeter and area of triangle *FGH*. *Note:* The heights of similar triangles have the same ratio as corresponding sides. Round to the nearest tenth when necessary.

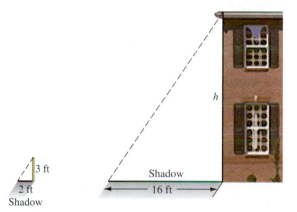

20. Triangles *JKL* and *MNO* are similar. Find the perimeter and area of triangle *MNO*.

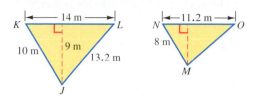

Solve each application problem. See Example 3.

21. The height of the house shown here can be found by comparing its shadow to the shadow cast by a 3 ft stick. Find the height of the house by writing a proportion and solving it.

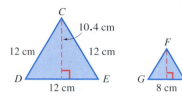

22. A fire lookout tower provides an excellent view of the surrounding countryside. The height of the tower can be found by lining up the top of the tower with the top of 2-meter stick. Use similar triangles to find the height of the tower.

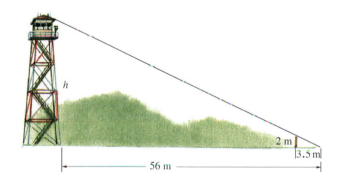

23. Refer to the building in Exercise 21. Later in the day, the same building had a shadow 6 ft long. How long would the stick's shadow be at that time?

24. Refer to the lookout tower in Exercise 22.

   (a) How far away from the tower was the 2 m stick?

   (b) To use a 1-meter stick and have it line up with the same endpoint, how far from the tower would it have to be?

25. Look up the word *similar* in a dictionary. What is the nonmathematical definition of this word? Describe two examples of similar objects at home, school, or work.

26. *Congruent* objects have the *same shape* and the *same size*. Sketch a pair of congruent triangles. Describe two examples of congruent objects at home, school, or work.

Find the unknown length in Exercises 27–30. Round your answers to the nearest tenth.
Note: When a line is drawn parallel to one side of a triangle, the smaller triangle that is
formed will be similar to the original triangle. In Exercises 27–28, the red segments are
parallel.

27.

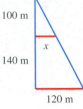

100 m

x

140 m

120 m

Hint: Redraw the two triangles and label the sides.

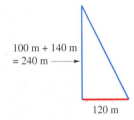

100 m

x

100 m + 140 m
= 240 m

120 m

28.

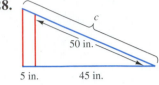

c

50 in.

5 in.   45 in.

Hint: Redraw the two triangles and label the sides.

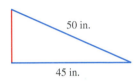

c

5 in. + 45 in. = 50 in.

50 in.

45 in.

29. Use similar triangles and a proportion to find the length of the lake shown here. (*Hint:* The side 100 m long in the smaller triangle corresponds to side of 100 m + 120 m = 220 m in the larger triangle.)

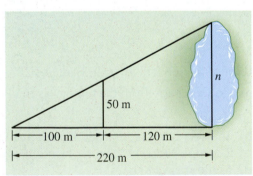

n

50 m

100 m   120 m

220 m

30. To find the height of the tree, find *y* and then add $5\frac{1}{2}$ ft for the distance from the ground to the person's eye level.

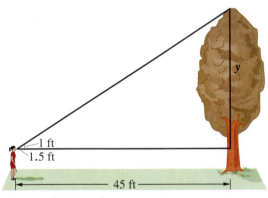

y

1 ft
1.5 ft

45 ft

Distance from person to tree.

Chapter 8

KEY TERMS

8.1	**point**	A point is a location in space. *Example:* Point *P* at the right.
	line	A line is a straight row of points that goes on forever in both directions. *Example:* Line *AB,* written $\overleftrightarrow{AB}$, at the right.
	line segment	A line segment is a piece of a line with two endpoints. *Example:* Line segment *PQ,* written $\overline{PQ}$, at the right.
	ray	A ray is a part of a line that has one endpoint and extends forever in one direction. *Example:* Ray *RS,* written $\overrightarrow{RS}$, at the right.
	angle	An angle is made up of two rays that have a common endpoint called the vertex. *Example:* Angle 1 at the right.
	degrees	Degrees are used to measure angles; a complete circle is 360 degrees, written 360°.
	right angle	A right angle is an angle that measures exactly 90°. *Example:* Angle *AOB* at the right.
	acute angle	An acute angle is an angle that measures less than 90°. *Example:* Angle *E* at the right.
	obtuse angle	An obtuse angle is an angle that measures more than 90° but less than 180°. *Example:* Angle *F* at the right.
	straight angle	A straight angle is an angle that measures exactly 180°; its sides form a straight line. *Example:* Angle *G* at the right.
	intersecting lines	Intersecting lines cross. *Example:* $\overleftrightarrow{RQ}$ intersects $\overleftrightarrow{AB}$ at point *P* at the right.
	perpendicular lines	Perpendicular lines are two lines that intersect to form a right angle. *Example:* $\overleftrightarrow{PQ}$ is perpendicular to $\overleftrightarrow{RS}$ at the right.
	parallel lines	Parallel lines are two lines in the same plane that never intersect (never cross). *Example:* $\overleftrightarrow{AB}$ is parallel to $\overleftrightarrow{ST}$ at the right.
8.2	**complementary angles**	Complementary angles are two angles whose measures add up to 90°.
	supplementary angles	Supplementary angles are two angles whose measures add up to 180°.
	congruent angles	Congruent angles are angles that measure the same number of degrees.

continued

	vertical angles	Vertical angles are two nonadjacent congruent angles formed by intersecting lines. *Example:* ∠*COA* and ∠*EOF* are vertical angles at the right.

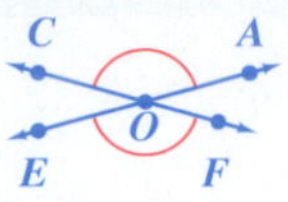

8.3–8.5   **perimeter**   Perimeter is the distance around the outside edges of a flat shape. It is measured in linear units such as ft, yd, cm, m, km, and so on.

8.3–8.6   **area**   Area is the surface inside a two-dimensional (flat) shape. It is measured by determining the number of squares of a certain size needed to cover the surface inside the shape. Some of the commonly used units for measuring area are square inches (in.^{2}), square feet (ft^2), square yards (yd^2), square centimeters (cm^2), and square meters (m^2).

8.3   **rectangle**   A rectangle is a four-sided figure with all sides meeting at 90° angles. The opposite sides are the same length. *Example:* The rectangle measuring 12 cm by 7 cm at the right.

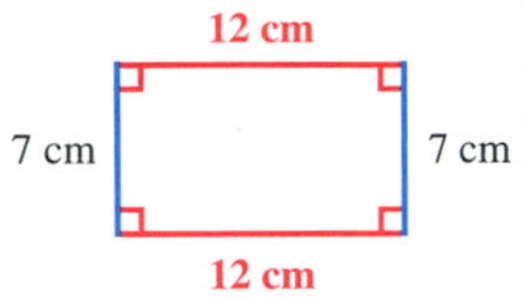

   square   A square is a rectangle with all four sides the same length. *Example:* The square with a side measurement of 20 inches at the right.

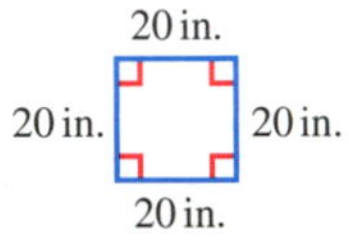

8.4   **parallelogram**   A parallelogram is a four-sided figure with both pairs of opposite sides parallel and equal in length. *Example:* See the parallelogram at the right with sides measuring 8 m and 12 m.

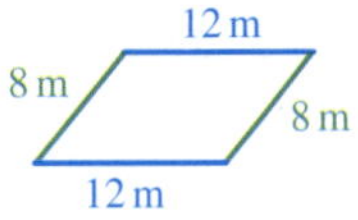

   trapezoid   A trapezoid is a four-sided figure with one pair of parallel sides. *Example:* Trapezoid *PQRS* at the right; $\overline{PQ}$ is parallel to $\overline{SR}$.

8.5   **triangle**   A triangle is a figure with exactly three sides. *Example:* Triangle *ABC* at the right.

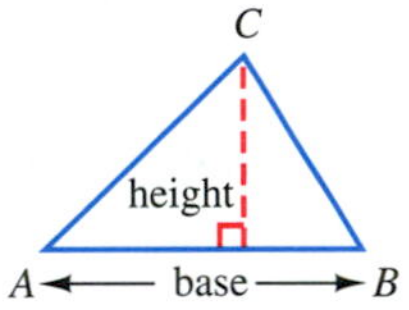

8.6   **circle**   A circle is a figure with all points the same distance from a fixed center point. *Example:* See figure at the right.

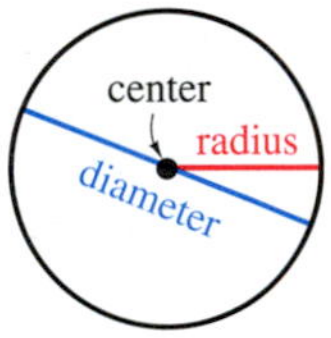

   radius   Radius is the distance from the center of a circle to any point on the circle. *Example:* See the red radius in the circle at the right.

   diameter   Diameter is the distance across a circle, passing through the center. *Example:* See the blue diameter in the circle at the right.

   circumference   Circumference is the distance around a circle.

   π (pi)   π is the ratio of the circumference to the diameter of any circle. It is approximately equal to 3.14.

8.7   **volume**   Volume is a measure of the space inside a three-dimensional (solid) shape. Volume is measured in cubic units such as in.^{3}, ft^3, yd^3, mm^3, cm^3, and so on.

8.8   **hypotenuse**   The hypotenuse is the side of a right triangle opposite the 90° angle; it is the longest side. *Example:* See the red side in the triangle at the right.

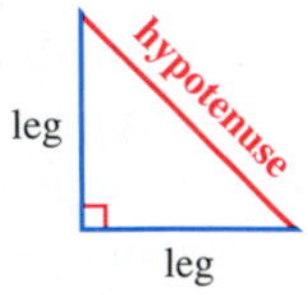

8.9   **similar triangles**   Similar triangles are triangles with the same shape but not necessarily the same size; corresponding angles measure the same number of degrees, and the *ratios* of the lengths of corresponding sides are equal.

NEW FORMULAS

Perimeter of a rectangle: $P = 2 \cdot l + 2 \cdot w$

Area of a rectangle: $A = l \cdot w$

Perimeter of a square: $P = 4 \cdot s$

Area of a square: $A = s^2$

Area of a parallelogram: $A = b \cdot h$

Area of a trapezoid: $A = \dfrac{1}{2} \cdot h \cdot (b + B)$

$\quad$ or $\quad A = 0.5 \cdot h \cdot (b + B)$

Area of a triangle: $A = \dfrac{1}{2} \cdot b \cdot h$

$\quad$ or $\quad A = 0.5 \cdot b \cdot h$

Diameter of a circle: $d = 2 \cdot r$

Radius of a circle: $r = \dfrac{d}{2}$

Circumference of a circle: $C = \pi \cdot d$

$\quad$ or $\quad C = 2 \cdot \pi \cdot r$

Area of a circle: $A = \pi \cdot r^2$

Area of semicircle: $A = \dfrac{\pi \cdot r^2}{2}$

Volume of rectangular solid: $V = l \cdot w \cdot h$

Volume of a sphere: $V = \dfrac{4}{3} \cdot \pi \cdot r^3$

$\quad$ or $\quad V = \dfrac{4 \cdot \pi \cdot r^3}{3}$

Volume of a hemisphere: $V = \dfrac{2}{3} \cdot \pi \cdot r^3$

$\quad$ or $\quad V = \dfrac{2 \cdot \pi \cdot r^3}{3}$

Volume of a cylinder: $V = \pi \cdot r^2 \cdot h$

Volume of a cone: $V = \dfrac{1}{3} \cdot B \cdot h$ $\quad$ or $\quad V = \dfrac{B \cdot h}{3}$

Volume of a pyramid: $V = \dfrac{1}{3} \cdot B \cdot h$ $\quad$ or $\quad V = \dfrac{B \cdot h}{3}$

Right triangle: hypotenuse $= \sqrt{(\text{leg})^2 + (\text{leg})^2}$

$\quad$ leg $= \sqrt{(\text{hypotenuse})^2 - (\text{leg})^2}$

TEST YOUR WORD POWER

See how well you have learned the vocabulary in this chapter. Answers follow the Quick Review.

1. Two angles that are **complementary**
   A. have measures that add up to 180°
   B. are always congruent
   C. form a straight angle
   D. have measures that add up to 90°.

2. The **perimeter** of a flat shape is
   A. measured in square units
   B. the distance around the outside edges
   C. the number of squares needed to cover the space inside the shape
   D. measured in cubic units.

3. An **obtuse angle**
   A. is formed by perpendicular lines
   B. is congruent to a right angle
   C. measures more than 90° but less than 180°
   D. measures less than 90°.

4. The **hypotenuse** is
   A. the long base in a trapezoid
   B. the height line in a parallelogram
   C. the longest side in a right triangle
   D. the distance across a circle, passing through the center.

5. π is the ratio of
   A. the diameter to the radius of a circle
   B. the circumference to the diameter of a circle
   C. the circumference to the radius of a circle
   D. the diameter to the circumference of a circle.

6. **Perpendicular lines**
   A. intersect to form a right angle
   B. intersect to form an acute angle
   C. never intersect
   D. have a common endpoint called the vertex.

7. In a pair of **similar triangles,**
   A. corresponding sides have the same length
   B. all the angles have the same measure
   C. the perimeters are equal
   D. the ratios of the lengths of corresponding sides are equal.

8. The **area of a rectangle** is found by
   A. using the formula $P = 2 \cdot l + 2 \cdot w$
   B. multiplying length times width
   C. adding the lengths of the sides
   D. using the formula $V = l \cdot w \cdot h$.

QUICK REVIEW

Concepts	Examples

8.1 Lines

A *line* is a straight row of points that goes on forever in both directions. If a line has one endpoint, it is a *ray.* If it has two endpoints, it is a *line segment.*

Identify each of the following as a line, line segment, or ray.

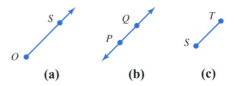

(a) (b) (c)

Figure **(a)** shows a ray, **(b)** shows a line, and **(c)** shows a line segment.

If two lines intersect at right angles, they are *perpendicular.*

If two lines in the same plane never intersect, they are *parallel.*

Label each pair of lines as appearing to be parallel or as perpendicular.

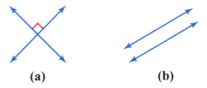

(a) (b)

Figure **(a)** shows perpendicular lines (they intersect at 90°).

Figure **(b)** shows lines that appear to be parallel (they never intersect).

8.2 Angles

If the sum of the measures of two angles is 90°, they are *complementary.*

If the sum of the measures of two angles is 180°, they are *supplementary.*

Find the complement and supplement of a 35° angle.

$$90° - 35° = 55° \text{ (the complement)}$$

$$180° - 35° = 145° \text{ (the supplement)}$$

If two angles measure the same number of degrees, the angles are *congruent.* The symbol for congruent is ≅.

Two nonadjacent angles formed by intersecting lines are called *vertical angles.* Vertical angles are congruent.

Identify the vertical angles in this figure. Which angles are congruent?

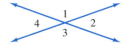

∠1 and ∠3 are vertical angles.

∠2 and ∠4 are vertical angles.

Vertical angles are congruent, so ∠1 ≅ ∠3 and ∠2 ≅ ∠4.

Concepts	**Examples**

8.3 *Rectangles and Squares*

Use this formula to find the perimeter of a *rectangle*.

$$P = 2 \cdot l + 2 \cdot w$$

Use this formula to find the area of a rectangle.

$$A = l \cdot w$$

Area is measured in **square units**.

Find the perimeter and area of this rectangle.

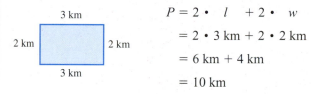

$$P = 2 \cdot l + 2 \cdot w$$
$$= 2 \cdot 3 \text{ km} + 2 \cdot 2 \text{ km}$$
$$= 6 \text{ km} + 4 \text{ km}$$
$$= 10 \text{ km}$$

$$A = l \cdot w = 3 \text{ km} \cdot 2 \text{ km} = 6 \textbf{ km}^2$$

Use these formulas to find the perimeter and area of a *square*.

$$P = 4 \cdot s$$
$$A = s^2$$

Area is measured in **square units**.

Find the perimeter and area of this square.

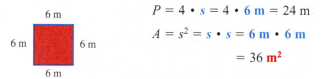

$$P = 4 \cdot s = 4 \cdot \textbf{6 m} = 24 \text{ m}$$
$$A = s^2 = s \cdot s = \textbf{6 m} \cdot \textbf{6 m}$$
$$= 36 \textbf{ m}^2$$

8.4 *Parallelograms*

Use these formulas to find the perimeter and area of a *parallelogram*.

$$P = \text{sum of the lengths of the sides}$$
$$A = b \cdot h$$

Area is measured in **square units**.

Find the perimeter and area of this parallelogram.

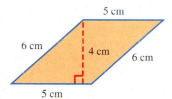

$$P = 5 \text{ cm} + 6 \text{ cm} + 5 \text{ cm} + 6 \text{ cm} = 22 \text{ cm}$$

$$A = 5 \text{ cm} \cdot 4 \text{ cm} = 20 \textbf{ cm}^2$$

8.4 *Trapezoids*

Use these formulas to find the perimeter and area of a *trapezoid*.

$$P = \text{sum of the lengths of the sides}$$
$$A = \frac{1}{2} \cdot h \cdot (b + B)$$

where *b* is the short base and *B* is the long base.

Area is measured in **square units**.

Find the perimeter and area of this trapezoid.

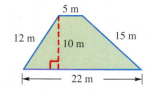

$$P = 5 \text{ m} + 15 \text{ m} + 22 \text{ m} + 12 \text{ m} = 54 \text{ m}$$

$$A = \frac{1}{\cancel{2}_1} \cdot \cancel{10}^{5} \text{ m} \cdot (5 \text{ m} + 22 \text{ m})$$

$$= 5 \text{ m} \cdot (27 \text{ m}) = 135 \textbf{ m}^2$$

8.5 *Triangles*

Use these formulas to find the perimeter and area of a *triangle*.

$$P = \text{sum of the lengths of the sides}$$
$$A = \frac{1}{2} \cdot b \cdot h$$
$$\downarrow$$
$$\text{or} \quad A = 0.5 \cdot b \cdot h$$

Area is measured in **square units**.

Find the perimeter and area of this triangle.

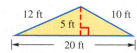

$$P = 12 \text{ ft} + 10 \text{ ft} + 20 \text{ ft} = 42 \text{ ft}$$

$$A = \frac{1}{2} \cdot b \cdot h$$

$$A = \frac{1}{\cancel{2}_1} \cdot \cancel{20}^{10} \text{ ft} \cdot 5 \text{ ft} = 50 \textbf{ ft}^2$$

$$\text{or} \quad A = 0.5 \cdot 20 \text{ ft} \cdot 5 \text{ ft} = 50 \textbf{ ft}^2$$

Concepts	Examples

8.6 Circles

Use this formula to find the *diameter* of a circle when you are given the radius.

$$d = 2 \cdot r$$

Find the diameter of a circle if the radius is 7 yd.

$$d = 2 \cdot r = 2 \cdot \mathbf{7\ yd} = 14\ yd$$

Use this formula to find the *radius* of a circle when you are given the diameter.

$$r = \frac{d}{2}$$

Find the radius of a circle if the diameter is 5 cm.

$$r = \frac{d}{2} = \frac{\mathbf{5\ cm}}{2} = 2.5\ cm$$

Use these formulas to find the *circumference* of a circle.

When you know the radius, use $C = 2 \cdot \pi \cdot r$

When you know the diameter, use $C = \pi \cdot d$

Use 3.14 as the approximate value for π.

Find the circumference of a circle with a radius of 3 cm.

$$C = 2 \cdot \pi \cdot r$$

$$C \approx 2 \cdot \mathbf{3.14} \cdot \mathbf{3\ cm} \approx 18.8\ cm \qquad \text{Rounded}$$

Use this formula to find the *area* of a circle.

$$A = \pi \cdot r^2$$

Area is measured in **square units**.

Find the area of this circle.

$$A = \pi \cdot r^2$$

$$A \approx \mathbf{3.14} \cdot \mathbf{3\ cm} \cdot \mathbf{3\ cm}$$

$$A \approx 28.3\ \mathbf{cm^2} \qquad \text{Rounded}$$

8.7 Volume of a Rectangular Solid

Use this formula to find the volume of *rectangular solids* (box-like solids).

$$V = l \cdot w \cdot h$$

Volume is measured in **cubic units**.

Find the volume of this box.

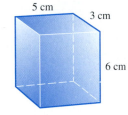

$$V = l \cdot w \cdot h$$

$$V = \mathbf{5\ cm} \cdot \mathbf{3\ cm} \cdot \mathbf{6\ cm}$$

$$V = 90\ \mathbf{cm^3}$$

8.7 Volume of a Sphere and Hemisphere

Use this formula to find the volume of a *sphere* (a ball-shaped solid).

$$V = \frac{4}{3} \cdot \pi \cdot r^3$$

or $V = \dfrac{4 \cdot \pi \cdot r^3}{3}$

where r is the radius of the sphere.

Volume is measured in **cubic units**.

Find the volume of a sphere with a radius of 5 m.

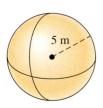

$$V = \frac{4 \cdot \pi \cdot (\mathbf{radius})^3}{3}$$

$$V \approx \frac{4 \cdot \mathbf{3.14} \cdot \mathbf{5\ m} \cdot \mathbf{5\ m} \cdot \mathbf{5\ m}}{3}$$

$$V \approx 523.3\ \mathbf{m^3} \qquad \text{Rounded}$$

Concepts	**Examples**

8.7 *Volume of a Sphere and Hemisphere (continued)*

Use this formula to find the volume of a *hemisphere* (half of a sphere).

$$V = \frac{2}{3} \cdot \pi \cdot r^3$$

$$\text{or} \quad V = \frac{2 \cdot \pi \cdot r^3}{3}$$

where *r* is the radius of the hemisphere.

Volume is measured in **cubic units**.

Find the volume of a hemisphere with a radius of 20 cm.

$$V = \frac{2 \cdot \pi \cdot (\text{radius})^3}{3}$$

$$V \approx \frac{2 \cdot 3.14 \cdot 20 \text{ cm} \cdot 20 \text{ cm} \cdot 20 \text{ cm}}{3}$$

$$V \approx 16,746.7 \text{ cm}^3 \quad \text{Rounded}$$

8.7 *Volume of a Cylinder*

Use this formula to find the volume of a *cylinder*.

$$V = \pi \cdot r^2 \cdot h$$

where *r* is the radius of the circular base and *h* is the height of the cylinder.

Volume is measured in **cubic units**.

Find the volume of this cylinder.

First, find the radius. $\quad r = \dfrac{8 \text{ m}}{2} = 4 \text{ m}$

$$V = \pi \cdot r^2 \cdot h$$

$$V \approx 3.14 \cdot 4 \text{ m} \cdot 4 \text{ m} \cdot 10 \text{ m}$$

$$V \approx 502.4 \text{ m}^3$$

8.7 *Volume of a Cone*

Use this formula to find the volume of a *cone*.

$$V = \frac{1}{3} \cdot B \cdot h$$

$$\text{or} \quad V = \frac{B \cdot h}{3}$$

where *B* is the area of the circular base and *h* is the height of the cone.

Volume is measured in **cubic units**.

Find the volume of this cone.

area of circular **Base** $\approx 3.14 \cdot 4$ in. $\cdot 4$ in.

$$\boldsymbol{B} \approx 50.24 \text{ in.}^2$$

$$V = \frac{\boldsymbol{B} \cdot \boldsymbol{h}}{3}$$

$$V \approx \frac{50.24 \text{ in.}^2 \cdot 9 \text{ in.}}{3}$$

$$V \approx 150.7 \text{ in.}^3 \quad \text{Rounded}$$

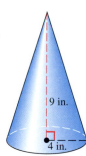

8.7 *Volume of a Pyramid*

Use this formula to find the volume of a *pyramid*.

$$V = \frac{1}{3} \cdot B \cdot h$$

$$\text{or} \quad V = \frac{B \cdot h}{3}$$

where *B* is the area of the square or rectangular base and *h* is the height of the pyramid.

Volume is measured in **cubic units**.

Find the volume of this pyramid.

area of square **Base** $= 2 \text{ cm} \cdot 2 \text{ cm}$

$$\boldsymbol{B} = 4 \text{ cm}^2$$

$$V = \frac{\boldsymbol{B} \cdot \boldsymbol{h}}{3}$$

$$V = \frac{4 \text{ cm}^2 \cdot 6 \text{ cm}}{3}$$

$$V = 8 \text{ cm}^3$$

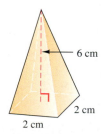

8.8 *Finding the Square Root of a Number*

Use the square root key on a calculator, $\boxed{\sqrt{}}$ or $\boxed{\sqrt{x}}$.

Round to the nearest thousandth if necessary.

$$\sqrt{64} = 8 \qquad \text{A perfect square}$$

$$\sqrt{43} \approx 6.557 \qquad \text{6.557438524 is rounded to nearest thousandth.}$$

Concepts	Examples

8.8 Finding the Unknown Length in a Right Triangle

To find the *hypotenuse,* use this formula.

$$\text{hypotenuse} = \sqrt{(\text{leg})^2 + (\text{leg})^2}$$

The hypotenuse is the side opposite the right angle; it is the longest side in a right triangle.

Find the unknown length in this right triangle. Round to the nearest tenth.

6 m
5 m

$$\text{hypotenuse} = \sqrt{(6)^2 + (5)^2}$$
$$= \sqrt{36 + 25}$$
$$= \sqrt{61} \approx 7.8 \text{ m}$$

To find a *leg,* use this formula.

$$\text{leg} = \sqrt{(\text{hypotenuse})^2 - (\text{leg})^2}$$

The legs are the sides that form the right angle.

Find the unknown length in this right triangle. Round to the nearest tenth.

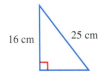

16 cm   25 cm

$$\text{leg} = \sqrt{(25)^2 - (16)^2}$$
$$= \sqrt{625 - 256}$$
$$= \sqrt{369} \approx 19.2 \text{ cm}$$

8.9 Finding the Unknown Lengths in Similar Triangles

Use the fact that in similar triangles, the *ratios* of the lengths of corresponding sides are equal. Write a proportion. Then find the cross products and show that they are equivalent. Finish solving for the unknown length.

Find *x* and *y* if the triangles are similar.

$$\frac{x}{8} = \frac{5}{10} \qquad\qquad \frac{y}{12} = \frac{5}{10}$$

$$x \cdot 10 = 8 \cdot 5 \qquad\qquad y \cdot 10 = 12 \cdot 5$$

$$\frac{x \cdot \cancel{10}^{1}}{\cancel{10}_{1}} = \frac{40}{10} \qquad\qquad \frac{y \cdot \cancel{10}^{1}}{\cancel{10}_{1}} = \frac{60}{10}$$

$$x = 4 \text{ m} \qquad\qquad y = 6 \text{ m}$$

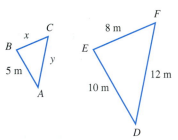

B   x   C
5 m   y
A

E   8 m   F
10 m   12 m
D

---

1. D; *Example:* If $\angle 1$ measures 35° and $\angle 2$ measures 55°, the angles are complementary because 35° + 55° = 90°.
2. B; *Example:* If a square measures 5 ft on each side, then the perimeter is 5 ft + 5 ft + 5 ft + 5 ft = 20 ft.
3. C; *Examples:* Angles that measure 91°, 120°, and 175° are all obtuse angles.
4. C; *Example:* In triangle *ABC* at the right, side *AC* is the hypotenuse; sides *AB* and *BC* are the legs.

C
A   B

5. B; *Example:* The ratio of a circumference of 12.57 cm to a diameter of 4 cm is $\dfrac{12.57}{4} \approx 3.14$ (rounded).

6. A; *Example:* $\overleftrightarrow{EF}$ is perpendicular to $\overleftrightarrow{GH}$, at the right.

G   F
E   H

7. D; *Example:* Triangle *ABC* is similar to triangle *DEF,* so the ratios of corresponding sides are equal.

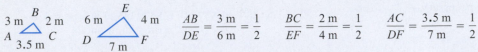

B
3 m   2 m
A   C
3.5 m

E
6 m   4 m
D   F
7 m

$$\frac{AB}{DE} = \frac{3 \text{ m}}{6 \text{ m}} = \frac{1}{2} \qquad \frac{BC}{EF} = \frac{2 \text{ m}}{4 \text{ m}} = \frac{1}{2} \qquad \frac{AC}{DF} = \frac{3.5 \text{ m}}{7 \text{ m}} = \frac{1}{2}$$

8. B; *Example:* in a rectangle with a length of 8 in. and a width of 5 in., Area = 8 in. • 5 in. = 40 in.²

Chapter 8

REVIEW EXERCISES

[8.1] *Name each line, line segment, or ray.*

1.

2.

3.

Label each pair of lines as appearing to be parallel, as perpendicular, or as intersecting.

4.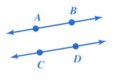

5.

6.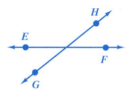

Label each angle as acute, right, obtuse, or straight. For right and straight angles, indicate the number of degrees in the angle.

7.

8.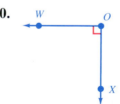

9.

10.

[8.2] *In each figure you are given the measures of two of the angles. Find the measure of each of the other angles.*

11.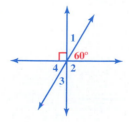

12.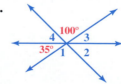

Name the pairs of supplementary angles in each figure.

13.

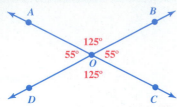

14.

Find the complement or supplement of each angle.

15. Find the complement of:

   (a) 80°

   (b) 45°

   (c) 7°

16. Find the supplement of:

   (a) 155°

   (b) 90°

   (c) 33°

[8.3] *Find the perimeter of each rectangle or square.*

17.

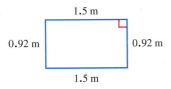

18.

19. A square-shaped pillow measures 38 cm along each side. How much lace is needed to trim all the edges?

20. A rectangular garden plot is $8\frac{1}{2}$ ft wide and 12 ft long. How much fencing is needed to surround the garden?

Find the area of each rectangle or square. Round your answers to the nearest tenth when necessary.

21.

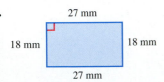

22.

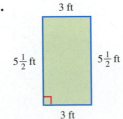

23.

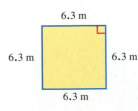

[8.4] *Find the perimeter and area of each parallelogram or trapezoid. Round your answers to the nearest tenth when necessary.*

24.
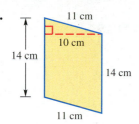
11 cm
10 cm
14 cm
14 cm
11 cm

25.
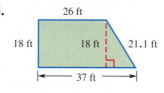
26 ft
18 ft   18 ft   21.1 ft
37 ft

26.
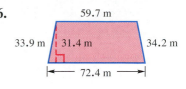
59.7 m
33.9 m   31.4 m   34.2 m
72.4 m

[8.5] *Find the perimeter and area of each triangle.*

27.
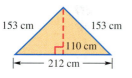
153 cm   153 cm
110 cm
212 cm

28.
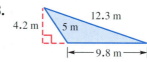
12.3 m
4.2 m   5 m
9.8 m

29.
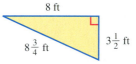
8 ft
$8\frac{3}{4}$ ft $3\frac{1}{2}$ ft

Find the number of degrees in the third angle of each triangle.

30.
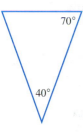
70°
40°

31.
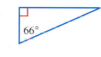
66°

[8.6] *Find the unknown length.*

32. The radius of a circular irrigation field is 68.9 m. What is the diameter of the field?

33. The diameter of a juice can is 3 in. What is the radius of the can?

Find the circumference and area of each circle. Use 3.14 as the approximate value for π. Round your answers to the nearest tenth.

34.
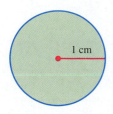
1 cm

35.
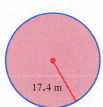
17.4 m

36.
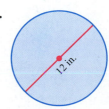
12 in.

[8.3–8.6] *Find each shaded area. Note that Exercises 44 and 45 contain semicircles. Use 3.14 as the approximate value for π. Round your answers to the nearest tenth when necessary.*

37.

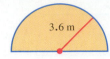

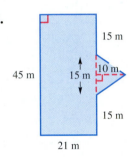

3.6 m

38.

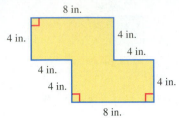

8 in.

4 in.

4 in.

4 in.

4 in.

4 in.

4 in.

8 in.

39.

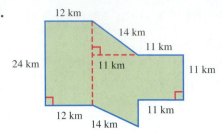

12 km

14 km

11 km

24 km

11 km

11 km

12 km

14 km

11 km

40.

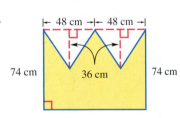

15 m

45 m

15 m

10 m

15 m

21 m

41.

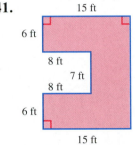

15 ft

6 ft

8 ft

7 ft

8 ft

6 ft

15 ft

42.

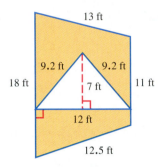

13 ft

9.2 ft

9.2 ft

18 ft

7 ft

11 ft

12 ft

12.5 ft

43.

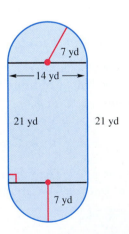

48 cm

48 cm

74 cm

74 cm

36 cm

44.

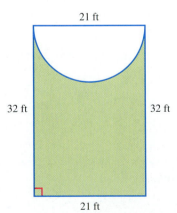

21 ft

32 ft

32 ft

21 ft

45.

7 yd

14 yd

21 yd

21 yd

7 yd

[8.7] *Name each solid and find its volume. Use 3.14 as the approximate value for π. Round your answers to the nearest tenth when necessary.*

46.

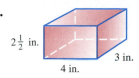

$2\frac{1}{2}$ in.

4 in.

3 in.

47.

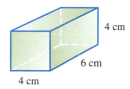

4 cm

6 cm

4 cm

48.

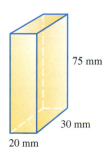

75 mm

30 mm

20 mm

49.

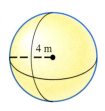

4 m

50.

6 ft

51.

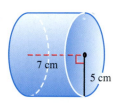

7 cm

5 cm

52.

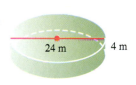

24 m

4 m

53.

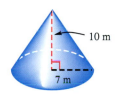

10 m

7 m

54.

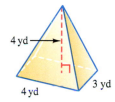

4 yd

4 yd

3 yd

[8.8] *Find each square root. Round your answers to the nearest thousandth when necessary.*

55. $\sqrt{49}$

56. $\sqrt{8}$

57. $\sqrt{3000}$

58. $\sqrt{144}$

59. $\sqrt{58}$

60. $\sqrt{625}$

61. $\sqrt{105}$

62. $\sqrt{80}$

Find the unknown length in each right triangle. Use a calculator to find square roots. Round your answers to the nearest tenth when necessary.

63.

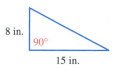

8 in.

90°

15 in.

64.

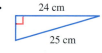

24 cm

25 cm

65.

15 cm

90°

11 cm

66.

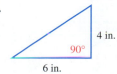

4 in.

90°

6 in.

67.

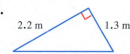

2.2 m

1.3 m

68.

12 km

8.5 km

[8.9] *Find the unknown lengths in each pair of similar triangles. Then find the perimeter of the larger triangle in each pair.*

69.

70.

71.

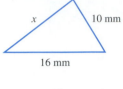

MIXED REVIEW EXERCISES

Name each figure and find its perimeter (or circumference) and area. Use **3.14** *as the approximate value for* π. *Round your answers to the nearest tenth when necessary.*

72.

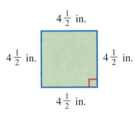

$4\frac{1}{2}$ in. $4\frac{1}{2}$ in. $4\frac{1}{2}$ in. $4\frac{1}{2}$ in.

73.

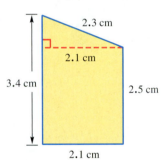

2.3 cm
2.1 cm
3.4 cm
2.5 cm
2.1 cm

74.

13 m

75.

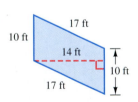

17 ft
10 ft
14 ft
17 ft
10 ft

76.

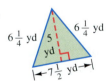

$6\frac{1}{4}$ yd 5 yd $6\frac{1}{4}$ yd
$7\frac{1}{2}$ yd

77.

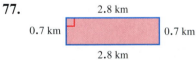

2.8 km
0.7 km   0.7 km
2.8 km

78.

8.5 m

79.

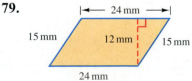

24 mm
15 mm   12 mm   15 mm
24 mm

80.

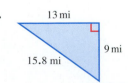

13 mi
9 mi
15.8 mi

Label each figure. Choose from these labels: line segment, ray, parallel lines, perpendicular lines, intersecting lines, acute angle, right angle, straight angle, obtuse angle. Indicate the number of degrees in the right angle and the straight angle.

81.

82.

83.

84.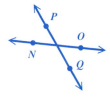

85.

86.

87.

88.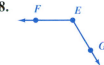

89.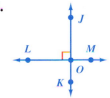

90. What is the complement of an angle measuring 9°?

91. What is the supplement of an angle measuring 42°?

Find the perimeter and area of each figure. In Exercise 92, assume that all angles are 90°.

92.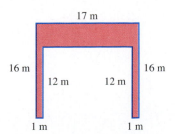

93.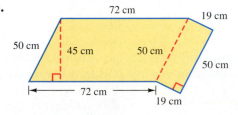

Find the volume of each solid. Use 3.14 as the approximate value for π. Round your answers to the nearest tenth when necessary.

94.

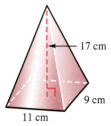

8 ft — 2 ft

95.

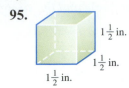

$1\frac{1}{2}$ in.

$1\frac{1}{2}$ in.

$1\frac{1}{2}$ in.

96.

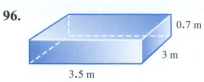

0.7 m

3 m

3.5 m

97.

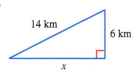

17 cm

9 cm

11 cm

98.

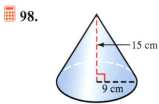

15 cm

9 cm

99.

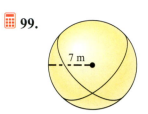

7 m

Find the unknown angle or side measurement. Round your answers to the nearest tenth when necessary.

100.

14 km

6 km

x

101.

C

60°

D

48°

E

102. similar triangles

x

18 mm   16.8 mm

10 mm

15 mm   *y*

103. Explain how you could use the information about prefixes from **Section 8.6** to solve a problem that asks, "How many decades are in two centuries?"

Chapter 8

T E S T

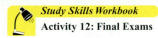
Study Skills Workbook
Activity 12: Final Exams

Choose the figure that matches each label. For right and straight angles, indicate the number of degrees in the angle.

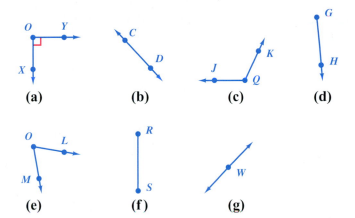

(a)   (b)   (c)   (d)

(e)   (f)   (g)

1. Acute angle is figure _____.

2. Right angle is figure _____ and its measure is _____.

3. Ray is figure _____.

4. Straight angle is figure _____ and its measure is _____.

5. Write a definition of parallel lines and a definition of perpendicular lines. Make a sketch to illustrate each definition.

6. Find the complement of an 81° angle.

7. Find the supplement of a 20° angle.

8. Find the measure of each unlabeled angle in the figure at the right.

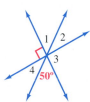

Name each figure and find its perimeter and area.

9.
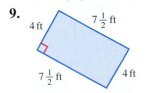

4 ft $7\frac{1}{2}$ ft

$7\frac{1}{2}$ ft   4 ft

10.
18 mm

18 mm   18 mm

18 mm

11.
7.2 m

5.9 m   4.6 m   5.9 m

|← 7.2 m →|

12.
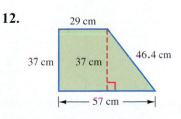

29 cm

37 cm   37 cm   46.4 cm

|← 57 cm →|

1. _____

2. _____

3. _____

4. _____

5. _____

6. _____

7. _____

8. _____

9. _____

_____

10. _____

_____

11. _____

_____

12. _____

_____

13. _____

_____

14. _____

_____

15. _____

16. _____

17. _____

18. _____

19. _____

20. _____

21. _____

22. _____

23. _____

24. _____

25. _____

Find the perimeter and area of each triangle.

13.

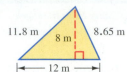

14.

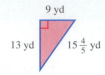

15. A triangle has angles that measure 90° and 35°. What does the third angle measure?

In Problems 16–22, use 3.14 as the approximate value for π. Round your answers to the nearest tenth when necessary.

16. Find the radius

17. Find the circumference

▦ *Find the area of each figure.*

18.

19.

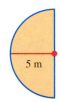

Name each solid and find its volume.

20.

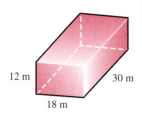

▦ **21.**

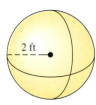

▦ **22.**

Find the unknown lengths. Round your answers to the nearest tenth when necessary.

▦ **23.**

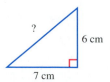

24. Similar triangles

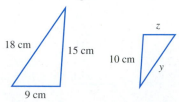

25. Explain the difference between cm, cm², and cm³. In what types of geometry problems might you use each of these units?

Cumulative Review Exercises

CHAPTERS 1–8

≈ *First use front end rounding to round each number and estimate the answer. Then find the exact answer.*

1. Estimate: Exact:

$$\begin{array}{r} 319 \\ 58{,}028 \\ +\ 6\ 227 \\ \hline \end{array}$$

$$+ \underline{\hspace{1.5cm}}$$

2. Estimate: Exact:

$$\begin{array}{r} 20.07 \\ -\ 9.828 \\ \hline \end{array}$$

$$- \underline{\hspace{1.5cm}}$$

3. Estimate: Exact:

$$\begin{array}{r} 3.664 \\ \times\ 7.3 \\ \hline \end{array}$$

$$\times \underline{\hspace{1.5cm}}$$

4. Estimate: Exact:

$$\begin{array}{r} 28{,}419 \\ \times\ \ \ \ 73 \\ \hline \end{array}$$

$$\times \underline{\hspace{1.5cm}}$$

5. Estimate: Exact:

$$2.8\overline{)562.24}$$

$$\underline{\hspace{1cm}}\,\overline{)\hspace{1.5cm}}$$

6. Estimate: Exact:

$$52\overline{)4888}$$

$$\underline{\hspace{1cm}}\,\overline{)\hspace{1.5cm}}$$

7. Estimate: Exact:

$$\begin{array}{r} 4\dfrac{1}{2} \\ +\ 4\dfrac{9}{10} \\ \hline \end{array}$$

$$+ \underline{\hspace{1.5cm}}$$

8. Estimate: Exact:

$$\begin{array}{r} 3\dfrac{1}{6} \\ -\ 1\dfrac{7}{8} \\ \hline \end{array}$$

$$- \underline{\hspace{1.5cm}}$$

9. Exact:

$$3\dfrac{1}{9} \cdot 1\dfrac{5}{7} = \underline{\hspace{1.5cm}}$$

Estimate:

$$\underline{\hspace{0.6cm}} \cdot \underline{\hspace{0.6cm}} = \underline{\hspace{0.6cm}}$$

Add, subtract, multiply, or divide as indicated. Write answers to fraction problems in lowest terms and as whole or mixed numbers when possible.

10. $3\dfrac{3}{5} \div 8$

11. $1 - 0.0868$

12. Write your answer using **R** for the remainder.

$$81\overline{)5749}$$

13. $10 \div \dfrac{5}{16}$

14. $(0.006)(0.013)$

15. $40{,}020 - 915$

16. $0.7 \div 0.036$ Round answer to the nearest hundredth.

17. $6\dfrac{1}{6} - 1\dfrac{3}{4}$

18. $752.6 + 83 + 0.485$

Use the order of operations to simplify each expression.

19. $16 - (10 - 2) \div 2(3) + 5$

20. $2^4 \div \sqrt{64} + 6^2$

21. Write 0.0208 in words.

22. Write six hundred sixty and five hundredths in numbers.

23. Arrange in order from smallest to largest.

2.55   2.505   2.055   2.5005

24. Explain how you could use the information on prefixes and root words in **Section 8.6** to remember the way to change a percent to a decimal.

Complete this chart.

Fraction/Mixed Number	Decimal	Percent
25. _____	0.02	**26.** _____
$1\frac{3}{4}$	**27.** _____	**28.** _____
29. _____	**30.** _____	40%

Write each rate or ratio as a fraction in lowest terms. Change to the same units when necessary.

31. 4 ft to 6 in.

32. Last month there were 9 cloudy days and 21 sunny days. What was the ratio of sunny days to cloudy days?

Find the unknown number in each proportion. Round your answer to hundredths if necessary.

33. $\dfrac{5}{13} = \dfrac{x}{91}$

34. $\dfrac{207}{69} = \dfrac{300}{x}$

35. $\dfrac{4.5}{x} = \dfrac{6.7}{3}$

Solve each percent problem.

36. 72 patients is what percent of 45 patients?

37. $18 is 3% of what number of dollars?

Convert each measurement.

38. $2\frac{1}{4}$ hours to minutes

39. 40 oz to pounds

40. 8 cm to meters

41. 1.8 L to milliliters

Write the most reasonable metric unit in each blank. Choose from km, m, cm, mm, L, mL, kg, g, and mg.

42. Her wristwatch strap is 15 _____ wide.

43. Jon added 2 _____ of oil to his car.

44. The child weighs 15 _____.

45. The bookcase is 90 _____ high.

46. List the metric temperatures at which water freezes and water boils.

Find the perimeter (or circumference) and area of each figure. Use 3.14 as the approximate value of π.
Also, in Exercises 47–51, name the figure.

47.

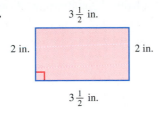

$3\frac{1}{2}$ in.

2 in.   2 in.

$3\frac{1}{2}$ in.

48.

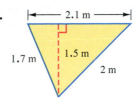

← 2.1 m →

1.7 m   1.5 m

2 m

49.

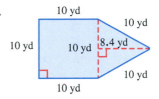

5 ft

50.

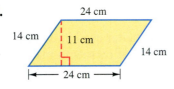

24 cm

14 cm   11 cm

14 cm

← 24 cm →

51.

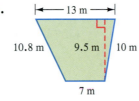

← 13 m →

10.8 m   9.5 m   10 m

7 m

52.

10 yd

10 yd

10 yd   10 yd   8.4 yd

10 yd

10 yd

Find the unknown length in each figure. Round your answers to the nearest tenth. In Exercise 54, also find the
perimeter of the smaller triangle.

53.

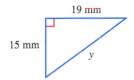

19 mm

15 mm

y

54. Similar triangles

15 ft

5.5 ft

x 22 ft

5.5 ft

15 ft

Solve each application problem.

55. Mei Ling must earn 90 credits
to receive an associate of arts
degree. She has 53 credits.
What percent of the necessary
credits does she have? Round
to the nearest whole percent.

56. Which bag of chips is the best buy: Brand T is
$15\frac{1}{2}$ oz for $2.99, Brand F is 14 oz for $2.49, and
Brand H is 18 oz for $3.89. You have a coupon for
40¢ off Brand H and another for 30¢ off Brand T.

57. A Folger's coffee can has a diameter of 13 cm and a height of 17 cm. Find the volume of the can. Use 3.14 for π and round your answer to the nearest tenth. (*Source:* Folger's.)

├──13 cm──┤

17 cm

58. Post-it® Notes introduced a new, larger size, as shown in the advertisement below. Find the area of each size of notes. Then check the statement about 75% more writing space. Is it true? If not, what is the correct answer to the nearest whole percent? (*Source:* 3M Company.)

Extra Large
4" × 4"

Original
3" × 3"

75% more writing space than original 3 × 3 notes.

XL

59. Steven bought $4\frac{1}{2}$ yd of canvas material to repair the tents used by the scout troop. He used $1\frac{2}{3}$ yd on one tent and $1\frac{3}{4}$ yd on another. How much material is left?

60. The cooks at a homeless shelter used 25 lb of meat, 35 lb of potatoes, and 15 lb of carrots to make stew for 140 people. At that rate, how much of each item is needed for stew to feed 200 people? Round to the nearest tenth.

61. Graciela needs 85 cm of yarn to make a tassel for one corner of a pillow. How many meters of yarn does she need to put a tassel on each corner of a square-shaped pillow?

62. Swimsuits are on sale in August at 65% off the regular price. How much will Lanece pay for a suit that has a regular price of $64?

The bar graph below shows the average number of vacation days received each year by workers in various countries. Use the information in the graph to answer Exercises 63–66. Round to the nearest tenth when necessary.

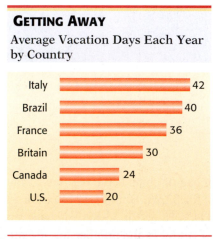

GETTING AWAY
Average Vacation Days Each Year by Country

Italy	42
Brazil	40
France	36
Britain	30
Canada	24
U.S.	20

Source: World Tourism Organization.

63. Assuming that a workweek is 5 days long, which country has an average of:

(a) six weeks of vacation;

(b) eight weeks of vacation.

64. Find the ratio of vacation days when comparing:

(a) Brazil to the United States;

(b) Canada to France.

65. (a) U.S. vacation days are what percent of Italy's vacation days?

(b) France's vacation days are what percent of U.S. vacation days?

66. By what percent would U.S. vacation days need to increase to match the vacation days in:

(a) Canada;   **(b)** Britain;

(c) Brazil.

Basic Algebra

9

The weather—we all talk about it and often complain about it. When reporting temperatures, we need both *positive* and *negative* numbers. If you live in Chicago, the highest temperature ever recorded is 104 °F and the lowest is −27 °F, a difference of 131 degrees! In Atlanta, the record high and low are 105 °F and −8 °F. And in Barrow, Alaska, the range of extreme temperatures is 79 °F to −56 °F, a difference of 135 degrees. To make things even more uncomfortable, humidity can make hot temperatures feel hotter and wind can make cold temperatures feel colder. Learn more about *windchill* as you work Exercises 69–72 in **Section 9.2.** (*Source:* National Climatic Data Center.)

9.1 Signed Numbers

OBJECTIVES

1. Write negative numbers.
2. Graph signed numbers on a number line.
3. Use the < and > symbols.
4. Find absolute value.
5. Find the opposite of a number.

All the numbers you have studied so far in this book have been either 0 or greater than 0. Numbers *greater* than 0 are called *positive numbers.* For example, you have worked with these positive numbers.

Salary of $45,000

Temperature of 98.6 °F

Length of $3\frac{1}{2}$ feet

OBJECTIVE 1 Write negative numbers. Not all numbers are positive. For example, "15 degrees below 0" or "a loss of $500" is expressed with a number *less* than 0. Numbers less than 0 are called **negative numbers.** Zero is neither positive nor negative.

1 Write each number.

(a) The temperature at the North Pole is 70 degrees below 0.

> **Writing a Negative Number**
>
> To write a negative number, put a *negative sign,* **−**, in front of it.

(b) Your checking account is overdrawn by 15 dollars.

For example, "15 degrees below 0" is written with a negative sign, as −15°. And "a loss of $500" is written −$500.

⫷⫷⫷ **Work Problem 1 at the Side.**

(c) A submarine dived to 284 ft below sea level.

OBJECTIVE 2 Graph signed numbers on a number line. In **Section 3.5** you graphed positive numbers on a number line. Negative numbers can also be shown on a number line. Zero separates the positive numbers from the negative numbers on the number line. The number −5 is read "negative five."

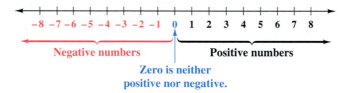

2 Write *positive, negative,* or *neither* for each number.

(a) −8

> **NOTE**
>
> For every *positive* number on a number line, there is a corresponding *negative* number on the *opposite* side of 0.

(b) $-\dfrac{3}{4}$

(c) 1

When you work with both positive and negative numbers (and 0), we say you are working with **signed numbers.**

(d) 0

> **Writing a Positive Number**
>
> A positive number can be written in two ways.
> 1. Use a "+" sign. For example, +2 is "positive two."
> 2. Do not write any sign. For example, when you write 3, it is assumed to be "positive three."

ANSWERS

1. (a) −70° (b) −$15 (c) −284 ft
2. (a) negative (b) negative
   (c) positive (d) neither

⫷⫷⫷ **Work Problem 2 at the Side.**

The next example shows you how to graph signed numbers on a number line.

EXAMPLE 1   **Graphing Signed Numbers**

Graph these numbers on the number line.

(a) -4 **(b)** 3 **(c)** -1 **(d)** 0 **(e)** $1\frac{1}{4}$

Place a dot at the correct location for each number.

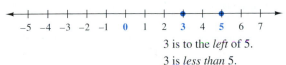

Work Problem 3 at the Side. ▶▶▶

OBJECTIVE **3**   **Use the $<$ and $>$ symbols.**   As shown on the number line below, 3 is to the *left* of 5.

3 is to the *left* of 5.

3 is *less than* 5.

Recall the following symbols for comparing two numbers.

$<$ means **"is less than."**

$>$ means **"is greater than."**

Use these symbols to write "3 is less than 5" as follows.

$$3 \quad < \quad 5$$
$$\downarrow \qquad \downarrow \qquad \downarrow$$

3 **is less than** 5

This example suggests the following.

> The *lesser* of two numbers is the one farther to the **left** on a number line. The *greater* of two numbers is the one farther to the **right** on a number line.

EXAMPLE 2   **Using the Symbols $<$ and $>$**

Use this number line to compare each pair of numbers. Then write $>$ or $<$ to make each statement true.

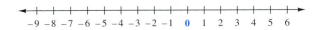

(a) $2 < 6$ (read "2 is less than 6") because 2 is to the *left* of 6 on the number line.

(b) $-9 < -4$ because -9 is to the *left* of -4.

(c) $2 > -1$ because 2 is to the *right* of -1.

(d) $-4 < 0$ because -4 is to the *left* of 0.

Continued on Next Page

---

3 Graph each set of numbers.

(a) $-1, 1, -3, 3$

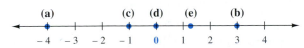

(b) $-2, 4, 0, -1, -4$

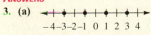

4 Write $<$ or $>$ in each blank to make a true statement.

(a) 4 _____ 0

(b) -1 _____ 0

(c) -3 _____ -1

(d) -8 _____ -9

(e) 0 _____ -3

5 Simplify each absolute value expression.

(a) $|5|$

(b) $|-5|$

(c) $|-17|$

(d) $-|-9|$

(e) $-|2|$

ANSWERS
4. (a) $>$ (b) $<$ (c) $<$ (d) $>$ (e) $>$
5. (a) 5 (b) 5 (c) 17 (d) -9 (e) -2

> **NOTE**
> When using $>$ and $<$, the *smaller* pointed end of the symbol points to the *smaller* (lesser) number.

◀◀◀ **Work Problem 4 at the Side.**

OBJECTIVE 4 Find absolute value. In order to graph a number on the number line, you need to answer two questions:

1. Which *direction* is it from 0? It can be in a *positive* direction or a *negative* direction. You can tell the direction by looking for a positive sign or a negative sign (or no sign, which is positive).

2. How *far* is it from 0? The *distance* from 0 is the **absolute value** of a number.

Absolute value is indicated by two vertical bars. For example, $|6|$ is read "the **absolute value** of 6."

> **NOTE**
> Absolute value is never negative because it is the *distance* from 0. A distance is never negative.

EXAMPLE 3 Finding Absolute Values

Simplify each absolute value expression.

(a) $|8|$ The *distance* from 0 to 8 is 8, so $|8| = 8$.

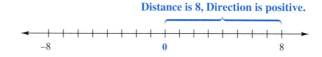

Distance is 8, Direction is positive.

(b) $|-8|$ The *distance* from 0 to -8 is also 8, so $|-8| = 8$.

Distance is 8, Direction is negative.

(c) $|0| = 0$

(d) $-|-3|$ First, $|-3|$ is 3. But there is also a negative sign outside the absolute value bars. So, -3 is the simplified expression.

> **CAUTION**
> A negative sign *outside* the absolute value bars is *not* affected by the absolute value bars. Therefore, your final answer is negative, as in Example 3(d) above.

◀◀◀ **Work Problem 5 at the Side.**

OBJECTIVE 5 **Find the opposite of a number.** Two numbers that are the *same distance* from 0 on a number line but on *opposite sides* of 0 are called **opposites** of each other. As this number line shows, −3 and 3 are opposites of each other.

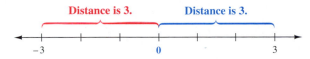

Distance is 3.   **Distance is 3.**

To indicate the opposite of a number, write a negative sign in front of the number.

EXAMPLE 4 **Finding Opposites**

Find the opposite of each number.

Number	Opposite
5	$-(5) = \, -5$ ── Write a negative sign
9	$-(9) = \, -9$
$\dfrac{4}{5}$	$-\left(\dfrac{4}{5}\right) = -\dfrac{4}{5}$
0	$-(0) = \quad 0$ ── No negative sign

The opposite of 0 is 0. Zero is neither positive nor negative.

Work Problem 6 at the Side. ▶▶▶

Some numbers have two negative signs. An example is shown below.

$$-(-3)$$

The negative sign in front of (-3) means the *opposite* of −3. The opposite of −3 is 3.

$$-(-3) = 3$$

Use the following rule to find the opposite of a negative number.

Double Negative Rule

The opposite of a negative number is positive.

For example, $-(-3) = 3$ and $-(-10) = 10$.

EXAMPLE 5 **Finding Opposites**

Find the opposite of each number.

Number	Opposite	
−2	$-(-2) = 2$	By the double negative rule
−9	$-(-9) = 9$	
$-\dfrac{1}{2}$	$-\left(-\dfrac{1}{2}\right) = \dfrac{1}{2}$	

Work Problem 7 at the Side. ▶▶▶

6 Find the opposite of each number.

(a) 4

(b) 10

(c) 49

(d) $\dfrac{2}{5}$

(e) 0

7 Find the opposite of each number.

(a) −4

(b) −10

(c) −25

(d) −1.9

(e) −0.85

(f) $-\dfrac{3}{4}$

Auto Aide, Part 1

The *Auto Aide Service* is a hypothetical business located in downtown Houston, Texas. It specializes in assisting motorists who have minor mechanical problems, such as a flat tire or a dead battery. The I-10 corridor in the greater Houston area is a major highway with constant heavy traffic, so the company assigns five vans to concentrate on I-10 assistance needs. The owner is a retired mathematics teacher who believes in using integers to track the actions of the auto aides along their service routes.

A schematic map of the I-10 corridor is shown below. Each tick mark represents 1 mile. The home office is located at 0. Locations to the west are represented by negative integers, and locations to the east are represented by positive integers.

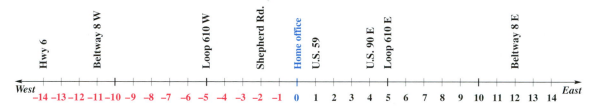

A radio dispatcher notifies the aide to assist a motorist at a given location. (Only a few locations are identified in this hypothetical problem, for simplicity.) Each aide must include four calculations on the daily report. First, the aide must translate the route instructions into an integer sentence that gives the direction and miles between locations. Second, the aide must record the displacement at the end of the route (location relative to home office). Third, the aide must record the distance from the home office at the end of the route. Fourth, the aide must record the total distance traveled on the route.

Aide Anne is shown as an example. Complete the table for the remaining aides.

Aide	Route Instructions	Integer Sentence	Displacement (final location)	Distance from Home Office	Total Distance
Anne	Home office to U.S. 90 E; to Loop 610 W; to Hwy 6; to Shepherd Rd.	$4 + (-9) + (-9) + 12$	-2 (Shepherd) (*Hint:* Find the integer sum.)	2 miles (*Hint:* Find the absolute value of displacement.)	34 miles (*Hint:* Find sum of absolute values of trip segments.)
Bill	Home office to Hwy 6; to U.S. 59; to Shepherd Rd.; to Loop 610 W				
Carlos	Home office to Shepherd Rd.; to Loop 610 W; to Beltway 8 W; to Hwy 6				
Dylan	Home office to Beltway 8 E; to U.S. 90 E; to Home office				
Ellen	Home office to U.S. 59; to Shepherd Rd.; to U.S. 59; to Loop 610 W; to U.S. 59				

9.1 Exercises

 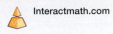
Write a signed number for each situation.

1. Water freezes at 32 degrees above 0 on the Fahrenheit temperature scale.

2. She made a profit of $920.

3. The price of the stock fell $12.

4. His checking account is overdrawn by $30.79.

5. Keith lost $6\frac{1}{2}$ lb while he was sick with the flu.

6. The river is 20 ft above flood stage.

7. Mount McKinley, the tallest mountain in the United States, rises 20,320 ft above sea level. (*Source: World Almanac.*)

8. The bottom of Lake Baykal in central Asia is 5315 ft below the surface of the water. (*Source: World Almanac.*)

Graph each set of numbers on the number line. See Example 1.

9. $4, -1, 2, 0, -5$

10. $-2, 1, -3, 5, 0$

11. $-\frac{1}{2}, -3, \frac{7}{4}, -4\frac{1}{2}, 3\frac{1}{4}$

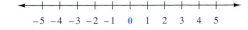

12. $-4, -\frac{3}{4}, 1, -1\frac{1}{4}, \frac{5}{2}$

13. $3, 4.5, -1.5, 2.2, -0.5$

14. $3.25, -1, -4.5, 1.25, 2$

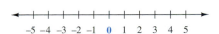

Write $<$ or $>$ in each blank to make a true statement. See Example 2.

15. 9 _____ 14

16. 6 _____ 11

17. 0 _____ -2

18. 0 _____ 2

19. -6 _____ 3

20. -9 _____ 9

21. 1 _____ -1

22. -1 _____ 0

23. -11 _____ -2

24. -5 _____ -1

25. -72 _____ -75

26. -50 _____ -60

RELATING CONCEPTS (EXERCISES 27–30) For Individual or Group Work

An ultrasound of a person's heel bone is a quick way to screen patients for osteoporosis (brittle bone disease). Use the information on the Patient Report Form to **work Exercises 27–30 in order.**

> **Patient Report Form**
>
> Your T score measures your bone density compared to that of a young, healthy woman when peak bone mass is achieved.
>
> | T Score | Interpretation |
> |---------|----------------|
> | Above 0 | Above normal |
> | 0 to –1.0 | Normal |
> | Below –1 | You may be at risk; further discussion with your health provider is recommended. |

Source: Health Partners, Inc.

27. Here are the T scores for four patients. Draw a number line and graph the four scores.

Patient A: −1.2

Patient B: 0.6

Patient C: −0.5

Patient D: 0

28. List the patients' scores in order from lowest to highest.

29. What is the interpretation of each patient's score?

30. (a) Explain what could happen if patient A did not understand the importance of a negative sign.

(b) For which patient does the sign of the score make no difference? Explain your answer.

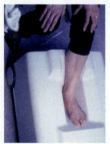

Taking an ultrasound of a patient's heel bone.

Simplify each absolute value expression. See Example 3.

31. $|3|$ **32.** $|9|$ **33.** $|-10|$ **34.** $|-2|$ **35.** $|0|$

36. $\left|-\dfrac{1}{2}\right|$ **37.** $-|-18|$ **38.** $-|-5|$ **39.** $-|32|$ **40.** $-|20|$

Find the opposite of each number. See Examples 4 and 5.

41. 7 **42.** 1 **43.** −14 **44.** −5 **45.** $\dfrac{2}{3}$

46. 0 **47.** −8.3 **48.** 0.2 **49.** $-\dfrac{1}{6}$ **50.** $-\dfrac{3}{10}$

Write true *or* false *for each statement.*

51. $|-5| > 0$ **52.** $|-12| > |-15|$ **53.** $0 < -(-6)$

54. $-9 < -(-9)$ **55.** $-|-4| < -|-7|$ **56.** $-|-0| > 0$

9.2 Adding and Subtracting Signed Numbers

You can show a *positive* number on a number line by drawing an arrow pointing to the *right*. On the number lines below, both arrows represent positive 4 units.

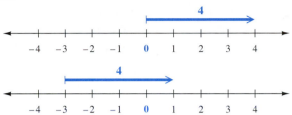

Draw arrows pointing to the *left* to show *negative* numbers. Both of the arrows below represent -3 units.

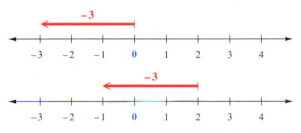

> **Work Problem 1 at the Side.**

OBJECTIVE 1 Add signed numbers by using a number line. You can use a number line to add signed numbers. For example, this number line shows how to add 2 and 3.

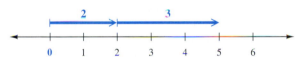

Add 2 and 3 by starting at 0 and drawing an arrow 2 units to the right. From the end of this arrow, draw another arrow 3 units to the right. This second arrow ends at 5, showing that the sum of $2 + 3$ is 5.

$$2 + 3 = 5$$

> **EXAMPLE 1** Adding Signed Numbers Using a Number Line

Add using a number line.

(a) $4 + (-1)$

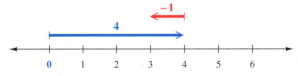

   Start at 0 and draw an arrow 4 units to the right. From the end of this arrow, draw an arrow 1 unit to the *left*. (Remember to go to the left for a negative number.) This second arrow ends at 3, so

$$4 + (-1) = 3$$

> **CAUTION**
> Always start at 0 when adding on the number line.

― **Continued on Next Page**

1 Complete each arrow so it represents the indicated number of units.

(a)

(b)
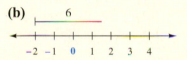

(c)
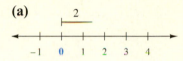

(d)
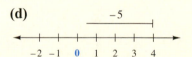

ANSWERS

1. **(a)**
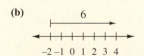

(b)
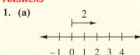

(c)

(d)
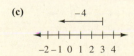

2 Draw arrows to find each sum.

(a) $3 + (-2)$

(b) $-4 + 1$

(c) $-3 + 7$

(d) $-1 + (-4)$

2. (a) $3 + (-2) = 1$

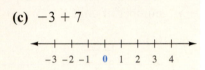

(b) $-4 + 1 = -3$

(c) $-3 + 7 = 4$

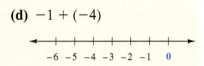

(d) $-1 + (-4) = -5$

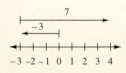

(b) $-6 + 2$

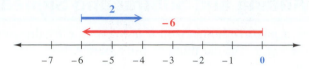

Draw an arrow from 0 going 6 units to the left. From the end of this arrow, draw an arrow 2 units to the right. This second arrow ends at -4, so

$$-6 + 2 = -4$$

(c) $-3 + (-5)$

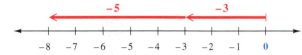

As the arrows along the number line show,

$$-3 + (-5) = -8$$

> **NOTE**
> When a negative number *follows* an operation sign, we usually write parentheses around the negative number to prevent confusion. You can see this in Example 1(c) above.
>
> $$-3 + (-5) = -8$$
>
> Operation sign ——↑   ↑—— Parentheses around the negative number that follows

◀◀ Work Problem 2 at the Side.

OBJECTIVE 2 Add signed numbers without using a number line. After working with number lines for a while, you will see ways to add signed numbers without drawing arrows. You already know how to add two positive numbers (from **Chapter 1**). Here are the steps for adding two negative numbers.

Adding Two Negative Numbers

Step 1   Add the absolute values of the numbers.

Step 2   Write a negative sign in front of the sum.

EXAMPLE 2 Adding Two Negative Numbers

Add (without using number lines).

(a) $-4 + (-12)$

The absolute value of -4 is **4.**

The absolute value of -12 is **12.**

Add the absolute values.

$$4 + 12 = 16$$

Write a negative sign in front of the sum.

— Write a negative sign in front of 16

$$-4 + (-12) = -16$$

(b) $-5 + (-25) = -30$ ← Sum of absolute values, with a negative sign written in front of 30

(c) $-11 + (-46) = -57$

— **Continued on Next Page**

(d) $-\dfrac{3}{4} + \left(-\dfrac{1}{2}\right)$

The absolute value of $-\frac{3}{4}$ is $\frac{3}{4}$, and the absolute value of $-\frac{1}{2}$ is $\frac{1}{2}$. Add the absolute values. Check that the sum is in lowest terms.

$$\frac{3}{4} + \frac{1}{2} = \frac{3}{4} + \frac{2}{4} = \frac{5}{4} \leftarrow \text{Lowest terms}$$

Write a negative sign in front of the sum.

$$-\frac{3}{4} + \left(-\frac{1}{2}\right) = -\frac{5}{4} \quad \text{Write a negative sign.}$$

> ### NOTE
> In algebra, we always write fractions in lowest terms, but usually do *not* change improper fractions to mixed numbers, because improper fractions are easier to work with. In Example 2(d) above, we checked that $-\frac{5}{4}$ was in lowest terms but did *not* rewrite it as $-1\frac{1}{4}$.

Work Problem 3 at the Side. ▶▶▶

Use the following steps to add two numbers with *different* signs.

Adding Two Numbers with Different Signs
Step 1 **Subtract** the smaller absolute value from the larger absolute value.

Step 2 Write the sign of the number with the **larger** absolute value in front of the answer.

EXAMPLE 3 Adding Two Numbers with Different Signs

Find each sum.

(a) $8 + (-3)$

First find this sum with a number line.

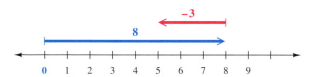

Because the second arrow ends at 5,

$$8 + (-3) = 5$$

Now find the sum by using the above rule. First, find the absolute value of each number.

$$|8| = 8 \quad \text{and} \quad |-3| = 3$$

Subtract the smaller absolute value from the larger absolute value.

$$8 - 3 = 5$$

The *positive* number 8 has the larger absolute value, so the answer is *positive*.

$$8 + (-3) = 5 \leftarrow \text{Positive answer}$$

Continued on Next Page

---

3 Add.

(a) $-4 + (-4)$

(b) $-3 + (-20)$

(c) $-31 + (-5)$

(d) $-10 + (-8)$

(e) $-\dfrac{9}{10} + \left(-\dfrac{3}{5}\right)$

ANSWERS
3. **(a)** -8 **(b)** -23 **(c)** -36
   (d) -18 **(e)** $-\dfrac{3}{2}$

4 Add.

(a) $10 + (-2)$

(b) $-7 + 8$

(c) $-11 + 11$

(d) $23 + (-32)$

(e) $-\dfrac{7}{8} + \dfrac{1}{4}$

(b) $4 + (-12)$

First, find the absolute values.

$$|4| = \mathbf{4} \quad \text{and} \quad |-12| = \mathbf{12}$$

Subtract the smaller absolute value from the larger absolute value.

$$\mathbf{12 - 4 = 8}$$

The *negative* number -12 has the larger absolute value, so the answer is *negative*.

$$4 + (-12) = \mathbf{-8}$$

— Write a negative sign in front of the answer because -12 has the larger absolute value.

(c) $-21 + 15 = \mathbf{-6}$

— Write a negative sign in front of the answer because -21 has the larger absolute value.

(d) $13 + (-9) = 4 \leftarrow$ Positive answer because the positive number 13 has the larger absolute value

(e) $-\dfrac{1}{2} + \dfrac{2}{3}$

The absolute value of $-\frac{1}{2}$ is $\frac{1}{2}$, and the absolute value of $\frac{2}{3}$ is $\frac{2}{3}$. Subtract the smaller absolute value from the larger absolute value.

$$\frac{2}{3} - \frac{1}{2} = \frac{4}{6} - \frac{3}{6} = \frac{1}{6}$$

Because the *positive* number $\frac{2}{3}$ has the larger absolute value, the answer is *positive*.

$$-\frac{1}{2} + \frac{2}{3} = \frac{1}{6} \leftarrow \text{Positive answer}$$

◀◀◀ Work Problem 4 at the Side.

OBJECTIVE 3 Find the additive inverse of a number. Recall that the opposite of 9 is -9, and the opposite of -4 is $-(-4)$, or 4. Adding opposites gives the results shown below.

$$9 + (-9) = 0 \quad \text{and} \quad -4 + 4 = 0$$

The sum of a number and its opposite is always 0. For this reason, opposites are also called *additive inverses* of each other.

Additive Inverse

The opposite of a number is called its **additive inverse.**
The sum of a number and its opposite is 0.

EXAMPLE 4 Finding Additive Inverses

This chart shows you several numbers and the additive inverse of each.

Number	Additive Inverse	Sum of Number and Inverse
6	-6	$6 + (-6) = 0$
-8	$-(-8)$ or 8	$(-8) + 8 = 0$
4	-4	$4 + (-4) = 0$
-3	$-(-3)$ or 3	$-3 + 3 = 0$
$\dfrac{5}{8}$	$-\dfrac{5}{8}$	$\dfrac{5}{8} + \left(-\dfrac{5}{8}\right) = 0$
0	0	$0 + 0 = 0$

Work Problem 5 at the Side. ▶▶▶

OBJECTIVE 4 Subtract signed numbers. You may have noticed that negative numbers are often written with parentheses, like (-8). This is especially helpful when subtracting because the $-$ sign is used both to indicate a **negative number** and to indicate **subtraction**.

Example	How to Say It
(-5)	**negative five**
8	positive eight
$-3 - 2$	**negative three** minus positive two
$6 - (-4)$	positive six **minus negative four**
$-7 - (-1)$	**negative seven** minus **negative one**
$8 - 3$	positive eight **minus** positive three

CAUTION
Be sure you understand when the "$-$" sign means "subtract," and when it means "negative number."

Work Problem 6 at the Side. ▶▶▶

When working with signed numbers, it is helpful to write a subtraction problem as an addition problem. For example, you know that $6 - 4 = 2$. But you get the same result by *adding* 6 and the *opposite* of 4, that is, $6 + (-4)$.

$$6 - 4 = 2$$
$$6 + (-4) = 2$$
Same result

This suggests the following definition of subtraction.

Defining Subtraction
The difference of two numbers, a and b, is
$$a - b = a + (-b).$$
In other words, subtract two numbers by adding the first number and the opposite (additive inverse) of the second.

---

5 Give the additive inverse of each number. Then find the sum of the number and its inverse.

(a) 12

(b) -9

(c) 3.5

(d) $-\dfrac{7}{10}$

(e) 0

6 Write each example in words.

(a) $-7 - 2$

(b) -10

(c) $3 - (-5)$

(d) 4

(e) $-8 - (-6)$

(f) $2 - 9$

7 Subtract.

(a) $-6 - 5$

(b) $3 - (-10)$

(c) $-8 - (-2)$

(d) $4 - 9$

(e) $-7 - (-15)$

(f) $-\dfrac{2}{3} - \left(-\dfrac{5}{12}\right)$

Subtracting Signed Numbers

To subtract two signed numbers, add the opposite of the second number to the first number. Use these steps.

Step 1 Change the subtraction sign to addition.

Step 2 Change the sign of the second number to its opposite.

Step 3 Proceed as in addition.

NOTE
The pattern in a subtraction problem is shown below.

$$\begin{array}{ccccccc} \text{1st} & & \text{2nd} & & \text{1st} & & \text{opposite of} \\ \text{number} & - & \text{number} & = & \text{number} & + & \text{2nd number} \end{array}$$

EXAMPLE 5 Subtracting Signed Numbers

Subtract.

(a) $8 - 11$

The first number, 8, stays the same. Change the subtraction sign to addition. Change the sign of the second number to its opposite.

Positive 8 stays the same. $8 - 11$ Positive 11 is changed to its opposite, -11
$8 + (-11)$

Subtraction is changed to addition.

Now add.

$$8 + (-11) = -3$$
$$\text{So} \quad 8 - 11 = -3 \quad \text{also.}$$

(b) $-9 - 15$ Subtraction is changed to addition.
Positive 15 is changed to its opposite, (-15).
$-9 + (-15) = -24$

(c) $-5 - (-7)$ Subtraction is changed to addition.
Negative 7 is changed to its opposite, $(+7)$.
$-5 + (+7) = 2$

(d) $7.6 - (-8.3)$ Subtraction is changed to addition.
Negative 8.3 is changed to its opposite, $(+8.3)$.
$7.6 + (+8.3) = 15.9$

(e) $\dfrac{5}{8} - \dfrac{1}{2}$ Subtraction is changed to addition.
Positive $\frac{1}{2}$ is changed to its opposite, $\left(-\frac{1}{2}\right)$.

$$\dfrac{5}{8} + \left(-\dfrac{1}{2}\right) = \dfrac{5}{8} + \left(-\dfrac{4}{8}\right) = \dfrac{1}{8}$$

Work Problem 7 at the Side.

OBJECTIVE 5 **Add or subtract a series of signed numbers.** If a problem involves both addition and subtraction, use the order of operations, from **Section 1.8.**

EXAMPLE 6 **Combining Addition and Subtraction of Signed Numbers**

According to the last step in the order of operations, perform addition and subtraction from left to right.

(a) $-6 + (-11) - \quad 5$

$\quad\quad -17 \quad - \quad 5$ Change subtraction to addition;
change positive 5 to its opposite, (-5).

$\quad\quad -17 \quad + \quad (-5)$

$\quad\quad\quad\quad -22$

(b) $4 - (-3) + (-9)$ Change subtraction to addition;
change -3 to its opposite, $(+3)$.

$\quad 4 + (+3) + (-9)$

$\quad\quad 7 \quad + (-9)$

$\quad\quad\quad -2$

> **Work Problem 8 at the Side.** ▶▶▶

EXAMPLE 7 **Using the Order of Operations to Combine More Than Two Numbers**

Find each sum.

(a) $-7 + 12 + (-3)$

$\quad\quad 5 \quad + (-3)$

$\quad\quad\quad 2$

(b) $14 + (-9) - (-8) + 10$

$\quad\quad 5 \quad - (-8) + 10$ Change subtraction to addition;
change -8 to its opposite, $(+8)$.

$\quad\quad 5 \quad + (+8) + 10$

$\quad\quad\quad 13 \quad + 10$

$\quad\quad\quad\quad 23$

(c) $\quad -6.3$
$\quad\quad -14.9$
$\quad\quad\quad 8.5$
$\quad\quad -7.4$
$\quad\quad\quad 5.2$

Start at the top.

$\begin{array}{rrrr} -6.3 \\ -14.9 \end{array}\Big\} \;\; -21.2 \;\Big\}$
$\quad\quad 8.5 \quad\quad\quad 8.5 \;\Big\} \;\; -12.7 \;\Big\}$
$\quad -7.4 \quad\quad -7.4 \quad\quad -7.4 \;\Big\} \;\; -20.1$
$\quad\quad 5.2 \quad\quad\quad 5.2 \quad\quad\quad 5.2 \quad\quad\quad 5.2$
$\quad\quad\quad\quad\quad\quad\quad\quad\quad\quad\quad\quad\quad\quad\quad -14.9$

> **Work Problem 9 at the Side.** ▶▶▶

8 Perform the addition and subtraction from left to right.

(a) $6 - 7 + (-3)$

(b) $-2 + (-3) - (-5)$

(c) $-3 - (-9) - (-5)$

(d) $8 - (-2) + (-6)$

9 Add.

(a) $-1 - 2 + 3 - 4$

(b) $7 - 6 - 5 + (-4)$

(c) $-6 + (-15) - (-19) + (-25)$

(d) $\quad -19.2$
$\quad\quad -6.7$
$\quad\quad 15.8$
$\quad\quad 17.1$
$\quad\quad -5.4$

ANSWERS
8. (a) -4 (b) 0 (c) 11 (d) 4
9. (a) -4 (b) -8 (c) -27 (d) 1.6

Real-Data Applications

Auto Aide, Part 2

A computer failure results in the loss of data for one day at the *Auto Aide Service* in Houston, Texas (see Auto Aide, Part 1 at the end of **Section 9.1**). The owner, the retired mathematics teacher, knows that the aides used integers to track their actions along their service routes. A schematic map of the I-10 corridor is shown below. Each tick mark represents one mile. The Home office is located at 0. Locations to the west are represented by negative integers, and locations to the east are represented by positive integers.

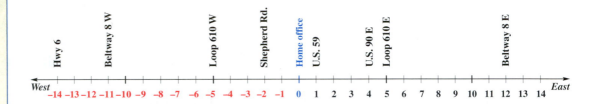

The owner asks each aide to re-create the route instructions for that day based on their written records. Each aide must translate the integer sentence into Route Instructions, by location. Also, the owner asks the aide to report the displacement (location relative to the home office at the end of the route) and the total distance traveled.

1. Aide Anne is given as an example. Complete the table for the remaining aides.

Aide	Integer Sentence	Route Instructions (start from the home office)	Displacement (end location)	Total Distance
Anne	$12 + (-11) + (-6) + 3$	Beltway 8 E; to U.S. 59; to Loop 610 W; to Shepherd Rd.	-2	32 miles
Bill	$-5 + 6 + 4 + (-16)$			
Carlos	$1 + 11 + (-8) + 1 + (-4)$			
Dylan	$-11 + (-3) + 15 + (-3) + 2$			
Ellen	$4 + 1 + (-7) + (-9) + 6$			

2. Suppose x is the aide's displacement from the home office in miles. List the aides whose journeys met the conditions of the following absolute value statements.
   (a) $|x| < 5$
   (b) $|x| = 1$
   (c) $|x| \leq 5$
   (d) $|x| > 3$

9.2 Exercises

FOR EXTRA HELP   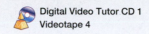 MathXL   Digital Video Tutor CD 1 Videotape 4   Student's Solutions Manual   MyMathLab    Interactmath.com

Addison-Wesley Math Tutor Center

Add by using the number line. See Example 1.

1. $-2 + 5$

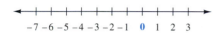

3. $-5 + (-2)$

5. $3 + (-4)$

2. $-3 + 4$

4. $-2 + (-2)$

6. $5 + (-1)$

Add. See Examples 2 and 3.

7. $-8 + 5$ **8.** $-3 + 2$ **9.** $-1 + 8$ **10.** $-4 + 10$

11. $-2 + (-5)$ **12.** $-7 + (-3)$ **13.** $6 + (-5)$ **14.** $11 + (-3)$

15. $4 + (-12)$ **16.** $9 + (-10)$ **17.** $-10 + (-10)$ **18.** $-5 + (-20)$

Write and solve an addition problem for each situation.

19. The football team gained 13 yd on the first play and lost 17 yd on the second play.

20. At penguin breeding grounds on Antarctic islands, winter temperatures routinely drop to -15 °C; but at the interior of the continent, temperatures may easily drop another 60° below that.

21. Nicole's checking account was overdrawn by $52.50. She deposited $50 in the account.

22. $48.40 was stolen from Jay's car. He got $30 of it back.

23. Use the score sheet to find each player's point total after three rounds in a card game.

	Jeff	Terry
Round 1	Lost 20 pts	Won 42 pts
Round 2	Won 75 pts	Lost 15 pts
Round 3	Lost 55 pts	Won 20 pts

24. Use the information in the table on flood water depths to find the new flood level for each river.

	Red River	Mississippi
Monday	Rose 8 ft	Rose 4 ft
Tuesday	Fell 3 ft	Rose 7 ft
Wednesday	Fell 5 ft	Fell 13 ft

Add.

25. $7.8 + (-14.6)$

26. $4.9 + (-8.1)$

27. $-\dfrac{1}{2} + \dfrac{3}{4}$

28. $-\dfrac{2}{3} + \dfrac{5}{6}$

29. $-\dfrac{7}{10} + \dfrac{2}{5}$

30. $-\dfrac{3}{4} + \dfrac{3}{8}$

31. $-\dfrac{7}{3} + \left(-\dfrac{5}{9}\right)$

32. $-\dfrac{8}{5} + \left(-\dfrac{3}{10}\right)$

Give the additive inverse of each number. See Example 4.

33. 3

34. 4

35. -9

36. -14

37. $\dfrac{1}{2}$

38. $\dfrac{7}{8}$

39. -6.2

40. -0.5

Subtract by changing subtraction to addition. See Example 5.

41. $19 - 5$

42. $24 - 11$

43. $10 - 12$

44. $1 - 8$

45. $7 - 19$

46. $2 - 17$

47. $-15 - 10$

48. $-10 - 4$

49. $-9 - 14$

50. $-3 - 11$

51. $-3 - (-8)$

52. $-1 - (-4)$

53. $6 - (-14)$

54. $8 - (-1)$

55. $1 - (-10)$

56. $6 - (-1)$

57. $-30 - 30$

58. $-25 - 25$

59. $-16 - (-16)$

60. $-20 - (-20)$

61. $-\dfrac{7}{10} - \dfrac{4}{5}$

62. $-\dfrac{8}{15} - \dfrac{3}{10}$

63. $\dfrac{1}{2} - \dfrac{9}{10}$

64. $\dfrac{2}{3} - \dfrac{11}{12}$

65. $-8.3 - (-9)$

66. $-2 - (-3.9)$

67. Explain and correct the mistakes in these subtraction problems.

(a) $-6 - 6$

$$-6 + 6 = 0$$

(b) $-9 - 5$

$$9 + (-5) = 4$$

68. Explain the purpose of the "−" sign in each of these examples.

(a) $6 - 9$ (b) (-9) (c) $-(-2)$

The windchill table below shows how wind increases a person's heat loss in cold weather. For example, suppose the temperature is 15 °F (along the top of the table) and the wind speed is 20 mph (along the left side of the table). This column and row intersect at −2 °F, the "windchill" temperature. The actual temperature is 15 °F but the wind makes it feel like −2 °F. Use the table to find the windchill temperature under each set of conditions in Exercises 69–72. Then write and solve a subtraction problem to calculate actual temperature minus windchill temperature.

WINDCHILL
Temperature (degrees Fahrenheit)

Calm	40	35	30	25	20	15	10	5	0	−5	−10	−15	−20	−25	−30
5	36	31	25	19	13	7	1	−5	−11	−16	−22	−28	−34	−40	−46
10	34	27	21	15	9	3	−4	−10	−16	−22	−28	−35	−41	−47	−53
15	32	25	19	13	6	0	−7	−13	−19	−26	−32	−39	−45	−51	−58
20	30	24	17	11	4	−2	−9	−15	−22	−29	−35	−42	−48	−55	−61
25	29	23	16	9	3	−4	−11	−17	−24	−31	−37	−44	−51	−58	−64
30	28	22	15	8	1	−5	−12	−19	−26	−33	−39	−46	−53	−60	−67
35	28	21	14	7	0	−7	−14	−21	−27	−34	−41	−48	−55	−62	−69
40	27	20	13	6	−1	−8	−15	−22	−29	−36	−43	−50	−57	−64	−71

Wind Speed (miles per hour)

Shaded area: Frostbite occurs in 15 minutes or less.

Source: National Weather Service

69. (a) 30 °F; 10 mph wind

    (b) 15 °F; 15 mph wind

70. (a) 40 °F; 20 mph wind

    (b) 20 °F; 35 mph wind

71. (a) 5 °F; 25 mph wind

    (b) −10 °F; 35 mph wind

72. (a) 10 °F; 15 mph wind

    (b) −5 °F; 30 mph wind

Follow the order of operations to work each problem. See Examples 6 and 7.

73. $-2 + (-11) - (-3)$

74. $-5 - (-2) + (-6)$

75. $4 - (-13) + (-5)$

76. $6 - (-1) + (-10)$

77. $-12 - (-3) - (-2)$

78. $-1 - (-7) - (-4)$

79. $4 - (-4) - 3$

80. $5 - (-2) - 8$

81. $\dfrac{1}{2} - \dfrac{2}{3} + \left(-\dfrac{5}{6}\right)$

82. $\dfrac{2}{5} - \dfrac{7}{10} + \left(-\dfrac{3}{2}\right)$

83. $-5.7 - (-9.4) - 8.1$

84. $-6.5 - (-11.2) - 1.4$

85. $-2 + (-11) + |-2|$

86. $|-7 + 2| + (-2) + 4$

   (*Hint:* Work inside the absolute value bars first.)

87. $-3 - (-2 + 4) + (-5)$

(*Hint:* Work inside parentheses first.)

88. $5 - 8 - (6 - 7) + 1$

Find the balance in each checking account.

89. LaVerle had $37 in her checking account. She wrote a $689 check for tuition, deposited a $908 paycheck, and got $60 in cash at an ATM machine.

90. Yvonne had $478 in her checking account. She deposited a $212 tax refund and wrote an $89 check for electricity and a $605 check for rent.

91. Rod's account was overdrawn by $89.62. He wrote checks for $110.70 and $99.68 before depositing three $100 bills into his account.

92. Dwayne's checking account was overdrawn by $23.77, so the bank charged a $16 fee. He deposited his $583.29 paycheck before withdrawing $50 in cash.

RELATING CONCEPTS (EXERCISES 93–96) For Individual or Group Work

*Use your knowledge of addition and subtraction of signed numbers to **work Exercises 93–96 in order.***

93. Work each pair of examples.

   (a) $-5 + 3 \ = \ \underline{\quad}$ $3 + (-5) = \underline{\quad}$

   (b) $-2 + (-6) = \underline{\quad}$ $-6 + (-2) = \underline{\quad}$

   (c) $17 + (-7) = \underline{\quad}$ $-7 + 17 \ = \ \underline{\quad}$

   Explain why the answers to each pair are the same.

94. Work each pair of examples

   (a) $-3 - 5 \ = \ \underline{\quad}$ $5 - (-3) = \underline{\quad}$

   (b) $-4 - (-6) = \underline{\quad}$ $-6 - (-4) = \underline{\quad}$

   (c) $3 - 10 \ = \ \underline{\quad}$ $10 - 3 \ = \ \underline{\quad}$

   Explain what happens when you try to apply the commutative property to subtraction.

95. Look back at your answers in Exercises 94.

   (a) Describe how the two answers for each pair are similar, and how they are different.

   (b) Write a rule that explains what happens when you switch the order of the numbers in a subtraction problem.

96. Work each set of exercises to see how 0 functions in addition and subtraction.

   (a) Describe the pattern in these addition answers.

   $-18 + 0 \ = \ \underline{\quad}$ $20 + 0 = \underline{\quad}$

   $0 + (-5) = \underline{\quad}$ $0 + 4 = \underline{\quad}$

   (b) Describe the pattern in these subtraction answers.

   $2 - 0 = \underline{\quad}$ $0 - 10 \ = \ \underline{\quad}$

   $-3 - 0 = \underline{\quad}$ $0 - (-7) = \underline{\quad}$

9.3 Multiplying and Dividing Signed Numbers

In mathematics, the rules or patterns must be consistent. We can use this idea to see how to multiply two numbers with *different* signs. Look for a pattern in this list of products.

$$4 \cdot 2 = 8$$

Blue numbers decrease by 1 $\quad 3 \cdot 2 = 6 \quad$ Red numbers decrease by 2

$$2 \cdot 2 = 4$$
$$1 \cdot 2 = 2$$
$$0 \cdot 2 = 0$$
$$-1 \cdot 2 = ?$$

As the numbers in blue decrease by 1, the numbers in red decrease by 2. To continue the red pattern, replace **?** with a number 2 *less than* 0, which is **−2**. Therefore,

$$-1 \cdot 2 = -2$$

OBJECTIVE 1 Multiply or divide two numbers with opposite signs. The pattern above suggests a rule for multiplying two numbers with different signs.

Multiplying Two Numbers with Different Signs

The product of two numbers with *different* signs is *negative*.

EXAMPLE 1 Multiplying Numbers with Different Signs

Multiply.

(a) $-8 \cdot 4 = -32$ Factors have *different* signs, so the product is *negative*.

Negative   Positive

(b) $6(-3) = -18$ Factors have *different* signs, so the product is *negative*.

Positive   Negative

(c) $(-5)(11) = -55$

(d) $12(-7) = -84$

Work Problem 1 at the Side. ▶▶▶

For two numbers with the *same* sign, look at this pattern.

$$4 \cdot (-2) = -8$$

Blue numbers decrease by 1 $\quad 3 \cdot (-2) = -6 \quad$ Red numbers increase by 2

$$2 \cdot (-2) = -4$$
$$1 \cdot (-2) = -2$$
$$0 \cdot (-2) = 0$$
$$-1 \cdot (-2) = ?$$

This time, as the numbers in blue decrease by 1, the products *increase* by 2. To continue the red pattern, replace **?** with a number 2 *greater than* 0, which is **positive 2**. Therefore,

$$-1 \cdot (-2) = 2$$

OBJECTIVE 2 Multiply or divide two numbers with the same sign. In the pattern above, a negative number times a negative number gave a positive result.

1 Multiply.

(a) $5 \cdot (-4)$

(b) $(-9)(15)$

(c) $12(-1)$

(d) $-6(6)$

(e) $\left(-\dfrac{7}{8}\right)\left(\dfrac{4}{3}\right)$

2 Multiply.

(a) $(-5)(-5)$

(b) $(-14)(-1)$

(c) $-7(-8)$

(d) $3(12)$

(e) $\left(-\dfrac{2}{3}\right)\left(-\dfrac{6}{5}\right)$

3 Divide.

(a) $\dfrac{-20}{4}$

(b) $\dfrac{-50}{-5}$

(c) $\dfrac{44}{2}$

(d) $\dfrac{6}{-6}$

(e) $\dfrac{-15}{-1}$

(f) $\dfrac{-\dfrac{3}{5}}{\dfrac{9}{10}}$

(g) $\dfrac{-35}{0}$

> **Multiplying Two Numbers with the Same Sign**
>
> The product of two numbers with the *same* sign is *positive*.

EXAMPLE 2 Multiplying Two Numbers with the Same Sign

Multiply.

(a) $(-9)(-2)$ The factors have the same sign (both are negative).

     $(-9)(-2) = 18 \leftarrow$ The product is positive.

(b) $-7(-4) = 28$ (c) $(-6)(-2) = 12$

(d) $(-10)(-5) = 50$ (e) $7(5) = 35$

> **Work Problem 2 at the Side.**

You can use the same rules for dividing signed numbers as you use for multiplying signed numbers.

> **Dividing Signed Numbers**
>
> When two nonzero numbers with *different* signs are divided, the result is *negative*.
>
> When two nonzero numbers with the *same* sign are divided, the result is *positive*.

Division involving 0 works the same as it did for whole numbers (see **Section 1.5**). Division by 0 cannot be done; we say it is *undefined*. But 0 divided by any other number is 0.

EXAMPLE 3 Dividing Signed Numbers

Divide.

(a) $\dfrac{-15}{5}$ Numbers have *different* signs, so the quotient is negative.

     $\dfrac{-15}{5} = -3$

(b) $\dfrac{-8}{-4}$ Numbers have the *same* sign (both negative), so the quotient is positive.

     $\dfrac{-8}{-4} = 2$

(c) $\dfrac{-75}{-25} = 3$

(d) $\dfrac{-6}{0}$ is undefined.   Division by 0 cannot be done.

(e) $\dfrac{0}{-5} = 0$

(f) $\dfrac{90}{-9} = -10$

(g) $\dfrac{-\dfrac{2}{3}}{\dfrac{5}{9}} = \left(-\dfrac{2}{3}\right)\left(-\dfrac{9}{5}\right)$ Use the reciprocal of $-\dfrac{5}{9}$, which is $-\dfrac{9}{5}$

$= \left(-\dfrac{2}{\overset{}{\underset{1}{3}}}\right)\left(-\dfrac{\overset{3}{9}}{5}\right)$ Divide out common factors. Then multiply.

$= \dfrac{6}{5}$ Both numbers were negative, so the quotient is positive.

> **Work Problem 3 at the Side.**

9.3 Exercises

 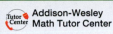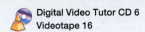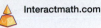
Multiply. See Examples 1 and 2.

1. $-5 \cdot 7$

2. $-10 \cdot 2$

3. $(-5)(9)$

4. $(-9)(4)$

5. $3(-6)$

6. $8(-6)$

7. $10 \cdot (-5)$

8. $5 \cdot (-11)$

9. $(-1)(40)$

10. $(75)(-1)$

11. $-8(-4)$

12. $-3(-9)$

13. $11(7)$

14. $4(25)$

15. $-19 \cdot (-7)$

16. $-21 \cdot (-3)$

17. $-13(-1)$

18. $-1(-31)$

19. $(0)(-25)$

20. $(-50)(0)$

21. $-\dfrac{1}{2} \cdot (-8)$

22. $\dfrac{1}{3} \cdot (-15)$

23. $-10\left(\dfrac{2}{5}\right)$

24. $-25\left(-\dfrac{7}{10}\right)$

25. $\left(\dfrac{3}{5}\right)\left(-\dfrac{1}{6}\right)$

26. $\left(-\dfrac{7}{9}\right)\left(-\dfrac{3}{4}\right)$

27. $-\dfrac{7}{5}\left(-\dfrac{10}{3}\right)$

28. $-\dfrac{9}{10}\left(\dfrac{5}{4}\right)$

29. $-\dfrac{7}{15} \cdot \dfrac{25}{14}$

30. $-\dfrac{5}{9} \cdot \dfrac{18}{25}$

31. $-\dfrac{5}{2}\left(-\dfrac{7}{10}\right)$

32. $-\dfrac{8}{5}\left(-\dfrac{15}{16}\right)$

33. $9(-4.7)$

34. $15(-6.3)$

35. $(-0.5)(-12)$

36. $(-3.15)(-5)$

37. $(-6.2)(5.1)$

38. $(-4.3)(9.7)$

39. $-1.25(-3.6)$

40. $6.33(0.2)$

41. $(-8.23)(-1)$

42. $(-1)(-0.69)$

43. $0(-58.6)$

44. $-91.3(0)$

Divide. See Example 3.

45. $\dfrac{-14}{7}$

46. $\dfrac{-8}{2}$

47. $\dfrac{30}{-6}$

48. $\dfrac{21}{-7}$

49. $\dfrac{-28}{0}$

50. $\dfrac{-40}{0}$

51. $\dfrac{14}{-1}$

52. $\dfrac{25}{-1}$

53. $\dfrac{-20}{-2}$

54. $\dfrac{-80}{-4}$

55. $\dfrac{-48}{-12}$

56. $\dfrac{-30}{-15}$

57. $\dfrac{-18}{18}$

58. $\dfrac{50}{-50}$

59. $\dfrac{-573}{-3}$

60. $\dfrac{-580}{-5}$

61. $\dfrac{0}{-9}$

62. $\dfrac{0}{-4}$

63. $\dfrac{-30}{-30}$

64. $\dfrac{-25}{-25}$

65. $\dfrac{-\dfrac{5}{7}}{-\dfrac{15}{14}}$

66. $\dfrac{-\dfrac{3}{4}}{-\dfrac{9}{16}}$

67. $-\dfrac{2}{3} \div (-2)$

68. $-\dfrac{3}{4} \div (-9)$

69. $5 \div \left(-\dfrac{5}{8}\right)$

70. $7 \div \left(-\dfrac{14}{15}\right)$

71. $-\dfrac{7}{5} \div \dfrac{3}{10}$

72. $-\dfrac{4}{9} \div \dfrac{8}{3}$

73. $\dfrac{-18.92}{-4}$

74. $\dfrac{-22.75}{-7}$

75. $\dfrac{-7.05}{1.5}$

76. $\dfrac{-17.02}{7.4}$

77. $\dfrac{45.58}{-8.6}$

78. $\dfrac{6.27}{-0.3}$

Following the order of operations, work from left to right in each exercise.

79. $(-4)(-6)\dfrac{1}{2}$

80. $(-9)(-3)\dfrac{2}{3}$

81. $(-0.6)(-0.2)(-3)$

82. $(-4)(-1.2)(-0.7)$

83. $\left(-\dfrac{1}{2}\right)\left(\dfrac{2}{5}\right)\left(\dfrac{7}{8}\right)$

84. $\left(\dfrac{3}{4}\right)\left(-\dfrac{5}{6}\right)\left(\dfrac{2}{3}\right)$

85. $-36 \div (-2) \div (-3) \div (-3) \div (-1)$

86. $-48 \div (-8) \cdot (-4) \div (-4) \div (-3)$

87. $|-8| \div (-4) \cdot |-5|$

88. $-6 \cdot |-3| \div |9| \cdot (-2)$

Solve each application problem. Be sure to indicate whether the answer is a positive or negative number.

89. A new computer-software store had losses of $9950 during each month of its first year. What was the total loss for the year?

90. A college's enrollment dropped by 3245 students over the last 11 years. What was the average drop in enrollment each year?

91. The greatest ocean depth is 36,198 ft below sea level. If an unmanned research sub dives to that depth in 18 equal steps, how far does it dive in each step?

92. Pat ate a dozen crackers as a snack. Each cracker had 17 calories. How many calories did Pat eat? (*Source:* Nabisco, Inc.)

93. Tuition at the state university is $182 per credit for undergraduates. How much tuition will Wei Chen pay for 13 credits?

94. A long-distance phone company estimates that it is losing 95 customers each week. How many customers will it lose in a year?

95. There is a 3-degree drop in temperature for every thousand feet that an airplane climbs into the sky. If the temperature on the ground is 50 degrees, what will be the temperature when the plane reaches an altitude of 24,000 ft? (*Source:* Lands' End.)

96. An unmanned submarine descends to 150 ft below the surface of the ocean. Then it continues to go deeper, taking a water sample every 25 ft. What is its depth when it takes the 15th sample?

RELATING CONCEPTS (EXERCISES 97–98)   For Individual or Group Work

Use your knowledge of multiplying and dividing signed numbers and look for patterns as you **work Exercises 97 and 98 in order.**

97. Write three numerical examples for each situation.

   (a) A positive number multiplied by -1

   (b) A negative number multiplied by -1

   Now write a rule that explains what happens when you multiply a signed number by -1.

98. Write three numerical examples for each situation.

   (a) A negative number divided by -1

   (b) A positive number divided by -1

   (c) A negative number divided by itself

   Now write a rule that explains what happens when you divide a signed number by -1. Write another rule for a negative number divided by itself.

9.4 Order of Operations

In the last two sections, you worked examples that mixed either addition and subtraction or multiplication and division. In those situations, you worked from left to right. Here are two more examples as a review.

Work additions and subtractions from left to right.

$$-8 - (-6) + (-11)$$
$$\underbrace{-2} + (-11)$$
$$-13$$

Work multiplications and divisions from left to right.

$$(-15) \div (-3)(6) \qquad \text{Divide.}$$
$$5(6) \qquad \text{Multiply.}$$
$$30$$

Work Problem 1 at the Side. ▶▶▶

OBJECTIVE 1 Use the order of operations. Before working examples that mix division with addition or include parentheses, let's review the order of operations from **Section 1.8.**

Order of Operations

1. Do all operations inside *parentheses* or *other grouping symbols.*

2. Simplify any expressions with *exponents* and find any *square roots.*

3. *Multiply* or *divide,* proceeding from left to right.

4. *Add* or *subtract,* proceeding from left to right.

EXAMPLE 1 Using the Order of Operations

Use the order of operations to simplify this expression.

$$4 - 10 \div 2 + 7 \qquad \text{Check for parentheses: none.}$$

Check for exponents and square roots: none.

Move from left to right, checking for multiplying and dividing.

Yes, here is dividing. Use the number on each side of the ÷ sign.

$$4 - \mathbf{10 \div 2} + 7 \qquad 10 \div 2 \text{ is 5. Bring down the other}$$
$$\downarrow\downarrow \quad \downarrow \quad \downarrow\downarrow \qquad \text{numbers and signs you haven't used.}$$
$$4 - \quad 5 \quad + 7 \qquad \text{Move from left to right, checking}$$
for adding and subtracting.

Yes, here is subtracting.

$$4 - \quad 5 \quad + 7 \qquad \text{Change subtraction to addition;}$$
$$\downarrow \qquad \downarrow \qquad \text{change 5 to its opposite, } (-5).$$
$$\mathbf{4 + (-5)} \quad + 7 \qquad \text{Add } 4 + (-5) \text{ to get } -1$$
$$\mathbf{-1} \quad + 7 \qquad \text{Add } -1 + 7 \text{ to get 6}$$
$$\mathbf{6}$$

Work Problem 2 at the Side. ▶▶▶

1 Simplify.

(a) $-9 + (-15) + (-3)$

(b) $-8 - (-2) + (-6)$

(c) $-2 - (-7) - (-4)$

(d) $3(-4) \div (-6)$

(e) $-18 \div 9(-4)$

2 Use the order of operations to simplify each expression.

(a) $10 + 8 \div 2$ Divide first!

(b) $4 - 6(-2)$ Multiply first!

(c) $-3 + (-5) \cdot 2 - 1$

(d) $-6 \div 2 + 3(-2)$

(e) $7 - 6(2) \div (-3)$

❸ Simplify.

(a) $2 + 40 \div (-5 + 3)$

(b) $5 - 3(2 + 4)$

(c) $(-24 \div 2) + (15 - 3)$

(d) $-3(2 - 8) - 5(4 - 3)$

(e) $3(3) - 10(3) \div 5$

(f) $6 - (2 + 7) \div (-4 + 1)$

ANSWERS
3. (a) -18 **(b)** -13 **(c)** 0
   (d) 13 **(e)** 3 **(f)** 9

EXAMPLE 2   Parentheses and the Order of Operations

Use the order of operations to simplify each expression.

(a) $-8(7 - 5) - 9$ — Work inside parentheses first.
Bring down the other numbers and signs you haven't used yet.

$-8(2) \quad - 9$ — Check for exponents and square roots: none.
Move from left to right, looking for multiplying and dividing. Here is multiplying.

$-8(2) \quad - 9$

$-16 \quad - 9$ — Move from left to right, doing any adding and subtracting. Change subtraction to addition.

$-16 \quad + (-9)$ — Add $-16 + (-9)$ to get -25

-25

(b) $3 + 2(6 - 8)(15 \div 3)$ — Work inside first set of parentheses; change $6 - 8$ to $6 + (-8)$ to get (-2).

$3 + 2(-2)(15 \div 3)$ — Work inside second set of parentheses; $15 \div 3$ is 5

$3 + 2(-2) \quad (5)$ — Multiply and divide from left to right. First multiply $2(-2)$ to get -4

$3 + (-4) \quad (5)$ — Then multiply $-4 \cdot 5$ to get -20

$3 + (-20)$ — Add last; $3 + (-20)$ gives -17

-17

◀◀◀ Work Problem 3 at the Side.

OBJECTIVE 2   Use the order of operations with exponents.   Recall from **Section 1.8** that 2^3 means 2 is used as a factor 3 times.

$$2^3 = 2 \cdot 2 \cdot 2 = 8$$

The small raised 3 in 2^3 is called an *exponent*. Exponents are also used with signed numbers. Here are three examples.

$$(-3)^2 = (-3) \cdot (-3) = 9$$

$$(-4)^3 = (-4) \cdot (-4) \cdot (-4)$$ — Be careful! Multiply two numbers at a time. Watch the signs.
$$= 16 \cdot (-4)$$
$$= -64$$ ← Negative product

$$\left(-\frac{1}{2}\right)^4 = \left(-\frac{1}{2}\right) \cdot \left(-\frac{1}{2}\right) \cdot \left(-\frac{1}{2}\right) \cdot \left(-\frac{1}{2}\right)$$
$$= \frac{1}{4} \cdot \left(-\frac{1}{2}\right) \cdot \left(-\frac{1}{2}\right)$$
$$= -\frac{1}{8} \cdot \left(-\frac{1}{2}\right)$$
$$= \frac{1}{16}$$

Be very careful with exponents and signed numbers. For example,

$$(-3)^2 = (-3) \cdot (-3) = 9 \leftarrow \text{Positive 9}$$

But the expression -3^2, with *no parentheses,* is different.

$$-3^2 = -(3 \cdot 3) = -9 \leftarrow \text{Negative 9}$$

CAUTION
$(-3)^2$ is *not* the same as -3^2.

$$(-3)^2 = (-3) \cdot (-3) = 9 \quad but \quad -3^2 = -(3 \cdot 3) = -9$$

You will need this information as you take more algebra classes.

EXAMPLE 3 Exponents and the Order of Operations

Simplify.

(a) $4^2 - (-3)^2$ There are parentheses around (-3) but no work can be done inside these parentheses.

Apply the exponents: $4^2 = 4 \cdot 4 = 16$ and $(-3)^2 = (-3) \cdot (-3) = 9$

$16 - 9$ Subtract: $16 - 9 = 7$

7

(b) $(-5)^2 - (4 - 6)^2(-3)$ Work inside parentheses first.

$(-5)^2 - (-2)^2(-3)$ Apply the exponents next.

$25 - 4(-3)$ Multiply.

$25 - (-12)$ Change subtraction to addition.

$25 + (+12)$ Add.

37

(c) $\left(\dfrac{2}{3} - \dfrac{1}{6}\right)^2 \div \left(-\dfrac{3}{8}\right)$ Inside parentheses: $\frac{2}{3} - \frac{1}{6} = \frac{4}{6} - \frac{1}{6} = \frac{3}{6} = \frac{1}{2}$

$\left(\dfrac{1}{2}\right)^2 \div \left(-\dfrac{3}{8}\right)$ Apply the exponent: $(\frac{1}{2})^2 = \frac{1}{2} \cdot \frac{1}{2} = \frac{1}{4}$

$\dfrac{1}{4} \div \left(-\dfrac{3}{8}\right)$ Divide by using the reciprocal of the divisor: $-\frac{3}{8}$ becomes $-\frac{8}{3}$

$\dfrac{1}{4} \cdot \left(-\dfrac{8}{3}\right)$ Divide out common factors, then multiply: $\frac{1}{4} \cdot -\frac{\overset{2}{\cancel{8}}}{3} = -\frac{2}{3}$

$-\dfrac{2}{3}$

Work Problem 4 at the Side. ▶▶▶

NOTE
Parentheses can be used in several different ways.

To indicate multiplication: $(4)(-3) = -12$

To separate a negative number from a minus sign: $8 - (-2) = 10$

To indicate which operation to do first: $35 + (6 - 2)$

$35 + 4$

39

4 Simplify.

(a) $2^3 - 3^2$

(b) $4^2 - 3^2(5 - 2)$

(c) $-18 \div (-3)(2^3)$

(d) $(-3)^3 + (3 - 8)^2$

(e) $\dfrac{3}{8} + \left(-\dfrac{1}{2}\right)^2 \div \dfrac{1}{4}$

ANSWERS
4. **(a)** -1 **(b)** -11 **(c)** 48 **(d)** -2
   (e) $\dfrac{11}{8}$

5 Simplify.

(a) $\dfrac{-3(2^3)}{-10 - 6 + 8}$

(b) $\dfrac{(-10)(-5)}{-6 \div 3(5)}$

(c) $\dfrac{6 + 18 \div (-2)}{(1 - 10) \div 3}$

(d) $\dfrac{6^2 - 3^2(4)}{5 + (3 - 7)^2}$

OBJECTIVE 3 Use the order of operations with fraction bars.
A fraction bar indicates division, as in $\frac{-6}{2}$, which means $-6 \div 2$. In an expression such as

$$\frac{-5 + 3^2}{16 - 7(2)}$$

the fraction bar also acts as a grouping symbol, like parentheses. It tells us to do the work in the numerator, and then the work in the denominator. The last step is to divide the results.

$$\frac{-5 + 3^2}{16 - 7(2)} \rightarrow \frac{-5 + 9}{16 - 14} \rightarrow \frac{4}{2} \rightarrow \text{Now divide.} \quad 4 \div 2 = 2$$

The final simplified result is 2.

EXAMPLE 4 Fraction Bars and the Order of Operations

Simplify.

$$\frac{-8 + 5(4 - 6)}{4 - 4^2 \div 8}$$

First do the work in the numerator.

$-8 + 5(4 - 6)$ Work inside parentheses.

$-8 + 5(-2)$ Multiply.

$-8 + (-10)$ Add.

Numerator $\rightarrow -18$

Now do the work in the denominator.

$4 - 4^2 \div 8$ No parentheses; apply the exponent.

$4 - 16 \div 8$ Divide.

$4 - 2$ Subtract.

Denominator $\rightarrow 2$

The last step is the division.

$$\frac{\text{Numerator} \rightarrow -18}{\text{Denominator} \rightarrow 2} = -9$$

Work Problem 5 at the Side.

ANSWERS

5. (a) $\dfrac{-24}{-8} = 3$ (b) $\dfrac{50}{-10} = -5$

   (c) $\dfrac{-3}{-3} = 1$ (d) $\dfrac{0}{21} = 0$

9.4 **Exercises**

FOR EXTRA HELP

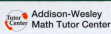 Addison-Wesley Math Tutor Center

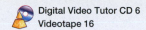 MathXL

 Digital Video Tutor CD 6 Videotape 16

Student's Solutions Manual

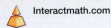 MyMathLab

Interactmath.com

Simplify. See Examples 1–3.

1. $6 + 3(-4)$

2. $10 - 30 \div 2$

3. $-1 + 15 + 2(-7)$

4. $9 + (-5) + 2(-2)$

5. $6^2 + 4^2$

6. $3^2 + 8^2$

7. $10 - 7^2$

8. $5 - 5^2$

9. $(-2)^5 + 2$

10. $(-2)^4 - 7$

11. $4^2 + 3^2 + (-8)$

12. $5^2 + 2^2 + (-12)$

13. $2 - (-5) + 3^2$

14. $6 - (-9) + 2^3$

15. $(-4)^2 + (-3)^2 + 5$

16. $(-5)^2 + (-6)^2 + 12$

17. $3 + 5(6 - 2)$

18. $4 + 3(8 - 3)$

19. $-7 + 6(8 - 14)$

20. $-3 + 5(9 - 12)$

21. $-6 + (-5)(9 - 14)$

22. $-5 + (-3)(6 - 7)$

23. $(-5)(7 - 13) \div (-10)$

24. $(-4)(9 - 17) \div (-8)$

25. $9 \div (-3)^2 + (-1)$

26. $-48 \div (-4)^2 + 3$

27. $2 - (-5)(-3)^2$

28. $1 - (-10)(-2)^3$

29. $(-2)(-7) + 3(9)$

30. $4(-2) + (-3)(-5)$

31. $30 \div (-5) - 36 \div (-9)$

32. $8 \div (-4) - 42 \div (-7)$

33. $2(5) - 3(4) + 5(3)$

34. $9(3) - 6(4) + 3(7)$

35. $4(3^2) + 7(3 + 9) - (-6)$

36. $5(4^2) - 6(1 + 4) - (-3)$

Simplify. See Example 4.

37. $\dfrac{-1 + 5^2 - (-3)}{-6 - 9 + 12}$

38. $\dfrac{-6 + 3^2 - (-7)}{7 - 9 - 3}$

39. $\dfrac{-2(4^2) - 4(6 - 2)}{-4(8 - 13) \div (-5)}$

40. $\dfrac{3(3^2) - 5(9 - 2)}{8(6 - 9) \div (-3)}$

41. $\dfrac{2^3(-2 - 5) + 4(-1)}{4 + 5(-6 \cdot 2) + (5 \cdot 11)}$

42. $\dfrac{3^3 + 4(-1 - 2) - 25}{-4 + 4(3 \cdot 5) + (-6 \cdot 9)}$

Simplify each expression.

43. $(-4)^2(7-9)^2 \div 2^3$

44. $(-5)^2(9-17)^2 \div (-10)^2$

45. $(-0.3)^2 + (-0.5)^2 + 0.9$

46. $(0.2)^3 - (-0.4)^2 + 3.02$

47. $(-0.75)(3.6-5)^2$

48. $(-0.3)(4-6.8)^2$

49. $(0.5)^2(-8) - (0.31)$

50. $(0.3)^3(-5) - (-2.8)$

51. $\dfrac{2}{3} \div \left(-\dfrac{5}{6}\right) - \dfrac{1}{2}$

52. $\dfrac{5}{8} \div \left(-\dfrac{10}{3}\right) - \dfrac{3}{4}$

53. $\left(-\dfrac{1}{2}\right)^2 - \left(\dfrac{3}{4} - \dfrac{7}{4}\right)$

54. $\left(-\dfrac{2}{3}\right)^2 - \left(\dfrac{1}{6} - \dfrac{11}{6}\right)$

55. $\dfrac{3}{5}\left(-\dfrac{7}{6}\right) - \left(\dfrac{1}{6} - \dfrac{5}{3}\right)$

56. $\dfrac{2}{7}\left(-\dfrac{14}{5}\right) - \left(\dfrac{4}{3} - \dfrac{13}{9}\right)$

57. $5^2(9-11)(-3)(-2)^3$

58. $4^2(13-17)(-2)(-3)^2$

59. $1.6(-0.8) \div (-0.32) \div 2^2$

60. $6.5(-4.8) \div (-0.3) \div (-2)^3$

61. Simplify.

$(-2)^2 =$ _____ $(-2)^6 =$ _____

$(-2)^3 =$ _____ $(-2)^7 =$ _____

$(-2)^4 =$ _____ $(-2)^8 =$ _____

$(-2)^5 =$ _____ $(-2)^9 =$ _____

(a) Describe the pattern you see in the sign of the answers.

(b) What would be the sign of $(-2)^{10}$? of $(-2)^{15}$? of $(-2)^{24}$?

62. Explain the difference between -5^2 and $(-5)^2$.

Simplify.

63. $\dfrac{-9 + 18 \div (-3)(-6)}{5 - 4(12) \div 3(2)}$

64. $\dfrac{-20 - 15(-4) - (-40)}{4 + 27 \div 3(-2) - 6}$

65. $-7\left(6 - \dfrac{5}{8} \cdot 24 + 3 \cdot \dfrac{8}{3}\right)$

66. $(-0.3)^2(-5)(3) + (6 \div 2)(0.4)$

67. $|-12| \div 4 + 2(3^2) \div 6$

68. $6 - (2 - 12) + 4^2 \div (-2)\left(\dfrac{5}{2}\right) + (2)^2$

Summary Exercises on Operations with Signed Numbers

Simplify each expression.

1. $2 - 8$

2. $(-16)(0)$

3. $-14 - (-7)$

4. $\dfrac{-42}{6}$

5. $-9(-7)$

6. $\dfrac{-12}{12}$

7. $(1)(-56)$

8. $1 + (-23)$

9. $5 - (-7)$

10. $-\dfrac{8}{3} \div \left(-\dfrac{4}{9}\right)$

11. $-18 + 5$

12. $\dfrac{0}{-10}$

13. $-40 - 40$

14. $-17 + 0$

15. $8(-6)$

16. $-\dfrac{1}{10} - \dfrac{9}{10}$

17. $\left(-\dfrac{5}{6}\right)\left(\dfrac{1}{5}\right)$

18. $\dfrac{30}{0}$

19. $0 - 14.6$

20. $\dfrac{1.8}{-3}$

21. $-4(-6)(2)$

22. $-2 + (-12) + (-5)$

23. $-60 \div 10 \div (-3)$

24. $-8 - 4 - 8$

25. $64(0) \div (-8)$

26. $2 - (-5) + 3^2$

27. $-9 + 8 + (-2)$

28. $(-6)(-2)(-3)$

29. $8 + 6 + (-8)$

30. $-72 \div (-9) \div (-4)$

31. $-7 + 28 + (-56) + 3$

32. $9 - 6 - 3 - 5$

33. $-6(-8) \div (-5 - 7)$

34. $-1(9732)(-1)(-1)$

35. $-80 \div 4(-5)$

36. $-10 - 4 + 0 + 18$

37. $-7 \cdot |7| \cdot |-7|$

38. $5 - |-3| + 3$

39. $-2(-3)(7) \div (-7)$

40. $-3 - (-2 + 4) - 5$

41. $0 - |-7 + 2|$

42. $(-4)^2 (7 - 9)^2 \div 2^3$

43. $12 \div 4 + 2(-2)^2 \div (-4)$

44. $\dfrac{-9 + 24 \div (-4)(-6)}{32 - 4(12) \div 3(2)}$

45. $\dfrac{5 - |2 - 4 \cdot 4| + (-5)^2 \div 5^2}{-9 \div 3(2 - 2) - (-8)}$

Solve each application problem. Be sure to indicate whether the final answer is a positive or negative number.

46. When Ashwini discovered that her checking account was overdrawn by $238, she quickly transferred $450 from her savings to her checking account. What is the balance in her checking account?

47. A discount store found that 174 items were lost to shoplifting last month. The total loss was $4176. What was the average loss on each item?

48. In a laboratory experiment, a mixture started at a temperature of −102 degrees. First the temperature was raised 37 degrees and then raised 52 degrees. What was the final temperature?

49. A plane ascended an average of 730 ft each minute during a 37-minute takeoff. How far did the plane ascend during the takeoff?

50. The Tigers gained a total of 148 yd during the first half of the football game. During the second half they lost 191 yd. How many yards did they gain or lose during the entire game?

51. A scuba diver was photographing fish at 65 ft below the surface of the lagoon. She swam up 24 ft and then swam down 49 ft. What was her final depth?

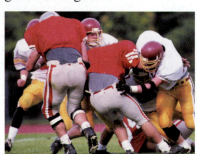

Elena Sanchez opened a shop that does alterations and designs custom clothing. Use the table of her income and expenses to answer Exercises 52–55.

Month	Income	Expenses	Profit or Loss
January	$2400	$3100	
February	$1900	$2000	
March	$2500	$1800	
April	$2300	$1400	
May	$1600	$1600	
June	$1900	$1200	

52. Complete the table by finding Elena's profit or loss for each month.

53. Which month had the greatest loss? Which month had the greatest profit?

54. What was Elena's average monthly income?

55. What was the average monthly amount of expenses?

56. Explain in your own words how to add two numbers with different signs. Include two examples in your explanation, one that has a positive answer and one that has a negative answer.

57. Explain what is different and what is similar between multiplying and dividing signed numbers.

9.5 Evaluating Expressions and Formulas

In formulas, you have seen that numbers can be represented by letters. For example, you used this formula for finding simple interest in **Section 6.7.**

$$I = p \cdot r \cdot t$$

In this formula, p (principal) represents the amount of money borrowed, r is the rate of interest, and t is the time in years. In algebra, we often write multiplication without the multiplication dots. If there is no operation sign written between two letters, or between a letter and a number, you assume it is multiplication.

OBJECTIVES

1 Define variable and expression.

2 Find the value of an expression when values of the variables are given.

Showing Multiplication in Algebra

If there is no operation sign, it is understood to be multiplication. Here are some examples.

$I = p \cdot r \cdot t$	is written	$I = prt$
$2 \cdot r$	is written	$2r$
$3 \cdot x + 4 \cdot y$	is written	$3x + 4y$

OBJECTIVE 1 **Define variable and expression.** Letters (such as the I, p, r, and t used above) that represent numbers are called **variables.** A combination of operations on letters and numbers is an **expression.** Three examples of expressions are shown here.

$$9 + p \qquad 8r \qquad 7k - 2m$$

OBJECTIVE 2 **Find the value of an expression when values of the variables are given.** The value of an expression changes depending on the value of each variable. To find the value of an expression, replace the variables with their values. It is helpful to write each value inside parentheses when multiplication is involved.

1 Find the value of $5x - 3y$, if:

(a) $x = 5$, $y = 2$

(b) $x = -3$, $y = 4$

(c) $x = 0$, $y = 6$

EXAMPLE 1 Finding the Value of an Expression

Find the value of $5x - 3y$, if $x = 2$ and $y = 7$.

Replace x with 2. Replace y with 7.

Using the order of operations, do all multiplications first. Then subtract.

$$5x - 3y$$
$$5(2) - 3(7)$$
$$10 - 21$$
$$-11$$

Work Problem 1 at the Side.

EXAMPLE 2 Finding the Value of an Expression

What is the value of $\dfrac{6k + 2r}{5s}$, if $k = -2$, $r = 5$, and $s = -1$?

Continued on Next Page

ANSWERS
1. **(a)** 19 **(b)** -27 **(c)** -18

② Find the value of $\dfrac{3k + r}{2s}$ if:

(a) $k = 1$, $r = 1$, $s = 2$

(b) $k = 8$, $r = -2$, $s = -4$

(c) $k = -3$, $r = 1$, $s = -2$

③ Find the value of each expression.

(a) $-x - 6y$, if $x = -1$ and $y = -4$

(b) $-4a - b$, if $a = 4$ and $b = -2$

(c) $-w - 3x - y$, if $w = 10$, $x = -5$, and $y = -6$

④ Using the given values, evaluate each formula.

(a) $A = \dfrac{1}{2}bh$
$b = 6$ yd, $h = 12$ yd

(b) $P = 2l + 2w$
$l = 10$ cm, $w = 8$ cm

(c) $C = 2\pi r$
$\pi \approx 3.14$, $r = 6$ ft

ANSWERS

2. (a) $\dfrac{4}{4} = 1$ (b) $\dfrac{22}{-8} = -\dfrac{11}{4}$ (c) $\dfrac{-8}{-4} = 2$

3. (a) 25 (b) -14 (c) 11

4. (a) $A = 36$ yd^2 (b) $P = 36$ cm
(c) $C \approx 37.68$ ft

Replace k with -2, r with 5, and s with -1.

$$\dfrac{6k + 2r}{5s} = \dfrac{6(-2) + 2(5)}{5(-1)} \qquad \text{Do all the multiplications first.}$$

$$= \dfrac{-12 + 10}{-5} \qquad \text{Add in the numerator.}$$

$$= \dfrac{-2}{-5} \qquad \text{Dividing two numbers with the same sign gives a positive answer.}$$

$$= \dfrac{2}{5}$$

Work Problem 2 at the Side.

The next example shows how to evaluate an expression with negative signs and/or subtraction, when the value of the variable is also negative.

EXAMPLE 3 Evaluating an Expression with Negative Signs and Negative Values

Find the value of $-c - 5b$ when $c = -2$ and $b = -3$.

Replace c with -2.
Replace b with -3.

$-(-2)$ is the opposite of (-2), which is $(+2)$.

$$-(-2) - 5(-3) \qquad \text{Multiply } 5 \cdot (-3).$$
$$+2 \;-\; (-15) \qquad \text{Change subtraction to addition; change } (-15)$$
$$2 \;+\; (+15) \qquad \text{to its opposite, } (+15).$$
$$17$$

CAUTION
Watch the signs carefully when there are negative signs in the expression and the value of a variable is also negative, as in Example 3 above.

Work Problem 3 at the Side.

EXAMPLE 4 Evaluating a Formula

The formula you used in **Chapter 8** for the area of a triangle can now be written without the multiplication dots.

$$A = \dfrac{1}{2} \cdot b \cdot h \quad \text{is written} \quad A = \dfrac{1}{2}bh$$

In this formula, b is the length of the base of the triangle and h is the height of the triangle. What is the area if $b = 9$ cm and $h = 24$ cm?

$$A = \dfrac{1}{2} \; b \; h$$

$$A = \dfrac{1}{2}(9\text{ cm})(24\text{ cm}) \qquad \text{Replace } b \text{ with 9 cm and } h \text{ with 24 cm.}$$

$$A = \dfrac{1}{\overset{1}{\cancel{2}}}(9\text{ cm})(\overset{12}{\cancel{24}}\text{ cm}) \qquad \text{Divide out any common factors.}$$

$$A = 108\text{ cm}^2 \quad \longleftarrow \quad \text{Recall that cm}^2 \text{ means } square\ centimeters.$$

The area of the triangle is 108 cm^2.

Work Problem 4 at the Side.

9.5 Exercises

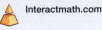

Find the value of the expression $2r + 4s$ for each set of values of r and s. See Example 1.

1. $r = 2, \quad s = 6$

2. $r = 6, \quad s = 1$

3. $r = 1, \quad s = -3$

4. $r = 7, \quad s = -2$

5. $r = -4, \quad s = 4$

6. $r = -3, \quad s = 5$

7. $r = -1, \quad s = -7$

8. $r = -3, \quad s = -5$

9. $r = 0, \quad s = -2$

10. $r = -7, \quad s = 0$

Use the given values of the variables to find the value of each expression. See Examples 1 and 2.

11. $8x - y$
$x = 1, \quad y = 8$

12. $a - 5b$
$a = 10, \quad b = 2$

13. $6k + 2s$
$k = 1, \quad s = -2$

14. $7p + 7q$
$p = -4, \quad q = 1$

15. $\dfrac{-m + 5n}{2s + 2}$
$m = 4, \quad n = -8, \quad s = 0$

16. $\dfrac{2y - z}{x - 2}$
$y = 0, \quad z = 5, \quad x = 1$

17. $-m - 3n$
$m = \dfrac{1}{2}, \quad n = \dfrac{3}{8}$

18. $7k - 3r$
$k = \dfrac{2}{3}, \quad r = \dfrac{1}{3}$

Be careful when an expression has a negative sign and the value of the variable is also negative.
Use the given values to find the value of each expression. See Example 3.

19. $-c - 5b$
$c = -8, \quad b = -4$

20. $-c - 5b$
$c = -1, \quad b = -2$

21. $-4x - y$
$x = 5, \quad y = -15$

22. $-4x - y$
$x = 3, \quad y = -8$

23. $-k - m - 8n$
$k = 6, \quad m = -9, \quad n = 0$

24. $-k - m - 8n$
$k = 0, \quad m = -7, \quad n = -1$

25. $\dfrac{-3s - t - 4}{-s + 6 + t}$
$s = -1, \quad t = -13$

26. $\dfrac{-3s - t - 4}{-s - 20 - t}$
$s = -3, \quad t = -6$

Using the given values, evaluate each formula. See Example 4.

27. $P = 4s; \quad s = 7.5$

28. $P = 4s; \quad s = 0.8$

29. $P = 2l + 2w; \quad l = 9, \quad w = 5$

30. $P = 2l + 2w; \quad l = 12, \quad w = 2$

31. $A = \pi r^2$; $\pi \approx 3.14$, $r = 5$

32. $A = \pi r^2$; $\pi \approx 3.14$, $r = 10$

33. $A = \dfrac{1}{2}bh$; $b = 15$, $h = 3$

34. $A = \dfrac{1}{2}bh$; $b = 5$, $h = 11$

35. $V = \dfrac{1}{3}Bh$; $B = 30$, $h = 60$

36. $V = \dfrac{1}{3}Bh$; $B = 105$, $h = 5$

37. $d = rt$; $r = 53$, $t = 6$

38. $d = rt$; $r = 180$, $t = 5$

39. $C = 2\pi r$; $\pi \approx 3.14$, $r = 4$

40. $C = 2\pi r$; $\pi \approx 3.14$, $r = 18$

Solve each application problem.

41. The expression for finding the perimeter of a triangle with sides of equal length is $3s$, where s is the length of one side. Evaluate the expression when

    (a) the length of one side is 11 in.

    (b) the length of one side is 3 ft.

42. The expression for finding the perimeter of a pentagon with sides of equal length is $5s$, where s is the length of one side. Evaluate the expression when

    (a) the length of one side is 25 cm

    (b) the length of one side is 8 in.

43. The expression for figuring a student's average test score is $\dfrac{p}{t}$, where p is the total points earned on all the tests and t is the number of tests. Evaluate the expression when

    (a) 332 points were earned on 4 tests

    (b) there were 7 tests and 637 points were earned.

44. The expression for deciding how many buses are needed for a group trip is $\dfrac{p}{b}$, where p is the total number of people and b is the number of people that one bus will hold. Evaluate the expression when

    (a) 176 people are going on a trip and one bus holds 44 people

    (b) a bus holds 36 people and 72 people are going on a trip.

45. Find and correct the error made by the student who solved this example:

Find the value of $-x - 4y$ if $x = -3$ and $y = -1$.

$$-x - 4y$$
$$-3 - 4(-1)$$
$$-3 - (-4)$$
$$-3 + (+4)$$
$$1$$

After stating the error, rework the problem and write a sentence next to each step, explaining what is being done in that step.

46. Go back to **Chapter 8** and find each of the formulas listed below. Pick values for the variables and then find the value of A or V. Then pick different values for the variables and again find the value of A or V.

(a) Area of a trapezoid: pick values for h, B, and b.

(b) Volume of a rectangular solid: pick values for l, w, and h.

RELATING CONCEPTS (EXERCISES 47–52)   For Individual or Group Work

Work Exercises 47–52 in order. Use the given formula and values of the variables to find the value of the remaining variable. If you studied **Chapters 7 and 8,** write a sentence telling when you would use each formula.

47. $F = \dfrac{9C}{5} + 32;$ $C = -40$

48. $C = \dfrac{5(F - 32)}{9};$ $F = -4$

49. $V = \dfrac{4\pi r^3}{3};$ $\pi \approx 3.14,$ $r = 3$

50. $c^2 = a^2 + b^2;$ $a = 3,$ $b = 4$

51. $A = \dfrac{1}{2}h(b + B);$ $h = 7,$ $b = 4,$ $B = 12$

52. $V = \dfrac{\pi r^2 h}{3};$ $\pi \approx 3.14,$ $r = 6,$ $h = 10$

9.6 Solving Equations

An **equation** is a statement that says two expressions are equal. Examples of equations are shown here.

$$x + 1 = 9 \qquad 20 = 5k \qquad 6r - 1 = 17$$

The **equal sign** in an equation divides the equation into two parts, the *left side* and the *right side*. In $6r - 1 = 17$, the left side is $6r - 1$, and the right side is 17. The equal sign tells us that the two sides are equivalent.

$$6r - 1 = 17$$

Left side **=** Right side

You solve an equation by finding all numbers that can be substituted for the variable to make the equation true. These numbers are called **solutions** of the equation.

OBJECTIVE 1 Determine whether a number is a solution of an equation. To decide whether a number is a solution of an equation, substitute the number in the equation to see whether the result is true.

EXAMPLE 1 Determining Whether a Number Is a Solution of an Equation

Is 7 a solution of either one of these equations?

(a) $12 = x + 5$
  Replace x with 7.

$$12 = x + 5 \qquad \text{Replace } x \text{ with } 7$$
$$12 = 7 + 5$$
$$12 = 12 \qquad \text{True}$$

Because the statement is true, 7 is a solution of the equation $12 = x + 5$.

(b) $2y + 1 = 16$
  Replace y with 7.

$$2y + 1 = 16$$
$$2(7) + 1 = 16$$
$$14 + 1 = 16$$
$$15 = 16 \qquad \text{False}$$

The *false* statement shows that 7 is *not* a solution of $2y + 1 = 16$.

Work Problem 1 at the Side.

OBJECTIVE 2 Solve equations using the addition property of equations. If the equation $a = b$ is true, and if a number c is added to both a and b, the new equation is also true. This rule, called the **addition property of equations,** means that you can add the *same* number to *both* sides of an equation and still have a true equation.

Addition Property of Equations

If $a = b$, then $a + c = b + c$.

In other words, you may add the *same* number to *both* sides of an equation.

OBJECTIVES

1 Determine whether a number is a solution of an equation.

2 Solve equations using the addition property of equations.

3 Solve equations using the multiplication property of equations.

1 Decide whether the given number is a solution of the equation.

(a) $p + 1 = 8$; 7

(b) $30 = 5r$; 6

(c) $3k - 2 = 4$; 3

(d) $23 = 4y + 3$; 5

You can use the addition property to solve equations. The idea is to get the variable (the letter) by itself on one side of the equal sign and a number by itself on the other side.

> ### EXAMPLE 2 Solving Equations Using the Addition Property
>
> Solve each equation.
>
> **(a)** $k - 4 = 6$
>
> To get k by itself on the left side, add 4 to the left side, because $k - 4 + 4$ gives $k + 0$. You must then add 4 to the right side also.
>
> $$k - 4 = 6 \qquad \leftarrow \text{Original equation}$$
> $$k - 4 + 4 = 6 + 4 \qquad \text{Add 4 to both sides.}$$
> $$k + 0 = 10 \qquad \text{On the left side, } -4 + 4 \text{ is } 0$$
> $$k = 10 \qquad \text{On the left side, } k + 0 \text{ is } k.$$
>
> The solution is 10. To check the solution, replace k with 10 in the original equation.
>
> $$k - 4 = 6 \leftarrow \text{Original equation}$$
> $$10 - 4 = 6 \qquad \text{Replace } k \text{ with } 10$$
> $$6 = 6 \qquad \text{True, so 10 is the correct solution.}$$
>
> This result is true, so **10 is the solution.**

> **CAUTION**
> When checking the solution in Example 2(a) above, we ended up with $6 = 6$. Notice that 6 is *not* the solution. The solution is 10, the number used to replace k in the original equation.

> **(b)** $2 = z + 8$
>
> To get z by itself on the right side, add -8 to both sides.
>
> $$2 = z + 8 \qquad \leftarrow \text{Original equation}$$
> $$2 + (-8) = z + 8 + (-8) \qquad \text{Add } (-8) \text{ to both sides.}$$
> $$-6 = z + 0$$
> $$-6 = z \qquad \leftarrow \text{The solution is } -6$$

> **NOTE**
> Notice that we *added* -8 to both sides of the equation to get z by itself. We can accomplish the same thing by *subtracting 8* from both sides. Recall from **Section 9.2** that subtraction is defined in terms of addition. On the left side of the equation above, $2 - 8$ gives the same result as $2 + (-8)$.

Check the solution by replacing z with -6 in the original equation.

$$2 = z + 8 \qquad \leftarrow \text{Original equation}$$
$$2 = -6 + 8 \qquad \text{Replace } z \text{ with } -6$$
$$2 = 2 \qquad \text{True, so } -6 \text{ is the correct solution.}$$

The result is true, so -6 is the solution (*not* 2).

Continued on Next Page

Here is a summary of the rules you can use to solve equations using the addition property. In these rules, x is the variable and a and b represent numbers.

> ### Solving an Equation Using the Addition Property
> Solve $x - a = b$ by adding a to *both* sides.
> Solve $x + a = b$ by subtracting a from *both* sides.

Work Problem 2 at the Side. ▶▶▶

OBJECTIVE 3 **Solve equations using the multiplication property of equations.** As long as you do the *same* thing to *both* sides of an equation, it will still be a true equation. So far you have added or subtracted on both sides. Now we will multiply or divide on both sides.

> ### Multiplication Property of Equations
> If $a = b$ and c does not equal 0, then
> $$a \cdot c = b \cdot c \quad \text{and} \quad \frac{a}{c} = \frac{b}{c}.$$
> In other words, you may multiply or divide *both* sides of an equation by the *same* number. (The only exception is you cannot divide by 0.)

EXAMPLE 3 **Solving Equations Using the Multiplication Property**

Solve each equation.

(a) $9p = 63$

You want to get the variable, p, by itself on the left side. The expression $9p$ means $9 \cdot p$. To undo the multiplication and get p by itself, *divide* both sides by 9.

$$9p = 63$$

$$\frac{\overset{1}{\cancel{9}} \cdot p}{\underset{1}{\cancel{9}}} = \frac{63}{9} \qquad \text{Divide both sides by 9}$$

$$p = 7 \quad \leftarrow \text{The solution is 7}$$

Check:

$$9\boldsymbol{p} = 63 \quad \leftarrow \text{Original equation}$$

$$9 \cdot 7 = 63 \qquad \text{Replace } p \text{ with 7}$$

$$63 = 63 \qquad \text{True}$$

The result is true, so **7** is the solution (***not*** 63).

Continued on Next Page

---

2 • Solve each equation. Check each solution.

(a) $n - 5 = 8$

(b) $5 = r - 10$

(c) $3 = z + 1$

(d) $k + 9 = 0$

(e) $-2 = y + 9$

(f) $x - 2 = -6$

❸ Solve each equation. Check each solution.

(a) $2y = 14$

(b) $42 = 7p$

(c) $-8a = 32$

(d) $-3r = -15$

(e) $-60 = -6k$

(f) $10x = 0$

(b) $-4r = 24$

Divide *both* sides by -4 to get r by itself on the left side.

$$\frac{-\overset{1}{\cancel{4}} \cdot r}{-\underset{1}{\cancel{4}}} = \frac{24}{-4} \qquad \text{Divide both sides by } -4$$

$$r = -6 \; \leftarrow \text{The solution is } -6$$

Check:

$$-4r = 24 \; \leftarrow \text{Original equation}$$
$$-4(-6) = 24 \qquad \text{Replace } r \text{ with } -6$$
$$24 = 24 \qquad \text{True}$$

The result is true, so -6 is the solution (**not** 24).

(c) $-55 = -11m$

Divide *both* sides by -11 to get m by itself on the right side.

$$\frac{-55}{-11} = \frac{-\overset{1}{\cancel{11}} \cdot m}{-\underset{1}{\cancel{11}}}$$

$$5 = m$$

Check:

$$-55 = -11m \; \leftarrow \text{Original equation}$$
$$-55 = -11(5) \qquad \text{Replace } m \text{ with } 5$$
$$-55 = -55 \qquad \text{True}$$

The result is true, so **5** is the solution (**not** -55).

▶▶▶ **Work Problem 3 at the Side.**

EXAMPLE 4 Solving Equations Using the Multiplication Property

Solve each equation.

(a) $\dfrac{x}{2} = 9$

Replace $\dfrac{x}{2}$ with $\dfrac{1}{2}x$, because dividing x by 2 is the same as multiplying x by $\dfrac{1}{2}$. Then, to get x by itself, multiply each side by the reciprocal of $\dfrac{1}{2}$, which is $\dfrac{2}{1}$. (Recall from **Section 2.7** that when two numbers are reciprocals, their product is 1.)

$$\frac{1}{2}x = 9$$

$$\frac{\overset{1}{\cancel{2}}}{1} \cdot \frac{1}{\underset{1}{\cancel{2}}}x = 2 \cdot 9 \qquad \text{Multiply both sides by } \tfrac{2}{1} \text{ (which equals 2).}$$

$$1x = 18$$

$$x = 18 \; \leftarrow \text{The solution is 18}$$

Continued on Next Page

Check: $\dfrac{x}{2} = 9$ ← Original equation

$\dfrac{18}{2} = 9$ Replace x with 18

$9 = 9$ True, so 18 is the correct solution.

18 is the correct solution (*not* 9).

(b) $-\dfrac{2}{3}r = 4$

Multiply both sides by the reciprocal of $-\frac{2}{3}$, which is $-\frac{3}{2}$.

$$-\dfrac{2}{3}r = 4$$

$$-\dfrac{\cancel{3}^{1}}{\cancel{2}_{1}} \cdot \left(-\dfrac{\cancel{2}^{1}}{\cancel{3}_{1}}r\right) = -\dfrac{3}{\cancel{2}_{1}} \cdot \dfrac{\cancel{4}^{2}}{1}$$ Multiply both sides by $-\frac{3}{2}$

$$r = -6$$ ← The solution is -6

Check by replacing r with -6 in the original equation. Write -6 as $\frac{-6}{1}$.

$$-\dfrac{2}{3}r = 4$$ ← Original equation

$$-\dfrac{2}{\cancel{3}_{1}} \cdot \dfrac{\cancel{-6}^{-2}}{1} = 4$$ Replace r with -6

$$4 = 4$$ True, so -6 is the correct solution.

-6 is the correct solution (*not* 4).

Here is a summary of the rules for using the multiplication property. In these rules, x is the variable and a, b, and c represent numbers.

> **Solving Equations Using the Multiplication Property**
>
> Solve the equation $ax = b$ by dividing *both* sides by a.
>
> Solve the equation $\dfrac{a}{b}x = c$ by multiplying *both* sides by $\dfrac{b}{a}$.

Work Problem 4 at the Side. ▶▶▶

4 Solve each equation. Check each solution.

(a) $\dfrac{a}{4} = 2$

(b) $\dfrac{y}{7} = -3$

(c) $-8 = \dfrac{k}{6}$

(d) $8 = -\dfrac{4}{5}z$

(e) $-\dfrac{5}{8}p = -10$

Expressions

A college with four campuses uses an auditorium for graduation that has 1500 seats. Each campus hosts its own graduation ceremony. Students are allocated a whole number of tickets. Round each result **down** to the nearest whole number.

1. How many tickets are allocated to each of the 458 graduates at the **North** Campus graduation?

2. How many tickets are allocated to each of the 297 graduates at the **East** Campus graduation?

3. How many tickets are allocated to each of the 315 graduates at the **South** Campus graduation?

4. How many tickets are allocated to each of the 186 graduates at the **West** Campus graduation?

5. Write an *expression* that represents the number of tickets that are allocated to each of *g* graduates.

6. What recommendations do you have for unallocated tickets?

Traffic engineers have to decide how long to have the red, yellow, and green lights showing on a traffic signal. To decide the number of seconds that a yellow light should be on, the engineers use the expression

$$\frac{5v}{100} + 1$$

where *v* is the speed limit in miles per hour (mph).

7. How many seconds should the yellow light be on if the speed limit is 20 mph? 40 mph? 60 mph?

8. Based on the answers you just calculated, how could you estimate the time for a yellow light if the speed limit is 30 mph? 50 mph?

9. Use the given expression to find the number of seconds that the yellow light should be on if the speed limit is 30 mph and 50 mph. Did you get the same result as in Problem 8?

To estimate the number of words in a child's vocabulary, a pediatrician uses the expression $60A - 900$, where A is the child's age in months.

10. Estimate the number of words that a child aged 20 months knows.

11. Estimate the number of words that a child aged 2 years knows. (*Hint:* How many months are in two years?)

12. How many words does a child learn between the ages of 20 months and 2 years?

13. Estimate the number of words in the vocabulary of a 3-year-old child.

14. How many words does a child learn between the ages of 2 years and 3 years?

15. Evaluate the expression for a child who is 15 months old. Do you think that the answer is reasonable? Explain why or why not.

16. Evaluate the expression for a child who is 12 months old. Do you think that the answer makes sense? Explain why or why not.

Here is a *rule of thumb* to estimate the distance in miles to a thunderstorm: "Count the number of seconds from the time you see a lightning flash until you hear the thunder. Divide by 5."

17. How far away is the storm if you count 15 *seconds* between the lightning flash and the thunder? 10 *seconds?* 5 *seconds?*

18. Write an expression that represents the distance in miles to a thunderstorm if *s* seconds elapse between seeing lightning and hearing thunder.

19. Use the expression to estimate the distance to a storm if the time lapse is $2\frac{1}{2}$ seconds.

9.6 Exercises

FOR EXTRA HELP 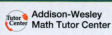 Addison-Wesley Math Tutor Center 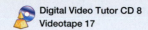 MathXL Digital Video Tutor CD 8 Videotape 17 📖 Student's Solutions Manual *MyMathLab* MyMathLab 🔺 Interactmath.com

Determine whether the given number is a solution of the equation. See Example 1.

1. $x + 7 = 11;$ 4

2. $k - 2 = 7;$ 9

3. $4y = 28;$ 7

4. $5p = 30;$ 6

5. $2z - 1 = -15;$ -8

6. $6r - 3 = -14;$ -2

Solve each equation by using the addition property. Check each solution. See Example 2.

7. $p + 5 = 9$

8. $a + 3 = 12$

9. $k + 15 = 0$

10. $y + 6 = 0$

11. $z - 5 = 3$

12. $x - 9 = 4$

13. $8 = r - 2$

14. $3 = b - 5$

15. $-5 = n + 3$

16. $-1 = a + 8$

17. $7 = r + 13$

18. $12 = z + 7$

19. $-4 + k = 14$

20. $-9 + y = 7$

21. $-12 + x = -1$

22. $-3 + m = -9$

23. $-5 = -2 + r$

24. $-1 = -10 + y$

25. $d + \dfrac{2}{3} = 3$

26. $x + \dfrac{1}{2} = 4$

27. $z - \dfrac{7}{8} = 10$

28. $m - \dfrac{3}{4} = 6$

29. $\dfrac{1}{2} = k - 2$

30. $\dfrac{3}{5} = t - 1$

31. $m - \dfrac{7}{5} = \dfrac{11}{4}$

32. $z - \dfrac{7}{3} = \dfrac{32}{9}$

33. $x - 0.8 = 5.07$

34. $a - 3.82 = 7.9$

35. $3.25 = 4.76 + r$

36. $8.9 = 10.5 + b$

Solve each equation. Check each solution. See Example 3.

37. $6z = 12$

38. $8k = 24$

39. $48 = 12r$

40. $99 = 11m$

41. $3y = 0$

42. $5a = 0$

43. $-6k = 36$

44. $-7y = 70$

45. $-36 = -4p$ **46.** $-54 = -9r$ **47.** $-1.2m = 8.4$

48. $-5.4z = 27$ **49.** $-8.4p = -9.24$ **50.** $-3.2y = -16.64$

Solve each equation. Check each solution. See Example 4.

51. $\dfrac{k}{2} = 17$ **52.** $\dfrac{y}{3} = 5$ **53.** $11 = \dfrac{a}{6}$

54. $5 = \dfrac{m}{8}$ **55.** $\dfrac{r}{3} = -12$ **56.** $\dfrac{z}{9} = -3$

57. $-\dfrac{2}{5}p = 8$ **58.** $-\dfrac{5}{6}k = 15$ **59.** $-\dfrac{3}{4}m = -3$

60. $-\dfrac{9}{10}b = -18$ **61.** $6 = \dfrac{3}{8}x$ **62.** $4 = \dfrac{2}{3}a$

63. $\dfrac{y}{2.6} = 0.5$ **64.** $\dfrac{k}{0.7} = 3.2$ **65.** $\dfrac{z}{-3.8} = 1.3$ **66.** $\dfrac{m}{-5.2} = 2.1$

67. Explain the addition property of equations. Then show an example of an equation where you would use the addition property to solve it. Make the equation so it has -3 as the solution.

68. Explain the multiplication property of equations. Then show an example of an equation where you would use the multiplication property to solve it. Make the equation so it has $+6$ as the solution.

Solve each equation.

69. $x - 17 = 5 - 3$

70. $y + 4 = 10 - 9$

71. $3 = x + 9 - 15$

72. $-1 = y + 7 - 9$

73. $\dfrac{7}{2}x = \dfrac{4}{3}$

74. $\dfrac{3}{4}x = \dfrac{5}{3}$

75. $\dfrac{1}{2} - \dfrac{3}{4} = \dfrac{a}{5}$

76. $\dfrac{2}{3} - \dfrac{8}{9} = \dfrac{c}{6}$

77. $m - 2 + 18 = |-3 - 4| + 5$

78. $10 - |0 - 8| = n + 1 - 4$

9.7 Solving Equations with Several Steps

OBJECTIVE **1** **Solve equations with several steps.** You cannot solve the equation $5m + 1 = 16$ by just adding the same number to both sides, nor by just dividing both sides by the same number. Instead, you use a combination of operations. Here are the steps.

OBJECTIVES

1 Solve equations with several steps.

2 Use the distributive property.

3 Combine like terms.

4 Solve more difficult equations.

> **Solving an Equation Using the Addition and Multiplication Properties**
>
> *Step 1* Add or subtract the *same* amount on *both* sides of the equation so that the variable term ends up by itself on one side.
>
> *Step 2* Multiply or divide *both* sides by the *same* number to find the solution.
>
> *Step 3* Check the solution in the original equation.

EXAMPLE 1 **Solving an Equation with Several Steps**

Solve $5m + 1 = 16$.

Step 1 Subtract 1 from *both* sides so that $5m$ will be by itself on the left side.

$$5m + 1 - 1 = 16 - 1$$
$$5m = 15$$

Step 2 Divide *both* sides by 5.

$$\frac{\overset{1}{\cancel{5}} \cdot m}{\underset{1}{\cancel{5}}} = \frac{15}{5}$$

$$m = 3 \leftarrow \text{The solution is 3}$$

Step 3 Check the solution.

$$5\mathbf{m} + 1 = 16 \leftarrow \text{Original equation}$$
$$5(\mathbf{3}) + 1 = 16 \qquad \text{Replace } m \text{ with 3}$$
$$15 + 1 = 16$$
$$16 = 16 \qquad \text{True, so 3 is the correct solution.}$$

3 is the correct solution (**not** 16).

Work Problem 1 at the Side. ▶▶▶

OBJECTIVE **2** **Use the distributive property.** We can use the order of operations to simplify these two expressions.

$$2(6 + 8) \qquad \text{and} \qquad 2 \cdot 6 + 2 \cdot 8$$
$$2(14) \qquad\qquad\qquad 12 \ + \ 16$$
$$28 \qquad\qquad\qquad\qquad 28$$

Because both answers are the same, the two expressions are equivalent.

$$2(6 + 8) = 2 \cdot 6 + 2 \cdot 8$$

This is an example of the **distributive property.**

> **Distributive Property**
>
> $$a(b + c) = ab + ac$$

1 Solve each equation. Check each solution.

(a) $2r + 7 = 13$

(b) $20 = 6y - 4$

(c) $7m + 9 = 9$

(d) $-2 = 4p + 10$

(e) $-10z - 9 = 11$

ANSWERS

1. **(a)** $r = 3$ **(b)** $y = 4$ **(c)** $m = 0$
   (d) $p = -3$ **(e)** $z = -2$

2 Use the distributive property.

(a) $3(2 + 6)$

(b) $8(k - 3)$

(c) $-6(r + 5)$

(d) $-9(s - 8)$

3 Combine like terms.

(a) $5y + 11y$

(b) $10a - 28a$

(c) $3x + 3x - 9x$

(d) $k + k$

(e) $6b - b - 7b$

EXAMPLE 2 Using the Distributive Property

Simplify each expression by using the distributive property.

(a) $9(4 + 2) = 9 \cdot 4 + 9 \cdot 2 = 36 + 18 = 54$

The 9 outside the parentheses is *distributed* over the 4 and the 2 inside the parentheses. That means that *every* number inside the parentheses is multiplied by 9.

(b) $-3(k + 9) = -3 \cdot k + (-3) \cdot 9 = -3k + (-27) = -3k - 27$

(c) $6(y - 5) = 6 \cdot y - 6 \cdot 5 = 6y - 30$

(d) $-2(x - 3) = -2 \cdot x - (-2) \cdot 3 = -2x - (-6) = -2x + 6$

CAUTION

Notice how the final step in Example 2(b) above uses the definition of subtraction "in reverse."

$$-3k + (-27) \quad \text{Change addition to subtraction;}$$
$$\downarrow \quad \downarrow \quad \text{change } (-27) \text{ to its opposite, } 27$$
$$-3k - \quad 27$$

▶▶▶ **Work Problem 2 at the Side.**

OBJECTIVE 3 Combine like terms. A single letter or number, or the product of a variable and a number, makes up a *term*. Here are six examples of terms.

$$3y \quad 5 \quad -9 \quad 8r \quad 10r^2 \quad x$$

Terms with exactly the same variable and the same exponent are called **like terms.**

$5x$	and	$3x$	like terms
$5x$	and	$3m$	*not* like terms; variables are different
$5x^2$	and	$5x^3$	*not* like terms; exponents are different
$5x^4$	and	$3x^4$	like terms

The distributive property can be used to simplify a sum of like terms such as $6r + 3r$:

$$6r + 3r = (6 + 3)r = 9r$$

This process of *adding* the like terms is called *combining like terms.*

EXAMPLE 3 Combining Like Terms

Use the distributive property to combine like terms.

(a) $5k + 11k = (5 + 11)k = 16k$

(b) $10m - 14m + 2m = (10 - 14 + 2)m = -2m$

(c) $-5x + x$ can be written $-5x + 1x = (-5 + 1)x = -4x$

▶▶▶ **Work Problem 3 at the Side.**

OBJECTIVE **4** **Solve more difficult equations.** The next examples show you how to solve more difficult equations using the addition, multiplication, and distributive properties.

EXAMPLE 4 **Solving Equations**

Solve each equation. Check each solution.

(a) $6r + 3r = 36$

You can combine $6r$ and $3r$ because they are *like terms*. $6r + 3r$ is $9r$, so the equation becomes

$$9r = 36$$

Next, divide both sides by 9.

$$\frac{\overset{1}{\cancel{9}} \cdot r}{\underset{1}{\cancel{9}}} = \frac{36}{9}$$

$$r = 4 \leftarrow \text{The solution is 4}$$

Check:

$$6r + 3r = 36 \leftarrow \text{Original equation}$$
$$6(4) + 3(4) = 36 \quad \text{Replace } r \text{ with 4}$$
$$24 + 12 = 36$$
$$36 = 36 \quad \text{True, so 4 is the correct solution.}$$

4 is the correct solution (*not* 36).

(b) $2k - 2 = 5k - 11$

One way to get the variable term by itself on one side is to subtract $5k$ from both sides.

$$2k - 2 - 5k = 5k - 11 - 5k$$
$$2k - 5k - 2 = 5k - 5k - 11$$
$$-3k - 2 = -11$$

Next, add 2 to both sides.

$$-3k - 2 + 2 = -11 + 2$$
$$-3k = -9$$

Finally, divide both sides by -3.

$$\frac{-\overset{1}{\cancel{3}} \cdot k}{\underset{1}{-\cancel{3}}} = \frac{-9}{-3}$$

$$k = 3 \quad \text{The solution is 3}$$

Check:

$$2k - 2 = 5k - 11 \quad \text{Original equation}$$
$$2(3) - 2 = 5(3) - 11 \quad \text{Replace } k \text{ with 3}$$
$$6 - 2 = 15 - 11$$
$$4 = 4 \quad \text{True, so 3 is the correct solution.}$$

3 is the correct solution (*not* 4).

4 Solve each equation. Check each solution.

(a) $3y - 1 = 2y + 7$

(b) $5a + 7 = 3a - 9$

(c) $3p - 2 = p - 6$

Work Problem 4 at the Side. ▶▶▶

5 Solve each equation. Check each solution.

(a) $-12 = 4(y - 1)$

(b) $5(m + 4) = 20$

(c) $6(t - 2) = 18$

Now that you know about the distributive property and combining like terms, here is a summary of all the steps you can use to solve an equation.

> **Solving an Equation**
>
> *Step 1* If possible, use the **distributive property** to remove parentheses.
>
> *Step 2* **Combine any like terms** on the left side of the equation. Combine any like terms on the right side of the equation.
>
> *Step 3* **Add or subtract** the same amount on *both* sides of the equation so that the variable term ends up by itself on one side.
>
> *Step 4* **Multiply or divide** *both* sides by the same number to find the solution.
>
> *Step 5* **Check** your solution by going back to the *original equation*. Replace the variable with your solution. Follow the order of operations to complete the calculations. If the two sides of the equation are equal, your solution is correct.

EXAMPLE 5 Solving Equations Using the Distributive Property

Solve $-6 = 3(y - 2)$

Step 1 Use the distributive property on the right side of the equation.

$$3(y - 2) \quad \text{becomes} \quad 3 \cdot y - 3 \cdot 2 \quad \text{or} \quad 3y - 6$$

Now the equation looks like this.

$$-6 = 3y - 6$$

Step 2 Combine like terms. Check the left side of the equation. There are no like terms. Check the right side. No like terms there either, so go on to Step 3.

Step 3 Add 6 to both sides to get the variable term by itself on the right side.

$$-6 + 6 = 3y - 6 + 6$$
$$0 = 3y$$

Step 4 Divide both sides by 3.

$$\frac{0}{3} = \frac{\overset{1}{\cancel{3}} \cdot y}{\underset{1}{\cancel{3}}}$$

$$0 = y \leftarrow \text{The solution is 0}$$

Step 5 Check. Go back to the original equation.

$$-6 = 3(\boldsymbol{y} - 2) \quad \leftarrow \text{Original equation}$$
$$-6 = 3(\boldsymbol{0} - 2) \qquad \text{Replace } y \text{ with 0}$$
$$-6 = 3(-2)$$
$$-6 = -6 \qquad \text{True, so 0 is the correct solution.}$$

0 is the correct solution (*not* -6).

◄◄◄ Work Problem 5 at the Side.

9.7 Exercises

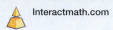

Solve each equation. Check each solution. See Example 1.

1. $7p + 5 = 12$

2. $6k + 3 = 15$

3. $2 = 8y - 6$

4. $10 = 11p - 12$

5. $-3m + 1 = 1$

6. $-4k + 5 = 5$

7. $28 = -9a + 10$

8. $5 = -10p + 25$

9. $-5x - 4 = 16$

10. $-12a - 3 = 21$

11. $-\dfrac{1}{2}z + 2 = -1$

12. $-\dfrac{5}{8}r + 4 = -6$

13. $-0.7 = 5b - 5.2$

14. $0.25 = -3c + 0.85$

Use the distributive property to simplify. See Example 2.

15. $6(x + 4)$

16. $8(k + 5)$

17. $7(p - 8)$

18. $9(t - 4)$

19. $-3(m + 6)$

20. $-5(a + 2)$

21. $-2(y - 3)$

22. $-4(r - 7)$

23. $-8(c + 8)$

24. $-6(n + 5)$

25. $-10(w - 9)$

26. $-11(x - 11)$

Combine like terms. See Example 3.

27. $11r + 6r$ **28.** $2m + 5m$ **29.** $8z - 7z$ **30.** $10x - 2x$

31. $y - 3y$ **32.** $-10a + a$ **33.** $-4t + t - 4t$ **34.** $3y - y - 4y$

35. $7p - 9p + 2p$ **36.** $-6c - c + 7c$ **37.** $\dfrac{5}{2}b - \dfrac{11}{2}b$ **38.** $\dfrac{3}{8}d - \dfrac{9}{8}d$

Solve each equation. Check each solution. See Example 4.

39. $4k + 6k = 50$ **40.** $3a + 2a = 15$ **41.** $54 = 10m - m$

42. $28 = x + 6x$ **43.** $2b - 6b = 24$ **44.** $3r - 9r = 18$

45. $-12 = 6y - 18y$ **46.** $-5 = 10z - 15z$ **47.** $6p - 2 = 4p + 6$

48. $5y - 5 = 2y + 10$ **49.** $9 + 7z = 9z + 13$ **50.** $8 + 4a = 2a + 2$

51. $-2y + 6 = 6y - 10$ **52.** $5x - 4 = -3x + 4$ **53.** $b + 3.05 = 2$

54. $t + 0.8 = -1.7$ **55.** $2.5r + 9 = -1$ **56.** $0.5x - 6 = 2$

Solve each equation by using the distributive property. Check each solution. See Example 5.

57. $-10 = 2(y + 4)$ **58.** $-3 = 3(x + 6)$ **59.** $-4(t + 2) = 12$

60. $-5(k + 3) = 25$ **61.** $6(x - 5) = -30$ **62.** $7(r - 5) = -35$

63. Solve $-2t - 10 = 3t + 5$. Show each step you take to solve it. Next to each step, write a sentence that explains what you did in that step. Be sure to tell when you used the addition property of equations and when you used the multiplication property of equations.

64. Here is one student's solution to an equation.
$$3(2x + 5) = -7$$
$$6x + 5 = -7$$
$$6x + 5 - 5 = -7 - 5$$
$$6x = -12$$
$$x = -2$$

Show how to check the solution. If the solution does not check, find and correct the error.

Solve each equation.

65. $30 - 40 = -2x + 7x - 4x$

66. $-6 - 5 + 14 = -50a + 51a$

67. $0 = -2(y - 2)$

68. $0 = -9(b - 1)$

69. $\dfrac{y}{2} - 2 = \dfrac{y}{4} + 3$

70. $\dfrac{z}{3} + 1 = \dfrac{z}{2} - 3$

71. $-3(w - 2) = |0 - 13| + 4w$

72. $-5b - |47 - 7| = -8(b + 8)$

73. $2(a + 0.3) = 1.2(a - 4)$

74. $-0.5(c - 4) = -3(c - 2.5)$

9.8 Using Equations to Solve Application Problems

It is rare for an application problem to be presented as an equation. Usually, the problem is given in words. You need to *translate* these words into an equation that you can solve.

OBJECTIVE 1 Translate word phrases into expressions with variables. Examples 1 and 2 show you how to translate word phrases into algebraic expressions.

OBJECTIVES

1. Translate word phrases into expressions with variables.
2. Translate sentences into equations.
3. Solve application problems.

EXAMPLE 1 Translating Word Phrases into Expressions with Variables

Write each word phrase in symbols, using x as the variable.

Words	Algebraic Expression
A number **plus** 2	$x + 2$ or $2 + x$
The **sum** of 8 and a number	$8 + x$ or $x + 8$
5 **more than** a number	$x + 5$ or $5 + x$
-35 **added to** a number	$-35 + x$ or $x + (-35)$
A number **increased by** 6	$x + 6$ or $6 + x$
9 **less than** a number	$x - 9$
A number **subtracted from** 3	$3 - x$
A number **decreased by** 4	$x - 4$
10 **minus** a number	$10 - x$

CAUTION
Recall that addition can be done in any order, so $x + 2$ gives the same result as $2 + x$. This is **not** true in subtraction, so be careful. $10 - x$ does **not** give the same result as $x - 10$.

Work Problem 1 at the Side. ▶▶▶

EXAMPLE 2 Translating Word Phrases into Expressions with Variables

Write each word phrase in symbols, using x as the variable.

Words	Algebraic Expression
8 **times** a number	$8x$
The **product** of 12 and a number	$12x$
Double a number (meaning "2 times")	$2x$
Twice a number (meaning "2 times")	$2x$
The **quotient** of 6 and a number	$\dfrac{6}{x}$
A number **divided by** 10	$\dfrac{x}{10}$
One-third of a number	$\dfrac{1}{3}x$ or $\dfrac{x}{3}$
The result **is**	$=$

Work Problem 2 at the Side. ▶▶▶

---

1 Write each word phrase in symbols, using x as the variable.

(a) 15 less than a number

(b) 12 more than a number

(c) A number increased by 13

(d) A number minus 8

(e) -10 plus a number

(f) A number subtracted from 6

2 Write each word phrase in symbols, using x as the variable.

(a) Double a number

(b) The product of -8 and a number

(c) The quotient of 15 and a number

(d) One-half of a number

ANSWERS
1. (a) $x - 15$ (b) $x + 12$ or $12 + x$
   (c) $x + 13$ or $13 + x$ (d) $x - 8$
   (e) $-10 + x$ or $x + (-10)$ (f) $6 - x$
2. (a) $2x$ (b) $-8x$ (c) $\dfrac{15}{x}$ (d) $\dfrac{1}{2}x$ or $\dfrac{x}{2}$

3 Translate each sentence into an equation and solve it. Check your solution by going back to the words in the original problem.

(a) If 3 times a number is added to 4, the result is 19. Find the number.

(b) If −6 times a number is added to 5, the result is −13. Find the number.

(c) If twice a number is subtracted from 65, the result is −21.

OBJECTIVE **2** **Translate sentences into equations.** The next example shows you how to translate a sentence into an equation that you can solve.

EXAMPLE 3 **Translating a Sentence into an Equation**

If 5 times a number is added to 11, the result is 26. Find the number.

Let x represent the unknown number.
Use the information in the problem to write an equation.

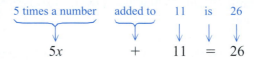

5 times a number	added to	11	is	26
$5x$	$+$	11	$=$	26

> **NOTE**
> The phrase "the result is" translates to "=."

Next, solve the equation.

$$5x + 11 - 11 = 26 - 11 \quad \text{Subtract 11 from both sides.}$$
$$5x = 15$$
$$\frac{5x}{5} = \frac{15}{5} \quad \text{Divide both sides by 5}$$
$$x = 3 \longleftarrow \text{The solution is 3}$$

The unknown number is 3.
To check the solution, go back to the words of the *original* problem.

If 5 times	a number	is added to	11,	the result is	26
$5 \cdot$	3	$+$	11	$=$	26

Does $5 \cdot 3 + 11$ really equal 26? Yes, $15 + 11 = 26$. So 3 is the correct solution because it "works" when you put it back into the original problem.

◀◀◀ **Work Problem 3 at the Side.**

OBJECTIVE **3** **Solve application problems.** In **Section 1.10,** you learned how to solve application problems using six steps. Now that you know how to use variables and solve equations, we can include those tools in the problem-solving process. Here is a revised list of the six steps we will use.

Solving an Application Problem Using Algebra

Step 1 **Read** the problem carefully to see what it is about.

Step 2 **Assign a variable** by identifying the unknown(s). Write down what the variable represents. If there are several unknowns, let the variable represent the one you know the least about.

Step 3 **Write an equation** using the variable expression(s).

Step 4 **Solve** the equation.

Step 5 **State the answer.**

Step 6 **Check** the solution in the words of the original problem.

Notice that Steps 1, 5, and 6 are nearly identical to what you have been using. Steps 2, 3, and 4 introduce the use of variables and equations.

ANSWERS
3. **(a)** $3x + 4 = 19$
     $x = 5$
   (b) $-6x + 5 = -13$
     $x = 3$
   (c) $65 - 2x = -21$
     $x = 43$

EXAMPLE 4 **Solving an Application Problem with One Unknown**

Michael has 5 less than three times as many lab experiments completed as David. If Michael has completed 13 experiments, how many lab experiments has David completed?

Step 1 **Read.** The problem asks for the number of lab experiments completed by David.

Step 2 **Assign a variable.** There is only *one* unknown: David's number of experiments.
Let x represent David's number of experiments.

Step 3 **Write an equation.**

$$13 \quad = \quad 3x - 5$$

Step 4 **Solve.**

$$13 = 3x - 5$$

$$13 + 5 = 3x - 5 + 5 \qquad \text{Add 5 to both sides.}$$

$$18 = 3x$$

$$\frac{18}{3} = \frac{\overset{1}{\cancel{3}}x}{\underset{1}{\cancel{3}}} \qquad \text{Divide both sides by 3}$$

$$6 = x \longleftarrow \text{The solution is 6}$$

Step 5 **State the answer.** David completed 6 lab experiments.

Step 6 **Check.** Use the words of the *original problem*.

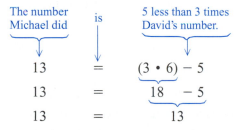

$$13 \quad = \quad (3 \cdot 6) - 5$$
$$13 \quad = \quad 18 - 5$$
$$13 \quad = \quad 13$$

So 6 is the correct solution because it "works" in the original problem.

NOTE

In *Step 3* above, you may also write the equation as $3x - 5 = 13$, with the two sides switched. The solution will be the same.

Work Problem 4 at the Side. ▶▶▶

④ Susan donated $10 more than twice what LuAnn donated. If Susan donated $22, how much did LuAnn donate?
  Show your work for each of the six problem-solving steps.

ANSWERS
4. *Step 1* Asks for amount of LuAnn's donation.
   Step 2 Let d be LuAnn's donation.
   Step 3 $2d + 10 = 22$ (or, $10 + 2d = 22$)
   Step 4

$$2d + 10 - 10 = 22 - 10$$
$$2d = 12$$
$$\frac{2d}{2} = \frac{12}{2}$$
$$d = 6$$

   Step 5 LuAnn donated $6.
   Step 6 Susan donated $10 more than twice $6, which is $10 + 2 \cdot \$6 = \$10 + \$12 = \$22$. That matches amount given in problem for Susan.

⑤ Solve each application using the six problem-solving steps.

(a) In a day of work, Keonda made $12 more than her daughter. Together they made $182. Find the amount made by each person. (*Hint:* Which amount do you know the least about, Keonda's or her daughter's? Let x be that amount.)

(b) A rope is 21 yd long. Marcos cut it into two pieces, so that one piece is 3 yd longer than the other. Find the length of each piece.

EXAMPLE 5 Solving an Application Problem with Two Unknowns

During the day, Sheila drove 72 km more than Russell. The total distance traveled by them both was 232 km. Find the distance traveled by each person.

Step 1 **Read.** The problem asks how far Sheila drove and how far Russell drove.

Step 2 **Assign a variable.** There are *two* unknowns: Sheila's distance and Russell's distance.

Let x be the distance traveled by Russell, because you know less about his distance than Sheila's distance.

Since Sheila drove 72 km *more* than Russell, the distance she traveled is $x + 72$ km, that is, Russell's distance (x) plus 72 km.

Step 3 **Write an equation.**

Distance for Russell	plus	distance for Sheila	is	total distance.
x	$+$	$x + 72$	$=$	232

Step 4 **Solve.**

$$x + x + 72 = 232$$ On the left side, $x + x$ is $1x + 1x$, which is $2x$.

$$2x + 72 = 232$$

$$2x + 72 - 72 = 232 - 72$$ Subtract 72 from both sides.

$$2x = 160$$

$$\frac{\overset{1}{\cancel{2}}x}{\cancel{2}_1} = \frac{160}{2}$$ Divide both sides by 2

$$x = 80 \longleftarrow \text{The solution is 80}$$

Step 5 **State the answer.**

Russell's distance is x, so Russell traveled 80 km.

Sheila's distance is $x + 72$, so Sheila traveled $80 + 72 = 152$ km.

Step 6 **Check.** Use the words of the *original* problem.

"Sheila drove 72 km more than Russell."

Sheila's 152 km is 72 km more than Russell's 80 km, so that checks.

"The total distance traveled by them both was 232 km."

Sheila's 152 km + Russell's 80 km = 232 km, so that checks.

◀◀◀ Work Problem 5 at the Side.

EXAMPLE 6 **Solving a Geometry Application Problem**

The length of a rectangle is 2 cm more than the width. The perimeter is 68 cm. Find the length and width.

Step 1   **Read.** The problem asks for the length and the width of a rectangle.

Step 2   **Assign a variable.** There are *two* unknowns, length and width. You know the least about the width, so let x represent the **width.**
    Since the length is 2 cm *more* than the width, the **length** is $x + 2$. A drawing of the rectangle will help you see these relationships.

width is x
length is $x + 2$
perimeter $= 68$

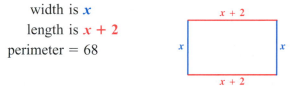

Step 3   **Write an equation.**

Use the formula for perimeter of a rectangle, $P = 2l + 2w$.

$$P = 2 \cdot l \quad + 2 \cdot w$$

Step 4   **Solve.**

$68 = \underbrace{2(x + 2)} + \underbrace{2 \cdot x}$ Use the distributive property.

$68 = 2x + 4 \; + \; 2x$ Combine like terms.

$68 = 4x + 4$

$68 - 4 = 4x + 4 - 4$ Subtract 4 from both sides.

$64 = 4x$

$$\frac{64}{4} = \frac{\overset{1}{\cancel{4}} \cdot x}{\underset{1}{\cancel{4}}}$$ Divide both sides by 4

$16 = x \longleftarrow$ The solution is 16

Step 5   **State the answer.**

The width is x, so the width is 16 cm.

The length is $x + 2$, so the length is $16 + 2$ or 18 cm.

The width is 16 cm and the length is 18 cm.

Step 6   **Check.** Use the words of the original problem. It says the length is 2 cm more than the width. 18 cm is 2 cm more than 16 cm, so that part checks.
    The original problem also says the perimeter is 68 cm. Use 18 cm and 16 cm to find the perimeter.

$$P = 2 \cdot 18 \text{ cm} + 2 \cdot 16 \text{ cm}$$

$$P = \underbrace{36 \text{ cm}} \; + \; \underbrace{32 \text{ cm}} \quad = 68 \text{ cm} \longleftarrow \text{Checks}$$

Work Problem 6 at the Side. ▶▶▶

6 Make a drawing and use the six steps to solve this problem. The length of Ann's rectangular garden plot is 3 yd more than the width. She used 22 yd of fencing around the edge. Find the length and the width of the garden.

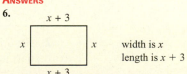

Real-Data Applications

Connections: Arithmetic to Algebra

As you begin to learn algebra, you may be frustrated when asked to work problems algebraically that are simple enough to be solved using arithmetic. Why bother with algebra? This activity shows that the *process* you use when solving a problem using arithmetic becomes *automated* when you use an **algebraic equation.** As the applications become more complicated and you find it more and more difficult to solve a problem using just arithmetic, the algebraic process becomes a powerful tool. In the example below, compare the steps of the arithmetic process to the steps involved in solving the algebraic equation. Note that the step-by-step *operations* are identical.

Problem: Kevin rented a chain saw for a one-time $15 sharpening fee plus $18-a-day rental. His total bill was $105. For how many days did Kevin rent the chain saw?

Arithmetic Solution	Algebraic Solution
To solve the problem, we have to first find the amount spent on rental fee alone. Then we have to find the number of days that gives that rental fee. The process is as follows.	First, let n = number of days of rental. Write the equation. The daily rental fee times the number of days, plus the sharpening fee, equals the total charge: $18n + 15 = 105$. Solve the equation.
1. First subtract the sharpening fee from the total charge to find the rental fee: $105 - 15 = 90$.	1. Subtract 15 from both sides. $$18n + 15 - 15 = 105 - 15$$ $$18n = 90$$
2. Divide the rental fee by the daily rental charge to find the number of days. $90 \div 18 = 5$	2. Divide both sides by 18. $$\frac{18n}{18} = \frac{90}{18}$$ $$n = 5$$
3. *Answer:* Kevin rented the saw for 5 days.	3. *Answer:* Kevin rented the saw for 5 days.

In the following problems, the first one can be solved easily using both arithmetic and algebraic methods. Show that the processes are the same, in a manner similar to the above example. The second problem should be much easier to work using an algebraic equation. In fact, the arithmetic solution will be a trial-and-error process. Show your work on separate paper, using a format similar to the one shown in the example.

1. Jose made a $300 down payment on a used car. His monthly payment was $150, and he paid $3900 in all. For how many months did he make payments?

Arithmetic Solution	Algebraic Solution
	Let x = number of months of payments

2. An 8 ft long board is to be cut into three parts. Two parts are the same length, and the third part is 2.75 ft longer than either of the other two parts. Find the length of each part.

Arithmetic Solution	Algebraic Solution
	Let x = length of Part 1 and also of Part 2. _____ = length of Part 3.

9.8 Exercises

  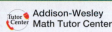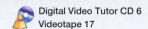

Write each word phrase in symbols, using x as the variable. See Examples 1 and 2.

1. 14 plus a number

2. The sum of a number and -8

3. -5 added to a number

4. 16 more than a number

5. 20 minus a number

6. A number decreased by 25

7. 9 less than a number

8. A number subtracted from -7

9. Subtract 4 from a number

10. 3 fewer than a number

11. Six times a number

12. The product of -3 and a number

13. Double a number

14. Half a number

15. A number divided by 2

16. 4 divided by a number

17. Twice a number added to 8

18. Five times a number plus 5

19. 10 fewer than seven times a number

20. 12 less than six times a number

21. The sum of twice a number and the number

22. Triple a number subtracted from the number

23. In your own words, write a definition for each of these words: **(a)** variable; **(b)** expression; **(c)** equation. Give three examples to illustrate each definition.

24. "You can use any letter to represent the unknown in an application problem." Is this statement true or false? Explain your answer.

Translate each sentence into an equation and solve it. See Example 3.

25. If four times a number is decreased by 2, the result is 26. Find the number.

26. The sum of 8 and five times a number is 53. Find the number.

27. If twice a number is added to the number, the result is −15. What is the number?

28. If a number is subtracted from three times the number, the result is −8. What is the number?

29. If the product of some number and 5 is increased by 12, the result is seven times the number. Find the number.

30. If eight times a number is subtracted from eleven times the number, the result is −9. Find the number.

31. When three times a number is subtracted from 30, the result is 2 plus the number. What is the number?

32. When twice a number is decreased by 8, the result is the number increased by 7. Find the number.

33. If half a number is added to twice the number, the answer is 50. Find the number.

34. If one-third of a number is added to three times the number, the result is 30. Find the number.

Solve each application problem by using the six problem-solving steps you learned in this section. See Examples 4 and 5.

35. When 75 is subtracted from four times Tamu's age, the result is Tamu's age. How old is Tamu?

36. If three times Linda's age is decreased by 36, the result is twice Linda's age. How old is Linda?

37. While shopping for clothes, Consuelo spent $3 less than twice what Brenda spent. Consuelo spent $81. How much did Brenda spend?

38. Dennis weighs 184 lb. His weight is 2 lb less than six times his child's weight. How much does his child weigh?

39. My sister is 9 years older than I am. The sum of our ages is 51. Find our ages.

40. Ed and Marge were candidates for city council. Marge won, with 93 more votes than Ed. The total number of votes cast in the election was 587. Find the number of votes received by each candidate.

41. Last year, Lien earned $1500 more than her husband. Together they earned $37,500. How much did each of them earn?

42. A $149,000 estate is to be divided between two charities so that one charity receives $18,000 less than the other. How much will each charity receive?

43. Jason paid five times as much for his computer as he did for his printer. He paid a total of $1320 for both items. What did each item cost?

44. The attendance at the Saturday night baseball game was three times the attendance at Sunday's game. In all, 56,000 fans attended the games. How many fans were at each game?

45. A board is 78 cm long. Rosa cut the board into two pieces, with one piece 10 cm longer than the other. Find the length of both pieces. (*Hint:* Make a sketch of the board.)

longer piece   shorter piece

46. A rope is 50 ft long. Juan cut it into two pieces so that one piece is 8 ft longer than the other. Find the length of each piece (*Hint:* Make a sketch of the rope.)

longer piece   shorter piece

47. A wire is cut into two pieces, with one piece 7 ft shorter than the other. The wire was 31 ft long before it was cut. How long was each piece?

48. A 90 cm pipe is cut into two pieces so that one piece is 6 cm shorter than the other. Find the length of each piece.

In Exercises 49–54, use the formula for the perimeter of a rectangle, P = 2l + 2w. Make a drawing to help you solve each problem, and use the six problem-solving steps. See Example 6.

49. The perimeter of a rectangle is 48 yd. The width is 5 yd. Find the length.

50. The length of a rectangle is 27 cm, and the perimeter is 74 cm. Find the width of the rectangle.

51. A rectangular dog pen is twice as long as it is wide. The perimeter of the pen is 36 ft. Find the length and the width of the pen.

52. A new city park has a rectangular shape. The length is triple the width. It will take 240 yd of fencing to go around the park. Find the length and width of the park.

53. The length of a rectangular jewelry box is 3 inches more than twice the width. The perimeter is 36 inches. Find the length and the width.

54. The perimeter of a rectangular house is 122 ft. The width is 5 ft less than the length. Find the length and the width.

9.1	**negative numbers**	Negative numbers are numbers that are less than 0.
	signed numbers	Signed numbers include positive numbers, negative numbers, and 0.
	absolute value	Absolute value is the distance of a number from 0 on a number line. Absolute value is never negative.
	opposite of a number	The opposite of a number is a number the same distance from 0 on a number line, but on the opposite side of 0.
9.2	**additive inverse**	The additive inverse is the opposite of a number. The sum of a number and its additive inverse is always 0.
9.5	**variables**	Variables are letters that represent numbers.
	expression	An expression is a combination of operations on variables and numbers.
9.6	**equation**	An equation is a statement that says two expressions are equal. An equation contains an equal sign.
	solution	The solution of an equation is a number that can replace the variable so that the equation is true.
	addition property of equations	The addition property of equations states that the same number can be added or subtracted on both sides of an equation.
	multiplication property of equations	The multiplication property of equations states that both sides of an equation can be multiplied or divided by the same number, except division by 0 is not allowed.
9.7	**distributive property**	If a, b, and c are three numbers, the distributive property says that $a(b + c) = ab + ac$.
	like terms	Like terms are terms with exactly the same variables and the same exponents.

See how well you have learned the vocabulary in this chapter. Answers follow the Quick Review.

1. An **expression**
   A. has an equal sign
   B. contains like terms with the same variables and exponents
   C. follows the order of operations
   D. is a combination of operations on variables and numbers.

2. An **equation**
   A. always has more than one solution
   B. can be graphed on a number line
   C. is evaluated by adding the same number to both sides
   D. contains an equal sign.

3. The **absolute value** of a number is
   A. its distance from 0
   B. never positive
   C. used when multiplying
   D. less than 0.

4. **Like terms**
   A. are always positive
   B. are the same distance from 0 but in opposite directions
   C. have matching variables and exponents
   D. can be multiplied but not added.

5. The **opposite** of a number is
   A. never negative
   B. called the additive inverse
   C. called the absolute value
   D. at the same point on the number line.

6. A **variable** is
   A. a solution of an equation
   B. a negative number
   C. a letter representing a number
   D. one of several like terms.

QUICK REVIEW

Concepts	Examples

9.1 *Graphing Signed Numbers*
Place a dot at the correct location on the number line. Positive numbers are to the right of 0; negative numbers are to the left of 0.

Graph $-2, 1, 0,$ and $2\frac{1}{2}$.

9.1 *Using the $<$ and $>$ Symbols*
Place the symbols $<$ (less than) or $>$ (greater than) between two numbers to make the statement true. The smaller pointed end of the symbol points to the smaller number.

Use the symbol $<$ or $>$ to make each statement true.

$2 \underline{\quad > \quad} 1$

$-3 \underline{\quad > \quad} -5$

$-6 \underline{\quad < \quad} 2$

9.1 *Finding the Absolute Value of a Number*
Determine the distance from 0 to the given number on the number line. Because the absolute value is a distance, it is never negative.

Simplify each absolute value expression.

(a) $|8|$ **(b)** $|-7|$ **(c)** $-|-5|$

$|8| = 8$ $|-7| = 7$ -5

9.1 *Finding the Opposite of a Number*
Find the number that is the same distance from 0 as the given number, but on the opposite side of 0 on the number line.

Find the opposite of each number.

(a) -6 **(b)** $+9$

$-(-6) = 6$ $-(+9) = -9$

9.2 *Adding Two Signed Numbers*

Adding Two Positive Numbers
Add the numerical values. The sum is positive.

Adding Two Negative Numbers
Add the absolute values and write a negative sign in front of the sum.

Add.

(a) $8 + 6 = 14$

(b) $-8 + (-6)$

Find absolute values. $\quad |-8| = 8 \quad\quad |-6| = 6$

Add absolute values: $8 + 6 = 14$.
Write a negative sign in front of the sum.

$$-8 + (-6) = -14$$

Adding Two Numbers with Different Signs
Subtract the smaller absolute value from the larger absolute value. Write the sign of the number with the larger absolute value in front of the answer.

(c) $5 + (-7)$

Find absolute values. $\quad |5| = 5 \quad\quad |-7| = 7$

Subtract the smaller absolute value from the larger:
$7 - 5 = 2$.

The number with the larger absolute value is -7. Its sign is negative, so write a negative sign in front of the answer.

$$5 + (-7) = -2$$

9.2 *Subtracting Two Signed Numbers*
To subtract two numbers, add the first number to the opposite of the second. Follow these steps.
Step 1 Change the subtraction sign to addition.
Step 2 Change the sign of the second number to its opposite.
Step 3 Proceed as in addition.

Subtract.

(a) $-6 - 5$

$-6 + (-5) = -11$

(b) $5 - (-8)$

$5 + (+8) = 13$

Concepts	Examples

9.3 *Multiplying Signed Numbers*

The product of two numbers with the *same* sign is *positive.*
The product of two numbers with *different* signs is *negative.*

Multiply.

(a) $7 \cdot 3 = 21$ (b) $(-3) \cdot 4 = -12$

(c) $5(-8) = -40$ (d) $(-9)(-6) = 54$

9.3 *Dividing Signed Numbers*

Use the same rules as for multiplying signed numbers.
When two numbers have the *same* sign, the quotient is *positive.*
When two numbers have *different* signs, the quotient is *negative.*

Divide

(a) $\dfrac{8}{4} = 2$ (b) $\dfrac{-20}{5} = -4$

(c) $\dfrac{50}{-5} = -10$ (d) $\dfrac{-12}{-6} = 2$

9.4 *Using the Order of Operations to Evaluate Numerical Expressions*

Use the order of operations to evaluate numerical expressions.

1. Do all operations inside **parentheses** or **other grouping symbols.**
2. Simplify any expressions with **exponents** and find any **square roots.**
3. **Multiply** or **divide,** proceeding from left to right.
4. **Add** or **subtract,** proceeding from left to right.

Simplify.

(a) $-4 + 6 \div (-2)$

$\quad -4 + \;\;(-3)$

$\qquad\quad -7$

(b) $3^2 + 3(8 \div 2)$

$\quad 3^2 + 3\;\;(4)$

$\quad\; 9\; + 3(4)$

$\quad\; 9\; + \;12$

$\qquad\quad 21$

9.5 *Evaluating Expressions*

Replace each variable in the expression with the specified number.
Use the order of operations to simplify the expression.
Remember that when no operation sign is written between a number and a letter, you assume it is multiplication. For example, $7x$ means $7 \cdot x$.

What is the value of $6p - 5s$, if $p = -3$ and $s = -4$?

$$6\;p\; - 5\;s$$
$$6(-3) - 5(-4)$$
$$-18\; - (-20)$$
$$2$$

9.6 *Determining Whether a Number Is a Solution of an Equation*

Substitute the number for the variable in the equation.
If the resulting statement is *true,* the number is a solution of the equation.
If the resulting statement is *false,* the number is *not* a solution of the equation.

Is 4 a solution of the equation $3x - 5 = 7$?

Replace x with 4.

$$3(4) - 5 = 7$$
$$12 - 5 = 7$$
$$7 = 7 \quad \text{True}$$

The result is true, so 4 is a solution of $3x - 5 = 7$.

9.6 *Using the Addition Property of Equations to Solve an Equation*

Add or subtract the same number on both sides of the equation so that you get the variable by itself on one side.

Solve each equation.

(a) $\quad x - 6 = 9$

$\quad x - 6 + 6 = 9 + 6 \quad$ Add 6 to both sides.

$\qquad x + 0 = 15$

$\qquad\qquad x = 15 \quad \leftarrow$ The solution is 15

(b) $\quad -7 = x + 9$

$\;\; -7 - 9 = x + 9 - 9 \quad$ Subtract 9 from both sides.

$\quad\; -16 = x + 0$

$\quad\; -16 = x \qquad\quad \leftarrow$ The solution is -16

Concepts	Examples

9.6 *Using the Multiplication Property of Equations to Solve an Equation*

Multiply or divide both sides of the original equation by the same number so that you get the variable by itself on one side. (Do not divide by 0.)

Solve each equation.

(a) $-54 = 6x$

$$\frac{-54}{6} = \frac{\overset{1}{\cancel{6}} \cdot x}{\underset{1}{\cancel{6}}}$$

$$-9 = x$$

(b) $\frac{1}{3}x = 8$

$$\frac{\cancel{3}}{1} \cdot \frac{1}{\cancel{3}}x = 3 \cdot 8$$

$$1x = 24$$
$$x = 24$$

9.7 *Solving Equations with Several Steps*

Use the following steps.

Step 1   Add or subtract the same amount on both sides of the equation so that the variable ends up by itself on one side.

Step 2   Multiply or divide both sides by the same number to find the solution.

Step 3   Check the solution by going back to the original equation.

Solve: $2p - 3 = 9$.

$$2p - 3 + 3 = 9 + 3 \qquad \text{Add 3 to both sides.}$$
$$2p = 12$$

$$\frac{\overset{1}{\cancel{2}} \cdot p}{\underset{1}{\cancel{2}}} = \frac{12}{2} \qquad \text{Divide both sides by 2}$$

$$p = 6 \qquad \leftarrow \text{The solution is 6}$$

Check: $2p - 3 = 9 \leftarrow$ Original equation

$$2(6) - 3 = 9 \qquad \text{Replace } p \text{ with 6}$$
$$12 - 3 = 9$$
$$9 = 9 \qquad \text{True, so 6 is the solution.}$$

6 is the solution (***not*** 9).

9.7 *Using the Distributive Property*

To simplify expressions, use the distributive property.

$$a(b + c) = ab + ac$$

Simplify: $-2(x + 4)$

$$= -2 \cdot x + (-2) \cdot 4$$
$$= -2x + (-8) = -2x - 8$$

9.7 *Combining Like Terms*

To combine like terms, add or subtract the number parts of the terms. The variable part stays the same.

Combine like terms.
(a) $6p + 7p = (6 + 7)p = 13p$
(b) $8m - 11m = (8 - 11)m = -3m$

9.8 *Translating Word Phrases into Expressions with Variables*

Use x (or any other letter) as a variable and symbolize the operations described by the words.

Write each word phrase in symbols, using x as the variable.
(a) Two more than a number $x + 2$
(b) 8 subtracted from a number $x - 8$
(c) The product of a number and 15 $15x$

(d) A number divided by 9 $\dfrac{x}{9}$

ANSWERS TO TEST YOUR WORD POWER

1. D; *Examples:* $3x - 5$; $2y^2$; $4a + b$.
2. D; *Examples:* $2x + 6 = -4$; $-8 + y = 7y - 14$.
3. A; *Example:* $|-3| = 3$ and $|3| = 3$ because the distance from 0 for both -3 and 3 is 3.
4. C; *Examples:* $2y$ and $-7y$ are like terms; $4x^2$ and x^2 are like terms; $3y^2$ and $3y^3$ are not like terms.
5. B; *Example:* The opposite of 5 is -5; it is the additive inverse because $5 + (-5) = 0$.
6. C; *Examples:* x, y, and d are variables.

Chapter 9
REVIEW EXERCISES

[9.1] *Graph each set of numbers on the number line.*

1. $2, -3, 4, 1, 0, -5$

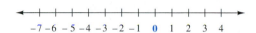

2. $-2, 5, -4, -1, 3, -6$

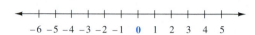

3. $-1\dfrac{1}{4}, -\dfrac{5}{8}, -3\dfrac{3}{4}, 2\dfrac{1}{8}, \dfrac{3}{2}, -2\dfrac{1}{8}$

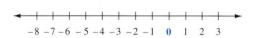

4. $0, -\dfrac{3}{4}, \dfrac{5}{4}, -4\dfrac{1}{2}, \dfrac{7}{8}, -7\dfrac{2}{3}$

Write $<$ *or* $>$ *in each blank to make a true statement.*

5. $0 \rule{1.5cm}{0.4pt} -2$

6. $-5 \rule{1.5cm}{0.4pt} 0$

7. $-1 \rule{1.5cm}{0.4pt} -4$

8. $-9 \rule{1.5cm}{0.4pt} -6$

Simplify each absolute value expression.

9. $|8|$

10. $|-19|$

11. $-|-7|$

12. $-|15|$

[9.2] *Add.*

13. $-4 + 6$

14. $-10 + 3$

15. $-11 + (-8)$

16. $-9 + (-24)$

17. $12 + (-11)$

18. $1 + (-20)$

19. $\dfrac{9}{10} + \left(-\dfrac{3}{5}\right)$

20. $-\dfrac{7}{8} + \dfrac{1}{2}$

21. $-6.7 + 1.5$

22. $-0.8 + (-0.7)$

Give the additive inverse (opposite) of each number

23. 6

24. -14

25. $-\dfrac{5}{8}$

26. 3.75

Subtract.

27. $4 - 10$

28. $7 - 15$

29. $-6 - 1$

30. $-12 - 5$

31. $8 - (-3)$

32. $2 - (-9)$

33. $-1 - (-14)$

34. $-10 - (-4)$

35. $-40 - 40$

36. $-15 - (-15)$

37. $\dfrac{1}{3} - \dfrac{5}{6}$

38. $2.8 - (-6.2)$

[9.3] *Multiply or divide.*

39. $-4(6)$

40. $5 \cdot (-4)$

41. $-3(-5)$

42. $-8(-8)$

43. $\dfrac{80}{-10}$

44. $\dfrac{-9}{3}$

45. $\dfrac{-25}{-5}$

46. $\dfrac{-120}{-6}$

47. $(-37)(0)$

48. $(-1)(81)$

49. $\dfrac{0}{-10}$

50. $\dfrac{-20}{0}$

51. $\left(\dfrac{2}{3}\right)\left(-\dfrac{6}{7}\right)$

52. $-\dfrac{4}{5} \div \left(-\dfrac{2}{15}\right)$

53. $(-0.5)(-2.8)$

54. $\dfrac{-5.28}{0.8}$

[9.4] *Use the order of operations to simplify each expression.*

55. $2 - 11(-5)$

56. $(-4)(-8) - 9$

57. $48 \div (-2)^3 - (-5)$

58. $-36 \div (-3)^2 - (-2)$

59. $5(4) - 7(6) + 3(-4)$

60. $2(8) - 4(9) + 2(-6)$

61. $3^3(-4) - 2(5 - 9)$

62. $6(-4)^2 - 3(7 - 14)$

63. $\dfrac{3 - (5^2 - 4^2)}{14 + 24 \div (-3)}$

64. $(-0.8)^2(0.2) - (-1.2)$

65. $\left(-\dfrac{1}{3}\right)^2 + \dfrac{1}{4}\left(-\dfrac{4}{9}\right)$

66. $\dfrac{12 \div (2 - 5) + 12(-1)}{2^3 - (-4)^2}$

[9.5] *Find the value of each expression using the given values of the variables.*

67. $3k + 5m$
    $k = 4, \quad m = 3$

68. $3k + 5m$
    $k = -6, \quad m = 2$

69. $2p - q$
    $p = -5, \quad q = -10$

70. $2p - q$
    $p = 6, \quad q = -7$

71. $\dfrac{5a - 7y}{2 + m}$
    $a = 1, \quad y = 4, \quad m = -3$

72. $\dfrac{5a - 7y}{2 + m}$
    $a = 2, \quad y = -2, \quad m = -26$

Using the given values, evaluate each formula.

73. $P = a + b + c; \quad a = 9, \quad b = 12, \quad c = 14$

74. $A = \dfrac{1}{2}bh; \quad b = 6, \quad h = 9$

[9.6–9.7] *Solve each equation. Check each solution.*

75. $y + 3 = 0$

76. $a - 8 = 8$

77. $-5 = z - 6$

78. $-8 = -9 + r$

79. $-\dfrac{3}{4} + x = -2$

80. $12.92 + k = 4.87$

81. $-8r = 56$

82. $3p = 24$

83. $\dfrac{z}{4} = 5$

84. $\dfrac{a}{5} = -11$

85. $20 = 3y - 7$

86. $-5 = 2b + 3$

Use the distributive property to simplify each expression.

87. $6(r - 5)$

88. $11(p + 7)$

89. $-9(z - 3)$

90. $-8(x + 4)$

Combine like terms.

91. $3r + 8r$

92. $10z - 15z$

93. $3p - 12p + p$

94. $-6x - x + 9x$

Solve each equation. Check each solution.

95. $-4z + 2z = 18$

96. $-35 = 9k - 2k$

97. $4y - 3 = 7y + 12$

98. $b + 6 = 3b - 8$

99. $-14 = 2(a - 3)$

100. $42 = 7(t + 6)$

[9.8] *Write each word phrase in symbols, using x to represent the variable.*

101. 18 plus a number

102. Half a number

103. −5 times a number

104. A number subtracted from 20

Translate each sentence into an equation and solve it.

105. The sum of four times a number and 6 is −14. Find the number.

106. If a number is subtracted from five times the number, the result is 100. Find the number.

Solve each application problem using the six problem-steps.

107. A $30,000 scholarship is being divided between two students so that one student receives three times as much as the other. How much will each student receive?

108. The perimeter of a rectangular game board is 48 in. The length is 4 in. more than the width. Find the length and width of the game board. (Draw a sketch of the board.)

MIXED REVIEW EXERCISES

Add, subtract, multiply, or divide as indicated.

109. $-6 - (-9)$

110. $-8 \cdot (-5)$

111. $-12 + 11$

112. $\dfrac{-70}{10}$

113. $-4(4)$

114. $5 - 14$

115. $\dfrac{-42}{-7}$

116. $16 + (-11)$

117. $-10 - 10$

118. $\dfrac{-5}{0}$

119. $-\dfrac{2}{3} + \dfrac{1}{9}$

120. $0.7(-0.5)$

121. $|-6| + 2 - 3(-8) - 5^2$

122. $9 \div |-3| + 6(-5) + 2^3$

Solve each equation.

123. $-45 = -5y$

124. $b - 8 = -12$

125. $6z - 3 = 3z + 9$

126. $-5 = r + 5$

127. $-3x = 33$

128. $2z - 7z = -15$

129. $3(k - 6) = 6 - 12$

130. $6(t + 3) = -2 + 20$

131. $-10 = \dfrac{a}{5} - 2$

132. $4 + 8p = 4p + 16$

Solve each application problem using the six problem-solving steps.

133. The recommended daily intake of iron for an adult female is 3 mg more than twice the recommended amount for a newborn infant. The amount for an adult female is 15 mg. How much should the infant receive? (*Source:* Food and Nutrition Board.)

134. In the U.S. Congress, the number of representatives is 65 less than five times the number of senators. There are a total of 535 members of Congress. Find the number of senators and the number of representatives. (*Source:* World Almanac.)

135. A cheetah's sprinting speed is 25 miles per hour (mph) faster than a zebra can run. The sum of their running speeds is 111 mph. How fast can each animal run? (*Source:* Grolier Multimedia Encyclopedia.)

136. A rectangular park has a perimeter of 1320 yd. The length of the park is five times the width. Find the length and width of the park.

Chapter 9

TEST

Work each problem.

1. Graph -4, -1, $1\frac{1}{2}$, 3, and 0 on the number line at the right.

2. Write $<$ or $>$ in each blank to make a true statement.

   -3 _____ 0 -4 _____ -8

3. Find each absolute value: $|-7|$ and $|15|$.

Add, subtract, multiply, or divide.

4. $-8 + 7$

5. $-11 + (-2)$

6. $6.7 + (-1.4)$

7. $8 - 15$

8. $4 - (-12)$

9. $-\dfrac{1}{2} - \left(-\dfrac{3}{4}\right)$

10. $8(-4)$

11. $-7(-12)$

12. $(-16)(0)$

13. $\dfrac{-100}{4}$

14. $\dfrac{-24}{-3}$

15. $-\dfrac{1}{4} \div \dfrac{5}{12}$

Use the order of operations to simplify each expression.

16. $-5 + 3(-2) - (-12)$

17. $2 - (6 - 8) - (-5)^2$

Find the value of $8k - 3m$, given each set of values.

18. $k = -4$, $m = 2$

19. $k = 3$, $m = -1$

20. In Exercises 18 and 19, you were evaluating an expression. Explain the difference between evaluating an expression and solving an equation.

1.
   $-5\ -4\ -3\ -2\ -1\ \ 0\ \ 1\ \ 2\ \ 3\ \ 4$

2. _____

3. _____

4. _____

5. _____

6. _____

7. _____

8. _____

9. _____

10. _____

11. _____

12. _____

13. _____

14. _____

15. _____

16. _____

17. _____

18. _____

19. _____

20. _____

21. _____

22. _____

23. _____

24. _____

25. _____

26. _____

27. _____

28. _____

29. _____

30. _____

21. The formula for the area of a triangle is $A = \frac{1}{2}bh$. Find A, if $b = 20$ and $h = 11$.

Solve each equation. Show your work.

22. $x - 9 = -4$

23. $30 = -1 + r$

24. $3t - 8t = -25$

25. $\frac{p}{5} = -3$

26. $-15 = 3(a - 2)$

27. $3m - 5 = 7m - 13$

Solve each application problem using the six problem-solving steps.

28. A board is 118 cm long. Karin cut it into two pieces, with one piece 4 cm longer than the other. Find the length of both pieces.

29. The perimeter of a rectangular building is 420 ft. The length is four times as long as the width. Find the length and the width. Make a drawing to help solve this problem.

30. Marcella and her husband, Tim, spent a total of 19 hours redecorating their living room. Tim spent 3 hours less time than Marcella. How long did each person work on the room?

Cumulative Review Exercises

CHAPTERS 1–9

≈ *First use front end rounding to round each number and estimate an answer. Then find the exact answer. Write answers to fraction problems in lowest terms and as whole or mixed numbers when possible.*

1. *Estimate:*   *Exact:*

$$
\begin{array}{r}
8.7 \\
0.902 \\
+41 \\
\hline
\end{array}
$$

$+\underline{}$

2. *Estimate:*   *Exact:*

$$
\begin{array}{r}
6.27 \\
\times\ 49.2 \\
\hline
\end{array}
$$

$\times\underline{}$

3. *Estimate:*   *Exact:*

$\underline{})\overline{}$ $78)\overline{39{,}234}$

4. *Estimate:*   *Exact:*

$$
\begin{array}{r}
3\dfrac{3}{5} \\
-2\dfrac{3}{4} \\
\hline
\end{array}
$$

$-\underline{}$

5. *Exact:*

$5\dfrac{5}{6} \cdot \dfrac{9}{10} = \underline{}$

Estimate:

$\underline{} \cdot \underline{} = \underline{}$

6. *Exact:*

$4\dfrac{1}{6} \div 1\dfrac{2}{3} = \underline{}$

Estimate:

$\underline{} \div \underline{} = \underline{}$

Add, subtract, multiply, or divide as indicated.

7. $17 - 8.094$

8. $(1309)(408)$

9. $4.06 \div 0.072$
Round to nearest tenth.

10. $-12 + 7$

11. $-5(-8)$

12. $-3 - (-7)$

13. $3.2 + (-4.5)$

14. $\dfrac{30}{-6}$

15. $\dfrac{1}{4} - \dfrac{3}{4}$

Use the order of operations to simplify each expression.

16. $45 \div \sqrt{25} - 2(3) + (10 \div 5)$

17. $-6 - (4 - 5) + (-3)^2$

Write < or > in each blank to make a true statement.

18. $\dfrac{3}{10} \underline{} \dfrac{4}{15}$

19. $0.7072 \underline{} 0.72$

20. $-5 \underline{} -2$

21. Write 8% as a decimal and as a fraction in lowest terms.

22. Write $4\frac{1}{2}$ as a decimal and as a percent.

The bar graph shows the average number of tornadoes each year in selected states.
Use the graph to answer Exercise 23.

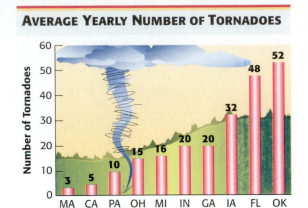

AVERAGE YEARLY NUMBER OF TORNADOES

Number of Tornadoes

Source: The Weather Book.

23. Write the ratio that compares the numbers of tornadoes in each pair of states. Write all ratios as fractions in lowest terms.

(a) Iowa (IA) to Michigan (MI)

(b) California (CA) to Ohio (OH)

(c) Georgia (GA) to Indiana (IN)

(d) Florida (FL) to Oklahoma (OK)

Find the unknown number in each proportion. Round your answer to hundredths if necessary.

24. $\dfrac{x}{12} = \dfrac{1.5}{45}$

25. $\dfrac{350}{x} = \dfrac{3}{2}$

26. $\dfrac{38}{190} = \dfrac{9}{x}$

Solve each percent problem.

27. 0.5% of 3000 students is how many students?

28. What percent of 12.5 miles is 6.8 miles?

29. 90 cars is 180% of what number of cars?

30. $5.80 is what percent of $145?

Convert these measurements.

31. $3\frac{1}{2}$ gal to quarts

32. 72 hr to days

33. 3.7 L to milliliters

34. 40 cm to meters

Write the most reasonable metric unit in each blank. Choose from km, m, cm, mm,
L, mL, kg, g, and mg.

35. The building is 15 _____ high.

36. Rita took 15 _____ of cough syrup.

37. Bruce walked 2 _____ to work.

38. The robin weighs 100 _____.

39. Identify which of these temperatures is normal body temperature and which is comfortable room temperature: 52 °C, 70 °C, 20 °C, 10 °C, 37 °C, 98 °C.

Find the perimeter (or circumference) and the area of each shape. Use **3.14** *as the approximate value of* π. *In Exercises 40–44, name the shape.*

40.

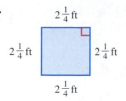

$2\frac{1}{4}$ ft

$2\frac{1}{4}$ ft $2\frac{1}{4}$ ft

$2\frac{1}{4}$ ft

41.

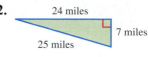

9 mm

42.

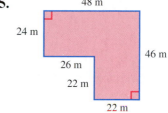

24 miles

7 miles

25 miles

43.

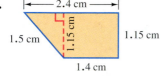

2.4 cm

1.5 cm   1.15 cm   1.15 cm

1.4 cm

44.

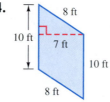

8 ft

10 ft   7 ft

10 ft

8 ft

45.

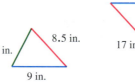

48 m

24 m

26 m   46 m

22 m

22 m

Find the unknown length in each figure. Round your answers to the nearest tenth if necessary.

46.

20 yd   y

15 yd

47. Similar triangles

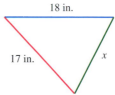

18 in.

7 in.   8.5 in.

9 in.

17 in.   x

Solve each equation.

48. $-20 = 6 + y$

49. $-2t - 6t = 40$

50. $3x + 5 = 5x - 11$

51. $6(p + 3) = -6$

In Exercise 52, write an equation and solve it. Then solve Exercises 53 and 54 using the six problem-solving steps.

52. If 40 is added to four times a number, the result is zero. Find the number.

53. $1000 in prize money is being split between Reggie and Donald. Donald should get $300 more than Reggie. How much will each man receive?

54. Make a drawing to help solve this problem. The length of a photograph is 5 cm more than the width. The perimeter of the photograph is 82 cm. Find the length and the width.

Solve each application problem. Round money answers to the nearest cent.

55. Portia bought two CDs at $14.98 each. The sales tax rate is $6\frac{1}{2}$%. Find the total amount charged to Portia's credit card.

56. Brian's spaghetti sauce recipe calls for $3\frac{1}{3}$ cups of tomato sauce. He wants to make $2\frac{1}{2}$ times the usual amount. How much tomato sauce does he need?

57. The local food pantry received 2480 lb of food this month. Their goal was 2000 lb. What percent of their goal was received?

58. Sayoko bought 720 g of chicken priced at $5.97 per kilogram. How much did she pay for the chicken?

59. A packing crate measures 2.4 m long, 1.2 m wide, and 1.2 m high. A trucking company wants crates that hold 4 m^3. The crate's volume is how much more or less than 4 m^3?

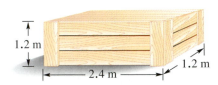

1.2 m

2.4 m

1.2 m

60. The Mercado family needs 35 ft of fencing to put around a rectangular garden plot. If the plot is $6\frac{1}{2}$ ft wide, find its length.

$6\frac{1}{2}$ ft

61. Rich spent 25 minutes reading 14 pages in his sociology textbook. At that rate, how long will it take him to read 30 pages? Round to the nearest whole number of minutes.

62. Jackie drove her car 364 miles on 14.5 gallons of gas. Maya used 16.3 gallons to drive 406 miles. Naomi drove 300 miles on 11.9 gallons. Which car had the highest number of miles per gallon? How many miles per gallon did that car get, rounded to the nearest tenth?

Statistics

10

10.1 **Circle Graphs**

10.2 **Bar Graphs and Line Graphs**

10.3 **Frequency Distributions and Histograms**

10.4 **Mean, Median, and Mode**

The old saying, "A picture is worth a thousand words" was never more true than when applied to the understanding of data. As an information society, we are constantly being bombarded with facts and numbers. The ability to understand and interpret the many types of information and the many ways it is presented has become essential. For example, if you own or manage a business, you need to understand and interpret the sales history of the business and the performance of employees to make wise managerial decisions. (See **Section 10.2**, Exercises 25–32 and 35–42, and **Section 10.3**, Exercises 29–36.)

10.1 Circle Graphs

OBJECTIVES

1 Read and understand a circle graph.

2 Use a circle graph.

3 Draw a circle graph.

1 Use the circle graph to answer each question.

(a) The greatest number of hours is spent in which activity?

(b) How many more hours are spent working than studying?

(c) Find the total number of hours spent studying, working, and attending classes.

2 Use the circle graph to find each ratio. Write the ratios as fractions in lowest terms.

(a) Hours spent driving to whole day

(b) Hours spent sleeping to whole day

(c) Hours spent attending class and studying to whole day

(d) Hours spent driving and working to whole day

The word *statistics* originally came from words that mean *state numbers*. State numbers refer to numerical information, or *data,* gathered by the government such as the number of births, deaths, or marriages in a population. Today, the word *statistics* has a much broader application; data from the fields of economics, social science, and business can all be organized and studied under the branch of mathematics called *statistics.*

OBJECTIVE 1 Read and understand a circle graph.   It can be hard to understand a large collection of data. The graphs described in this section can be used to help you make sense of such data. The **circle graph** is used to show how a total amount is divided into parts. The circle graph below shows you how 24 hours in the life of a college student are divided among different activities.

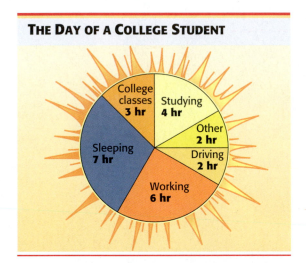

THE DAY OF A COLLEGE STUDENT

◀◀◀ **Work Problem 1 at the Side.**

OBJECTIVE 2 Use a circle graph.   The above circle graph uses pie-shaped pieces called *sectors* to show the amount of time spent on each activity (the total must be 24 hours); a circle graph can therefore be used to compare the time spent on one activity to the total number of hours in the day.

EXAMPLE 1 Using a Circle Graph

Find the ratio of time spent in college classes to the total number of hours in a day. Write the ratio as a fraction in lowest terms. (See **Section 5.1.**)

The circle graph shows that 3 of the 24 hours in a day are spent in class. The ratio of class time to the hours in a day is shown below.

$$\frac{3 \text{ hours (college classes)}}{24 \text{ hours (whole day)}} = \frac{3 \text{ hours}}{24 \text{ hours}} = \frac{3 \div 3}{24 \div 3} = \frac{1}{8} \longleftarrow \text{Lowest terms}$$

◀◀◀ **Work Problem 2 at the Side.**

The circle graph above can also be used to find the ratio of the time spent on one activity to the time spent on any other activity.

EXAMPLE 2 Finding a Ratio from a Circle Graph

Use the circle graph on a student's day to find the ratio of study time to class time. Write the ratio as a fraction in lowest terms.

The circle graph shows 4 hours spent studying and 3 hours spent in class. The ratio of study time to class time is shown below.

$$\frac{4 \text{ hours (study)}}{3 \text{ hours (class)}} = \frac{4 \text{ hours}}{3 \text{ hours}} = \frac{4}{3}$$

Work Problem 3 at the Side. ▶▶▶

A circle graph often shows data as percents. For example, suppose that the yearly vending machine snack food sales in the United States were $36 billion. The circle graph below shows how sales were divided among various types of snack foods. The entire circle represents $36 billion in sales. Each sector represents the sales of one snack item as a percent of the total sales (the total must be 100%).

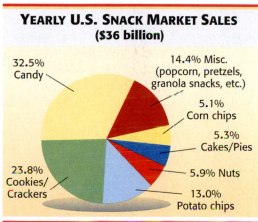

YEARLY U.S. SNACK MARKET SALES
($36 billion)

32.5% Candy

14.4% Misc. (popcorn, pretzels, granola snacks, etc.)

5.1% Corn chips

5.3% Cakes/Pies

5.9% Nuts

13.0% Potato chips

23.8% Cookies/ Crackers

Source: Natural Choice—USA.

EXAMPLE 3 Calculating an Amount Using a Circle Graph

Use the circle graph above on vending machine snack sales to find the amount spent on candy for the year.

Recall the percent equation.

part = percent • whole

The total sales are $36 billion, so the whole is $36 billion. The percent is 32.5, or as a decimal, 0.325. Find the part.

$$\begin{array}{ccc} \text{part} = \text{percent} \cdot & \text{whole} \\ \downarrow & \downarrow & \downarrow \end{array}$$

$$x = 0.325 \cdot 36 \text{ billion}$$

$$x = 11.7 \text{ billion}$$

The amount spent on candy was $11.7 billion or $11,700,000,000.

Work Problem 4 at the Side. ▶▶▶

3 Use the circle graph on a student's day to find the following ratios. Write the ratios as fractions in lowest terms.

(a) Hours spent in class to hours spent studying

(b) Hours spent working to hours spent sleeping

(c) Hours spent driving to hours spent working

(d) Hours spent in class to hours spent for "Other"

4 Use the circle graph on vending machine snack sales to find the following.

(a) The amount spent on corn chips

(b) The amount spent on miscellaneous (popcorn, pretzels, granola snacks, etc.)

(c) The amount spent on cakes/pies

(d) The amount spent on cookies/crackers

ANSWERS

3. (a) $\frac{3}{4}$ (b) $\frac{6}{7}$ (c) $\frac{1}{3}$ (d) $\frac{3}{2}$

4. (a) $1.836 billion or $1,836,000,000
   (b) $5.184 billion or $5,184,000,000
   (c) $1.908 billion or $1,908,000,000
   (d) $8.568 billion or $8,568,000,000

OBJECTIVE 3   Draw a circle graph.   The coordinator of the Fair Oaks Youth Soccer League organizes teams in five age groups. She places the players in various age groups as follows.

Age Group	Percent of Total
Under 8 years	20%
Ages 8–9	15%
Ages 10–11	25%
Ages 12–13	25%
Ages 14–15	15%
Total	100%

You can show these percents by using a circle graph. A circle has 360 degrees (written 360°). The 360° represents the entire league, or 100% of the soccer players.

EXAMPLE 4   **Drawing a Circle Graph**

Using the data on *age groups,* find the number of degrees in the sector that would represent the "Under 8" group, and begin constructing a circle graph.

Recall that a complete circle has 360° (see **Section 8.1**). Because the "Under 8" group makes up 20% of the total number of players, the number of degrees needed for the "Under 8" sector of the circle graph is 20% of 360°.

$$(360°)(\mathbf{20\%}) = (360°)(\mathbf{0.2}) = 72°$$

Use a tool called a **protractor** to make a circle graph. First, using a straightedge, draw a line from the center of a circle to the left edge. Place the hole in the protractor over the center of the circle, making sure that 0 on the protractor lines up with the line that was drawn. Find 72° and make a mark as shown in the illustration. Then remove the protractor and use the straightedge to draw a line from the center of the circle to the 72° mark at the edge of the circle. This sector is 72° and represents the "Under 8" group.

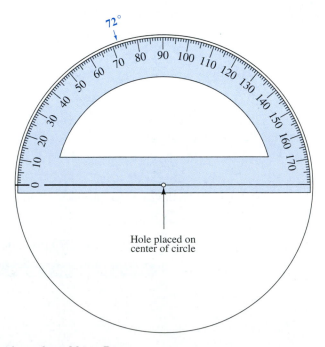

Hole placed on center of circle

Continued on Next Page

To draw the "Ages 8–9" sector, begin by finding the number of degrees in the sector.

$$(360°)(15\%) = (360°)(0.15) = 54°$$

Again, place the hole of the protractor over the center of the circle, but this time align 0 on the second line that was drawn. Make a mark at 54° and draw a line as before. This sector is 54° and represents the "Ages 8–9" group.

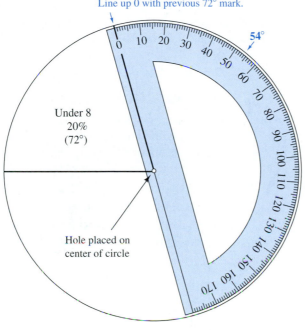

CAUTION
You must be certain that the hole in the protractor is placed over the exact center of the circle each time you measure the size of a sector.

Work Problem 5 at the Side.)))

Use this circle for Problem 5 at the side.

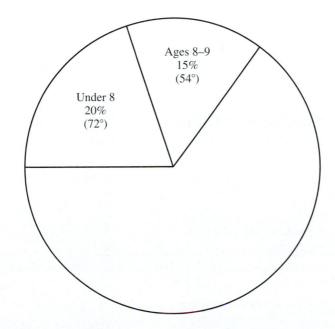

5 Using the information on the soccer age groups in the table, find the number of degrees needed for each sector and complete the circle graph at the bottom left.

(a) Ages 10–11

(b) Ages 12–13

(c) Ages 14–15

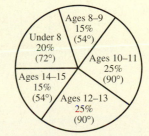

Real-Data Applications

Just in Time

Antique clocks usually chime either on the hour or both on the hour and the half hour. The mechanism that controls the chimes is a set of gears called the count plate and hammer wheel, and a lever called the count hook.

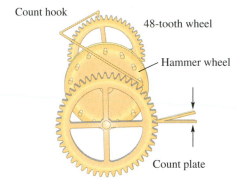

Count hook

48-tooth wheel

Hammer wheel

Count plate

Hour-Striking Clocks

1. If a clock chimes only on the hour, what will be the total number of chimes in a 12-hour period? (*Hint:* For example, it will chime 6 times at 6:00.)

2. If the count plate has one gear tooth for each chime, what fractional part of the count plate is one tooth?

3. How many degrees correspond to one gear tooth?

4. The mechanism designed to move the count plate has two wheels. The hammer wheel has 13 pins, which is combined with a wheel with 48 teeth and a pinion that fits 8 teeth. That gives a 6:1 ratio. What is significant about the combination of 13 and 6 that causes the count plate to move one gear tooth?

Hour- and Half-Hour-Striking Clocks

5. If a clock chimes on the hour and also one time on each half hour, what will be the total number of chimes in a 12-hour period?

6. If the count plate has one tooth for each chime, what fractional part of the count plate makes one tooth?

7. How many degrees correspond to one gear tooth?

8. If the count plate has a diameter of 2 in., what is the circumference of the count plate, rounded to the nearest hundredth? What is the width of each gear tooth, rounded to the nearest hundredth? (*Note:* Use $\pi \approx 3.14$.)

10.1 Exercises

FOR EXTRA HELP

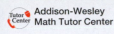 Addison-Wesley Math Tutor Center

 MathXL

 Digital Video Tutor CD 6 Videotape 18

Student's Solutions Manual

 MyMathLab

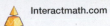 Interactmath.com

This circle graph shows the cost of adding an art studio to an existing home. Use this circle graph to answer Exercises 1–6. Write ratios as fractions in lowest terms. See Examples 1 and 2.

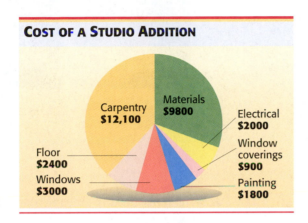

COST OF A STUDIO ADDITION

Carpentry $12,100
Materials $9800
Electrical $2000
Window coverings $900
Floor $2400
Windows $3000
Painting $1800

1. Find the total cost of adding the art studio

2. What is the largest single expense in adding the studio?

3. Find the ratio of the cost of materials to the total remodeling cost.

4. Find the ratio of the cost of painting to the total remodeling cost.

5. Find the ratio of the cost of carpentry to the cost of window coverings.

6. Find the ratio of the cost of windows to the cost of the floor.

This circle graph, adapted from USA Today, *shows the number of people in a survey who gave various reasons for eating dinner at restaurants. Use this circle graph to answer Exercises 7–14. Write ratios as fractions in lowest terms. See Examples 1 and 2.*

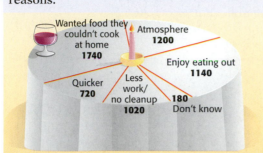

ON THE TOWN
When asked in a survey why they ate dinner in restaurants, a group of people gave these reasons.

Wanted food they couldn't cook at home **1740**

Atmosphere **1200**

Enjoy eating out **1140**

Quicker **720**

Less work/no cleanup **1020**

180 Don't know

Source: Market Facts for Tyson Foods.

7. Which reason was given by the least number of people?

8. Which reason was given by the second-highest number of people?

9. Those who said dining out is "Quicker" to total people in the survey

10. Those who said "Enjoy eating out" to the total people in the survey

11. Those who said "Less work/no cleanup" to those who said "Atmosphere"

12. Those who said "Don't know" to those who said "Quicker"

13. Those who said "Wanted food they couldn't cook at home" to those who said "Less work/no cleanup"

14. Those who said "Atmosphere" to those who said "Enjoy eating out"

This circle graph shows the favorite hot dog toppings in the United States. Each topping is expressed as a percent of the 3200 people in the survey. Use the graph to find the number of people in the survey who favored each of the toppings in Exercises 15–20.

15. Onions

16. Ketchup

17. Sauerkraut

18. Relish

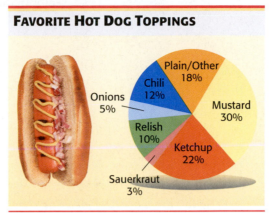

FAVORITE HOT DOG TOPPINGS

Plain/Other 18%
Chili 12%
Onions 5%
Mustard 30%
Relish 10%
Ketchup 22%
Sauerkraut 3%

Source: National Hot Dog and Sausage Council.

19. Mustard

20. Chili

The National Academy of Sciences recommends that adults get 1000 mg of calcium daily, or about three 8-ounce glasses of milk. The circle graph, adapted from USA Today, *shows the daily consumption of milk products by adults. If 5540 adults were surveyed in this study, find the number of people giving each response in Exercises 21–26. Round to the nearest whole number. See Example 3.*

21. Consume none or very few milk products

GOT MILK

Consume two servings 27%
Consume none or very few milk products 22%
Consume recommended servings 13%
Consume one serving 33%
Don't know 5%

Source: Market Facts for Milk Mustache Mobile 100-City Cruise for Calcium.

22. Consume recommended servings

23. Consume one serving

24. Consume two servings

25. Don't know

26. Consume less than the recommended servings (Do not include those who don't know.)

27. Describe the procedure for determining how large each sector must be to represent each of the items in a circle graph.

28. A protractor is the tool used to draw a circle graph. Give a brief explanation of what the protractor does and how you would use it to measure and draw each sector in the circle graph.

During one semester Kara Diano spent $5460 for school expenses as shown in the following chart. Find all numbers missing from the chart.

Item	Dollar Amount	Percent of Total	Degrees of a Circle
29. Rent	$1365	25%	_____
30. Food	$1092	_____	72°
31. Clothing	$546	_____	_____
32. Books	$546	10%	_____
33. Tuition	$819	15%	_____
34. Savings	$273	_____	_____
35. Entertainment	$819	_____	54°

36. Draw a circle graph by using the above information. See Example 4.

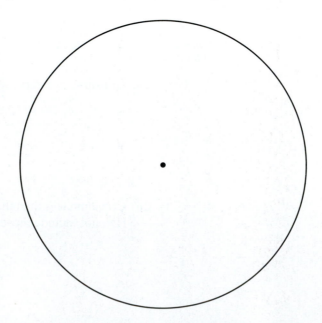

37. White Water Rafting Company divides its annual sales into five categories as follows.

Category	Annual Sales
Adventure classes	$12,500
Grocery and provision sales	$40,000
Equipment rentals	$60,000
Rafting tours	$50,000
Equipment sales	$37,500

(a) Find the total sales for the year.

(b) Find the number of degrees in a circle graph for each item.

(c) Make a circle graph showing this information.

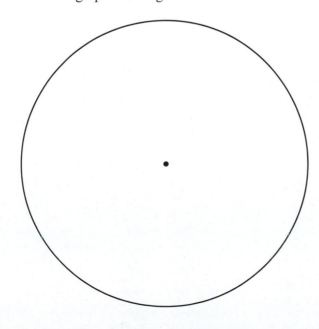

38. A book publisher had 25% of total sales in mysteries, 10% in biographies, 15% in cookbooks, 15% in romance novels, 20% in science, and the rest in business books.

(a) Find the number of degrees in a circle graph for each type of book.

(b) Draw a circle graph.

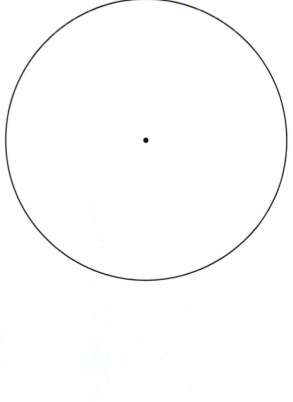

39. The Pathfinder Research Group asked 4488 Americans how they fall asleep, and the results are shown in the figure on the right.

(a) Use this information to complete the chart and draw a circle graph. Round to the nearest whole percent and to the nearest degree.

Sleeping Position	Number of Americans	Percent of Total	Number of Degrees
Side	_____	_____	_____
Not sure	_____	_____	_____
Stomach	_____	_____	_____
Varies	_____	_____	_____
Back	_____	_____	_____

SET TO SLEEP

Number of Americans surveyed who fall asleep on their:

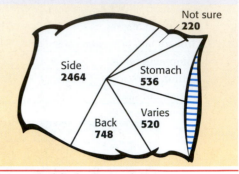

Source: Pathfinder Research Group.

(b) Add up the percents. Is the total 100%? Explain why or why not.

(c) Add up the degrees. Is the total 360°? Explain why or why not.

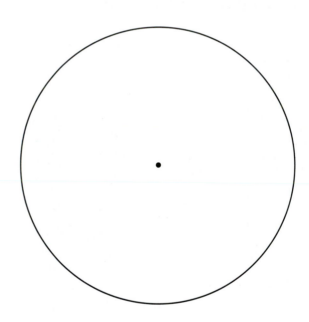

10.2 Bar Graphs and Line Graphs

OBJECTIVE 1 Read and understand a bar graph. Bar graphs are useful when showing comparisons. For example, the bar graph below compares the number of college graduates who continued taking advanced courses in their major field during each of 5 years.

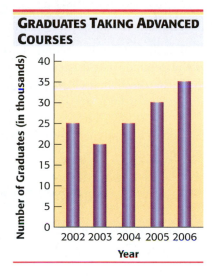

GRADUATES TAKING ADVANCED COURSES

1 Use the bar graph at the left to find the number of college graduates who took advanced classes in their major field in each of these years.

(a) 2002

EXAMPLE 1 Using a Bar Graph

How many college graduates took advanced classes in their major field in 2004?

The bar for 2004 rises to 25. Notice the label along the left side of the graph that says "Number of Graduates (in thousands)." The phrase *in thousands* means you have to multiply 25 by 1000 to get 25,000. So, 25,000 (not 25) graduates took advanced classes in their major field in 2004.

Work Problem 1 at the Side. ▶▶▶

(b) 2003

OBJECTIVE 2 Read and understand a double-bar graph. A double-bar graph can be used to compare two sets of data. The graph below shows the number of DSL (digital subscriber line) installations each quarter for two different years.

(c) 2005

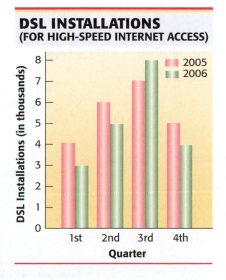

DSL INSTALLATIONS (FOR HIGH-SPEED INTERNET ACCESS)

■ 2005
■ 2006

(d) 2006

2 Use the double-bar graph to find the number of DSL installations in 2005 and 2006 for each quarter.

(a) 1st quarter

(b) 3rd quarter

(c) 4th quarter

(d) Find the greatest number of installations. Identify the quarter and the year in which they occurred.

3 Use the line graph at the right to find the number of trout stocked in each month.

(a) June

(b) May

(c) April

(d) July

EXAMPLE 2 Reading a Double-Bar Graph

Use the double-bar graph on the previous page to find the following.

(a) The number of DSL installations in the second quarter of 2005

There are two bars for the second quarter. The color code in the upper right-hand corner of the graph tells you that the **red bars** represent 2005. So the **red bar** on the *left* is for the 2nd quarter of 2005. It rises to 6. Multiply 6 by 1000 because the label on the left side of the graph says *in thousands*. So there were 6000 DSL installations for the second quarter in 2005.

(b) The number of DSL installations in the second quarter of 2006

The **green bar** for the second quarter rises to 5 and 5 times 1000 is 5000. So, in the second quarter of 2006, there were 5000 DSL installations.

> **CAUTION**
> Use a ruler or straightedge to line up the top of the bar with the number on the left side of the graph.

◀◀ Work Problem 2 at the Side.

OBJECTIVE 3 Read and understand a line graph. A **line graph** is often useful for showing a trend. The line graph below shows the number of trout stocked along the Feather River over a 5-month period. Each dot indicates the number of trout stocked during the month directly below that dot.

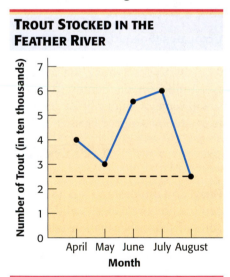

TROUT STOCKED IN THE FEATHER RIVER

EXAMPLE 3 Understanding a Line Graph

Use the line graph to find the following.

(a) In which month were the least number of trout stocked?

The lowest point on the graph is the dot directly over August, so the least number of trout were stocked in August.

(b) How many trout were stocked in August?

Use a ruler or straightedge to line up the August dot with the numbers along the left edge of the graph. The August dot is halfway between the 2 and the 3. Notice that the label on the left side says *in ten thousands*. So August is halfway between (2 • 10,000) and (3 • 10,000). It is halfway between 20,000 and 30,000. That means 25,000 trout were stocked in August.

◀◀ Work Problem 3 at the Side.

OBJECTIVE **4** **Read and understand a comparison line graph.** Two sets of data can also be compared by drawing two line graphs together as a **comparison line graph.** For example, the line graph below compares the number of minivans and the number of sport-utility vehicles (SUVs) sold during each of 5 years.

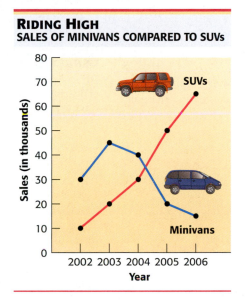

RIDING HIGH
SALES OF MINIVANS COMPARED TO SUVs

4 Use the comparison line graph at the left to find the following.

(a) The number of minivans sold in 2002, 2004, 2005, and 2006.

(b) The number of SUVs sold in 2002, 2003, 2004, and 2005

EXAMPLE 4 **Interpreting a Comparison Line Graph**

Use the comparison line graph above to find the following.

(a) The number of minivans sold in 2003
    Find the dot on the **blue line** above 2003. Use a ruler or straightedge to line up the dot with the numbers along the left edge. The dot is halfway between 40 and 50, which is 45. Then, 45 times 1000 is 45,000 minivans sold in 2003.

(b) The number of SUVs sold in 2006
    The **red line** on the graph shows that 65,000 SUVs were sold in 2006.

(c) The first full year in which the number of SUVs sold was greater than the number of minivans sold

NOTE
Both the double-bar graph and the comparison line graph are used to compare two or more sets of data.

Work Problem 4 at the Side.

ANSWERS
4. **(a)** 30,000; 40,000; 20,000;
       15,000 minivans
    (b) 10,000; 20,000; 30,000; 50,000 SUVs
    (c) 2005

Real-Data Applications

Grocery Shopping

1. The graph at the right is a double-bar graph. What is it about? Write a brief paragraph describing the general purpose of the graph.

2. Which two strategies do women use most often?

3. Which two strategies do men use most often?

4. Which two strategies show the greatest difference in use between the men and women surveyed?

5. Which two strategies show the least difference?

6. The sum of the percents for the women's responses and the men's responses do not add to 100%. Why?

7. Is it possible to decide how many of the people in the survey use *none* of the strategies listed? Why?

8. Sometimes people "jump to conclusions" without enough evidence. Which of these conclusions are appropriate, based on the information in the graph?

   (a) Men and women use a variety of saving strategies when grocery shopping.
   (b) Men spend more for groceries than women.
   (c) Men eat more groceries than women.
   (d) Women are better grocery shoppers than men.
   (e) There are some differences in the grocery shopping strategies used by men and women.

9. Conduct a survey of your class members. Find out how many of them regularly use each of the saving strategies shown in the graph. Complete the table below.

10. Make a double-bar graph showing your survey data. How is your data similar to the graph shown above? How is it different?

GENDER BUYS

On average, each person spends $32 weekly on groceries. Men average $35 and women $32. Saving strategies include:

Stock up on bargains — 26% / 28%
Stick to list — 23% / 15%
Check ads for specials — 21% / 33%
Mail/newspaper coupons — 21% / 30%
Buy store brands — 21% / 21%
In-store coupons — 20% / 25%

Men ▬ Women ▬

Source: Food Marketing Institute.

Strategy	Number of Women Using Strategy	Percent of Women Using Strategy	Number of Men Using Strategy	Percent of Men Using Strategy
Stock up on bargains				
Stick to list				
Check ads for specials				
Mail/newspaper coupons				
Buy store brands				
In-store coupons				
Total				

10.2 Exercises

| FOR EXTRA HELP | 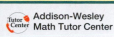 Addison-Wesley Math Tutor Center |  MathXL | Digital Video Tutor CD 6 Videotape 18 | Student's Solutions Manual | 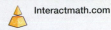 MyMathLab | Interactmath.com |

The American Farm Bureau Federation reports that the average adult in the United States will work 40 days (rounded to the nearest day) to earn enough to pay the annual household food bill. This was found by multiplying the average percent of household income spent on food by 365 (the number of days in a year). This bar graph shows the percent of income spent in various countries of the world. Use this graph to answer Exercises 1–6. See Example 1.

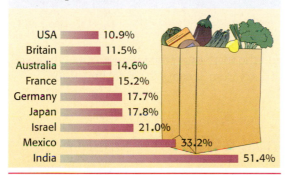

TAKING A BITE OUT OF HOUSEHOLD INCOME

The average American adult will work 40 days each year to earn enough to pay the household food bill. Percent of household income spent on food in:

USA 10.9%
Britain 11.5%
Australia 14.6%
France 15.2%
Germany 17.7%
Japan 17.8%
Israel 21.0%
Mexico 33.2%
India 51.4%

Source: American Farm Bureau Federation.

1. In which country is the highest percent of income spent on food? What percent is this?

2. In which country is the lowest percent of income spent on food? What percent is this?

3. List all countries in the graph in which less than 15% of household income is spent, on average, for food.

4. List all countries in the graph in which more than 20% of household income is spent, on average, for food.

5. How many days each year will the average adult have to work to earn enough to pay for food in Mexico? Round to the nearest day.

6. How many days each year will the average adult have to work to earn enough to pay for food in Israel? Round to the nearest day.

This double-bar graph shows the number of workers who were unemployed in a city during the first six months of 2005 and 2006. Use this graph to answer Exercises 7–12. See Example 2.

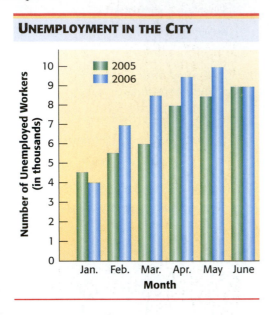

UNEMPLOYMENT IN THE CITY

7. In which month in 2006 were the greatest number of workers unemployed? What was the total number unemployed in that month?

8. How many workers were unemployed in January of 2005?

9. How many more workers were unemployed in February of 2006 than in February of 2005?

10. How many fewer workers were unemployed in March of 2005 than in March of 2006?

11. Find the increase in the number of unemployed workers from February 2005 to April 2006.

12. Find the increase in the number of unemployed workers from January 2006 to June 2006.

This double-bar graph shows sales of super unleaded and supreme unleaded gasoline at a service station for each of 5 years. Use this graph to answer Exercises 13–18. See Example 2.

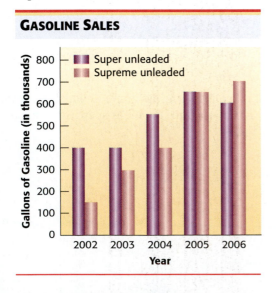

GASOLINE SALES

13. How many gallons of supreme unleaded gasoline were sold in 2002?

14. How many gallons of super unleaded gasoline were sold in 2005?

15. In which year did the greatest difference in sales between super unleaded and supreme unleaded gasoline occur? Find the difference.

16. In which year did the sales of supreme unleaded gasoline surpass the sales of super unleaded gasoline?

17. Find the increase in supreme unleaded gasoline sales from 2002 to 2006.

18. Find the increase in super unleaded gasoline sales from 2002 to 2006.

This line graph shows how the personal computer (PC) has evolved over the two decades of its existence. What began as a technician's dream is now a common tool of business and home life. Use this line graph to answer Exercises 19–24. See Example 3.

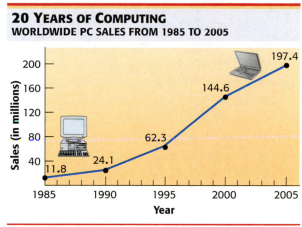

20 YEARS OF COMPUTING
WORLDWIDE PC SALES FROM 1985 TO 2005

Source: Gartner Dataquest.

19. Find the number of PCs shipped in 1990.

20. What was the number of PCs shipped in 1995?

21. Find the increase in the number of PCs shipped in 2005 from the number shipped in 1995.

22. How many more PCs were shipped in 2000 than in 1990?

23. Give two possible explanations for the increase in the number of PCs shipped.

24. Give two possible conditions that could result in a decrease in PC shipments in the future.

This comparison line graph shows the number of compact discs (CDs) sold by two different chain stores during each of 5 years. Use this graph to find the annual number of CDs sold in each year shown in Exercises 25–30. See Example 4.

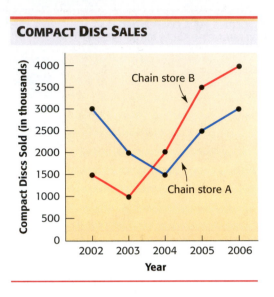

COMPACT DISC SALES

25. Chain store A in 2006

26. Chain store A in 2005

27. Chain store A in 2004

28. Chain store B in 2006

29. Chain store B in 2005

30. Chain store B in 2004

31. Looking at the comparison line graph above, which store would you like to own? Explain why. Based on the graph, what amount of sales would you predict for your store in 2007?

32. In the comparison line graph above, store B used to have lower sales than store A. What might have happened to cause this change? Give two possible explanations.

33. Explain in your own words why a bar graph or a line graph (not a double-bar graph or comparison line graph) can be used to show only one set of data.

34. The double-bar graph and the comparison line graph are both useful for comparing two sets of data. Explain how this works and give your own example.

This comparison line graph shows the sales and profits of Tacos-to-Go for each of 4 years. Use the graph to answer Exercises 35–42. See Example 4.

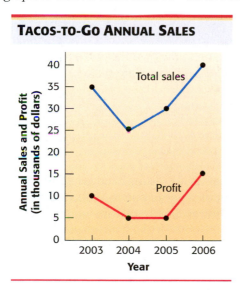

TACOS-TO-GO ANNUAL SALES

35. Total sales in 2006

36. Total sales in 2005

37. Total sales in 2004

38. Profit in 2006

39. Profit in 2005

40. Profit in 2004

41. Give two possible explanations for the decrease in sales from 2003 to 2004 and two possible explanations for the increase in sales from 2004 to 2006.

42. Based on the graph, what conclusion can you make about the relationship between sales and profits?

RELATING CONCEPTS (EXERCISES 43–48)   For Individual or Group Work

This newspaper clipping, adapted from USA Today, *shows some statistics regarding LifeSavers. Use this information to* **work Exercises 43–48 in order.**

ROLL WITH IT

Sweet-toothed fans recently voted to change three of the original flavors. The new five-flavor roll will include cherry, watermelon, pineapple, raspberry, and blackberry. The orange, lemon, and lime flavors have been replaced.

LifeSavers by the numbers:

Year first flavor invented: **1912 (Pep-O-Mint)**

Year five-flavor roll invented: **1935**

Total number of flavors today: **25**

Number of candies per roll: **14**

LifeSavers produced daily: **3 million**

Pounds of sugar used per day: **250,000**

Number of miniature rolls given out at Halloween: **88 million**

43. The first LifeSaver flavor was Pep-O-Mint. How long after the Pep-O-Mint LifeSaver was invented was the five-flavor roll invented?

44. In addition to the five flavors, how many more flavors of LifeSavers are there?

45. Find the number of rolls of LifeSavers produced daily. Round to the nearest whole number.

46. Use your answer from Exercise 45 to find the amount of sugar in one roll of LifeSavers. Round to the nearest hundredth of a pound.

47. Check your work in Exercise 46. Is the answer reasonable? Explain why or why not.

48. Name three possible causes of errors in statistics.

10.3 Frequency Distributions and Histograms

The owner of Towne Insurance Agency has kept track of her personal phone sales call activity over the past 50 weeks. The number of sales calls made for each of the weeks is given below. Read down the columns, beginning with the left column, for successive weeks of the year.

75	65	40	50	45	30	30	35	45	25
75	70	60	55	30	25	44	30	35	30
75	70	50	30	50	20	30	30	20	25
60	62	45	45	48	40	35	25	20	25
75	45	50	40	35	40	40	30	27	40

OBJECTIVE 1 Understand a frequency distribution. A long list of numbers can be confusing. You can make the data easier to read by putting it in a special type of table called a **frequency distribution.**

EXAMPLE 1 Preparing a Frequency Distribution

Using the data above, construct a table that shows each possible number of sales calls. Then go through the original data and place a *tally* mark (I) in the tally column next to each corresponding value. Total the tally marks and place the totals in the third column. The result is a frequency distribution table.

Number of Sales Calls	Tally	Frequency	Number of Sales Calls	Tally	Frequency
20	III	3	48	I	1
25	IIII	5	50	IIII	4
27	I	1	55	I	1
30	IIII IIII	9	60	II	2
35	IIII	4	62	I	1
40	IIII I	6	65	I	1
44	I	1	70	II	2
45	IIII	5	75	IIII	4

Work Problem 1 at the Side.

OBJECTIVE 2 Arrange data in class intervals. The frequency distribution given in Example 1 above contains a great deal of information—perhaps too much to digest. It can be simplified by combining the number of sales calls into groups, forming the class intervals shown below.

GROUPED DATA

Class Intervals (Number of Sales Calls)	Class Frequency (Number of Weeks)
20–29	9
30–39	13
40–49	13
50–59	5
60–69	4
70–79	6

OBJECTIVES

1. Understand a frequency distribution.
2. Arrange data in class intervals.
3. Read and understand a histogram.

1 Use the frequency distribution table at the left to find the following.

(a) The least number of sales calls made in a week

(b) The most common number of sales calls made in a week

(c) The number of weeks in which 35 calls were made

(d) The number of weeks in which 45 calls were made

2 Use the grouped data for the insurance agency on the previous page to answer each question.

(a) During how many weeks were fewer than 50 calls made?

(b) During how many weeks were 50 or more calls made?

> **NOTE**
> The number of class intervals in the left column of the grouped data table is arbitrary. Grouped data usually has between 5 and 15 class intervals.

EXAMPLE 2 Analyzing a Frequency Distribution

Use the grouped data for the insurance agency (on the preceding page) to answer the following questions.

(a) During how many weeks were fewer than 30 calls made?
The first class in the grouped data table (20–29) is the number of weeks during which fewer than 30 calls were made. Therefore, the owner made fewer than 30 calls during 9 weeks out of the 50 weeks shown.

(b) During how many weeks were 40 or more calls made?
The last four classes in the grouped data table are the number of weeks during which 40 or more calls were made.

$$13 + 5 + 4 + 6 = 28 \text{ weeks}$$

◀◀◀ Work Problem 2 at the Side.

OBJECTIVE 3 Read and understand a histogram. The results in the grouped data table have been used to draw the special bar graph below, which is called a **histogram.** In a histogram, the width of each bar represents a range of numbers (*class interval*). The height of each bar in a histogram gives the *class frequency,* that is, the number of occurrences in each class interval.

3 Use the histogram at the right to answer each question.

(a) During how many weeks were fewer than 60 calls made?

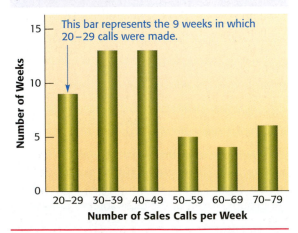

SALES CALL DATA FOR THE PAST 50 WEEKS (GROUPED DATA)

This bar represents the 9 weeks in which 20–29 calls were made.

Number of Weeks / Number of Sales Calls per Week

(b) During how many weeks were 60 or more calls made?

EXAMPLE 3 Using a Histogram

Use the histogram to find the number of weeks in which fewer than 40 calls were made.
Because 20–29 calls were made during 9 of the weeks and 30–39 calls were made during 13 of the weeks, the number of weeks in which fewer than 40 calls were made is $9 + 13 = 22$ weeks.

◀◀◀ Work Problem 3 at the Side.

ANSWERS
2. (a) 35 weeks (b) 15 weeks
3. (a) 40 weeks (b) 10 weeks

10.3 Exercises

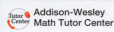
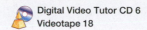

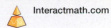

The Quilters Club of America recorded the ages of its members and used the results to construct this histogram. Use the histogram to answer Exercises 1–6. See Example 3.

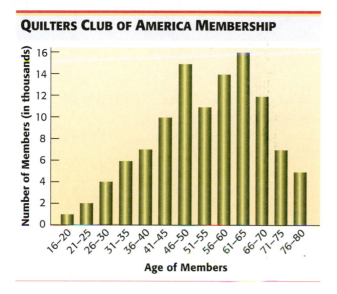

1. The greatest number of members are in which age group? How many members are in that group?

2. The least number of members are in which age group? How many members are in that group?

3. Find the number of members 35 years of age and under.

4. Find the number of members ages 61 to 80.

5. How many members are 41 to 60 years of age?

6. How many members are 46 to 55 years of age?

This histogram shows the annual earnings for the part-time employees of Wally World Amusement Park. Use this histogram to answer Exercises 7–12. See Example 3.

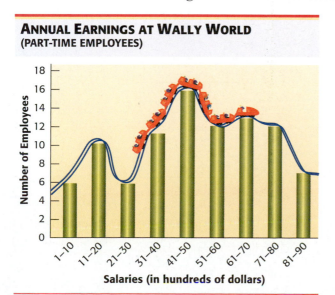

7. The greatest number of employees are in which earnings group? How many are in that group?

8. The fewest number of employees are in which earnings groups? How many are in each group?

9. Find the number of employees who earn $3100 to $4000.

10. Find the number of employees who earn $1100 to $2000.

11. How many employees earn $5000 or less?

12. How many employees earn $6100 or more?

13. Describe class interval and class frequency. How are they used when preparing a histogram?

14. What might be a problem of using two few or too many class intervals?

This list shows the number of new accounts opened annually by the employees of the Schools Credit Union. Use it to complete the table. See Example 1.

186	191	144	198	147	158	174
193	142	155	174	162	151	178
145	151	199	182	147	195	146

	Class Intervals (Number of New Accounts)	**Tally**	**Class Frequency (Number of Employees)**
15.	140–149	_____	_____
16.	150–159	_____	_____
17.	160–169	_____	_____
18.	170–179	_____	_____
19.	180–189	_____	_____
20.	190–199	_____	_____

As part of a new college-assistance program, the manager of Zapp Software asked her 30 employees how many college credits each had completed. Use her list of responses to complete the following table. See Examples 1–3.

74	133	0	127	20	30
103	27	158	118	138	121
149	132	64	141	130	76
42	50	95	56	65	104
4	140	12	88	119	64

	Class Intervals (Number of Credits)	**Tally**	**Class Frequency (Number of Employees)**
21.	0–24	_____	_____
22.	25–49	_____	_____
23.	50–74	_____	_____
24.	75–99	_____	_____
25.	100–124	_____	_____
26.	125–149	_____	_____
27.	150–174	_____	_____

28. Construct a histogram by using the data in Exercises 21–27.

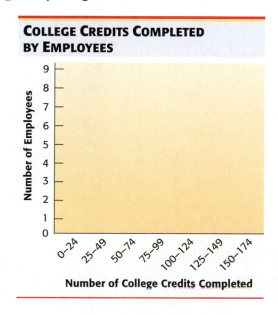

COLLEGE CREDITS COMPLETED BY EMPLOYEES

Number of Employees

9
8
7
6
5
4
3
2
1
0

0–24 25–49 50–74 75–99 100–124 125–149 150–174

Number of College Credits Completed

Century 21 All Professional Realty has 80 salespeople spread over its five offices. The number of homes sold by each of these salespeople during the past year is shown below. Use these numbers to complete the following table. See Example 1.

9	33	14	8	17	10	25	11	4	16	3	9	5	7	14	18
15	24	19	30	16	31	21	20	30	2	6	6	27	17	3	32
3	8	5	11	15	26	7	18	29	10	7	3	12	9	25	15
11	6	10	4	2	35	10	25	5	19	34	2	4	14	11	28
8	13	25	15	23	26	12	4	22	12	21	12	22	10	18	21

	Class Intervals (New Homes Sold)	Tally	Class Frequency (Number of Salespeople)
29.	1–5	_____	_____
30.	6–10	_____	_____
31.	11–15	_____	_____
32.	16–20	_____	_____
33.	21–25	_____	_____
34.	26–30	_____	_____
35.	31–35	_____	_____

36. Make a histogram showing the results from Exercises 29–35.

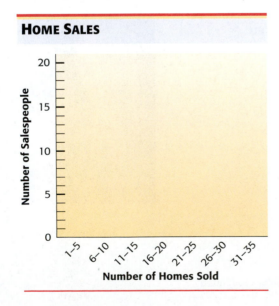

10.4 Mean, Median, and Mode

Businesses, governments, colleges, and others working with lists of numbers are often faced with the problem of analyzing a mass of raw data. The measures discussed in this section are helpful for such analyses.

OBJECTIVE 1   Find the mean of a list of numbers.   When analyzing data, one of the first things to look for is a *measure of central tendency*— a single number that we can use to represent the entire list of numbers. One such measure is the *average* or **mean.** The mean can be found with the following formula.

Finding the Mean (Average)

$$\text{mean} = \frac{\text{sum of all values}}{\text{number of values}}$$

EXAMPLE 1   Finding the Mean

David had test scores of 84, 90, 95, 98, and 88. Find the mean (average) of these scores.

Use the formula for finding the mean. Add up all the test scores and then divide by the number of tests.

$$\text{mean} = \frac{84 + 90 + 95 + 98 + 88}{5} \quad \begin{array}{l} \leftarrow \text{Sum of test scores} \\ \\ \leftarrow \text{Number of tests} \end{array}$$

$$\text{mean} = \frac{455}{5} \quad \text{Divide.}$$

$$\text{mean} = 91$$

David has a mean score of 91.

Work Problem 1 at the Side. ▷▷▷

EXAMPLE 2   Applying the Average or Mean

Milk sales at a local 7-Eleven for each of the days last week were

$$\$86, \$103, \$118, \$117, \$126, \$158, \text{ and } \$149$$

For the local 7-Eleven, find the mean (rounded to the nearest cent) as shown below.

$$\text{mean} = \frac{\$86 + \$103 + \$118 + \$117 + \$126 + \$158 + \$149}{7}$$

$$\text{mean} = \frac{\$857}{7}$$

$$\text{mean} \approx \$122.43$$

The mean daily sales amount for milk was $122.43.

Work Problem 2 at the Side. ▷▷▷

OBJECTIVES

1 Find the mean of a list of numbers.
2 Find a weighted mean.
3 Find the median.
4 Find the mode.

1 Tanya has test scores of 96, 98, 84, 88, 82, and 92. Find her mean (average) score.

2 Find the mean for each list of numbers. Round to the nearest cent if necessary.

(a) Monthly gasoline expenses of $50.28, $85.16, $110.50, $78, $120.70, $58.64, $73.80, $86.24, $67.85, $96.56, $138.65, $48.90

(b) A list of the sales for one year at eight different office supply stores: $749,820; $765,480; $643,744; $824,222; $485,886; $668,178; $702,294; $525,800

ANSWERS

1. 90

2. **(a)** $\frac{\$1015.28}{12} \approx \$84.61$ (rounded)

   (b) $\frac{\$5,365,424}{8} = \$670,678$

❸ The numbers below show the amount that Heather Hall spent for lottery tickets and the number of days that she spent that amount. Find the weighted mean.

Value	Frequency
$ 2	4
$ 4	6
$ 6	5
$ 8	6
$10	12
$12	5
$14	8
$16	4

OBJECTIVE 2 Find a weighted mean. Some items in a list might appear more than once. In this case, we find a **weighted mean,** in which each value is "weighted" by multiplying it by the number of times it occurs.

EXAMPLE 3 Understanding the Weighted Mean

The following table shows the amount of contribution and the number of times (frequency) the amount was given to a food pantry. Find the weighted mean.

Contribution Value	Frequency
$ 3	4 ← 4 people each contributed $3
$ 5	2
$ 7	1
$ 8	5
$ 9	3
$10	2
$12	1
$13	2

The same amount was given by more than one person: for example, $5 was given twice and $8 was given five times. Other amounts, such as $12, were given once. To find the mean, multiply each contribution value by its frequency. Then add the products. Next, add the numbers in the *frequency* column to find the total number of values.

Value	Frequency	Product
$ 3	4	$(3 \cdot 4) = \$12$
$ 5	2	$(5 \cdot 2) = \$10$
$ 7	1	$(7 \cdot 1) = \$ 7$
$ 8	5	$(8 \cdot 5) = \$40$
$ 9	3	$(9 \cdot 3) = \$27$
$10	2	$(10 \cdot 2) = \$20$
$12	1	$(12 \cdot 1) = \$12$
$13	2	$(13 \cdot 2) = \$26$
Totals	**20**	**$154**

Finally, divide the totals.

$$\text{mean} = \frac{\$154}{20} = \$7.70$$

The mean contribution to the food pantry was $7.70.

Work Problem 3 at the Side.

A common use of the weighted mean is to find a student's *grade point average* (GPA), as shown by the next example.

EXAMPLE 4 Applying the Weighted Mean

Find the grade point average for a student earning the following grades. Assume A = 4, B = 3, C = 2, D = 1, and F = 0. The number of credits determines how many times the grade is counted (the frequency).

Course	Credits	Grade	Credits • Grade
Mathematics	4	A (= 4)	4 • 4 = 16
Speech	3	C (= 2)	3 • 2 = 6
English	3	B (= 3)	3 • 3 = 9
Computer science	2	A (= 4)	2 • 4 = 8
Art history	2	D (= 1)	2 • 1 = 2
Totals	**14**		**41**

It is common to round grade point averages to the nearest hundredth. So the grade point average for this student is rounded to 2.93.

$$\text{GPA} = \frac{41}{14} \approx 2.93$$

Work Problem 4 at the Side. ▶▶▶

OBJECTIVE 3 Find the median. Because it can be affected by extremely high or low numbers, the mean is often a poor indicator of central tendency for a list of numbers. In cases like this, another measure of central tendency, called the *median,* can be used. The **median** divides a group of numbers in half; half the numbers lie above the median, and half lie below the median.

Find the median by listing the numbers *in order* from *smallest* to *largest.* If the list contains an *odd* number of items, the median is the *middle number.*

EXAMPLE 5 Finding the Median

Find the median for the following list of prices for women's T-shirts.

$9, $23, $15, $8, $18, $12, $24

First arrange the numbers in numerical order from smallest to largest.

Smallest → 8, 9, 12, 15, 18, 23, 24 ← Largest

Next, find the middle number in the list.

8, 9, 12, **15**, 18, 23, 24

Three are below. ↓ Three are above.
Middle number.

The median price is $15.

Work Problem 5 at the Side. ▶▶▶

---

4 Find the grade point average (GPA) for Greg Barnes, who earned the following grades last semester. Round to the nearest hundredth.

Course	Credits	Grade
Mathematics	3	A (= 4)
P.E.	1	C (= 2)
English	3	C (= 2)
Keyboarding	2	B (= 3)
Recreation	4	B (= 3)

5 Find the median for the following weights of bagged groceries.

14 lb, 18 lb, 10 lb, 17 lb, 15 lb, 19 lb, 20 lb

6 Find the median for the following list of measurements.

125 m, 87 m, 96 m, 108 m, 136 m, 74 m

If a list contains an *even* number of items, there is no single middle number. In this case, the median is defined as the mean (average) of the *middle two* numbers.

EXAMPLE 6   **Finding the Median for an Even Number of Items**

Find the median for the following list of ages.

$$74, 7, 15, 13, 25, 28, 47, 59, 32, 68$$

First arrange the numbers in numerical order from smallest to largest. Then find the middle two numbers.

Smallest → 7, 13, 15, 25, **28, 32**, 47, 59, 68, 74 ← Largest

Middle two numbers

The median age is the mean of the two middle numbers.

$$\text{median} = \frac{\mathbf{28 + 32}}{2} = \frac{\mathbf{60}}{2} = \mathbf{30 \text{ years}}$$

◀◀◀ **Work Problem 6 at the Side.**

7 Find the mode for each list of numbers.

(a) Ages of summer work applicants (in years): 19, 18, 22, 20, 18

OBJECTIVE **4** **Find the mode.**   The last important statistical measure is the **mode,** the number that occurs *most often* in a list of numbers. For example, if the test scores for 10 students were

↓   ↓   ↓

74, 81, 39, **74**, 82, 80, 100, 92, **74**, and 85,

then the mode is 74. Three students earned a score of 74, so 74 appears more times on the list than any other score. (It is not necessary to place the numbers in numerical order when looking for the mode.)

A list can have two modes; such a list is sometimes called **bimodal.** If no number occurs more frequently than any other number in a list, the list has *no mode*.

(b) Total points on a screening exam: 312, 219, 782, 312, 219, 426

EXAMPLE 7   **Finding the Mode**

Find the mode for each list of numbers.

(a) 51, 32, 49, 73, 49, 90

The number **49** occurs more often than any other number; therefore, 49 is the mode.

(b) 482, 485, 483, 485, 487, 487, 489

Because both **485** and **487** occur twice, each is a mode. This list is *bimodal.*

(c) Monthly commissions of sales people: $1706, $1289, $1653, $1892, $1301, $1782

(c) $10,708; $11,519; $10,972; $12,546; $13,905; $12,182

No number occurs more than once. This list has *no mode*.

◀◀◀ **Work Problem 7 at the Side.**

Measures of Central Tendency

The **mean** is the sum of all the values divided by the number of values. It is the mathematical average.

6. $\dfrac{96 + 108}{2} = \dfrac{204}{2} = 102$ m

7. **(a)** 18 yr
   (b) bimodal, 219 points and 312 points (this list has two modes)
   (c) no mode (no number occurs more than once)

The **median** is the middle number in a group of values that are listed from smallest to largest. It divides a group of numbers in half.

The **mode** is the value that occurs most often in a group of values.

10.4 Exercises

 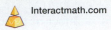

Find the mean for each list of numbers. Round answers to the nearest tenth if necessary. See Example 1.

1. Shopping center ages (in years) of 6, 22, 15, 2, 8, 13

2. Minutes of cell phone use each day: 53, 77, 38, 29, 46, 48, 52

3. Inches of rain per month of 3.1, 1.5, 2.8, 0.8, 4.1

4. Algebra quiz scores of 32, 26, 30, 19, 51, 46, 38, 39

5. Annual salaries of $38,500; $39,720; $42,183; $21,982; $43,250

6. Numbers of people attending baseball games: 27,500; 18,250; 17,357; 14,298; 33,110

Solve each application problem. See Example 2.

7. The Athletic Shoe Store sold shoes at the following prices: $75.52, $36.15, $58.24, $21.86, $47.68, $106.57, $82.72, $52.14, $28.60, $72.92. Find the average (mean) shoe sales amount.

8. In one evening, a waitress collected the following checks from her dinner customers: $30.10, $42.80, $91.60, $51.20, $88.30, $21.90, $43.70, $51.20. Find the average (mean) dinner check amount.

Find the weighted mean. Round answers to the nearest tenth. See Example 3.

9.

Customers Each Hour	Frequency
8	2
11	12
15	5
26	1

10.

Deliveries Each Week	Frequency
4	1
8	3
16	5
20	1

11.

Fish per Boat	Frequency
12	4
13	2
15	5
19	3
22	1
23	5

12.

Patients per Clinic	Frequency
25	1
26	2
29	5
30	4
32	3
33	5

Solve each application problem. See Example 4.

13. The table below shows the face value (policy amount) of life insurance policies sold and the number of policies sold for each amount by the New World Life Company during one week. Find the weighted mean amount for the policies sold.

Policy Amount	Number of Policies Sold
$ 10,000	6
$ 20,000	24
$ 25,000	12
$ 30,000	8
$ 50,000	5
$100,000	3
$250,000	2

14. Detroit Metro-Sales Company prepared the following table showing the gasoline mileage obtained by each of the cars in their automobile fleet. Find the weighted mean to determine the miles per gallon for the fleet of cars.

Miles per Gallon	Number of Autos
15	5
20	6
24	10
30	14
32	5
35	6
40	4

Find the median for each list of numbers. See Example 5.

15. Number of books loaned: 125, 100, 150, 135, 114, 172

16. Number of hits for a World Wide Web site (in thousands): 140, 85, 122, 114, 98

17. Calories in fast-food menu items: 501, 412, 521, 515, 298, 621, 346, 528

18. Number of cars in the parking lot each day: 520, 523, 513, 1283, 338, 509, 290, 420

Find the mode or modes for each list of numbers. See Example 7.

19. Porosity of soil samples: 21%, 18%, 21%, 28%, 22%, 21%, 25%

20. Low daily temperatures: 21, 32, 46, 32, 49, 32, 49

21. Ages of residents (in years) at Leisure Village: 74, 68, 68, 68, 75, 75, 74, 74, 70

22. Number of pages read: 86, 84, 79, 75, 88, 66, 72, 85, 71

23. When can the mean be a poor measure of central tendency? What might you do in such a situation?

24. What is the purpose of the weighted mean? Give an example of where it is used.

25. When is the median a better measure of central tendency than the mean to describe a set of data? Make up a list of numbers to illustrate your explanation. Calculate both the mean and the median.

26. Suppose you own a hat shop and can order a certain hat in only one size. You look at last year's sales to decide on the size to order. Should you find the mean, median, or mode for these sales? Explain your answer.

Find the grade point average for students earning the following grades. Assume A = 4, B = 3, C = 2, D = 1, *and* F = 0. *Round answers to the nearest hundredth.*

27.

Credits	Grade
4	B
2	A
5	C
1	F
3	B

28.

Credits	Grade
3	A
3	B
4	B
3	C
3	C

29.

Credits	Grade
2	A
3	C
4	A
1	C
4	B

30.

Credits	Grade
3	A
2	A
5	B
4	A
1	A

RELATING CONCEPTS (EXERCISES 31–40) For Individual or Group Work

Gluco Industries manufactures and sells glucose monitors and other diabetes-related products. The number of sales calls made over an 8-week period by two sales representatives, Scott Samuels and Rob Stricker, is shown below. Use this information to **work Exercises 31–40 in order.**

| | Number of Calls | |
Week	Samuels	Stricker
1	39	21
2	15	22
3	40	20
4	22	23
5	13	19
6	22	24
7	17	25
8	8	22

31. Find the total number of sales calls made by each of the sales representatives.

32. Find the mean number of sales calls made by Samuels and by Stricker.

33. Find the median number of sales calls made by Samuels and by Stricker.

34. Find the mode for the number of sales calls for each of the sales representatives.

35. How do the mean, median, and mode for the two sales representatives compare?

36. Explain how the number of weekly sales calls made by each of the sales representatives compare to each other.

To show the variation *or spread of the number of sales calls made by each of the sales representatives requires some* **measure of the dispersion,** *or spread of the numbers around the mean. A common measure of the dispersion is the* **range.** *The range is the* difference between the largest value and the smallest value in the set of numbers.

37. Find the range for the number of sales calls made by Samuels.

38. Find the range for the number of sales calls made by Stricker.

39. Is the performance data for these two sales representatives sufficient to accurately determine which is the best? What else might you want to know?

40. List three possible explanations for the wide variation (range) in the number of sales calls for Samuels.

KEY TERMS

10.1	**circle graph**	A circle graph shows how a total amount is divided into parts or sectors. It is based on percents of 360°.
	protractor	A protractor is a device (usually in the shape of a half-circle) used to measure the number of degrees in angles or parts of a circle.
10.2	**bar graph**	A bar graph uses bars of various heights or lengths to show quantity or frequency.
	double-bar graph	A double-bar graph compares two sets of data by showing two sets of bars.
	line graph	A line graph uses dots connected by lines to show trends.
	comparison line graph	A comparison line graph shows how two sets of data relate to each other by showing a line graph for each item.
10.3	**frequency distribution**	A frequency distribution is a table that includes a column showing each possible number in the data collected. The original data is then entered in another column using a tally mark for each corresponding value. The tally marks are totaled and the totals are placed in a third column.
	histogram	A histogram is a bar graph in which the width of each bar represents a range of numbers (class interval) and the height represents the quantity or frequency of items that fall within the interval.
10.4	**mean**	The mean is the sum of all the values divided by the number of values. It is often called the *average*.
	weighted mean	The weighted mean is a mean calculated so that each value is multiplied by its frequency.
	median	The median is the middle number in a group of values that are listed from smallest to largest. It divides a group of values in half. If there are an even number of values, the median is the mean (average) of the two middle values.
	mode	The mode is the value that occurs most often in a group of values.
	bimodal	A list of numbers is bimodal when it has two modes. The two values occur the same number of times.
	dispersion	The dispersion is the variation or spread of the numbers around the mean.
	range	The range is a common measure of the dispersion of numbers. It is the difference between the largest value and the smallest value in the set of numbers.

Mean or average: $\text{mean} = \dfrac{\text{sum of all values}}{\text{number of values}}$

See how well you have learned the vocabulary in this chapter. Answers follow the Quick Review.

1. A **circle graph**
   A. uses bars of various heights to show quantity or frequency
   B. shows how a total amount is divided into parts or sectors
   C. uses dots connected by lines to show trends
   D. uses bars of various widths to represent a range of numbers.

2. A **bar graph**
   A. uses bars of various heights to show quantity or frequency
   B. shows how a total amount is divided into parts or sectors
   C. uses dots connected by lines to show trends
   D. uses bars of various widths to represent a range of numbers.

3. A **histogram** is a graph in which
   A. tally marks are used to record original data
   B. two sets of data are compared using two sets of bars
   C. dots are connected by lines to show trends
   D. the width of each bar represents a range of numbers and the height represents the frequency of items within that range.

4. A **protractor** is a device used to
   A. construct a histogram
   B. calculate measures of central tendency
   C. measure the number of degrees in angles or parts of a circle
   D. compare two sets of data.

5. The **mean** is
   A. calculated so that each value is multiplied by its frequency
   B. the sum of all values divided by the number of values
   C. the middle number in a group of values that are listed from smallest to largest
   D. the value that occurs most often in a group of values.

6. The **mode** is
   A. calculated so that each value is multiplied by its frequency
   B. the sum of all values divided by the number of values
   C. the middle number in a group of values that are listed from smallest to largest
   D. the value that occurs most often in a group of values.

Concepts

Examples

10.1 *Constructing a Circle Graph*

Step 1 Determine the percent of the total for each item.
Step 2 Find the number of degrees out of 360° that each
percent represents.
Step 3 Use a protractor to measure the number of degrees
for each item in the circle.

Construct a circle graph for the following table, which lists
expenses for a business trip.

Item	Amount
Transportation	$200
Lodging	$300
Food	$250
Entertainment	$150
Other	$100
Total	**$1000**

Item	Amount	Percent of Total	Sector Size
Transportation	$200	$\frac{\$200}{\$1000} = \frac{1}{5} =$ **20%** so $360° \cdot 20\%$ $= 360 \cdot 0.20$	$= 72°$
Lodging	$300	$\frac{\$300}{\$1000} = \frac{3}{10} =$ **30%** so $360° \cdot 30\%$ $= 360 \cdot 0.30$	$= 108°$
Food	$250	$\frac{\$250}{\$1000} = \frac{1}{4} =$ **25%** so $360° \cdot 25\%$ $= 360 \cdot 0.25$	$= 90°$
Entertainment	$150	$\frac{\$150}{\$1000} = \frac{3}{20} =$ **15%** so $360° \cdot 15\%$ $= 360 \cdot 0.15$	$= 54°$
Other	$100	$\frac{\$100}{\$1000} = \frac{1}{10} =$ **10%** so $360° \cdot 10\%$ $= 360 \cdot 0.10$	$= 36°$

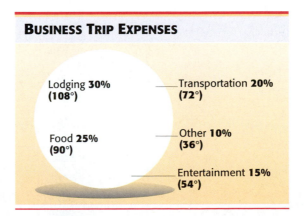

BUSINESS TRIP EXPENSES

Lodging **30%**
(108°)

Transportation **20%**
(72°)

Food **25%**
(90°)

Other **10%**
(36°)

Entertainment **15%**
(54°)

Concepts

Examples

10.2 *Reading a Bar Graph*
The height of the bar is used to show the quantity or frequency (number) in a specific category. Use a ruler or straightedge to line up the top of each bar with the numbers on the left side of the graph.

Use the bar graph below to determine the number of students who earned each letter grade.

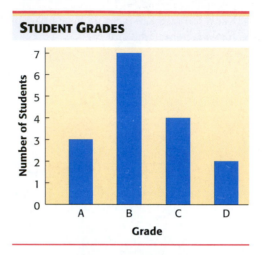

Grade	Number of Students
A	3
B	7
C	4
D	2

10.2 *Reading a Line Graph*
A dot is used to show the number or quantity in a specific class. The dots are connected with lines. This kind of graph is used to show a trend.

The line graph below shows the annual sales for the Fabric Supply Center for each of 4 years.

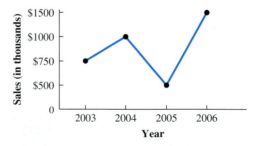

Find the sales in each year.

Year	Total Sales
2003	$750 \cdot 1000 = \$\ \ 750{,}000$
2004	$1000 \cdot 1000 = \$1{,}000{,}000$
2005	$500 \cdot 1000 = \$\ \ 500{,}000$
2006	$1500 \cdot 1000 = \$1{,}500{,}000$

Concepts

10.3 *Preparing a Frequency Distribution and a Histogram from Raw Data*

Step 1 Construct a table listing each value, and the number of times this value occurs.

Step 2 Divide the data into groups, categories, or classes.

Step 3 Draw bars representing these groups to make a histogram.

Examples

Draw a histogram for the following list of student quiz scores.

12	15	15	14
13	20	10	12
11	9	10	12
17	20	16	17
14	18	19	13

Quiz Score	Tally	Frequency	
9	I	1	1st
10	II	2	class
11	I	1	interval
12	III	3	2nd
13	II	2	class
14	II	2	interval
15	II	2	3rd
16	I	1	class
17	II	2	interval
18	I	1	4th
19	I	1	class
20	II	2	interval

Class Interval (Quiz Scores)	Frequency (Number of Students)
9–11	4
12–14	7
15–17	5
18–20	4

STUDENT QUIZ SCORES

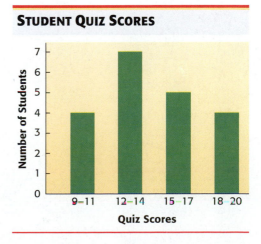

Concepts	*Examples*

10.4 *Finding the Mean (Average) of a Set of Numbers*

Step 1 Add all values to obtain a total.

Step 2 Divide the total by the number of values.

$$\text{mean (average)} = \frac{\text{sum of all values}}{\text{number of values}}$$

The test scores for Keith Zagorin in his algebra course were as follows:

80	92	92	94
76	88	84	93

Find Keith's mean (average) test score to the nearest tenth.

$$\text{mean} = \frac{80 + 92 + 92 + 94 + 76 + 88 + 84 + 93}{8}$$

$$= \frac{699}{8} \approx 87.4$$

Keith's mean test score is approximately 87.4.

10.4 *Finding the Weighted Mean*

Step 1 Multiply frequency by value.

Step 2 Add all the products from Step 1.

Step 3 Divide the sum in Step 2 by the total number of pieces of data.

This table shows the distribution of the number of school-age children in a survey of 30 families.

Number of School-Age Children	Frequency (Number of Families)
0	12
1	6
2	7
3	3
4	2
Total of 30 families	

Find the mean number of school-age children per family. Round to the nearest hundredth.

Value	Frequency	Product
0	12	$(0 \cdot 12) = 0$
1	6	$(1 \cdot 6) = 6$
2	7	$(2 \cdot 7) = 14$
3	3	$(3 \cdot 3) = 9$
4	2	$(4 \cdot 2) = 8$
Totals	**30**	**37**

$$\text{mean} = \frac{37}{30} \approx 1.23$$

The mean number of school-age children per family is approximately 1.23.

Concepts

Examples

10.4 *Finding the Median of a Set of Numbers*

Step 1   Arrange the data from smallest to largest.

Step 2   Select the middle value, or, if there is an even number of values, find the average of the two middle values.

Find the median for Keith Zagorin's test scores from the previous page.

The data arranged from smallest to largest is as follows:

$$76 \quad 80 \quad 84 \quad \underbrace{88 \quad 92}_{\text{Middle values}} \quad 92 \quad 93 \quad 94$$

The middle two values are 88 and 92. The average of these two values is

$$\frac{88 + 92}{2} = 90$$

Keith's median test score is 90.

10.4 *Finding the Mode of a Set of Values*

Find the value that appears most often in the list of values. If no value appears more than once, there is no mode. If two different values appear the same number of times, the list is bimodal.

Find the mode for Keith's test scores shown above.

The most frequently occurring score is 92 (it occurs twice). Therefore, the mode is 92.

ANSWERS TO TEST YOUR WORD POWER

1. (B) *Example:*

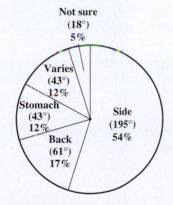

2. (A) *Example:*

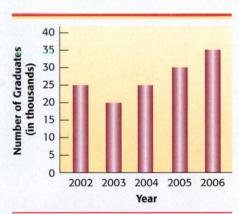

3. (D) *Example:*

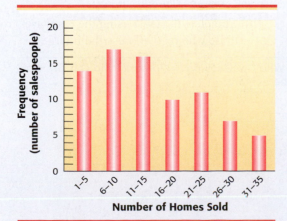

4. (C) *Example:*

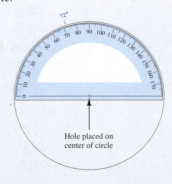

Hole placed on center of circle

5. (B) *Example:* The mean of the values $5, $9, $7, $5, $2, and $8, is

$$\frac{\$5 + \$9 + \$7 + \$5 + \$2 + \$8}{6} = \frac{\$36}{6} = \$6.$$

6. (D) *Example:* The mode of the values $5, $9, $7, $5, $2, and $8 is $5 because $5 appears twice in the list.

Real-Data Applications

Surfing the Net

1. Look at the "Source" information at the bottom of the graph. How were the numbers in the graph obtained?

2. A researcher seeks information about members of a *population*. The individuals who are polled must be representative of the *population*. Describe the population that was targeted by this survey.

3. How many people in the poll said they cut back on television viewing to find time to use the Internet?

4. Find the number of people in the poll who cut back on each of the other activities listed in the graph.

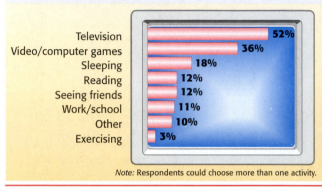

CAUGHT IN THE NET

Web users have cut back on the following activities to get more online time:

Television 52%
Video/computer games 36%
Sleeping 18%
Reading 12%
Seeing friends 12%
Work/school 11%
Other 10%
Exercising 3%

Note: Respondents could choose more than one activity.

Source: NUKE InterNETWORK poll of 500 regular users.

5. Add up all the responses to the poll from Problems 3 and 4. Why is the total more than the 500 people that were in the poll?

6. Suppose you took a similar poll of 100 students at your school. Would you expect the results to be similar to those shown in the graph? Why or why not?

7. Suppose you first asked students if they regularly used the Internet, and then took a similar poll of 100 of those students. Would you expect the results to be similar to those shown in the graph? Why or why not?

8. Conduct a survey of your class members. First find out if they regularly use the Internet. Ask those who regularly use the Internet which of the activities they cut back on to have more surfing time. Each person polled can select more than one activity.

   (a) How many students are in your class poll?

   (b) Complete the table using the responses from those who regularly use the Internet.

Activity	Number Who Cut Back on the Activity	Percent Who Cut Back on the Activity
Television		
Video/computer games		
Sleeping		
Reading		
Seeing friends		
Work/school		
Other		
Exercising		

9. Make a bar graph showing your survey data. How is your data similar to the graph shown above? How is it different?

Chapter 10

REVIEW EXERCISES

[10.1]

1. This circle graph shows the cost of a family vacation. What is the largest single expense of the vacation? How much is that item?

COST OF A FAMILY VACATION

Other $160

Food $400

Sightseeing $280

Gasoline $300

Lodging $560

Using the circle graph in Exercise 1, find each ratio. Write the ratios as fractions in lowest terms.

2. Cost of the food to the total cost of the vacation

3. Cost of gasoline to the total cost of the vacation

4. Cost of sightseeing to the total cost of the vacation

5. Cost of gasoline to the cost of the Other category

6. Cost of the lodging to the cost of the food

[10.2] *This bar graph shows recent trends in employee benefits. It includes the most frequently offered "work perks" and the percent of the responding companies offering them. The survey was conducted on-line and included 4800 companies ranging in size from 2 to 5000 employees. Use this graph to find the number of companies offering each work perk listed in Exercises 7–10 and to answer Exercises 11 and 12. (Source: Work Perks Survey, Ceridian Employer Services, www.ces.ceridian.com)*

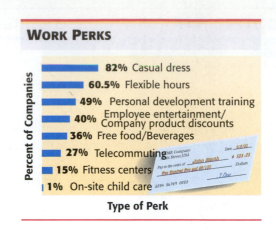

7. Casual dress

8. Free food/Beverages

9. Fitness centers

10. Flexible hours

11. Which two work perks do companies offer least often? Give one possible explanation why these work perks are not offered.

12. Which two work perks do companies offer most often? Give one possible explanation why these work perks are so popular.

This double-bar graph shows the number of acre-feet of water in Lake Natoma for each of the first six months of 2005 and 2006. Use this graph to answer Exercises 13–18.

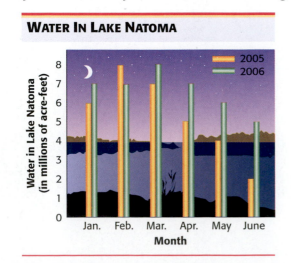

13. During which month in 2006 was the greatest amount of water in the lake? How much was there?

14. During which month in 2005 was the least amount of water in the lake? How much was there?

15. How many acre-feet of water were in the lake in June of 2006?

16. How many acre-feet of water were in the lake in May of 2005?

17. Find the decrease in the amount of water in the lake from March 2005 to June 2005.

18. Find the decrease in the amount of water in the lake from April 2006 to June 2006.

This comparison line graph shows the annual floor-covering sales of two different home improvement centers during each of 5 years. Use this graph to find the amount of annual floor-covering sales in each year shown in Exercises 19–22 and to answer Exercises 23 and 24.

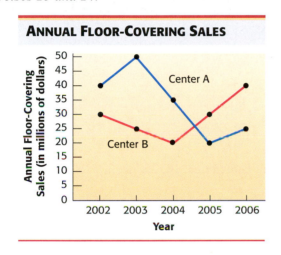

19. Center A in 2003

20. Center A in 2005

21. Center B in 2004

22. Center B in 2006

23. What trend do you see in center A's sales from 2003 to 2006? Why might this have happened?

24. What trend do you see in center B's sales starting in 2004? Why might this have happened?

[10.4] *Find the mean for each list of numbers. Round answers to the nearest tenth if necessary.*

25. Digital cameras sold: 18, 12, 15, 24, 9, 42, 54, 87, 21, 3

26. Number of harassment complaints filed: 31, 9, 8, 22, 46, 51, 48, 42, 53, 42

Find the weighted mean for each list. Round to the nearest tenth if necessary.

27.

Dollar Value	Frequency
$42	3
$47	7
$53	2
$55	3
$59	5

28.

Total Points	Frequency
243	1
247	3
251	5
255	7
263	4
271	2
279	2

Find the median for each list of numbers.

29. The number of accident forms filed: 43, 37, 13, 68, 54, 75, 28, 35, 39

30. Commissions of $576, $578, $542, $151, $559, $565, $525, $590

Find the mode or modes for each list of numbers.

31. Running shoes priced at $79, $56, $110, $79, $72, $86, $79

32. Boat launchings: 18, 25, 63, 32, 28, 37, 32, 26, 18

MIXED REVIEW EXERCISES

The Broadway Hair Salon spent $22,400 to open a new shop. This amount was spent as shown below. Find all the missing numbers in Exercises 33–37.

Item	Dollar Amount	Percent of Total	Degrees of Circle
33. Plumbing and electrical changes	$2240	10%	_____
34. Work stations	$7840	_____	126°
35. Small appliances	$4480	_____	_____
36. Interior decoration	$5600	_____	_____
37. Supplies	$2240	_____	_____

38. Draw a circle graph using the information in Exercises 33–37.

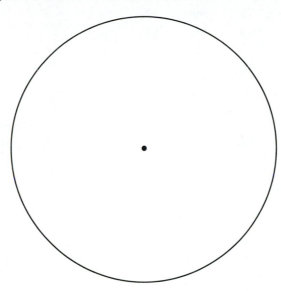

Find the mean for each list of numbers. Round answers to the nearest tenth if necessary.

39. Number of volunteers for the project: 48, 72, 52, 148, 180

40. Number of flu vaccinations in a day: 122, 135, 146, 159, 128, 147, 168, 139, 158

Find the mode or modes for each list of numbers.

41. Job applicants meeting the qualifications: 48, 43, 46, 47, 48, 48, 43

42. Number of two-bedroom apartments in each building: 26, 31, 31, 37, 43, 51, 31, 43, 43

Find the median for each list of numbers.

43. Hours worked: 4.7, 3.2, 2.9, 5.3, 7.1, 8.2, 9.4, 1.0

44. Number of e-mails each day: 35, 51, 9, 2, 17, 12, 46, 23, 3, 19, 39, 27

Here are the scores of 40 students on a computer science exam. Complete the table.

78	89	36	59	78	99	92	86
73	78	85	57	99	95	82	76
63	93	53	76	92	79	72	62
74	81	77	76	59	84	76	94
58	37	76	54	80	30	45	38

	Class Intervals (Scores)	Tally	Class Frequency (Number of Students)
45.	30–39	_____	_____
46.	40–49	_____	_____
47.	50–59	_____	_____
48.	60–69	_____	_____
49.	70–79	_____	_____
50.	80–89	_____	_____
51.	90–99	_____	_____

52. Construct a histogram by using the data in Exercises 45–51.

COMPUTER SCIENCE EXAM SCORES

Number of Students (vertical axis): 0, 5, 10, 15
Exam Scores (horizontal axis): 30–39, 40–49, 50–59, 60–69, 70–79, 80–89, 90–99

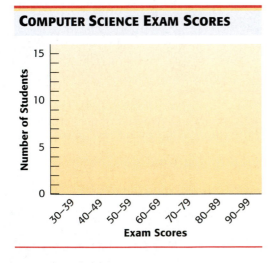

Find each weighted mean. Round answers to the nearest tenth if necessary.

53.

Test Score	Frequency
46	4
54	10
62	8
70	12
78	10

54.

Units Sold	Frequency
104	6
112	14
115	21
119	13
123	22
127	6
132	9

Chapter 10

TEST

This circle graph shows the advertising budget for Lakeland Amusement Park. Find the dollar amount budgeted for each category. The total advertising budget of Lakeland Amusement Park is $2,800,000.

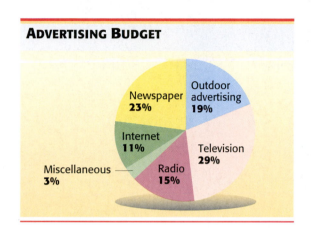

ADVERTISING BUDGET

Newspaper 23%
Outdoor advertising 19%
Internet 11%
Miscellaneous 3%
Radio 15%
Television 29%

1. Newspaper

1. _____

2. Outdoor advertising

2. _____

3. Television

3. _____

4. Radio

4. _____

5. Miscellaneous

5. _____

6. Internet

6. _____

During a one-year period, Whitings Oak Furniture Sales had the following expenses. Find all numbers missing from the chart.

Item	Dollar Amount	Percent of Total	Degrees of a Circle
7. Salaries	$144,000	30%	_____
8. Delivery expense	$48,000	10%	_____
9. Advertising	$96,000	20%	_____
10. Rent	$144,000	30%	_____
11. Other	$48,000	_____	36°

7. _____

8. _____

9. _____

10. _____

11. _____

12.

12. Draw a circle graph using the information in Problems 7–11.

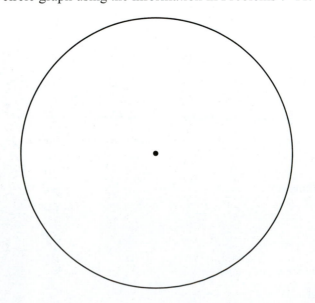

Here are the profits for each of the past 20 weeks from Alan's Snack Bar vending machines. Complete the table.

$142 $137 $125 $132 $147 $129 $151 $172 $175 $129

$159 $148 $173 $160 $152 $174 $169 $163 $149 $173

Profit	Number of Weeks
13. $120–129	_____
14. $130–139	_____
15. $140–149	_____
16. $150–159	_____
17. $160–169	_____
18. $170–179	_____

19. Use the information in Problems 13–18 to draw a histogram.

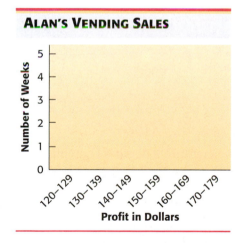

ALAN'S VENDING SALES

Find the mean for each list of numbers. Round answers to the nearest tenth if necessary.

20. Number of miles run each week while training: 52, 61, 68, 69, 73, 75, 79, 84, 91, 98

21. Weight in pounds for the largest bass caught in the lake: 11, 14, 12, 14, 20, 16, 17, 18

22. Airplane speeds in miles per hour: 458, 432, 496, 491, 500, 508, 512, 396, 492, 504

---

13. _____

14. _____

15. _____

16. _____

17. _____

18. _____

19. See histogram at left.

20. _____

21. _____

22. _____

23. Explain why a weighted mean must be used to determine a student's grade point average. Calculate your own grade point average for last semester or quarter. If you are a new student, make up a grade point average problem of your own and solve it. Round to the nearest hundredth.

23. _____

24. Explain in your own words the procedure for finding the median when there are an odd number of values in a list. Make up a problem with a list of five numbers and solve for the median.

24. _____

Find the weighted mean for the following. Round answers to the nearest tenth if necessary.

25. _____

25.

Cost	Frequency
12	5
20	6
22	8
28	4
38	6
48	2

26.

Value	Frequency
150	15
160	17
170	21
180	28
190	19
200	7

26. _____

Find the median for each list of numbers.

27. _____

27. Lowest daily temperatures in degrees Fahrenheit: 32, 41, 28, 28, 37, 35, 16, 31

28. _____

28. The length of steel beams in meters: 7.6, 11.4, 6.2, 12.5, 31.7, 22.8, 9.1, 10.0, 9.5

Find the mode or modes for each list of numbers.

29. _____

29. Blood sample amounts in milliliters: 72, 46, 52, 37, 28, 18, 52, 61

30. _____

30. Hot tub temperatures in degrees Fahrenheit of 96, 104, 103, 104, 103, 104, 91, 74, 103

Cumulative Review Exercises

Round each number to the place shown.

1. $65.236 to the nearest cent

2. $781.499 to the nearest dollar

3. 93,367 to the nearest ten

4. 854,279 to the nearest ten thousand

Simplify each expression by using the order of operations.

5. $4 + 10 \div 2 + 7(2)$

6. $\sqrt{81} - 4(2) + 9$

Simplify.

7. $2^2 \cdot 3^3$

8. $6^2 \cdot 3^2$

≈ *First use front end rounding to round each number and estimate the answer. Then find the exact answer.*

9. *Estimate:*   *Exact:*

$$\begin{array}{r} 54,289 \\ 171,352 \\ 48 \\ +\ 18,208 \\ \hline \end{array}$$

$+$ _____

10. *Estimate:*   *Exact:*

$$\begin{array}{r} 2.607 \\ 796.2 \\ 37.96 \\ 53.72 \\ +\ 8.06 \\ \hline \end{array}$$

$+$ ____

11. *Estimate:*   *Exact:*

$$\begin{array}{r} 445,306 \\ -\ 234,867 \\ \hline \end{array}$$

$-$ _____

12. *Estimate:*   *Exact:*

$$\begin{array}{r} 875.62 \\ -\ 63.757 \\ \hline \end{array}$$

$-$ ____

13. *Estimate:*   *Exact:*

$$\begin{array}{r} 7064 \\ \times\ \ \ \ 635 \\ \hline \end{array}$$

$\times$ _____

14. *Estimate:*   *Exact:*

$$\begin{array}{r} 62.75 \\ \times\ \ 2.644 \\ \hline \end{array}$$

$\times$ ____

15. *Estimate:*   *Exact:*

$\overline{})\overline{}$

$18\overline{)11,556}$

16. *Estimate:*   *Exact:*

$\overline{})\overline{}$

$4.25\overline{)62.56}$

Add, subtract, multiply, or divide as indicated. Write answers in lowest terms and as whole or mixed numbers when possible.

17. $\dfrac{7}{8} + \dfrac{3}{4}$

18. $\dfrac{3}{4} + \dfrac{5}{8} + \dfrac{1}{2}$

19. $\begin{aligned} 4\dfrac{3}{5} \\ + 5\dfrac{2}{3} \\ \hline \end{aligned}$

20. $\dfrac{5}{6} - \dfrac{1}{3}$

21. $\begin{aligned} 6\dfrac{2}{3} \\ - 4\dfrac{3}{4} \\ \hline \end{aligned}$

22. $\begin{aligned} 8 \\ - 5\dfrac{2}{5} \\ \hline \end{aligned}$

23. $\dfrac{3}{5} \cdot \dfrac{5}{8}$

24. $\left(9\dfrac{3}{5}\right)\left(4\dfrac{5}{8}\right)$

25. $22\left(\dfrac{2}{5}\right)$

26. $\dfrac{5}{6} \div \dfrac{5}{8}$

27. $18 \div \dfrac{3}{4}$

28. $3\dfrac{1}{3} \div 8\dfrac{3}{4}$

Use the order of operations to simplify each expression.

29. $\dfrac{2}{3}\left(\dfrac{7}{8} - \dfrac{3}{4}\right)$

30. $\left(\dfrac{5}{6} - \dfrac{1}{3}\right) + \left(\dfrac{1}{2}\right)^2 \cdot \dfrac{3}{4}$

Write each fraction in decimal form. Round to the nearest thousandth if necessary.

31. $\dfrac{3}{8}$

32. $\dfrac{4}{5}$

33. $\dfrac{1}{6}$

34. $\dfrac{17}{20}$

Write in order, from smallest to largest.

35. $0.218, 0.22, 0.199, 0.207, 0.2215$

36. $0.6319, \dfrac{5}{8}, 0.608, \dfrac{13}{20}, 0.58$

Write each ratio in lowest terms. Be sure to make all necessary conversions.

37. $5\frac{1}{2}$ in. to 44 in.

38. 3 hr to 45 min

Find cross products to determine whether each proportion is true *or* false.

39. $\dfrac{6}{15} = \dfrac{18}{45}$

40. $\dfrac{52}{180} \quad \dfrac{36}{120}$

Find the unknown number in each proportion.

41. $\dfrac{1}{5} = \dfrac{x}{30}$

42. $\dfrac{15}{x} = \dfrac{390}{156}$

43. $\dfrac{200}{135} = \dfrac{24}{x}$

44. $\dfrac{x}{208} = \dfrac{6.5}{26}$

Write each percent as a decimal. Write each decimal as a percent.

45. 65%

46. 0.035

47. 380%

48. 7.75%

Write each percent as a fraction or mixed number in lowest terms. Write each fraction or mixed number as a percent.

49. 4%

50. $87\frac{1}{2}\%$

51. $\dfrac{7}{20}$

52. $5\dfrac{3}{4}$

Solve each percent problem.

53. 65% of $780 is how much?

54. Find 5.4% of 6000 homes.

55. $8\frac{1}{2}\%$ of what number of people is 238 people?

56. 72 DVDs is 18% of what number of DVDs?

57. What percent of $1040 is $468?

58. 13 weeks is what percent of 52 weeks?

Convert each measurement using unit fractions.

59. _____ ft = 3 yd

60. 28 qt = _____ gal

61. 5 days = _____ hr

62. _____ lb = 6 T

Convert each measurement using unit fractions or the metric conversion line.

63. 10 km to m

64. 3815 mm to m

65. 8.3 g to mg

66. 230 g to kg

67. 72 mL to L

68. 0.28 L to mL

Write the most reasonable metric unit in each blank. Choose from L, mL, kg, g, mg, km, cm, m, and mm.

69. The fuel tank on the chain saw has a capacity of 750 _____ of fuel.

70. A nickel weighs 5 _____.

71. The distance of the run this Saturday is 10 _____.

72. The heaviest player on the team weighs 108 _____.

Name each shape and find its area. Use 3.14 as the approximate value of π. Round answers to the nearest tenth.

73.

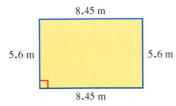

8.45 m
5.6 m   5.6 m
8.45 m

74.

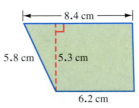

8.4 cm
5.8 cm   5.3 cm
6.2 cm

75.

5.75 ft
4.5 ft
4.25 ft

76.

13 cm

Name each solid and find its volume. Use 3.14 as the approximate value for π.
Round answers to the nearest tenth if necessary.

77.

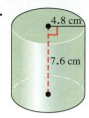
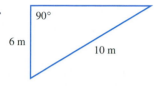

4.8 cm

7.6 cm

78.

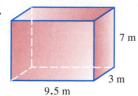
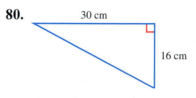

7 m

3 m

9.5 m

Find the unknown length in each right triangle.

79.

90°

6 m

10 m

80.

30 cm

16 cm

Add, subtract, multiply, or divide as indicated.

81. $-12 + (-10)$

82. $-5.7 - (-12.6)$

83. $9(-5)$

84. $(-14.6)(-5.7)$

85. $\dfrac{-42}{-7}$

86. $\dfrac{-34.04}{14.8}$

Solve each equation. Show your work.

87. $3x - 5 = 16$

88. $-12 = 3(x + 2)$

89. $15x - 11x = 12$

90. $3.4x + 6 = 1.4x - 8$

Find the mean, the median, and the mode for each list of numbers. Round to the nearest tenth if necessary.

91. Cable hookups per installer: 16, 37, 27, 31, 19, 25, 15, 38, 43, 19

92. Number of acres plowed each hour: 10.3, 4.3, 1.65, 2.85, 5.3, 5.7, 2.3, 4.35, 2.85

Solve each application problem.

93. Trader Joe's sold 3620 bags of tortilla chips in a recent week. If 1267 of these bags were baked tortilla chips (fat-free), find the percent that were baked.

94. In one state the sales tax is 7%. On a recent purchase, the amount of sales tax Sue paid was $78.68. Find the cost of the item she purchased.

95. A gasoline additive is used at the rate of $2\frac{3}{4}$ gallons for each storage tank. If $280\frac{1}{2}$ gallons of additive are available, how many storage tanks can receive the additive?

96. A survey found that 34 out of every 50 hotel rooms are for nonsmokers. If a hotel has 1200 rooms, how many would you expect to be for nonsmokers?

97. Breathe Right™ nasal strips are sold in four package sizes.

> 12 nasal strips $6.50
> 24 nasal strips $7.50
> 30 nasal strips $8.95
> 38 nasal strips $9.95

You have a $2-off coupon for the 12-strip size and a $1-off coupon for the 30-strip size. Which choice is the best buy?

98. The sketch below shows the plans for a lobby in a large commercial complex. What is the cost of carpeting the lobby, excluding the atrium, if the contractor charges $43.50 per square yard? Use 3.14 for π.

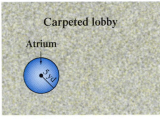

99. The Spa Service Center services 140 spas. If each spa needs 125 mL of muriatic acid, how many liters of acid are needed for all the spas?

100. Jonas will pay back a loan of $9400 with $8\frac{1}{2}$% interest at the end of 9 months. Find the total amount due.

Appendix A:
An Introduction
to Calculators

Appendix A Scientific Calculators

Calculators are among the more popular inventions of the last four decades. Each year better calculators are developed and costs drop. The first all-transitor desktop calculator was introduced to the market in 1966; it weighed 55 lb, cost $2500, and was slow. Today, these same calculations are performed quite well on a calculator costing less than $10. And today's $100 calculators have more ability to solve problems than some of the early computers.

Many colleges allow students to use calculators in basic mathematics courses. Although you can still purchase a basic four-function calculator, you're probably better off spending $10 to $20 on a **scientific calculator** that allows you to do a lot more. A **graphing calculator** allows you to graph functions and data, but it is beyond the scope of this text. The discussion here is confined to the common scientific calculator with the percent key, reciprocal key, exponent key, square root key, memory function, order of operations, and parentheses keys.

> **NOTE**
> For explanations of your specific calculator model and its special function keys, refer to the instruction booklet supplied with your calculator.

OBJECTIVE 1 Learn the basic calculator keys.   Most calculators use **algebraic logic.** Some problems can be solved by entering number and function keys in the same order as you would solve problems by hand, but many others require a knowledge of the order of operations when entering the problem. The problem $14 + 28$ would be entered as

$$14 \;\boxed{+}\; 28 \;\boxed{=}$$

and 42 would appear as the answer. Enter $387 - 62$ as

$$387 \;\boxed{-}\; 62 \;\boxed{=}$$

and 325 appears as the answer. If your calculator does not work problems in this way, check its instruction book to see how to proceed.

OBJECTIVES

1. Learn the basic calculator keys.
2. Understand the $\boxed{C}$, $\boxed{CE}$, and $\boxed{ON/C}$ or $\boxed{ON/AC}$ keys.
3. Understand the floating decimal point.
4. Use the $\boxed{\%}$ key.
5. Use the $\boxed{x^2}$ and the $\boxed{x^3}$ keys.
6. Use the $\boxed{y^x}$ and $\boxed{\sqrt{x}}$ keys.
7. Use the $\boxed{a^{b/c}}$ key.
8. Solve problems with negative numbers.
9. Use the calculator memory function.
10. Solve chain calculations using the order of operations.
11. Use the parentheses keys.

OBJECTIVE **2** **Understand the** ⓒ **,** (CE) **, and** (ON/C) **or** (ON/AC) **keys.**
All calculators have a

ⓒ ,   (ON/C) ,   or   (ON/AC)

key. Pressing this key erases everything in the calculator and prepares the
calculator to begin a new problem. Some calculators also have a

(CE)

key. Pressing this key erases *only* the number displayed and allows the person using the calculator to correct a mistake without having to start the problem over.

Many calculators combine the ⓒ key and the (CE) key and use an
(ON/C) key. This key turns the calculator on and is also used to erase the
calculator display. If the (ON/C) key is pressed after the (=) or one of the
operation keys ((+), (−), (×), (÷)), everything in the calculator is erased. If
the wrong operation key is pressed, you press the correct key and the error is
corrected. For example, 7 (+) (−) 3 (=) 4. Pressing the (−) key cancels out
the previous (+) key entry.

> **CAUTION**
> Be sure to look at the directions that come with your calculator to see
> how to clear the memory.

OBJECTIVE **3** **Understand the floating decimal point.**   Most calculators have a **floating decimal,** which locates the decimal point in the final
result. For example, to find the cost of 55.75 square yards of vinyl floor
covering at a cost of $18.99 per square yard, proceed as follows.

55.75 (×) 18.99 (=) 1058.6925

The decimal point is automatically placed in the answer. You should
round money answers to the nearest cent. Draw a cut-off line after the
hundredths place.

Look only at the first digit being cut off.
↓
1058.69 | 25
↑
Cent position (hundredths)

Because the first digit being cut off is *4 or less,* the part you are keeping
remains the same. The answer is rounded to $1058.69. If the first digit being
cut off had been *5 or more,* you would have rounded up by adding 1 to the
cent position.

When using a calculator with a floating decimal, enter the decimal point
as needed. For example, enter $47 as

47

with no decimal point, but enter 95¢ as

(·) 95

One problem in using a floating decimal is shown by the following example (adding $21.38 and $1.22).

$$21.38 \; ⊕ \; 1.22 \; ⊜ \; 22.6$$

The calculator does not show the final 0. You must remember that the problem dealt with money and write the final 0, making the answer $22.6**0**.

OBJECTIVE 4 **Use the ⊛ key.** The ⊛ key moves the decimal point two places to the left when pressed following multiplication or division. The problem 8% of $4205 is solved as follows.

$$4205 \; ⊗ \; 8 \; ⊛ \; ⊜ \; 336.4$$

Because the problem involved money, write the answer as $336.40.

OBJECTIVE 5 **Use the x^2 and the x^3 keys.** The squaring key, x^2, allows you to square the number in the display (multiply the number by itself). The square of 7 is found as follows.

$$7 \; x^2 \; 49$$

If your calculator has the x^3 key, it allows you to find the cube of a number (the number is multiplied by itself three times). To find the cube of 6.8 (that is, $6.8 \times 6.8 \times 6.8$), follow these keystrokes.

$$6.8 \; x^3 \; 314.432$$

OBJECTIVE 6 **Use the y^x and $\sqrt{x}$ keys.** Use an exponent to write the product of $3 \times 3 \times 3 \times 3 \times 3$ as shown below.

Exponent

$$3^5$$

Base

The exponent, 5, shows how many times the base is multiplied by itself (multiply 3 by itself five times). The y^x key raises a base to any desired power. Find 3^5 as shown below.

$$3 \; y^x \; 5 = 243$$

Since $3^2 = 9$, the number 3 is called the square root of 9. Square roots are written with the symbol $\sqrt{}$. Use the $\sqrt{x}$ key to find the square root of 144, $\sqrt{144}$, as follows.

$$144 \; \sqrt{x} \; 12$$

Find $\sqrt{20}$ as follows.

$$20 \; \sqrt{x} \; 4.472135955$$

The result may be rounded to the desired position.

OBJECTIVE 7 **Use the $\boxed{a^{b/c}}$ key.**   The $\boxed{a^{b/c}}$ key is used when solving problems containing fractions and mixed numbers.

Solve $\dfrac{3}{4} + \dfrac{6}{11}$ as shown below.

$$3 \;\boxed{a^{b/c}}\; 4 \;\boxed{+}\; 6 \;\boxed{a^{b/c}}\; 11 \;\boxed{=}\; 1\_13\lrcorner 44$$

The answer is $1\dfrac{13}{44}$.

Solve the mixed number problem $4\dfrac{7}{8} \div 3\dfrac{4}{7}$ as follows.

$$4 \;\boxed{a^{b/c}}\; 7 \;\boxed{a^{b/c}}\; 8 \;\boxed{\div}\; 3 \;\boxed{a^{b/c}}\; 4 \;\boxed{a^{b/c}}\; 7 \;\boxed{=}\; 1\_73\lrcorner 200$$

The answer is $1\dfrac{73}{200}$.

> **NOTE**
> Many calculators with a fraction key automatically show fractions in lowest terms and as mixed numbers when possible.

OBJECTIVE 8 **Solve problems with negative numbers.**   Negative numbers may be entered by first entering the number and then using the $\boxed{+/-}$ key. This changes the number entered to a negative number. For example, solve $-10 + 6 - 8$ as follows.

$$10 \;\boxed{+/-}\; \boxed{+}\; 6 \;\boxed{-}\; 8 \;\boxed{=}\; -12$$

OBJECTIVE 9 **Use the calculator memory function.**   Many calculators feature memory keys, which are a sort of electronic scratch paper. These memory keys are used to store intermediate steps in a calculation. On some basic calculators, a key labeled $\boxed{M}$ is used to store the numbers in the display, with $\boxed{MR}$ used to recall the numbers from memory.

Some calculators have $\boxed{M+}$ and $\boxed{M-}$ keys. The $\boxed{M+}$ key adds the number displayed to the number already in memory. For example, if the memory contains the number 0 at the beginning of a problem, and the calculator display contains the number 29.4, then pushing $\boxed{M+}$ will cause 29.4 to be stored in the memory (the result of adding 0 and 29.4). If 57.8 is then entered into the display, pushing $\boxed{M+}$ will cause

$$29.4 + 57.8 = 87.2$$

to be stored. If 11.9 is then entered into the display, with $\boxed{M-}$ pushed, the memory will contain the following.

$$87.2 - 11.9 = 75.3$$

The $\boxed{MR}$ key is used to recall the number in memory as needed, with $\boxed{MC}$ used to clear the memory.

Scientific calculators typically have one or more storage registers in which to store numbers. These memory keys are usually labeled as (STO) for store and (RCL) for recall. For example, you can store 25.6 in register 1 by pressing

$$25.6 \; (\text{STO}) \; 1$$

or you can store it in register 2 by pressing 25.6 (STO) 2 and so forth for other registers. Values are retrieved from a particular memory register by using the (RCL) key followed by the number of the register; for example, (RCL) 2 recalls the contents of memory in register 2.

With a scientific calculator, a number stays in memory until it is replaced by another number or until the memory is cleared. With some calculators, the contents of the memory are saved even when the calculator is turned off.

Here is an example of a problem that uses the memory keys. Suppose an elevator technician wants to find the average weight of a person using an elevator. To do this, she counts the number of people entering an elevator and also measures the weight of each group of people.

Number of People	Total Weight (in pounds)
6	839
8	1184
4	640

First, find the total weight of all three groups and store this in memory register 1.

$$839 \; (+) \; 1184 \; (+) \; 640 \; (=) \; 2663 \; (\text{STO}) \; 1$$

Then, find the total number of people.

$$6 \; (+) \; 8 \; (+) \; 4 \; (=) \; 18 \; (\text{STO}) \; 2$$

Finally, divide the contents of memory register 1 (total weight) by the contents of memory register 2 (total people).

$$(\text{RCL}) \; 1 \; (\div) \; (\text{RCL}) \; 2 \; (=) \; 147.9444444 \text{ pounds}$$

This answer can be rounded as needed.

OBJECTIVE 10 **Solve chain calculations using the order of operations.** Long calculations involving several different operations are called **chain calculations.** They must be done in a specific sequence called the **order of operations.** The logic of the following order of operations is built into most scientific calculators and can help you work problems without having to store or write down a lot of intermediate steps.

Order of Operations

Step 1 Do all operations inside parentheses or other grouping symbols.

Step 2 Simplify any expressions with exponents and find any square roots.

Step 3 Multiply and divide, proceeding from left to right.

Step 4 Add and subtract, proceeding from left to right.

Your scientific calculator can be used to solve the problem $3 + 7 \times 9\frac{3}{4}$.

$$3 \; \oplus \; 7 \; \otimes \; 9 \; \fbox{$a^{b}/_c$} \; 3 \; \fbox{$a^{b}/_c$} \; 4 \; \ominus \; 71\frac{1}{4}$$

The calculator automatically multiplies $7 \times 9\frac{3}{4}$ *before* adding 3.

The problem $42.1 \times 5 - 90 \div 4$ is solved as follows.

$$42.1 \; \otimes \; 5 \; \ominus \; 90 \; \oplus \; 4 \; \ominus \; 188$$

The calculator automatically does the multiplication (42.1×5) and the division ($90 \div 4$) *before* doing the subtraction.

> **CAUTION**
> Scientific calculators keep track of the order of operations for us. All we have to do is enter the problem correctly into the calculator and the calculator does the rest. However, the basic four-function calculator is *not* programmed to observe the order of operations and will calculate correctly *only* if you enter numbers in the proper order.

OBJECTIVE 11 Use the parentheses keys. The parentheses keys allow you to group numbers in a complex chain calculation. For example, $\frac{4}{5+7}$ can be written as $\frac{4}{(5+7)}$, which can be solved as follows.

Left parentheses key

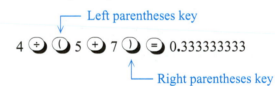

$$4 \; \oplus \; (\; 5 \; \oplus \; 7 \;) \; \ominus \; 0.333333333$$

Right parentheses key

Without the parentheses, the calculator would have automatically divided 4 by 5 before adding 7, giving an *incorrect* answer of 7.8.

To solve the problem

$$\frac{16 \div 2.5}{39.2 - 29.8 \times 0.6}$$

you must think of the problem as

$$\frac{(16 \div 2.5)}{(39.2 - 29.8 \times 0.6)}$$

Use parentheses to set off the numerator and the denominator.

$$(\; 16 \; \oplus \; 2.5 \;) \; \oplus \; (\; 39.2 \; \ominus \; 29.8 \; \otimes \; .6 \;) \; \ominus \; 0.300187617$$

Appendix B: Inductive and Deductive Reasoning

Appendix B Inductive and Deductive Reasoning

OBJECTIVE 1 Use inductive reasoning to analyze patterns. In many scientific experiments, conclusions are drawn from specific outcomes. After many repetitions and similar outcomes, the findings are generalized into statements that appear to be true. When general conclusions are drawn from specific observations, we are using a type of reasoning called **inductive reasoning.** The next several examples illustrate this type of reasoning.

EXAMPLE 1 Using Inductive Reasoning

Find the next number in the sequence 3, 7, 11, 15, ...
 To discover a pattern, calculate the difference between each pair of successive numbers.

$$7 - 3 = 4$$
$$11 - 7 = 4$$
$$15 - 11 = 4$$

Notice that the difference is always 4. Each number is 4 greater than the previous one. Thus, the next number in the pattern is $15 + 4$, or 19.

> **Work Problem 1 at the Side.** ▶▶▶

EXAMPLE 2 Using Inductive Reasoning

Find the next number in this sequence.

$$7, 11, 8, 12, 9, 13, \ldots$$

The pattern in this example involves addition and subtraction.

$$7 + 4 = 11$$
$$11 - 3 = 8$$
$$8 + 4 = 12$$
$$12 - 3 = 9$$
$$9 + 4 = 13$$

To get the second number, we add 4 to the first number. To get the third number, we subtract 3 from the second number. To obtain subsequent numbers, we continue the pattern. The next number is $13 - 3 = 10$.

> **Work Problem 2 at the Side.** ▶▶▶

OBJECTIVES

1. Use inductive reasoning to analyze patterns.
2. Use deductive reasoning to analyze arguments.
3. Use deductive reasoning to solve problems.

① Find the next number in the sequence 2, 8, 14, 20, . . . Describe the pattern.

② Find the next number in the sequence 6, 11, 7, 12, 8, 13, . . . Describe the pattern.

ANSWERS
1. 26; add 6 each time.
2. 9; add 5, subtract 4.

③ Find the next number in the sequence 2, 6, 18, 54, ... Describe the pattern.

EXAMPLE 3 Using Inductive Reasoning

Find the next number in the sequence 1, 2, 4, 8, 16, ...

Each number after the first is obtained by multiplying the previous number by 2. So the next number would be 16 • **2** = 32.

◀◀◀ Work Problem 3 at the Side.

EXAMPLE 4 Using Inductive Reasoning

(a) Find the next geometric shape in this sequence.

The figures alternate between a circle and a triangle. Also, the number of dots increases by 1 in each subsequent figure. Thus, the next figure should be a circle with five dots inside it.

(b) Find the next geometric shape in this sequence.

The first two shapes consist of vertical lines with horizontal lines at the bottom extending first *left* and then *right*. The third shape is a vertical line with a horizontal line at the top extending to the *left*. Therefore, the next shape should be a vertical line with a horizontal line at the top extending to the *right*.

④ Find the next shape in this sequence.

◀◀◀ Work Problem 4 at the Side.

OBJECTIVE **2** **Use deductive reasoning to analyze arguments.**   In the previous discussion, specific cases were used to find patterns and predict the next event. There is another type of reasoning called **deductive reasoning,** which moves from general cases to specific conclusions.

EXAMPLE 5 Using Deductive Reasoning

Does the conclusion follow from the premises in this argument?

All Buicks are automobiles. ← Premise

All automobiles have horns. ← Premise

∴   All Buicks have horns. ← Conclusion

In this example, the first two statements are called *premises* and the third statement (below the line) is called a *conclusion*. The symbol ∴ is a mathematical symbol meaning "**therefore**." The entire set of statements is called an *argument*.

ANSWERS
3. 162; multiply by 3.

4.

Continued on Next Page

The focus of deductive reasoning is to determine whether the conclusion follows (is valid) from the premises. A set of circles called **Euler circles** is used to analyze the argument.

In Example 5, the statement "All Buicks are automobiles" can be represented by two circles, one for Buicks and one for automobiles. Note that the circle representing Buicks is totally inside the circle representing automobiles because the first premise states that *all* Buicks are automobiles.

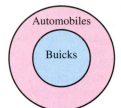

Now, a circle is added to represent the second statement, vehicles with horns. This circle must completely surround the circle representing automobiles because the second premise states that *all* automobiles have horns.

To analyze the conclusion, notice that the circle representing Buicks is *completely* inside the circle representing vehicles with horns. Therefore, it must follow that all Buicks have horns. ***The conclusion is valid.***

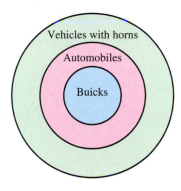

Work Problem 5 at the Side. ▶▶▶

EXAMPLE 6 **Using Deductive Reasoning**

Does the conclusion follow from the premises in this argument?

All tables are round.

All glasses are round.
∴   All glasses are tables.

Use Euler circles. Draw a circle representing tables *inside* a circle representing round objects, because the first premise states that *all* tables are round.

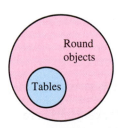

The second statement requires that a circle representing glasses must now be drawn inside the circle representing round objects, but not necessarily inside the circle representing tables. Therefore, the conclusion does *not* follow from the premises. This means that ***the conclusion is invalid.***

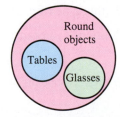

Work Problem 6 at the Side. ▶▶▶

---

5 Does the conclusion follow from the premises in the following argument?

All cars have four wheels.
All Fords are cars.
∴   All Fords have four wheels.

6 Does each conclusion follow from the premises?

(a) All animals are wild.
All cats are animals
∴   All cats are wild.

(b) All students use math.
All adults use math.
∴   All adults are students.

7 In a college class of 100 students, 35 take both math and history, 50 take history, and 40 take math. How many take neither math nor history? Draw a Venn diagram.

8 A Chevy, BMW, Cadillac, and Ford are parked side by side. The known facts are:

(a) The Ford is on the right end.

(b) The BMW is next to the Cadillac.

(c) The Chevy is between the Ford and the Cadillac.

Which car is parked on the left end?

OBJECTIVE 3   **Use deductive reasoning to solve problems.**   Another type of deductive reasoning problem occurs when a set of facts is given in a problem and a conclusion must be drawn using these facts.

EXAMPLE 7   **Using Deductive Reasoning**

There were 25 students enrolled in a ceramics class. During the class, 10 of the students made a bowl and 8 students made a birdbath. Three students made both a bowl and a birdbath. How many students did not make either a bowl or a birdbath?

This type of problem is best solved by organizing the data using a drawing called a **Venn diagram.** Two overlapping circles are drawn, with each circle representing one item made by the students, as shown below.

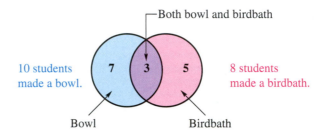

In the region where the circles overlap, write the number of students who made *both* items, namely, 3. In the remaining portion of the birdbath circle, write the number 5, which when added to 3 will give the total number of students who made a birdbath, namely, 8. In a similar manner, write 7 in the remaining portion of the bowl circle, since 7 + 3 = 10, the total number of students who made a bowl. The total of all three numbers written in the circles is 15. Since there were 25 students in the class, this means 25 − 15 or 10 students did not make either a birdbath or a bowl.

◀◀◀ Work Problem 7 at the Side.

EXAMPLE 8   **Using Deductive Reasoning**

Four cars in a race finish first, second, third, and fourth. The following facts are known.

(a) Car A beat Car C.

(b) Car D finished between Cars C and B.

(c) Car C beat Car B.

In which order did the cars finish?

To solve this type of problem, it is helpful to use a line diagram.

1. *Write A before C,* because Car A beat Car C (fact **a**).

$$A \qquad C$$

2. *Write B after C,* because Car C beat Car B (fact **c**).

$$A \qquad C \qquad B$$

3. *Write D between C and B,* because Car D finished between Car C and Car B (fact **b**).

The correct order of finish is shown below.

$$A \qquad C \qquad D \qquad B$$

◀◀◀ Work Problem 8 at the Side.

ANSWERS

7.

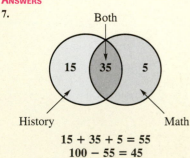

$$15 + 35 + 5 = 55$$
$$100 - 55 = 45$$

45 students take neither math nor history.

8. BMW

Appendix **B** Exercises

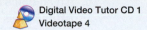

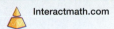
Find the next number in each sequence. Describe the pattern in each sequence. See Examples 1–3.

1. 2, 9, 16, 23, 30, . . .

2. 5, 8, 11, 14, 17, . . .

3. 0, 10, 8, 18, 16, . . .

4. 3, 9, 7, 13, 11, . . .

5. 1, 2, 4, 8, . . .

6. 1, 4, 16, 64, . . .

7. 1, 3, 9, 27, 81, . . .

8. 3, 6, 12, 24, 48, . . .

9. 1, 4, 9, 16, 25, . . .

10. 6, 7, 9, 12, 16, . . .

Find the next shape in each sequence. See Example 4.

11.

12.

13.

14.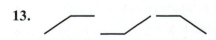

For each argument, draw Euler circles and then state whether or not the conclusion follows from the premises. See Examples 5 and 6.

15. All animals are wild.
All lions are animals.

∴   All lions are wild.

16. All students are hard workers.
All business majors are students.

∴   All business majors are hard workers.

17. All teachers are serious.
All mathematicians are serious.

∴   All mathematicians are teachers.

18. All boys ride bikes.
All Americans ride bikes.

∴   All Americans are boys.

Solve each application problem. See Examples 7 and 8.

19. In a given 30-day period, a husband watched television 20 days and his wife watched television 25 days. If they watched television together 18 days, how many days did neither watch television? Draw a Venn diagram.

20. In a class of 40 students, 21 students take both calculus and physics. If 30 students take calculus and 25 students take physics, how many do not take either calculus or physics? Draw a Venn diagram.

21. Tom, Dick, Mary, and Joan all work for the same company. One is a secretary, one is a computer operator, one is a receptionist, and one is a mail clerk.

(a) Tom and Joan eat dinner with the computer operator.

(b) Dick and Mary carpool with the secretary.

(c) Mary works on the same floor as the computer operator and the mail clerk.

Who is the computer operator?

22. Four cars—a Ford, a Buick, a Mercedes, and an Audi—are parked in a garage in four spaces.

(a) The Ford is in the last space.

(b) The Buick and Mercedes are next to each other.

(c) The Audi is next to the Ford but not next to the Buick.

Which car is in the first space?

Answers to Selected Exercises

In this section we provide the answers that we think most students will obtain when they work the exercises using the methods explained in the text. If your answer does not look exactly like the one given here, it is not necessarily wrong. In many cases there are equivalent forms of the answer that are correct. For example, if the answer section shows $\frac{3}{4}$ and your answer is 0.75, you have obtained the right answer but written it in a different (yet equivalent) form. Unless the directions specify otherwise, 0.75 is just as valid an answer as $\frac{3}{4}$.

In general, if your answer does not agree with the one given in the text, see whether it can be transformed into the other form. If it can, then it is the correct answer. If you still have doubts, talk with your instructor.

DIAGNOSTIC PRETEST

(page xxiii)

1. 89,023,507 **2.** 4331 **3.** 697 **4.** 90,000 **5.** $29\frac{3}{8}$

6. $2^3 \cdot 7^2$ **7.** $11\frac{1}{4}$ cups **8.** *Estimate:* $6 \cdot 2 = 12$;

Exact: $12\frac{7}{32}$ **9.** 120 **10.** $\frac{5}{6}$ **11.** *Estimate:* $8 + 13 = 21$;

Exact: $21\frac{1}{18}$ **12.** $\frac{35}{48}$ **13.** \$1.39 **14.** 6.556 (rounded)

15. 24.25 **16.** \$8.43 (rounded) **17.** $\frac{5}{24}$ **18.** false **19.** $26\frac{2}{3}$

20. \$340 **21.** 58.2% **22.** 875% **23.** \$135.68 **24.** 7%
25. (a) 9 gal (b) 76 oz **26.** (a) 67.5 cm (b) 4.528 kg
27. 2.92 kg **28.** (a) mg (b) cm **29.** 27 in. **30.** 24.5 cm²
31. 75.36 ft³ **32.** 15 cm **33.** 25 **34.** -4.6 **35.** 11
36. $r = -2$ **37.** 1632 parents **38.** 41.3 years **39.** 7.8 (rounded)
40. \$8.95

CHAPTER 1

Section 1.1 (page 7)

1. 3; 6 **3.** 1; 0 **5.** 8; 2 **7.** 3; 561; 435 **9.** 60; 0; 502; 109
11. Evidence suggests that this is true. It is common to count using fingers. **13.** twenty-three thousand, one hundred fifteen
15. three hundred forty-six thousand, nine **17.** twenty-five million, seven hundred fifty-six thousand, six hundred sixty-five
19. 63,163 **21.** 10,000,223 **23.** 3,200,000 **25.** 2,000,000,000
27. 854,795 **29.** 800,000,621,020,215 **31.** public transportation; six million, sixty-nine thousand, five hundred eighty-nine
33. seven million, eight hundred ninety-four thousand, nine hundred eleven

Section 1.2 (page 15)

1. 97 **3.** 89 **5.** 889 **7.** 889 **9.** 7785 **11.** 1589 **13.** 7676
15. 78,446 **17.** 8928 **19.** 59,224 **21.** 150 **23.** 155
25. 121 **27.** 145 **29.** 102 **31.** 1651 **33.** 1154 **35.** 413
37. 1771 **39.** 1410 **41.** 6391 **43.** 11,624 **45.** 17,611
47. 15,954 **49.** 10,648 **51.** 15,594 **53.** 11,557 **55.** 12,078
57. 4250 **59.** 12,268 **61.** correct **63.** incorrect; should be 769
65. correct **67.** incorrect; should be 11,577 **69.** correct
71. Changing the order in which numbers are added does not change the sum. You can add from bottom to top when checking addition. **73.** 33 miles **75.** 38 miles **77.** \$16,342
79. 699 people **81.** \$212 **83.** 550 ft **85.** 72 ft **87.** 9421

88. 1249 **89.** 77,762 **90.** 22,267 **91.** 9,994,433
92. 3,334,499 **93.** Write the largest digits on the left, using the smaller digits as you move right.

Section 1.3 (page 25)

1. 16 **3.** 33 **5.** 17 **7.** 213 **9.** 101 **11.** 7111 **13.** 3412
15. 2111 **17.** 13,160 **19.** 41,110 **21.** correct **23.** incorrect; should be 62 **25.** incorrect; should be 121 **27.** correct
29. incorrect; should be 7222 **31.** 38 **33.** 45 **35.** 19
37. 281 **39.** 519 **41.** 7059 **43.** 7589 **45.** 8859 **47.** 3
49. 23 **51.** 19 **53.** 2833 **55.** 7775 **57.** 503 **59.** 156
61. 2184 **63.** 5687 **65.** 19,038 **67.** 31,556 **69.** 6584
71. correct **73.** correct **75.** correct **77.** correct
79. Possible answers are 1. $3 + 2 = 5$ could be changed to $5 - 2 = 3$ or $5 - 3 = 2$ 2. $6 - 4 = 2$ could be changed to $2 + 4 = 6$ or $4 + 2 = 6$. **81.** 47 calories **83.** 367 ft
85. 467 passengers **87.** 2172 more people on Friday
89. 9539 flags **91.** \$263 **93.** 758 more people visited on Tuesday **95.** \$57,500 **97.** 48 deliveries **99.** 284 deliveries

Section 1.4 (page 35)

1. 24 **3.** 48 **5.** 0 **7.** 24 **9.** 40 **11.** 0 **13.** Factors may be multiplied in any order to get the same answer. They are the same; you may add or multiply numbers in any order. **15.** 210 **17.** 238
19. 3210 **21.** 1872 **23.** 8612 **25.** 10,084 **27.** 20,488
29. 258,447 **31.** 280 **33.** 480 **35.** 2220 **37.** 3600
39. 3750 **41.** 65,400 **43.** 270,000 **45.** 86,000,000
47. 48,500 **49.** 720,000 **51.** 1,940,000 **53.** 476 **55.** 2400
57. 3735 **59.** 2378 **61.** 6164 **63.** 15,792 **65.** 21,665
67. 15,730 **69.** 82,320 **71.** 183,996 **73.** 2,468,928
75. 66,005 **77.** 86,028 **79.** 19,422,180 **81.** 2,278,410
83. To multiply by 10, 100, or 1000, just add one, two, or three zeros to the number you are multiplying and that's your answer.
85. 3000 balls **87.** 216 plants **89.** 352 miles **91.** \$600
93. \$1560 **95.** \$112,888 **97.** 50,568 **99.** 38,250 trees
101. 3175 miles **103.** \$7390 **105.** (a) 452 (b) 452
106. commutative **107.** (a) 281 (b) 281 **108.** associative
109. (a) 15,840 (b) 15,840 **110.** commutative
111. (a) 6552 (b) 6552 **112.** associative **113.** No. Some examples are 1. $7 - 5 = 2$, but $5 - 7$ does not equal 2
2. $12 - 6 = 6$, but $6 - 12$ does not equal 6 3. $(8 - 2) - 5 = 1$, but $8 - (2 - 5)$ does not equal 1. **114.** No. Some examples are 1. $10 \div 2 = 5$, but $2 \div 10$ does not equal 5 2. $(16 \div 8) \div 2 = 1$, but $16 \div (8 \div 2)$ does not equal 1.

Section 1.5 (page 49)

1. $4)\overline{24}$ $\frac{24}{4} = 6$ **3.** $9)\overline{45}$ $45 \div 9 = 5$ **5.** $16 \div 2 = 8$ $\frac{16}{2} = 8$
7. 1 **9.** 7 **11.** undefined **13.** 24 **15.** 0 **17.** undefined
19. 15 **21.** 8 **23.** 25 **25.** 18 **27.** 304 **29.** 627 R1
31. 1522 R5 **33.** 309 **35.** 3005 **37.** 5006 **39.** 811 R1
41. 2589 R2 **43.** 7324 R2 **45.** 3157 R2 **47.** 2630
49. 12,458 R3 **51.** 10,253 R5 **53.** 18,377 R6 **55.** correct
57. incorrect; should be 1908 R1 **59.** incorrect; should be 670 R2
61. incorrect; should be 3568 R1 **63.** correct **65.** correct
67. incorrect; should be 9628 R3 **69.** correct **71.** Multiply the quotient by the divisor and add any remainder. The result should be the dividend. **73.** 328 tables **75.** 11,200 people each day
77. \$48,500 **79.** 135 acres **81.** \$1,137,500 **83.** \$9135

85. ✓ ✓ ✓ ✓   **87.** ✓ X X X   **89.** X X ✓ X
91. X ✓ X X   **93.** ✓ ✓ X X   **95.** X X X X

Section 1.6 (page 59)

1. 53   **3.** 250   **5.** 120 R7   **7.** 1105 R5   **9.** 7134 R12
11. 900 R100   **13.** 73 R5   **15.** 476 R15   **17.** 2407 R1
19. 1146 R15   **21.** 3331 R82   **23.** 850   **25.** incorrect;
should be 101 R14   **27.** incorrect; should be 658   **29.** incorrect;
should be 62   **31.** When dividing by 10, 100, or 1000, drop the
same number of zeros from the dividend to get the quotient. One
example is 2500 ÷ 100 = 25.   **33.** 35 days   **35.** 56 floor clocks
37. $355   **39.** 176,000 wagons   **41.** $39 per week   **43.** $0
44. 0   **45.** undefined   **46.** impossible; if you have 6 cookies, it is
not possible to divide them among 0 people.   **47. (a)** 14   **(b)** 17
(c) 38   **48.** Yes. Some examples are 18 • 1 = 18; 26 • 1 = 26;
43 • 1 = 43.   **49. (a)** 3200   **(b)** 320   **(c)** 32   **50.** Drop the
same number of zeros that appear in the divisor. The result is the
quotient. With the divisor 10, drop one 0; with 100, drop two zeros;
with 1000, drop three zeros.

Summary Exercises on Whole Numbers (page 63)

1. 3; 4   **2.** 6; 0   **3.** 1; 6   **4.** eighty-six thousand, two
5. four hundred twenty-five million, two hundred eight thousand,
seven hundred thirty-three   **6.** 97   **7.** 905   **8.** 21   **9.** 409
10. 17,573   **11.** 82,164   **12.** 677   **13.** 37,674   **14.** 35,889
15. 560   **16.** 5600   **17.** 350,000   **18.** 252,000   **19.** 8,642,180
20. 6,783,430   **21.** 1   **22.** 0   **23.** undefined   **24.** 15   **25.** 56
26. 0   **27.** 96   **28.** 304   **29.** 2750 R2   **30.** 761 R3   **31.** 3380
32. 220,545   **33.** 2016   **34.** 1476   **35.** 78   **36.** 210
37. 18,038,816   **38.** 506 R28   **39.** 52   **40.** 1208 R3

Section 1.7 (page 71)

1. 620   **3.** 860   **5.** 6800   **7.** 86,800   **9.** 28,500   **11.** 6000
13. 16,000   **15.** 78,000   **17.** 8000   **19.** 10,000   **21.** 600,000
23. 5,000,000   **25.** 4480; 4500; 4000   **27.** 3370; 3400; 3000
29. 6050; 6000; 6000   **31.** 5340; 5300; 5000   **33.** 19,540;
19,500; 20,000   **35.** 26,290; 26,300; 26,000   **37.** 93,710; 93,700;
94,000   **39.** 1. Locate the place to be rounded and underline it.
2. Look only at the next digit to the right. If this digit is 5 or more,
increase the underlined digit by 1.   3. Change all digits to the right
of the underlined place to zeros.   **41.** 30 60 50 80 220; 219
43. 80 40 40; 35   **45.** 70 30 2100; 2278   **47.** 900 700 400 800
2800; 2828   **49.** 900 400 500; 435   **51.** 800 400 320,000;
282,000   **53.** 8000 60 700 4000 12,760; 12,605   **55.** 700 500
200; 158   **57.** 900 30 27,000; 27,231   **59.** Perhaps the best explanation is that 3492 is closer to 3500 than 3400, but 3492 is closer to
3000 than 4000.   **61.** 80 million people; 300 million people
63. 50 yr; 80 yr   **65.** 1,670,000 pounds; 1,700,000 pounds;
2,000,000 pounds   **67.** $25,765,500,000; $25,800,000,000;
$26,000,000,000   **69.** 71,500   **70.** 72,499   **71.** 7500   **72.** 8499
73. 1160; 1000; 980; 940; 790   **74.** 1000; 1000; 1000; 900; 800
75. (a) When using front end rounding, all digits are 0 except the
first digit. These numbers are easier to work with when estimating
answers.   **(b)** When using front end rounding to estimate an answer, the estimated answer can vary greatly from the exact answer.

Section 1.8 (page 77)

1. 2; 3; 9   **3.** 2; 5; 25   **5.** 2; 8; 64   **7.** 2; 15; 225   **9.** 4   **11.** 8
13. 10   **15.** 12   **17.** 36; 36   **19.** 400; 400   **21.** 1225; 1225
23. 625; 625   **25.** 10,000; 10,000   **27.** A perfect square is the
square of a whole number. The number 25 is the square of 5
because 5 • 5 = 25. The number 50 is not a perfect square. There is
no whole number that can be squared to get 50.   **29.** 12   **31.** 15
33. 4   **35.** 20   **37.** 45   **39.** 63   **41.** 118   **43.** 22   **45.** 30
47. 102   **49.** 9   **51.** 63   **53.** 33   **55.** 70   **57.** 7   **59.** 17

61. 55   **63.** 108   **65.** 26   **67.** 26   **69.** 27   **71.** 16   **73.** 16
75. 21   **77.** 7   **79.** 20   **81.** 14   **83.** 25   **85.** 16   **87.** 23
89. 233

Section 1.9 (page 83)

1. 4500 stores   **3.** Dollar General   **5.** 1250 stores   **7.** 9 people
9. (a) Saw ad   **(b)** 25 people   **11.** 9 people   **13.** 2006; 7000
homes sales   **15.** 4500 home sales   **17.** Possible answers are
1. shortage of homes for sale   2. lack of qualified buyers
3. poor economy   4. high interest rates on home loans.
19. (7 − 2) • 3 − 6   **20.** (4 + 2) • (5 + 1)
21. 36 ÷ (3 • 3) • 4   **22.** 56 ÷ (2 • 2 • 2) + $\frac{0}{6}$
23. (a) 100 + 50 + 65 + (50 − 15 + (100 − 65) + 15)
(b) 3600 ft

Section 1.10 (page 91)

1. *Estimate:* 80 + 80 + 100 + 40 + 50 = 350 invoices;
Exact: 382 invoices   **3.** *Estimate:* $3000 − $2000 = $1000;
Exact: $880   **5.** *Estimate:* 200 × 20 = 4000 kits; *Exact:* 5664 kits
7. *Estimate:* 3000 ÷ 700 ≈ 4 toys; *Exact:* 4 toys
9. *Estimate:* 8000 − 4000 = 4000 people; *Exact:* 4174 people
11. *Estimate:* $30 × 5 = $150; *Exact:* $170
13. *Estimate:* $50,000 − $50,000 = $0; *Exact:* $9300
15. (a) *Estimate:* $30,000 + $40,000 = $70,000 Garrett;
$30,000 + $50,000 = $80,000 Harcos; *Exact:* $68,000 Garrett;
$82,900 Harcos; Mr. and Mrs. Harcos   **(b)** *Estimate:* $80,000 −
$70,000 = $10,000; *Exact:* $14,900   **17.** *Estimate:* $2000 −
$700 − $300 − $400 − $200 − $200 = $200; *Exact:* $350
19. *Estimate:* 40,000 × 100 = 4,000,000 ft²; *Exact:* 6,011,280 ft²
21. *Estimate:* $200 + $400 + $600 + $200 + $200 = $1800;
Exact: $1720   **23.** *Estimate:* $200 + $600 + $200 + $400 =
$1400; $1400 − $1200 = $200; *Exact:* $120   **25.** *Estimate:*
($1000 × 6) + ($900 × 20) = $24,000; *Exact:* $20,961   **27.** Possible
answers are   Addition: more; total; gain of   Subtraction: less; loss
of; decreased by   Multiplication: twice; of; product   Division:
divided by; goes into; per   Equals: is; are   **29.** Estimating the
answer can help you avoid careless mistakes like decimal or calculation errors. Examples of reasonable answers in daily life might be a
$25 bag of groceries, $20 to fill the gas tank, or $45 for a phone bill.
31. $165   **33.** 2477 pounds   **35.** $500   **37.** $375   **39.** 20 seats

Chapter 1 Review Exercises (page 99)

1. 6; 573   **2.** 36; 215   **3.** 105; 724   **4.** 1; 768; 710; 618
5. seven hundred thirty-five   **6.** fifteen thousand, three hundred ten
7. three hundred nineteen thousand, two hundred fifteen
8. sixty-two million, five hundred thousand, five   **9.** 10,008
10. 200,000,455   **11.** 110   **12.** 121   **13.** 5464   **14.** 15,657
15. 10,986   **16.** 9845   **17.** 40,602   **18.** 49,855   **19.** 36
20. 27   **21.** 189   **22.** 184   **23.** 6849   **24.** 4327   **25.** 224
26. 25,866   **27.** 49   **28.** 0   **29.** 32   **30.** 64   **31.** 45   **32.** 42
33. 56   **34.** 81   **35.** 40   **36.** 45   **37.** 48   **38.** 8   **39.** 0
40. 42   **41.** 48   **42.** 0   **43.** 84   **44.** 368   **45.** 522   **46.** 98
47. 5000   **48.** 2992   **49.** 5396   **50.** 45,815   **51.** 14,912
52. 20,160   **53.** 465,525   **54.** 174,984   **55.** 875   **56.** 2368
57. 1176   **58.** 5100   **59.** 15,576   **60.** 30,184   **61.** 887,169
62. 500,856   **63.** $360   **64.** $1064   **65.** $20,352   **66.** $684
67. 14,000   **68.** 23,800   **69.** 206,800   **70.** 318,500
71. 128,000,000   **72.** 90,300,000   **73.** 5   **74.** 7   **75.** 6   **76.** 2
77. 6   **78.** 4   **79.** 7   **80.** 0   **81.** undefined   **82.** 0   **83.** 8
84. 9   **85.** 82   **86.** 98   **87.** 4422   **88.** 352   **89.** 150 R4
90. 124 R25   **91.** 820   **92.** 15,200   **93.** 21,000   **94.** 70,000
95. 3490; 3500; 3000   **96.** 20,070; 20,100; 20,000
97. 98,200; 98,200; 98,000   **98.** 352,120; 352,100; 352,000
99. 4   **100.** 7   **101.** 12   **102.** 14   **103.** 3; 7; 343

104. 6; 3; 729 **105.** 3; 5; 125 **106.** 5; 4; 1024 **107.** 34
108. 26 **109.** 9 **110.** 4 **111.** 9 **112.** 6 **113.** 8 parents
114. 5 parents **115.** Keeping bedroom clean **116.** Hanging up
wet bath towels **117.** *Estimate:* 70 × $20 = $1400; *Exact:* $1170
118. *Estimate:* 1000 × 60 = 60,000 revolutions; *Exact:* 84,000
revolutions **119.** *Estimate:* 100 × 20 = 2000 forks; *Exact:* 2160
forks **120.** *Estimate:* 6000 × 30 = 180,000 brackets; *Exact:*
180,000 brackets **121.** *Estimate:* 2000 × 10 = 20,000 hr; *Exact:*
24,000 hr **122.** *Estimate:* 80 × 5 = 400 mi; *Exact:* 400 mi
123. *Estimate:* ($30 × 20) + ($20 × 30) = $1200; *Exact:* $1301
124. *Estimate:* (60 × $20) + (20 × $7) = $1340; *Exact:* $1139
125. *Estimate:* $400 − $40 = $360; *Exact:* $349
126. *Estimate:* 30,000 ÷ 1000 = 30 hr; *Exact:* 33 hr
127. *Estimate:* 9000 ÷ 200 = 45 pounds; *Exact:* 50 pounds
128. *Estimate:* $2000 − $500 − $400 = $1100; *Exact:* $1019
129. *Estimate:* 30,000 ÷ 600 = 50 acres; *Exact:* 52 acres
130. *Estimate:* 6000 ÷ 200 = 30 homes; *Exact:* 32 homes
131. 332 **132.** 448 **133.** 253 **134.** 588 **135.** 1041
136. 1661 **137.** 32,062 **138.** 24,947 **139.** 3 **140.** 7
141. 93,635 **142.** 83,178 **143.** undefined **144.** 7 **145.** 6900
146. 2310 **147.** 1,079,040 **148.** 130,212 **149.** 108
150. 207 **151.** three hundred seventy-six thousand, eight hundred
fifty-three **152.** four hundred eight thousand, six hundred ten
153. 8700 **154.** 401,000 **155.** 8 **156.** 9 **157.** $5544
158. $31,080 **159.** $2288 **160.** $15,782 **161.** 468 cards
162. 3600 textbooks **163.** $280 **164.** $114,635 **165.** $1905
166. $12,420 **167.** 497 ft **168.** 269 ft **169.** (a) 5108 ft
(b) less than a mile by 172 ft **170.** $100,600

Chapter 1 Test (page 106)

1. nine thousand, two hundred five **2.** seventy-five thousand,
sixty-five **3.** 426,005 **4.** 8530 **5.** 112,630 **6.** 1045
7. 6206 **8.** 168 **9.** 171,000 **10.** 1615 **11.** 4,450,743
12. 7047 **13.** undefined **14.** 458 R5 **15.** 160 **16.** 6350
17. 77,000 **18.** 41 **19.** 28 **20.** *Estimate:* $500 + $500 +
$500 + $400 − $800 = $1100; *Exact:* $1140 **21.** *Estimate:*
70,000 ÷ 500 = 140 days; *Exact:* 125 days **22.** *Estimate:*
$2000 − $500 − $200 − $200 = $1100; *Exact:* $948 **23.** *Estimate:*
(200 × 4) + (200 × 4) = 1600 cameras; *Exact:* 1784 cameras
24. 1. Locate the place to which you are rounding and underline it.
2. Look only at the next digit to the right. If this digit is a 4 or less,
do not change the underlined digit. If the digit is 5 or more, increase
the underlined digit by 1. 3. Change all digits to the right of the
underlined place to zeros. Each person's rounding example will
vary. **25.** 1. Read the problem carefully. 2. Work out a plan.
3. Estimate a reasonable answer. 4. Solve the problem.
5. State the answer. 6. Check your work.

CHAPTER 2

Section 2.1 (page 113)

1. $\frac{3}{4}, \frac{1}{4}$ **3.** $\frac{1}{3}, \frac{2}{3}$ **5.** $\frac{7}{5}, \frac{3}{5}$ **7.** $\frac{5}{6}$ **9.** $\frac{8}{25}$ **11.** $\frac{155}{218}$ **13.** 4; 5

15. 9; 8 **17.** Proper $\frac{1}{3}, \frac{5}{8}, \frac{7}{16}$ Improper $\frac{8}{5}, \frac{6}{6}, \frac{12}{2}$

19. Proper $\frac{3}{4}, \frac{9}{11}, \frac{7}{15}$ Improper $\frac{3}{2}, \frac{5}{5}, \frac{19}{18}$

21. One possibility is

$\frac{3}{4}$ ← Numerator / ← Denominator

The denominator shows the number of equal parts in the whole
and the numerator shows how many of the parts are being
considered.

23. 3; 8 **25.** 5; 24

Section 2.2 (page 119)

1. $\frac{5}{4}$ **3.** $\frac{23}{5}$ **5.** $\frac{13}{2}$ **7.** $\frac{33}{4}$ **9.** $\frac{18}{11}$ **11.** $\frac{19}{3}$ **13.** $\frac{81}{8}$ **15.** $\frac{43}{4}$

17. $\frac{27}{8}$ **19.** $\frac{43}{5}$ **21.** $\frac{54}{11}$ **23.** $\frac{131}{4}$ **25.** $\frac{221}{12}$ **27.** $\frac{269}{15}$

29. $\frac{187}{24}$ **31.** $1\frac{1}{3}$ **33.** $2\frac{1}{4}$ **35.** 8 **37.** $7\frac{3}{5}$ **39.** $4\frac{7}{8}$ **41.** 9

43. $15\frac{3}{4}$ **45.** $5\frac{2}{9}$ **47.** $8\frac{1}{8}$ **49.** $16\frac{4}{5}$ **51.** 28 **53.** $26\frac{1}{7}$

55. Multiply the denominator by the whole number and add the numerator. The result becomes the new numerator, which is placed
over the original denominator.

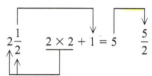

$2\frac{1}{2}$ $2 \times 2 + 1 = 5$ $\frac{5}{2}$

57. $\frac{501}{2}$ **59.** $\frac{1000}{3}$ **61.** $\frac{4179}{8}$ **63.** $154\frac{1}{4}$ **65.** 171

67. $122\frac{13}{32}$ **69.** $\frac{2}{3}, \frac{4}{5}, \frac{3}{4}, \frac{7}{10}$ **70.** (a) numerator; denominator

(b)

(c) less **71.** $\frac{5}{5}, \frac{10}{3}, \frac{6}{5}$ **72.** (a) numerator; denominator

(b)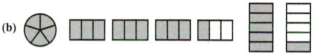

(c) greater **73.** $\frac{5}{3} = 1\frac{2}{3}; \frac{7}{7} = 1; \frac{11}{6} = 1\frac{5}{6}$

74. (a) improper; greater than or equal to

(b)

(c) Divide the numerator by the denominator and place the remainder over the denominator.

Section 2.3 (page 127)

1. 1, 2, 4, 8 **3.** 1, 3, 5, 15 **5.** 1, 2, 3, 4, 6, 8, 12, 16, 24, 48
7. 1, 2, 3, 4, 6, 9, 12, 18, 36 **9.** 1, 2, 4, 5, 8, 10, 20, 40
11. 1, 2, 4, 8, 16, 32, 64 **13.** composite **15.** prime
17. composite **19.** prime **21.** prime **23.** composite
25. composite **27.** composite **29.** 2^3 **31.** $2^2 \cdot 5$ **33.** $2^2 \cdot 3^2$
35. 5^2 **37.** $2^2 \cdot 17$ **39.** $2^3 \cdot 3^2$ **41.** $2^2 \cdot 11$ **43.** $2^2 \cdot 5^2$
45. 5^3 **47.** $2^2 \cdot 3^2 \cdot 5$ **49.** $2^6 \cdot 5$ **51.** $2^3 \cdot 3^2 \cdot 5$
53. A composite number has a factor(s) other than itself or 1.
Examples include 4, 6, 8, 9, 10. A prime number is a whole number
that has exactly two *different* factors, itself and 1. Examples include
2, 3, 5, 7, 11. The numbers 0 and 1 are neither prime nor composite.
55. All the possible factors of 24 are 1, 2, 3, 4, 6, 8, 12, and 24.
This list includes both prime numbers and composite numbers.
The prime factors of 24 include only prime numbers. The prime
factorization of 24 is $2 \cdot 2 \cdot 2 \cdot 3 = 2^3 \cdot 3$ **57.** $2 \cdot 5^2 \cdot 7$
59. $2^6 \cdot 3 \cdot 5$ **61.** $2^3 \cdot 3 \cdot 5 \cdot 13$ **63.** $2^2 \cdot 3^2 \cdot 5 \cdot 7$
65. 2, 3, 5, 7, 11, 13, 17, 19, 23, 29, 31, 37, 41, 43, 47 **66.** A
prime number is a whole number that is evenly divisible by itself
and 1 only. **67.** It is true because any even number is divisible by
the number 2 in addition to itself and 1. **68.** No. A multiple of a
prime number can never be prime because it will always be divisible

evenly by the prime number.   **69.** $2 \cdot 2 \cdot 3 \cdot 5 \cdot 5 \cdot 7$
70. $2^2 \cdot 3 \cdot 5^2 \cdot 7$

Section 2.4 (page 133)

1. ✓ ✓ ✓ ✓   **3.** ✓ ✓ X X   **5.** ✓ X ✓ ✓   **7.** ✓ ✓ X X
9. $\dfrac{3}{4}$ **11.** $\dfrac{1}{4}$ **13.** $\dfrac{3}{5}$ **15.** $\dfrac{6}{7}$ **17.** $\dfrac{7}{8}$ **19.** $\dfrac{6}{7}$ **21.** $\dfrac{4}{7}$ **23.** $\dfrac{1}{50}$
25. $\dfrac{8}{11}$ **27.** $\dfrac{5}{9}$

29. $\dfrac{\cancel{2} \cdot \cancel{3} \cdot 3}{\cancel{2} \cdot 2 \cdot 2 \cdot \cancel{3}} = \dfrac{3}{4}$ **31.** $\dfrac{\cancel{3} \cdot 7}{2 \cdot 2 \cdot 2 \cdot \cancel{3}} = \dfrac{7}{8}$

33. $\dfrac{\cancel{2} \cdot \cancel{3} \cdot \cancel{3} \cdot \cancel{3}}{\cancel{2} \cdot 2 \cdot \cancel{3} \cdot \cancel{3} \cdot \cancel{3}} = \dfrac{1}{2}$ **35.** $\dfrac{\cancel{2} \cdot \cancel{2} \cdot \cancel{3} \cdot 3}{\cancel{2} \cdot \cancel{2} \cdot \cancel{3}} = 3$

37. $\dfrac{2 \cdot 2 \cdot 2 \cdot \cancel{3} \cdot \cancel{3}}{\cancel{3} \cdot \cancel{3} \cdot 5 \cdot 5} = \dfrac{8}{25}$ **39.** equivalent   **41.** not equivalent

43. not equivalent   **45.** equivalent   **47.** not equivalent
49. equivalent   **51.** A fraction is in lowest terms when the numerator and the denominator have no common factors other than 1.
Some examples are $\dfrac{1}{2}, \dfrac{3}{8},$ and $\dfrac{2}{3}$.   **53.** $\dfrac{5}{8}$ **55.** $\dfrac{2}{1} = 2$

Summary Exercises on Fraction Basics (page 135)

1. $\dfrac{5}{6}; \dfrac{1}{6}$ **2.** $\dfrac{1}{3}; \dfrac{2}{3}$ **3.** $\dfrac{5}{8}; \dfrac{3}{8}$ **4.** 3; 4   **5.** 8; 5   **6.** Proper
$\dfrac{3}{5}, \dfrac{4}{25}, \dfrac{1}{32}$ Improper $\dfrac{8}{2}, \dfrac{16}{7}, \dfrac{8}{8}$ **7.** $\dfrac{6}{31}$ **8.** $\dfrac{19}{31}$ **9.** $2\dfrac{1}{2}$ **10.** $1\dfrac{3}{8}$
11. $1\dfrac{2}{7}$ **12.** $2\dfrac{2}{3}$ **13.** 8   **14.** 5   **15.** $7\dfrac{1}{5}$ **16.** $4\dfrac{7}{10}$ **17.** $\dfrac{10}{3}$
18. $\dfrac{43}{8}$ **19.** $\dfrac{34}{5}$ **20.** $\dfrac{53}{5}$ **21.** $\dfrac{51}{4}$ **22.** $\dfrac{62}{13}$ **23.** $\dfrac{71}{6}$ **24.** $\dfrac{189}{8}$
25. $2 \cdot 5$ **26.** $5 \cdot 11$ **27.** $2^2 \cdot 3^2$ **28.** 3^4 **29.** $2^3 \cdot 5 \cdot 7$
30. $2^3 \cdot 3^2 \cdot 5$ **31.** $\dfrac{1}{3}$ **32.** $\dfrac{1}{2}$ **33.** $\dfrac{1}{4}$ **34.** $\dfrac{3}{5}$ **35.** $\dfrac{3}{4}$ **36.** $\dfrac{5}{6}$
37. $\dfrac{7}{8}$ **38.** $\dfrac{1}{50}$ **39.** $\dfrac{1}{8}$ **40.** $\dfrac{5}{9}$ **41.** $\dfrac{4}{7}$ **42.** $\dfrac{5}{9}$
43. $\dfrac{\cancel{2} \cdot \cancel{2} \cdot 2 \cdot \cancel{3}}{\cancel{2} \cdot \cancel{2} \cdot 3 \cdot \cancel{3}} = \dfrac{2}{3}$ **44.** $\dfrac{\cancel{2} \cdot \cancel{2} \cdot \cancel{2} \cdot \cancel{2} \cdot \cancel{3}}{\cancel{2} \cdot \cancel{2} \cdot \cancel{2} \cdot \cancel{2} \cdot 2 \cdot \cancel{3}} = \dfrac{1}{2}$
45. $\dfrac{\cancel{2} \cdot \cancel{3} \cdot 3 \cdot \cancel{7}}{\cancel{2} \cdot \cancel{3} \cdot \cancel{7}} = 3$ **46.** $\dfrac{\cancel{2} \cdot \cancel{2} \cdot \cancel{2} \cdot \cancel{2} \cdot 2 \cdot 3}{\cancel{2} \cdot \cancel{2} \cdot \cancel{2} \cdot \cancel{2} \cdot 7} = \dfrac{6}{7}$

Section 2.5 (page 143)

1. $\dfrac{1}{4}$ **3.** $\dfrac{2}{35}$ **5.** $\dfrac{3}{4}$ **7.** $\dfrac{1}{4}$ **9.** $\dfrac{5}{12}$ **11.** $\dfrac{9}{32}$ **13.** $\dfrac{2}{5}$ **15.** $\dfrac{13}{32}$
17. $\dfrac{21}{128}$ **19.** 4   **21.** 40   **23.** 24   **25.** $13\dfrac{1}{2}$ **27.** $31\dfrac{1}{2}$ **29.** 80
31. $94\dfrac{2}{3}$ **33.** 400   **35.** 810   **37.** $\dfrac{1}{4}$ mi²   **39.** 9 m²   **41.** $\dfrac{3}{10}$ in.²
43. Multiply the numerators and multiply the denominators. An
example is $\dfrac{3}{4} \cdot \dfrac{1}{2} = \dfrac{3 \cdot 1}{4 \cdot 2} = \dfrac{3}{8}$ **45.** $1\dfrac{1}{2}$ yd²   **47.** $1\dfrac{1}{2}$ mi²
49. They are both the same size: $\dfrac{3}{64}$ mi²   **51.** 2,400,000 customers
52. 2,337,917 customers   **53.** 300,000; 281,421 premium channel
customers   **54.** 171,429; 154,638 movie channel customers

55. $\dfrac{3}{5} \cdot 450{,}000$ (multiple of 5) $= 270{,}000$ customers
56. $\dfrac{3}{7} \cdot 350{,}000$ (multiple of 7) $= 150{,}000$ customers

Section 2.6 (page 151)

1. $\dfrac{1}{2}$ yd²   **3.** $\dfrac{8}{9}$ ft²   **5.** $\dfrac{3}{10}$ yd²   **7.** \$2568   **9.** 375 students
11. 910 women   **13.** 0−24 years; 40 million or 40,000,000 books
15. 250 million or 250,000,000 books   **17.** Because everyone is
included and fractions are given for *all* age groups, the sum of the
fractions must be *1* or *all* of the people.   **19.** \$58,000
21. \$11,600   **23.** \$3625
25. The correct solution is
$$\dfrac{9}{10} \times \dfrac{20}{21} = \dfrac{\overset{3}{\cancel{9}}}{\cancel{10}} \times \dfrac{\overset{2}{\cancel{20}}}{\cancel{21}} = \dfrac{6}{7}$$
27. \$48   **29.** 9000 votes   **31.** $\dfrac{1}{32}$ of the estate

Section 2.7 (page 161)

1. $\dfrac{8}{3}$ **3.** $\dfrac{6}{5}$ **5.** $\dfrac{5}{8}$ **7.** $\dfrac{1}{4}$ **9.** $\dfrac{2}{3}$ **11.** $2\dfrac{5}{8}$ **13.** $\dfrac{9}{20}$ **15.** 4
17. 6   **19.** $\dfrac{13}{16}$ **21.** $\dfrac{4}{5}$ **23.** 18   **25.** 24   **27.** $\dfrac{1}{14}$ **29.** $\dfrac{2}{9}$ liter
31. 15 times   **33.** 88 dispensers   **35.** 60 trips   **37.** You can
divide two fractions by using the reciprocal of the second fraction
(divisor) and then multiplying   **39.** 12 pounds   **41.** 208 homes
43. 304 miles   **45.** 2432 towels   **47.** double, twice, times,
product, twice as much   **48.** goes into, divide, per, quotient,
divided by   **49.** reciprocal   **50.** $\dfrac{4}{3}; \dfrac{8}{7}; \dfrac{1}{5}; \dfrac{19}{12}$

51. $\dfrac{15}{\cancel{16}} \times \dfrac{\overset{1}{\cancel{4}}}{1} = \dfrac{15}{4} = 3\dfrac{3}{4}$ in. Multiply the length of one side by 3,

4, 5, or 6.   **52.** $\dfrac{225}{256}$ in.²; Multiply the length by the width.

Section 2.8 (page 171)

1. *Exact:* $7\dfrac{7}{8}$;   *Estimate:* $5 \cdot 2 = 10$ **3.** *Exact:* $4\dfrac{1}{2}$;
Estimate: $2 \cdot 3 = 6$ **5.** *Exact:* 4;   *Estimate:* $3 \cdot 1 = 3$
7. *Exact:* 50;   *Estimate:* $8 \cdot 6 = 48$ **9.** *Exact:* $49\dfrac{1}{2}$;
Estimate: $5 \cdot 2 \cdot 5 = 50$ **11.** *Exact:* 12;
Estimate: $3 \cdot 2 \cdot 3 = 18$ **13.** *Exact:* $\dfrac{1}{3}$;   *Estimate:* $1 \div 4 = \dfrac{1}{4}$
15. *Exact:* $\dfrac{5}{6}$;   *Estimate:* $3 \div 3 = 1$ **17.** *Exact:* $3\dfrac{3}{5}$;
Estimate: $9 \div 3 = 3$ **19.** *Exact:* $\dfrac{5}{12}$;   *Estimate:* $1 \div 2 = \dfrac{1}{2}$
21. *Exact:* $\dfrac{3}{10}$;   *Estimate:* $2 \div 6 = \dfrac{1}{3}$ **23.** *Exact:* $\dfrac{17}{18}$;   *Estimate:*
$6 \div 6 = 1$ **25. (a)** *Estimate:* $1 \cdot 3 = 3$ cups;   *Exact:* $1\dfrac{7}{8}$ cups of
applesauce   **(b)** *Estimate:* $1 \cdot 3 = 3$ teaspoons;
Exact: $1\dfrac{1}{4}$ teaspoons of salt   **(c)** *Estimate:* $2 \cdot 3 = 6$ cups;
Exact: $4\dfrac{3}{4}$ cups of flour   **27. (a)** *Estimate:* $1 \div 2 = \dfrac{1}{2}$ teaspoon;

Exact: $\frac{1}{4}$ teaspoon vanilla extract   **(b)** *Estimate:* $1 \div 2 = \frac{1}{2}$ cup;

Exact: $\frac{3}{8}$ cup applesauce   **(c)** Estimate: $2 \div 2 = 1$ cup;

Exact: $\frac{7}{8}$ cup of flour   **29.** *Estimate:* $1314 \div 110 = 12$ homes;

Exact: 12 homes   **31.** *Estimate:* $135 \cdot 20 = 2700$ in.;

Exact: $2632\frac{1}{2}$ in.   **33.** The answer should include include

Step 1 Change mixed numbers to improper fractions. *Step 2* Multiply the fractions. *Step 3* Write the answer in lowest terms, changing to mixed or whole numbers where possible.
35. *Estimate:* $51,460 \div 20 = 2573$ tires;   *Exact:* 2480 tires
37. *Estimate:* $10 \div 1 = 10$ spacers;   *Exact:* 13 spacers
39. *Estimate:* $7 \cdot 26 = 182$ gallons;   *Exact:* $172\frac{1}{8}$ gallons

Chapter 2 Review Exercises (page 179)

1. $\frac{1}{3}$ **2.** $\frac{5}{8}$ **3.** $\frac{2}{4}$ **4.** Proper $\frac{1}{8}, \frac{3}{4}, \frac{2}{3}$; Improper $\frac{4}{3}, \frac{5}{5}$
5. Proper $\frac{15}{16}, \frac{1}{8}$; Improper $\frac{6}{5}, \frac{16}{13}, \frac{5}{3}$ **6.** $\frac{17}{4}$ **7.** $\frac{59}{6}$ **8.** $3\frac{3}{8}$
9. $12\frac{3}{5}$ **10.** 1, 2, 3, 6   **11.** 1, 2, 3, 4, 6, 8, 12, 24
12. 1, 5, 11, 55   **13.** 1, 2, 3, 5, 6, 9, 10, 15, 18, 30, 45, 90
14. 3^3 **15.** $2 \cdot 3 \cdot 5^2$ **16.** $2^3 \cdot 3 \cdot 7$ **17.** 25   **18.** 288
19. 1728   **20.** 2048   **21.** $\frac{2}{3}$ **22.** $\frac{4}{5}$ **23.** $\frac{15}{16}$
24. $\dfrac{\cancel{3} \cdot 5}{2 \cdot 2 \cdot 3 \cdot \cancel{3}}; \dfrac{5}{12}$ **25.** $\dfrac{\cancel{2} \cdot \cancel{2} \cdot \cancel{2} \cdot \cancel{2} \cdot \cancel{2} \cdot 2 \cdot 2 \cdot \cancel{3}}{\cancel{2} \cdot \cancel{2} \cdot \cancel{2} \cdot \cancel{2} \cdot \cancel{2} \cdot \cancel{3}}; 4$
26. equivalent   **27.** not equivalent   **28.** $\frac{3}{5}$ **29.** $\frac{3}{16}$ **30.** $\frac{1}{7}$
31. $\frac{4}{21}$ **32.** 15   **33.** 625   **34.** $1\frac{1}{3}$ **35.** $\frac{5}{3} = 1\frac{2}{3}$ **36.** $\frac{5}{2} = 2\frac{1}{2}$
37. 2   **38.** 8   **39.** 24   **40.** $\frac{5}{24}$ **41.** $\frac{2}{15}$ **42.** $\frac{4}{13}$ **43.** $\frac{9}{40}$ ft²
44. $\frac{7}{12}$ in.²   **45.** $4\frac{1}{6}$ ft²   **46.** 276 yd²   **47.** *Exact:* $6\frac{7}{8}$;
Estimate: $6 \cdot 1 = 6$ **48.** *Exact:* $21\frac{3}{8}$;   *Estimate:* $2 \cdot 7 \cdot 1 = 14$
49. *Exact:* $5\frac{1}{6}$;   *Estimate:* $16 \div 3 = 5\frac{1}{3}$ **50.** *Exact:* $\frac{3}{4}$;
Estimate: $5 \div 6 = \frac{5}{6}$ **51.** 512 bins   **52.** $\frac{2}{15}$ of the estate
53. *Estimate:* $158 \div 4 = 40$ pull cords;   *Exact:* 36 pull cords
54. *Estimate:* $9 \cdot 38 = \$342$;   *Exact:* $323   **55.** 25 pounds
56. $510   **57.** $\frac{3}{32}$ of the budget   **58.** $\frac{4}{5}$ ton   **59.** $\frac{3}{8}$ **60.** $\frac{2}{5}$
61. $31\frac{1}{4}$ **62.** $28\frac{1}{8}$ **63.** $\frac{1}{10}$ **64.** $\frac{5}{32}$ **65.** 30   **66.** $2\frac{1}{6}$
67. $1\frac{3}{5}$ **68.** $38\frac{1}{4}$ **69.** $\frac{17}{3}$ **70.** $\frac{307}{8}$
71. $\dfrac{\cancel{2} \cdot \cancel{2} \cdot 2}{\cancel{2} \cdot \cancel{2} \cdot 3} = \dfrac{2}{3}$ **72.** $\dfrac{\cancel{2} \cdot 2 \cdot \cancel{3} \cdot 3 \cdot 3}{\cancel{2} \cdot \cancel{3} \cdot 5 \cdot 7} = \dfrac{18}{35}$
73. $\frac{5}{6}$ **74.** $\frac{2}{3}$ **75.** $\frac{2}{5}$ **76.** $\frac{1}{3}$

77. *Estimate:* $4 \cdot 44 = 176$ ounces;   *Exact:* $152\frac{4}{9}$ ounces
78. *Estimate:* $7 \cdot 26 = 182$ quarts;   *Exact:* $184\frac{7}{8}$ quarts
79. $1\frac{17}{32}$ in.²   **80.** $\frac{5}{16}$ yd²

Chapter 2 Test (page 183)

1. $\frac{5}{6}$ **2.** $\frac{3}{8}$ **3.** $\frac{2}{3}, \frac{6}{7}, \frac{1}{4}, \frac{5}{8}$ **4.** $\frac{27}{8}$ **5.** $30\frac{3}{4}$ **6.** 1, 2, 3, 6, 9, 18
7. $3^2 \cdot 5$ **8.** $2^4 \cdot 3^2$ **9.** $2^2 \cdot 5^3$ **10.** $\frac{3}{4}$ **11.** $\frac{5}{6}$
12. $\dfrac{56}{84} = \dfrac{\cancel{2} \cdot \cancel{2} \cdot 2 \cdot \cancel{7}}{\cancel{2} \cdot \cancel{2} \cdot 3 \cdot \cancel{7}} = \dfrac{2}{3}$
13. Multiply fractions by multiplying the numerators and multiplying the denominators. Divide two fractions by using the reciprocal of the second fraction (divisor) and multiplying.   **14.** $\frac{1}{3}$ **15.** 36
16. $\frac{21}{32}$ yd²   **17.** 1528 men   **18.** $\frac{9}{10}$ **19.** $15\frac{3}{4}$
20. 200 sports bottles   **21.** *Estimate:* $4 \cdot 4 = 16$;   *Exact:* $14\frac{7}{16}$
22. *Estimate:* $2 \cdot 4 = 8$;   *Exact:* $7\frac{17}{18}$ **23.** *Estimate:* $10 \div 2 = 5$;
Exact: $4\frac{4}{15}$ **24.** *Estimate:* $9 \div 2 = 4\frac{1}{2}$;   *Exact:* $5\frac{1}{10}$
25. *Estimate:* $3 \cdot 12 = 36$;   *Exact:* $30\frac{5}{8}$ g

Cumulative Review Exercises: Chapters 1–2 (page 185)

1. hundreds 7; tens 8   **2.** millions 8; ten thousands 2   **3.** 177
4. 149,199   **5.** 4452   **6.** 3,221,821   **7.** 747   **8.** 72
9. 2,168,232   **10.** 450,400   **11.** 9   **12.** 7581   **13 .** 8471 R2
14. 22 R26   **15.** 6580; 6600; 7000   **16.** 76,270; 76,300; 76,000
17. 8   **18.** 5   **19.** $992   **20.** 60 decibels   **21.** 47,000 hairs
22. 188 hours   **23.** $\frac{11}{16}$ ft²   **24.** 60¢   **25.** proper   **26.** improper
27. proper   **28.** $\frac{27}{8}$ **29.** $\frac{32}{5}$ **30.** 2   **31.** $12\frac{7}{8}$ **32.** $2^3 \cdot 3^2$
33. $2 \cdot 3^2 \cdot 7$ **34.** $2 \cdot 5^2 \cdot 7$ **35.** 64   **36.** 288   **37.** 640
38. $\frac{7}{8}$ **39.** $\frac{2}{3}$ **40.** $\frac{5}{9}$ **41.** $\frac{3}{8}$ **42.** 12   **43.** 25   **44** $\frac{24}{25}$
45. $\frac{7}{12}$ **46.** $2\frac{2}{5}$

CHAPTER 3

Section 3.1 (page 191)

1. $\frac{5}{8}$ **3.** $\frac{5}{6}$ **5.** $\frac{1}{2}$ **7.** $1\frac{1}{5}$ **9.** $\frac{1}{3}$ **11.** $\frac{13}{20}$ **13.** $\frac{11}{15}$ **15.** $1\frac{1}{2}$
17. $\frac{11}{27}$ **19.** $\frac{3}{8}$ **21.** $\frac{6}{11}$ **23.** $\frac{3}{5}$ **25.** $1\frac{1}{7}$ **27.** $\frac{1}{5}$ **29.** $1\frac{1}{6}$
31. $1\frac{1}{10}$ **33.** Three steps to add like fractions are: 1. Add the numerators of the fractions to find the numerator of the sum (the answer).   2. Use the denominator of the fractions as the denominator of the sum.   3. Write the answer in lowest terms.
35. $\frac{4}{5}$ **37.** $\frac{1}{3}$ **39.** $\frac{1}{2}$ acre

Section 3.2 (page 199)

1. 6   **3.** 15   **5.** 36   **7.** 14   **9.** 30   **11.** 100   **13.** 20   **15.** 60

17. 36   **19.** 120   **21.** 180   **23.** 144   **25.** $\frac{16}{24}$ **27.** $\frac{18}{24}$ **29.** $\frac{20}{24}$

31. 3   **33.** 12   **35.** 28   **37.** 12   **39.** 32   **41.** 72   **43.** 84

45. 96   **47.** 27   **49.** It probably depends on how large the numbers are. If the numbers are small, the method using multiples of the largest number seems best. If the numbers are larger, or there are more than two numbers, then the factorization method will be better.   **51.** 3600   **53.** 10,584   **55.** like; unlike

56. numerators; denominator; lowest   **57.** least; smallest

58. 40 is the least common multiple.   **59.** 70   **60.** 450

61. 240 is a common multiple but twice as large as the least common multiple; 120 is the LCM.   **62.** The least common multiple can be no smaller than the largest number in a group and the number 1760 is a multiple of 55.

Section 3.3 (page 207)

1. $\frac{7}{8}$ **3.** $\frac{8}{9}$ **5.** $\frac{3}{4}$ **7.** $\frac{39}{40}$ **9.** $\frac{23}{36}$ **11.** $\frac{14}{15}$ **13.** $\frac{29}{36}$ **15.** $\frac{17}{20}$

17. $\frac{23}{30}$ **19.** $\frac{3}{8}$ **21.** $\frac{23}{48}$ **23.** $\frac{1}{2}$ **25.** $\frac{1}{2}$ **27.** $\frac{7}{15}$ **29.** $\frac{1}{6}$

31. $\frac{19}{45}$ **33.** $\frac{3}{40}$ **35.** $\frac{17}{48}$ **37.** $\frac{1}{4}$ in.   **39.** $\frac{7}{12}$ acre   **41.** $\frac{31}{40}$ in.

43. $\frac{3}{8}$ gallon   **45.** You cannot add or subtract until all the fractional pieces are the same size. For example, halves are larger than fourths, so you cannot add $\frac{1}{2} + \frac{1}{4}$ until you rewrite $\frac{1}{2}$ as $\frac{2}{4}$.

47. $\frac{7}{24}$ **49.** work and travel; 8 hr   **51.** $\frac{3}{16}$ in.

Section 3.4 (page 215)

1. *Estimate:* $6 + 3 = 9$; *Exact:* $8\frac{5}{6}$

3. *Estimate:* $7 + 4 = 11$; *Exact:* $11\frac{1}{2}$

5. *Estimate:* $1 + 4 = 5$; *Exact:* $4\frac{5}{24}$

7. *Estimate:* $25 + 19 = 44$; *Exact:* $43\frac{2}{3}$

9. *Estimate:* $34 + 19 = 53$; *Exact:* $52\frac{1}{10}$

11. *Estimate:* $23 + 15 = 38$; *Exact:* $38\frac{5}{28}$

13. *Estimate:* $13 + 19 + 15 = 47$; *Exact:* $45\frac{5}{6}$

15. *Estimate:* $15 - 12 = 3$; *Exact:* $2\frac{5}{8}$

17. *Estimate:* $13 - 1 = 12$; *Exact:* $11\frac{7}{15}$

19. *Estimate:* $28 - 6 = 22$; *Exact:* $22\frac{7}{30}$

21. *Estimate:* $17 - 7 = 10$; *Exact:* $10\frac{3}{8}$

23. *Estimate:* $19 - 6 = 13$; *Exact:* $12\frac{19}{20}$

25. *Estimate:* $20 - 12 = 8$; *Exact:* $7\frac{11}{12}$

27. $9\frac{1}{8}$ **29.** $11\frac{1}{2}$ **31.** $3\frac{5}{6}$ **33.** $6\frac{11}{12}$ **35.** $8\frac{1}{8}$ **37.** $\frac{5}{6}$

39. $2\frac{7}{8}$ **41.** $2\frac{7}{12}$ **43.** $5\frac{9}{20}$ **45.** $3\frac{16}{21}$

47. Find the least common denominator. Change the fraction parts so that they have the same denominator. Add the fraction parts. Add the whole number parts. Write the answer as a mixed number.

49. *Estimate:* $25 - 15 = 10$ ft; *Exact:* $10\frac{3}{4}$ ft

51. *Estimate:* $23 - 19 = 4$ in.; *Exact:* $4\frac{1}{8}$ in.

53. *Estimate:* $23 + 22 + 21 = 66$ in.; *Exact:* 66 in.

55. *Estimate:* $3 + 6 + 5 + 3 + 6 = 23$ hr; *Exact:* $22\frac{7}{8}$ hr

57. *Estimate:* $24 + 35 + 24 + 35 = 118$ in.; *Exact:* $116\frac{1}{2}$ in.

59. *Estimate:* $9 - 3 - 3 - 2 = 1$ yd³; *Exact:* $1\frac{5}{8}$ yd³

61. *Estimate:* $527 - 108 - 151 - 139 = 129$ ft; *Exact:* 130 ft

63. *Estimate:* $3 + 7 + 2 + 3 + 7 = 22$ tons; *Exact:* $21\frac{7}{12}$ tons

65. $4\frac{11}{16}$ in.   **67.** $21\frac{3}{8}$ in.   **69. (a)** 30   **(b)** 28   **(c)** 25

(d) 264   **70.** least common denominator   **71. (a)** $\frac{23}{24}$ **(b)** $\frac{8}{15}$

(c) $\frac{43}{48}$ **(d)** $\frac{4}{21}$ **72.** fraction parts   **73.** improper

74. (a) $4\frac{5}{8} + 3\frac{2}{8} = 7\frac{7}{8}$; $\frac{37}{8} + \frac{26}{8} = \frac{63}{8} = 7\frac{7}{8}$

(b) $11\frac{56}{40} - 8\frac{35}{40} = 3\frac{21}{40}$; $\frac{496}{40} - \frac{355}{40} = \frac{141}{40} = 3\frac{21}{40}$

Section 3.5 (page 227)

1.–12.

2. 1.   **10.**   **4.**   **3.** 12.   **7.** 5.   **6.**   **11.** 9.   **8.**

13. >   **15.** <   **17.** >   **19.** <   **21.** >   **23.** >   **25.** $\frac{1}{9}$ **27.** $\frac{25}{64}$

29. $\frac{9}{16}$ **31.** $\frac{64}{125}$ **33.** $\frac{81}{16} = 5\frac{1}{16}$ **35.** $\frac{81}{256}$

37. A number line is a horizontal line with a range of numbers placed on it. The lowest number is on the left and the highest number is on the right. It can be used to compare the size or value of numbers.

39. 4   **41.** 10   **43.** 1   **45.** $\frac{3}{16}$ **47.** $\frac{4}{9}$ **49.** $\frac{1}{3}$ **51.** $\frac{1}{2}$ **53.** $\frac{3}{8}$

55. $\frac{1}{4}$ **57.** $1\frac{1}{2}$ **59.** $\frac{1}{12}$ **61.** 3   **63.** $\frac{5}{16}$ **65.** $\frac{1}{4}$ **67.** $\frac{1}{32}$

69. $\frac{9}{25}$ of the employees is greater.   **71.** <; >

72. (a) like; numerators; numerator   **(b)** Answers will vary.

73. parentheses; exponents; square; multiply; divide; add; subtract

74. $\frac{2}{45}$ **75.** $\frac{4}{9}$ **76.** $2\frac{1}{2}$ **77.** $\frac{27}{125}$ **78.** $1\frac{9}{16}$ **79.** 2   **80.** $2\frac{57}{64}$

Summary Exercises on Fractions (page 231)

1. proper   **2.** improper   **3.** improper   **4.** proper   **5.** $\frac{5}{6}$ **6.** $\frac{7}{8}$

7. $\frac{3}{7}$ **8.** $\frac{23}{47}$ **9.** $\frac{1}{2}$ **10.** $\frac{3}{8}$ **11.** 35   **12.** $\frac{5}{6}$ **13.** $1\frac{1}{6}$ **14.** 56

15. $1\frac{13}{24}$ **16.** $1\frac{13}{16}$ **17.** $2\frac{1}{12}$ **18.** $\frac{1}{12}$ **19.** $\frac{11}{24}$ **20.** $\frac{2}{15}$

21. *Exact:* $7\frac{7}{8}$; *Estimate:* $4 \cdot 2 = 8$

22. *Exact:* $17\frac{15}{32}$; *Estimate:* $5 \cdot 3 = 15$

23. *Exact:* $107\frac{2}{3}$; *Estimate:* $8 \cdot 6 \cdot 2 = 96$

24. *Exact:* $1\frac{1}{6}$; *Estimate:* $4 \div 4 = 1$

25. *Exact:* $3\frac{7}{16}$; *Estimate:* $7 \div 2 = 3\frac{1}{2}$

26. *Exact:* $6\frac{1}{6}$; *Estimate:* $5 \div 1 = 5$

27. *Estimate:* $6 + 4 = 10$; *Exact:* $9\frac{11}{12}$

28. *Estimate:* $18 + 10 = 28$; *Exact:* $28\frac{1}{6}$

29. *Estimate:* $15 + 11 = 26$; *Exact:* $25\frac{4}{15}$

30. *Estimate:* $9 - 4 = 5$; *Exact:* $4\frac{19}{20}$

31. *Estimate:* $14 - 7 = 7$; *Exact:* $6\frac{5}{8}$

32. *Estimate:* $32 - 23 = 9$; *Exact:* $9\frac{1}{4}$ **33.** $\frac{1}{12}$ **34.** $\frac{9}{10}$ **35.** $\frac{5}{18}$

36. 40 **37.** 72 **38.** 84 **39.** 35 **40.** 12 **41.** 55 **42.** $<$
43. $>$ **44.** $>$

Chapter 3 Review Exercises (page 237)

1. $\frac{6}{7}$ **2.** $\frac{7}{9}$ **3.** $\frac{3}{4}$ **4.** $\frac{1}{8}$ **5.** $\frac{4}{5}$ **6.** $\frac{1}{6}$ **7.** $\frac{13}{31}$ **8.** $\frac{1}{3}$

9. $\frac{3}{4}$ of her patients **10.** $\frac{1}{4}$ Web page less **11.** 10 **12.** 12

13. 60 **14.** 24 **15.** 120 **16.** 180 **17.** 8 **18.** 21 **19.** 10

20. 45 **21.** 32 **22.** 20 **23.** $\frac{5}{6}$ **24.** $\frac{7}{8}$ **25.** $\frac{5}{8}$ **26.** $\frac{5}{12}$

27. $\frac{13}{24}$ **28.** $\frac{17}{36}$ **29.** $\frac{23}{24}$ of the sack **30.** $\frac{59}{60}$ of the budget

31. *Estimate:* $19 + 14 = 33$; *Exact:* $32\frac{3}{8}$

32. *Estimate:* $23 + 15 = 38$; *Exact:* $38\frac{1}{9}$

33. *Estimate:* $13 + 9 + 10 = 32$; *Exact:* $31\frac{43}{80}$

34. *Estimate:* $32 - 15 = 17$; *Exact:* $17\frac{1}{12}$

35. *Estimate:* $34 - 16 = 18$; *Exact:* $18\frac{1}{3}$

36. *Estimate:* $215 - 136 = 79$; *Exact:* $79\frac{7}{16}$

37. $9\frac{1}{10}$ **38.** $10\frac{5}{12}$ **39.** $3\frac{1}{4}$ **40.** $1\frac{2}{3}$ **41.** $5\frac{1}{2}$ **42.** $2\frac{19}{24}$

43. *Estimate:* $15 - 6 - 7 = 2$ gallons; *Exact:* $2\frac{5}{12}$ gallons

44. *Estimate:* $29 + 25 = 54$ tons; *Exact:* $53\frac{5}{12}$ tons

45. *Estimate:* $9 + 9 + 7 = 25$ pounds; *Exact:* $24\frac{23}{24}$ pounds

46. *Estimate:* $9 - 2 - 3 = 4$ acres; *Exact:* $4\frac{1}{16}$ acres

47.–50.

47.   48. 49. 50.

51. $<$ **52.** $<$ **53.** $>$ **54.** $>$ **55.** $<$ **56.** $>$ **57.** $<$
58. $>$ **59.** $\frac{1}{4}$ **60.** $\frac{4}{9}$ **61.** $\frac{27}{1000}$ **62.** $\frac{81}{4096}$ **63.** $\frac{1}{2}$

64. $6\frac{3}{4}$ **65.** $\frac{1}{16}$ **66.** 1 **67.** $\frac{3}{16}$ **68.** $1\frac{25}{64}$ **69.** $\frac{3}{4}$ **70.** $\frac{2}{5}$

71. $\frac{19}{32}$ **72.** $\frac{11}{16}$ **73.** $2\frac{1}{6}$ **74.** $26\frac{1}{4}$ **75.** $5\frac{3}{8}$ **76.** $11\frac{43}{80}$

77. $15\frac{5}{12}$ **78.** $\frac{8}{11}$ **79.** $\frac{1}{250}$ **80.** $\frac{1}{2}$ **81.** $\frac{2}{9}$ **82.** $\frac{11}{27}$

83. $>$ **84.** $<$ **85.** $<$ **86.** $>$ **87.** 36 **88.** 120
89. 126 **90.** 18 **91.** 108 **92.** 60

93. *Estimate:* $93 - 14 - 22 = 57$ ft; *Exact:* $56\frac{7}{8}$ ft

94. *Estimate:* $45 - 11 - 26 = 8$ cubic yards; *Exact:* $8\frac{5}{8}$ cubic yards

Chapter 3 Test (page 243)

1. $\frac{3}{4}$ **2.** $\frac{1}{2}$ **3.** $\frac{2}{5}$ **4.** $\frac{1}{6}$ **5.** 12 **6.** 30 **7.** 108 **8.** $\frac{5}{8}$

9. $\frac{23}{36}$ **10.** $\frac{5}{24}$ **11.** $\frac{1}{40}$

12. *Estimate:* $8 + 5 = 13$; *Exact:* $12\frac{1}{2}$

13. *Estimate:* $16 - 12 = 4$; *Exact:* $4\frac{11}{15}$

14. *Estimate:* $19 + 9 + 12 = 40$; *Exact:* $40\frac{29}{60}$

15. *Estimate:* $24 - 18 = 6$; *Exact:* $5\frac{5}{8}$

16. Probably addition and subtraction of fractions is more difficult because you have to find the least common denominator and then change the fractions to the same denominator. **17.** Round mixed numbers to the nearest whole number. Then add, subtract, multiply, or divide to estimate the answer. The estimate may vary from the exact answer but it lets you know if your answer is reasonable.

18. *Estimate:* $6 + 5 + 4 + 7 + 5 = 27$ hr; *Exact:* $26\frac{3}{4}$

19. *Estimate:* $148 - 69 - 37 - 6 = 36$ gal; *Exact:* $35\frac{7}{8}$ gal

20. $>$ **21.** $>$ **22.** 2 **23.** $\frac{13}{48}$ **24.** $1\frac{3}{4}$ **25.** $1\frac{1}{3}$

Cumulative Review Exercises: Chapters 1–3 (page 245)

1. 8, 1 **2.** 5, 9 **3.** 1440; 1400; 1000 **4.** 59,800; 59,800; 60,000 **5.** *Estimate:* $2000 + 400 + 50,000 + 30,000 = 82,400$; *Exact:* 79,779 **6.** *Estimate:* $20,000 - 10,000 = 10,000$; *Exact:* 14,389 **7.** *Estimate:* $4000 \times 300 + 1,200,000$; *Exact:* 1,251,040 **8.** *Estimate:* $100,000 \div 40 = 2500$; *Exact:* 3211 **9.** 26 **10.** 1,255,609 **11.** 591 **12.** 2,801,695 **13.** 120 **14.** 126 **15.** 160 **16.** 456 **17.** 369,408 **18.** 17,000 **19.** 135 **20.** 2693 R2 **21.** 32 R166 **22.** *Estimate:* $20 + 9 + 5 + 20 + 9 + 5 = 68$ ft; *Exact:* 64 ft **23.** *Estimate:* $20 \cdot 10 = 200$ ft^2; *Exact:* 252 ft^2 **24.** *Estimate:* $40,000 \div 30 \approx 1333$ cartons; *Exact:* 1260 cartons **25.** *Estimate:* $4000 \times 60 = 240,000$ revolutions; *Exact:* 216,000 revolutions

26. *Estimate:* $2 \cdot 3 = 6$ yd^2; *Exact:* $4\frac{2}{3}$ yd^2

27. *Estimate:* $3 \cdot 5 = 15$ mi^2; *Exact:* $12\frac{1}{4}$ mi^2

28. *Estimate:* $5 \cdot 4 = 20$ cords; *Exact:* $18\frac{3}{8}$ cords

29. *Estimate:* $1537 - 83 = 1454$ ft; *Exact:* $1454\frac{3}{8}$ ft

30. $2 \cdot 5^2$ **31.** $2^3 \cdot 4^2$ **32.** $5^2 \cdot 7^2$ **33.** 144 **34.** 256

35. 972 **36.** 6 **37.** 9 **38.** 12 **39.** 9 **40.** 44 **41.** $\frac{4}{45}$

42. $\frac{9}{10}$ **43.** $1\frac{1}{16}$ **44.** proper **45.** improper **46.** improper

47. $\frac{5}{12}$ **48.** $\frac{7}{8}$ **49.** $\frac{9}{10}$ **50.** $\frac{1}{2}$ **51.** $\frac{5}{16}$ **52.** $36\frac{3}{4}$ **53.** $1\frac{2}{3}$

54. $2\frac{3}{16}$ **55.** $13\frac{1}{2}$ **56.** $\frac{25}{28}$ **57.** $\frac{13}{16}$ **58.** $\frac{7}{36}$

59. *Estimate:* $3 + 5 = 8$; *Exact:* $7\frac{7}{8}$

60. *Estimate:* $22 + 4 = 26$; *Exact:* $26\frac{7}{24}$

61. *Estimate:* $5 - 2 = 3$; *Exact:* $2\frac{5}{8}$

62. 24 **63.** 120 **64.** 144 **65.** 36 **66.** 56 **67.** 81 **68.** 60
69.–72. **73.** < **74.** > **75.** <

70. 71. 69. 72.

CHAPTER 4

Section 4.1 (page 255)

1. 7; 0; 4 **3.** 5; 1; 8 **5.** 4; 7; 0 **7.** 1; 6; 3 **9.** 1; 8; 9

11. 6; 2; 1 **13.** 410.25 **15.** 6.5432 **17.** 5406.045 **19.** $\frac{7}{10}$

21. $13\frac{2}{5}$ **23.** $\frac{1}{4}$ **25.** $\frac{33}{50}$ **27.** $10\frac{17}{100}$ **29.** $\frac{3}{50}$ **31.** $\frac{41}{200}$

33. $5\frac{1}{500}$ **35.** $\frac{343}{500}$ **37.** five tenths **39.** seventy-eight
hundredths **41.** one hundred five thousandths **43.** twelve and
four hundredths **45.** one and seventy-five thousandths **47.** 6.7
49. 0.32 **51.** 420.008 **53.** 0.0703 **55.** 75.030 **57.** Anne
should not say "and" because that denotes a decimal point.

59. ten thousandths inch; $\frac{10}{1000} = \frac{1}{100}$ inch **61.** 12 pounds

63. 3-C **65.** 4-A **67.** One and six hundred two thousandths
centimeters **69.** millionths, ten-millionths, hundred-millionths,
billionths; these match the words on the left side of the chart with
"ths" attached. **70.** The first place to the left of the decimal point
is ones, so the first place to the right could be one*ths,* like tens and
ten*ths.* But anything that is 1 or more is to the left of the decimal
point. **71.** Seventy-two million four hundred thirty-six thousand
nine hundred fifty-five hundred-millionths **72.** six hundred
seventy-eight thousand five hundred fifty-four billionths
73. eight thousand six and five hundred thousand one millionths
74. twenty thousand, sixty and five hundred five millionths
75. 0.0302040 **76.** 9,876,543,210.100200300

Section 4.2 (page 265)

1. 16.9 **3.** 0.956 **5.** 0.80 **7.** 3.661 **9.** 794.0 **11.** 0.0980
13. 49 **15.** 9.09 **17.** 82.0002 **19.** $0.82 **21.** $1.22

23. $0.50 **25.** $48,650 **27.** $310 **29.** $849 **31.** $500
33. $1.00 **35.** $1000 **37. (a)** 322 miles per hour **(b)** 107
miles per hour **39. (a)** 186.0 miles per hour **(b)** 763.0 miles
per hour **41.** Rounds to $0 (zero dollars) because $0.499 is closer
to $0 than to $1. **42.** Round amounts less than $1.00 to nearest
cent instead of nearest dollar. **43.** Rounds to $0.00 (zero cents)
because $0.0015 is closer to $0.00 than to $0.01. **44.** Both round
to $0.60. Rounding to nearest thousandth (tenth of a cent) would
allow you to identify $0.597 as less than $0.601.

Section 4.3 (page 271)

1. 17.48 **3.** 23.013 **5.** 7.763 **7.** 77.006 **9.** 20.104
11. 0.109 **13.** 330.86895 **15. (a)** 24.75 in. **(b)** 3.95 in.
17. (a) 62.27 in. **(b)** 0.39 in. **19.** 6 should be written 6.00;
sum is 46.22. **21.** *Estimate:* $20 - 7 = $13; *Exact:* $13.16
23. *Estimate:* $400 + 1 + 20 = 421$; *Exact:* 414.645
25. *Estimate:* $9 - 4 = 5$; *Exact:* 4.849 **27.** *Estimate:* $60 + 500$
$+ 6 = 566$; *Exact:* 608.4363 **29.** 0.275 **31.** 6.507 **33.** 1.81
35. 6056.7202 **37.** *Estimate:* $19 - 15 = 4$ million people;
Exact: 3.6 million people **39.** *Estimate:* $100 + 30 + 20 + 20 +$
$20 + 9 = 199$ million people; *Exact:* 234.4 million people
41. *Estimate:* $2 + 2 + 2 = 6$ meters; *Exact:* 6.1 meters, which is
less than the rhino by 0.3 meter **43.** *Estimate:* $11 - 10 =$
1 ounce; *Exact:* 0.65 ounce **45.** *Estimate:* $20 + 6 + 20 + 6 =$
52 in.; *Exact:* 52.1 in. **47.** *Estimate:* $5 - $5 = 0; *Exact:* $0.30
49. *Estimate:* $19 + 2 + 2 + 10 + 2 = 35; *Exact:* $35.25
51. $1939.36 **53.** $3.97 **55.** $598.22
57. $b = 1.39$ centimeters **59.** $q = 23.843$ ft

Section 4.4 (page 277)

1. 0.1344 **3.** 159.10 **5.** 15.5844 **7.** $34,500.20 **9.** 43.2
11. 0.432 **13.** 0.0432 **15.** 0.00432 **17.** 0.0000312
19. 0.000025 **21.** 59.6; 32; 4.76; 803.5; 7226; 9. Multiplying by
10, decimal point moves one place to the right; by 100, two places
to the right; by 1000, three places to the right. **22.** 5.96; 0.32;
0.0476; 8.035; 6.5; 52.3. Multiplying by 0.1, decimal point moves
one place to the left; by 0.01, two places to the left; by 0.001, three
places to the left. **23.** *Estimate:* $40 \times 5 = 200$; *Exact:* 190.08
25. *Estimate:* $40 \times 40 = 1600$; *Exact:* 1558.2
27. *Estimate:* $7 \times 5 = 35$; *Exact:* 30.038
29. *Estimate:* $3 \times 7 = 21$; *Exact:* 19.24165
31. unreasonable; $189.00 **33.** reasonable
35. unreasonable; $4.19 **37.** unreasonable; 9.5 pounds
39. $945.87 (rounded) **41.** $2.45 (rounded) **43.** $32.54
(rounded) **45.** $16,065 **47. (a)** Area before 1929 ≈ 23.2 in.^{2};
Area today ≈ 16.0 in.^{2} **(b)** 7.2 in.^{2} **49. (a)** 0.43 in.
(b) 4.3 in. **51.** $984.04; $2207.80 **53.** $76.50 **55.** $4.09
(rounded) **57.** $129.25 **59. (a)** $70.05 **(b)** $25.80

Section 4.5 (page 287)

1. 3.9 **3.** 0.47 **5.** 400.2 **7.** 36 **9.** 0.06 **11.** 6000
13. 25.3 **15.** 516.67 (rounded) **17.** 24.291 (rounded)
19. 10,082.647 (rounded) **21.** 0.377; 0.91; 0.0886; 3.019;
40.65; 662.57. **(a)** Dividing by 10, decimal point moves one place
to the left; by 100, two places to the left; by 1000, three places to the
left. **(b)** The decimal point moved to the *right* when *multiplying* by
10 or 100 or 1000. Here it moves to the *left* when *dividing* by those
numbers. **22.** 402; 71; 3.39; 157.7; 460; 8730. **(a)** Dividing by
0.1, decimal point moves one place to the right; by 0.01, two places
to the right; by 0.001, three places to the right. **(b)** The decimal
point moved to the *left* when multiplying by 0.1 or 0.01 or 0.001.
Here the decimal point moves to the *right* when *dividing* by those
numbers. **23.** unreasonable; $40 \div 8 = 5$; Correct answer is 4.725.
25. reasonable; $50 \div 50 = 1$ **27.** unreasonable; $300 \div 5 = 60$;

Correct answer is 60.2.   **29.** unreasonable; $9 \div 1 = 9$; Correct answer is 7.44.   **31.** $4.00 (rounded)   **33.** $67.08 (rounded)   **35.** $0.30   **37.** $11.92 per hour   **39.** 21.2 miles per gallon (rounded)   **41.** 7.37 meters (rounded)   **43.** 0.08 meter   **45.** 22.49 meters   **47.** 14.25   **49.** 73.4   **51.** 1.205   **53.** 0.334   **55.** $0.03 per can (rounded)   **57. (a)** 1,541,667 pieces (rounded)   **(b)** 25,694 pieces (rounded)   **(c)** 428 pieces (rounded).   **59.** 100,000 box tops   **61.** 2632 box tops (rounded)

Summary Exercises on Decimals (page 291)

1. $\frac{4}{5}$ **2.** $6\frac{1}{250}$ **3.** $\frac{7}{20}$ **4.** ninety-four and five tenths   **5.** two and three ten-thousandths   **6.** seven hundred six thousandths   **7.** 0.05   **8.** 0.0309   **9.** 10.7   **10.** 6.19   **11.** 1.0   **12.** 0.420   **13.** $0.89   **14.** $3.00   **15.** $100   **16.** 0.945   **17.** 49.6199   **18.** 50   **19.** 15.03   **20.** 0.00488   **21.** 2.15   **22.** 10.955   **23.** 18.4009   **24.** 5.005   **25.** 0.9   **26.** 8.08   **27.** 0.04 (rounded)   **28.** $P = 3.1$ in.; $A \approx 0.5$ in.^{2} **29.** $P = 3.25$ in.; $A \approx 0.66$ in.^{2} **30. (a)** Watermelon is heaviest; strawberry is lightest   **(b)** 261.49 pounds   **31.** 91.803 pounds   **32.** 40.5 pounds   **33.** 75 apples (rounded)   **34.** $2052.50   **35.** $63.28

Section 4.6 (page 297)

1. 0.5   **3.** 0.75   **5.** 0.3   **7.** 0.9   **9.** 0.6   **11.** 0.875   **13.** 2.25   **15.** 14.7   **17.** 3.625   **19.** 0.333 (rounded)   **21.** 0.833 (rounded)   **23.** 1.889 (rounded)   **25. (a)** A proper fraction is less than 1, so it cannot be equivalent to a mixed number.   **(b)** $\frac{5}{9}$ means $5 \div 9$ or $9\overline{)5}$ so correct answer is 0.556 (rounded). This makes sense because both the fraction and decimal are less than 1.

26. (a) $2.035 = 2\frac{35}{1000} = 2\frac{7}{200}$, not $2\frac{7}{20}$.   **(b)** Adding the whole number part gives $2 + 0.35$, which is 2.35, not 2.035. To check, $2.35 = 2\frac{35}{100} = 2\frac{7}{20}$.   **27.** Just add the whole number part to 0.375. So $1\frac{3}{8} = 1.375$; $3\frac{3}{8} = 3.375$; $295\frac{3}{8} = 295.375$.

28. It works only when the fraction part has a one-digit numerator and a denominator of 10, a two-digit numerator and a denominator of 100, and so on.   **29.** $\frac{2}{5}$ **31.** $\frac{5}{8}$ **33.** $\frac{7}{20}$ **35.** 0.35   **37.** $\frac{1}{25}$ **39.** $\frac{3}{20}$ **41.** 0.2   **43.** $\frac{9}{100}$ **45.** shorter; 0.72 inch   **47.** too much; 0.005 gram   **49.** 0.9991 cm, 1.0007 cm   **51.** more; 0.05 inch   **53.** 0.5399, 0.54, 0.5455   **55.** 5.0079, 5.79, 5.8, 5.804   **57.** 0.6009, 0.609, 0.628, 0.62812   **59.** 2.8902, 3.88, 4.876, 5.8751   **61.** 0.006, 0.043, $\frac{1}{20}$, 0.051   **63.** 0.37, $\frac{3}{8}$, $\frac{2}{5}$, 0.4001   **65.** red box   **67.** 0.01 in.   **69.** 1.4 in. (rounded)   **71.** 0.3 in. (rounded)   **73.** 0.4 in. (rounded)

Chapter 4 Review Exercises (page 305)

1. 0; 5   **2.** 0; 6   **3.** 8; 9   **4.** 5; 9   **5.** 7; 6   **6.** $\frac{1}{2}$ **7.** $\frac{3}{4}$ **8.** $4\frac{1}{20}$ **9.** $\frac{7}{8}$ **10.** $\frac{27}{1000}$ **11.** $27\frac{4}{5}$ **12.** eight tenths   **13.** four hundred and twenty-nine hundredths   **14.** twelve and seven thousandths   **15.** three hundred six ten-thousandths   **16.** 8.3   **17.** 0.205   **18.** 70.0066   **19.** 0.30   **20.** 275.6   **21.** 72.79   **22.** 0.160   **23.** 0.091   **24.** 1.0   **25.** $15.83

26. $0.70   **27.** $17,625.79   **28.** $350   **29.** $130   **30.** $100   **31.** $29   **32.** *Estimate:* $6 + 400 + 20 = 426$; *Exact:* 444.86   **33.** *Estimate:* $80 + 1 + 100 + 1 + 30 = 212$; *Exact:* 233.515   **34.** *Estimate:* $300 - 20 = 280$; *Exact:* 290.7   **35.** *Estimate:* $9 - 8 = 1$; *Exact:* 1.2684   **36.** *Estimate:* 80 million − 50 million = 30 million; *Exact:* 31.4 million people   **37.** *Estimate:* $300 - $200 - $40 = 60; *Exact:* $45.80   **38.** *Estimate:* $2 + $5 + $20 = 27; $30 - $27 = 3; *Exact:* $4.14   **39.** *Estimate:* $2 + 4 + 5 = 11$ kilometers; *Exact:* 11.55 kilometers   **40.** *Estimate:* $6 \times 4 = 24$; *Exact:* 22.7106   **41.** *Estimate:* $40 \times 3 = 120$; *Exact:* 141.57   **42.** 0.0112   **43.** 0.000355   **44.** reasonable; $700 \div 10 = 70$ **45.** unreasonable; $30 \div 3 = 10$; Correct answer is 9.5.   **46.** 14.467 (rounded)   **47.** 1200   **48.** 0.4   **49.** $708 (rounded)   **50.** $2.99 (rounded)   **51.** 133 shares (rounded)   **52.** $3.47 (rounded)   **53.** 29.215   **54.** 10.15   **55.** 3.8   **56.** 0.64   **57.** 1.875   **58.** 0.111 (rounded)   **59.** 3.6008, 3.68, 3.806   **60.** 0.209, 0.2102, 0.215, 0.22   **61.** $\frac{1}{8}$, $\frac{3}{20}$, 0.159, 0.17   **62.** 404.865   **63.** 254.8   **64.** 3583.261 (rounded)   **65.** 29.0898   **66.** 0.03066   **67.** 9.4   **68.** 175.675   **69.** 9.04   **70.** 19.50   **71.** 8.19   **72.** 0.928   **73.** 35   **74.** 0.259   **75.** 0.3   **76.** $3.00 (rounded)   **77.** $2.17 (rounded)   **78.** $35.96   **79.** $199.71   **80.** $78.50   **81. (a)** baked potato with skin   **(b)** cup of orange juice   **(c)** 0.59 milligram   **82. (a)** 2.06 milligrams   **(b)** more, by 0.06 milligram

Chapter 4 Test (page 309)

1. $18\frac{2}{5}$ **2.** $\frac{3}{40}$ **3.** sixty and seven thousandths   **4.** two hundred eight ten-thousandths   **5.** 725.6   **6.** 0.630   **7.** $1.49   **8.** $7860   **9.** *Estimate:* $8 + 80 + 40 = 128$; *Exact:* 129.2028   **10.** *Estimate:* $80 - 4 = 76$; *Exact:* 75.498   **11.** *Estimate:* $6(1) = 6$; *Exact:* 6.948   **12.** *Estimate:* $20 \div 5 = 4$; *Exact:* 4.175   **13.** 839.762   **14.** 669.004   **15.** 0.0000483   **16.** 480   **17.** 2.625   **18.** 0.44, $\frac{9}{20}$, 0.4506, 0.451   **19.** 35.49   **20.** $446.87   **21.** pintails, wigeons, gadwalls   **22.** $5.35 (rounded)   **23.** 2.8 degrees   **24.** $4.55 per meter   **25.** Answer varies.

Cumulative Review Exercises: Chapters 1–4 (page 311)

1. 5, 9, 2   **2.** 0, 5, 8   **3.** 500,000   **4.** 602.49   **5.** $710   **6.** $0.05   **7.** *Estimate:* $4000 + 600 + 9000 = 13,600$; *Exact:* 13,339   **8.** *Estimate:* $4 + 20 + 1 = 25$; *Exact:* 20.683   **9.** *Estimate:* $5000 - 2000 = 3000$; *Exact:* 3209   **10.** *Estimate:* $50 - 7 = 43$; *Exact:* 44.506   **11.** *Estimate:* $3000 \times 200 = 600,000$; *Exact:* 550,622   **12.** *Estimate:* $7 \times 7 = 49$; *Exact:* 49.786   **13.** *Estimate:* $100,000 \div 50 = 2000$; *Exact:* 2690   **14.** *Estimate:* $40 \div 8 = 5$; *Exact:* 4.5   **15.** *Estimate:* $2 \cdot 4 = 8$; *Exact:* $7\frac{1}{8}$ **16.** *Estimate:* $2 \div 1 = 2$; *Exact:* $2\frac{4}{5}$ **17.** *Estimate:* $2 + 2 = 4$; *Exact:* $3\frac{7}{15}$ **18.** *Estimate:* $5 - 2 = 3$; *Exact:* $2\frac{5}{8}$

19. 9.671   **20.** $1\frac{4}{9}$ **21.** 73,225   **22.** $1\frac{2}{5}$ **23.** 4914

24. 93.603   **25.** 404 R3   **26.** $1\frac{17}{24}$ **27.** 233,728   **28.** 0.03264

29. 8   **30.** 45   **31.** $\frac{4}{31}$ **32.** $\frac{2}{3}$ **33.** 0.51 (rounded)   **34.** 4

35. 14   **36.** $\frac{1}{4}$ **37.** 20.81   **38.** 576   **39.** 14   **40.** $2^3 \cdot 5^2$

41. forty and thirty-five thousandths   **42.** 0.0306   **43.** $\frac{1}{8}$

44. $3\frac{2}{25}$ **45.** 2.6   **46.** 0.636 (rounded)   **47.** >

48. 7.005, 7.5, 7.5005, 7.505   **49.** 0.8, 0.8015, $\frac{21}{25}, \frac{7}{8}$

50. size M (medium)

51. XXS $1\frac{1}{2}$ in.; XS $\frac{3}{4}$ in.; S $\frac{3}{4}$ in.; M $\frac{7}{8}$ in.; L $\frac{3}{4}$ in.

52. $21.125 - 20.25 = 0.875$ in.; $7 \div 8 = 0.875$ in.

or $\frac{875}{1000} = \frac{875 \div 125}{1000 \div 125} = \frac{7}{8}$ in.

53. Answers will vary. Many people prefer using decimals because you do not need to find a common denominator or rewrite answers in lowest terms.   **54.** *Estimate:* $40 - $18 - $1 = 21;

Exact: $21.17   **55.** *Estimate:* 50 − 47 = 3 in.; *Exact:* $3\frac{3}{8}$ in.

56. *Estimate:* $10 × 20 = $200; *Exact:* $191.90 (rounded)
57. *Estimate:* $(8 × 20) + (10 × 30) = 160 + 300 = 460$ students;

Exact: 488 students

58. *Estimate:* 2 + 4 = 6 yards; *Exact:* $6\frac{5}{24}$ yards

59. *Estimate:* $30 + $200 − $40 − $20 = $170; *Exact:* $191.50
60. *Estimate:* $6 ÷ 3 = $2 per pound; *Exact:* $2.29 per pound
(rounded)   **61.** *Estimate:* $80,000 ÷ 100 = $800; Exact $729
(rounded)   **62. (a)** 9 days (rounded)   **(b)** 15.6 pounds (rounded)
63. Average weight of food eaten by bee is 0.32 ounce

64. 0.112 ounce; $\frac{14}{125}$ ounce; $14 \div 125 = 0.112$

CHAPTER 5

Section 5.1 (page 321)

1. $\frac{8}{9}$ **3.** $\frac{2}{1}$ **5.** $\frac{1}{3}$ **7.** $\frac{8}{5}$ **9.** $\frac{3}{8}$ **11.** $\frac{9}{7}$ **13.** $\frac{6}{1}$ **15.** $\frac{5}{6}$ **17.** $\frac{8}{5}$

19. $\frac{1}{12}$ **21.** $\frac{5}{16}$ **23.** $\frac{4}{1}$ **25.** $\frac{1}{2}$ **27.** $\frac{36}{1}$ **29.** Answers will

vary. One possibility is stocking cards of various types in the same

ratios as those in the table.   **31.** $\frac{6}{5}; \frac{36}{17}$ **33.** *White Christmas* to

It's Now or Never; White Christmas to *I Will Always Love You;*

Candle in the Wind to *I Want to Hold Your Hand.*   **35. (a)** $\frac{6}{1}$

(b) $\frac{5}{2}$ **(c)** $\frac{7}{16}$ **37.** $\frac{7}{5}$ **39.** $\frac{6}{1}$ **41.** $\frac{38}{17}$ **43.** $\frac{1}{4}$ **45.** $\frac{34}{35}$

47. $\frac{1}{1}$; as long as the sides all have the same length, any

measurement you choose will maintain the ratio.
48. Answers will vary. Some possibilities are
$\frac{4}{5} = \frac{8}{10} = \frac{12}{15} = \frac{16}{20} = \frac{20}{25} = \frac{24}{30} = \frac{28}{35}$.   **49.** It is not possible.
Amelia would have to be older than her mother to have a ratio of
5 to 3.   **50.** Answers will vary, but a ratio of 3 to 1 means your
income is 3 times your friend's income.

Section 5.2 (page 329)

1. $\frac{5 \text{ cups}}{3 \text{ people}}$ **3.** $\frac{3 \text{ feet}}{7 \text{ seconds}}$ **5.** $\frac{18 \text{ miles}}{1 \text{ gallon}}$

7. $12 per hour or $12/hour   **9.** 1.25 pounds/person
11. $103.30/day   **13.** 325.9   21.0 (rounded)   **15.** 338.6   20.9
(rounded)   **17.** 4 ounces for $3.09   **19.** 15 ounces for $3.15
21. 18 ounces for $1.79   **23.** Answers will vary. For example, you
might choose Brand B because you like more chicken, so the cost
per chicken chunk may actually be the same or less than Brand A.
25. 1.75 pounds/week   **27.** $12.26/hour   **29. (a)** Radiant, $0.44;
IDT, $0.25; Access; $0.235   **(b)** Radiant, $0.088/min; IDT,
$0.05/min; Access, $0.047/min; Access America is the best buy.
31. For a 15-minute call: Radiant, $0.036/min; IDT, $0.031/min;
Access, $0.047/min; IDT is the best buy. For a 25-minute call: Ra-
diant, $0.026/min; IDT, $0.028/min; Access, $0.047/min; Radiant is
the best buy.   **33.** one battery for $1.79; like getting 3 batteries so
$1.79 ÷ 3 ≈ $0.597 per battery   **35.** Brand P with the 50¢ coupon
is the best buy. ($3.39 − $0.50 = $2.89 ÷ 16.5 ounces ≈ $0.175
per ounce)   **37.** $2.92 (rounded) per month for Verizon, T-Mobile,
and Nextel; $3 per month for Sprint.   **38.** Average number of
weekdays per month is about 21.7; Verizon ≈ 18 min/weekday;
T-Mobile ≈ 28 min/weekday; Nextel and Sprint ≈ 23 min.
39. Round to hundredths (nearest cent) to see that T-Mobile is best
buy at $0.07 per "anytime minute"; Verizon ≈ $0.16; Sprint ≈ $0.12;
Nextel ≈ $0.10.   **40.** Verizon ≈ $1.65 per "anytime minute";
T-Mobile ≈ $1.57; Nextel ≈ $1.63; Sprint = $1.48.

Section 5.3 (page 337)

1. $\frac{\$9}{12 \text{ cans}} = \frac{\$18}{24 \text{ cans}}$ **3.** $\frac{200 \text{ adults}}{450 \text{ children}} = \frac{4 \text{ adults}}{9 \text{ children}}$

5. $\frac{120}{150} = \frac{8}{10}$ **7.** $\frac{3}{5} = \frac{3}{5}$; true   **9.** $\frac{5}{8} = \frac{5}{8}$; true   **11.** $\frac{3}{4} \neq \frac{2}{3}$; false

13. $\frac{14}{5} = \frac{14}{5}$; true   **15.** $\frac{16}{9} = \frac{16}{9}$; true   **17.** $\frac{7}{6} \neq \frac{9}{8}$; false

19. 54 = 54; True   **21.** 336 ≠ 320; False   **23.** 2880 ≠ 2970; False
25. 28 = 28; True   **27.** 44.8 ≠ 45; False   **29.** 66 = 66; True

31. $68\frac{1}{4} = 68\frac{1}{4}$; True   **33.** $5\frac{2}{5} \neq 5\frac{1}{3}$; False

35. $2\frac{1}{80} \neq 2\frac{7}{100}$ or   2.0125 ≠ 2.07; False

37. $\frac{16 \text{ hits}}{50 \text{ at bats}} = \frac{128 \text{ hits}}{400 \text{ at bats}}$ $\begin{array}{c} 50 \cdot 128 = 6400 \\ 16 \cdot 400 = 6400 \end{array}$ Cross products

are *equal* so the proportion is *true;* they hit equally well.

Section 5.4 (page 343)

1. $x = 4$ **3.** $x = 2$ **5.** $x = 88$ **7.** $x = 91$ **9.** $x = 5$
11. $x = 10$ **13.** $x \approx 24.44$ (rounded)   **15.** $x = 50.4$

17. $x \approx 17.64$ (rounded)   **19.** $x = 1$ **21.** $x = 3\frac{1}{2}$

23. $x = 0.2$ or $x = \frac{1}{5}$ **25.** $x = 0.005$ or $x = \frac{1}{200}$

27. Find cross products: 20 ≠ 30, so the proportion is false.

$\dfrac{6\frac{2}{3}}{4} = \dfrac{5}{3}$ or $\dfrac{10}{6} = \dfrac{5}{3}$ or $\dfrac{10}{4} = \dfrac{7.5}{3}$ or $\dfrac{10}{4} = \dfrac{5}{2}$

28. Find cross products: 192 ≠ 180, so the proportion is false.

$\dfrac{6.4}{8} = \dfrac{24}{30}$ or $\dfrac{6}{7.5} = \dfrac{24}{30}$ or $\dfrac{6}{8} = \dfrac{22.5}{30}$ or $\dfrac{6}{8} = \dfrac{24}{32}$

Summary Exercises on Ratios, Rates, and Proportions (page 345)

1. $\frac{6}{5}$ 2. $\frac{2}{7}$ 3. $\frac{2}{1}$ 4. $\frac{13}{11}$ 5. Comparing the violin to piano, guitar, organ, clarinet, and drums gives ratios of $\frac{1}{11}, \frac{1}{10}, \frac{1}{3}, \frac{1}{2},$ and $\frac{2}{3}$ respectively.   6. (a) guitar to clarinet (b) organ to drums, or clarinet to violin   7. 2.1 points/min; 0.5 min/point
8. 1.6 points/min; 0.6 min/point   9. \$16.32/hour; \$24.48/hour for overtime   10. \$0.50/channel; \$0.39/channel; \$0.32/channel
11. 34 ounces, at \$0.16 per ounce   12. Brand P with the \$2 coupon is the best buy at \$0.57 per pound.
13. $\frac{4}{3} = \frac{4}{3}$ or $924 = 924$; true   14. $2.0125 \neq 2.07$; false

15. $68\frac{1}{4} = 68\frac{1}{4}$; true   16. $x = 28$ 17. $x = 3.2$ 18. $x = 182$
19. $x \approx 3.64$ (rounded)   20. $x \approx 0.93$ (rounded)   21. $x = 1.56$
22. $x \approx 0.05$ (rounded)   23. $x = 1$ 24. $x = \frac{3}{4}$

Section 5.5 (page 351)

1. 22.5 hours   3. \$7.20   5. 42 pounds   7. \$273.45
9. 10 ounces (rounded)   11. 5 quarts   13. 14 ft, 10 ft
15. 14 ft, 8 ft   17. 96 pieces chicken; 33.6 pounds lasagna;
10.8 pounds deli meats; $5\frac{3}{5}$ pounds cheese; 7.2 dozen (about 86)
buns; 14.4 pounds of salad   19. 2065 students (reasonable); about 4214 students with incorrect setup (only 2950 students in the group)
21. about 30 people (reasonable); about 1904 people with incorrect setup (only 238 people attended)   23. 105,350,000 households (reasonable); about 109,693,878 households with incorrect setup (only 107,500,000 U.S. households).   25. 625 stocks
27. 4.06 meters (rounded)   29. 311 calories (rounded)
31. 10.53 meters (rounded)   33. You cannot solve this problem using a proportion because the ratio of age to weight is not constant. As Jim's age increases, his weight may decrease, stay the same, or increase.   35. 4800 students use cream   37. 120 calories and 12 grams of fiber   39. $1\frac{3}{4}$ cups water, 3 Tbsp margarine,
$\frac{3}{4}$ cup milk, 2 cups flakes   40. $5\frac{1}{4}$ cups water, 9 Tbsp margarine,
$2\frac{1}{4}$ cups milk, 6 cups flakes   41. $\frac{7}{8}$ cup water, $1\frac{1}{2}$ Tbsp margarine,
$\frac{3}{8}$ cup milk, 1 cup flakes   42. $2\frac{5}{8}$ cups water, $4\frac{1}{2}$ Tbsp margarine,
$1\frac{1}{8}$ cups milk, 3 cups flakes

Chapter 5 Review Exercises (page 361)

1. $\frac{3}{4}$ 2. $\frac{4}{1}$ 3. great white shark to whale shark; whale shark to blue whale   4. $\frac{2}{1}$ 5. $\frac{2}{3}$ 6. $\frac{5}{2}$ 7. $\frac{1}{6}$ 8. $\frac{3}{1}$ 9. $\frac{3}{8}$ 10. $\frac{4}{3}$
11. $\frac{1}{9}$ 12. $\frac{10}{7}$ 13. $\frac{7}{5}$ 14. $\frac{5}{6}$ 15. $\frac{\$11}{1\text{ dozen}}$ 16. $\frac{12\text{ children}}{5\text{ families}}$
17. 0.2 page/minute or $\frac{1}{5}$ page/minute; 5 minutes/page
18. \$8/hour; 0.125 hour/dollar or $\frac{1}{8}$ hour/dollar
19. 13 ounces for \$2.29   20. 25 pounds for \$10.40 − \$1 coupon
21. $\frac{3}{5} = \frac{3}{5}$ or $90 = 90$; true   22. $\frac{1}{8} \neq \frac{1}{4}$ or $432 \neq 216$; false

23. $\frac{47}{10} \neq \frac{49}{10}$ or $980 \neq 940$; false   24. $\frac{16}{9} = \frac{16}{9}$ or $3456 = 3456$; true   25. $4.8 = 4.8$; true   26. $14 = 14$; true   27. $x = 1575$
28. $x = 20$ 29. $x = 400$ 30. $x = 12.5$ 31. $x \approx 14.67$
(rounded)   32. $x \approx 8.17$ (rounded)   33. $x = 50.4$ 34. $x \approx 0.57$
(rounded)   35. $x \approx 2.47$ (rounded)   36. 27 cats   37. 46 hits
38. \$15.63 (rounded)   39. 3299 students (rounded)   40. 68 feet

41. $27\frac{1}{2}$ hours or 27.5 hours   42. 511 calories (rounded)

43. 14.7 milligrams   44. $x = 105$ 45. $x = 0$ 46. $x = 128$
47. $x \approx 23.08$ (rounded)   48. $x = 6.5$ 49. $x \approx 117.36$ (rounded)
50. $1440 \neq 1485$; False   51. $10.8 \neq 10.864$; False
52. $2\frac{1}{3} = 2\frac{1}{3}$; True   53. $\frac{8}{5}$ 54. $\frac{33}{80}$ 55. $\frac{15}{4}$ 56. $\frac{4}{1}$ 57. $\frac{4}{5}$

58. $\frac{37}{7}$ 59. $\frac{3}{8}$ 60. $\frac{1}{12}$ 61. $\frac{45}{13}$ 62. 24,900 fans (rounded)

63. $\frac{8}{3}$ 64. 75 ft for \$1.99 − \$0.50 coupon   65. 21 ft long;

15 ft wide   66. 7.5 hours or $7\frac{1}{2}$ hours   67. $\frac{1}{2}$ teaspoon or

0.5 teaspoon   68. 21 points (rounded)   69. Set up the proportion to compare teaspoons to pounds on both sides.

$$\frac{1.5 \text{ teaspoons}}{24 \text{ pounds}} = \frac{x \text{ teaspoons}}{8 \text{ pounds}}$$

Show that cross products are equal. $(24)(x) = (1.5)(8)$

Divide both sides by 24. $\dfrac{\overset{1}{\cancel{24}}(x)}{\underset{1}{\cancel{24}}} = \dfrac{12}{24}$

so $x = \frac{1}{2}$ teaspoon or 0.5 teaspoon   70. (a) 1400 milligrams

(b) 100 milligrams

Chapter 5 Test (page 365)

1. $\frac{4}{5}$ 2. $\frac{20 \text{ miles}}{1 \text{ gallon}}$ 3. $\frac{\$1}{5 \text{ minutes}}$ 4. $\frac{9}{2}$ 5. $\frac{15}{4}$
6. Best buy is 8 in. sub $\approx$ \$0.686 per inch.   7. 18 ounces for \$1.89 − \$0.25 coupon   8. You earned less this year. An example is: $\dfrac{\text{Last year} \rightarrow \$15,000}{\text{This year} \rightarrow \$10,000} = \dfrac{3}{2}$ 9. $\frac{3}{7} \neq \frac{2}{5}$ or $252 \neq 270$; false
10. $5.88 = 5.88$; true   11. $x = 25$ 12. $x \approx 2.67$ (rounded)
13. $x = 325$ 14. $x = 10\frac{1}{2}$ 15. 576 words   16. 3.6 ounces

17. 87 students (rounded)   18. No, 4875 cannot be correct because there are only 650 students in the whole school.
19. 23.8 grams (rounded)   20. 60 ft

Cumulative Review Exercises: Chapters 1–5 (page 367)

1. 5; 3; 6; 4   2. 9; 0; 5; 3   3. 9900   4. 617.1   5. \$100
6. \$3.06   7. *Estimate:* 30 + 5000 + 400 = 5430; *Exact:* 5585
8. *Estimate:* 60 − 6 = 54; *Exact:* 57.408
9. *Estimate:* 5000 × 800 = 4,000,000; *Exact:* 3,791,664
10. *Estimate:* 1 × 18 = 18; *Exact:* 17.4796
11. *Estimate:* 50,000 ÷ 50 = 1000; *Exact:* 907
12. *Estimate:* 2000 ÷ 5 = 400; *Exact:* 364
13. *Estimate:* 2 • 4 = 8; *Exact:* $6\frac{3}{5}$ 14. *Estimate:* 5 ÷ 1 = 5;
Exact: 6   15. *Estimate:* 3 − 2 = 1; *Exact:* $\frac{29}{30}$
16. *Estimate:* 3 + 11 = 14; *Exact:* $13\frac{2}{5}$

17. 374,416   **18.** 29.34   **19.** 610 R27   **20.** 0.0076   **21.** 2312
22. 68.381   **23.** 55.6   **24.** 35,852,728   **25.** 39   **26.** 18
27. 64   **28.** 0.95   **29.** one hundred five ten-thousandths
30. 60.071   **31.** $1\frac{9}{16} \approx 1.563$; $1\frac{3}{8} = 1.375$; $1\frac{1}{8} = 1.125$

32. $1\frac{1}{8}$ inches, 1.25 inches, $1\frac{3}{8}$ inches, 1.5 inches, $1\frac{9}{16}$ inches

33. $\frac{7}{16}$ inch   **34.** $\frac{1}{16}$ inch or 0.063 inch (rounded)   **35.** $\frac{11}{5}$ **36.** $\frac{1}{5}$

37. $\frac{4}{1}$ **38.** $\frac{1}{12}$ **39.** $\frac{3}{1}$ **40.** 36 servings for $3.24 − $0.50 coupon

41. $x = 21$ **42.** $x \approx 17.14$ (rounded)   **43.** $x = 11\frac{1}{4}$

44. $x \approx 0.98$ (rounded)   **45.** 250 pounds

46. 26.7 centimeters (rounded)   **47.** $7\frac{3}{20}$ miles

48. 18.0 miles per gallon (rounded)   **49.** 140 residents

50. $1\frac{1}{4}$ teaspoons   **51.** $0.08, $0.10, $0.104, $0.111, $0.14,

0.172, 0.333   **52.** 71 minutes (rounded)   **53.** 30 minutes
(rounded) per $10 card, so buy 4 cards for 120 minutes (2 hours).
54. $\frac{125}{100} = \frac{5}{4}$ **55.** (a) $0.024 (rounded) (b) $0.015
56. (a) $892.50 (b) $0.036 (rounded)

CHAPTER 6

Section 6.1 (page 377)

1. 0.12   **3.** 0.70 or 0.7   **5.** 0.25   **7.** 1.40 or 1.4   **9.** 0.055
11. 1.00 or 1   **13.** 0.005   **15.** 0.0035   **17.** 60%   **19.** 58%
21. 1%   **23.** 12.5%   **25.** 37.5%   **27.** 200%   **29.** 370%
31. 3.12%   **33.** 416.2%   **35.** 0.28%   **37.** Answers will vary.
Some possibilities are: No common denominators are needed with
percents. The denominator is always 100 with percent, which makes
comparisons easier to understand.   **39.** 0.45   **41.** 0.18
43. 3.5%   **45.** 200%   **47.** 0.5%   **49.** 1.536   **51.** 20 children
53. 420 employees   **55.** 270 chairs   **57.** $377.50
59. 820 commuters   **61.** 26 plants   **63.** (a) Since 100% means
100 parts out of 100 parts, 100% is all of the number. (b) Answers
will vary. For example, 100% of $72 is $72.   **65.** (a) Since 200%
is two times a number, find 200% of the number by multiplying the
number by 2 (double it). (b) Answers will vary. For example,
200% of $20 is 2 • $20 = $40.   **67.** (a) Since 10% means 10

parts out of 100 parts or $\frac{1}{10}$, the shortcut for finding 10% of a

number is to move the decimal point in the number one place to the
left. (b) Answers will vary. For example, 10% of $90 is $9.
69. 63%; 0.63   **71.** (a) landscaping and gardening
(b) 54%; 0.54   **73.** 26%; 0.26   **75.** (a) Marine Corps (b) 5%;
0.05   **77.** 16.8%; 0.168   (a) Germany; Bulgaria (b) 16.5%;
0.165   **81.** 95% shaded; 5% unshaded   **83.** 30% shaded;
70% unshaded   **85.** 55% shaded; 45% unshaded

Section 6.2 (page 389)

1. $\frac{1}{4}$ **3.** $\frac{3}{4}$ **5.** $\frac{17}{20}$ **7.** $\frac{5}{8}$ **9.** $\frac{1}{16}$ **11.** $\frac{1}{6}$ **13.** $\frac{1}{15}$ **15.** $\frac{1}{200}$

17. $1\frac{4}{5}$ **19.** $3\frac{3}{4}$ **21.** 50%   **23.** 80%   **25.** 70%   **27.** 37%

29. 62.5%   **31.** 87.5%   **33.** 48%   **35.** 46%   **37.** 35%
39. 83.3% (rounded)   **41.** 55.6% (rounded)

43. 14.3% (rounded)   **45.** $\frac{1}{2}$; 50%   **47.** $\frac{7}{8}$; 0.875   **49.** $\frac{4}{5}$; 80%

51. 0.167 (rounded); 16.7% (rounded)   **53.** $\frac{7}{10}$; 70%

55. $\frac{1}{8}$; 0.125   **57.** 0.667 (rounded); 66.7% (rounded)

59. 0.06; 6%   **61.** 0.08; 8%   **63.** 0.005; 0.5%   **65.** $2\frac{1}{2}$; 250%

67. 3.25; 325%   **69.** There are many possible answers. Examples
2 and 3 show the steps that students should include in their answers.

71. $\frac{9}{50}$; 0.18; 18%   **73.** $\frac{13}{100}$; 0.13; 13%   **75.** $\frac{3}{5}$; 0.6; 60%

77. $\frac{1}{5}$; 0.20; 20%   **79.** $\frac{1}{10}$; 0.1; 10%   **81.** $\frac{1}{4}$; 0.25; 25%

83. $\frac{1}{20}$; 0.05; 5%   **85.** 100; 100

86. (a) 765 workers (b) 96 letters (c) 21 videos   **87.** 50; 100;

half or $\frac{1}{2}$ **88.** 10; 100; 1; left   **89.** 1; 100; 2; left

90. (a) 525 homes (b) 37 printers (c) $0.08

91. Find 10% of 160, then add $\frac{1}{2}$ of 10%.

$$10\% \;+\; 5\% \;=\; 15\%$$
$$\downarrow \qquad \downarrow \qquad \downarrow$$
$$16 \;+\; 8 \;=\; 24$$

92. Find 100% of 160, then add 50% of 160.
$$100\% \;+\; 50\% \;=\; 150\%$$
$$\downarrow \qquad \downarrow \qquad \downarrow$$
$$160 \;+\; 80 \;=\; 240$$

93. From 100% of $450, subtract 10% of $450.
$$100\% \;-\; 10\% \;=\; 90\%$$
$$\downarrow \qquad \downarrow \qquad \downarrow$$
$$\$450 \;-\; \$45 \;=\; \$405$$

94. To 100% of $800, add 100% of $800, and then add 10% of $800.
$$100\% \;+\; 100\% \;+\; 10\% \;=\; 210\%$$
$$\downarrow \qquad \downarrow \qquad \downarrow \qquad \downarrow$$
$$\$800 \;+\; \$800 \;+\; \$80 \;=\; \$1680$$

Section 6.3 (page 399)

1. whole = 50   **3.** whole = 150   **5.** whole = 70   **7.** percent =
25%   **9.** percent = 150%   **11.** percent = 33.3% (rounded)
13. part = 26   **15.** part = 21.6   **17.** part = 26.5% (rounded)
19. 115   **21.** 0.4%   **23.** 2.5%   **25.** 0.3%

27. Part $\rightarrow$ $\dfrac{60}{\text{unknown}} = \dfrac{10}{100}$ $\leftarrow$ Percent
Whole $\rightarrow$ $\phantom{\dfrac{60}{\text{unknown}}}$ $\leftarrow$ Always 100

29. $\dfrac{600}{800} = \dfrac{75}{100}$ **31.** $\dfrac{\text{unknown}}{970} = \dfrac{25}{100}$ **33.** $\dfrac{12}{\text{unknown}} = \dfrac{20}{100}$

35. $\dfrac{34}{68} = \dfrac{50}{100}$ **37.** $\dfrac{177}{296} = \dfrac{\text{unknown}}{100}$ **39.** $\dfrac{54.34}{\text{unknown}} = \dfrac{3.25}{100}$

41. $\dfrac{\text{unknown}}{487} = \dfrac{0.68}{100}$

43. Percent—the ratio of the part to the whole. It appears with
the word *percent* or "%" after it. whole—the entire quantity.
Often appears after the word *of.* Part—the part being compared
with the whole.   **45.** $\dfrac{640}{810} = \dfrac{\text{unknown}}{100}$ **47.** $\dfrac{86}{142} = \dfrac{\text{unknown}}{100}$

49. $\dfrac{\text{unknown}}{610} = \dfrac{23}{100}$ **51.** $\dfrac{680}{2000} = \dfrac{\text{unknown}}{100}$

53. $\dfrac{\text{unknown}}{480} = \dfrac{55}{100}$ **55.** $\dfrac{\text{unknown}}{822} = \dfrac{49.5}{100}$

57. $\dfrac{504}{\text{unknown}} = \dfrac{16.8}{100}$ **59.** $\dfrac{\text{unknown}}{680} = \dfrac{45}{100}$

Section 6.4 (page 411)

1. 42 test tubes **3.** 1836 military personnel **5.** 4.8 ft
7. 315 files **9.** 819 trucks **11.** $3.28 **13.** 1530 tables
15. 182 homes **17.** $21.60 **19.** 320 e-mails **21.** 160 hay bales
23. 550 students **25.** 330 mountain bikes **27.** 2800 **29.** 50%
31. 52% **33.** 8% **35.** 1.5% **37.** 18.6% (rounded) **39.** 9.2%
41. 150% of $30 cannot be less than $30 because 150% is greater
than 1 (100%). The answer must be greater than $30. 25% of $16
cannot be greater than $16 because 25% is less than 1 (100%). The
answer must be less than $16. **43.** $52.80 **45.** 44 females
47. 220 students **49.** $204.97 **51.** 33% **53.** (a) 2074 children
55. 2% **57.** $3200 monthly earnings; $38,400 yearly earnings
59. March **61.** 52.5 million cans sold in January **63.** Hardee's
65. $18.06 billion **67.** $645 **69.** 2156 products **71.** whole; 100
72. whole **73.** 108 calories **74.** 300 grams **75.** 67 grams
76. Yes, since they would eat 28% × 4 servings = 112% of the
daily value. **77.** 17 packages (rounded). It may be possible but
would result in a diet that is high in total fat, saturated fat, sodium,
and total carbohydrates.

Section 6.5 (page 421)

1. 270 donors **3.** 1350 bath towels **5.** 83.2 quarts
7. 3500 airbags **9.** 1029.2 meters **11.** $4.16 **13.** 160 patients
15. 325 salads **17.** 680 circuits **19.** 1080 people
21. 300 gallons **23.** 50% **25.** 76% **27.** 125% **29.** 1.5%
31. 250% **33.** You must first change the fraction in the percent to
a decimal, then divide the percent by 100 to change it to a decimal.

$$2\frac{1}{2}\% = 2.5\% = 0.025 \leftarrow 2.5\% \text{ as a decimal}$$

Change $2\frac{1}{2}$ to 2.5.

35. 3.78 million or 3,780,000 office workers
37. 82.3 million homes (rounded) **39.** 924 employers
41. (a) 10%; (b) 90% **43.** 36.9% (rounded) **45.** 2106 people
47. 2916 people **49.** $1817.2 million (rounded)
51. 478,175 Mustangs (rounded) **53.** $510,390 **55.** $439.61

Section 6.6 (page 431)

1. $0.24; $6.24 **3.** 3%; $437.75 **5.** 5%; $298.20
7. $672.60 (rounded); $12,901.60 (rounded) **9.** $22.40
11. 20% **13.** $185.51 (rounded) **15.** $6615 **17.** $20 (rounded);
$179.99 (rounded) **19.** 30%; $126 **21.** $4.38 (rounded); $13.12
23. $8.76; $49.64 **25.** On the basis of commission alone you
would choose Company A. Other considerations might be:
reputation of the company; expense allowances; other fringe
benefits; travel; promotion and training, to name a few.
27. $199.89 (rounded) **29.** $87.58 (rounded) **31.** 5%
33. 14.0% (rounded) **35.** 22.5% (rounded) **37.** $134 **39.** $285
41. 4% **43.** $106.20; $483.80 **45.** 5.6% (rounded) **47.** $18.43
(rounded) **49.** $4276.97 (rounded) **51.** $9007.47 (rounded)
53. percent; whole **54.** rate (percent) of tax; cost of item
55. $146.25 (rounded) **56.** Yes, the same; (12.4% • $123) +
(6.5% • $123) = $23.25 (rounded) and (12.4% + 6.5%) • ($123) =
$23.25 **57.** $1358.12 **58.** No. In Exercise 57 the excise tax of
$12.20 cannot be added to the sales tax rate of $7\frac{3}{4}$% because the
excise tax is an amount, not a percent.

Section 6.7 (page 437)

1. $6 **3.** $105 **5.** $28.80 **7.** $488.75 **9.** $2227.50
11. $10 **13.** $49.20 **15.** $42.30 **17.** $16.84 (rounded)
19. $647.06 (rounded) **21.** $210 **23.** $773.30 **25.** $2043
27. $3412.50 **29.** $17,797.81 (rounded) **31.** The answer

should include: amount of principal—This is the amount of
money borrowed or loaned. Interest rate—This is the percent used
to calculate the interest. Time of loan—The length of time that
money is loaned or borrowed is an important factor in determining
interest. **33.** $41.25 **35.** $34,200 **37.** $1240 **39.** $277.50
41. $159.50 **43.** $12,254.69 (rounded) **45.** $2165.63 (rounded)

Summary Exercises on Percent (page 441)

1. 0.0625 **2.** 3.80 or 3.8 **3.** 37.5% **4.** 0.6% **5.** $\frac{7}{8}$ **6.** $1\frac{3}{5}$
7. 62.5% **8.** 0.8% **9.** $46.80 **10.** 500 rolls **11.** 55%
12. 28 screening exams **13.** 363 acres **14.** 2.7% (rounded)
15. $9.68; $185.68 **16.** 7%; $990.82 **17.** $301.20 **18.** 6.5%
or $6\frac{1}{2}$% **19.** $8.99 (rounded); $40.96 (rounded)
20. 35% $444.60 **21.** $30 $1030 **22.** $535.50 $2915.50
23. $44.10 $1514.10 **24.** $664.95 $7484.95 **25.** 139.1%
increase (rounded) **26.** 10.5% decrease (rounded)

Section 6.8 (page 447)

1. $540.80 **3.** $1966.91 (rounded) **5.** $4587.79 (rounded)
7. $1102.50 **9.** $1873.52 (rounded) **11.** $2027.46 (rounded)
13. $17,360.41 (rounded) **15.** $1216.70; $216.70
17. $14,326.40; $6326.40 **19.** $10,976.01 (rounded);
$2547.84 (rounded) **21.** Interest paid on past interest as well as
on the principal. Many people describe compound interest as
"interest on interest." **23.** $13,467.61 **25.** (a) $128,402
(b) $52,402 **27.** (a) $87,786 (b) $17,786 **29.** principal;
rate; time p; r; t **30.** principal; interest **31.** principal; interest
32. principal; interest **33.** (a) $254.48 (b) No; it has more
than doubled (about five times). (c) examples will vary
34. (a) the compound interest account; $1091.20 more
(b) Greater amounts of interest are earned with compound interest
than simple interest because interest is earned on both principal and
past interest.

Chapter 6 Review Exercises (page 457)

1. 0.35 **2.** 1.5 **3.** 0.9944 **4.** 0.00085 **5.** 315% **6.** 2%
7. 87.5% **8.** 0.2% **9.** $\frac{3}{20}$ **10.** $\frac{3}{8}$ **11.** $1\frac{3}{4}$ **12.** $\frac{1}{400}$
13. 75% **14.** 62.5% or $62\frac{1}{2}$% **15.** 325% **16.** 0.5%
17. 0.125 **18.** 12.5% **19.** $\frac{1}{4}$ **20.** 25% **21.** $1\frac{4}{5}$ **22.** 1.8
23. whole = 250 **24.** part = 24
25. Part $\rightarrow$ $\frac{287}{820} = \frac{35}{100}$ $\leftarrow$ Percent, Whole $\rightarrow$ $\leftarrow$ Always 100
26. $\frac{73}{90} = \frac{\text{unknown}}{100}$ **27.** $\frac{\text{unknown}}{160} = \frac{14}{100}$ **28.** $\frac{418}{\text{unknown}} = \frac{16}{100}$
29. $\frac{3}{8} = \frac{\text{unknown}}{100}$ **30.** $\frac{\text{unknown}}{1280} = \frac{88}{100}$ **31.** 171 programs
32. 870 reference books **33.** 31.2 acres **34.** 2.8 kilograms
35. 750 crates **36.** 2320 test tubes **37.** 484 miles
38. 17,000 cases **39.** 55% **40.** 5.2% (rounded)
41. 9.5% (rounded) **42.** 30.8% (rounded) **43.** $2.3 million
(rounded) **44.** 10.4% (rounded) **45.** $145.28 **46.** 186 trucks
47. 0.4% **48.** 175% **49.** 120 miles **50.** $575 **51.** $31.50;
$661.50 **52.** $7\frac{1}{2}$%; $838.50 **53.** $276 **54.** 5% **55.** $33.75;
$78.75 **56.** 25%; $189 **57.** $8 **58.** $67.50 **59.** $7
60. $152.10 **61.** $832.50 **62.** $1598.85 **63.** $5375.60;
$1375.60 **64.** $2187.71 (rounded); $317.71 (rounded)
65. $4534.92; $934.92 **66.** $16,337.50; $3837.50 **67.** part = 12

68. whole = 1640 **69.** 23.28 meters **70.** 150% **71.** $0.51
72. 1980 employees **73.** 40% **74.** 498.8 liters **75.** 0.55
76. 3 **77.** 500% **78.** 471% **79.** 0.086 **80.** 62.1%
81. 0.00375 **82.** 0.06% **83.** 75% **84.** $\frac{21}{50}$ **85.** $\frac{7}{8}$
86. 37.5% or $37\frac{1}{2}$% **87.** $\frac{13}{40}$ **88.** 60% **89.** $\frac{1}{400}$ **90.** 375%
91. $1326.94 (rounded) **92.** $16,520 **93.** 32.4% (rounded)
94. (a) $15,780.96 (rounded) **(b)** $3280.96 (rounded)
95. $6225 **96.** 8.1% (rounded) **97.** $848.40 (rounded)
98. 1055 people **99.** $9758 **100.** 13.1% (rounded)

Chapter 6 Test (page 463)

1. 0.65 **2.** 80% **3.** 175% **4.** 87.5% or $87\frac{1}{2}$% **5.** 3.00 or 3
6. 0.0005 **7.** $\frac{1}{8}$ **8.** $\frac{1}{400}$ **9.** 60% **10.** 62.5% or $62\frac{1}{2}$%
11. 250% **12.** 800 sacks **13.** 20% **14.** $18,500
15. $3450.60 **16.** 8% **17.** 25%
18. $\dfrac{\text{amount of increase}}{\text{last year's salary}} = \dfrac{p}{100}$
A possible answer is: Part is the increase in salary.
$\qquad$ this year − last year = increase
Whole is last year's salary. Percent of increase is unknown.
19. $\qquad I = p \cdot r \cdot t$
$$I = (1000)(0.05)\left(\frac{9}{12}\right) = \$37.50$$
$$I = (1000)(0.05)(2.5) = \$125$$
The interest formula is $I = p \cdot r \cdot t$. If time is in months, it is expressed as a fraction with 12 as the denominator. If time is expressed in years, it is placed over 1 or shown as a decimal number.
20. $11.52; $84.48 **21.** $91; $189 **22.** $378 **23.** $240
24. $5657.75 **25. (a)** $10,668 (rounded) **(b)** $1668

Cumulative Review Exercises: Chapters 1–6 (page 465)

1. *Estimate:* 6000 + 90 + 700 = 6790; *Exact:* 6441
2. *Estimate:* 1 + 50 + 8 = 59; *Exact:* 58.574
3. *Estimate:* 80,000 − 50,000 = 30,000; *Exact:* 28,988
4. *Estimate:* 8 − 4 = 4; *Exact:* 4.2971
5. *Estimate:* 7000 × 700 = 4,900,000; *Exact:* 4,628,904
6. *Estimate:* 70 × 9 = 630; *Exact:* 568.11
7. *Estimate:* 40,000 ÷ 40 = 1000; *Exact:* 902
8. *Estimate:* 2000 ÷ 8 = 250; *Exact:* 320
9. *Estimate:* 7 ÷ 1 = 7; *Exact:* 8.45 **10.** 18 **11.** 19 **12.** 39
13. 4680 **14.** 7,600,000 **15.** $513 **16.** $362.74 **17.** $1\frac{3}{8}$
18. $1\frac{1}{12}$ **19.** $13\frac{3}{8}$ **20.** $\frac{1}{8}$ **21.** $3\frac{7}{8}$ **22.** $8\frac{8}{15}$ **23.** $\frac{7}{12}$
24. $26\frac{5}{32}$ **25.** $28\frac{4}{5}$ **26.** $\frac{7}{8}$ **27.** 16 **28.** $\frac{11}{30}$ **29.** <
30. < **31.** > **32.** $\frac{1}{4}$ **33.** 1 **34.** $\frac{1}{4}$ **35.** 0.75 **36.** 0.375
37. 0.583 (rounded) **38.** 0.55 **39.** $\frac{3}{1}$ **40.** $\frac{4}{3}$ **41.** $\frac{1}{8}$
42. 360 = 360; True **43.** 6912 = 6912; True **44.** $x = 6$
45. $x = 4$ **46.** $x = 16$ **47.** $x = 30$ **48.** 0.78 **49.** 0.03
50. 2.00 or 2 **51.** 0.005 **52.** 87% **53.** 380% **54.** 2.3%
55. $\frac{2}{25}$ **56.** $\frac{5}{8}$ **57.** $1\frac{3}{4}$ **58.** 87.5% or $87\frac{1}{2}$% **59.** 30%
60. 420% **61.** 2132 DVDs **62.** $39.60 **63.** 180 tires
64. 1700 miles **65.** 50% **66.** 40% **67.** $0.34; $8.84

68. 6.5% or $6\frac{1}{2}$%; $489.90 **69.** $1003.04 **70.** 2.5% or $2\frac{1}{2}$%
71. $205.20; $250.80 **72.** 15%; $922.25 **73.** $1115.50
74. $23,028.75 **75.** 33 hearing tests **76.** 255 ounces
77. 12.5% **78.** 2% **79.** 5.5% (rounded) **80.** 8.1% (rounded)
81. $107,520 **82. (a)** $19,604.40 (rounded)
(b) $4604.40 (rounded)

CHAPTER 7

Section 7.1 (page 479)

1. 3 **3.** 8 **5.** 5280 **7.** 2000 **9.** 60 **11.** 2 **13.** 2
15. 14,000 to 16,000 lb **17.** 27 **19.** 112 **21.** 10
23. $1\frac{1}{2}$ or 1.5 **25.** $\frac{1}{4}$ or 0.25 **27.** $1\frac{1}{2}$ or 1.5 **29.** $2\frac{2}{2}$ or 2.5
31. snowmobile/ATV; person walking **33.** 5000 **35.** 17
37. 4 to 8 in. **39.** 216 **41.** 28 **43.** 518,400 **45.** 48,000
47. (a) pound/ounces **(b)** quarts/pints or pints/cups
(c) minutes/hours or seconds/minutes **(d)** feet/inches
(e) pounds/tons **(f)** days/weeks **49.** 174,240 **51.** 800
53. 0.75 or $\frac{3}{4}$ **55.** $1.83 (rounded) **57.** $140 **59. (a)** 1056 sec;
(b) 17.6 min **61. (a)** $12\frac{1}{2}$ qt **(b)** 4 jugs, because you can't buy
part of a jug **63. (a)** 1391 times around Earth (rounded)
(b) 12,436 times from Los Angeles to New York (rounded)
64. (a) 35,000,000 ÷ 25,000 becomes 35,000 ÷ 25 = 1400 times
around Earth. **(b)** 12,500 times from Los Angeles to New York
(c) They are very close and just as useful for making comparisons
of such large numbers.

Section 7.2 (page 489)

1. 1000; 1000 **3.** $\frac{1}{1000}$ or 0.001; $\frac{1}{1000}$ or 0.001 **5.** $\frac{1}{100}$ or 0.01;
$\frac{1}{100}$ or 0.01 **7.** Answers will vary; about 8 to 10 cm.
9. Answers will vary; about 20 to 25 mm. **11.** cm **13.** m
15. km **17.** mm **19.** cm **21.** m **23.** Some possible answers
are: 35 mm film for cameras, track and field events, metric auto
parts, and lead refills for mechanical pencils. **25.** 700 cm
27. 0.040 m or 0.04 m **29.** 9400 m **31.** 5.09 m **33.** 40 cm
35. 910 mm **37.** less; 18 cm or 0.18 m
39. $\qquad$ 5 mm = 0.5 cm
$\qquad$ 1 mm = 0.1 cm
41. 0.018 km **43.** 164 cm; 1640 mm **45.** 0.0000056 km

Section 7.3 (page 497)

1. mL **3.** L **5.** kg **7.** g **9.** mL **11.** mg **13.** L
15. kg **17.** unreasonable; too much **19.** unreasonable; too much
21. reasonable **23.** reasonable **25.** Some capacity examples
are 2 L bottles of soda and shampoo bottles marked in mL; weight
examples are grams of fat listed on cereal boxes and vitamin doses
in milligrams. **27.** Unit for your answer (g) is in numerator; unit
being changed (kg) is in denominator so it will divide out. The unit
fraction is $\dfrac{1000 \text{ g}}{1 \text{ kg}}$. **29.** 15,000 mL **31.** 3 L **33.** 0.925 L
35. 0.008 L **37.** 4150 mL **39.** 8 kg **41.** 5200 g **43.** 850 mg
45. 30 g **47.** 0.598 g **49.** 0.06 L **51.** 0.003 kg **53.** 990 mL
55. mm **57.** mL **59.** cm **61.** mg **63.** 0.3 L **65.** 1340 g
67. 0.9 L **69.** 3 kg to 4 kg **71.** greater; 5 mg or 0.005 g
73. 200 nickels **75. (a)** 1,000,000
(b) $\dfrac{3.5 \text{ Mm}}{1} \cdot \dfrac{1,000,000 \text{ m}}{1 \text{ Mm}} = 3,500,000 \text{ m}$

76. (a) 1,000,000,000

(b) $\dfrac{2500 \cancel{m}}{1} \cdot \dfrac{1 \text{ Gm}}{1,000,000,000 \cancel{m}} = 0.0000025 \text{ Gm}$

77. (a) 1,000,000,000,000   **(b)** 1000; 1,000,000
78. 1,000,000; 1,000,000,000; $2^{20} = 1,048,576$; $2^{30} = 1,073,741,824$

Section 7.4 (page 503)

1. $1.33 (rounded)   **3.** 89.5 kg   **5.** 71 beats (rounded)
7. 180 cm; $0.02/cm (rounded)   **9.** 5.03 m   **11.** 4 bottles
13. 1.89 g   **15. (a)** 10 km   **(b)** 600 km   **(c)** 36,000 km
17. 215 g; 4.3 g; 4300 mg   **18.** 330 g; 0.33 g; 330 mg
19. 1550 g   or   1.55 kg; 1500 g; 3 g   **20.** 55 g; 50 g; 0.5 g

Section 7.5 (page 509)

1. 21.8 yd   **3.** 262.4 ft   **5.** 4.8 m   **7.** 5.3 oz   **9.** 111.6 kg
11. 30.3 qt   **13. (a)** about 0.2 oz   **(b)** probably not
15. about 31.8 L   **17.** about 1.3 cm   **19.** 28 °F   **21.** 40 °C
23. 150 °C   **25.** More. There are 180 degrees between freezing
and boiling on the Fahrenheit scale, but only 100 degrees on the
Celsius scale, so each Celsius degree is a greater change in temperature.   **27.** 16 °C (rounded)   **29.** 40 °C   **31.** 46 °F (rounded)
33. 95 °F   **35.** 58 °C (rounded)   **37. (a)** pleasant weather, above
freezing but not hot   **(b)** 75 °F to 39 °F (rounded)
(c) Answers will vary. In Minnesota, it's 0 °C to −40 °C; in
California, 24 °C to 0 °C.   **39.** about 1.8 m   **40.** about 5.0 kg
41. about 3040.2 km   **42.** 38 hr 23 min   **43.** about 300 m
44. less than 3780 mL   **45.** about 59.4 mL   **46.** about 1.6%

Chapter 7 Review Exercises (page 519)

1. 16   **2.** 3   **3.** 2000   **4.** 4   **5.** 60   **6.** 8   **7.** 60   **8.** 5280
9. 12   **10.** 48   **11.** 3   **12.** 4   **13.** $\dfrac{3}{4}$ or 0.75   **14.** $2\dfrac{1}{2}$ or 2.5
15. 28   **16.** 78   **17.** 112   **18.** 345,600   **19. (a)** $4153\dfrac{1}{3}$ yd
(exact) or $4153.\overline{3}$ yd (rounded)   **(b)** 2.4 mi (rounded)
20. $2465.20   **21.** mm   **22.** cm   **23.** km   **24.** m   **25.** cm
26. mm   **27.** 500 cm   **28.** 8500 m   **29.** 8.5 cm   **30.** 3.7 m
31. 0.07 km   **32.** 930 mm   **33.** mL   **34.** L   **35.** g   **36.** kg
37. L   **38.** mL   **39.** mg   **40.** g   **41.** 5 L   **42.** 8000 mL
43. 4580 mg   **44.** 700 g   **45.** 0.006 g   **46.** 0.035 L   **47.** 31.5 L
48. 357 g (rounded)   **49.** 87.25 kg   **50.** $3.01 (rounded)
51. 6.5 yd (rounded)   **52.** 11.7 in. (rounded)
53. 67.0 mi (rounded)   **54.** 1288 km   **55.** 21.9 L (rounded)
56. 44.0 qt (rounded)   **57.** 0 °C   **58.** 100 °C   **59.** 37 °C
60. 20 °C   **61.** 25 °C   **62.** 33 °C (rounded)   **63.** 43 °F (rounded)
64. 120 °F (rounded)   **65.** L   **66.** kg   **67.** cm   **68.** mL
69. mm   **70.** km   **71.** g   **72.** m   **73.** mL   **74.** g   **75.** mg
76. L   **77.** 105 mm   **78.** $\dfrac{3}{4}$ hr or 0.75 hr   **79.** $7\dfrac{1}{2}$ ft or 7.5 ft
80. 130 cm   **81.** 77 °F   **82.** 14 qt   **83.** 0.7 g   **84.** 810 mL
85. 80 oz   **86.** 60,000 g   **87.** 1800 mL   **88.** 30 °C   **89.** 36 cm
90. 0.055 L   **91.** 1.26 m   **92.** 18 tons   **93.** 113 g; 177 °C
(both rounded)   **94.** 178.0 lb; 6.0 ft (both rounded)   **95.** 13.2 m
96. 22.0 ft   **97.** 3.6 m   **98.** 39.8 in.   **99.** 484,000 lb
100. 18.9 to 22.7 L   **101. (a)** 242 tons   **(b)** 144 in.
102. (a) 1.02 m   **(b)** 670 cm

Chapter 7 Test (page 523)

1. 36 qt   **2.** 15 yd   **3.** 2.25 hr   or $2\dfrac{1}{4}$ hr   **4.** 0.75 ft or $\dfrac{3}{4}$ ft
5. 56 oz   **6.** 7200 min   **7.** kg   **8.** km   **9.** mL   **10.** g   **11.** cm
12. mm   **13.** L   **14.** cm   **15.** 2.5 m   **16.** 4600 m   **17.** 0.5 cm
18. 0.325 g   **19.** 16,000 mL   **20.** 400 g   **21.** 1055 cm
22. 0.095 L   **23.** 3.2 ft (rounded)   **24. (a)** 2310 mg
(b) 500 mg or 0.5 g more   **25.** 95 °C   **26.** 0 °C
27. 1.8 m (rounded)   **28.** 56.3 kg (rounded)   **29.** 13 gal

30. 5.0 mi (rounded)   **31.** 23 °C (rounded)   **32.** 36 °F (rounded)
33. 6 m ≈ 6.54 yd; $26.03 (rounded)   **34.** Possible answers: Use
same system as rest of the world; easier system for children to
learn; less use of fractional numbers; compete internationally.

Cumulative Review Exercises: Chapters 1–7 (page 525)

1. *Estimate:* $100 + 3 + 70 = 173$; *Exact:* 178.838
2. *Estimate:* $30,000 − 800 = 29,200$; *Exact:* 30,178
3. *Estimate:* $90,000 \times 200 = 18,000,000$; *Exact:* 16,849,725
4. *Estimate:* $60 \times 5 = 300$; *Exact:* 265.644
5. *Estimate:* $20,000 \div 40 = 500$; *Exact:* 530
6. *Estimate:* $40 \div 8 = 5$; *Exact:* 4.6
7. *Estimate:* $2 + 4 = 6$; *Exact:* $5\dfrac{1}{2}$
8. *Estimate:* $6 − 1 = 5$; *Exact:* $4\dfrac{3}{14}$
9. *Exact:* $14\dfrac{7}{9}$; *Estimate:* $3 \cdot 5 = 15$
10. *Exact:* $2\dfrac{1}{2}$; *Estimate:* $2 \div 1 = 2$
11. $\dfrac{11}{16}$ **12.** 488,045   **13.** $26\dfrac{2}{3}$ **14.** 13.0 (rounded)   **15.** $1\dfrac{5}{24}$
16. 7.0793   **17.** $\dfrac{5}{8}$ **18.** 307 R38   **19.** 0.00762   **20.** 8   **21.** 265
22. three hundred seven and nineteen hundredths   **23.** 0.0082
24. 0.067, 0.6, 0.6007, 0.67   **25.** 16 diapers for $3.87, about $0.242
per diaper   **26.** 0.875   **27.** 87.5% or $87\dfrac{1}{2}$%   **28.** $\dfrac{1}{20}$ **29.** 5%
30. $3\dfrac{1}{2}$ **31.** 3.5   **32.** $\dfrac{13}{18}$ **33.** $\dfrac{6}{1}$ **34.** salmon to light chicken;
light chicken to cod   **35.** $x = 12$ **36.** $x ≈ 1.67$ (rounded)
37. 5%   **38.** 30 hours   **39.** 30 in.   **40.** $1\dfrac{3}{4}$ or 1.75 min
41. 280 cm   **42.** 0.065 g   **43.** 122.76 mi   **44.** 10 °C   **45.** g
46. cm   **47.** m   **48.** kg   **49.** mg   **50.** mL   **51.** 38 °C
52. 12 °C   **53.** $18.99 (rounded); $170.95   **54.** $0.46 (rounded)
55. $12\dfrac{3}{4}$ or 12.75 lb   **56. (a)** 93.6 km   **(b)** 58.0 mi (rounded)
57. $2756.25; $11,506.25   **58.** 88.6% (rounded)
59. $9.74 (rounded)   **60.** $\dfrac{1}{6}$ cup more than the amount needed
61. 120 rows   **62.** 2100 students   **63.** $0.15   **64.** 37.5% decrease
65. (a) 76.5% (rounded)   **(b)** 5.9% (rounded)   **66. (a)** $\dfrac{1}{2}$
(b) 0.5   **(c)** 50%

CHAPTER 8

Section 8.1 (page 535)

1. line named $\overleftrightarrow{CD}$ or $\overleftrightarrow{DC}$ **3.** line segment named $\overline{GF}$ or $\overline{FG}$
5. ray named $\overrightarrow{PQ}$ **7.** perpendicular   **9.** parallel
11. intersecting   **13.** ∠AOS or ∠SOA   **15.** ∠CRT or ∠TRC
17. ∠AQC or ∠CQA   **19.** right (90°)   **21.** acute   **23.** straight
(180°)   **25. (a)** The car turned around in a complete circle.
(b) The governor took the opposite view, for example, having once
opposed taxes but now supporting them.   **27.** True, because $\overrightarrow{UQ}$ is
perpendicular to $\overleftrightarrow{ST}$.   **28.** True, because they form a 90° angle, as
indicated by the small red square.   **29.** False; the angles have the
same measure (both are 180°).   **30.** False; $\overleftrightarrow{ST}$ and $\overleftrightarrow{PR}$ are parallel.
31. False; $\overleftrightarrow{QU}$ and $\overleftrightarrow{TS}$ are perpendicular.   **32.** True; both angles
are formed by perpendicular lines, so they both measure 90°.

Section 8.2 (page 541)

1. $\angle EOD$ and $\angle COD$; $\angle AOB$ and $\angle BOC$ **3.** $\angle HNE$ and $\angle ENF$; $\angle HNG$ and $\angle GNF$; $\angle HNE$ and $\angle HNG$; $\angle ENF$ and $\angle GNF$ **5.** 50°
7. 4° **9.** 50° **11.** 90° **13.** $\angle SON \cong \angle TOM$; $\angle TOS \cong \angle MON$
15. $\angle GOH$ measures 63°; $\angle EOF$ measures 37°; $\angle AOC$ and $\angle GOF$ both measure 80°. **17.** Two angles are complementary if their sum is 90°. Two angles are supplementary if their sum is 180°. Drawings will vary; examples are:

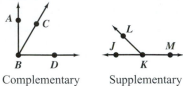

Complementary   Supplementary

19. $\angle ABF \cong \angle ECD$; Both are 138°. $\angle ABC \cong \angle BCD$; Both are 42°.
21. No; obtuse angles are >90°, so their sum would be >180°.

Section 8.3 (page 549)

1. $P = 28$ yd; $A = 48$ yd² **3.** $P = 3.6$ km; $A = 0.81$ km²
5. $P = 40$ ft; $A = 100$ ft² **7.** $P = 29$ m; $A = 7$ m²
9. $P = 196.2$ ft; $A = 1674.2$ ft² **11.** $P = 12$ mi; $A = 9$ mi²
13. $P = 38$ m; $A = 39$ m² **15.** $P = 98$ m; $A = 492$ m²
17. unlabeled side = 12 in.; $P = 78$ in.; $A = 234$ in.²
19. $P = 48$ m; $A = 144$ m² **21.** $460 **23.** $94.81
25. Panoramic: $P = 36$ in., $A = 56$ in.²; 4 in. × 6 in.: $P = 20$ in., $A = 24$ in.²; 4 in. × 7 in.: $P = 22$ in., $A = 28$ in.² **27.** 53 yd
29. $A = 6528$ ft²

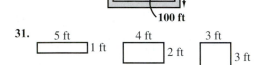

31.

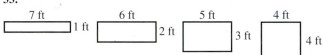

32. (a) 5 ft by 1 ft has area of 5 ft²; 4 ft by 2 ft has area of 8 ft²; 3 ft by 3 ft has area of 9 ft² **(b)** The square plot 3 ft by 3 ft has greatest area.
33.

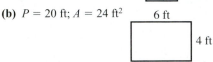

34. (a) 7 ft²; 12 ft²; 15 ft²; 16 ft² **(b)** Square plots have the greatest area.
35. (a) $P = 10$ ft; $A = 6$ ft²

(b) $P = 20$ ft; $A = 24$ ft²

(c) Perimeter is twice the original; area is four times the original.
36. (a) $P = 30$ ft; $A = 54$ ft²

(b) Perimeter is three times the original; area is nine times the original. **(c)** Perimeter will be four times the original; area will be 16 times the original.

Section 8.4 (page 557)

1. $P = 208$ m **3.** $P = 207.2$ m **5.** $P = 5.78$ km
7. $A = 775$ mm² **9.** $A = 19.25$ ft² **11.** $A = 3099.6$ cm²
13. $1410.75

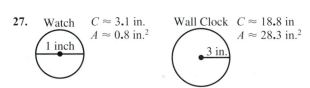

15. $A = 437.5$ in.²

17. Height is not part of perimeter; square units are used for area, not perimeter. $P = 2.5$ cm + 2.5 cm + 2.5 cm + 2.5 cm = 10 cm
19. $A = 3.02$ m² **21.** $A = 25,344$ ft²

Section 8.5 (page 563)

1. $P = 202$ m; $A = 1914$ m² **3.** $P = 58.9$ cm; $A = 139.15$ cm²
5. $P = 26\frac{1}{4}$ yd or 26.25 yd; $A = 30\frac{3}{4}$ yd² or 30.75 yd²
7. $P = 85.2$ cm; $A = 302.46$ cm² **9.** $A = 198$ m²
11. $A = 1664$ m² **13.** 32° **15.** 48° **17.** No. Right angles are 90°, so two right angles are 180°, and the sum of all *three* angles in a triangle equals 180°. **19.** $7\frac{7}{8}$ ft² or 7.875 ft² **21. (a)** 126.8 m of curb **(b)** 672 m² of sod **23. (a)** $A = 32$ m² **(b)** $A = 13.41$ m²
25. (a) 175 yd of frontage **(b)** $A = 8750$ yd²

Section 8.6 (page 573)

1. $d = 18$ mm **3.** $r = 0.35$ km **5.** $C \approx 69.1$ ft; $A \approx 379.9$ ft²
7. $C \approx 8.2$ m; $A \approx 5.3$ m² **9.** $C \approx 47.1$ cm; $A \approx 176.6$ cm²
11. $C \approx 23.6$ ft; $A \approx 44.2$ ft² **13.** $C \approx 27.2$ km; $A \approx 58.7$ km²
15. $A \approx 57$ cm² **17.** $A \approx 197.8$ cm² **19.** π is the ratio of circumference of a circle to its diameter. If you divide the circumference of any circle by its diameter, the answer is always a little more than 3. The approximate value is 3.14, which we call π (pi). Your test question could involve finding the circumference or the area of a circle. **21.** $A \approx 7850$ yd² **23.** $C \approx 91.4$ in.; Bonus: 693 revolutions/mile (rounded) **25.**

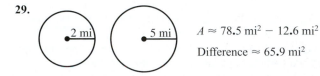

$A \approx 70,650$ mi²

27.

Watch $C \approx 3.1$ in.
$A \approx 0.8$ in.²

Wall Clock $C \approx 18.8$ in
$A \approx 28.3$ in.²

29.

$A \approx 78.5$ mi² − 12.6 mi²
Difference ≈ 65.9 mi²

31. (a) $d \approx 45.9$ cm **(b)** Divide the circumference by π (3.14).
33. $1170.33 (rounded). **35.** The prefix *rad* tells you that radius is a ray from the center of the circle. The prefix *dia* means the diameter goes through the circle, and the prefix *circum* means the circumference is the distance around. **37.** $A \approx 44.2$ in.²
38. $A \approx 132.7$ in.² **39.** $A \approx 201.0$ in.² **40.** small: $0.063 (rounded); medium: $0.049 (rounded); large: $0.046 (rounded) Best Buy **41.** small: $0.084 (rounded); medium: $0.067 (rounded) Best Buy; large: $0.071 (rounded) **42.** small: $0.077 (rounded) Best Buy; medium: $0.083 (rounded); large: $0.078 (rounded)

Summary Exercises on Perimeter, Circumference, and Area (page 577)

1. (a) **(b)**

All sides have the same length.

Opposite sides have the same length.

(c) 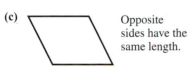 Opposite sides have the same length.

2. (a) **(b)**

(c) The diameter is twice the radius, or the radius is half the diameter. **3.** Add up the lengths of all the sides to find the perimeter. **4.** Perimeter is the total distance around the outside edges of a shape. Area is the number of square units needed to cover the space inside the shape. **5.** f, e, d, b, c, a **6. (a)** $C = 2 \cdot \pi \cdot r$ **(b)** $C = \pi \cdot d$ **(c)** Divide the diameter by 2 to find the radius. **7.** $P = 63$ ft; $A = 180$ ft^2 $P = 32.4$ cm; $A \approx 45.1$ cm^2 **8. (a)** $A = 12$ cm^2 **(b)** $P = 6\frac{1}{2}$ ft **(c)** $C \approx 28.5$ m **(d)** $A = 307$ in.^{2} **9.** Rectangle; $P = 27$ in.; $A = 31\frac{1}{2}$ in.^{2} or 31.5 in.^{2} **10.** Square; $P = 48$ yd; $A = 144$ yd^2 **11.** Parallelogram; $P = 6.4$ m; $A \approx 2.2$ m^2 **12.** Trapezoid; $P = 33.5$ mm; $A = 64$ mm^2 **13.** $d = 12$ cm; $C \approx 37.7$ cm; $A \approx 113.0$ cm^2 **14.** $r = 15$ mi; $C \approx 94.2$ mi; $A \approx 706.5$ mi^2 **15.** $r = 4.5$ ft; $C \approx 28.3$ ft; $A \approx 63.6$ ft^2 **16.** $A \approx 4.5$ m^2 **17.** $A = 217.5$ yd^2 **18.** $A \approx 86$ in.^{2} **19.** $C \approx 14.6$ ft;

Bonus: About 361.6 revolutions; the Mormons counted 360 revolutions, which is the answer you get using $2\frac{1}{3}$ ft instead of 2.33 ft as the radius. **20.** \$4.47 (rounded)

Section 8.7 (page 585)

1. Rectangular solid; $V = 550$ cm^3 **3.** Sphere; $V \approx 44{,}579.6$ m^3 **5.** Hemisphere; $V \approx 3617.3$ in.^{3} **7.** Cylinder; $V \approx 471$ ft^3 **9.** Cone; $V \approx 418.7$ m^3 **11.** Pyramid; $V = 800$ cm^3 **13.** $V = 18$ in.^{3} **15.** $V \approx 2481.5$ cm^3 **17.** $V = 651{,}775$ m^3 **19.** $V \approx 3925$ ft^3 **21.** Student used diameter of 7 cm; should use radius of 3.5 cm in formula. Units for volume are cm^3, not cm^2. Correct answer is $V \approx 192.3$ cm^3. **23.** $V = 513$ cm^3

Section 8.8 (page 591)

1. 4 **3.** 8 **5.** 3.317 **7.** 2.236 **9.** 8.544 **11.** 10.050 **13.** 13.784 **15.** 31.623 **17.** 30 is about halfway between 25 and 36, so $\sqrt{30}$ should be about halfway between 5 and 6, or about 5.5. Using a calculator, $\sqrt{30} \approx 5.477$. Similarly, $\sqrt{26}$ should be a little more than $\sqrt{25}$; by calculator $\sqrt{26} \approx 5.099$. And $\sqrt{35}$ should be a little less than $\sqrt{36}$; by calculator $\sqrt{35} \approx 5.916$. **19.** $\sqrt{1521} = 39$ ft **21.** $\sqrt{289} = 17$ in. **23.** $\sqrt{144} = 12$ mm **25.** $\sqrt{73} \approx 8.5$ in. **27.** $\sqrt{65} \approx 8.1$ yd **29.** $\sqrt{195} \approx 14.0$ cm **31.** $\sqrt{7.94} \approx 2.8$ m **33.** $\sqrt{65.01} \approx 8.1$ cm **35.** $\sqrt{292.32} \approx 17.1$ km **37.** $\sqrt{65} \approx 8.1$ ft **39.** $\sqrt{360{,}000} = 600$ m **41.** $\sqrt{32.5} \approx 5.7$ ft

43. $\sqrt{135} \approx 11.6$ ft

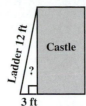

Ladder 12 ft · Castle · ? · 3 ft

45. The student did not square the numbers correctly: 9^2 is 81 and 7^2 is 49. Also, the final answer is rounded to thousandths instead of tenths. Correct answer is $\sqrt{130} \approx 11.4$ in. **47.** $\sqrt{16200} \approx 127.3$ ft **48. (a)**

Second · 60 ft · Third · First · 60 ft · Home

(b) $\sqrt{7200} \approx 84.9$ ft

49. The distance from third to first is the same as the distance from home to second because the baseball diamond is a square. **50.** One possibility is:

major league $\left\{ \dfrac{90 \text{ ft}}{127.3 \text{ ft}} = \dfrac{60 \text{ ft}}{x} \right\}$ softball

$x \approx 84.9$ ft

Section 8.9 (page 599)

1. similar **3.** not similar **5.** similar **7.** $\angle 1$ and $\angle 4$; $\angle 2$ and $\angle 5$; $\angle 3$ and $\angle 6$; $\overline{AB}$ and $\overline{PQ}$; $\overline{BC}$ and $\overline{QR}$; $\overline{AC}$ and $\overline{PR}$ **9.** $\angle 1$ and $\angle 6$; $\angle 2$ and $\angle 5$; $\angle 3$ and $\angle 4$; $\overline{MP}$ and $\overline{QS}$; $\overline{MN}$ and $\overline{QR}$; $\overline{NP}$ and $\overline{RS}$ **11.** $\dfrac{3}{2}; \dfrac{3}{2}; \dfrac{3}{2}$ **13.** $a = 5$ mm; $b = 3$ mm **15.** $a = 6$ cm; $b = 15$ cm **17.** $x = 24.8$ m; $P = 72.8$ m; $y = 15$ m; $P = 54.6$ m **19.** $P = 8$ cm + 8 cm + 8 cm = 24 cm; $A = 0.5 \cdot 8$ cm $\cdot$ 6.9 cm = 27.6 cm^2 **21.** $h = 24$ ft **23.** $\dfrac{3}{4}$ ft or 0.75 ft **25.** One dictionary definition is "resembling, but not identical." Examples of similar objects are sets of different-size pots or measuring cups; small- and- large size cans of beans; child's tennis shoe and adult's tennis shoe. **27.** $x = 50$ m **29.** $n = 110$ m

Chapter 8 Review Exercises (page 611)

1. line segment named $\overline{AB}$ or $\overline{BA}$ **2.** line named $\overleftrightarrow{CD}$ or $\overleftrightarrow{DC}$ **3.** ray named $\overrightarrow{OP}$ **4.** parallel **5.** perpendicular **6.** intersecting **7.** acute **8.** obtuse **9.** straight; 180° **10.** right; 90° **11.** $\angle 1$ and $\angle 3$ measure 30°; $\angle 2$ measures 90°; $\angle 4$ measures 60° **12.** $\angle 1$ measures 100°; $\angle 2$ and $\angle 4$ measure 45°; $\angle 3$ measures 35° **13.** $\angle AOB$ and $\angle BOC$; $\angle BOC$ and $\angle COD$; $\angle COD$ and $\angle DOA$; $\angle DOA$ and $\angle AOB$ **14.** $\angle ERH$ and $\angle HRG$; $\angle HRG$ and $\angle GRF$; $\angle FRG$ and $\angle FRE$; $\angle FRE$ and $\angle ERH$ **15. (a)** 10° **(b)** 45° **(c)** 83° **16. (a)** 25° **(b)** 90° **(c)** 147° **17.** $P = 4.84$ m **18.** $P = 128$ in. **19.** 152 cm **20.** 41 ft **21.** $A = 486$ mm^2 **22.** $A = 16.5$ ft^2 or $16\frac{1}{2}$ ft^2 **23.** $A \approx 39.7$ m^2 **24.** $P = 50$ cm; $A = 140$ cm^2 **25.** $P = 102.1$ ft; $A = 567$ ft^2 **26.** $P = 200.2$ m; $A \approx 2074.0$ m^2 **27.** $P = 518$ cm; $A = 11{,}660$ cm^2 **28.** $P = 27.1$ m; $A = 20.58$ m^2 **29.** $P = 20\frac{1}{4}$ ft or 20.25 ft; $A = 14$ ft^2 **30.** 70° **31.** 24° **32.** $d = 137.8$ m **33.** $r = 1\frac{1}{2}$ in. or 1.5 in. **34.** $C \approx 6.3$ cm; $A \approx 3.1$ cm^2 **35.** $C \approx 109.3$ m; $A \approx 950.7$ m^2 **36.** $C \approx 37.7$ in.; $A \approx 113.0$ in.^{2} **37.** $A \approx 20.3$ m^2 **38.** $A = 64$ in.^{2} **39.** $A = 673$ km^2 **40.** $A = 1020$ m^2 **41.** $A = 229$ ft^2 **42.** $A = 132$ ft^2 **43.** $A = 5376$ cm^2 **44.** $A \approx 498.9$ ft^2 **45.** $A \approx 447.9$ yd^2

46. Rectangular solid; $V = 30$ in.^{3} **47.** Rectangular solid; $V = 96$ cm^3 **48.** Rectangular solid; $V = 45,000$ mm^3
49. Sphere; $V \approx 267.9$ m^3 **50.** Hemisphere; $V \approx 452.2$ ft^3
51. Cylinder; $V \approx 549.5$ cm^3 **52.** Cylinder; $V \approx 1808.6$ m^3
53. Cone; $V \approx 512.9$ m^3 **54.** Pyramid; $V = 16$ yd^3 **55.** 7
56. 2.828 (rounded)   **57.** 54.772 (rounded)   **58.** 12
59. 7.616 (rounded)   **60.** 25   **61.** 10.247 (rounded)
62. 8.944 (rounded)   **63.** $\sqrt{289} = 17$ in.   **64.** $\sqrt{49} = 7$ cm
65. $\sqrt{104} \approx 10.2$ cm   **66.** $\sqrt{52} \approx 7.2$ in.   **67.** $\sqrt{6.53} \approx 2.6$ m
68. $\sqrt{71.75} \approx 8.5$ km   **69.** $y = 30$ ft; $x = 34$ ft; $P = 104$ ft
70. $y = 7.5$ m; $x = 9$ m; $P = 22.5$ m   **71.** $x = 12$ mm;
$y = 7.5$ mm; $P = 38$ mm   **72.** Square; $P = 18$ in.; $A \approx 20.3$ in.^{2}
or $A = 20\frac{1}{4}$ in.^{2} **73.** Trapezoid; $P = 10.3$ cm; $A \approx 6.2$ cm^2
74. Circle; $C \approx 40.8$ m; $A \approx 132.7$ m^2 **75.** Parallelogram;
$P = 54$ ft; $A = 140$ ft^2 **76.** Triangle; $P = 20$ yd; $A = 18\frac{3}{4}$ yd^2 or
18.8 yd^2 (rounded)   **77.** Rectangle; $P = 7$ km; $A \approx 2.0$ km^2
78. Circle; $C \approx 53.4$ m; $A \approx 226.9$ m^2 **79.** Parallelogram;
$P = 78$ mm; $A = 288$ mm^2 **80.** Triangle; $P = 37.8$ mi;
$A = 58.5$ mi^2 **81.** parallel lines   **82.** line segment
83. acute angle   **84.** intersecting lines   **85.** right angle; 90°
86. ray   **87.** straight angle; 180°   **88.** obtuse angle
89. perpendicular lines   **90.** 81°   **91.** 138°   **92.** $P = 90$ m;
$A = 92$ m^2 **93.** $P = 282$ cm; $A = 4190$ cm^2 **94.** $V \approx 100.5$ ft^3
95. $V \approx 3.4$ in.^{3} or $V = 3\frac{3}{8}$ in.^{3} **96.** $V \approx 7.4$ m^3
97. $V = 561$ cm^3 **98.** $V \approx 1271.7$ cm^3 **99.** $V \approx 1436.0$ m^3
100. $x \approx 12.6$ km   **101.** $\angle D = 72°$ **102.** $x = 12$ mm;
$y = 14$ mm   **103.** The prefix *dec* in *dec*ade means 10 and the
prefix *cent* in *cent*ury means 100, so divide 200 (two centuries)
by 10. The answers is 20 decades.

Chapter 8 Test (page 619)

1. (e)   **2.** (a); 90°   **3.** (d)   **4.** (g); 180°   **5.** Parallel lines are
lines in the same plane that never intersect. Perpendicular lines
intersect to form a right angle.

Parallel   Perpendicular
lines   lines

6. 9°   **7.** 160°   **8.** $\angle 1$ measures 50°; $\angle 3$ measures 90°; $\angle 2$ and
$\angle 4$ measure 40°   **9.** Rectangle; $P = 23$ ft; $A = 30$ ft^2
10. Square; $P = 72$ mm; $A = 324$ mm^2 **11.** Parallelogram;
$P = 26.2$ m; $A = 33.12$ m^2 **12.** Trapezoid; $P = 169.4$ cm;
$A = 1591$ cm^2 **13.** $P = 32.45$ m; $A = 48$ m^2 **14.** $P = 37.8$ yd or
$37\frac{4}{5}$ yd; $A = 58.5$ yd^2 **15.** 55°   **16.** $r = 12.5$ in. or $12\frac{1}{2}$ in.
17. $C \approx 5.7$ km (rounded)   **18.** $A \approx 206.0$ cm^2 (rounded)
19. $A \approx 39.3$ m^2 (rounded)   **20.** Rectangular solid; $V = 6480$ m^3
21. Sphere; $V \approx 33.5$ ft^3 (rounded)   **22.** Cylinder; $V \approx 5086.8$ ft^3
23. $\sqrt{85} \approx 9.2$ cm (rounded)   **24.** $y = 12$ cm; $z = 6$ cm
25. Linear units like cm are used to measure perimeter, radius,
diameter, and circumference. Area is measured in square units like
cm^2 (squares that measure 1 cm on each side). Volume is measured
in cubic units like cm^3.

Cumulative Review Exercises: Chapters 1–8 (page 621)

1. *Estimate:* $300 + 60,000 + 6000 = 66,300$; *Exact:* 64,574
2. *Estimate:* $20 - 10 = 10$; *Exact:* 10.242
3. *Estimate:* $4 \times 7 = 28$; *Exact:* 26.7472

4. *Estimate:* $30,000 \times 70 = 2,100,000$; *Exact:* 2,074,587
5. *Estimate:* $600 \div 3 = 200$; *Exact:* 200.8
6. *Estimate:* $5000 \div 50 = 100$; *Exact:* 94
7. *Estimate:* $5 + 5 = 10$; *Exact:* $9\frac{2}{5}$
8. *Estimate:* $3 - 2 = 1$; *Exact:* $1\frac{7}{24}$
9. *Estimate:* $3 \cdot 2 = 6$; *Exact:* $5\frac{1}{3}$ **10.** $\frac{9}{20}$ **11.** 0.9132
12. 70 R79   **13.** 32   **14.** 0.000078   **15.** 39,105
16. 19.44 (rounded)   **17.** $4\frac{5}{12}$ **18.** 836.085   **19.** 9   **20.** 38
21. two hundred eight ten-thousandths   **22.** 660.05   **23.** 2.055;
2.5005; 2.505; 2.55   **24.** *Per* means divide and *cent* means 100,
so divide by 100 to change a percent to a decimal.   **25.** $\frac{1}{50}$
26. 2%   **27.** 1.75   **28.** 175%   **29.** $\frac{2}{5}$ **30.** 0.4   **31.** $\frac{8}{1}$ **32.** $\frac{7}{3}$
33. $x = 35$ **34.** $x = 100$ **35.** $x \approx 2.01$ (rounded)   **36.** 160%
37. \$600   **38.** 135 min   **39.** $2\frac{1}{2}$ or 2.5 lb   **40.** 0.08 m
41. 1800 mL   **42.** mm   **43.** L   **44.** kg   **45.** cm
46. water freezes at 0 °C and boils at 100 °C   **47.** Rectangle;
$P = 11$ in.; $A = 7$ in.^{2} **48.** Triangle; $P = 5.8$ m; $A = 1.575$ m^2
49. Circle; $C \approx 31.4$ ft; $A \approx 78.5$ ft^2 **50.** Parallelogram;
$P = 76$ cm; $A = 264$ cm^2 **51.** Trapezoid; $P = 40.8$ m; $A = 95$ m^2
52. $P = 50$ yd; $A = 142$ yd^2 **53.** $y \approx 24.2$ mm   **54.** $x \approx 8.1$ ft;
$P \approx 19.1$ ft   **55.** 59% (rounded)   **56.** Brand T at 15.5 oz for
\$2.99 $-$ \$0.30 coupon   **57.** $V \approx 2255.3$ cm^3
58. Original Area $= 9$ in.^{2}; XL Area $= 16$ in.^{2}; XL is 78% larger
than original (rounded).   **59.** $1\frac{1}{12}$ yd   **60.** 35.7 lb meat (rounded);
50 lb potatoes; 21.4 lb carrots (rounded)   **61.** 3.4 m   **62.** \$22.40
63. (a) Britain   (b) Brazil   **64.** (a) $\frac{2}{1}$ (b) $\frac{2}{3}$
65. (a) 47.6% (rounded)   (b) 180%   **66.** (a) 20%   (b) 50%
(c) 100%

CHAPTER 9

Section 9.1 (page 631)

1. $+32$ or 32 degrees   **3.** $-\$12$ **5.** $-6\frac{1}{2}$ lb   **7.** $+20,320$ ft or
20,320 ft   **9.**

```
← + + + + ● + + ● + + + →
 -5 -4 -3 -2 -1 0 1 2 3 4 5
```

11.

```
← +●+ ● + +●+ +● ●+ +→
 -5 -4 -3 -2 -1 0 1 2 3 4 5
```

13.

```
← + + + +●+● + +●● +●+ →
 -5 -4 -3 -2 -1 0 1 2 3 4 5
```

15. $<$ **17.** $>$ **19.** $<$ **21.** $>$ **23.** $<$ **25.** $>$
27.

```
        A  C D  B
        ↓  ↓ ↓  ↓
   ←—+——●—+—●—●—+—●—+—→
     -2  -1   0    1
```

28. $-1.2, -0.5, 0, 0.6$ **29.** A: may be at risk; B: above normal;
C: normal; D: normal   **30.** (a) The patient would think the
interpretation was "above normal" and wouldn't get treatment.
(b) Patient D's score of 0; because 0 is neither positive nor negative.
31. 3   **33.** 10   **35.** 0   **37.** -18 **39.** -32 **41.** -7 **43.** 14
45. $-\frac{2}{3}$ **47.** 8.3   **49.** $\frac{1}{6}$ **51.** true   **53.** true   **55.** false

Section 9.2 (page 641)

1. 3

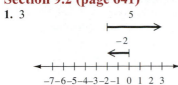

3. -7

5. -1

7. -3 9. 7   11. -7 13. 1   15. -8 17. -20
19. $13 + (-17) = -4$ yd   21. $-\$52.50 + \$50 = -\$2.50$
23. Jeff: $-20 + 75 + (-55) = 0$ pts; Terry: $42 + (-15) + 20 =$
47 pts   25. -6.8 27. $\dfrac{1}{4}$ 29. $-\dfrac{3}{10}$ 31. $-\dfrac{26}{9}$ 33. -3 35. 9

37. $-\dfrac{1}{2}$ 39. 6.2   41. 14   43. -2 45. -12 47. -25

49. -23 51. 5   53. 20   55. 11   57. -60 59. 0   61. $-\dfrac{3}{2}$

63. $-\dfrac{2}{5}$ 65. 0.7   67. (a) Change 6 to its opposite, -6.

Correct answer is $-6 + (-6) = -12$.   (b) Do *not* change 9 to its
opposite. Correct answer is $-9 + (-5) = -14$.   69. (a) 21 °F;
$30 - 21 = 9$ degrees difference   (b) 0 °F; $15 - 0 = 15$ degrees
difference   71. (a) -17 °F; $5 - (-17) = 22$ degrees difference
(b) -41 °F; $-10 - (-41) = 31$ degrees difference   73. -10
75. 12   77. -7 79. 5   81. -1 83. -4.4 85. -11
87. -10 89. $\$196$ 91. $\$0$ 93. (a) $-2; -2$ (b) $-8; -8$
(c) 10; 10 Addition is commutative, so changing the order of the
addends does *not* change the sum.   94. (a) $-8; 8$ (b) $2; -2$
(c) $-7; 7$ Changing the order of the numbers in subtraction
does change the result.   95. (a) The answers have the same
absolute value but opposite signs.   (b) When the order of the
numbers in a subtraction problem is switched, change the sign on
the answer to its opposite.   96. (a) $-18; 20; -5; 4$. Adding 0 to
any number leaves the number unchanged.   (b) $2; -10; -3; 7$.
Subtracting 0 from a number leaves the number unchanged, but
subtracting a *number from 0* changes the sign of the number.

Section 9.3 (page 647)

1. -35 35. -45 5. -18 7. -50 9. -40 11. 32

13. 77   15. 133   17. 13   19. 0   21. 4   23. -4 25. $-\dfrac{1}{10}$

27. $\dfrac{14}{3}$ 29. $-\dfrac{5}{6}$ 31. $\dfrac{7}{4}$ 33. -42.3 35. 6   37. -31.62

39. 4.5   41. 8.23   43. 0   45. -2 47. -5 49. undefined
51. -14 53. 10   55. 4   57. -1 59. 191   61. 0   63. 1

65. $\dfrac{2}{3}$ 67. $\dfrac{1}{3}$ 69. -8 71. $-\dfrac{14}{3}$ 73. 4.73   75. -4.7

77. -5.3 79. 12   81. -0.36 83. $-\dfrac{7}{40}$ 85. -2 87. -10

89. $-\$119,400$ 91. -2011 ft   93. $\$2366$ 95. -22 degrees
97. (a) Examples will vary. Some possibilities: $6(-1) = -6$;
$2(-1) = -2; 15(-1) = -15$ (b) Examples will vary. Some
possibilities: $-6(-1) = 6; -2(-1) = 2; -15(-1) = 15$. The
result of multiplying any nonzero number times -1 is the number
with the opposite sign.   98. (a) Examples will vary. Some

possibilities: $\dfrac{-6}{-1} = 6; \dfrac{-2}{-1} = 2; \dfrac{-15}{-1} = 15$

(b) Some possibilities: $\dfrac{6}{-1} = -6; \dfrac{2}{-1} = -2; \dfrac{15}{-1} = -15$

(c) Some possibilities: $\dfrac{-6}{-6} = 1; \dfrac{-2}{-2} = 1; \dfrac{-15}{-15} = 1$

When dividing by -1, the sign of the number changes to its opposite. A negative number divided by itself gives a result of 1.

Section 9.4 (page 655)

1. -6 3. 0   5. 52   7. -39 9. -30 11. 17   13. 16
15. 30   17. 23   19. -43 21. 19   23. -3 25. 0   27. 47

29. 41   31. -2 33. 13   35. 126   37. $\dfrac{27}{-3} = -9$

39. $\dfrac{-48}{-4} = 12$ 41. $\dfrac{-60}{-1} = 60$ 43. 8   45. 1.24   47. -1.47

49. -2.31 51. $-\dfrac{13}{10}$ 53. $\dfrac{5}{4}$ 55. $\dfrac{4}{5}$ 57. -1200 59. 1

61. $4, -8, 16, -32, 64, -128, 256, -512$.   (a) When a negative
number is raised to an even power, the answer is positive; when
raised to an odd power, the answer is negative.

(b) positive; negative; positive   63. $\dfrac{27}{-27} = -1$ 65. 7   67. 6

Summary Exercises on Operations with Signed Numbers (page 659)

1. -6 2. 0   3. -7 4. -7 5. 63   6. -1 7. -56
8. -22 9. 12   10. 6   11. -13 12. 0   13. -80 14. -17

15. -48 16. -1 17. $-\dfrac{1}{6}$ 18. undefined   19. -14.6

20. -0.6 21. 48   22. -19 23. 2   24. -20 25. 0   26. 16
27. -3 28. -36 29. 6   30. -2 31. -32 32. -5
33. -4 34. -9732 35. 100   36. 4   37. -343 38. 5
39. -6 40. -10 41. -5 42. 8   43. 1

44. $\dfrac{27}{0}$ is undefined.   45. $\dfrac{-8}{8} = -1$ 46. $+\$212$ or $\$212$

47. $-\$24$ 48. -13 degrees   49. $+27,010$ ft or $27,010$ ft
50. -43 yd   51. -90 ft   52. January: $-\$700$; February: $-\$100$;
March: $\$700$; April: $\$900$; May: $\$0$; June: $\$700$ 53. January;
April   54. $\$2100$ 55. $-\$1850$ 56. Subtract the smaller
absolute value from the larger absolute value. The answer has the
same sign as the addend with the larger absolute value. Examples:
$-6 + 2 = -4$ and $6 + (-2) = 4$ 57. Similar: If the signs match,
the result is positive. If the signs are different, the result is negative.
Different: Multiplication is commutative, division is not. You can
multiply by 0, but dividing by 0 is not allowed.

Section 9.5 (page 663)

1. 28   3. -10 5. 8   7. -30 9. -8 11. 0   13. 2

15. -22 17. $-\dfrac{13}{8}$ 19. 28   21. -5 23. 3   25. $\dfrac{12}{-6} = -2$

27. $P = 30$ 29. $P = 28$ 31. $A \approx 78.5$ 33. $A = \dfrac{45}{2}$ or $22\dfrac{1}{2}$

35. $V = 600$ 37. $d = 318$ 39. $C \approx 25.12$
41. (a) $P = 33$ in.   (b) $P = 9$ ft   43. (a) average score = 83
(b) average score = 91
45. The error was made when replacing x with -3; should be
$-(-3)$, not -3.

$-x - 4y$	Replace x with (-3) and y with (-1).
$-(-3) - 4(-1)$	Opposite of (-3) is $+3$.
$(+3) - (-4)$	Change subtraction to addition.
$3 + (+4)$	Add.
7	

47. $F = -40$; convert a Celsius temperature to Fahrenheit
48. $C = -20$; convert a Fahrenheit temperature to Celsius
49. $V \approx 113.04$; find the volume of a sphere
50. $c = 5$; Pythagorean Theorem for finding the hypotenuse or a leg in a right triangle   **51.** $A = 56$; find the area of a trapezoid
52. $V \approx 376.8$; find the volume of a cone

Section 9.6 (page 673)

1. yes   **3.** yes   **5.** no   **7.** $p = 4$ **9.** $k = -15$ **11.** $z = 8$
13. $r = 10$ **15.** $n = -8$ **17.** $r = -6$ **19.** $k = 18$ **21.** $x = 11$
23. $r = -3$ **25.** $d = \dfrac{7}{3}$ or $2\dfrac{1}{3}$ **27.** $z = \dfrac{87}{8}$ or $10\dfrac{7}{8}$ **29.** $k = \dfrac{5}{2}$
or $2\dfrac{1}{2}$ **31.** $m = \dfrac{83}{20}$ **33.** $x = 5.87$ **35.** $r = -1.51$ **37.** $z = 2$
39. $r = 4$ **41.** $y = 0$ **43.** $k = -6$ **45.** $p = 9$ **47.** $m = -7$
49. $p = 1.1$ **51.** $k = 34$ **53.** $a = 66$ **55.** $r = -36$
57. $p = -20$ **59.** $m = 4$ **61.** $x = 16$ **63.** $y = 1.3$
65. $z = -4.94$ **67.** You may add or subtract the same number on both sides of an equation. Many different equations could have -3 as the solution. One possibility is:

$$x + 5 = 2$$
$$x + 5 - 5 = 2 - 5$$
$$x = -3$$

69. $x = 19$ **71.** $x = 9$ **73.** $x = \dfrac{8}{21}$ **75.** $a = -\dfrac{5}{4}$ **77.** $m = -4$

Section 9.7 (page 681)

1. $p = 1$ **3.** $y = 1$ **5.** $m = 0$ **7.** $a = -2$ **9.** $x = -4$
11. $z = 6$ **13.** $b = 0.9$ **15.** $6x + 24$ **17.** $7p - 56$
19. $-3m - 18$ **21.** $-2y + 6$ **23.** $-8c - 64$ **25.** $-10w + 90$
27. $17r$ **29.** $1z$ or z **31.** $-2y$ **33.** $-7t$ **35.** $0p = 0$
37. $-3b$ **39.** $k = 5$ **41.** $m = 6$ **43.** $b = -6$ **45.** $y = 1$
47. $p = 4$ **49.** $z = -2$ **51.** $y = 2$ **53.** $b = -1.05$
55. $r = -4$ **57.** $y = -9$ **59.** $t = -5$ **61.** $x = 0$
63.
$$-2t - 10 = 3t + 5$$

$-2t - 3t - 10 = 3t - 3t + 5$	Subtract $3t$ from both sides (addition property).
$-5t - 10 = 5$	
$-5t - 10 + 10 = 5 + 10$	Add 10 to both sides (addition property).
$-5t = 15$	
$\dfrac{-5 \cdot t}{-5} = \dfrac{15}{-5}$	Divide both sides by -5 (multiplication property).
$t = -3$	

65. $x = -10$ **67.** $y = 2$ **69.** $y = 20$ **71.** $w = -1$
73. $a = -6.75$

Section 9.8 (page 691)

1. $14 + x$ or $x + 14$ **3.** $-5 + x$ or $x + (-5)$ **5.** $20 - x$
7. $x - 9$ **9.** $x - 4$ **11.** $6x$ **13.** $2x$ **15.** $\dfrac{x}{2}$ **17.** $8 + 2x$ or
$2x + 8$ **19.** $7x - 10$ **21.** $2x + x$ or $x + 2x$
23. (a) A variable is a letter that represents an unknown quantity. Examples: x, w, p (b) An expression is a combination of operations on variables and numbers. Examples: $6x, w - 5, 2p + 3x$
(c) An equation has an $=$ sign and shows that two expressions are equal. Examples: $2y = 14; x + 5 = 2x; 8p - 10 = 54$
25. $4n - 2 = 26; n = 7$ **27.** $2n + n = -15; n = -5$
29. $5n + 12 = 7n; n = 6$ **31.** $30 - 3n = 2 + n; n = 7$
33. $\dfrac{1}{2}n + 2n = 50; n = 20$ **35.** Let x be Tamu's age. $4x - 75 = x$.

Tamu is 25.   **37.** Let x be the amount Brenda spent. $81 = 2x - 3$. Brenda spent \$42.   **39.** Let x be my age; $x + 9$ is my sister's age. $x + x + 9 = 51$. I am 21; my sister is 30.   **41.** Let x be husband's earnings; $x + 1500$ is Lien's earnings. $x + x + 1500 = 37,500$. Husband earned \$18,000; Lien earned \$19,500.

43. Let x be printer's cost; $5x$ is computer's cost. $x + 5x = 1320$. The printer cost \$220; the computer cost \$1100.
45. Let x be length of shorter piece; $x + 10$ is length of longer piece. $x + x + 10 = 78$. Shorter piece is 34 cm; longer piece is 44 cm.   **47.** Let x be one piece; $x - 7$ is the other piece. $x + x - 7 = 31$. One piece is 19 ft; the other piece is 12 ft.
49. $48 = 2(5) + 2x$; length is 19 m.

51. $36 = 2(2w) + 2w$; length is 12 ft; width is 6 ft.

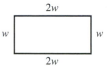

53. $36 = 2(2x + 3) + 2x$; length is 13 in.; width is 5 in.

Chapter 9 Review Exercises (page 699)

1.
$-7\,-6\,-5\,-4\,-3\,-2\,-1\ 0\ 1\ 2\ 3\ 4$

2.
$-6\,-5\,-4\,-3\,-2\,-1\ 0\ 1\ 2\ 3\ 4\ 5$

3.
$-7\,-6\,-5\,-4\,-3\,-2\,-1\ 0\ 1\ 2\ 3$

4.
$-8\,-7\,-6\,-5\,-4\,-3\,-2\,-1\ 0\ 1\ 2$

5. $>$ **6.** $<$ **7.** $>$ **8.** $<$ **9.** 8   **10.** 19   **11.** -7 **12.** -15
13. 2   **14.** -7 **15.** -19 **16.** -33 **17.** 1   **18.** -19 **19.** $\dfrac{3}{10}$
20. $-\dfrac{3}{8}$ **21.** -5.2 **22.** -1.5 **23.** -6 **24.** 14   **25.** $\dfrac{5}{8}$
26. -3.75 **27.** -6 **28.** -8 **29.** -7 **30.** -17 **31.** 11
32. 11   **33.** 13   **34.** -6 **35.** -80 **36.** 0   **37.** $-\dfrac{1}{2}$ **38.** 9
39. -24 **40.** -20 **41.** 15   **42.** 64   **43.** -8 **44.** -3 **45.** 5
46. 20   **47.** 0   **48.** -81 **49.** 0   **50.** undefined   **51.** $-\dfrac{4}{7}$
52. 6   **53.** 1.4   **54.** -6.6 **55.** 57   **56.** 23   **57.** -1 **58.** -2
59. -34 **60.** -32 **61.** -100 **62.** 117   **63.** -1 **64.** 1.328
65. 0   **66.** 2   **67.** 27   **68.** -8 **69.** 0   **70.** 19   **71.** 23
72. -1 **73.** $P = 35$ **74.** $A = 27$ **75.** $y = -3$ **76.** $a = 16$
77. $z = 1$ **78.** $r = 1$ **79.** $x = -\dfrac{5}{4}$ or $-1\dfrac{1}{4}$ **80.** $k = -8.05$
81. $r = -7$ **82.** $p = 8$ **83.** $z = 20$ **84.** $a = -55$ **85.** $y = 9$
86. $b = -4$ **87.** $6r - 30$ **88.** $11p + 77$ **89.** $-9z + 27$
90. $-8x - 32$ **91.** $11r$ **92.** $-5z$ **93.** $-8p$ **94.** $2x$
95. $z = -9$ **96.** $k = -5$ **97.** $y = -5$ **98.** $b = 7$
99. $a = -4$ **100.** $t = 0$ **101.** $18 + x$ or $x + 18$
102. $\dfrac{1}{2}x$ or $\dfrac{x}{2}$ **103.** $-5x$ **104.** $20 - x$
105. $4x + 6 = -14; x = -5$ **106.** $5x - x = 100; x = 25$
107. Let x be one student's scholarship; $3x$ is the other student's scholarship. $x + 3x = 30,000$. One student receives \$7500; the

other student receives $22,500.
108. $48 = 2(w + 4) + 2w$; width is 10 in.; length is 14 in.

109. 3 **110.** 40 **111.** -1 **112.** -7 **113.** -16 **114.** -9
115. 6 **116.** 5 **117.** -20 **118.** undefined **119.** $-\dfrac{5}{9}$
120. -0.35 **121.** 7 **122.** -19 **123.** $y = 9$ **124.** $b = -4$
125. $z = 4$ **126.** $r = -10$ **127.** $x = -11$ **128.** $z = 3$
129. $k = 4$ **130.** $t = 0$ **131.** $a = -40$ **132.** $p = 3$
133. 6 mg **134.** 100 senators; 435 representatives **135.** zebra:
43 mph; cheetah: 68 mph **136.** width is 110 yd; length is 550 yd

Chapter 9 Test (page 705)

1.

$-5 -4 -3 -2 -1 \ 0 \ 1 \ 2 \ 3 \ 4 \ 5$

2. $<, >$ **3.** 7, 15 **4.** -1 **5.** -13 **6.** 5.3 **7.** -7 **8.** 16
9. $\dfrac{1}{4}$ **10.** -32 **11.** 84 **12.** 0 **13.** -25 **14.** 8 **15.** $-\dfrac{3}{5}$
16. 1 **17.** -21 **18.** -38 **19.** 27 **20.** When evaluating, you
are given specific values to replace each variable. When solving an
equation, you are not given the value of the variable. You must find
a value that "works"; that is, when your solution is substituted for
the variable, the two sides of the equation are equal. **21.** 110
22. $x = 5$ **23.** $r = 31$ **24.** $t = 5$ **25.** $p = -15$ **26.** $a = -3$
27. $m = 2$ **28.** Let x be the shorter piece; $x + 4$ is longer piece.
$x + x + 4 = 118$. Shorter piece is 57 cm; longer piece is 61 cm.
29. $420 = 2(4w) + 2w$; width is 42 ft; length is 168 ft.

30. Let x be Marcella's time; $x - 3$ is Tim's time. $x + x - 3 = 19$.
Marcella spent 11 hr; Tim spent 8 hr.

Cumulative Review Exercises: Chapters 1–9 (page 707)

1. *Estimate:* $9 + 1 + 40 = 50$; *Exact:* 50.602
2. *Estimate:* $6 \times 50 = 300$; *Exact:* 308.484
3. *Estimate:* $40,000 \div 80 = 500$; *Exact:* 503
4. *Estimate:* $4 - 3 = 1$; *Exact:* $\dfrac{17}{20}$
5. *Exact:* $5\dfrac{1}{4}$; *Estimate:* $6 \cdot 1 = 6$
6. *Exact:* $2\dfrac{1}{2}$; *Estimate:* $4 \div 2 = 2$ **7.** 8.906 **8.** 534,072
9. 56.4 (rounded) **10.** -5 **11.** 40 **12.** 4 **13.** -1.3
14. -5 **15.** $-\dfrac{1}{2}$ **16.** 5 **17.** 4 **18.** $>$ **19.** $<$ **20.** $<$
21. $0.08; \dfrac{2}{25}$ **22.** 4.5; 450% **23.** (a) $\dfrac{2}{1}$ (b) $\dfrac{1}{3}$ (c) $\dfrac{1}{1}$
(d) $\dfrac{12}{13}$ **24.** $x = 0.4$ **25.** $x \approx 233.33$ (rounded) **26.** $x = 45$
27. 15 students **28.** 54.4% **29.** 50 cars **30.** 4% **31.** 14 qt
32. 3 days **33.** 3700 mL **34.** 0.4 m **35.** m **36.** mL
37. km **38.** g **39.** 37 °C; 20 °C **40.** Square; $P = 9$ ft;
$A = 5\dfrac{1}{16}$ or 5.0625 ft² **41.** Circle; $C \approx 28.26$ mm;

$A \approx 63.585$ mm² **42.** Right triangle; $P = 56$ mi; $A = 84$ mi²
43. Trapezoid; $P = 6.45$ cm; $A = 2.185$ cm² **44.** Parallelogram;
$P = 36$ ft; $A = 70$ ft² **45.** $P = 188$ m; $A = 1636$ m²
46. $y \approx 13.2$ yd **47.** $x = 14$ in. **48.** $y = -26$ **49.** $t = -5$
50. $x = 8$ **51.** $p = -4$ **52.** $4x + 40 = 0; x = -10$
53. $m + m + 300 = 1000$; $350 for Reggie; $650 for Donald
54. $82 = 2(w + 5) + 2w$; length is 23 cm; width is 18 cm

55. $31.91 (rounded) **56.** $8\dfrac{1}{3}$ cups **57.** 124%
58. $4.30 (rounded) **59.** 0.544 m³ less **60.** 11 ft
61. 54 min (rounded) **62.** Naomi's car; 25.2 miles per gallon
(rounded)

CHAPTER 10

Section 10.1 (page 717)

1. $32,000 **3.** $\dfrac{9800}{32,000} = \dfrac{49}{160}$ **5.** $\dfrac{12,100}{900} = \dfrac{121}{9}$
7. Don't know **9.** $\dfrac{720}{6000} = \dfrac{3}{25}$ **11.** $\dfrac{1020}{1200} = \dfrac{17}{20}$
13. $\dfrac{1740}{1020} = \dfrac{29}{17}$ **15.** 160 people **17.** 96 people **19.** 960 people
21. 1219 people (rounded) **23.** 1828 people (rounded)
25. 277 people (rounded) **27.** First find the percent of the total
that is to be represented by each item. Next, multiply the percent by
360° to find the size of each sector. Finally, use a protractor to draw
each sector. **29.** 90° **31.** 10%; 36° **33.** 54° **35.** 15%
37. (a) $200,000 (b) 22.5°; 72°; 108°; 90°; 67.5°
(c)

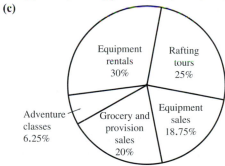

39. (a) 2464; 55% (rounded); 198°; 220; 5% (rounded); 18°;
536; 12% (rounded); 43° (rounded); 520; 12% (rounded);
43° (rounded); 748; 17% (rounded); 61° (rounded)

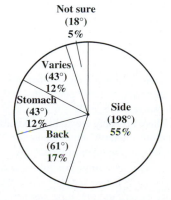

(b) No. The total is 101% due to rounding.
(c) No. The total is 363° due to rounding.

Section 10.2 (page 727)

1. India; 51.4% **3.** USA, Britain, and Australia **5.** 121 days (rounded) **7.** May; 10,000 unemployed **9.** 1500 workers
11. 4000 workers **13.** 150,000 gal **15.** 2002; 250,000 gal
17. 550,000 gal **19.** 24.1 million or 24,100,000 PCs
21. 135.1 million or 135,100,000 PCs **23.** Answers will vary. Some possibilities are: greater demand as a result of lower price; more uses and applications for the owner; improved technology **25.** 3,000,000 CDs **27.** 1,500,000 CDs **29.** 3,500,000 CDs
31. Probably Store B with greater sales. Predicted sales might be 4,500,000 CDs to 5,000,000 CDs in 2007. **33.** A single bar or a single line must be used for each set of data. To show multiple sets of data, multiple sets of bars or lines must be used. **35.** $40,000
37. $25,000 **39.** $5000 **41.** Answers will vary. Some possibilities are that the decrease in sales may have resulted from poor service or greater competition. The increase in sales may have been a result of more advertising or better service. **43.** 23 yr
44. 20 additional flavors **45.** 214,286 rolls each day (rounded)
46. 1.17 lb (rounded) **47.** The weight of one roll of LifeSavers is much less than 1.17 lb. The answer is correct using the information given, but some of the data given must not be accurate.
48. Answers will vary. Possible answers are: Misprints or typographical errors; careless reporting of data; math errors.

Section 10.3 (page 735)

1. 61–65 years; 16,000 members **3.** 13,000 members
5. 50,000 members **7.** $4100 to $5000; 16 employees
9. 11 employees **11.** 49 employees **13.** Class intervals are the result of combining data into groupings. Class frequency is the number of data items that fit in each class interval. These are used to group data and to have multiple responses (frequency) in a class interval—this makes the data easier to interpret. **15.** 𝄂𝄂 l; 6
17. l; 1 **19.** ll; 2 **21.** llll; 4 **23.** 𝄂𝄂 l; 6 **25.** 𝄂𝄂; 5 **27.** l; 1
29. 𝄂𝄂 𝄂𝄂 llll; 14 **31.** 𝄂𝄂 𝄂𝄂 𝄂𝄂 l;16 **33.** 𝄂𝄂 𝄂𝄂 l; 11 **35.** 𝄂𝄂; 5

Section 10.4 (page 743)

1. 11 yr **3.** 2.5 in. of rain (rounded) **5.** $37,127 **7.** $58.24
9. 12.5 customers (rounded) **11.** 17.2 fish (rounded)
13. $35,500 **15.** 130 books **17.** 508 calories **19.** 21%
21. 68 and 74 yr **23.** When the data contains a few very low or a few very high values, the mean will give a poor indication of the average. Consider using the median or mode instead.
25. The median is a better measure of central tendency when the list contains one or more extreme values. Find the mean and the median of the following home values. $182,000; $164,000; $191,000; $115,000; $982,000; mean home value = $326,800; median home value = $782,000 **27.** 2.60 **29.** 3.14 (rounded)
31. 176 sales calls **32.** mean Samuels: 22; mean Stricker: 22
33. median Samuels: 19.5; median Stricker: 22
34. mode Samuels: 22; mode Stricker: 22 **35.** The means and mode are identical for both sales representatives and the medians are close. **36.** The number of weekly sales calls made by Samuels varies greatly from week to week while the number of weekly sales calls made by Stricker remains fairly constant.
37. range = 40 − 8 = 32 **38.** range = 25 − 19 = 6
39. No, not with any certainty. There probably are additional questions that need to be answered before a determination could be made, such as the dollar amount of sales, number of repeat customers, and so on. **40.** Answers will vary. Some possible answers are: He works hard one week, then takes it easy the next week; the characteristics of the sales territories vary greatly; illness or personal problems may be affecting performance.

Chapter 10 Review Exercises (page 755)

1. lodging: $560 **2.** $\frac{400}{1700} = \frac{4}{17}$ **3.** $\frac{300}{1700} = \frac{3}{17}$ **4.** $\frac{280}{1700} = \frac{14}{85}$
5. $\frac{300}{160} = \frac{15}{8}$ **6.** $\frac{560}{400} = \frac{7}{5}$ **7.** 3936 companies
8. 1728 companies **9.** 720 companies **10.** 2904 companies
11. On-site child care and Fitness centers   Answers will vary. Perhaps employers feel that they are not needed or would not be used. Or, it may be that they would be too expensive for the benefit derived.   **12.** Flexible hours and Casual dress   Answers will vary. Perhaps employees request them and appreciate them. Or, it may be that neither of them cost the employer anything to offer.
13. March; 8,000,000 acre-feet **14.** June; 2,000,000 acre-feet
15. 5,000,000 acre-feet **16.** 4,000,000 acre-feet
17. 5,000,000 acre-feet **18.** 2,000,000 acre-feet
19. $50,000,000 **20.** $20,000,000 **21.** $20,000,000
22. $40,000,000 **23.** The floor-covering sales decreased for 2 years and then moved up slightly. Answers will vary. Perhaps there is less new construction, remodeling, and home improvement in the area near Center A, or a better product selection and service have reversed the decline in sales.   **24.** The floor-covering sales are increasing. Answers will vary. Perhaps new construction and home remodeling have increased in the area near Center B, or greater advertising has attracted more customers.   **25.** 28.5 digital cameras
26. 35.2 complaints **27.** $51.05 **28.** 257.3 points (rounded)
29. 39 forms **30.** $562 **31.** $79 **32.** 18 and 32 launchings (bimodal) **33.** 36° **34.** 35% **35.** 20%; 72° **36.** 25%; 90°
37. 10%; 36° **38.**

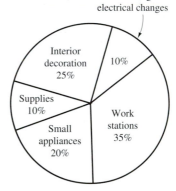

39. 100 volunteers **40.** 144.7 vaccinations (rounded)
41. 48 applicants **42.** 31 and 43 two-bedroom apartments (bimodal) **43.** 5.0 hr **44.** 21 e-mails **45.** llll; 4 **46.** l; 1
47. 𝄂𝄂 l; 6 **48.** ll; 2 **49.** 𝄂𝄂 𝄂𝄂 lll; 13 **50.** 𝄂𝄂 ll; 7 **51.** 𝄂𝄂 ll; 7
52.

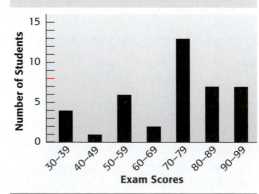

53. 64.5 (rounded) **54.** 118.8 units (rounded)